Failure Analysis of Engineering Materials

Failure Analysis of Engineering Materials

Charlie R. Brooks
Materials Science and Engineering Department
University of Tennessee
Knoxville, Tennessee

Ashok Choudhury
Oak Ridge National Laboratory
Oak Ridge, Tennessee

McGraw-Hill

New York Chicago San Francisco Lisbon London Madrid
Mexico City Milan New Delhi San Juan Seoul
Singapore Sydney Toronto

Library of Congress Cataloging-in-Publication Data

Brooks, Charlie R., date
Failure analysis of engineering materials / Charlie R. Brooks, Ashok Choudhury.
p. cm.
ISBN 0-07-135758-0
1. Fracture mechanics. 2. Materials—Fatigue. I. Choudhury, A. (Ashok) II. Title.
TA409.B775 2002
620.1'126—dc21 2001044238

McGraw-Hill

A Division of The ***McGraw-Hill*** *Companies*

7 8 9 10 11 IBT/IBT 1 9 8 7 6 5 4 3 2 1

ISBN 0-07-135758-0

The sponsoring editor for this book was Kenneth P. McCombs, the editing supervisor was Stephen M. Smith, and the production supervisor was Pamela A. Pelton. It was set in Century Schoolbook following the MHT 6x9 design by Joanne Morbit of McGraw-Hill Professional's Hightstown, N.J., composition unit.

Printed and bound by IBT Global.

Portions of this book were previously published in *Metallurgical Failure Analysis,* © 1993.

A.C. dedicates this book to his rays of sunshine, Sona and Puchun.

Contents

Contents

Preface

The predecessor (*Metallurgical Failure Analysis*) to the current book was limited in scope to examination of the principles and practices of failure analysis of metallic materials. However, with the increasing use of polymeric and ceramic materials as structural components, examination of the failure processes in these substances has become increasingly important. Thus, we cover in this book the failure mechanisms of ceramics and polymers. Another area of failure analysis that has become prominent concerns electronic materials and components. These solid-state devices are constructed of metallic, polymeric, and ceramic materials, and hence their failures involve the same mechanisms and modes of failure as the individual materials. However, electronic systems have additional features that may be involved in failures and thus are important to understand. Another critical property of many of these devices is the semiconducting nature of the materials used. Failures of electronic devices can be caused by processes or events that disrupt or skew this property. For example, physical and electronic damage owing to electrical arcing may induce failure; it is important to be able to detect this phenomenon when making a failure analysis. We have therefore added a new chapter on the failure analysis of electronic materials.

The rationale for this book on materials failure analysis is the same as that given in the preface to the preceding book. It should be pointed out that failure analysis is now established as a career in its own right. The number of firms and people involved in this industrial segment continues to increase. This book is designed to be of benefit to materials engineers and materials scientists, and other engineers involved in the design of components, specification of materials, and fabrication of components. It serves as an introduction to failure analysis for the novice, and as a refresher and a source book for those already familiar with the subject. The inclusion of the electronic failure analysis chapter hopefully extends the book's usefulness to electrical and electronics engineers.

Charlie R. Brooks
Ashok Choudhury

Failure Analysis of Engineering Materials

Chapter

1

Introduction

The general conclusion is this—Frost does not make either iron (cast or wrought) or steel brittle, and accidents arise from neglect to submit wheels, axles, and all other parts of the rolling stock to a practical and sufficient test before using them.

JAMES PRESCOTT JOULE
Philosophical Magazine, *1871*

1.1 Objectives

The analysis of failures of engineering components is an extremely important aspect of engineering. Establishing the causes of failures provides information for improvements in design, operating procedures, and the use of components. Also, determining the cause of a failure can play a pivotal role in establishing liability in litigation. Failure analysis is often difficult and frustrating, but understanding how to approach an analysis and how to interpret observations provides a basis for assuring meaningful results.

The objective of this book is to introduce the important aspects associated with the failure analysis of engineering components. Emphasis is placed on the analysis of broken components, where observations of the fracture surface play a key role. Thus a treatment of both macroscopic and microscopic observations of fracture surfaces is given. Since loading conditions are often an important aspect of the possible causes of failures, a simplified treatment of the mechanics involved is presented. It is to be noted that some information about prior loading conditions can often be gleaned from a careful observation of the general macroscopic orientation of the fracture surface. Also included is a section which reviews the common experimental methods used in failure analysis of materials. The failure of metallic, ceramic, and polymeric materials is examined, and there is special treatment of electronic materials in Chap. 6. Finally, some case studies of metallurgical failure analyses are introduced, and the approaches taken are related to the information presented in preceding chapters.

1.2 Approach to Failure Analysis of Materials

Failure analysis deals with the determination of the causes of the failure of engineering parts or components. In the broad, and correct, sense, *failure* can be defined as the inability of a component to function properly, and this definition does not imply fracture. *Failure analysis* can be defined as the examination of a failed component and of the failure situation in order to determine the causes of the failure. The purpose of a failure analysis is to define the mechanism and causes of the failure and usually to recommend a solution to the problem.

The causes of failures can be broken down into the following categories:

1. *Misuse.* The component is placed under conditions for which it was not designed. This is a common cause of failure, and its establishment sometimes relies on determining that the assembly of the component and the design were correct, leaving misuse as a suspected cause.
2. *Assembly errors and improper maintenance.* Assembly errors involve such factors as leaving off a bolt or using incorrect lubricant. Maintenance of equipment ranges from painting surfaces to cleaning and lubrication, and its neglect may lead to failure. It is also pointed out that a failure may be caused by some other part of the system not functioning properly, thereby placing the component which failed under conditions for which it was not designed. Thus failure of a component may point to a problem elsewhere in the system.
3. *Design errors.* This is a very common cause of failure. In this category the following items are considered to be specified by the design process:
   a. *Size and shape of the part.* This is usually determined by stress analysis or geometric constraints.
   b. *Material.* This refers to the chemical composition *and* the treatment (for example, heat treatment) necessary to achieve the required properties.
   c. *Properties.* This is related to stress analysis, but other properties such as corrosion resistance must also be considered.

It is interesting to examine some information about the causes of failures and compare it to the preceding list.

1. *Improper material selection.* Table 1.1 shows that improper material selection is a common problem.

TABLE 1.1 Frequency of Causes of Failure in Some Engineering Industry Investigations

Origin	%
Improper material selection	38
Fabrication defects	15
Faulty heat treatments	15
Mechanical design fault	11
Unforeseen operating conditions	8
Inadequate environment control	6
Improper or lack of inspection and quality control	5
Material mixup	2

SOURCE: Adapted from Davies.[1]

TABLE 1.2 Frequency of Causes of Failure of Aircraft Components (Laboratory Data)

Origin	%
Improper maintenance	44
Fabrication defects	17
Design deficiencies	16
Abnormal service damage	10
Defective material	7
Undetermined cause	6

SOURCE: Adapted from Davies.[1]

2. *Improper maintenance.* The data in Table 1.2 show that improper maintenance is the main problem in failed aircraft components.
3. *Faulty design considerations.* Causes of failures due to faulty design considerations or misapplication of material include the following (adapted from Dolan[2]):
   a. Ductile failure (excess deformation, elastic or plastic; tearing or shear fracture)
   b. Brittle fracture (from flaw or stress raiser of critical size)
   c. Fatigue failure (load cycling, strain cycling, thermal cycling, corrosion fatigue, rolling contact fatigue, fretting fatigue)
   d. High-temperature failure (creep, oxidation, local melting, warping)
   e. Static delayed fractures (hydrogen embrittlement, caustic embrittlement, environmentally stimulated slow growth of flaws)
   f. Excessively severe stress raisers inherent in the design
   g. Inadequate stress analysis, or impossibility of a rational stress calculation in a complex part
   h. Mistake in designing on the basis of static tensile properties, instead of the significant material properties that measure the resistance of the material to each possible failure mode

4. *Faulty processing.* Causes of failures due to faulty processing include the following (adapted from Dolan[2]):
   a. Flaws due to faulty composition (inclusions, embrittling impurities, wrong material)
   b. Defects originating in ingot making and casting (segregation, unsoundness, porosity, pipes, nonmetallic inclusions)
   c. Defects due to working (laps, seams, shatter cracks, hot-short splits, delamination, excess local plastic deformation)
   d. Irregulatities and mistakes due to machining, grinding, or stamping (gouges, burns, tearing, fins, cracks, embrittlement)
   e. Defects due to welding (porosity, undercuts, cracks, residual stress, lack of penetration, underbead cracking, heat-affected zone)
   f. Abnormalities due to heat treating (overheating, burning, quench cracking, grain growth, excessive retained austenite, decarburization, precipitation)
   g. Flaws due to case hardening (intergranular carbides, soft core, wrong heat cycles)
   h. Careless assembly (such as mismatch of mating parts, entrained dirt or abrasive, residual stress, gouges or injury to parts)
   i. Parting-line failures in forging due to poor transverse properties
5. *Deterioration in service.* Causes of failures due to deterioration during service conditions include the following (adapted from Dolan[2]):
   a. Overload or unforeseen loading conditions
   b. Wear (erosion, galling, seizing, gouging, cavitation)
   c. Corrosion (including chemical attack, stress corrosion, corrosion fatigue, dezincification, graphitization of cast iron, contamination by atmosphere)
   d. Inadequate or misdirected maintenance or improper repair (such as welding, grinding, punching holes, cold straightening)
   e. Disintegration due to chemical attack or attack by liquid metals or platings at elevated temperatures
   f. Radiation damage (sometimes must decontaminate for examination, which may destroy vital evidence of cause of failure); varies with time, temperature, environment, and dosage
   g. Accidental conditions (such as abnormal operating temperatures, severe vibration, sonic vibrations, impact or unforeseen collisions, ablation, thermal shock)

Most failures involve fracture of the component, and thus most failure analyses involve examination of the mechanical loading situation. In this book, the term *mode* of fracture will be used to reflect the type of loading involved, such as tensile overload, fatigue, or creep, and is covered in Chap. 4. The term fracture *mechanism* will be used to

TABLE 1.3 Frequency of Causes of Failure in Some Engineering Industry Investigations

Cause	%
Corrosion	29
Fatigue	25
Brittle fracture	16
Overload	11
High-temperature corrosion	7
Stress corrosion/corrosion fatigue/hydrogen embrittlement	6
Creep	3
Wear, abrasion, and erosion	3

SOURCE: Adapted from Davies.[1]

TABLE 1.4 Frequency of Causes of Failure of Aircraft Components

Cause	%
Fatigue	61
Overload	18
Stress corrosion	8
Excessive wear	7
Corrosion	3
High-temperature oxidation	2
Stress rupture	1

SOURCE: Adapted from Davies.[1]

define the type of microscopic process whereby the material fractured. This refers to processes such as cleavage or void coalesence, which are covered in detail in Chap. 3. The frequencies of the various types of fracture modes which have been identified are illustrated by the data in Tables 1.3 and 1.4.

The steps involved in conducting a failure analysis and their sequence depend upon the failure. One sequence to do this includes the following eight steps:

1. *Description of the failure situation.* Here the history of the failure should be documented. Any information pertaining to the failure, such as the design of the component (including the material and properties), and how the component was being used, is important to obtain. Especially useful are photographs of the part and of associated components.
2. *Visual examination.* Here the general appearance of the part should be documented. Care should be exercised in handling the part so as not to damage any of the fracture surfaces or other important features.

3. *Mechanical design analysis (stress analysis).* When the part clearly involved mechanical design as a major design component, a stress analysis should be carried out. This will help to establish whether the part was of sufficient size and of proper shape, and what mechanical properties were required. In some cases this analysis may establish the cause of failure. For example, if the load on a part can be determined and estimates of the mechanical properties made, then it may be possible to establish that the part is too small for this load.
4. *Chemical design analysis.* This step refers to an examination of the suitability of the material from the standpoint of corrosion resistance.
5. *Fractography.* Examination of the fracture surface with the unaided eye, with optical microscopes, and with electron microscopes should be carried out in order to establish the mechanism of fracture.
6. *Metallographic examination.* This requires sectioning and metallographic preparation. It may require agreement between all parties involved before sectioning. This step will help to establish such facts as whether the part had the correct heat treatment.
7. *Properties.* The properties pertinent to the design should be determined. This is not always possible because the test to determine a property may destroy the part. In terms of mechanical properties, hardness is especially important. Hardness will frequently correlate with many other mechanical properties (such as yield strength). It is a simple test to perform, and it usually will not damage the part.
8. *Failure simulation.* A very useful approach is to take an identical (supposedly) part and subject it to the exact condition under which it is designed to operate. This may be too expensive to carry out and is not done frequently.

An alternative procedure for metallurgical failure analysis, as adapted from Ryder et al.,[3] follows these steps:

1. Collection of background data and selection of samples
2. Preliminary examination of the failed part (visual examination and record keeping)
3. Nondestructive testing
4. Mechanical testing (including hardness and toughness testing)
5. Selection, identification, preservation, and/or cleaning of all specimens

6. Macroscopic examination and analysis (fracture surfaces, secondary cracks, and other surface phenomena)
7. Microscopic examination and analysis
8. Selection and preparation of metallographic sections
9. Examination and analysis of metallographic sections
10. Determination of failure mechanism
11. Chemical analyses (bulk, local, surface corrosion products, deposits or coatings, and microprobe analysis)
12. Analysis of fracture mechanics
13. Testing under simulated service conditions (special tests)
14. Analysis of all the evidence, formulation of conclusions, and writing the report (including recommendations)

The importance of obtaining background information cannot be overemphasized. Also, prior to the physical destruction of any broken components and the assembly of which they are a part, it is important to document, usually in pictures, their external features. In cases that may involve litigation it is recommended that all parties involved agree on any testing in which physical destruction may occur. Determination of the mechanical properties frequently plays a prominent role in failure analysis, and these properties can often be estimated from hardness measurements. This procedure is essentially nondestructive. Thus consideration of hardness measurements usually occurs as an early step.

A critical step is examination of the fracture surface. This usually is best made by sectioning the broken component for ease of handling. However, it is possible to reproduce the fracture surface topology by preparing a replica and making the observations on the replica. This can be very useful, and perhaps necessary, since it is nondestructive. Another critical step is usually microstructural analysis, which gives information about the processing and properties of the material. However, this nearly always requires sectioning.

A very useful step is failure simulation. Here a part identical to the one that broke is subjected to an operation that simulates what occurred during service. However, such a simulation may be too difficult or expensive to conduct.

Another feature of failure analysis is interaction with people. It is important to be able to obtain information from those involved in the failure, and to be able to interpret their statements correctly.

The type of background information that may be required for a failure analysis is depicted in Fig. 1.1. The results of an analysis usually

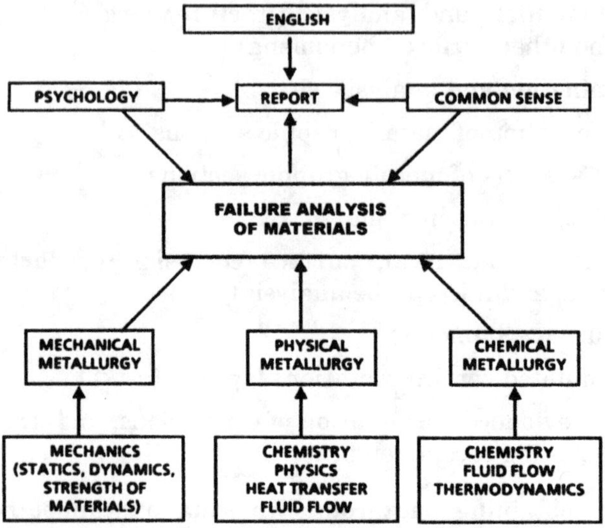

Figure 1.1 Disciplines and subjects involved in failure analysis of materials.

culminate in a report, which contains the findings and recommendations. Thus careful writing is very important.

1.3 Tools of Failure Analysis

Characterization of the microstructure and of the fracture surface topology plays a prominent role in failure analysis. The most common tools for this are the eye and the optical microscope, the scanning electron microscope, and the transmission electron microscope. Their utilization is reviewed in this section. Less commonly used techniques are covered in the volume in the Ninth Edition of the *Metals Handbook* on materials characterization.[4]

1.3.1 Optical microscopy

In an optical microscope (OM) utilized in fractography (and in metallography), light passes through an objective lens which, if at the proper distance from the surface, will form an image of the surface. To further magnify this image, this light passes through another lens (the eyepiece) and is then focused on the retina of the eye. The general scheme for an inverted-stage metallurgical microscope is shown in Fig. 1.2. The light from the source passes through a condenser lens to collimate the beam. There is also an adjustable aperture to control the intensity and to limit off-axis rays. A filter can be placed in the beam

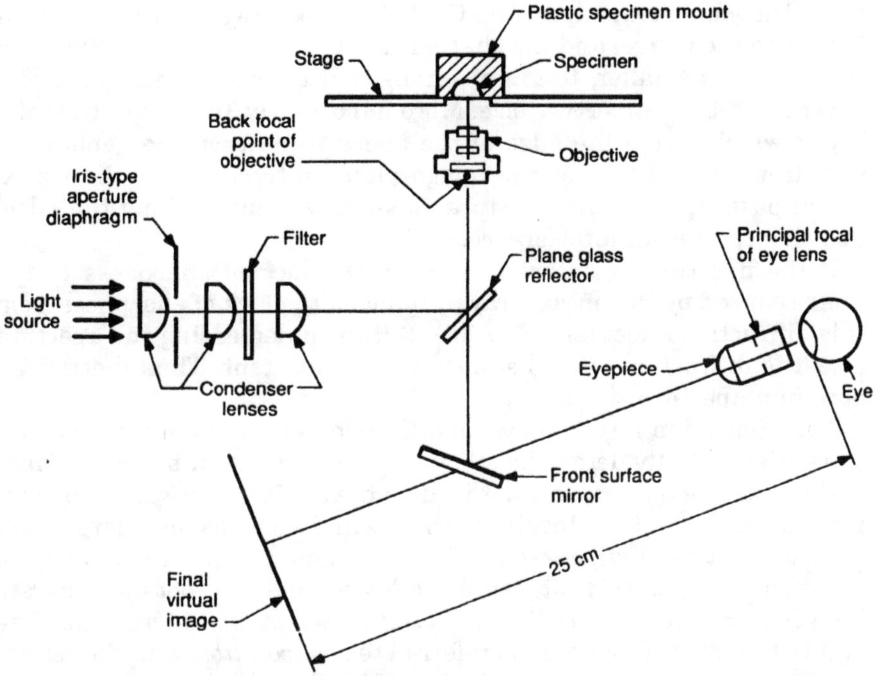

Figure 1.2 Schematic diagram of the optical system typical of that of an inverted-stage metallurgical microscope.

to create a monochromatic beam. The light impinges upon a plane glass reflector, is reflected up through the objective lens to the specimen surface, and is then reflected back, forming an image. These rays pass through the plane glass reflector onto a front surface mirror, to the eyepiece lens, and then to the eye.

The angle of refraction depends on the wavelength of the light. Thus two rays of different wavelengths, coming in the same direction from the same point on an object, will refract differently when striking the surface of the lens. This leads to a focusing error called *chromatic aberration.* To avoid it, a filter can be placed in the beam to generate a monochromatic beam.

Another lens error is *spherical aberration.* If a lens were ideal, two rays emitted at different angles from an object, upon passing through the lens, should image at the same point. However, the correct angle of refraction will be attained only for the correct surface geometry. Most lenses are ground to an approximately spherical surface, and this is not the correct geometry. Thus the two rays starting from the same point will not focus at the same point. This error can be minimized by placing an adjustable aperture near the objective lens to limit off-axis rays. The smaller the aperture, the less off-axis rays there are to contribute to the image, and the sharper the image. However, as the aperture becomes smaller, the light intensity decreases; hence there is a lower limit on the aperture size. More importantly, those rays from the object which strike the edge of the aperture diffract and generate a wavefront. This front, at the image plane, interacts with the waves which pass through the aperture opening without striking the edge, and this creates an interference.

If the aperture is sufficiently wide, the lack of sharpness of the image caused by the off-axis rays will mask the lack of sharpness from this diffraction effect (see Fig. 1.3). But upon decreasing the aperture size, the diffraction effect becomes more important. Thus there is an optimum aperture size to use.

This limitation imposed by the diffraction effect is quite important, since it essentially determines how fine the detail can be on a surface and yet still be discerned. That is, if on the surface there are two point protrusions too close together, they will image as one large spot instead of two distinct spots. The minimum separation distance between two points that can be allowed and yet obtain separate images is referred to as the *resolving power* of the microscope. The ability to resolve fine detail is referred to as *resolution,* and the resolving power is a measure of resolution. Note that the resolving power is a distance, and to see (or resolve) fine detail, the resolving power should be small. The resolving power (RP) depends on some optical parameters, and the mathematical relation can be stated as

$$\mathrm{RP} \cong \frac{0.6\lambda}{2\mathrm{NA}} \tag{1.1}$$

where λ is the wavelength of the light being used. NA is the numerical aperture, which is given by

$$\mathrm{NA} = \eta \sin \mu \tag{1.2}$$

where η is the index of refraction of the medium between the object and the objective lens and μ is the light-gathering angle (Fig. 1.4).

Note that there are three methods of improving resolution (reducing resolving power). One is to use light with a smaller wavelength. This is limited by the need to use visible light and by the fact that most lenses are corrected for chromatic aberration at a single wavelength. Another method is to increase the index of refraction of the medium between the object and the objective lens. This medium is normally air, with an index of essentially 1.0. An oil-immersion lens, with an oil having an index of refraction of up to 1.4, can be used. This will improve

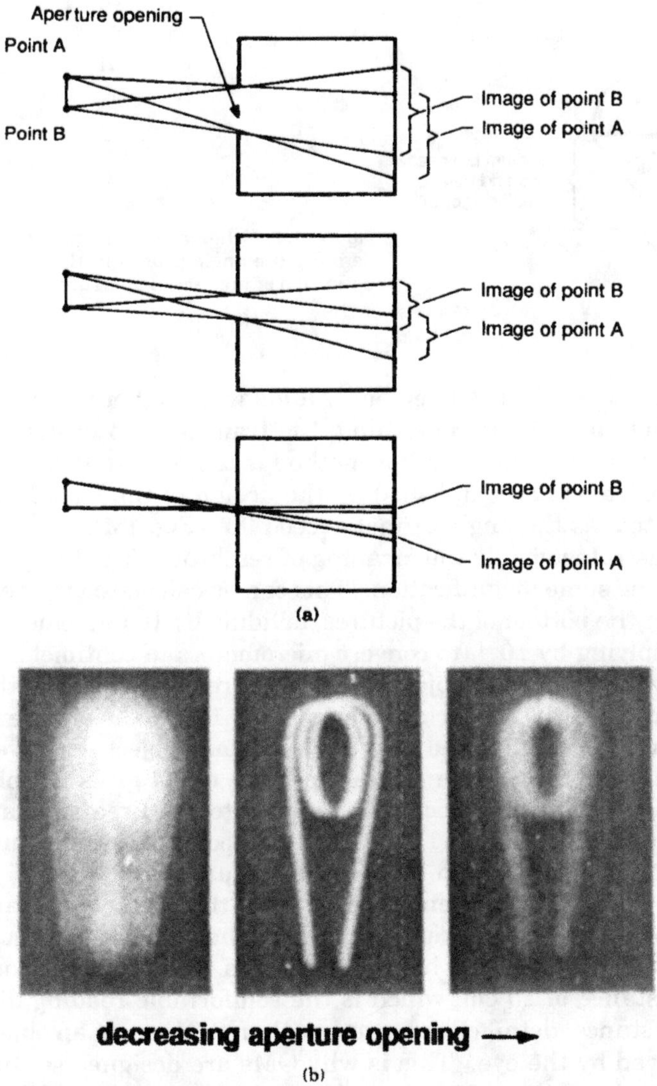

Figure 1.3 (*a*) Schematic illustration of the effect of reduction of aperture size on image sharpness. However, at too small an aperture opening, diffraction from the edges limits the resolution. (*b*) Pinhole photographs of incandescent filament, where image sharpness improves, then deteriorates. (*From Ruechardt.*[5])

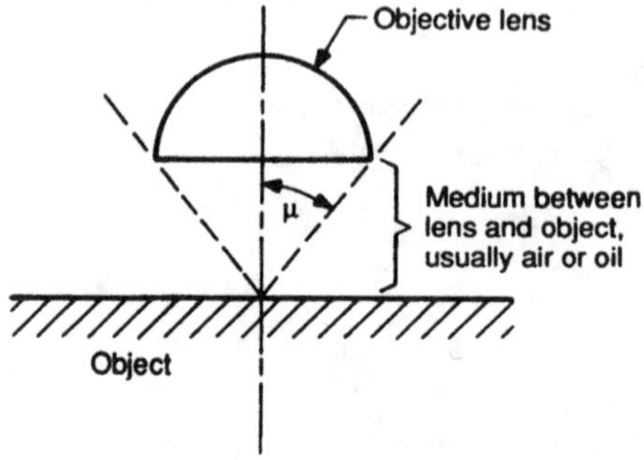

Figure 1.4 Schematic diagram defining the angle μ used in the definition of numerical aperture.

the resolving power by about 40 percent. Fluids with higher indices of refraction are not suitable; for one thing, the transmission of light is reduced to an unusable level. Another method is to increase the light-gathering angle. This is accomplished by the design of the lens. Note that this is limited, as the angle cannot exceed 90° (Fig. 1.4).

Figure 1.5 illustrates vividly the meaning of resolution. The two photographs are at the same magnification. [This can be calculated by measuring the bar at the bottom of the pictures, dividing by 10 micrometers (μm), and multiplying by 10^{-4} (to convert micrometers to centimeters); see App. F.] However, different objective lenses were used to obtain the pictures. The lens with the numerical aperture of 0.1 gave the picture in Fig. 1.5*b*, which appears to be the image of an oblong object. However, using an objective lens with a numerical aperture of 1.4 gives the picture in Fig. 1.5*a*. The same object now appears to be three separate rhomboidal objects. Using Eq. (1.1), the resolving power for the picture on the right is 1.6 μm, for that on the left it is 0.1 μm.

The example just described emphasizes that the most important characteristic of microscopy is resolution. Now, how does magnification enter into the picture? The human eye has a resolution of about 0.01 cm at a distance of 25 cm, which is the comfortable reading distance. At this distance, detail closer together than 0.01 cm on an object cannot be resolved by the eye. (This is why OMs are designed so that the virtual image is at about 25 cm; see Fig. 1.2.) The eye can see finer detail than this only by magnification. For example, consider Table 1.5, which lists the eyepieces and objective lenses for a typical metallurgical microscope. The resolving power was calculated from Eq. (1.1) using the wavelength of green light [5460 angstroms (Å)]. If an object being observed has detail separated by 10,000 Å (1 μm), then using the 5× objective will not resolve this detail, no matter what the magnification. However, using the 10× objective should allow imaging the detail, as the resolving power of this lens is 5460 Å, that is, less than 10,000 Å. If a 20× eyepiece is used with this 10× objective lens, the

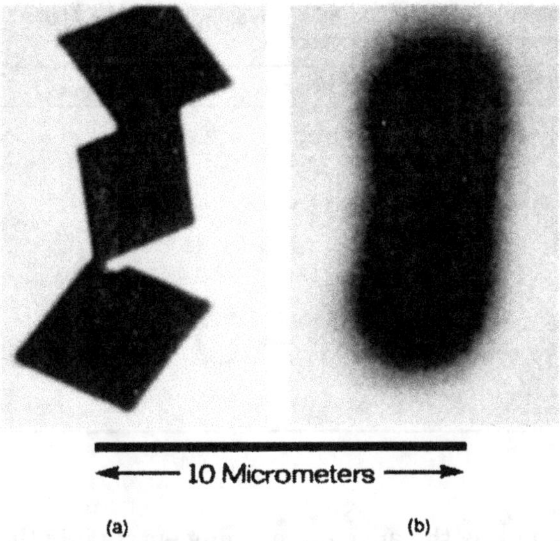

Figure 1.5 Photomicrograph of silver chromate crystals taken with objectives of different numerical apertures. Wavelength was 546 nm (green light). (*a*) NA = 1.4. (*b*) NA = 0.1. (*From Dyer.*[6])

total magnification is 200×, and the detail which is separated by 10,000 Å will appear to the eye to be separated by 10,000 × 200 = 2,000,000 Å, or 0.02 cm. This value is greater than the resolving power of the eye, and the detail will be resolved.

Thus we see that microscope lens systems must be able to magnify the resolvable detail to about 0.01 cm for the eye to see it. In optical microscopy, the wavelength is approximately fixed (it must remain in the visible range), the numerical aperture is limited by the angle μ, which cannot be greater than 90° (see Fig. 1.4), and the index of refraction is limited to about 1.4. Therefore there is an approximate limit on the resolving power, and there is no need to provide a lens system that is capable of greater magnification than that required to allow the eye to see the finest detail resolvable. This magnification is approximately 1500×, and that is why OMs, even with photographic bellows attachments, do not magnify above approximately 3000×.

We have seen that one of the most important characteristics in microscopy is resolution, and we have seen how magnification is related to resolution. In fractography, another important characteristic is *depth of field*. This term refers to the allowable variation of the height of the object surface topology while retaining common focus on the image plane (that is, objects at different levels should be focused on the same plane). This range is related to the aperture opening and to

TABLE 1.5 Magnification and Resolving Power (Using a Wavelength of 5460 Å) of a Set of Lenses Typical of That for a Metallurgical Microscope

Eyepiece	Objective	Total magnification	Resolving power, Å
5×	5× (NA* = 0.10)	25×	16,380
10×	5×	50×	
15×	5×	75×	
20×	5×	100×	
5×	10× (NA = 0.30)	50×	5,460
10×	10×	100×	
15×	10×	150×	
20×	10×	200×	
5×	40× (NA = 0.65)	200×	2,520
10×	40×	400×	
15×	40×	600×	
20×	40×	800×	

*NA = numerical aperture.

the wavelength of light. The way the aperture opening enters into the picture is shown clearly in Fig. 1.6. It appears that the smallest aperture, consistent with sufficient intensity to see the image, would be the optimum aperture. However, again the diffraction effect from the edge of the aperture must be considered, and this is why the wavelength affects the depth of field. The smaller the wavelength, the less this interference effect, and the greater the depth of field.

To place the importance of depth of field in perspective, consider using an OM to observe detail at the limit of resolution of the objective lens, which we will assume to be about 2000 Å (see Table 1.5). In this case the depth of field is about 4000 Å, or 0.4 μm. Thus if this microscope were being used to observe a fracture surface, any irregularities varying in height more than 0.4 μm would not be in simultaneous focus. If the lens has a resolving power of about 16,000 Å [a lens one might use for low-magnification (such as 50×) observation], the depth of field is about 10 μm. For the unaided eye, the depth of field is about 0.01 cm.

1.3.2 Transmission electron microscopy

The operation of the transmission electron microscope (TEM) is very similar in principle to that of the OM. Figure 1.7 compares the general optical path for the two instruments. There are, of course, important practical differences between the two microscopes. By definition, the sample is viewed in transmission, so it must be sufficiently thin for the beam to penetrate completely. This creates a problem of sample preparation (see Sec. 1.4.2). Due to the ease of scattering of electrons by air, the sample must be in a vacuum, although this is not a serious practi-

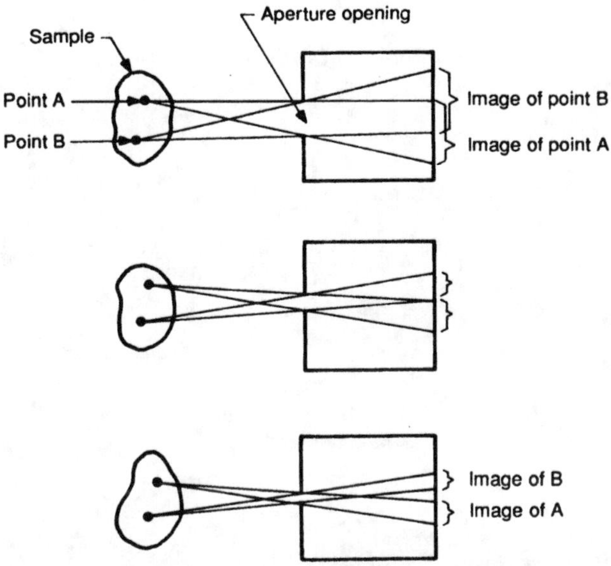

Figure 1.6 Effect of aperture opening on image sharpness from two points located at different levels in the sample.

cal problem. The area that can be viewed is small, so that the examination of a large sample is difficult and time-consuming. Since the eye cannot see electrons, the image is generated by the action of the electron beam upon a phosphorescent screen, on which the observations are made.

The disadvantages of the TEM are far outweighed by its advantages, so that it has developed into a widely used tool in failure analysis. The principal advantages of the TEM compared to the OM are caused by the shorter wavelength of the electron compared to visible light. Equation (1.1), used to calculate resolving power, clearly shows that the resolution can be improved by lowering the wavelength. The wavelength of visible light is around 5000 Å, whereas that typical of electrons used in electron microscopy is about 0.04 Å. This should give an improvement in resolution by a factor of about 10^5. However, to minimize spherical aberration, small apertures are used, so that the angle μ [Eq. (1.2)] is quite small, on the order of 0.1°. Although this will degrade the resolution, the smaller wavelength of the electrons is the dominant factor, and the resolution of the TEM is considerably better than that of the optical microscope. It is this better resolution that makes the TEM so widely used.

Using Eq. (1.1), the resolution of the TEM will be found to be about 10 Å, compared to about 1500 Å for the OM. Now we can see why a TEM has a projector lens system which can magnify so greatly. To be

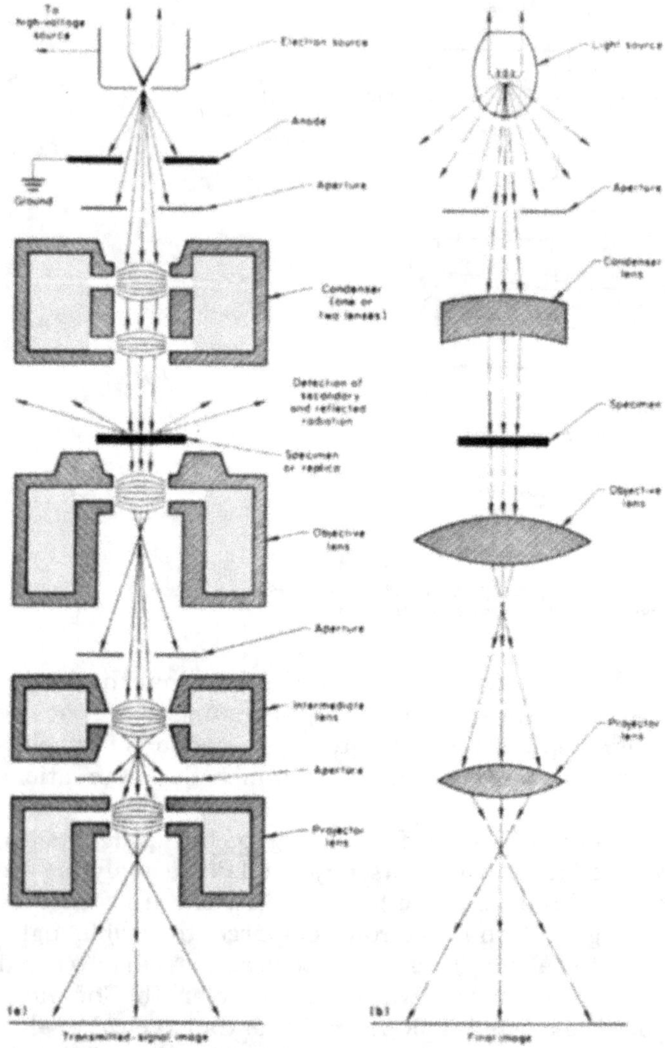

Figure 1.7 Comparison of the optical system for a transmission electron microscope and an optical microscope. (*From* Metals Handbook.[7])

able to see with the eye a detail of 10 Å, the magnification required to raise this to 0.01 cm is 100,000×. Thus the high magnification capability associated with TEM is required to magnify the fine detail so that the eye can observe it.

Also the depth of field is improved greatly by using electrons. For example, if the resolving power is 10 Å, typically the depth of field will be about 5000 Å. This is generally the maximum thickness allowable, which will transmit sufficient electrons to obtain an image. This means that the entire sample will be in focus. If the TEM is operated so that the resolving power is reduced to 1000 Å (such as by using larger apertures), then the depth of field increases to about 500,000 Å (0.005 cm).

Thus we see that, compared to the OM, the TEM offers greatly improved resolution and depth of field. The two major disadvantages are sample preparation problems and the small-size sample that must be used, making the area of observation limited. Actually, the TEM and the OM are complementary tools, both of which are used in fractography.

1.3.3 Scanning electron microscopy

The scanning electron microscope (SEM) operates somewhat differently from the TEM. The electron beam is collimated by the condenser lenses, then focused by the objective lens into a small-diameter beam. This beam strikes the surface of the sample, and the interaction of the beam with the sample generates emitted electrons whose quantity is especially sensitive to the surface topography. A scanning coil causes the beam to raster the surface of the specimen. The quantity of electrons emitted from any point on the surface controls the intensity of a synchronized cathode-ray tube (CRT) display. Hence as the electron beam rasters the surface, an image is generated on the CRT which is essentially a picture of the surface. Figure 1.8 compares the optical systems of the OM, TEM, and SEM.

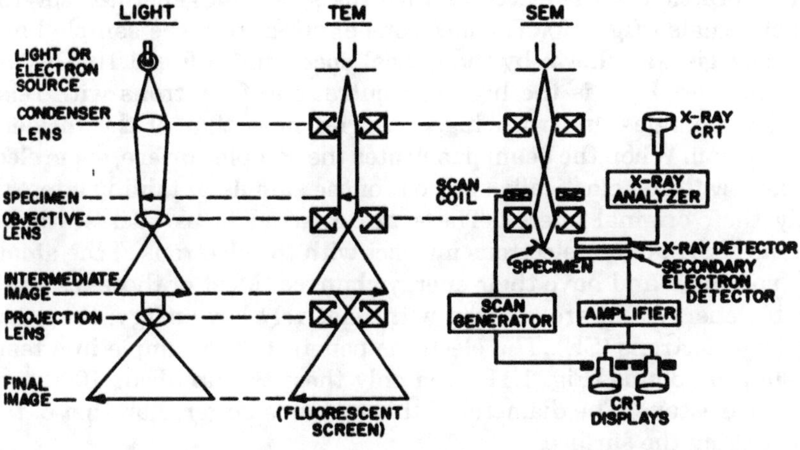

Figure 1.8 Comparison of the optical paths of the optical light microscope, the transmission electron microscope, and the scanning electron microscope. (*From Becker.*[8])

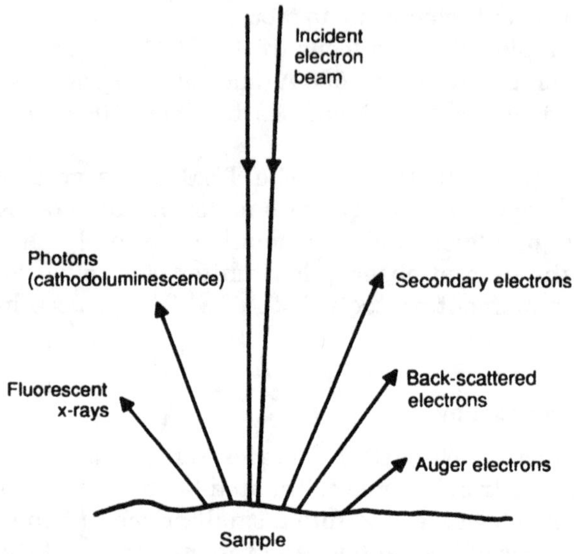

Figure 1.9 Types of signals generated by the interaction of an electron beam with a specimen. Back-scattered and secondary electrons are used to obtain an image of the surface in the scanning electron microscope. The Auger electrons are very weak and are emitted from the atoms in the first two or three layers of the surface; thus they are used for surface analysis. However, an ultrahigh vacuum system is required to prevent significant electron absorption. The fluorescent x-rays are used to obtain chemical analysis (see Sec. 1.3.5).

The interaction of the electrons with the specimen generates several types of signals (Fig. 1.9). The electrons emitted from the sample have varied energies, as shown by the typical spectrum in Fig. 1.10. Of particular interest here is the high concentration of electrons with relatively low energies and with high energies near that of the incident electron beam. When the beam penetrates the sample surface, some electrons are scattered elastically back out of the sample, retaining approximately their original energy. These are referred to as back-scattered electrons (BSE). Other electrons interact with the electrons of the atoms of the specimen and have their energy changed. Most of these lose considerable energy and are emitted with relatively low energy; these are secondary electrons (SE). The electrons penetrate the sample in a teardrop-shaped volume (Fig. 1.11), and only those within about 1000 Å of the surface escape. The diameter of this region is larger than that of the beam striking the surface.

To collect all of the electrons, a rather simple device can be used (Fig. 1.12). The electrons emitted from the surface are drawn toward

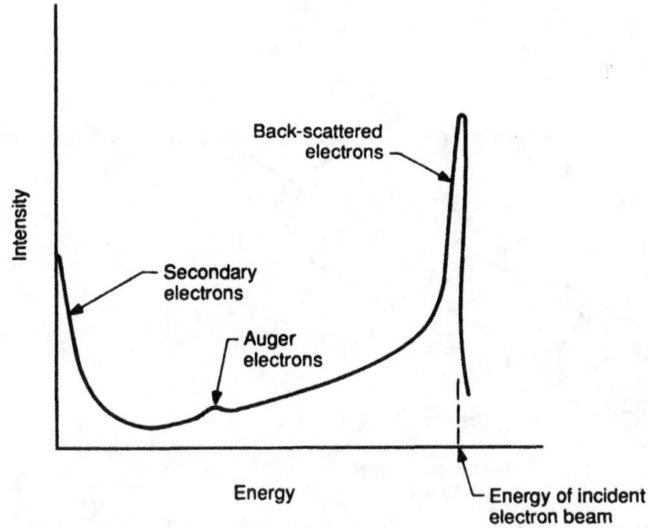

Figure 1.10 Typical electron spectrum from a sample surface in a scanning electron microscope.

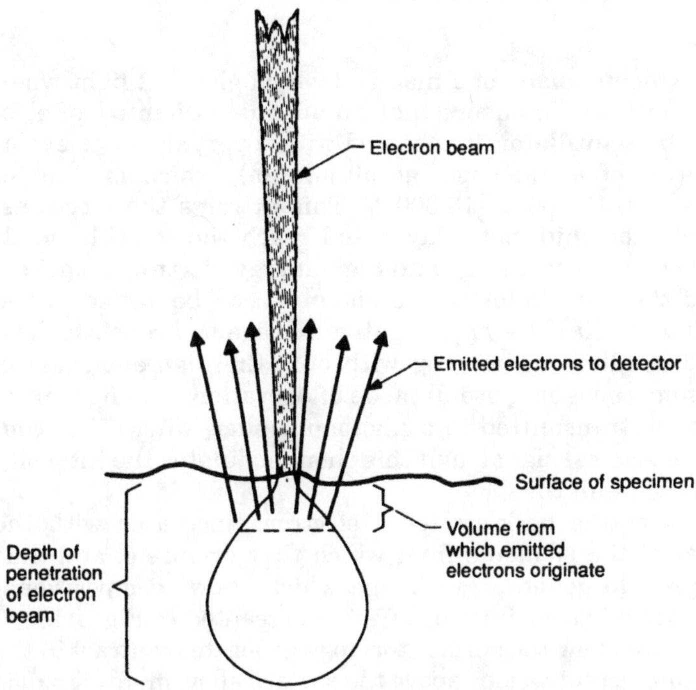

Figure 1.11 Schematic diagram to illustrate that the region from which the electrons are emitted is larger than that of the beam.

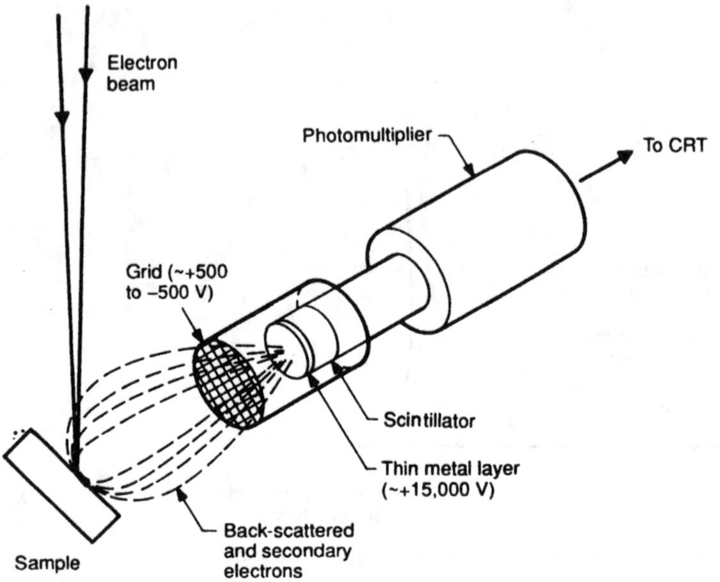

Figure 1.12 Schematic representation of the operation of the electron detector in a scanning electron microscope.

a scintillator, which is made of a material which gives off light when struck by the electron. To ensure that an adequate quantity of electrons reaches the scintillator, it is coated with a very thin (for example, 200 Å) layer of metal (such as aluminum), which is charged positively to about 10,000 to 15,000 V. This attracts the electrons, which penetrate the thin metal layer and reach the scintillator. To discriminate between low-energy and high-energy electrons, a grid is placed around the scintillator whose potential can be varied in the range of −500 to +500 V. Using a negative voltage will repel the low-energy electrons, allowing imaging with only the high-energy electrons, which sometimes is a useful mode of operation. The light from the scintillator is transmitted to a photomultiplier, where it is converted into an electrical signal, and this then modulates the intensity of the electron beam in the CRT.

The intensity of the back-scattered electrons increases with the atomic number of the elements from which they originate, and thus they can be used to generate an image which shows compositional variation. An example showing this effect is presented in Fig. 1.13.

Figure 1.14 shows how the surface topology generates contrast in the image. For example, protrusions above the surface allow more electrons to be emitted as a greater volume of material excited by the electron beam is within the escape distance of the surface. Note that the reso-

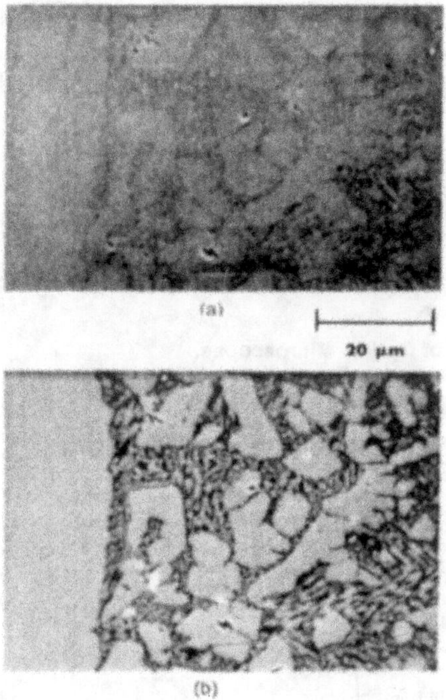

Figure 1.13 Micrographs of an as-polished copper-tin diffusion couple in a scanning electron microscope. (*a*) Using a secondary electron detector. (*b*) Using a back-scattered electron detector. Note the enhanced contrast using the back-scattered electron detector due to atomic number differences. (*From Verhoeven.*[9])

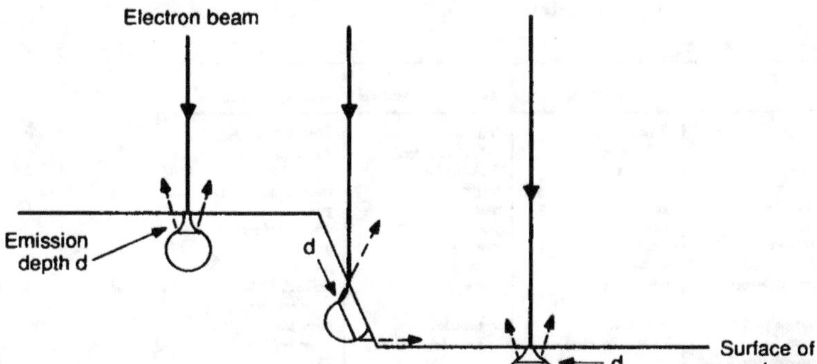

Figure 1.14 Schematic illustration of how the surface topology influences the volume of material within the emission distance d from the surface and enhances the intensity of the emitted electrons.

lution is basically fixed by the beam diameter and not the wavelength. Thus a SEM may have a resolution of 100 Å, but not the 10 Å obtained by the TEM. However, a high depth of field is still retained. Even if it is not desired to resolve very fine detail, good depth of field makes the SEM attractive for low-magnification observations.

Unlike the TEM, in the SEM a relatively large sample can be used and rather large areas observed. Even if the sample is too large to be placed in the SEM, a replica of the surface can be prepared and placed in the SEM.

1.3.4 Comparison of OM, TEM, and SEM

The information in Table 1.6 compares the features of these microscope types. In the use of these instruments for fractography, it should

TABLE 1.6 Comparison of Characteristics of Optical Microscopes, Transmission Electron Microscopes, and Scanning Electron Microscopes

Based on a 10-cm² screen viewed at 25 cm

Magnification		Resolution	Field	Depth of focus OM	Depth of focus SEM
OM	1×	0·2 mm	100 mm		
OM	10×	0·02 mm	10 mm	0·1 mm	10 mm
OM, SEM	100×	2 μm	1 mm	1 μm	1 mm
SEM, EM	1000×	0·2 μm	0·1 mm		10 μm
SEM, EM	10,000×	20 nm (200 Å)	10 μm		1 μm
EM	100,000×	2 nm (20 Å)	1 μm		
EM	10⁶×	0·2 nm (2 Å)	0·1 μm		

(a)

	Optical	Scanning electron	Direct electron
Resolution—easy	*5 μm*	0·2 μm	**100 Å (10 nm)**
—skilled	*0·2 μm*	100 Å (10 nm)	**10 Å (1 nm)**
—special	*0·1 μm*	5 Å (0·5 nm)	**2 Å (0·2 nm)**
Depth of focus	*poor*	**high**	moderate
Mode—transmission	yes	yes	yes
—reflection	yes	yes	*not satisfactory*
—diffraction	yes	yes	yes
—other	some	**many**	*no*
Specimen—preparation	usually easy	**easy**	*skilled, liable to artefacts*
—range and type	versatile real or replica	versatile real or replica	*only thin, or replica*
—maximum thickness for transmission	**thick**	medium	*very thin*
—environment	**versatile**	usually vacuum but can be modified	*vacuum*
—available space	small	**large**	small
Field of view	large enough	large enough	*limited*
Signal	only as image	**available for processing**	only as image
Cost	**low**	high	high

Advantages over others are indicated in bold type; disadvantages in italics.

(b)

SOURCE: From Hearle et al.[10]

be kept in mind that they complement each other, each having advantages over the others under certain circumstances.

1.3.5 Related tools and techniques

There are a number of tools and techniques related to microscopy which are useful in fractography. Here three are described briefly: x-ray fluorescence, x-ray diffraction, and stereomicroscopy.

X-ray fluorescence chemical analysis. This method of elemental chemical analysis is used widely. The description here will be given in terms of its use on the SEM, but it can be used in several other ways. When an atom is sufficiently excited, an inner-shell electron may be removed. When an electron from a neighboring orbital falls into this vacancy, the energy decrease is manifested as a fluorescent x-ray (see Fig. 1.15). The energy of the radiation from transitions in the outer (valence) electron shells depends on the type of atom bonding, but that from the transitions in the inner shells does not. Hence for a given element, the energy of the

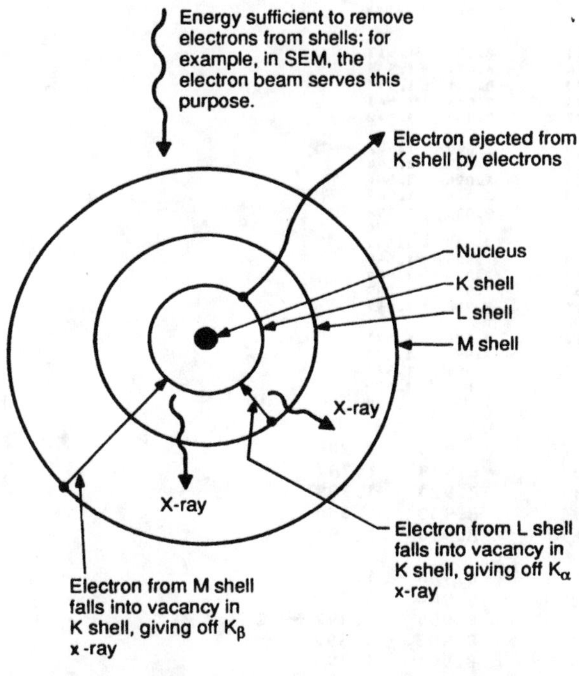

Figure 1.15 Schematic illustration of the origin of fluorescent x-rays and of the nomenclature used to describe the energy transitions.

radiation is characteristic of the element from which it was emitted. Thus the detection of these x-rays is a method for identifying the type of atoms present. For example, the x-ray spectrum from pure iron will show strong lines or peaks at specific energies. If the spectrum of iron carbide or iron carbonate is obtained, peaks or lines in the spectrum reveal what elements are present, but not how the elements are combined.

The designation of the x-ray emissions is determined by the shell to which the electron falls and from which it comes. The scheme is shown in Fig. 1.15. For example, K_α is an x-ray photon caused by an electron from the L shell falling into a K-shell vacancy. K_β is an x-ray photon caused by an electron from the M-shell falling into a K-shell vacancy.

The energies of these electron transitions have been measured carefully, and the values can be found tabulated in several publications. Figure 1.16 shows one such listing. The energy is usually given in keV

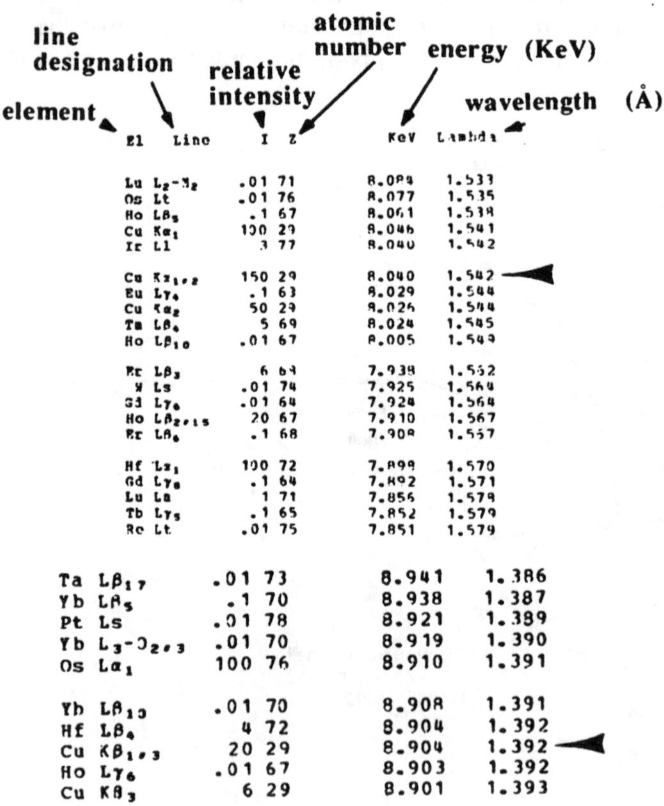

El	Line	I	Z	KeV	Lambda
Lu	L_2-M_2	.01	71	8.084	1.533
Os	Lt	.01	76	8.077	1.535
Ho	$L\beta_5$	.1	67	8.061	1.538
Cu	$K\alpha_1$	100	29	8.046	1.541
Ir	Ll	3	77	8.040	1.542
Cu	$K\alpha_{1,2}$	150	29	8.040	1.542
Eu	$L\gamma_4$	.1	63	8.029	1.544
Cu	$K\alpha_2$	50	29	8.026	1.544
Tm	$L\beta_4$	5	69	8.024	1.545
Ho	$L\beta_{10}$	.01	67	8.005	1.549
Er	$L\beta_3$	6	68	7.938	1.562
W	Ls	.01	74	7.925	1.564
Gd	$L\gamma_6$	.01	64	7.924	1.564
Ho	$L\beta_{2,15}$	20	67	7.910	1.567
Er	$L\beta_6$	.1	68	7.908	1.567
Hf	$L\alpha_1$	100	72	7.899	1.570
Gd	$L\gamma_8$	.1	64	7.892	1.571
Lu	La	1	71	7.856	1.579
Tb	$L\gamma_5$	.1	65	7.852	1.579
Re	Lt	.01	75	7.851	1.579
Ta	$L\beta_{17}$	.01	73	8.941	1.386
Yb	$L\beta_5$	.1	70	8.938	1.387
Pt	Ls	.01	78	8.921	1.389
Yb	L_3-$O_{2,3}$	.01	70	8.919	1.390
Os	$L\alpha_1$	100	76	8.910	1.391
Yb	$L\beta_{10}$	.01	70	8.908	1.391
Hf	$L\beta_4$	4	72	8.904	1.392
Cu	$K\beta_{1,3}$	20	29	8.904	1.392
Ho	$L\gamma_6$	.01	67	8.903	1.392
Cu	$K\beta_3$	6	29	8.901	1.393

Figure 1.16 Section of a table listing the energies of x-ray emissions of the elements. CuK_α and CuK_β are noted. (*Adapted from Johnson and White.*[11])

(1000 electronvolts). Also listed are the corresponding wavelengths in angstroms. The relation connecting these two quantities is

$$E = h\nu = \frac{hc}{\lambda} \tag{1.3}$$

where E is the energy, λ is the wavelength, ν is the frequency, h is Planck's constant, and c is the speed of light. Thus as the energy increases, the wavelength decreases (see Fig. 1.16). The other listing of importance in Fig. 1.16 is I, which is the relative intensity to be expected for a transition. These values are based on experimental data and reflect the fact that all transitions do not occur with equal probability. Note that the $K_{\alpha 1-2}$ transition for copper has a relative intensity of 150 (150 percent) and that of $K_{\beta 1-3}$ a value of 20. Thus in the spectrum from copper, the peak at 8.040 keV should be only about $^1/_7$ as strong as that at 8.904 keV. When the I value decreases to below 5, it generally will be difficult to detect the x-ray.

There are two methods used in scanning electron microscopy to measure or record the x-ray spectrum: One method uses a diffracting crystal, the other a solid-state detector. The method using the diffracting crystal is referred to as wavelength dispersive, or sometimes wavelength-dispersive x-ray analysis (WDXA) or wavelength-dispersive spectrometry (WDS). In this technique, the fluorescent x-rays are allowed to strike a single crystal. The x-rays will either pass through the crystal, will be absorbed, will be generally scattered, or will be diffracted. Those which satisfy Bragg's law

$$\lambda = 2d \sin \theta \tag{1.4}$$

will diffract, as shown in Fig. 1.17. Here d is the spacing between the planes of the crystal which are diffracting the x-rays to the detector, and its value is known for a given crystal. Thus the intensity for this specific wavelength is measured. The crystal can then be set to a different value of θ, always maintaining the θ–2θ relation shown in Fig. 1.18. Then the intensity at this wavelength is obtained. The process can be automated. It is customary to obtain a plot of intensity versus 2θ, not intensity versus wavelength, but the wavelength can be calculated from the values of 2θ by using Eq. (1.4). Figure 1.19 shows a typical spectrum. Note that for this stainless-steel sample it is quite obvious which peak is caused by which element.

There are two important disadvantages of this method of obtaining the x-ray spectrum. One is that the instrumentation may be difficult to place in the proper location for use on a SEM. Unlike the electrons, which can be focused or attracted by the high positive charge on the

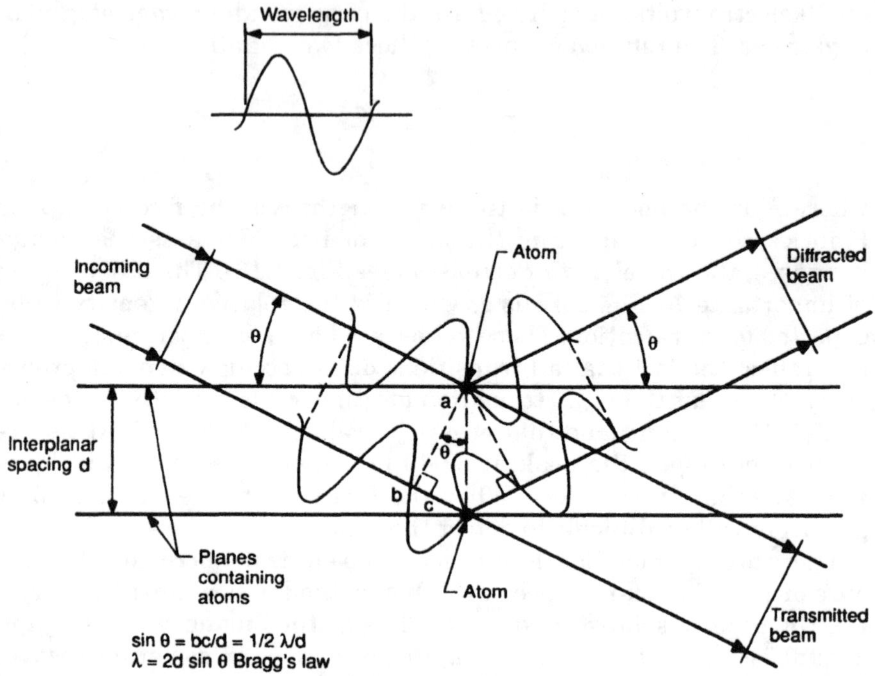

Figure 1.17 Simplified derivation of Bragg's law of x-ray or electron diffraction. If angle θ is such that the distance *bc* is ½ the wavelength λ, then Bragg's law is satisfied, and constructive interference occurs along the direction of angle θ, causing diffraction.

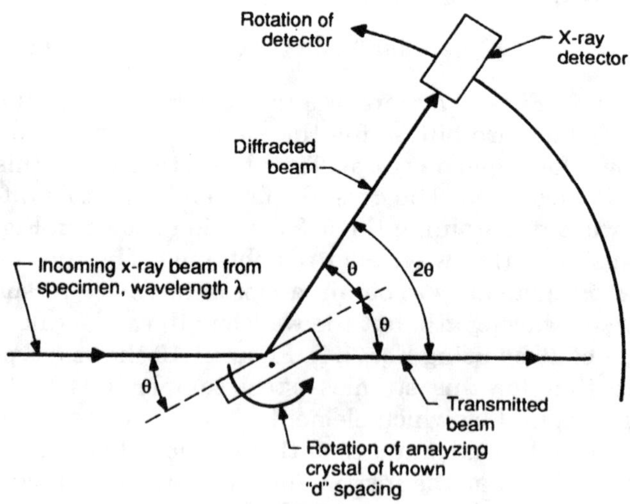

Figure 1.18 Schematic illustration of a diffractometer used to analyze the spectrum from an x-ray beam.

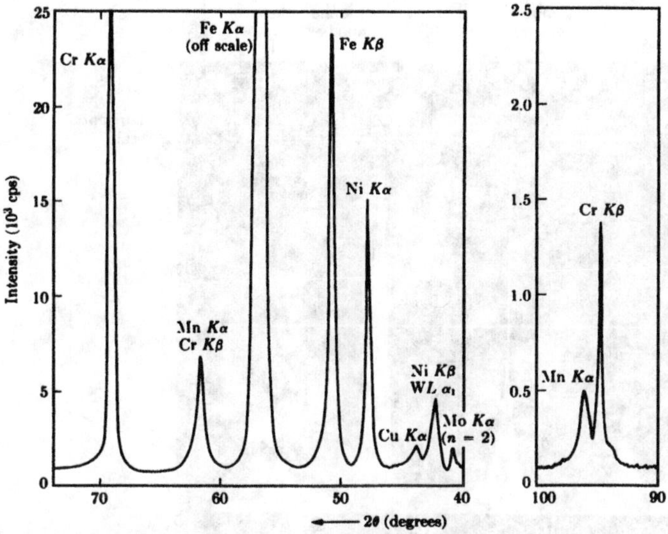

Figure 1.19 X-ray spectrum obtained by using an x-ray diffractometer (wavelength-dispersive spectrometry). The sample was stainless steel containing 19.4% Cr, 9.5% Ni, 1.5% Mo, 1.4% W, and 1.0% Mn, balance mainly Fe. A flat LiF crystal was used as the analyzer. A platinum target x-ray tube was used, operated at 50 kV and 30 mA. (*Adapted from Cullity*[12]; *courtesy of Diano Corporation.*)

electron detector (see Fig. 1.12), x-rays cannot be focused so easily. Thus only those x-rays which have a direct path to the detecting crystal will be analyzed. The other limitation is that it takes from 30 to 60 min to obtain a spectrum. On the other hand, this method can detect elements as low in atomic number as carbon.

The method using a solid-state detector is referred to as the nondispersive method, as the x-rays are not dispersed by an analyzing crystal. It is commonly referred to as energy-dispersive x-ray analysis (EDXA) or energy-dispersive spectrometry (EDS). This method uses a Si (Li) counter and a FET preamplifier, both of which must be maintained at liquid nitrogen temperature. The signal from the detector is separated into a spectrum by a multichannel analyzer, so that a spectrum of intensity versus energy is displayed on a CRT. Figure 1.20 shows a typical spectrum.

This method of analysis has two important advantages in a SEM, namely, the detector can be placed close to the sample, inside the vacuum chamber, and the time to obtain a spectrum is short. Obtaining a spectrum like the one shown in Fig. 1.20 by the wavelength-dispersive spectrometry method would take perhaps 60 min. The spectrum using energy-dispersive spectrometry would require about 1 min. An important limitation, however, is that the energy-dispersive detector

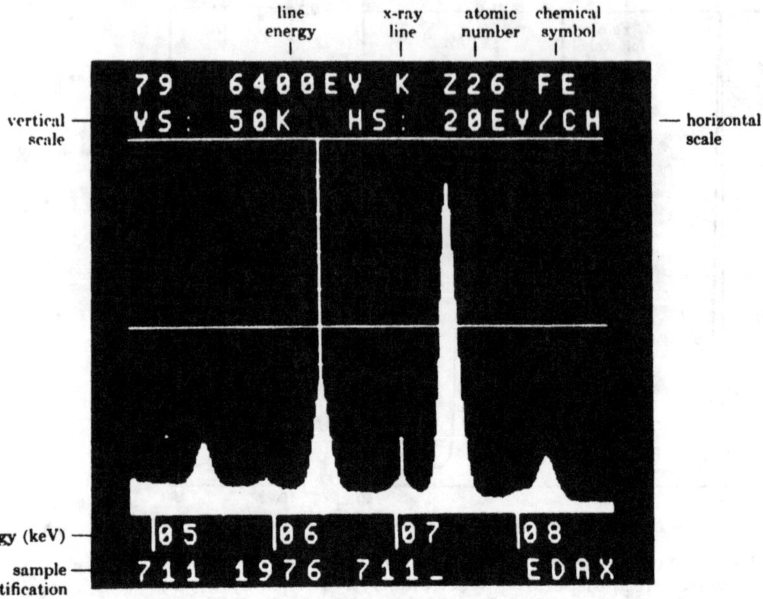

Figure 1.20 X-ray spectrum obtained by an energy-dispersive spectrometry system. (*Adapted from Cullity*[12]; *courtesy of Philips Electronic Instruments, Inc.*)

is not as sensitive to light elements (such as carbon) as the wavelength-dispersive system.

In terms of the use of fluorescent x-ray analysis in the SEM, it is emphasized that usually the analysis is qualitative or semiquantitative. It is particularly useful in determining what elements are present and their relative amounts, but not in determining the exact composition.

The strength of x-ray fluorescence in fractrography lies in the microchemical analysis which can be performed with the SEM. For example, in operating the SEM, it is possible to stop the raster of the beam and move the beam spot to any location in the field of view. Thus if the chemical analysis of a particle on the surface is desired, it can be obtained by locating the beam on the particle and compiling the x-ray spectrum. This is illustrated in Fig. 1.21, which shows a fracture surface with a small particle on it. The diameter of the particle is about 4000 Å, and the electron beam was perhaps 200 Å in diameter. Recalling that this beam will blossom into a tear-dropped shape, the actual diameter of the excited region is approximately 1000 Å. The depth of penetration is approximately 1000 Å. This gives a volume of the material from which the x-rays are emitted of approximately 10^{-15} cm^3, or a mass of material of about 10^{-14} g. Hence the use of the term microchemical analysis. The spectrum from this particle is shown in

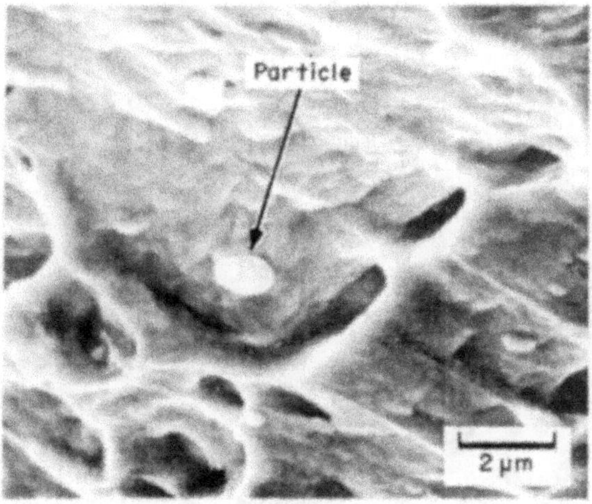

(a)

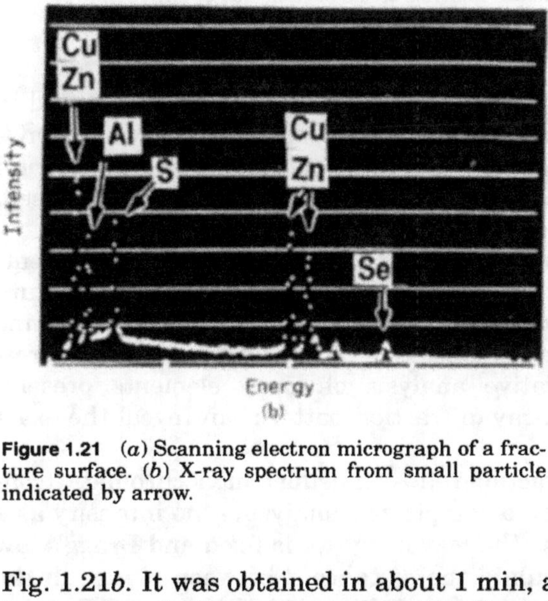

(b)

Figure 1.21 (*a*) Scanning electron micrograph of a fracture surface. (*b*) X-ray spectrum from small particle indicated by arrow.

Fig. 1.21*b*. It was obtained in about 1 min, and it took about a minute to photograph the spectrum from the face of the CRT.

Some precautions must be taken in obtaining x-ray spectra in the SEM. It must be remembered that the electron beam is considerably smaller than the diameter of the region from which x-rays are emitted. Thus if a particle of approximately 500 Å is to be analyzed and the beam diameter is 200 Å, the x-ray spectrum may contain emissions from material outside of the particle. It is equally important to realize that secondary fluorescence can occur. This is a constant hazard in

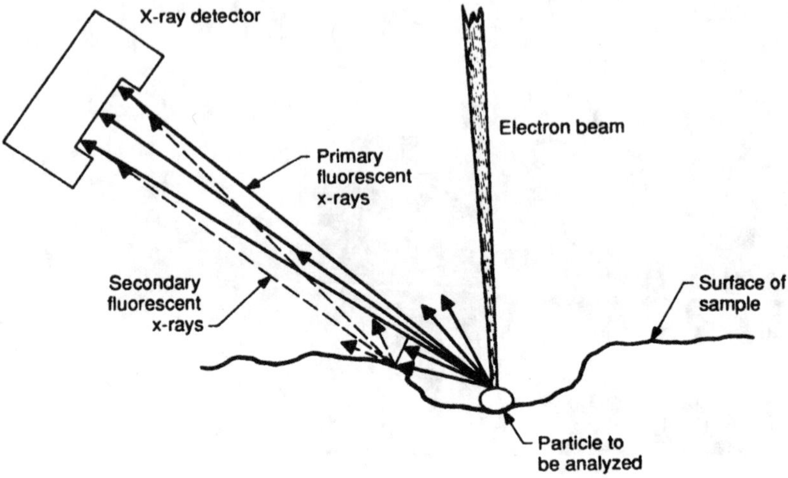

Figure 1.22 Schematic illustration of the origin of secondary fluorescence, causing a false indication of the elemental analysis of the particle due to x-rays from the edge of the matrix.

analyzing rough surfaces typical of fracture surfaces, as illustrated in Fig. 1.22. The primary x-ray emissions may be absorbed by surrounding higher regions, causing them to emit x-rays and giving a spectrum not actually characteristic of the region on which the beam is focused.

X-ray and electron diffraction. X-ray and electron diffraction is frequently used in conjunction with x-ray fluorescence to determine the nature of crystalline materials. For example, it may be important to determine the chemical and structural composition of wear debris. X-ray fluorescence can give a qualitative analysis of many elements present. However, analysis of an x-ray diffraction pattern can reveal the exact nature of the debris.

X-ray and electron diffraction relies on using a monochromatic (constant-wavelength) beam on a sample and analyzing the intensity as a function of 2θ. In this case the wavelength λ is fixed and Bragg's law [Eq. (1.4)] is used to calculate the distance d between planes in the crystals. These values can be compared to the ASTM x-ray diffraction data file to see if a match reveals what material is present. If this approach is not fruitful, then, in principle, the data can be analyzed to determine the crystal structure, although this may be quite difficult.

Electron diffraction in the TEM can be useful to determine the crystal structure of extremely fine individual particles as the diffraction pattern can be obtained by the equivalence of limiting the beam to the illumination of very small regions.

Stereomicroscopy. Stereomicroscopy refers to the observation of an object such that the entire field of view remains in focus and at the same time a three-dimensional effect is obtained. This is a particularly valuable tool in fractography. The amount of additional information obtained by viewing fracture surfaces in stereo cannot be overestimated, neither can the importance be proven in words. One has to view a surface under this condition to properly appreciate its significance.

The viewing itself is rather easy, but it may take practice. Low-magnification stereooptical microscopes are available. Where the resolution of finer detail is sought, the OM is not suitable because the high magnification required will have associated with it a poor depth of field. However, the SEM is especially appropriate for this. It is simple to obtain two pictures which, when viewed with a stereo viewer, will give the required three-dimensional effect. A photograph is taken of the area of interest, then the sample is tilted 6 to 12°, and the same area is photographed. These photographs are then properly positioned under a stereo viewer until the three-dimensional effect is observed. The same procedure can be used for samples being observed in the TEM. Volume 9 of the *Metals Handbook*[7] has a stereo viewer in the back, and several stereo fractographs are in the book.

A proper method of obtaining stereopairs in the SEM and of determining the variation in the height of the surface topology is outlined in App. 1A.

1.4 Sample Preparation

1.4.1 Cleaning of surfaces

A common problem is that the fracture surface is dirty and contaminated. There are several methods of cleaning fracture surfaces. In general, the surface is washed carefully, rinsed in alcohol or acetone, then dried. The washing is best done in an ultrasonic cleaner, which will remove most loose debris. Further removal of debris and surface reaction products such as rust can be accomplished by replicating the surface several times. In this procedure the surface is wet with acetone, and one side of a piece of cellulose acetate tape (about 0.01 cm thick) is moistened with acetone. Then, with the fracture surface still wet, the wet side of the tape is pressed carefully but firmly onto the fracture surface. It is rubbed to force the sticky tape to be embedded into the surface. After several minutes of drying, the tape is peeled off, giving a replica of the surface and removing debris. (Such a replica can be observed in the SEM and the surface debris can be analyzed.) This process can be repeated several times to assist in cleaning the surface.

A major problem in surface preparation is the removal of rust on ferrous samples. How to treat the surface to remove the rust without destroying the underlying surface which may reflect the true fracture surface topology before rusting occurred is a touchy decision. Most methods of cleaning involve the use of chemicals which are designed to dissolve the oxides without attacking the underlying metal. Appendixes 1B through 1G give information on methods of cleaning surfaces. Additional sources of information can be traced from the references in the sources cited there.

1.4.2 Preparation of replicas for the TEM

In using the TEM in fractography, the actual fracture surface usually cannot be observed, of course, as it is extremely difficult to prepare the required small and thin specimen. Thus the surface topology is examined indirectly by preparing a suitable replica of the surface. One method is the one just described for cleaning the surface, namely, to use cellulose acetate tape to make the replica. However, this tape is too thick for use directly in the TEM, and in addition such polymeric materials are subject to heating by the electron beam and, hence, damage. Instead it is common to take the tape replica and put a thin (200 Å) layer of carbon on it in a vacuum coater. To enhance contrast of this carbon replica, a thin layer (200 Å) of a heavy metal, such as chromium or platinum, is deposited, or shadowed, onto the surface. Then the acetate tape is dissolved in a suitable solvent (such as acetone), freeing the thin and fragile replica into the solvent. This replica is removed onto a screen or grid, which is of suitable size (about 0.3 cm in diameter) to fit into the TEM. The replica is viewed through the openings in the grid. The process of making such a replica is depicted in Fig. 1.23.

It is possible to prepare a replica of the surface by depositing the carbon directly and not using the tape. The carbon then is removed by adhesive tape, which is dissolved to free the replica. This gives a replica of greatest fidelity, having a resolution of about 20 Å. The two-stage tape-carbon replica has a resolution of about 100 Å. The artifacts which may be present in replicas have been quite well documented (see Sec. 3.18).

If it is desired to analyze particles on the surface, after the carbon layer is deposited, the underlying metal (original specimen) can be dissolved chemically, freeing the replica with the particles embedded in it. This is an extraction replica. It allows analysis of the particles in the TEM using electron diffraction and x-ray fluorescence. However, this process does destroy the fracture surface. The preparation of an extraction replica is illustrated in Fig. 1.24.

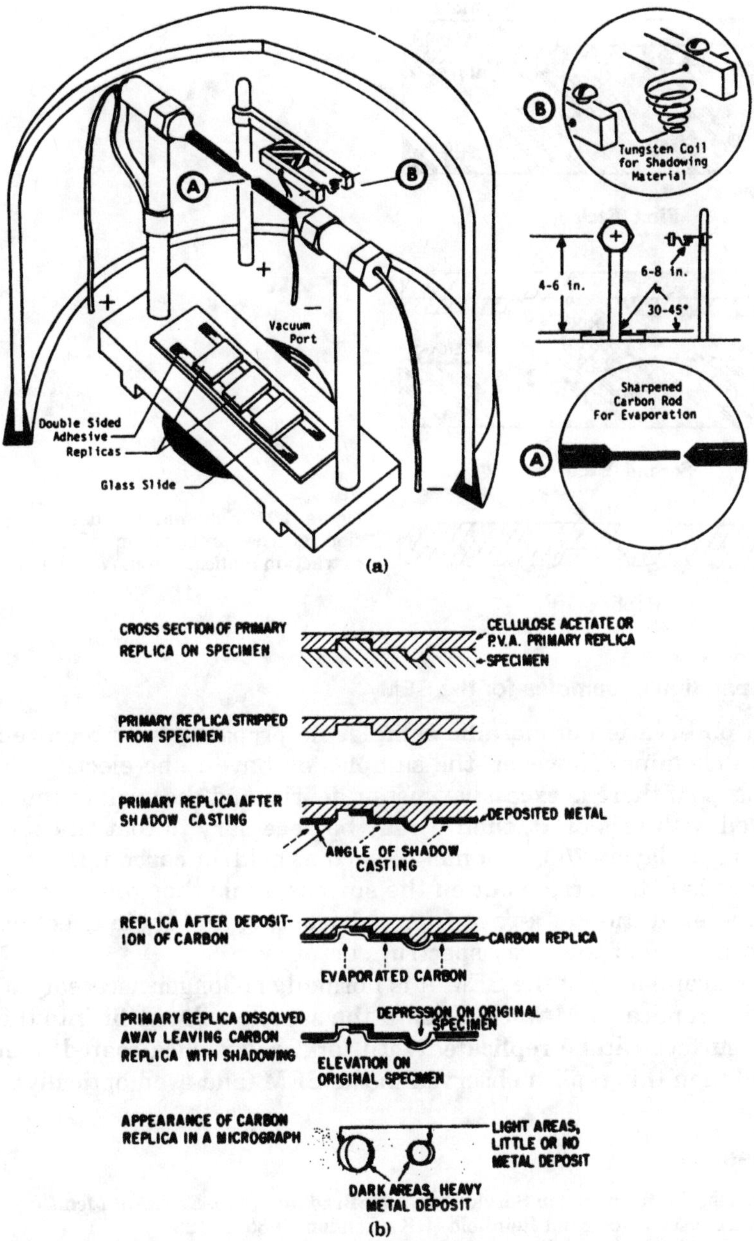

Figure 1.23 Schematic representation of a method of preparing replicas for transmission electron microscopy. (*a*) Vacuum deposition technique. (*From Nail,*[13] *copyright ASTM; reprinted with permission.*) (*b*) Steps in preparing an indirect carbon replica using a cellulose acetate or polyvinyl alcohol (PVA) primary replica. (*Prepared by E. F. Koch; from Phillips*[14]; *courtesy of General Electric Company.*)

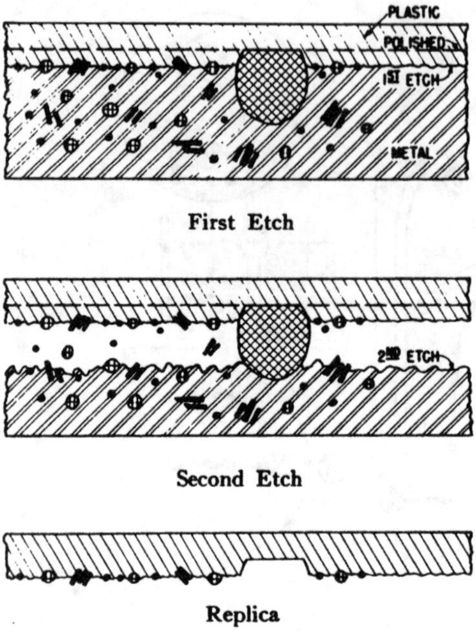

Figure 1.24 Schematic illustration of the preparation of an extraction replica. (*From Nail.*[13])

1.4.3 Preparation of samples for the SEM

For direct observation of metallic samples no preparation is required other than cleaning. However, the samples do have to be electrically conducting, so if there is excessive rust or debris, and the surface must be observed with this on it, then it may be necessary to coat the surface with a thin layer (200 Å) of metal (such as gold) or carbon. If x-ray fluorescence is to be carried out on the surface, some thought must be given to the excitation of x-rays from this coating and their effect on the interpretation of the x-ray spectrum obtained.

With the availability of the SEM it is normally no longer necessary to do extensive replica TEM work. Even if the sample will not fit into the SEM, the surface can be replicated with tape, the replica coated with metal, and then this replica observed in the SEM (and even optically).

References

1. G. J. Davies, "Performance in Service," in E. J. Bradbury (ed.), *Essential Metallurgy for Engineers,* Van Nostrand Reinhold (UK), London, 1985, p. 126.
2. T. J. Dolan, "Analyzing Failures of Metal Components," *Metals Eng. Quart.,* vol. 12(4), p. 32, 1972.
3. D. A. Ryder, T. J. Davies, I. Brough, and F. R. Hutchings, "General Practice in Failure Analysis," in *Metals Handbook,* 8th ed., vol. 10: *Failure Analysis and Prevention,* American Society for Metals, Metals Park, Ohio, 1975, p. 10.
4. *Metals Handbook,* 9th ed., vol. 10: *Materials Characterization,* American Society for Metals, Metals Park, Ohio, 1986.

5. E. Ruechardt, *Light,* University of Michigan Press, Ann Arbor, 1958.
6. D. L. Dyer, "Optical Limits in TV Microscopy," *Research / Develop.,* Sept. 1973.
7. *Metals Handbook,* 8th ed., vol. 9: *Fractography and Atlas of Fractographs,* American Society for Metals, Metals Park, Ohio, 1974.
8. H. C. Becker, "Scanning Electron Microscopy," *Lubrication,* vol. 61, p. 37, 1975.
9. J. D. Verhoeven, in *Metals Handbook,* 9th ed., vol. 10: *Materials Characterization,* American Society for Metals, Metals Park, Ohio, 1986.
10. J. W. S. Hearle, J. T. Sparrow, and P. M. Cross, *The Use of the Scanning Electron Microscope,* Pergamon, New York, 1972.
11. G. G. Johnson and E. W. White, "X-Ray Emission Wavelength and KeV Tables for Nondiffractive Analysis," ASTM Data Series DS 46, American Society for Testing and Materials, Philadelphia, Pa., 1970.
12. B. D. Cullity, *Elements of X-Ray Diffraction,* 2d ed., Addison Wesley, Reading, Mass., 1978.
13. D. A. Nail, "Procedures for Standard Replication Techniques for Electron Microscopy," in *Manual on Electron Metallography Techniques,* STP 547, American Society for Testing and Materials, Philadelphia, Pa., 1973, Chap. 1.
14. V. A. Phillips, *Modern Metallographic Techniques and Their Applications,* Wiley, New York, 1971.

Bibliography

The list of written material dealing, directly and indirectly, with failures of materials is extensive, and no attempt is made in this book to compile a bibliography. Instead, the references cited in this book can be used to trace additional information. Of special mention, though, are the different volumes of the *Metals Handbook* listed below and the article by Vander Voort that lists 531 citations.

Some of the books listed below deal specifically with failures in metallic components, and some with the more general aspects of component failures. Of special interest to illustrate the importance of failure analyses is the book by Petroski.

Barer, R. D., and B. F. Peters: *Why Metals Fail,* Gordon and Breach, New York, 1970.

Burke, J. J., and J. Weiss (eds.): *Risk and Failure Analysis for Improved Performance Reliability,* Plenum, New York, 1980.

Carper, K. L. (ed.): *Forensic Engineering,* Elsevier, New York, 1989.

Collins, J. A.: *Failure of Materials in Mechanical Design,* Wiley, New York, 1981.

Goel, V. S. (ed.): *Analyzing Failures: The Problems and the Solutions,* American Society for Metals, Metals Park, Ohio, 1986.

Lange, G. A. (ed.): *Systematic Analysis of Technical Failures,* DGM Informations Gesellschaft Verlag, Braunschweig, Germany, 1986.

Metals Handbook, 8th ed., vol. 10: *Failure Analysis and Prevention,* American Society for Metals, Metals Park, Ohio, 1975.

Metals Handbook, 9th ed., vol. 11: *Failure Analysis and Prevention,* American Society for Metals, Metals Park, Ohio, 1986.

Petroski, H.: *To Engineer Is Human: The Roles of Failure in Successful Design,* St. Martin's Press, New York, 1985.

Polushkin, E. P.: *Defects and Failures of Metals,* Elsevier, New York, 1956.

Smith, A. L.: *Reliability of Engineering Materials,* Butterworths, New York, 1983.

Sourcebook in Failure Analysis, American Society for Metals, Metals Park, Ohio, 1974.

Vander Voort, G. F.: in *Metals Handbook,* 9th ed., vol. 12: *Fractography,* American Society for Metals, Metals Park, Ohio, 1987.

Whyte, R. R. (ed.): *Engineering Progress through Trouble,* Institution of Mechanical Engineers, London, 1975.

Appendix 1A Stereomicroscopy*

Although SEM images appear three-dimensional, their very format reduces them to two-dimensional representations. Their multidimensional appearance is due to high depth of focus, but perspective distortions introduced by the geometry of the beam/specimen/detector invalidate spatial measurements (both height and lateral dimensions). Our subjective impressions are based on the direction of illumination, which cannot be adequately described in a single micrograph produced by a complex geometry. These problems are aggravated when rough-surfaced specimens are examined, because the exact angle of a field of view is both unknown and unmeasurable. This in turn implies that magnification varies within a field of view.

The phenomenon of perspective distortion may be observed by recording an image having two prominent features at 20°, recording a second micrograph of the same field at 45°, and then measuring the distances between the two features on each micrograph: the measurements will differ and neither is valid. *Stereo imaging* involves recording a given field of view twice at slightly different orientations, and simultaneously viewing the stereo pair such that a three-dimensional image is perceived. Perspective is restored, and valid spatial judgments or measurements replace subjective impressions.

Recording and viewing stereo images

The four methods used to record stereo pairs are (1) the tilt method, where an angle is applied between the two micrographs; (2) the lateral-shift method, where there is a horizontal displacement between the two micrographs; (3) the rotation method, where a specimen is rotated between exposures; and (4) electromagnetic deflection of the electron beam between images (Boyde, 1975; Chatfield, 1978; Wergin and Pawley, 1980). Methods 1 and 2 are readily applied in any SEM, method 3 is difficult, and method 4 requires special accessories for the SEM. Discussed below are the tilt and lateral-shift methods of stereo image recording.

The *tilt method* of stereo recording is a versatile technique that may be used in any type of SEM. It is desirable but not required to have a eucentric-tilt specimen stage (the tilt axis passes through the center of the specimen, not through the center of the stage). The tilt method is as follows. Select and record the desired field of view, noting the tilt

---

*From B. L. Gabriel, *SEM: A User's Manual for Materials Science,* American Society for Metals, Metals Park, Ohio, 1985.

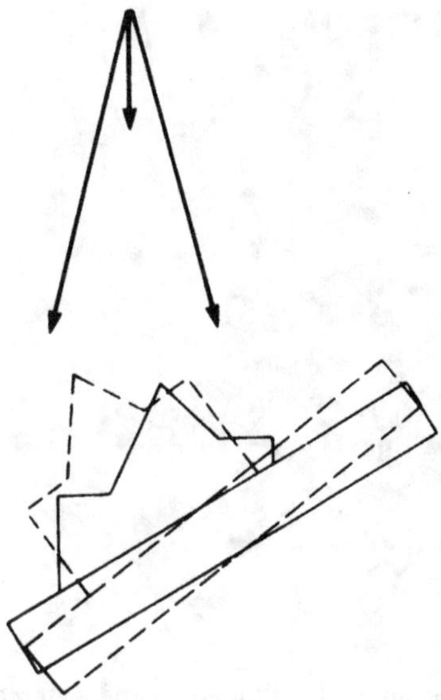

Figure 1A.1 Effect of a 7° stereo angle. Continuous line—specimen at 30°; dashed line—specimen tilted to 37°.

value of the specimen stage. With a wax pencil, mark the location of a prominent surface feature on the observation screen. Tilt the specimen approximately 7° (the stereo angle; see Figs. 1A.1 and 1A.2) while manipulating the stage X and Y axes to maintain the same field of view. Align the prominent surface feature beneath the wax pencil mark. Refocus the image using the Z-axis control; do not refocus with the objective lens controls. Adjust the contrast and brightness levels to match the first micrograph, and record the image.

The choice of the *stereo angle,* or the tilt difference between each micrograph of the stereo pair, is a function of the topography of the specimen. In general, smooth specimens require a stereo angle of 7 to 15°, while rough specimens require a stereo angle of 3 to 7°. The stereo angle determines *parallax* (synonym: horizontal displacement), which is a measure of the vertical position above or below a particular datum plane. With too large a stereo angle there is excessive displacement between features (i.e., excessive parallax), and the stereo image has "too much depth." With too small a stereo angle there is insufficient displacement, and the stereo image is not a true three-dimensional representation. If the microscopist is unsure of the optimal stereo angle for a given field of view, simply record several micrographs while changing the angle until the desired stereo image is obtained.

Another method useful for recording stereo pairs below 50× magnification is the *lateral-shift method* (synonyms: linear displacement or shift; translational shift). In this method, a micrograph is recorded,

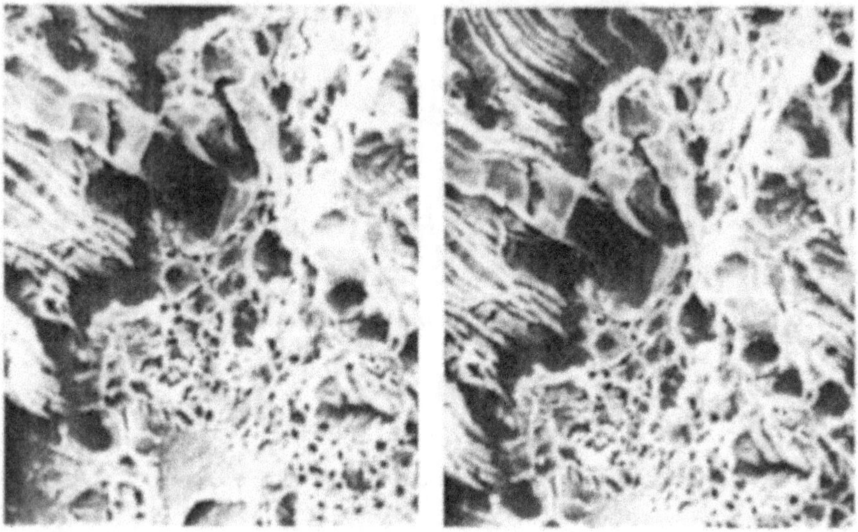

Figure 1A.2 Stereo pair of a ductile fracture prepared using the tilt method and a stereo angle of 7° (300×).

and the image is then moved horizontally while keeping the desired feature in the field of view. The second member of the stereo pair is then recorded. The distance shifted determines the depth of the stereo image; a very large lateral shift produces more depth than a small shift. Operating conditions must be the same when recording each half of the stereo pair; again, refocus (if necessary) by manipulating the Z axis, adjust contrast and brightness to match the first micrograph, and maintain the same tilt angle for both recordings. This method is appropriate only for low-magnification images, because at moderate or high levels lateral shift may displace the desired field of view before the optimal stereo image is visible (Fig. 1A.3).

Stereo pairs are viewed using simple pocket viewers, double-prism viewers, or a mirror stereoscope (Boyde, 1979). Pocket viewers are adequate for simple viewing of stereo pairs, but the more sophisticated prism and mirror stereoscopes offer significant advantages when stereo analysis is routinely conducted. The stereo effect (stereopsis) is perceived by positioning both micrographs within the viewer such that the tilt axis is vertical, i.e., rotate both micrographs 90°. The micrograph with the lower tilt value is placed to the left of the second half of the pair, and the distance between then is adjusted until the stereo image is prominent.

Because stereo viewers are designed for individual use, different methods have been developed to simultaneously project stereo pairs. The two most common projection techniques are the polarized and the

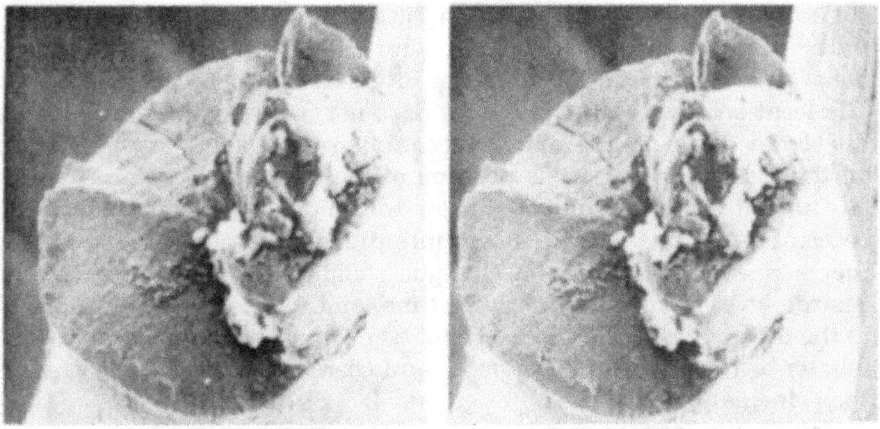

Figure 1A.3 Stereo pair of a fractured wire prepared with the lateral-shift method (75×).

anaglyph methods. The *polarized method* of stereo projection involves the simultaneous projection of each member of a stereo pair through adjacent slide projectors onto a lenticular silver screen. The projectors are equipped with filters that polarize the image from one projector at 45° to the vertical and from the other at an angle perpendicular to this. The audience must wear similarly polarized lenses to perceive the stereo effect. Wergin and Pawley (1980) thoroughly discuss the methods and equipment of the polarized method.

The *anaglyph method* of stereo projection applies a different color (usually red or green) to each half of the stereo pair. The simultaneous projection of the color-coded images produces a stereo image that is perceived by individuals wearing red-green lenses. This was the method used to project the 1950's 3D horror movies. This method is not as popular as the polarized method, because color breakthrough and other problems may degrade the stereo effect (Barber and Brett, 1982). Nemanic (1974) and Barber and Emerson (1980) review the preparation and presentation of anaglyphs.

Quantitative stereoscopy

In addition to restoring perspective, stereo images may be used for spatial measurements. Referred to as *photogrammetry* or *quantitative stereoscopy,* valid measurements derived from stereo pairs are used in the construction of three-dimensional models. Boyde and his colleagues have contributed extensively to SEM stereoscopy (Howell and Boyde, 1972; Boyde, 1974, 1981; Howell, 1975; and other references cited throughout this chapter). Summarized below are the more basic principles of quantitative stereoscopy.

The first stage of photogrammetry is establishment of the location of the *principal point,* defined as the point where the central ray (principal projector) of the scanning raster intersects the photographic plane. The location of the principal point does not necessarily correspond to the center of the micrograph; its location may be defined by switching off the SEM scan coils. The position of the stationary beam is visible as a bright spot. Alternatively, the position of the principal point can be established on a specimen by intentionally carbon-contaminating the surface: increase magnification roughly ten times above the desired level, permit a brief dwell time, and reduce the magnification to the desired level. A darkened spot (the site of contamination) is visible on beam-sensitive specimens, and corresponds to the location of the principal point.

Next, a *datum plane* which can be used as a baseline for measurements is identified with the tilt method. The common conventions use a datum plane parallel to one or the other halves of the stereo pair, or the midplane between the two micrographs. Using the parallel datum plane of one micrograph (Fig. 1A.4), Howell and Boyde (1972) defined the following spatial measurements, where M is magnification:

$$MZ_L = \frac{(MD)^2\,(X_L \cos\alpha - X_R) - MDX_LX_R \sin\alpha}{[(MD)^2 - X_LX_R]\sin\alpha + MD\,(X_L - X_R)\cos\alpha}$$

$$MX_L = \frac{X_L\,(MD - Z_L)}{MD}$$

$$MY_L = \frac{Y_L\,(MD - Z_L)}{MD}$$

Figure 1A.4 also shows the convention where an imaginary datum midplane between the stereo pair is located, and the following coordinates are shown:

$$MZ_C = \frac{(MD)^2\,(X_L - X_R)\cos(\alpha/2) + 2MDX_LX_R \sin(\alpha/2)}{[(MD)^2 + X_LX_R]\sin\alpha + MD\,(X_L - X_R)\cos\alpha}$$

$$MX_C = \frac{(MD)^2\,(X_L + X_R)\sin(\alpha/2)}{[(MD)^2 + X_LX_R]\sin\alpha + MD\,(X_L - X_R)\cos\alpha}$$

$$MY_C = \frac{Y_L}{MD}\left(MD + X_C \sin\frac{\alpha}{2} - Z_C \cos\frac{\alpha}{2}\right)$$

Individuals who regularly perform quantitative SEM measurements will be pleased to learn that modestly priced microcomputer programs are available for this purpose. Various systems are described by

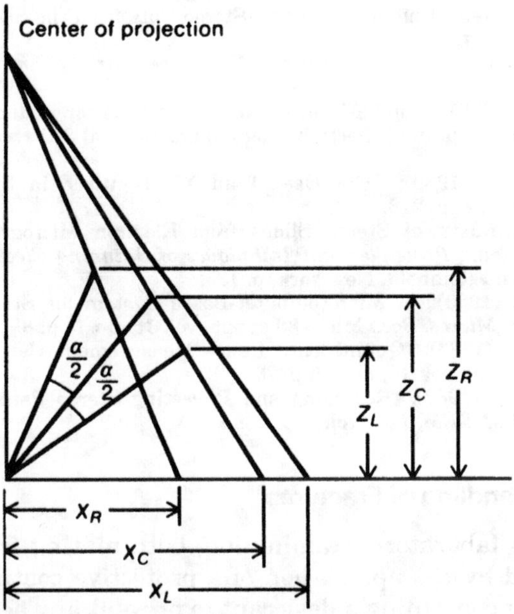

Figure 1A.4 Trigonometric relationship of the Howell and Boyde method. (*Courtesy of Howell and Boyde, 1972.*)

Howell and Boyde (1980), Roberts and Page (1980), Howell (1981), Boyde (1981), and Russ and Stewart (1983). Future developments in stereo SEM include real-time stereo imaging and image recording, computer control of the specimen stage, and continued acceptance and utilization by microscopists.

References to App. 1A

Barber, V. C., and D. A. L. Brett (1982): "'Colour Bombardment'—A Human Visual Problem that Interferes with the Viewing of Anaglyph Stereo Materials," *SEM, Inc.*, vol. 2, p. 495.

Barber, V. C., and C. J. Emerson (1980): "Preparation of SEM Anaglyph Stereo Material for Use in Teaching and Research," *Scanning,* vol. 3, p. 202.

Boyde, A. (1974): "Photogrammetry of Stereopair SEM Images Using Separate Images from the Two Images," *IITRI/SEM,* p. 101.

Boyde, A. (1975): "Measurement of Specimen Height Difference and Beam Tilt Angle in Anaglyph Real Time Stereo TV SEM System," *IITRI/SEM,* p. 189.

Boyde, A. (1979): "The Perception and Measurement of Depth in the SEM," *SEM, Inc.*, vol. 2, p. 67.

Boyde, A. (1981): "Recent Developments in Stereo SEM (1981 Update)," *SEM, Inc.*, vol. 1, p. 91.

Cannon, T. M., and B. R. Hunt (1981): "Image Processing by Computer," *Sci. Amer.*, vol. 245, no. 4, p. 214.

Chatfield, E. J. (1978): "Introduction to Stereo Scanning Electron Microscopy," in M. A. Hayat (ed.), *Principles and Techniques of Scanning Electron Microscopy,* vol. 6, Van Nostrand Reinhold, New York, p. 47.

Howell, P. G. T. (1975): "Taking, Presenting, and Treating Stereo Data from the SEM," *IITRI/SEM,* p. 697.
Howell, P. G. T. (1981): "Semi-Automatic Profiling from SEM Stereopairs," *Scanning,* vol. 4, p. 40.
Howell, P. G. T., and A. Boyde (1972): "Comparison of Various Methods for Reducing Measurements from Stereo-Pair Scanning Electron Micrographs to 'Real 3-D Data,'" *IITRI/SEM,* p. 233.
Howell, P. G. T., and A. Boyde (1980): "The Use of an XY Digitiser in SEM Photogrammetry," *Scanning,* vol. 3, p. 218.
Nemanic, M. K. (1974): "Preparation of Stereo Slides from Electron Micrograph Stereopairs," in M. A. Hayat (ed.), *Principles and Techniques of Scanning Electron Microscopy,* vol. 1, Van Nostrand Reinhold, New York, p. 135.
Roberts, S. G., and T. F. Page (1980): "A Microcomputer-Based System for Stereogrammetric Analysis," *Proc. R. Micros. Soc.,* Micro 80 suppl., vol. 15, no. 5, p. 8.
Russ, J. C., and W. D. Stewart (1983): "Quantitative Image Measurement Using a Microcomputer System," *Am. Lab.,* vol. 15, no. 12, p. 70.
Wergin, W. P., and J. B. Pawley (1980): "Recording and Projecting Stereo Pairs of Scanning Electron Micrographs," *SEM, Inc.,* vol. 1, p. 239.

Appendix 1B Care and Handling of Fractures*

When a fracture requires laboratory examination, both mating fractures should be preserved by the application of a protective coating† and sealed in a plastic bag containing a desiccant to prevent any accumulation of undue moisture until the examination can be made. Coatings used should be soluble in organic solvents so that they can be completely removed prior to replication. Handling the fracture faces with fingers, rubbing, or mating the fractures together can cause serious damage. Picking at the fracture with a sharp instrument should be avoided. Rough treatment or the formation of corrosion products on the fracture will obscure vital information (Fig. 1B.1). Education in the proper handling of specimens prior to any fractographic examination is strongly recommended for anyone dealing in fractures either in the field or in the laboratory.

Appendix 1C Preparation and Preservation of Fracture Specimens‡

Fracture surfaces are fragile and subject to mechanical and environmental damage that can destroy microstructural features. Consequently, fracture specimens must be carefully handled during all stages of analy-

*From A. Phillips, V. Kerlins, and B. V. Whiteson, *Electron Fractography Handbook,* Tech. Rep. ML-TDR-64-416, Air Force Materials Laboratory, Wright-Patterson Air Force Base, Ohio, 1965.

†For example, Krylon Crystal Clear spray coating no. 1302, Krylon, Inc., Norristown, Pa. Solvent: trichloroethylene.

‡From R. D. Zipp and E. P. Dahlberg, in *Metals Handbook,* 9th ed., vol. 12: *Fractography,* ASM International, Metals Park, Ohio, 1987, pp. 72–77.

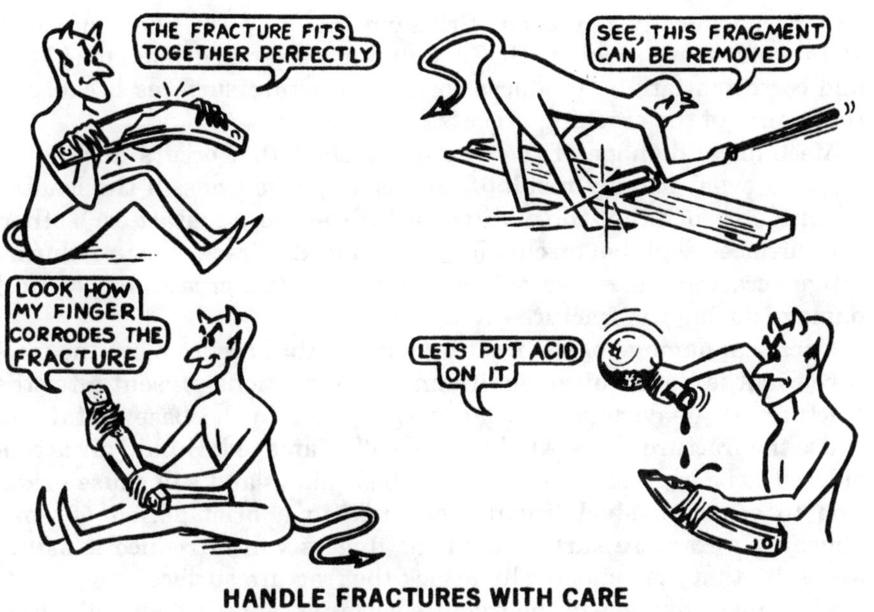

Figure 1B.1 Care and handling of fractures.

sis. This appendix discusses the importance of care and handling of fractures and what to look for during the preliminary visual examination, fracture-cleaning techniques, procedures for sectioning a fracture and opening secondary cracks, and the effect of nondestructive inspection on subsequent evaluation.

Care and handling of fractures[1]

Fracture interpretation is a function of the fracture surface condition. Because the fracture surface contains a wealth of information, it is important to understand the types of damage that can obscure or obliterate fracture features and obstruct interpretation. These types of damage are usually classified as chemical and mechanical damage. Chemical or mechanical damage of the fracture surface can occur during or after the fracture event. If damage occurs during the fracture event, very little can usually be done to minimize it. However, proper handling and care of fractures can minimize damage that can occur after the fracture.[2–4]

Chemical damage of the fracture surface that occurs during the fracture event is the result of environmental conditions. If the environment adjacent to an advancing crack front is corrosive to the base metal, the resultant fracture surface in contact with the environment will be chemically damaged. Cracking due to such phenomena as stress-corrosion cracking (SCC), liquid-metal embrittlement (LME), and corrosion fatigue produces corroded fracture surfaces because of the nature of the cracking process.

Mechanical damage of the fracture surface that occurs during the fracture event usually results from loading conditions. If the loading condition is such that the mating fracture surfaces contact each other, the surfaces will be mechanically damaged. Crack closure during fatigue cracking is an example of a condition that creates mechanical damage during the fracture event.

Chemical damage of the fracture surface that occurs after the fracture event is the result of environmental conditions present after the fracture. Any environment that is aggressive to the base metal will cause the fracture surface to be chemically damaged. Humid air is considered to be aggressive to most iron-base alloys and will cause oxidation to occur on steel fracture surfaces in a brief period of time. Touching a fracture surface with the fingers will introduce moisture and salts that may chemically attack the fracture surface.

Mechanical damage of the fracture surface that occurs after the fracture event usually results from handling or transporting of the fracture. It is easy to damage a fracture surface while opening primary cracks, sectioning the fracture from the total part, and transporting the fracture. Other common ways of introducing mechanical damage include fitting the two fracture halves together or picking at the fracture with a sharp instrument. Careful handling and transporting of the fracture are necessary to keep damage to a minimum.

Once mechanical damage occurs on the fracture surface, nothing can be done to remove its obliterating effect on the original fracture morphology. Corrosive attack, such as high-temperature oxidation, often precludes successful surface restoration. However, if chemical damage occurs and if it is not too severe, cleaning techniques can be implemented that will remove the oxidized or corroded surface layer and will restore the fracture surface to a state representative of its original condition.

Preliminary visual examination

The entire fracture surface should be visually inspected to identify the location of the fracture-initiating site or sites and to isolate the areas in the region of crack initiation that will be most fruitful for further microanalysis. The origin often contains the clue to the cause of fracture, and both low- and high-magnification analyses are critical to accurate failure analysis. Where the size of the failed part permits, visual examination should be conducted with a low-magnification wide-field stereomicroscope having an oblique source of illumination.

In addition to locating the failure origin, visual analysis is necessary to reveal stress concentrations, material imperfections, the presence of surface coatings, case-hardened regions, welds, and other structural details that contribute to cracking. The general level of stress, the relative ductility of the material, and the type of loading (torsion, shear, bending, and so on) can often be determined from visual analysis.

Finally, a careful macroexamination is necessary to characterize the condition of the fracture surface so that the subsequent microexamination strategy can be determined. Macroexamination can be used to identify areas of heavy burnishing in which opposite halves of the fracture have rubbed together and to identify regions covered with corrosion products. The regions least affected by this kind of damage should be selected for microanalysis. When stable crack growth has continued for an extended period, the region nearest the fast fracture is often the least damaged because it is the newest crack area. Corrodents often do not penetrate to the crack tip, and this region remains relatively clean.

The visual macroanalysis will often reveal secondary cracks that have propagated only partially through a cracked member. These part-through cracks can be opened in the laboratory and are often in much better condition than the main fracture. Areas for sectioning can be identified for subsequent metallography, chemical analysis, and mechanical-property determinations.

Preservation techniques[1]

Unless a fracture is evaluated immediately after it is produced, it should be preserved as soon as possible to prevent attack from the environment. The best way to preserve a fracture is to dry it with a gentle stream of dry compressed air, then store it in a desiccator, a vacuum storage vessel, or a sealed plastic bag containing a desiccant. However, such isolation of the fracture is often not practical. Therefore, corrosion-preventive surface coatings must be used to inhibit oxidation and corrosion of the fracture surface. The primary disadvantage of using these surface coatings is that fracture surface debris, which often provides clues to the cause of fracture, may be displaced during removal of the coating. However, it is still possible to recover the surface debris from the solvent used to remove these surface coatings by filtering the spent solvent and capturing the residue.

The main requirements for a surface coating are as follows:

- It should not react chemically with the base metal.
- It should prevent chemical attack of the fracture from the environment.

- It must be completely and easily removable without damaging the fracture.

Fractures in the field may be coated with fresh oil or axle grease if the coating does not contain substances that might attack the base metal. Clear acrylic lacquers or plastic coatings are sometimes sprayed on the fracture surfaces. These clear sprays are transparent to the fracture surface and can be removed with organic solvents. However, on rough fracture surfaces, it can be difficult to achieve complete coverage and to remove the coating completely.

Another type of plastic coating that has been successfully used to protect most fracture surfaces is cellulose acetate replicating tape. The tape is softened in acetone and applied to the fracture surface with finger pressure. As the tape dries, it adheres tightly to the fracture surface. The main advantage of using replicating tape is that it is available in various thicknesses. Rough fracture surfaces can be coated with relatively thick replicating tape to ensure complete coverage. The principal limitation of using replicating tape is that on rough fracture surfaces it is difficult to remove the tape completely.

Solvent-cutback petroleum-base compounds have been used by Boardman et al. to protect fracture surfaces and can be easily removed with organic solvents.[5] In this study, seven rust-inhibiting compounds were selected for screening as fracture surface coating materials. These inhibitor compounds were applied to fresh steel fracture surfaces and exposed to 100% relative humidity at 38°C (100°F) for 14 days. The coatings were removed by ultrasonic cleaning with the appropriate solvent, and the fracture surfaces were visually evaluated. Only the Tectyl 506 compound protected the fractures from rusting during the screening tests. Therefore, further studies were conducted with a scanning electron microscope to ensure that the Tectyl 506 compound would inhibit oxidation of the fracture surface and could be completely removed on the microscopic level without damaging the fracture surface.

Initially, steel Charpy samples and nodular iron samples were fractured in the laboratory by single-impact overload and fatigue, respectively. Representative fracture areas were photographed in the scanning electron microscope at various magnifications in the as-fractured condition. The fracture surfaces were then coated with Tectyl 506, exposed to 100% relative humidity at 38°C (100°F) for 14 days, and cleaned before scanning electron microscopy (SEM) evaluation by ultrasonically removing the coating in a naphtha solution. Figure 1C.1 shows a comparison of identical fracture areas in the steel at increasing magnifications in the as-fractured condition and after coating, exposing, and cleaning. These fractographs show that the solvent-cutback petroleum-

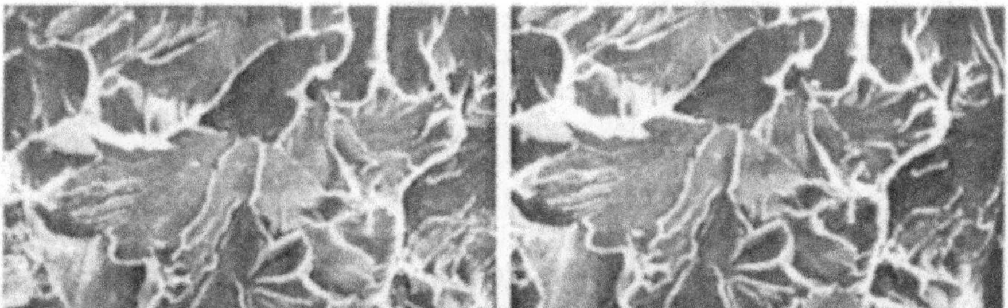

Figure 1C.1 Comparison of identical fracture areas of steel Charpy specimens at increasing magnifications. (*a*), (*c*) As-fractured surface. (*b*), (*d*) Same fracture surface after coating with Tectyl 506, exposing to 100% relative humidity for 14 days, and cleaning with naphtha.

base compound prevented chemical attack of the fracture surface from the environment and that the compound was completely removed in the appropriate solvent. It is interesting to note that Tectyl 506 is a rust-inhibiting compound that is commonly used to rustproof automobiles.

Fracture-cleaning techniques[1]

Fracture surfaces exposed to various environments generally contain unwanted surface debris, corrosion or oxidation products, and accumulated artifacts that must be removed before meaningful fractography can be performed. Before any cleaning procedures begin, the fracture surface should be surveyed with a low-power stereo binocular microscope, and the results should be documented with appropriate sketches or photographs. Low-power microscope viewing will also establish the severity of the cleaning problem and should also be used to monitor the effectiveness of each subsequent cleaning step. It is important to emphasize that the debris and deposits on the fracture surface can contain information that is vital to understanding the cause of fracture. Examples are fractures that initiate from such phenomena as SCC, LME, and corrosion fatigue. Often, knowing the nature of the surface debris and deposits, even when not essential to the fracture analysis, will be useful in determining the optimum cleaning technique. The most common techniques for cleaning fracture surfaces, in order of increasing aggressiveness, are:

- Dry air blast or soft organic-fiber brush cleaning
- Replica stripping
- Organic-solvent cleaning
- Water-base detergent cleaning
- Cathodic cleaning
- Chemical-etch cleaning

The mildest, least aggressive cleaning procedure should be tried first, and as previously mentioned, the results should be monitored with a stereo binocular microscope. If residue is still left on the fracture surface, more aggressive cleaning procedures should be implemented in order of increasing aggressiveness.

Air blast or brush cleaning. Loosely adhering particles and debris can be removed from the fracture surface with either a dry air blast or a soft organic-fiber brush. The dry air blast also dries the fracture surface. Only a soft organic-fiber brush, such as an artist's brush, should be used on the fracture surface because a hard-fiber brush or a metal wire brush will mechanically damage the fine details.

Replica-stripping cleaning. This technique is very similar to that described in the section "Preservation Techniques" above. However, instead of leaving the replica on the fracture surface to protect it from the environment, it is stripped off of the fracture surface, removing debris and deposits. Successive replicas are stripped until all the surface contaminants are removed. Figure 1C.2 shows successive replicas stripped from a rusted steel fracture surface and demonstrates that the first replicas stripped from the fracture surface contain the most contaminants and that the last replicas stripped contain the least. Capturing these contaminants on the plastic replicas, relative to their position on the fracture surface, can be a distinct advantage. The replicas can be retained, and the embedded contaminants can be chemically analyzed if the nature of these deposits is deemed important.

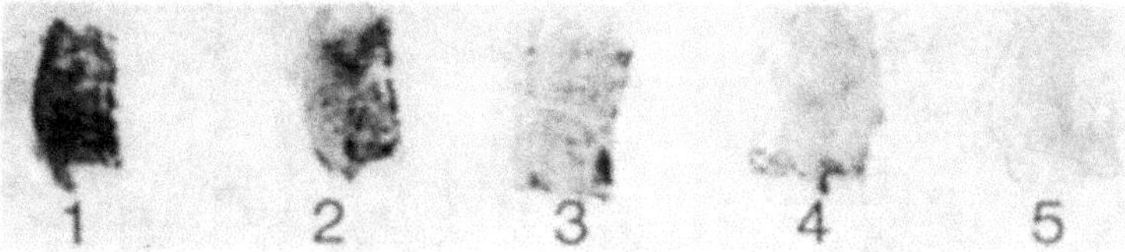

Figure 1C.2 Successive replicas (numbered 1 to 5) stripped from a rusted steel fracture surface. Note that the first replica stripped contains the most surface contaminants, while the last replica stripped is the cleanest. (Approximate size.)

The one disadvantage of using plastic replicas to clean a fracture surface is that on rough surfaces it is very difficult to remove the replicating material completely. However, if the fracture surface is ultrasonically cleaned in acetone after each successive replica is stripped from the fracture surface, removal of the residual replicating material is possible. Ultrasonic cleaning in acetone or the appropriate solvent should be mandatory when using the replica-stripping cleaning technique.

Organic solvents. Organic solvents, such as xylene, naphtha, toluene, freon TF, ketones, and alcohols, are primarily used to remove grease, oil, protective surface coatings, and crack-detecting fluids from the fracture surface. It is important to avoid use of the chlorinated organic solvents, such as trichloroethylene and carbon tetrachloride, because most of them have carcinogenic properties. The sample to be cleaned is usually soaked in the appropriate organic solvent for an extended period of time, immersed in a solvent bath where jets from a pump introduce fresh solvent to the fracture surface, or placed in a beaker containing the solvent and ultrasonically cleaned for a few minutes.

The ultrasonic cleaning method is probably the most popular of the three methods mentioned above, and the ultrasonic agitation will also remove any particles that adhere lightly to the fracture surface. However, if some of these particles are inclusions that are significant for fracture interpretation, the location of these inclusions relative to the fracture surface and the chemical composition of these inclusions should be investigated before their removal by ultrasonic cleaning.

Water-base detergent cleaning. This technique, assisted by ultrasonic agitation, is effective in removing debris and deposits from the fracture surface and, if proper solution concentrations and times are used, does not damage the surface. A particular detergent that has proved effective in cleaning ferrous and aluminum materials is Alconox. The cleaning solution is prepared by dissolving 15 g of Alconox powder in a beaker containing 350 mL of water. The beaker is placed

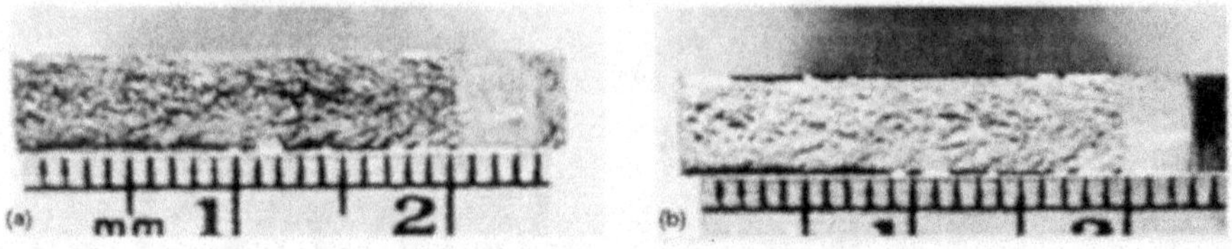

Figure 1C.3 Fracture toughness specimen that has been intentionally corroded in a 5% salt steam chamber for 6 h. (*a*) Before ultrasonic cleaning in a heated Alconox solution for 30 min. (*b*) After ultrasonic cleaning.

in an ultrasonic cleaner preheated to about 95°C (205°F). The fracture is then immersed in the solution for about 30 min, rinsed in water then alcohol, and air-dried.

Figure 1C.3*a* shows the condition of a laboratory-tested fracture toughness sample (AISI 1085 heat-treated steel) after it was intentionally corroded in a 5% salt steam spray chamber for 6 h. Figure 1C.3*b* shows the condition of this sample after cleaning in a heated Alconox solution for 30 min. The fatigue precrack region is the smoother fracture segment located to the right of the rougher single overload region. Figure 1C.4*a* and *b* shows identical views of an area in the fatigue precrack region before and after ultrasonic cleaning in a heated Alconox solution. Only corrosion products are visible, and the underlying fracture morphology is completely obscured in Fig. 1C.4*a*. Figure 1C.4*b* shows that the water-base detergent cleaning has removed the corrosion products on the fracture surface. The sharp edges on the fracture features indicate that cleaning has not damaged the surface, as evidenced by the fine and shallow fatigue striations clearly visible in Fig. 1C.4*b*.

The effect of prolonged ultrasonic cleaning in the Alconox solution is demonstrated in Fig. 1C.5*a* and *b*, which shows identical views of an area in the fatigue precrack region after cleaning for 30 min and 3.5 h, respectively. Figure 1C.5*b* reveals that the prolonged exposure has not only chemically etched the fracture surface but has also dislodged the originally embedded inclusions. Any surface corrosion products not completely removed within the first 30 min of water-base detergent cleaning are difficult to remove by further cleaning; therefore, prolonged cleaning provides no additional benefits.

Cathodic cleaning. This is an electrolytic process in which the sample to be cleaned is made the cathode, and hydrogen bubbles generated at the sample cause primarily mechanical removal of surface debris and deposits. An inert anode, such as carbon or platinum, is normally used to avoid contamination by plating upon the cathode. During cathodic cleaning, it is common practice to vibrate the electrolyte ultrasonically

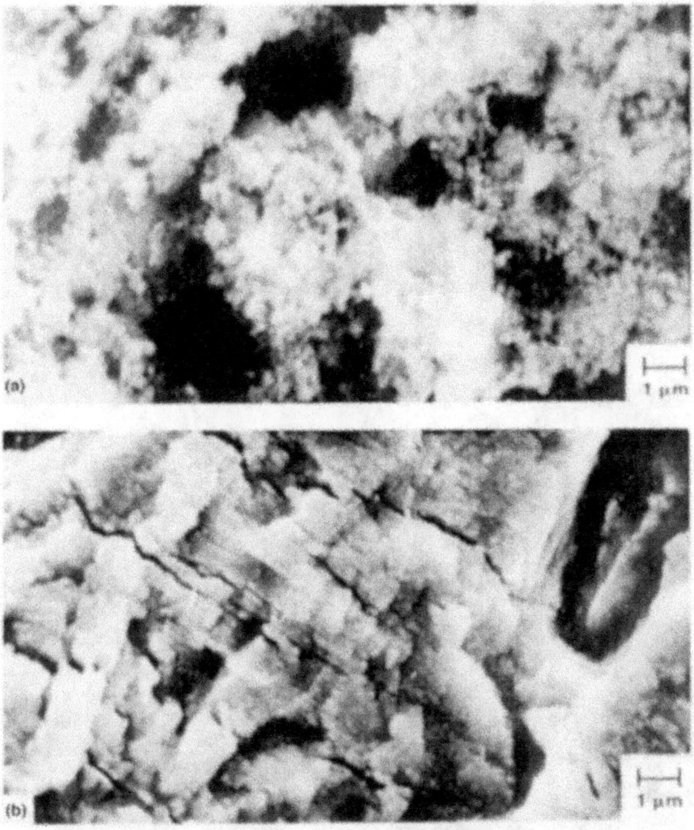

Figure 1C.4 Fatigue precrack region shown in Fig. 1C.3. (*a*) Before ultrasonic cleaning in a heated Alconox solution for 30 min. (*b*) After ultrasonic cleaning.

or to rotate the specimen (cathode) with a small motor. The electrolytes commonly used to clean ferrous fractures are sodium cyanide,[6,7] sodium carbonate, sodium hydroxide solutions, and inhibited sulfuric acid.[8] Because cathodic cleaning occurs primarily by the mechanical removal of deposits due to hydrogen liberation, the fracture surface should not be chemically damaged after elimination of the deposits.

The use of cathodic cleaning to remove rust from steel fracture surfaces has been successfully demonstrated.[9] In this study, AISI 1085 heat-treated steel and EX16 carburized steel fractures were exposed to a 100% humidity environment at 65°C (150°F) for 3 days. A commercially available sodium cyanide electrolyte, ultrasonically agitated, was used in conjunction with a platinum anode for cleaning. A 1-min cathodic cleaning cycle was applied to the rusted fractures, and the

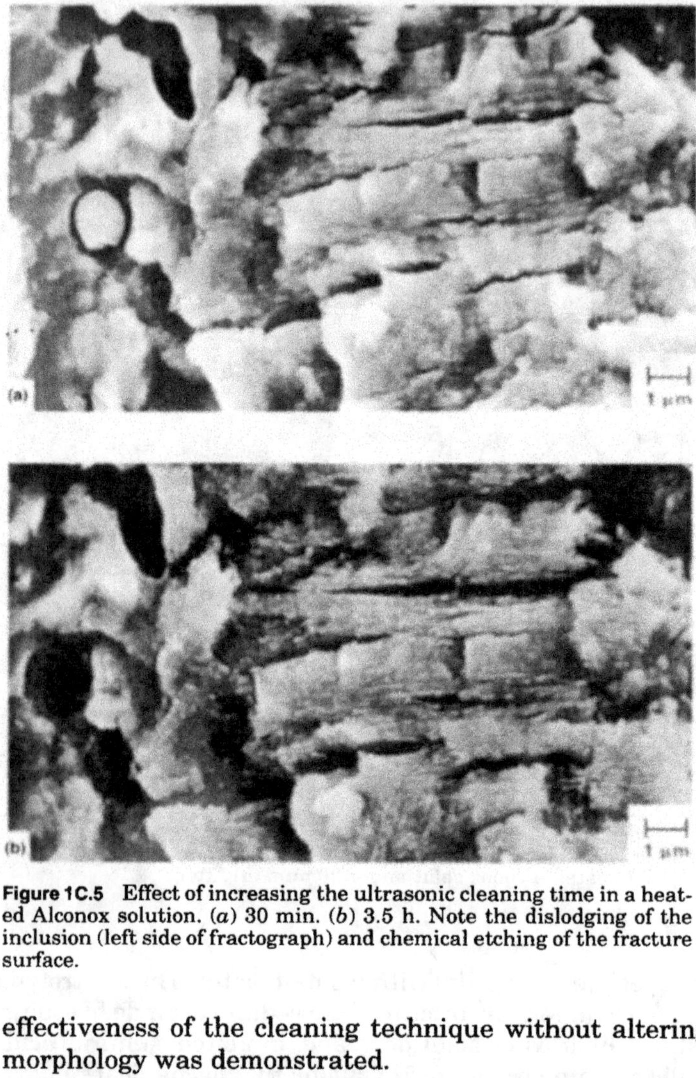

Figure 1C.5 Effect of increasing the ultrasonic cleaning time in a heated Alconox solution. (*a*) 30 min. (*b*) 3.5 h. Note the dislodging of the inclusion (left side of fractograph) and chemical etching of the fracture surface.

effectiveness of the cleaning technique without altering the fracture morphology was demonstrated.

Figure 1C.6 shows a comparison of an as-fractured surface with a corroded and cathodically stable ductile cracking region in a quenched-and-tempered 1085 carbon steel. The relatively low magnification (1000×) shows that the dimpled topography characteristic of ductile tearing was unchanged as a result of the corrosion and cathodic cleaning. High magnification (5000×) shows that the perimeters of the small interconnecting dimples were corroded away.

Chemical etching. If the above techniques are attempted and prove ineffective, the chemical-etch cleaning technique, which involves

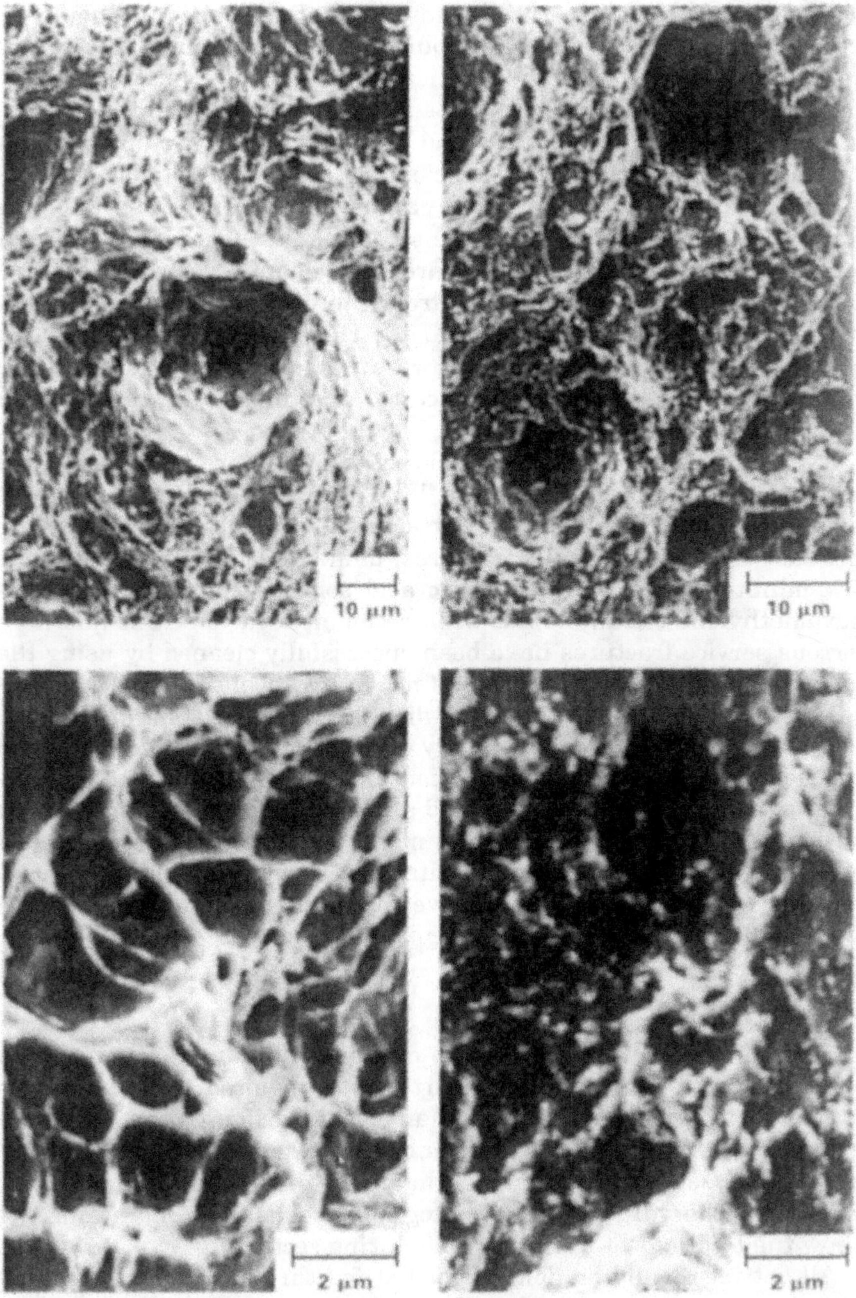

Figure 1C.6 Comparison of stable ductile crack growth areas from quenched-and-tempered 1085 carbon steel at increasing magnifications. The fractographs on the left show the as-fractured surface; those on the right show the fracture surface after corrosion exposure and cathodic cleaning.

treating the surface with mild acids or alkaline solutions, should be implemented. This technique should be used only as a last resort because it involves possible chemical attack of the fracture surface. In chemical-etch cleaning, the specimen is placed in a beaker containing the cleaning solution and is vibrated ultrasonically. It is sometimes necessary to heat the cleaning solution. Acetic acid, phosphoric acid, sodium hydroxide, ammonium citrate, ammonium oxalate solutions, and commercial solutions have been used to clean ferrous alloys.[8] Titanium alloys are best cleaned with nitric acid.[4] Oxide coatings can be removed from aluminum alloys by using a warmed solution containing 70 mL of orthophosphoric acid (85%), 32 g of chromic acid, and 130 mL of water.[10] However, it has also been recommended that fracture surfaces of aluminum alloys be cleaned only with organic solvents.[4]

Especially effective for chemical-etch cleaning are acids combined with organic corrosion inhibitors.[11,12] These inhibited acid solutions limit the chemical attack to the surface contaminants while protecting the base metal. For ferrous fractures, immersion of the samples for a few minutes in a 6*N* hydrochloric acid solution containing 2 g/L of hexamethylene tetramine has been recommended.[6] Ferrous and nonferrous service fractures have been successfully cleaned by using the following inhibited acid solution: 3 mL of hydrochloric acid (1.19 specific gravity), 4 mL of 2-butyne-1,4-diol (35% aqueous solution), and 50 mL of deionized water.[13] This study demonstrated the effectiveness of the cleaning solution in removing contaminants from the fracture surfaces of a low-carbon steel pipe and a Monel Alloy 400 expansion joint without damaging the underlying metal. Various fracture morphologies were not affected by the inhibited acid treatment when the cleaning time was appropriate to remove contaminants from these service fractures.

Sectioning a fracture

It is often necessary to remove the portion containing a fracture from the total part, because the total part is to be repaired, or to reduce the specimen to a convenient size. Many of the examination tools—for example, the scanning electron microscope and the electron microprobe analyzer—have specimen chambers that limit specimen size. Records, either drawings or photographs, should be maintained to show the locations of the cuts made during sectioning.

All cutting should be done such that fracture faces and their adjacent areas are not damaged or altered in any way; this includes keeping the fracture surface dry whenever possible. For large parts, the common method of specimen removal is flame cutting. Cutting must be done at a sufficient distance from the fracture so that the microstructure of the metal underlying the fracture surface is not altered by the heat of the flame and so that none of the molten metal from flame cutting is deposited on the fracture surface.

Saw cutting and abrasive cutoff wheel cutting can be used for a wide range of part sizes. Dry cutting is preferable because coolants may corrode the fracture or may wash away foreign matter from the fracture. A coolant may be required, however, if a dry cut cannot be made at a sufficient distance from the fracture to avoid heat damage to the fracture region. In such cases, the fracture surface should be solvent cleaned and dried immediately after cutting.

Some of the coating procedures mentioned above may be useful during cutting and sectioning. For example, the fracture can be protected during flame cutting by taping a cloth over it and can be protected during sawing by spraying or coating it with a lacquer or a rust-preventive compound.

Opening secondary cracks

When the primary fracture has been damaged or corroded to a degree that obscures information, it is desirable to open any secondary cracks to expose their fracture surfaces for examination and study. These cracks may provide more information than the primary fracture. If rather tightly closed, they may have been protected from corrosive conditions, and if they have existed for less time than the primary fracture, they may have corroded less. Also, primary cracks that have not propagated to total fracture may have to be opened.

In opening these types of cracks for examination, care must be exercised to prevent damage, primarily mechanical, to the fracture surface. This can usually be accomplished if opening is done such that the two faces of the fracture are moved in opposite directions, normal to the fracture plane. A saw cut can usually be made from the back of the fractured part to a point near the tip of the crack, using extreme care to avoid actually reaching the crack tip. This saw cut will reduce the amount of solid metal that must be broken. Final breaking of the specimen can be done by:

- Clamping the two sides of the fractured part in a tensile-testing machine, if the shape permits, and pulling
- Placing the specimen in a vise and bending one half away from the other half by striking it with a hammer in a way that will avoid damaging the crack surfaces
- Gripping the halves of the fracture in pliers or vise grips and bending or pulling them apart

It is desirable to be able to distinguish between a fracture surface produced during opening of a primary or secondary crack. This can be accomplished by ensuring that a different fracture mechanism is active in making the new break; for example, the opening can be performed at a very low temperature. During low-temperature fracture, care should be taken to avoid condensation of water, because this could corrode the fracture surface.

Crack separations and crack lengths should be measured before opening. The amount of strain that occurred in a specimen can often be determined by measuring the separation between the adjacent halves of a fracture. This should be done before preparation for opening a secondary crack has begun. The lengths of cracks may also be important for analyses of fatigue fractures or for fracture mechanics considerations.

Effect of nondestructive inspection

Many of the so-called nondestructive inspection methods are not entirely nondestructive. The liquid penetrants used for crack detection may corrode fractures in some metals, and they will deposit foreign compounds on the fracture surfaces; corrosion and the depositing of foreign compounds could lead to misinterpretation of the nature of the fracture. The surface of a part that contains, or is suspected to contain, a crack is often cleaned for more critical examination, and rather strong acids that can find their way into a tight crack are frequently used. Many detections of chlorine on a fracture surface of steel, for example, which were presumed to prove that the fracture mechanism was SCC, have later been found to have been derived from the hydrochloric acid used to clean the part.

Even magnetic-particle inspection, which is often used to locate cracks in ferrous parts, may affect subsequent examination. For example, the arcing that may occur across tight cracks can affect fracture surfaces. Magnetized parts that are to be examined by SEM will require demagnetization if scanning is to be done at magnifications above about 500×.

References to App. 1C

1. R. D. Zipp, "Preservation and Cleaning of Fractures for Fractography," *Scan. Elec. Microsc.*, no. 1, pp. 355–362, 1979.
2. A. Phillips et al., *Electron Fractography Handbook,* MCIC-HB-08, Metals and Ceramics Information Center, Battelle Columbus Laboratories, June 1976, pp. 4–5.
3. W. R. Warke et al., "Techniques for Electron Microscope Fractography," in *Electron Fractography,* STP 436, American Society for Testing and Materials, Philadelphia, Pa., 1968, pp. 212–230.
4. J. A. Fellows et al., in *Metals Handbook,* 8th ed., vol. 9: *Fractography and Atlas of Fractographs,* American Society for Metals, Metals Park, Ohio, 1974, pp. 9–10.

5. B. E. Boardman et al., "A Coating for the Preservation of Fracture Surfaces," Paper 750967, presented at SAE Automobile Engineering Meeting, Detroit, Mich., Society of Automotive Engineers, Oct. 13–17, 1975.
6. H. DeLeiris et al., "Techniques of De-Rusting Fractures of Steel Parts in Preparation for Electronic Micro-Fractography," *Mem. Sci. Rev. de Met.,* vol. 63, pp. 463–472, May 1966.
7. P. M. Yuzawich and C. W. Hughes, "An Improved Technique for Removal of Oxide Scale from Fractured Surfaces of Ferrous Materials," *Pract. Metallog.,* vol. 15, pp. 184–195, 1978.
8. B. B. Knapp, "Preparation and Cleaning of Specimens," in *The Corrosion Handbook,* Wiley, New York, 1948, pp. 1077–1083.
9. E. P. Dahlberg and R. D. Zipp, "Preservation and Cleaning of Fractures for Fractography—Update," *Scan. Elec. Microsc.,* no. 1, pp. 423–429, 1981.
10. G. F. Pittinato et al., *SEM/TEM Fractography Handbook,* MCIC-HB-06, Metals and Ceramics Information Center, Battelle Columbus Laboratories, Dec. 1975, pp. 4–5.
11. C. R. Brooks and C. D. Lundin, "Rust Removal from Steel Fractures—Effect on Fractographic Evaluation," *Microstruc. Sci.,* vol. 3, pp. 21–33, 1975.
12. G. G. Elibredge and J. C. Warner, "Inhibitors and Passivators," in *The Corrosion Handbook,* Wiley, New York, 1948, pp. 905–916.
13. E. P. Dahlberg, "Techniques for Cleaning Service Failures in Preparation for Scanning Electron Microscope and Microprobe Analysis," *Scan. Elec. Microsc.,* pp. 911–918, 1974.

Appendix 1D Cleaning of Fracture Surfaces*

A clean fracture surface is a prerequisite for the definition of the mode of failure. As with any other type of specimen, fracture surfaces must be clean for successful SEM imaging. Because fracture surfaces are fragile, their cleaning must be approached with caution and common sense. The cleaning methods are classified in Table 1D.1. As a rule, the least aggressive method must be attempted before proceeding to more aggressive techniques, because the latter are capable of damaging the fracture surface. Dahlberg (1974, 1976), Zipp (1979), and Dahlberg and Zipp (1981) comprehensively review and compare each method, and their views are summarized below.

The least aggressive cleaning method is capable of removing loosely adhering dust or debris from the fracture surface. Short-haired *soft brushes* (e.g., a trimmed artist's paintbrush) or bursts of *compressed gas* are useful for removing dust. This method alone is rarely sufficient to clean a surface; more often, organic films (oil, grease, etc.) also obscure the surface. An ultrasonic treatment with an *organic solvent* followed by blowing with compressed gas removes both organic films and debris. Zipp (1979) recommends the various solvents listed in Table 1D.1. For heavily contaminated surfaces, it may be necessary to pass through several changes or a series of solvents.

*From B. L. Gabriel, *SEM: A User's Manual for Materials Science,* American Society for Metals, Metals Park, Ohio, 1985.

TABLE 1D.1 Classification of Methods for Cleaning Fracture Surfaces according to Their Degree of Aggressiveness

Method	For removal of	Degree of aggressiveness
Soft fiber brush and dry air	Loosely adhering debris and dust	Least aggressive
Organic solvents and ultrasonic bath		↓
Toluene or xylene	Oil and grease	↓
Ketones	Varnish and gum	↓
Alcohol	Dyes and fatty acids	↓
Replica stripping	Insoluble debris and oxides	↓
Detergents (e.g., Alconox)	Corrosion products and oxides	↓
Cathodic cleaning	Deposits and oxides	↓
Corrosion-inhibited acids	Sulfides and oxides	↓
Acid etches	Oxides	Most aggressive

If the debris obscuring the surface is of interest, solvent cleaning should not be used immediately. For example, if the surface is corroded, it may be desirable to first analyze the composition of the corrosion products with energy-dispersive spectroscopy, then clean the specimen and examine the native fracture surface. Alternatively, the surface may be simultaneously cleaned and the reaction products preserved by preparing *extraction replicas* (synonymous with cleaning replicas). When a strip of cellulose acetate softened with acetone is placed over a fracture and firmly pressed into position (with one's thumb), the gel will encapsulate any material that is not part of the base metal. After the acetone has evaporated, the cellulose acetate retains the entrapped reaction products. The replica is then removed from the surface, and both the particles and their location relative to the fracture surface are preserved.

Heavily contaminated/oxidized surfaces may require sequential stripping of several replicas before the native surface is exposed. Those replicas applied first will contain heavier deposits than those applied subsequently, i.e., the first replica removes the most material (Fig. 1D.1). The fractured component is then cleaned in an ultrasonic bath with acetone followed by blowing with compressed air, and the effectiveness of cleaning is evaluated with a stereo microscope.

Although extraction replicas are very effective for cleaning oxidized surfaces, their main disadvantage is that, despite solvent cleaning, fragments of cellulose acetate adhere to very rough surfaces. Because this material is nonconductive, it will charge during SEM irradiation and degrade image quality. This problem is aggravated if the replica is stripped before it has completely dried; the cellulose acetate must be dry. If particle preservation is not an issue, this problem is minimized

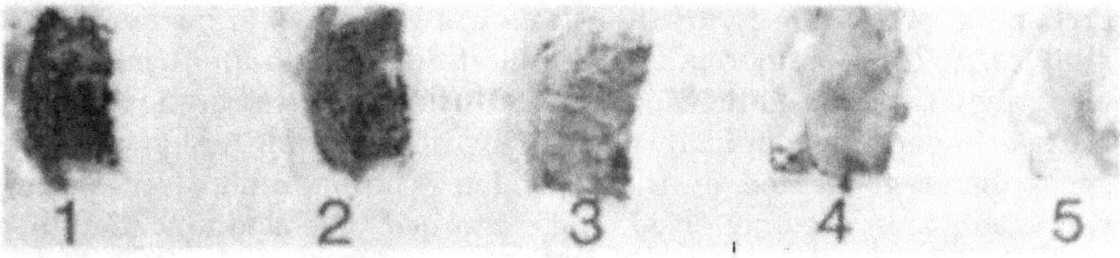

Figure 1D.1 Replicas stripped sequentially from an oxidized fracture surface. (Approximate size.) (*Courtesy of Richard Zipp.*)

by ultrasonic cleaning of the specimen with acetone between replicas. The final acetone rinse should be repeated two or three times with fresh acetone to ensure that all residual cellulose acetate has been removed.

Extraction replicas are effective for removing most oxidation products, but severely oxidized surfaces may require more rigorous cleaning. It must be understood that although cleaning will expose the metal surface beneath the oxide layer, oxidation itself has consumed some of the base metal, destroying the outermost layer of the native surface. Consequently the removal of oxide scale does not restore the fracture surface to its condition at the moment of failure. Under very harsh conditions, such as the environmental exposure of a failed component for months or years, the native fracture surface may be completely destroyed. Under less severe conditions, enough of the surface usually survives for definition of the fracture mode. Further, the very aggressive cleaning methods may themselves attack the base metal and erase all fine structural features. With these factors in mind, the analyst must be cautious and prepared to interrupt any of the following cleaning processes. One can readily resume the cleaning method, but the fracture surface cannot be restored once it has been obliterated by inappropriate cleaning methods.

Moderately aggressive cleaning with *water-based detergents* removes adherent oxides and corrosion products. Alconox is a popular detergent available from many laboratory suppliers. Zipp (1979) recommends a solution of 15 g Alconox/350 mL water; the solution is heated to 90°C and fractures are cleaned for 30 min. Simultaneous ultrasonic treatment is desirable. The specimen is then thoroughly rinsed with water followed by acetone and dried. Prolonging this treatment or increasing the concentration of the Alconox is ineffective and may cause attack of the base metal.

Cathodic cleaning is another moderately aggressive cleaning method for removal of oxides or heavy surface deposits. The specimen is made the cathode, an inert metal or graphite the anode, and both are submerged within an electrolytic bath of sodium cyanide, sodium carbonate, or sodium hydroxide (DeLeiris et al., 1966; Yuzawich and Hughes, 1978). Commercially available Endox 214 is another popular electrolyte (Dahlberg and Zipp, 1981). During the electrolytic reaction, the specimen is mechanically cleaned by the scrubbing action of hydrogen bubbles generated by the specimen. The specimen is rinsed in water and then acetone, dried, and examined. The cleaning should be periodically interrupted and its effectiveness evaluated with a stereo microscope; if required, cathodic cleaning may be repeated. High-strength steels or other alloys susceptible to hydrogen-induced cracking may be adversely affected by cathodic cleaning.

Sulfides and oxides are removed from fracture surfaces using *corrosion-inhibited acids*. The acid attacks and displaces the reaction products while the inhibitor protects the base metal from attack. However, the base metal will be attacked (etched) if the progression of cleaning is not carefully monitored. Ferrous alloys have been cleaned with 6*N* HCl containing 2 g/L of hexamethylene tetramine (DeLeiris et al., 1966; Lane and Ellis, 1971; Dahlberg, 1974). Kayafas (1980) used 1,3-di-*n*-butyl-2-thiourea to remove iron sulfide films. Both ferrous and nonferrous alloys may be cleaned with 2-butyne-1,4-diol inhibited HCl (Nathan, 1965; Farrar, 1974; Dahlberg, 1976). This corrosion-inhibited acid is prepared as follows:

HCl (1.190 specific gravity)	3 mL
2-Butyne-1,4-diol (35% aqueous)	4 mL
Distilled water	50 mL

The specimen is cleaned by immersion in an ultrasonic bath for 30 s, followed by rinsing with water and then acetone, and drying. Do not exceed a 30-s exposure to these solutions; a prolonged treatment increases the probability of base-metal attack.

If the corrosion-inhibited acid method fails, as a last resort the specimen may be cleaned in a *weak acid or base*. This extremely aggressive method will attack the base metal unless constantly monitored. Knapp (1948) recommends weak acetic acid, phosphoric acid, or sodium hydroxide for cleaning ferrous alloys. Titanium alloys may be cleaned with nitric acid (Zipp, 1979). Aluminum alloys are cleaned with a mixture of orthophosphoric acid (70 mL of 85% aqueous solution), chromic acid (32 g), and distilled water (130 mL), as described by Pittinato et al. (1975). Following the brief submersion in an acid or base, the specimen is rinsed in water and then acetone, and dried. Water washing must be thorough to stop the reaction; residual acid or base will consume the base metal, and vapors will attack stereo microscope objective lenses.

To summarize, the analyst should always choose the least aggressive cleaning method that effectively exposes the fracture surface. The objectives of the study should be known before the specimen is cleaned; if debris removed from the surface requires preservation, extraction replicas should be prepared, because with other methods the surface materials are lost. One should also evaluate the effectiveness of the treatment while it is in progress; cleaning will not be adversely affected if stopped and restarted. Always evaluate cleaning with a low-power stereo microscope, and when the cleaning appears adequate, examine the specimen in the SEM. By following these precautions, damage of the surface of interest is avoided.

References to App. 1D

Dahlberg, E. P. (1974): "Techniques for Cleaning Service Failures in Preparation for Scanning Electron Microscopy and Microprobe Analysis," *IITRI/SEM,* p. 911.

Dahlberg, E. P. (1976): "Failure Analysis by Examination of Fracture Surfaces. Analytical Procedures and Cleaning Techniques for Field Failures," *IITRI/SEM,* vol. 1, p. 715.

Dahlberg, E. P., and R. D. Zipp (1981): "Preservation and Cleaning of Fractures for Fractography—Update," *SEM, Inc.,* vol. 1, p. 423.

DeLeiris, H., et al. (1966): "Techniques of De-Rusting Fractures of Steel Parts in Preparation for Electronic Micro-Fractography," *Mem. Sci. Rev. de Met.,* vol. 63, p. 463.

Farrar, J. C. M. (1974): "The Role of the SEM in the Failure Analysis of Welded Structures," *IITRI/SEM,* p. 859.

Kayafas, I. (1980): "Corrosion Product Removal from Steel Fracture Surfaces for Metallographic Examination," *Corrosion,* vol. 36, no. 8, p. 443.

Knapp, B. B. (1948): "Preparation and Cleaning of Specimens," *The Corrosion Handbook,* Wiley, New York, p. 1077.

Lane, G. S., and J. Ellis (1971): "The Examination of Corroded Fracture Surfaces in the Scanning Electron Microscope," *Corr. Sci.,* vol. 11, p. 661.

Nathan, C. C. (1965): "Corrosion Inhibitors," in *Encyclopedia of Chemical Technology,* vol. 6, Wiley, New York, p. 317.

Pittinato, G. F., et al. (1975): *SEM/TEM Fractography Handbook,* Metals and Ceramics Information Center, Battelle Columbus Laboratories, Columbus, Ohio.

Yuzawich, P. M., and C. W. Hughes (1978): "An Improved Technique for Removal of Oxide Scale from Fractured Surfaces of Ferrous Material," *Pract. Metallog.,* vol. 15, p. 184.

Zipp, R. D. (1979): "Preservation and Cleaning of Fractures for Fractography," *SEM, Inc.,* vol. 1, p. 355.

Appendix 1E Examination of Cleaning Techniques for Postfailure Analysis*

The metallurgist has found fractographic examination of fracture surfaces to be a valuable tool in failure analysis. A meaningful examination,

---

*From R. S. Vecchio and R. W. Herzberg, in J. J. Mecholsky, Jr., and S. R. Powell, Jr. (eds.), *Fractography of Ceramic and Metal Failures,* STP 827, American Society for Testing and Materials, Philadelphia, Pa., 1982. (Copyright ASTM; reprinted with permission.)

however, may not be possible if the fracture surfaces become contaminated with grease, oil, debris, oxidation, and corrosion products. Typical cleaning practices include the use of a dry air blast, organic solvents (such as acetone, toluene, and alcohol), and repeated stripping of cellulose-acetate replicas. These procedures are useful in the removal of loosely adhering particles, grease, and oils, whereas more tightly bonded oxidation and corrosion products require more aggressive removal methods. Several previous studies[1,2] have demonstrated the effectiveness of acid-based corrosion removal techniques. For example, Kayafas[2] has shown that cleaning with an inhibited hydrochloric acid pickling solution yields adequate fractographic results on the internal surface of hydrogen blisters in pipe line steel. Lohberg et al.[3] have reported success in removing oxidation products from ferritic materials using a deoxidizing agent known as Endox. In addition, Zipp[4] has recommended that if the benign cleaning methods (air blast, organic solvents, and repeated replica stripping) are unsuccessful in the removal of tightly bonded oxidation and corrosion products, then the fracture surface could be cleaned in the following manner: submerge the sample in a water-based detergent (15 g Alconox powder* + 350 cm^3 water) heated to 90°C and agitated in an ultrasonic cleaner for 30 min.

The object of this study is to evaluate the effectiveness of the cleaning procedure recommended by Zipp. To this end, a comparison is made of fracture markings from surfaces of as-fractured samples and corroded samples which were subsequently cleaned with the benign techniques and with the Alconox solution.

Experimental procedures

The material chosen for this study was ASTM A36 bridge steel because of its wide commercial use in structures and its tendency for fracture surface deterioration in the presence of aggressive environments. Three different fracture mechanisms were generated and examined in an ETEC Autoscan scanning electron microscope (SEM) at an accelerating voltage of 20 keV. None of the surfaces was coated before viewing. Cleavage fracture was generated in simple tension at liquid nitrogen temperature. A fatigue fracture surface was generated by load cycling a compact tension specimen at an R ratio of 0.7 over a stress intensity factor range of 13 to 28 $Mpa \cdot m^{1/2}$. Unstable crack growth in the remaining ligament of the fatigue sample led to the formation of microvoid coalescence.

*Alconox powder is a detergent made by Alconox Inc., New York, N.Y. 10003; ingredients are not available.

The as-fractured samples were exposed to a corrosive environment of 5% salt water at 100°C for 2 h and then left exposed in room air (60% humidity) for 48 h. This resulted in the formation of corrosion products on the fracture surfaces. An energy dispersive x-ray analysis using a Tracor Northern 1710 system was performed to examine the nature of corrosion products resulting from the salt water exposure and reaction products which might have resulted from the Alconox cleaning treatment. In addition to the laboratory-generated fracture surfaces a polished metallographic section was treated in the Alconox solution for different lengths of time (15 and 30 min) and examined in the SEM.

Results and discussion

The typical appearance of the as-fractured samples is shown in Fig. 1E.1*a* to *c*, which reveals examples of microvoid coalescence, cleavage,

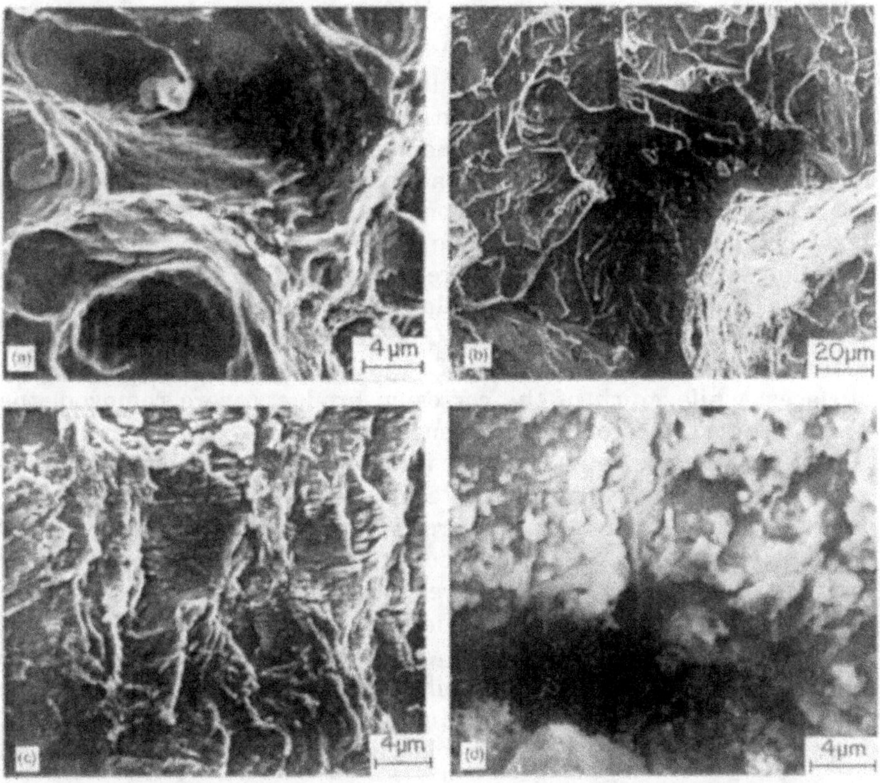

Figure 1E.1 Photomicrographs of as-fractured and as-corroded surfaces. (*a*) Microvoid coalescence. (*b*) Cleavage. (*c*) Fatigue striations. (*d*) Typical appearance of fracture surfaces after salt water exposure.

and fatigue striations, respectively, as developed in this material. Isolated regions of fracture through pearlite packets, as evidenced by the observed carbide and ferrite lamellae, were also noted on all fracture surfaces. A photomicrograph typical of all fracture regimes in the corroded condition is shown in Fig. 1E.1*d*. This photograph clearly illustrates how difficult it is to examine a corroded fracture surface for evidence of specific micromechanisms of failure. After use of the benign cleaning techniques some oxidation and corrosion product remained on the surface, though the microvoid coalescence and cleavage regions (Fig. 1E.2*a* and *b*) could be identified. The fatigue-damaged region did not lend itself to easy analysis owing to the persistent corrosion products found on the fracture surface (Fig. 1E.2*c* and *d*).

Figure 1E.3 shows the appearance of the fracture surfaces after being cleaned with Alconox solution; much of the corrosion debris remaining after the standard cleaning procedure was removed, though there is evidence that the fracture surfaces were etched by the Alconox. Note the considerable presence of pearlite colonies on the fracture surfaces (see especially Fig. 1E.3*b* and *c*). As a result, the interpretation of fatigue fracture markings is greatly complicated by the simultaneous presence of parallel lamellae within the etched pearlite regions and parallel fatigue striation markings. Some parallel fracture surface markings were verified as fatigue striations by matching their spacings with the striation spacings predicted from the prevailing stress intensity factor conditions associated with the fatigue test.[5] Fatigue striations were also differentiated from most pearlite lamellae, since the striations were generally parallel to the advancing crack front whereas pearlite lamellae assumed a random orientation relative to the crack front. Overall, however, it was difficult to make a reliable determination of striations on the fatigue fracture surfaces; while much of the corrosion products were removed, the etching introduced complicating surface artifacts.

In general, it was found that the fatigue surfaces exhibited the poorest response to corrosion removal techniques. Other experiences in our laboratory have shown similar results with failed turbine disks.* Here, intergranular failure was easily identified, while evidence for fatigue damage remained obscured after both fracture zones were cleaned with the benign techniques. It appears that these cleaning techniques are most effective in removing corrosion products from morphological features that exhibit long-range flatness (that is, 20 to 40 μm). The (100) cleavage facets in body-centered-cubic materials, the side walls of the larger microvoids, and the intergranular facets all

*Unpublished research, Lehigh University, Bethlehem, Pa. 18015.

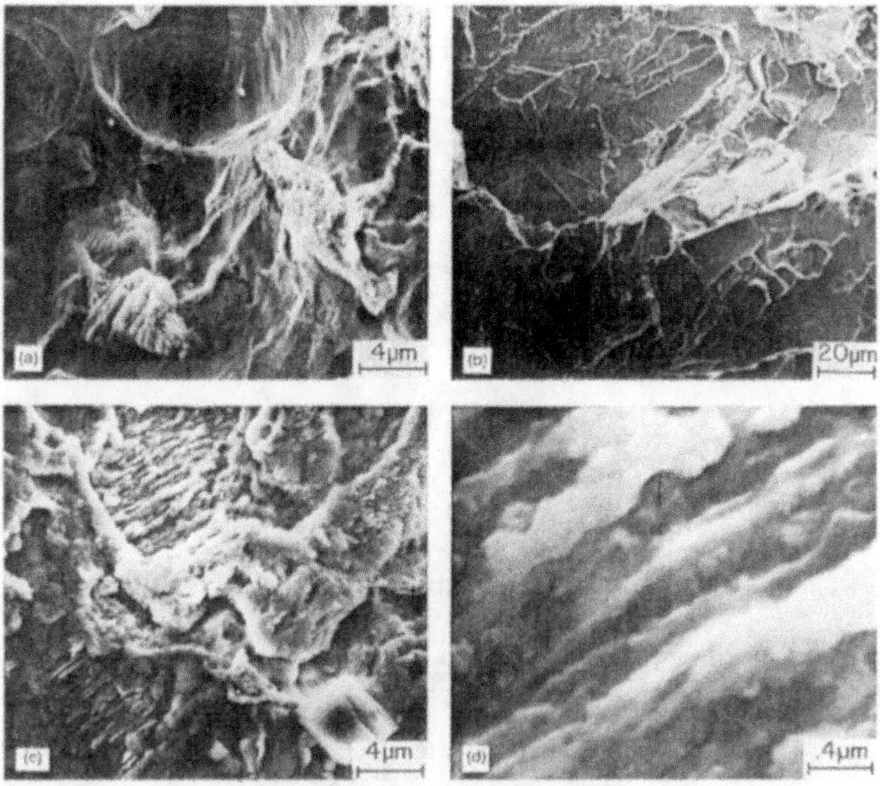

Figure 1E.2 Fracture surface appearance after cleaning by benign techniques. (*a*) Microvoid coalescence. (*b*) Cleavage. (*c*), (*d*) Fatigue striations.

exhibit such long-range flatness and appear to clean more easily than the fatigue fracture surfaces. By comparison, the sizes of fatigue striations are on the average one to two orders of magnitude smaller than these other fracture markings, and thus are more likely to be obscured by corrosion debris. If the benign cleaning techniques leave numerous corrosion-free regions on cleavage, microvoid coalescence, and intergranular fracture surfaces, then the absence of large corrosion-free regions may serve as an indication of the existence of fatigue damage. If this is the case, then these observations may aid in failure analysis. More studies are needed to clarify this point.

In many failure investigations an x-ray analysis is essential in determining the nature of fracture surface contaminants. Figure 1E.4*b* indicates the presence of large amounts of chlorine on the fracture surfaces of the as-corroded samples. If this were an actual service failure, the chlorine peak would serve as a clue in determining the nature and possible source of the corrosive medium.[6] The presence of

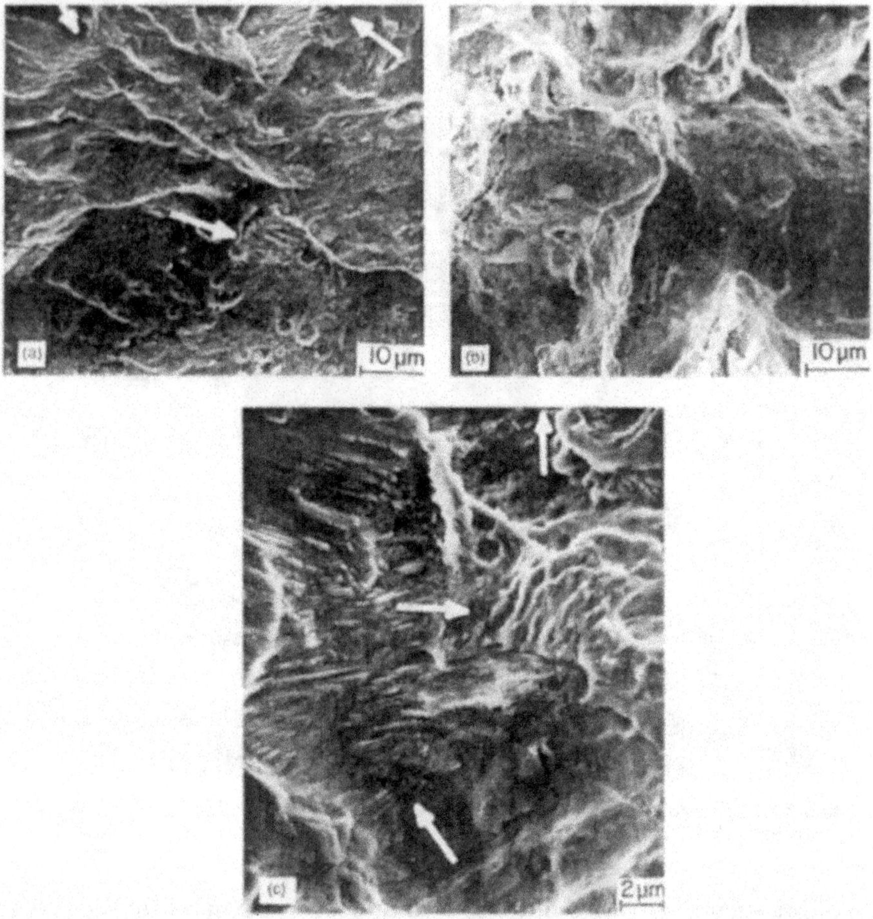

Figure 1E.3 Typical fracture surface appearance following cleaning in Alconox solution. (*a*) Microvoid coalescence. (*b*) Cleavage. (*c*) Fatigue fracture region. Note considerable cleavage and evidence of pearlite on fatigue fracture surfaces (see arrows).

the corrosive environment may have contributed to the failure (that is, stress corrosion cracking or corrosion-assisted fatigue). After cleaning the surface by the benign techniques, small amounts of chlorine were still detectable (Fig. 1E.4*c*). After use of the Alconox detergent, however, no indications remained of surface contaminants. Therefore it is suggested that if an x-ray analysis is indicated with failure examination, then use of the Alconox detergent should be avoided or deferred until after the analysis is completed.

To further study the nature of cleaning-induced damage, a metallographic section was examined after a 15-min detergent exposure. No pearlite colonies were revealed (Fig. 1E.5*b*), though the surface

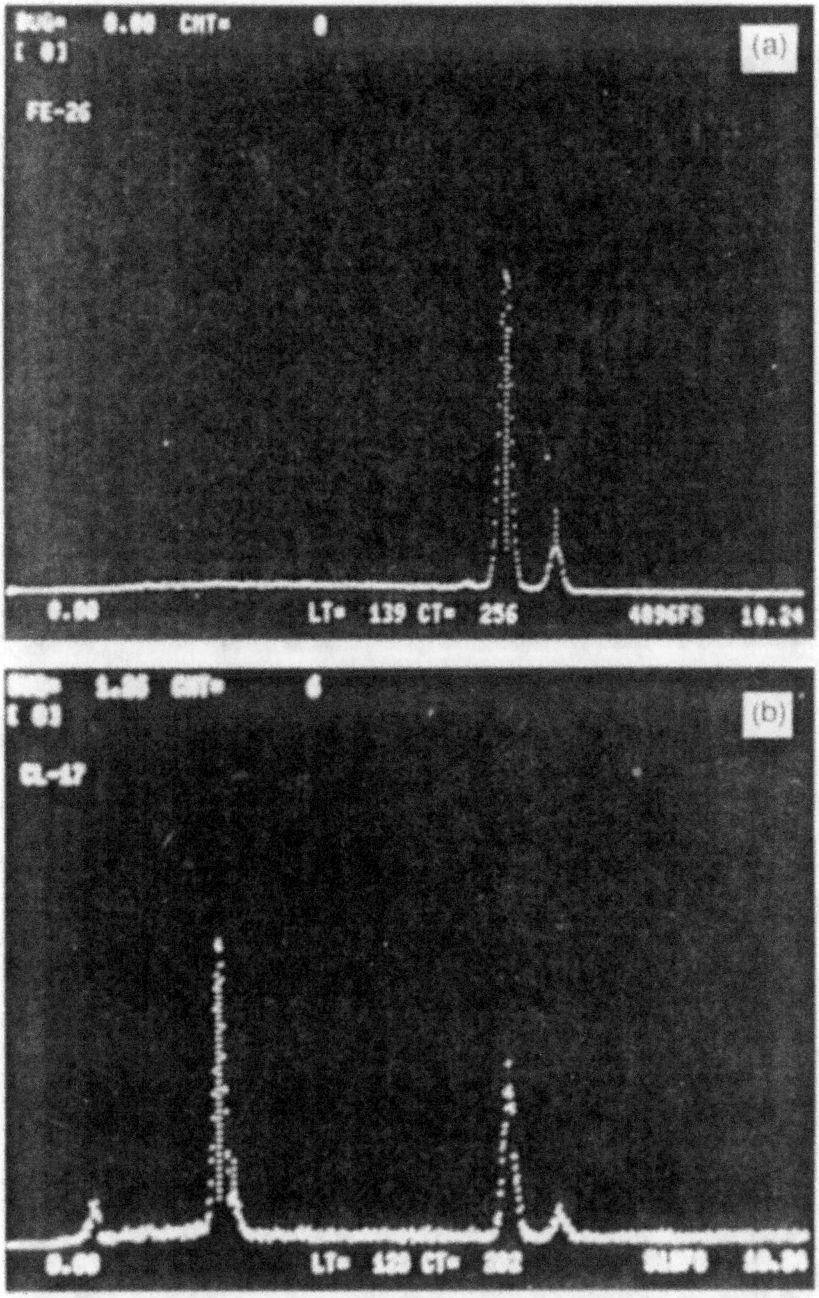

Figure 1E.4 Photospectragraphs of fracture surfaces. (*a*) As-fractured. (*b*) As-corroded. (*c*) After being cleaned with benign techniques. (*d*) After being cleaned in Alconox solution. Note the chlorine peak (at vertical dashed line) after cleaning by benign procedures and the absence of the peak following the cleaning in Alconox.

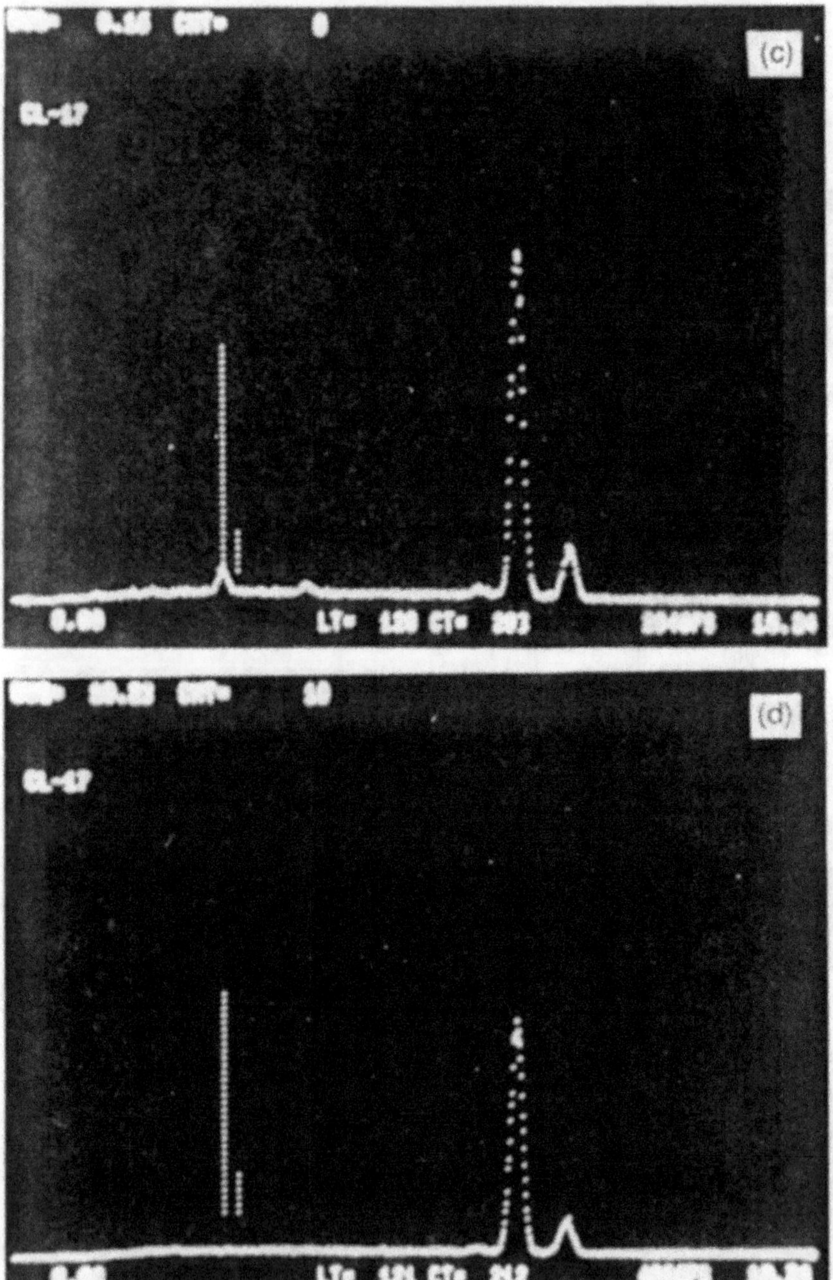

Figure 1E.4 (*Continued*)

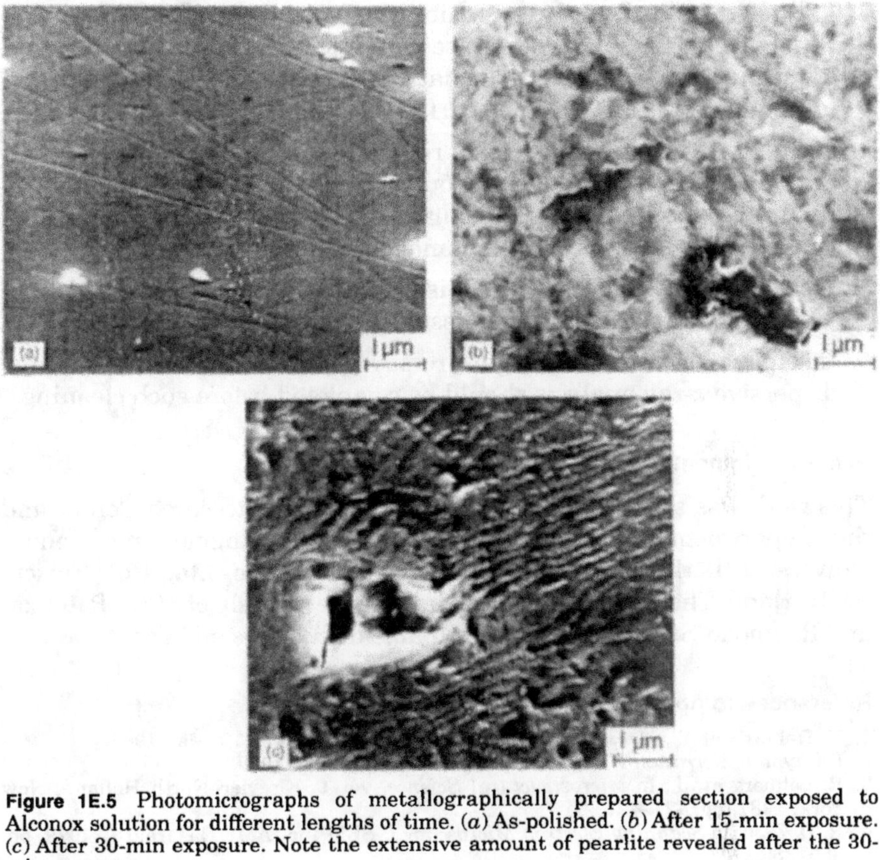

Figure 1E.5 Photomicrographs of metallographically prepared section exposed to Alconox solution for different lengths of time. (*a*) As-polished. (*b*) After 15-min exposure. (*c*) After 30-min exposure. Note the extensive amount of pearlite revealed after the 30-min exposure.

exhibited a moderate degree of degradation. After a 30-min exposure in Alconox, pearlite colonies were clearly evident. Figure 1E.5*c* shows dramatically how the surface of the metallographic section has been etched. We conclude that if cleaning in this detergent is necessary for the purpose of a fractographic examination, the exposure time should be limited to 15 min.

Conclusions

Based on the results of this investigation, the following conclusions have been reached:

1. Benign cleaning techniques are effective in the removal of loosely adhering particles, grease, and oils.

2. Fatigue-damaged surfaces exhibited the poorest response to the benign cleaning techniques. Based on this observation, identification of mechanisms responsible for fatigue damage may be difficult and unreliable when the fracture surface is obscured by corrosion debris.
3. Cleaning in Alconox detergent removes most tightly adhering oxidation and corrosion debris, though there is evidence that this procedure results in etching the surface. This degradation complicates the interpretation of fatigue-damaged regions.
4. If cleaning in Alconox detergent is necessary for the purpose of a fractographic examination, the exposure should be limited to 15 min.
5. Insomuch as Alconox detergent removes most surface contaminants, dispersive x-ray analysis should be completed before such cleaning.

Acknowledgments

This work was supported in part by the Materials Research Center and the Department of Metallurgy and Materials Engineering, Lehigh University, Bethlehem, Pa., and Swiss Aluminum Ltd. Co., Zurich, Switzerland. The authors wish to acknowledge the help of Steve Paterson and Raymond Stofanak for their valuable discussions and assistance.

References to App. 1E

1. H. DeLeiris et al., *Mem. Sci. Rev. de Met.,* vol. 63, pp. 463–472, May 1966.
2. I. Kayafas, *Corrosion NACE,* vol. 36, pp. 443–445, Aug. 1980.
3. R. Lohberg et al., in *Microstructural Science,* vol. 9, Elsevier–North Holland, New York, 1981, pp. 421–427.
4. R. D. Zipp, in *Scanning Electron Microscopy I,* SEM Inc., AMF O'Hare, Ill., 1979, pp. 355–362.
5. R. C. Bates and W. G. Clark, Jr., *Trans. Am. Soc. Met.,* vol. 62, p. 380, 1969.
6. D. B. Ballard and H. Yakowitz, in *Scanning Electron Microscopy 1970,* IITRI, Chicago, Ill., April 1970, p. 32.

Appendix 1F Recommended Cleaning Solutions for Metallic Fractures*

Solution	Purpose
Organic solvents	To remove oil, grease, or plastic coatings from fractured surfaces
Acetic acid Phosphoric acid Sodium hydroxide	Used either cold or warm to clean Fe-base alloy fractures

*From S. Battacharyya, V. E. Johnson, S. Agarwal, and M. A. H. Howes (eds.), *Failure Analysis of Metallic Materials by Scanning Electron Microscopy,* IIT Research Institute, Chicago, Ill., 1979.

Solution	Purpose
Ammonium citrate Ammonium oxalate Sulfamic acid Immersion dip for 1 to 15 min in 6*N* HCl containing 2 g per liter of hexamethylenetetramine	For cleaning rust, scale, or oxidation products from Fe-base alloy fractures
Nitric acid	Titanium alloy fractures
Orthophosphoric acid (70 mL) + chromic acid (32 g) + water (130 mL) or organic solvents	Used either cold or warm to clean aluminum alloy fractures

Appendix 1G A Scale and Rust Removal Solution*

For scale and rust removal without attack on iron, use either solution A at 150°F *or* solution B at room temperature.

Solution A	Solution B
Water 80 mL	Water 49 mL
HCl (conc.) 20 mL	HCl (conc.) 49 mL
Reilly Inhibitor #22, 4 drops	Rodine #50, 2 mL
Use at 150°F	Use at room temperature

- *Immerse* the iron parts until the rust and scale are removed.
- *Wash* the parts with water and with alcohol.
- *Coat* or *paint* the clean dry surfaces with Rustarest D14-50-S.

Note: Reilly Inhibitor #22 and probably Rodine #50 give off gas on standing. Leave the closure loose so that the gas can escape without building up an excessive pressure.

- Reilly Inhibitor #22 from Reilly Industries, Inc., 3201 Independence Road, Cleveland, OH 44105.
- Rodine #50 from Henkel Surface Technologies, 32100 Stephenson Highway, Madison Heights, MI 48071.
- Rustarest® from PPG Industries, One PPG Place, Pittsburgh, PA 15272.

*Origin unknown.

Chapter

2

Mechanical Aspects and Macroscopic Fracture-Surface Orientation

As is usual in a severe frost, we have recently heard of many severe accidents consequent upon the fracture of the tires of the wheel of railway-carriages. JAMES PRESCOTT JOULE
Philosophical Magazine, *1871*

2.1 Introduction

The overall macroscopic orientation of the fracture surface of a broken component is generally related to the loading conditions. The relationship between this orientation and the load may be complex and difficult to deduce, but in spite of this, in many cases considerable information may be gleaned from the fracture-surface orientation. For example, if the fracture clearly did not involve any plastic deformation (as deduced from the absence of significant geometric or dimensional change) and hence the fracture was brittle, and the macroscopic fracture surface was relatively flat, then fracture probably occurred by a stress normal to the fracture plane.

The purpose of this chapter is to review mechanics aspects which are related to the cause of the fracture-plane orientation. The principles are outlined only in sufficient detail to illustrate their application, and examples of macroscopic fracture surfaces are given to demonstrate the ideas. It is emphasized that it is the macroscopic fracture-surface orientation that is the subject of this chapter. The detailed fracture-surface topography is the subject of Chaps. 3 and 4.

2.2 Tensile Test

To introduce the concepts of loading and fracture, it is useful to examine the simple tensile test. This provides a definition of the common tensile mechanical properties and also allows distinguishing between elastic and plastic deformation.

Consider a cylinder loaded along its axis, and the tensile machine operated so that the cylinder elongates at an approximately constant rate. Let the load be increased so that the cylinder increases in length, and then let the load be reduced to zero. If the length is now the same as before the load, the material is said to have been deformed elastically. Thus *elastic* deformation can be defined as deformation such that any changes in dimensions or shape are recovered when the external load is removed. However, if the axial load is increased beyond a certain value and then reduced to zero, the length will be greater than that before loading, and thus the cylinder has deformed plastically. *Plastic* deformation can be defined as deformation in which there is permanent change in dimensions or shape after the removal of the external load. The load beyond which plastic deformation occurs is called the *elastic limit.* This behavior is shown schematically in Fig. 2.1.

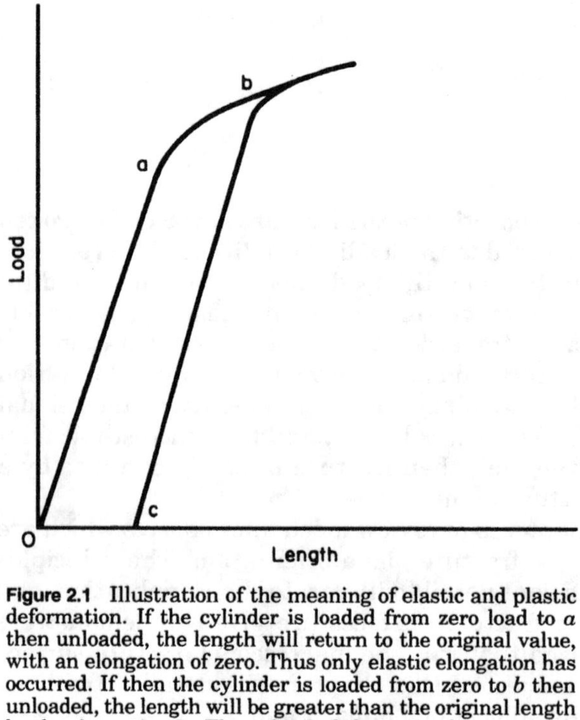

Figure 2.1 Illustration of the meaning of elastic and plastic deformation. If the cylinder is loaded from zero load to *a* then unloaded, the length will return to the original value, with an elongation of zero. Thus only elastic elongation has occurred. If then the cylinder is loaded from zero to *b* then unloaded, the length will be greater than the original length by the elongation 0c. Thus plastic deformation has occurred.

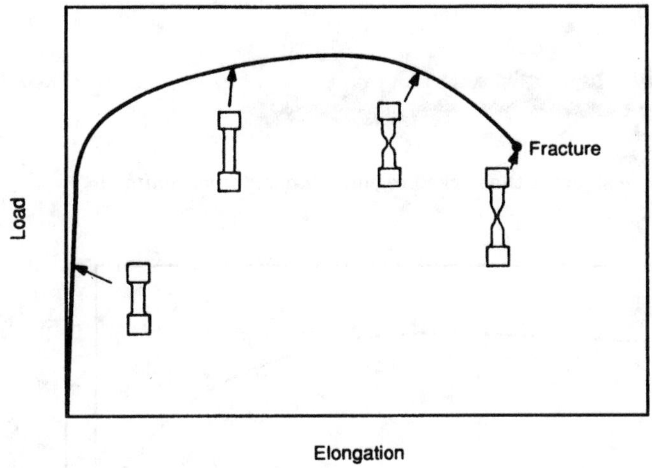

Figure 2.2 Typical load-elongation curve, showing the change in shape of the cylinder at various test stages.

For most metallic materials, to cause the cylinder to continue to elongate requires a continually increasing load. This is because the material becomes stronger as it is plastically deformed, and the material is said to *work* or *strain harden*. However, beyond a certain load and elongation, it is observed that subsequent plastic deformation occurs in a very localized region, which exhibits a decrease in cross-sectional area. This is called *necking*. Since the supporting cross-sectional area is now reduced, less load is required to continue to elongate the cylinder. Therefore the load-elongation curve passes through a maximum, and the load decreases until fracture occurs. A complete load-elongation curve is presented schematically in Fig. 2.2, which also shows the change in the profile of the cylinder. Figure 2.3 shows a tensile sample in the necking stage.

In a common standard tensile test, a specimen of specified dimensions (such as a gage length of 2.000 in and a diameter of 0.505 in) is used. The increase in the gage length is recorded as a function of load until fracture occurs. To make the results comparable to cases where the sample might have a different cross-sectional area, the load is divided by the original cross-sectional area to obtain the *nominal* or *engineering stress*. Also the elongation (change in length) is divided by the original length to obtain the *nominal* or *engineering strain*. It is customary to plot the engineering stress versus the engineering strain to obtain a stress-strain curve. A typical result is shown in Fig. 2.4. On the scale plotted, the linear elastic range is not well resolved. Figure 2.5 shows a stress-strain curve with the low-strain region expanded so that the elastic region is revealed.

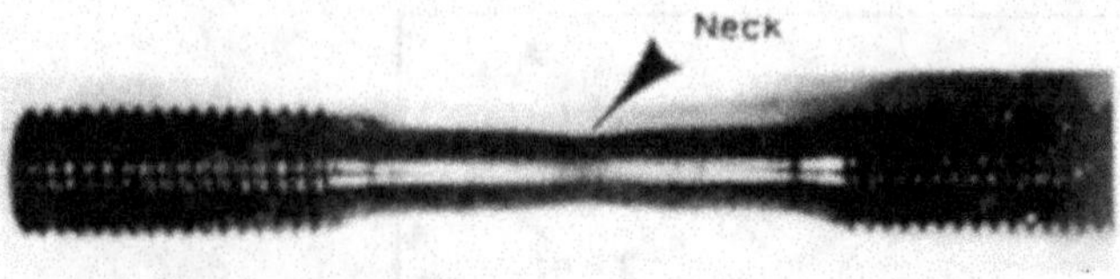

Figure 2.3 Tensile sample showing the necked region. (*Adapted from Smith.*[1])

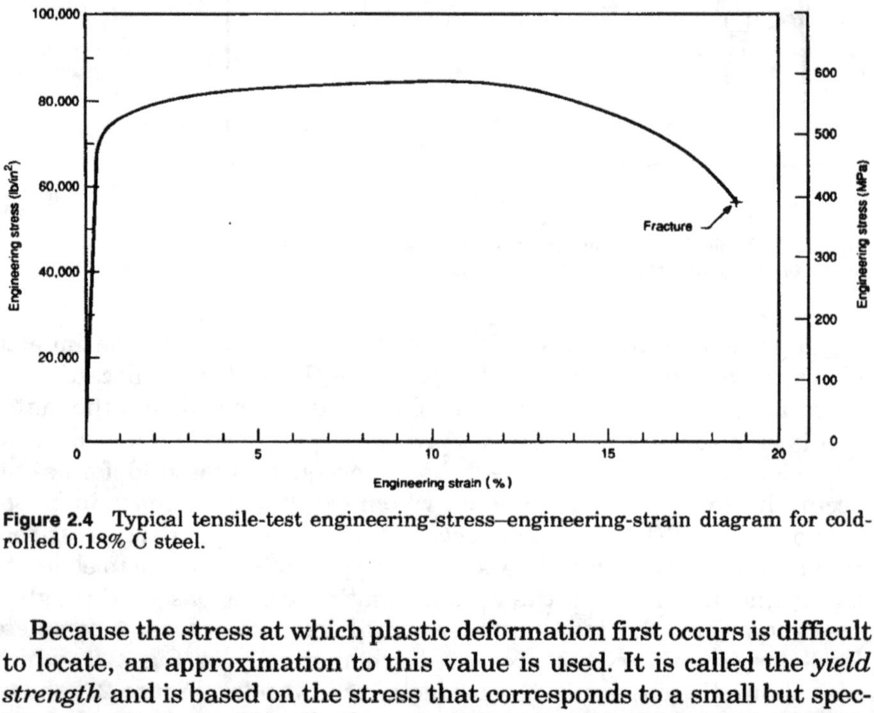

Figure 2.4 Typical tensile-test engineering-stress–engineering-strain diagram for cold-rolled 0.18% C steel.

Because the stress at which plastic deformation first occurs is difficult to locate, an approximation to this value is used. It is called the *yield strength* and is based on the stress that corresponds to a small but specified plastic strain. Usually the value of strain of 0.002 is used, which is 0.2 percent. The procedure for obtaining such a yield strength is illustrated in Fig. 2.5. A line parallel to the linear elastic line is drawn from a strain of 0.002 (0.2 percent), and its intersection with the stress-strain curve defines the 0.2 percent yield strength. In some materials (for example, normalized low and medium carbon steels), yielding is very prominent in the stress-strain curve. This is illustrated in Fig. 2.6. In this case, the *yield point* or *yield-point stress* is used as a measure of the stress to induce yielding, and an effective yield strength is not quoted.

A prominent feature of the engineering-stress–engineering-strain curve is the maximum, and the stress at the maximum is called the *ultimate strength, ultimate tensile strength,* or more commonly *tensile strength.* Note that it is an artifact of the way the test is made, associated with necking.

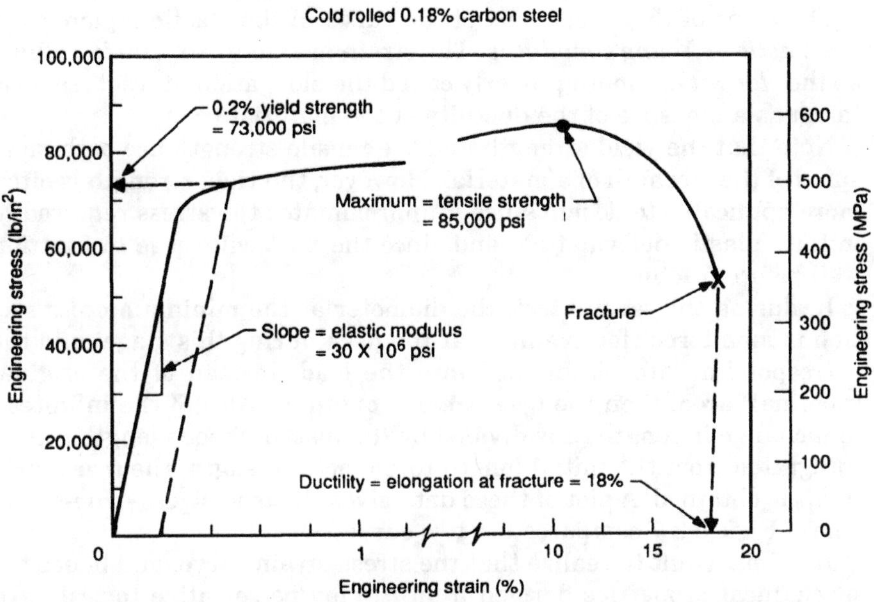

Figure 2.5 Engineering-stress–engineering-strain diagram for cold-rolled 0.18% C steel, showing how the 0.2 percent yield strength and other tensile mechanical properties are determined.

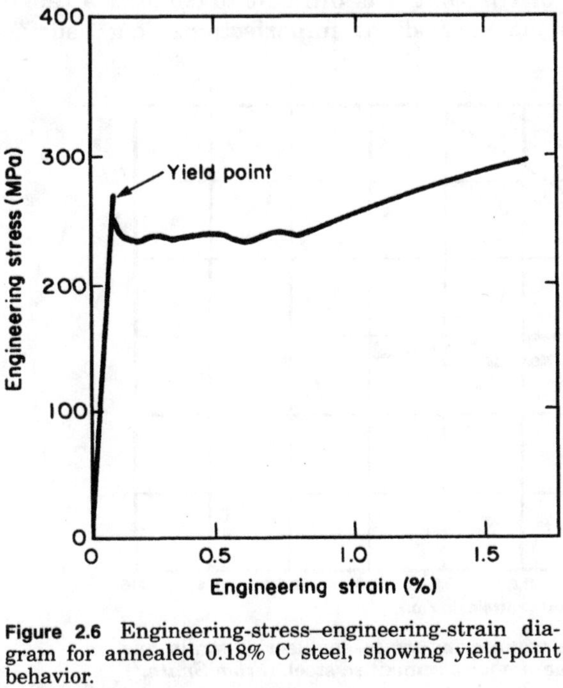

Figure 2.6 Engineering-stress–engineering-strain diagram for annealed 0.18% C steel, showing yield-point behavior.

The slope of the linear part of the curve in the elastic region gives the *elastic* or *Young's modulus.* The strain at fracture × 100 is defined as the *elongation,* more properly called the elongation at fracture. It is taken as a measure of the ductility of the material.

Note that the yield strength and the tensile strength are both measures of the strength of a material. However, the yield strength is often more applicable to design since it approximates the stress required to initiate plastic deformation, and since the part will cease to function well before fracture.

If, during the tensile test, the diameter at the minimum cross section is monitored (for example, at the neck during this stage) and the corresponding area is divided into the load, instead of the original (nominal) area, then the *true stress* is obtained. Also, if the infinitesimal change in length *dl* is divided by the instantaneous length *l,* then integrated from the initial length to the actual length, the *true strain* ln l/l_0 is obtained. A plot of these data gives the true-stress–true-strain curve. Figure 2.7 compares the two curves.

It is important to realize that the stress-strain curve, and hence the mechanical properties derived from it, may be sensitive to test variables such as test temperature and strain rate (see Sec. 2.9). Thus this behavior has to be taken into account in using tensile mechanical properties.

Most ceramics are so brittle that it is difficult to conduct a tensile test as cracks nucleate at very slight imperfections (e.g., surface

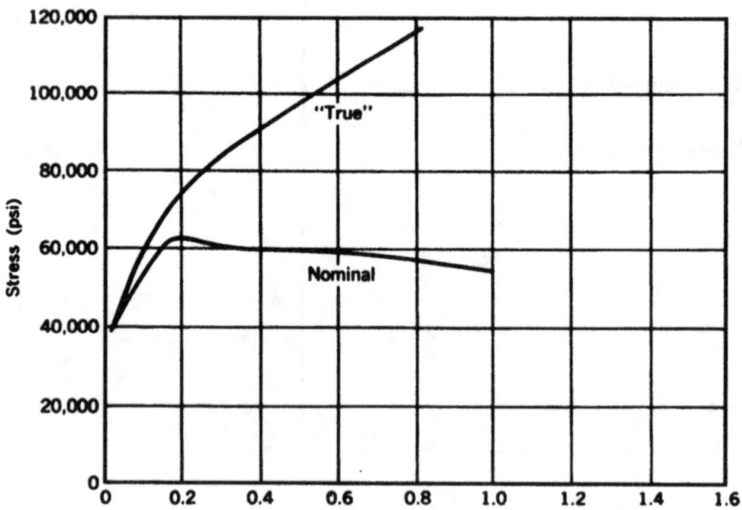

Figure 2.7 Comparison of the engineering-stress–engineering-strain and the true-stress–true-strain diagram for a semikilled steel. (*From Smith.*[2])

scratches) and propagate catastrophically. This requires data to be obtained in compression testing, but even then the elongation (negative) at which cracking is observed is very low. There are a few ceramics that exhibit some ductility, such as those in which elastic stresses initiate a phase transformation with a volume change to allow change in shape (e.g., transformation-induced toughened Zirconia).

The stress-strain behavior of polymers depends on the molecular structure (see Sec. 3.15). Those with relatively simple structure (e.g., polyethylene consisting of linear —C— chains) will consist of crystals separated by amorphous regions. These polymers show a marked yield point followed by great elongation. An example of the stress-strain curve is shown in Fig. 2.8. It is pointed out that the elastic modulus of polymers is considerably less than that of metallic materials and ceramics.

As the molecular arrangement of a polymer becomes more complicated (e.g., cross-linking, side chains), the material becomes more rigid and the ductility decreases. The structures of some polymers are so complicated that they occur only as amorphous solids. At room temperature and below, many of these polymeric materials are very brittle, and exhibit no or very little elongation at fracture.

The response of polymers to loading can be quite sensitive to strain rate and temperature at and near 25°C. One reason is that at these temperatures there can be considerable thermal activation. Diffusion in the sense of atoms moving from one lattice site to another, such as occurs in metallic and ceramic materials, does not occur in polymers. Instead, elevated temperature allows relaxation of molecular positions to occur. The sensitivity of the stress-strain curve to strain rate is illustrated by the data in Fig. 2.9.

2.3 Principal Stresses

The loading on most components is usually neither unidirectional nor uniform. Thus the three-dimensional nature of the loading must be considered. The loads are expressed in terms of stress, which is the load at any point divided by an area of specified orientation relative to the direction of the force, with the area reduced to an infinitesimal size. Two types of stresses are commonly used: a *normal stress* [which can be positive (tension) or negative (compression)], based on an area normal to the direction of the force, and a *shear stress,* based on an area parallel to the force. The concept of normal and shear stresses is illustrated in Fig. 2.10. Upon twisting the hose, the vertical markers across the longitudinal slit are offset (Fig. 2.10*a*), showing that a shear stress would be generated if the tube were not cut. If the tube is slit at 45°, then upon twisting, the tube separates normal to this

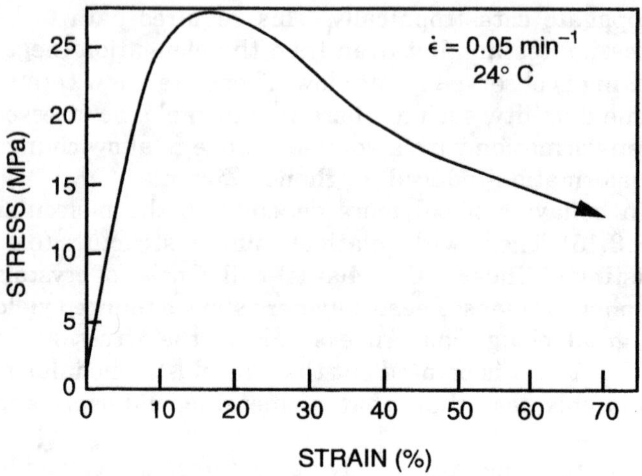

Figure 2.8 A stress-strain curve for linear polyethylene. (*From Rabinowitz and Brown.*[3])

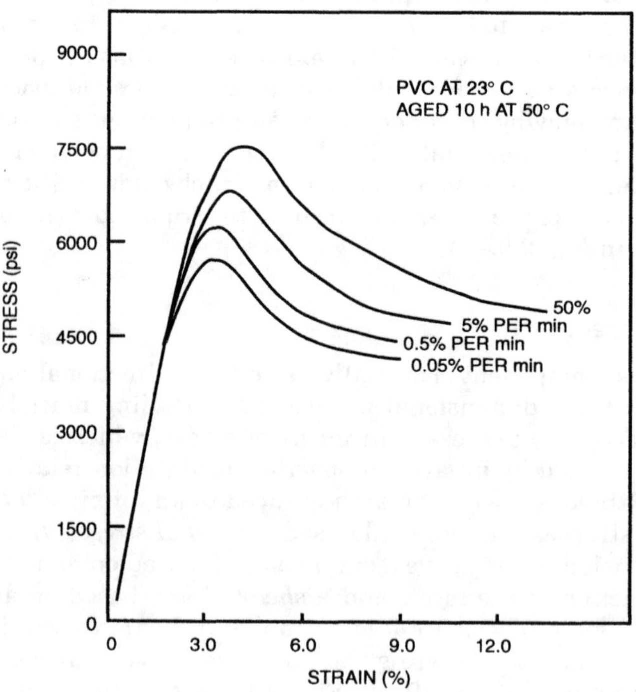

Figure 2.9 Effect of strain rate on the stress-strain curves for polyvinyl chloride. (*From Matsuoka.*[4])

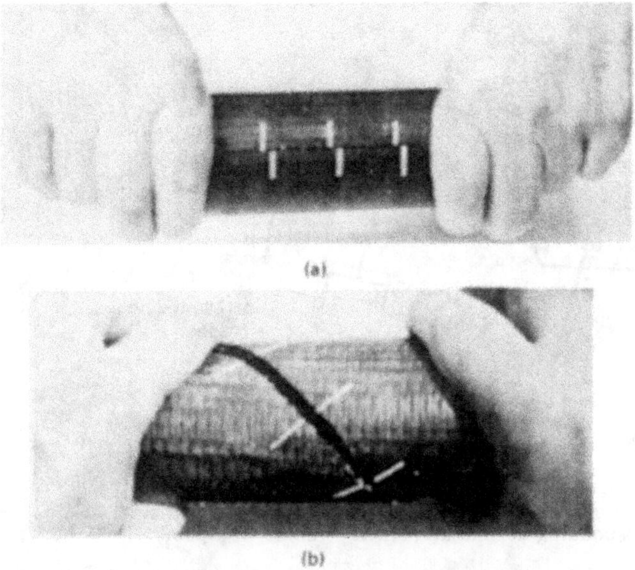

Figure 2.10 Illustration of the concept of normal and shear stresses. (*From Wulpi.*[5])

slit, showing that a normal stress would exist perpendicular to the line at 45° (Fig. 2.10*b*).

The external load can be resolved into three orthogonal components to produce three orthogonal stresses. This is easier to visualize in two dimensions, as illustrated in Fig. 2.11. The normal stress at the location shown is that of the resolved forces acting on an infinitesimal plane normal to the plane of the page and perpendicular to the force. Since the small region is constrained from rotating, the overall loading causes shear loads to act on adjacent faces to prevent this. The corresponding shear stresses are associated with the same areas (normal to the plane of the page) as the normal stresses.

A fundamental result of the application of mechanics to such a configuration is that there are orientations for which the normal stresses are a maximum and a minimum, and on these planes the shear stress is zero. This is illustrated in Fig. 2.12 for a simple two-dimensional loading case. The resolution of the forces onto different planes clearly gives a maximum and a minimum normal stress for the orientation in Fig. 2.12*a*. The maximum and minimum normal stresses are called the *principal normal stresses,* and the planes on which these extreme values of normal stress are obtained are called the *principal planes.* There also exists a plane on which the shear stress is a maximum and one on which it is a minimum, and these stresses are called *principal*

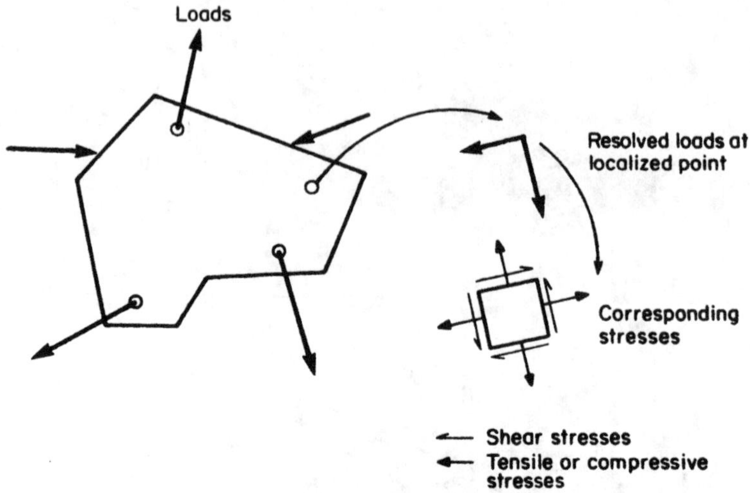

Figure 2.11 Resolution of external loads on a component into two orthogonal loads, and the corresponding normal and shear stresses, at a point.

shear stresses. (The normal stresses are not necessarily zero on these planes.) It is necessary to determine these principal stresses at all locations within the component in order to know where failure is likely to commence. These principal stresses are used in almost all modeling, stress analysis, and failure criteria.

In some loading configurations the orientation of the principal stresses is obvious, but in most cases an analytical treatment is required to determine this. The relation of the orientation of the external loading to that of the principal stresses is illustrated by a few examples, which were chosen for their simplicity and also because they involve common types of loading. Consider a round cylinder loaded axially, and the load uniformly distributed across the area normal to the loading direction (Fig. 2.13*a*); there is no other external load. The maximum stress σ_1 is the axial tensile stress. The other two normal stresses σ_2 and σ_3 are both zero, and hence are the minimum normal stresses. In this case the maximum shear stress is at 45° to the cylinder axis. In the case of simple bending of a beam of circular cross section (Fig. 2.13*b*), the upper surface is in tension in the longitudinal direction, and the lower surface is in compression (negative stress). The maximum normal stress is at the top surface, and the minimum normal stress is at the bottom (and negative). The maximum shear stress is on two orthogonal planes: perpendicular and parallel to the cylinder axis. It is a maximum at the axis. For the case of simple torsion of a cylinder (Fig. 2.13*c*), the maximum and minimum normal stresses are at 45° to the cylinder axis (the third normal stress is zero),

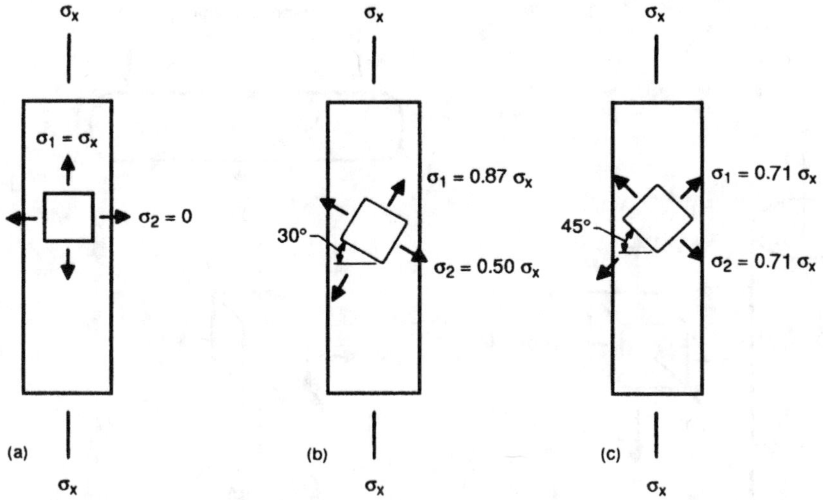

Figure 2.12 Illustration of the concept of maximum and minimum normal stress. The sample is a cylinder loaded axially with stress σ_x. (*a*) On the element shown, the maximum normal stress is σ_1 which is equal to σ_x. The minimum normal stress is zero. (*b*), (*c*) As the element is rotated, the maximum normal stress never equals the value of $\sigma_1 = \sigma_x$ in *a*.

and the maximum shear stress is parallel to the axis. This is illustrated in Fig. 2.10.

The effect of the principal stresses on the macroscopic fracture-surface plane can be visualized in the case of a material which fractures on a plane normal to the principal stress. Such a material would be considered brittle (but see Sec. 2.11). A ductile material tends to flow along the planes of maximum shear stress. Thus for the case of simple tensile loading, the maximum normal stress is parallel to the cylinder axis, and if the material is brittle, fracture occurs across the plane normal to the axis (Fig. 2.14). However, if the material is ductile, so that extensive plastic deformation can occur, then the fracture-plane orientation is more difficult to visualize, but should involve flow on planes of maximum shear stress; that is, on planes at 45° to the load. For simple torsional loading, the maximum normal stresses are at 45° to the cylinder axis, and the fracture-surface path is spiral and inclined at 45° to the cylinder axis (see also Fig. 2.13*c*). Cracking may also be found on a longitudinal plane which has the same shear stress acting on it. For simple bending, the maximum normal stress is parallel to the cylinder axis and varies from a tensile value on one surface to zero at the center and to a compressive value on the other surface. If the material is brittle, fracture then occurs on a plane normal to the cylinder axis. If the material is ductile, a topography similar to that of a tensile fracture of a ductile material occurs (Fig. 2.14*a*).

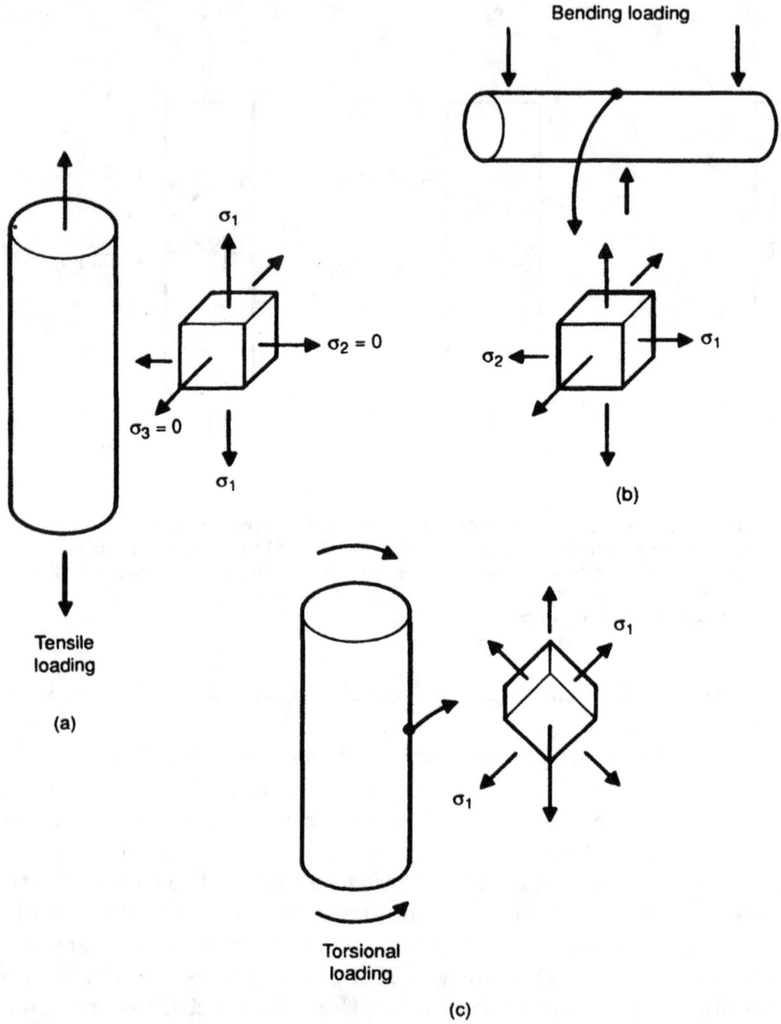

Figure 2.13 Illustration of the principal normal stresses and their orientation for cylinders loaded in tension, bending, and torsion. (The principal shear stresses are not shown.)

2.4 Stress Concentration

So far we have considered geometries with constant cross-sectional areas. Machine components, however, are usually not so simple; even at the surface there are irregularities from machining. Consider a cylinder of a cross-sectional area of 20 in^2 (129 cm^2) loaded axially with 2000 lb (8900 N). If the load is applied uniformly, then each unit of area has the same load on it. For simplicity we can think of each

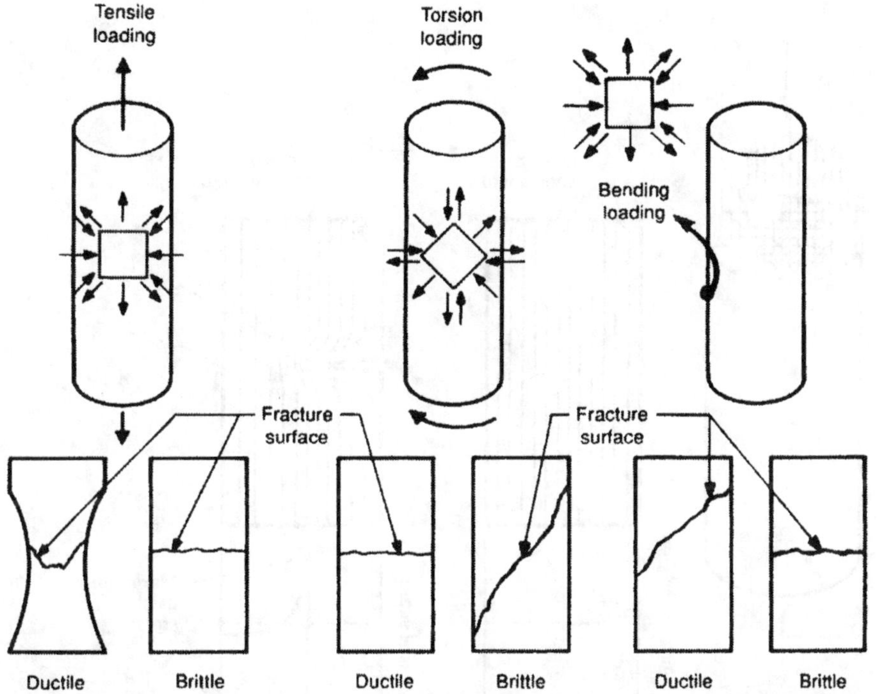

(a)

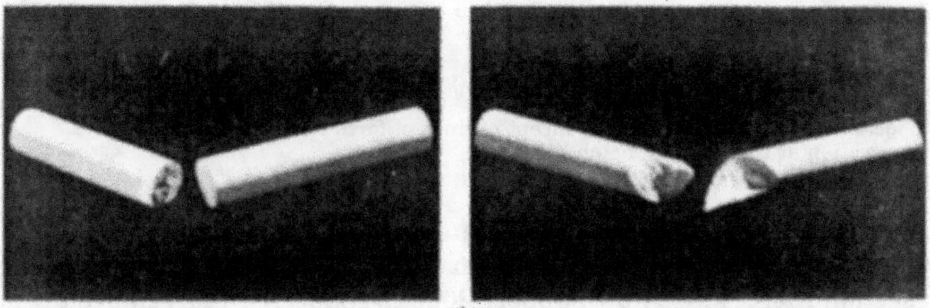

(b)

Figure 2.14 (*a*) Illustration of the orientation of the principal normal and shear stresses at the surface of cylinders loaded in simple tension, torsion, and bending, and the expected orientation of the fracture surface if the material is brittle and ductile. (*b*) Photographs of a piece of chalk broken by pulling axially (left) and by twisting (right). (*From Feynman et al.*[6])

square inch supporting 100 lb (445 N) to give a normal stress of 100 lb/in^2 (689,000 Pa). Also the lines centered on each square inch can be considered to be a line of force (or stress line), as depicted in Fig. 2.15*a*. Now consider a cylinder which has a circumferential notch in it so that the load-supporting area at the notch location is less than 20 in^2 (129

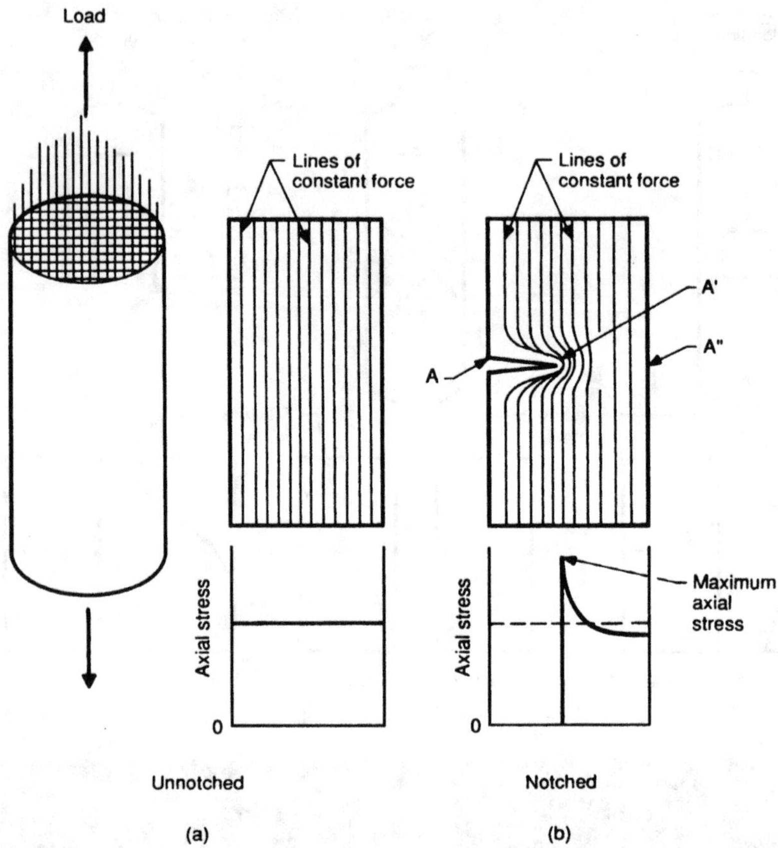

Figure 2.15 Illustration of the concept of lines of force and of the stress concentration at the root of a notch.

cm²) (Fig. 2.15*b*). There is no load applied to the surface *A–A′*, so the internal stress must go to zero on this surface. The remaining cross section *A′–A″* must now support 2000 lb (8900 N). Far removed from the notched surface, the load is still distributed essentially uniformly, but as the notch is approached, the lines of force become denser, attaining a maximum density right at the root of the notch. Thus the axial stress increases, as shown in Fig. 2.15*b*.

The stress distribution depends on the size and shape of the component and the change in geometry (such as hole, circumferential notch). Examples in terms of the lines of force are shown in Fig. 2.16. The locations at which the axial stress is high (higher density of lines of stress or force) are clear, and the increase in the stress intensity with increasing sharpness of the surface discontinuity is clear (compare Fig. 2.16*a*, *b*, and *c*).

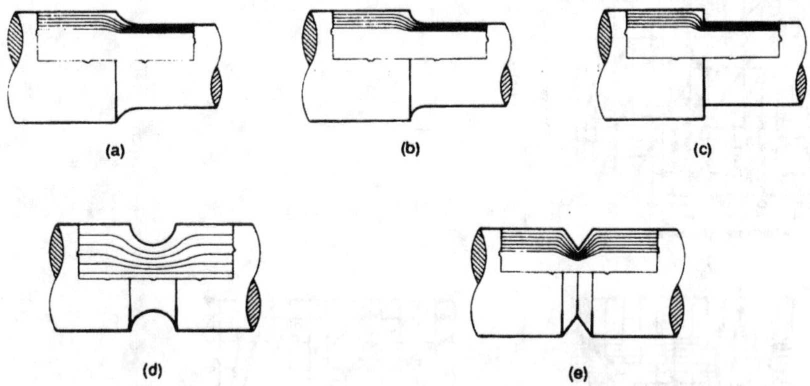

Figure 2.16 Illustration of the effect of geometry on the stress distribution in cylinders which are loaded axially. (*Adapted from* Metals Handbook.[7])

For simple cases the effect of such geometry changes can be given in terms of a stress concentration factor K_t. This factor is defined as the maximum normal stress (principal stress) divided by the nominal stress in the same direction, and its magnitude depends on the geometry and type of loading. The curves in Fig. 2.17 illustrate this. Consider the case for simple tension in Fig. 2.17*a*. The maximum axial stress occurs right at the root of the notch. It depends on the geometric factors shown in the figure and increases as the radius of the notch decreases relative to the diameter of the cylinder. For example, if the surface of a 1-in-diameter cylinder has a circumferential notch around it which is 0.1 in in radius, then $h/d = 0.1$ and $h/r = 1$. This gives $K_t = 1.6$. Thus the axial stress right at the root of the notch is about twice that calculated based on the overall area and the axial load. If the notch is about 0.001 in, the maximum stress is considerably higher. This points up the fact that surface irregularities act as *stress concentrators* or *stress risers,* and thus are common locations for fracture to commence. Internal geometry changes, such as holes, also act as stress concentrators, and thus the critical locations depend on the geometry of the component.

The curves in Fig. 2.17 indicate that fracture would be ensured for surface notches (or irregularities) sufficiently acute since K_t increases greatly as the notch size decreases. However, these curves are based on elasticity theory, and in even relatively brittle alloys the region at the root of the notch undergoes some plastic deformation to relieve the stress. Thus in real materials the stresses are modulated by the plastic deformation zone which develops around the notch root.

Another important consideration is the variation in K_t with the location in the component. This is illustrated in Fig. 2.18 for the axial loading of a cylinder and of a plate with a surface notch. It is seen that K_t is a maximum at the root of the notch, as demonstrated before. Thus

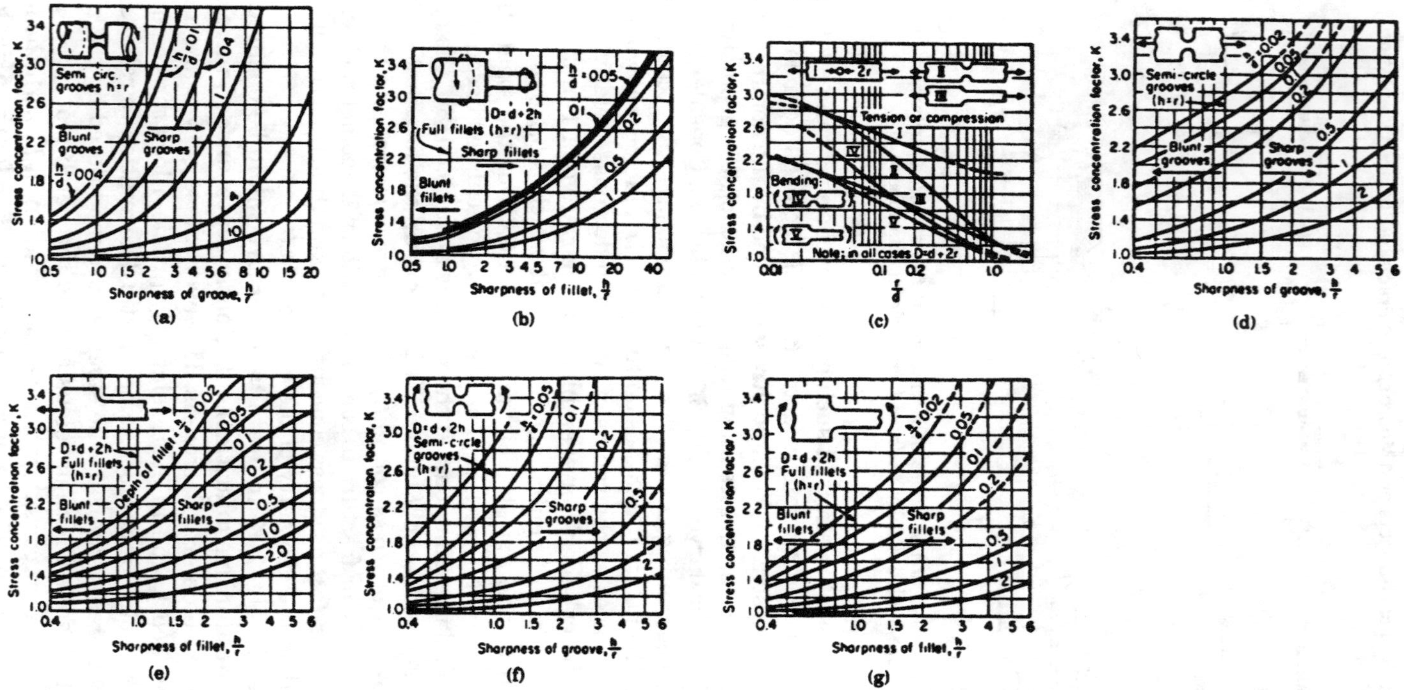

Figure 2.17 Calculated values of the stress concentration factor for various geometries. (*a*) Filleted shaft in tension. (*b*) Filleted shaft in torsion. (*c*) Flat plate with semicircular fillets and grooves or with holes in tension or compression. (*d*) Flat plate with grooves in tension. (*e*) Flat plate with fillets in tension. (*f*) Flat plate with grooves in bending. (*g*) Flat plate with fillets in bending. (*From Avallone and Baumester.*[8])

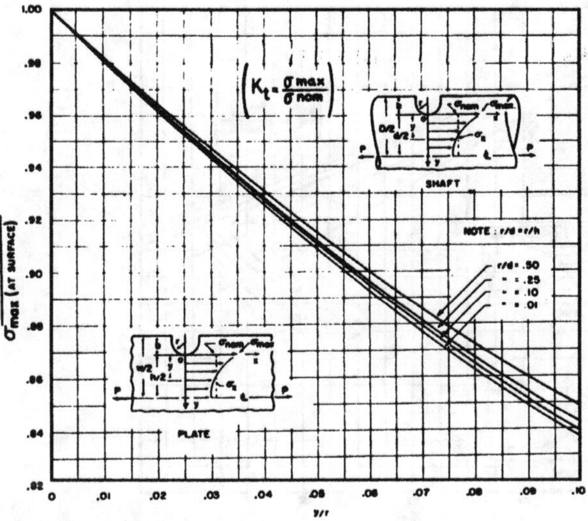

(a)

Figure 2.18 Calculated values of the variation of the maximum normal stress with depth in cylinders and plates loaded in (*a*) tension, (*b*) bending, and (*c*) torsion. (*From Lipson and Juvinall.*[9])

this is the location of expected initiation of fracture. However, consider the case of a cylinder made of an alloy for which the strength at the surface is sufficient to resist fracture. Then if the strength decreases with radius (such as a carburized steel), fracture may initiate below the surface at a location where the loading stress exceeds the material's strength. (Note that this possibility is present even if there is no surface discontinuity.)

2.5 Triaxial Stress and Constraint

It is important to recognize that a triaxial stress distribution can develop even if the external load is uniaxial. When a sample is loaded elastically in uniaxial tension (for example, a tensile test), the length of the sample increases. However, there is a concomitant reduction in radius. The ratio of the dimensional change (strain) in the axial direction to that in the radial direction is called *Poisson's ratio.* It is between 0.25 and 0.35 for most alloys.

Now consider a sample loaded in uniaxial tension with a notch present (Fig. 2.19*a*). We want to examine the stresses in the axial, width, and thickness directions (*Y, X,* and *Z*). In the width direction and at the surface of the notch, there is a bending stress due to the cantilever action of the notch, which decreases to zero at the root of

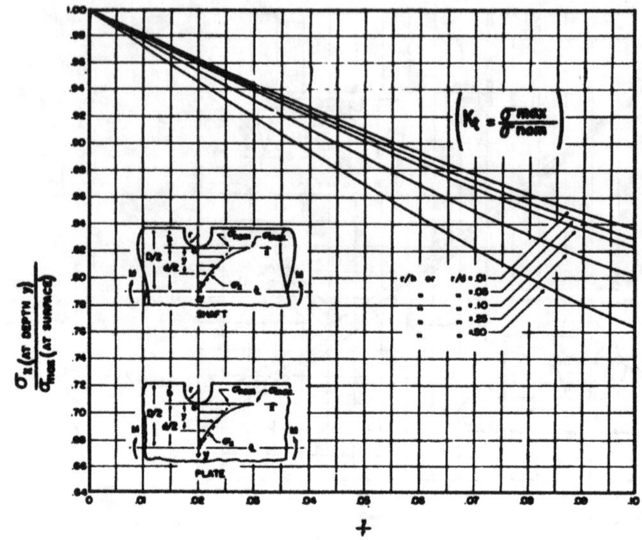

(b)

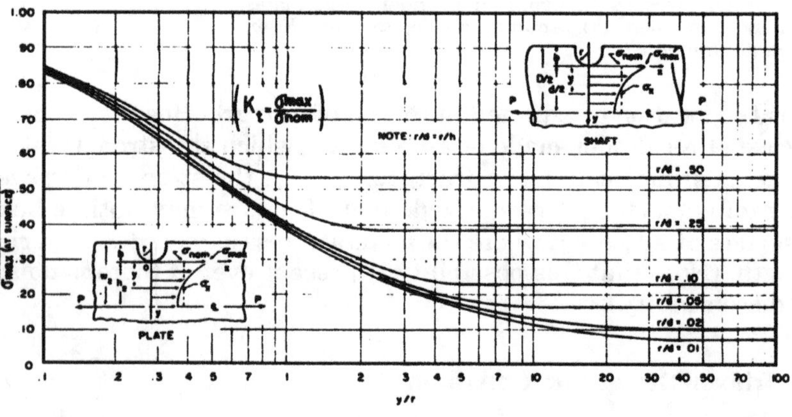

(c)

Figure 2.18 *(Continued)*

the notch. This is shown in Fig. 2.19*b*. Also, just inside the root of the notch, the axial stress is high and the sample tries to shrink in the X and Z directions. Just outside the notch there is no axial stress applied on the free surface and hence here the material experiences no strain in these two directions. However, these two locations are continuous, and thus the contraction of the region inside the notch is resisted by the lack of contraction outside the notch. This uneven

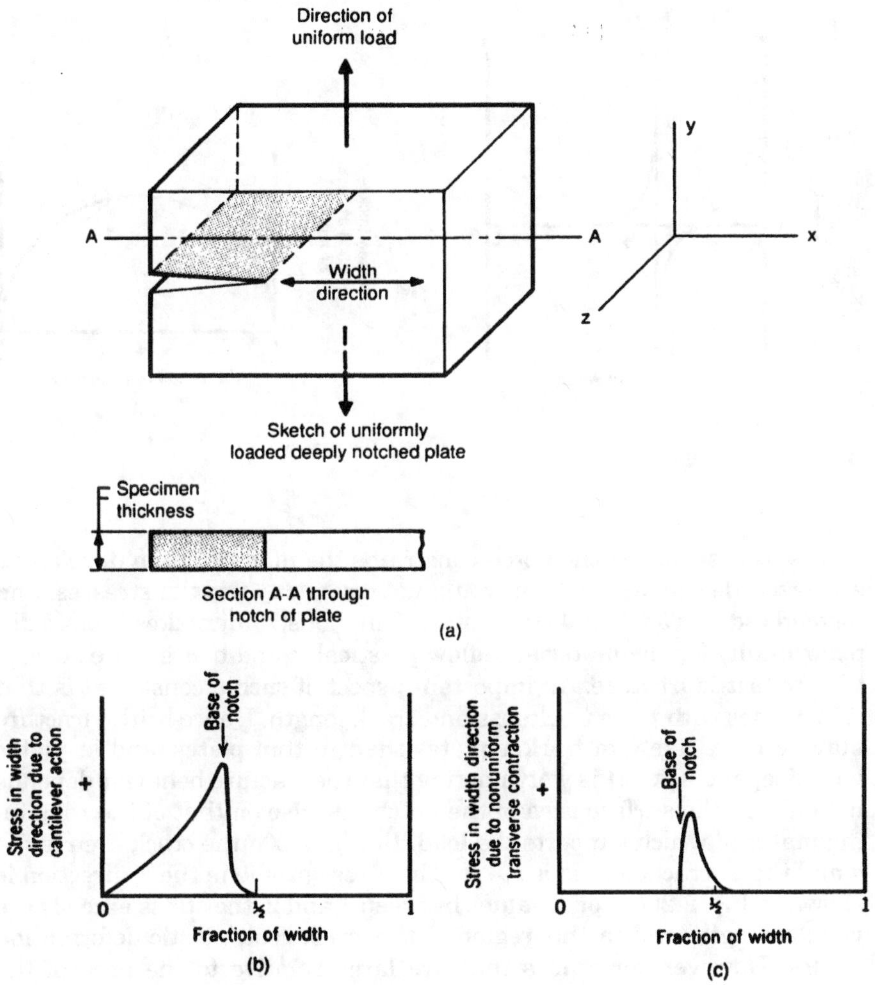

Figure 2.19 Illustration of the development of triaxial stresses due to the presence of a notch. (*Adapted from Parker.*[10])

strain in the X and Z directions induces stresses in these directions. Thus even though the external load is axial, the presence of the notch causes a triaxial stress state, namely, stresses along all three axes. The stress distribution in the width direction (X direction) is shown in Fig. 2.19*c,* and in the thickness direction (at midthickness) in Fig. 2.19*d*. The stress analysis also reveals that the stress in the thickness direction depends on the location across the thickness. At the root of the notch it varies from zero at the surface to a maximum at midthickness (Fig. 2.19*e*).

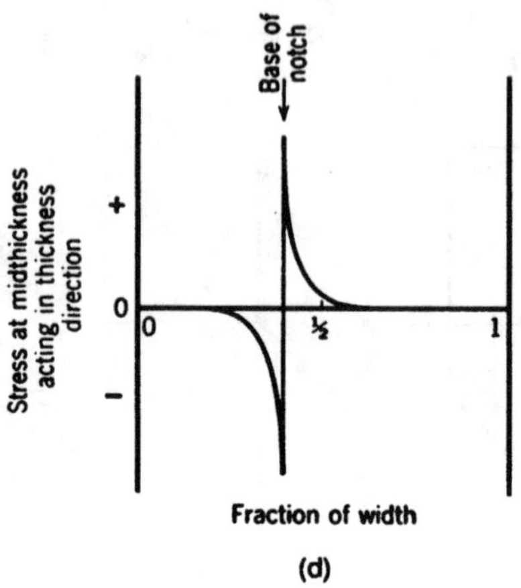

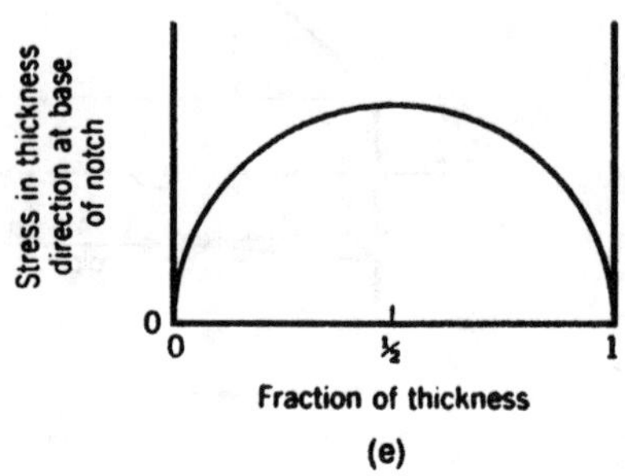

Figure 2.19 (*Continued*)

It is thus seen that the notch constrains the material from developing an elastic deformation which would not cause any triaxial stresses, and instead induces a triaxial stress state. This constraint makes it more difficult locally for the material to flow plastically, and hence more susceptible to brittle fracture. An important aspect of such a constraint is that it increases with plate thickness and crack length. Hence brittle fracture may be more likely in thicker plates than in thin plates, and in plates with deeper cracks. It is worth noting that the fracture behavior depends not only on the surface area of the notch but also on that of the remaining material which supports the load. In Fig. 2.20*a* the crack area is $t \times a$ and the uncracked area is $t \times w$. The stress profile in the Y direction is shown in Fig. 2.20*b*. For X values between 1 and 2, the stress exceeds the yield strength, and in this region of the crack root, plastic deformation occurs. However, for cracks that are large relative to the area of the uncracked material (as will occur as the crack grows), the uncracked area supporting the load ($t \times w$) is small, and the axial stress is high. Over the entire length of the remaining uncracked region, the axial stress may become greater than the yield strength, and hence gross plastic deformation will occur. This situation is shown in Fig. 2.20*c*.

2.6 Plane Stress

If a small region (or element) in a body is stressed, the stress state can be represented as shown in Fig. 2.21*a*. There are three principal normal stresses and also shear stresses on each face of the element (shown on only two faces in the figure). Now if the stress σ_Z is small relative to σ_X and σ_Y, then the state of stress is referred to as a *plane*

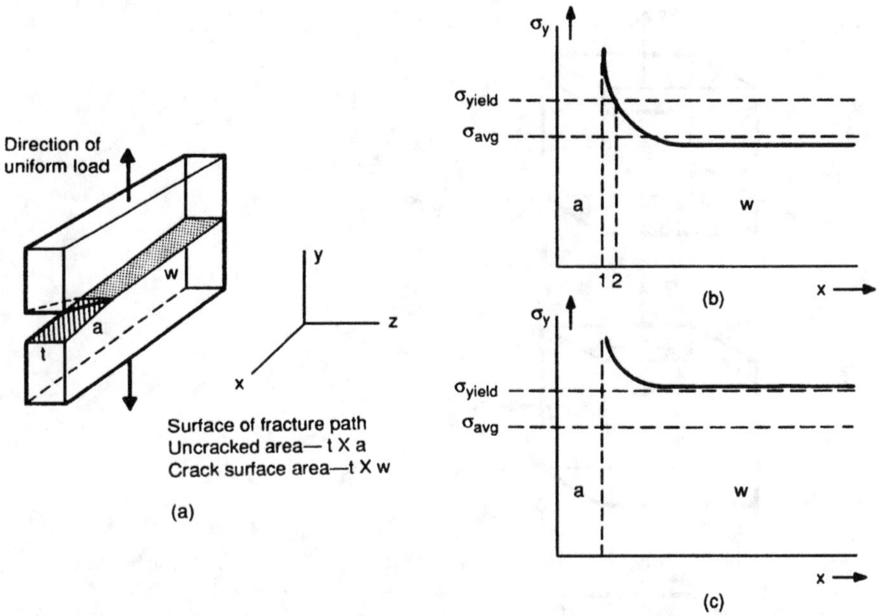

Figure 2.20 Illustration of the effect of reduction in the area supporting the load on the development of gross yielding.

stress, since the normal stresses of finite or significant magnitude lie only in the *X–Y* plane (Fig. 2.21*b*).

As pointed out in the preceding section, the constraint at the root of a notch in a uniaxially loaded sample produces a triaxial stress state. Consider a notched sample of rectangular cross section loaded in tension (Fig. 2.22), where the notch surface area is small compared to that of the remaining material supporting the load. At the surface and near the notch root, the stress in the *Z* direction is zero, since there is no load applied to this surface. This stress rises to a maximum value at the center. If the thickness in the *Z* direction is small, then the magnitude of this stress is low, relative to those in the *X* and *Y* directions. Thus a plane-stress condition exists. The plane-stress condition is common to loaded thin sheets or tubes.

For the plane-stress geometry there is no gross section yield because of the low normal stress and the high value of K_t, and this makes the value of the axial stress in the notch root relatively low. However, there is a considerably large maximum shear stress, which lies on planes at 45° to the axial load direction, as shown in Fig. 2.23*a*. Thus plastic deformation is limited to shear along these planes (Fig. 2.23*b*), leading to a *shear-lip* fracture surface. This fracture surface is also called a *slant fracture* (Fig. 2.24*a*).

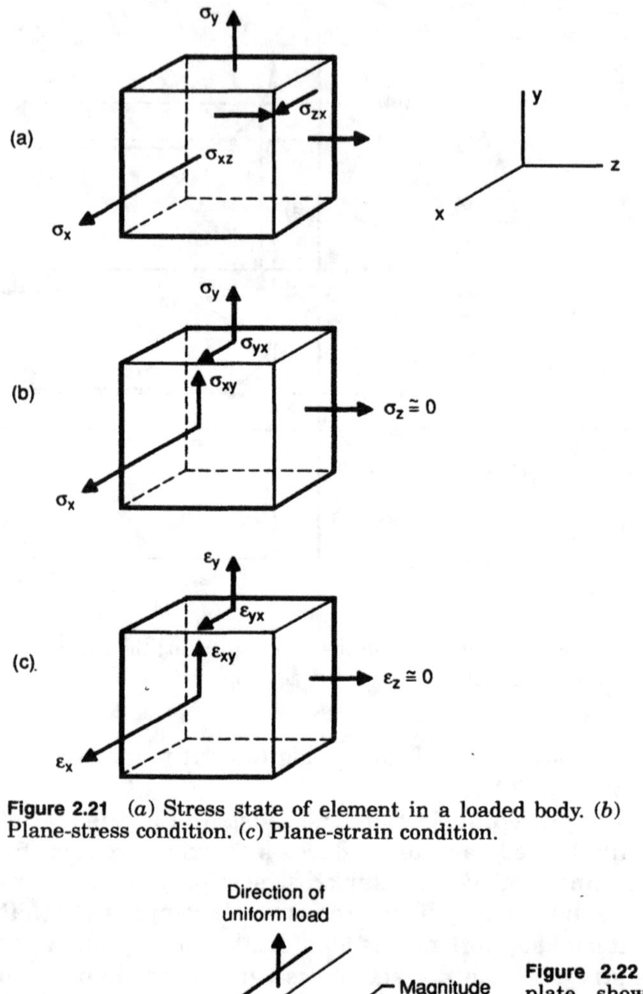

Figure 2.21 (*a*) Stress state of element in a loaded body. (*b*) Plane-stress condition. (*c*) Plane-strain condition.

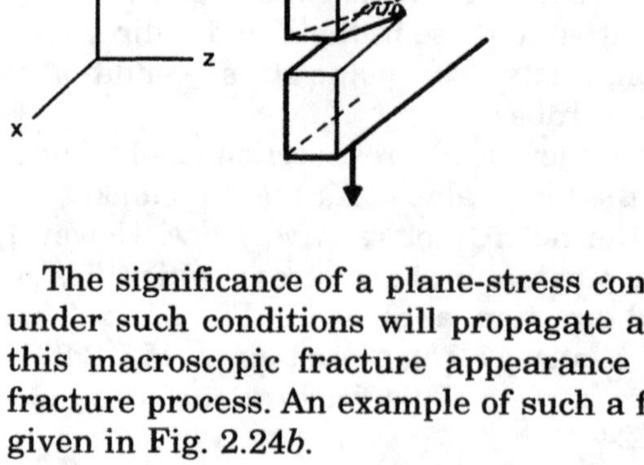

Figure 2.22 Notch in a thin plate, showing the low stress distribution in the thickness (Z) direction at the root of the notch corresponding to the plane-stress condition. The height of the region shows the magnitude of the stress in the Z direction. It is zero at both surfaces.

The significance of a plane-stress condition is that a crack present under such conditions will propagate as a slant fracture, and hence this macroscopic fracture appearance gives information about the fracture process. An example of such a fracture-surface orientation is given in Fig. 2.24*b*.

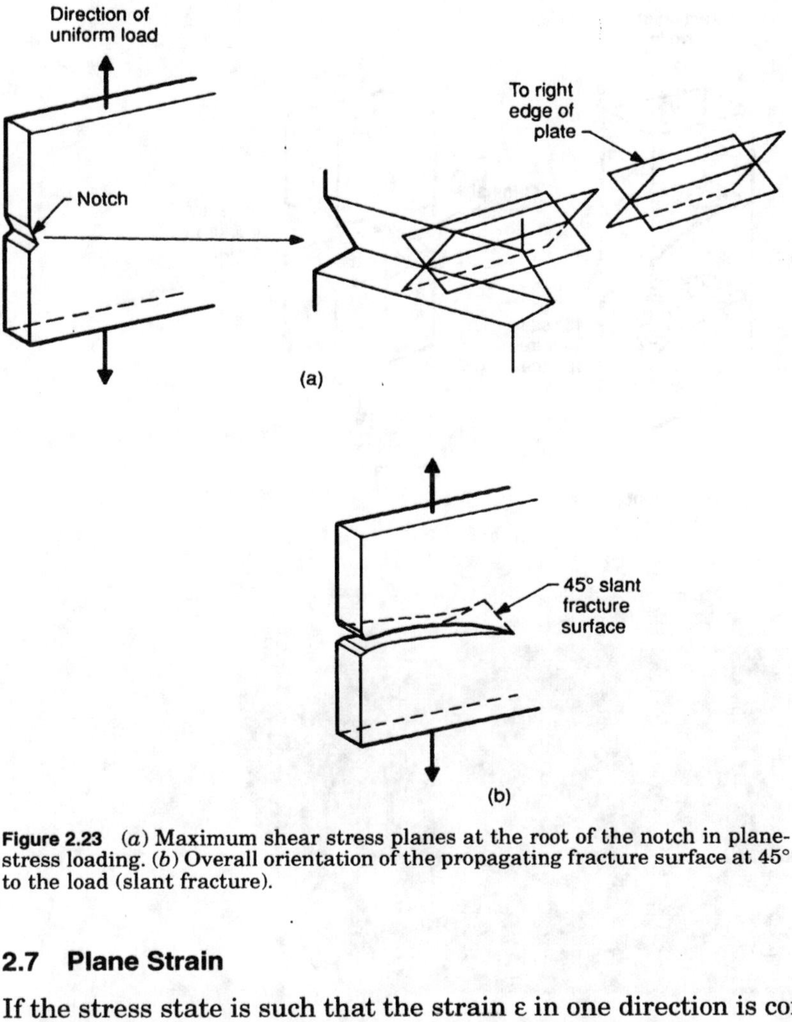

Figure 2.23 (*a*) Maximum shear stress planes at the root of the notch in plane-stress loading. (*b*) Overall orientation of the propagating fracture surface at 45° to the load (slant fracture).

2.7 Plane Strain

If the stress state is such that the strain ε in one direction is considerably less than that in the other two directions, the condition is referred to as *plane strain* (see Fig. 2.21*c*), since the change in dimensions is limited to two dimensions, defining a plane. When a notched plate loaded in uniaxial tension has significant thickness, the strain in the *Z* direction is constrained, leading to a plane-strain condition. If strain hardening is sufficiently high, the size of the plastic zone at the crack tip is smaller than in the plane-stress case, and hence the axial stress at the crack root is very large (compare Fig. 2.22 and Fig. 2.25*a*). Then the maximum shear stresses are relatively low, and the high normal axial stress controls the overall fracture path, which is thus relatively flat and normal to the axis of the external load. This is depicted in Fig. 2.25*c*, and an example is shown in Fig. 2.26.

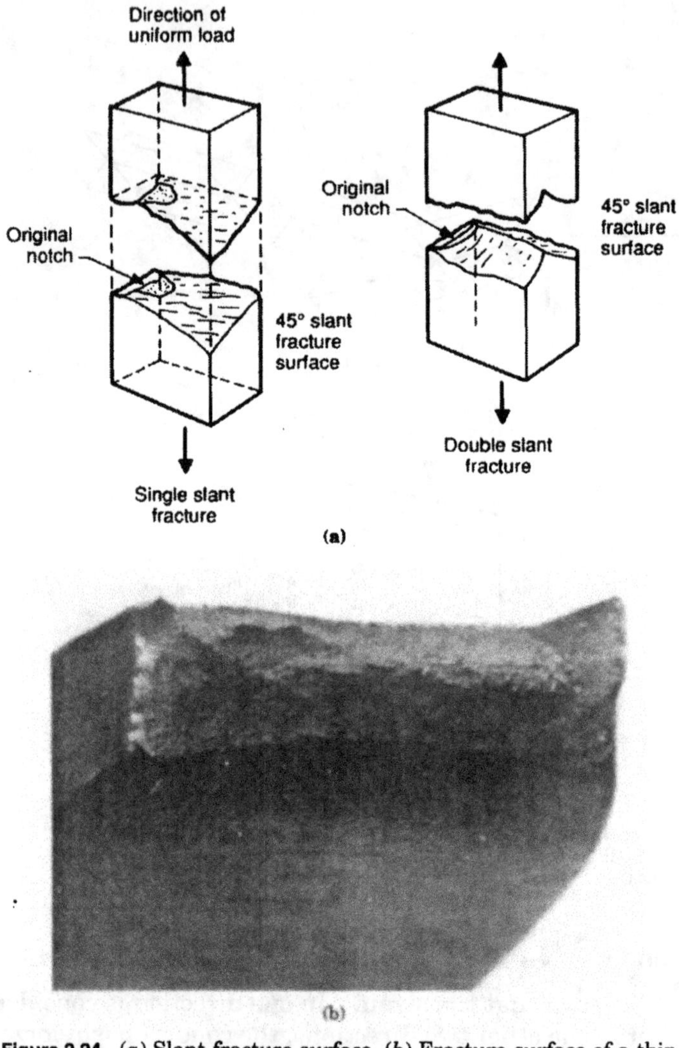

Figure 2.24 (*a*) Slant fracture surface. (*b*) Fracture surface of a thin plate. (*From Henry and Horstmann.*[11])

2.8 Fracture of Tensile Samples

In a tensile test of a ductile metal, the initial elastic (and essentially linear) deformation is followed by plastic deformation during which the length increases and the cross section decreases (see Sec. 2.2). In the plastic deformation region, usually strain (work) hardening occurs and the material gets stronger, and hence an increasing load is required to continue to strain (elongate) the sample at a constant (approximately) strain rate. Eventually a strain is attained where further plastic defor-

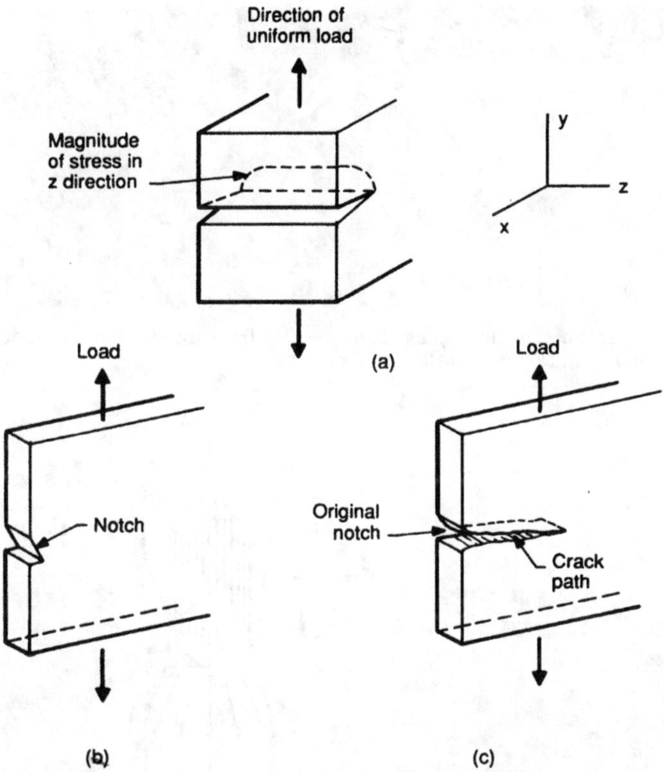

Figure 2.25 (*a*) Stress in the Z direction at the root of the notch in a thick plate, corresponding to plane-strain condition. (*b*) Notched plate. (*c*) Fracture surface propagating from notch has overall orientation perpendicular to the load.

mation becomes very localized. This may initiate at a region of slightly smaller cross section (such as a machining notch) or at a microstructural feature which has work-hardened less. In any case, at this location the cross section becomes locally reduced and hence serves as a surface notch. This can be shown to result in a triaxial state of stress, with the radial and axial tensile stresses being a maximum at the center (Fig. 2.27).

This stress distribution makes the center of the tensile bar susceptible to crack initiation, and in many alloys this is where fracture initiates. (However, this depends on the alloy and test conditions; the crack initiation mechanisms are discussed in Chap. 3.) If a crack initiates in the center, it then spreads radially, so that at this stage the tensile bar has an internal notch. As the crack approaches the surface, the remaining material is a thin shell, and hence the plane-stress condition develops. This leads to a change in crack propagation from flat

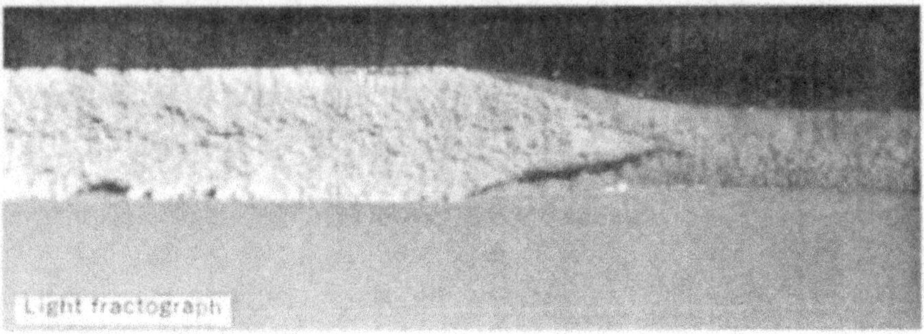

Figure 2.26 Light fractograph of a thick plate showing flat fracture surface developed from a plane-strain condition. (*From* Metals Handbook.[12])

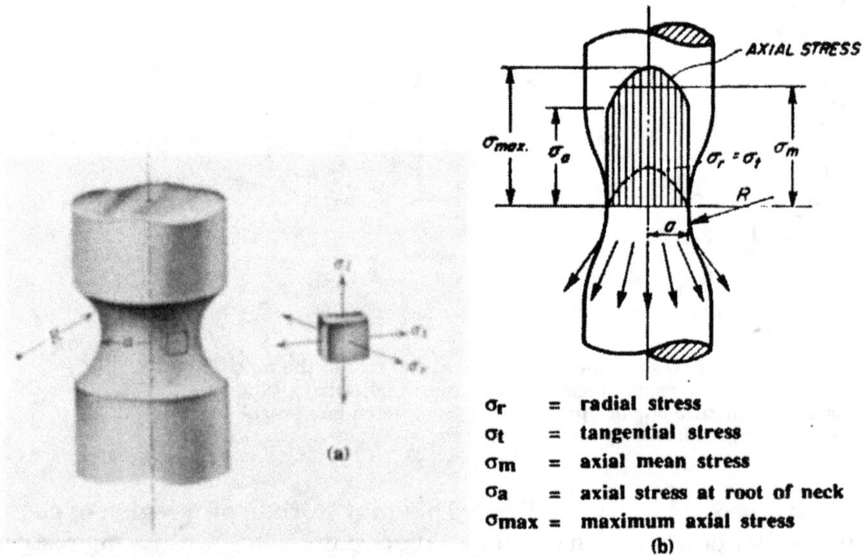

Figure 2.27 (*a*) Stress distribution at the surface of a notch in a cylinder loaded axially. (*From Hertzberg.*[13]) (*b*) Axial stress across the diameter of a cylindrical tensile sample after necking has occurred. (*Adapted from Nadai.*[14])

fracture to slant process. Thus the fracture surface consists of a central region, which is macroscopically relatively flat with its surface normal to the tensile load direction (plain-strain region), and a rim inclined at 45° (plane-stress region). This is illustrated in Fig. 2.28 and is called a *cup-and-cone fracture.* If fracture does not initiate in the center, then the material will pull down to a point, as shown in Fig. 2.29. Thus the fracture appearance depends on the effect of internal microstructural features on the initiation of internal cracks.

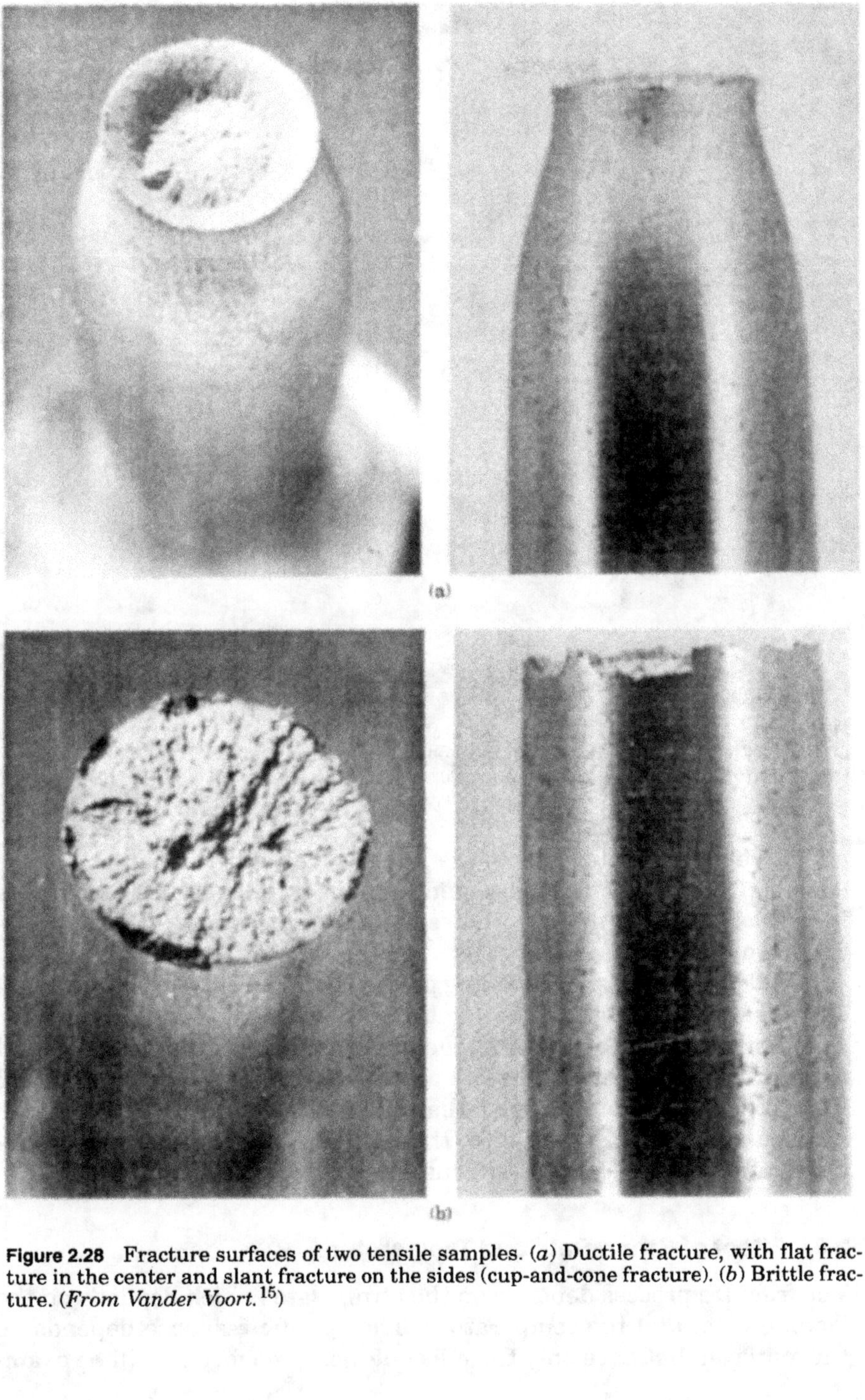

Figure 2.28 Fracture surfaces of two tensile samples. (*a*) Ductile fracture, with flat fracture in the center and slant fracture on the sides (cup-and-cone fracture). (*b*) Brittle fracture. (*From Vander Voort.*[15])

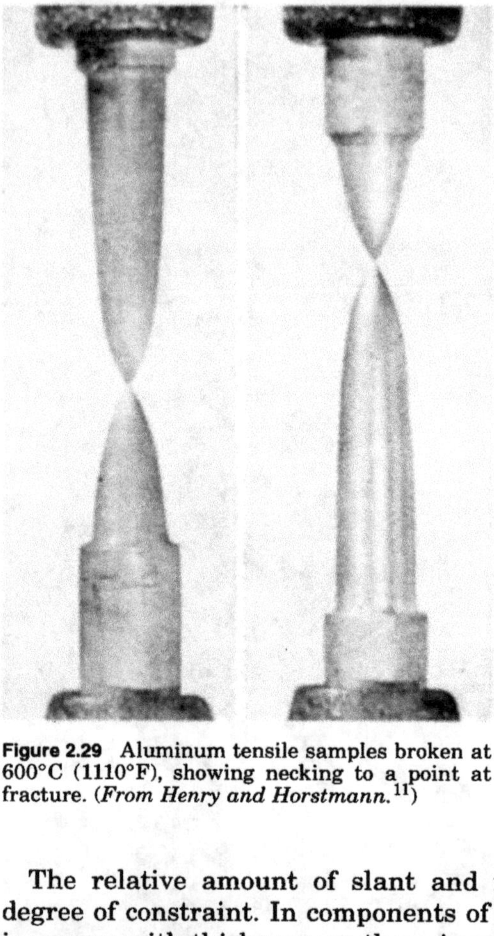

Figure 2.29 Aluminum tensile samples broken at 600°C (1110°F), showing necking to a point at fracture. (*From Henry and Horstmann.*[11])

The relative amount of slant and flat fracture depends on the degree of constraint. In components of rectangular cross section this increases with thickness, so there is more flat fracture. This is illustrated in Fig. 2.30.

In wide thin sheets, the necking process is different from that in cylinders. A large width-to-thickness ratio (such as 10) produces a constraint which prevents necking on a plane containing the width direction and perpendicular to the load direction. Instead, it is found (and predicted from a plain-strain analysis) that necking is restricted to a band inclined approximately at 55° to the load axis. This, then, is the orientation of the fracture plane, as illustrated schematically in Fig. 2.31.

2.9 Effect of Strain Rate and Temperature

The fracture process depends on the strain rate (which depends on the loading rate) and the temperature. The specific response depends on the material, but here only the effect of these two factors will be exam-

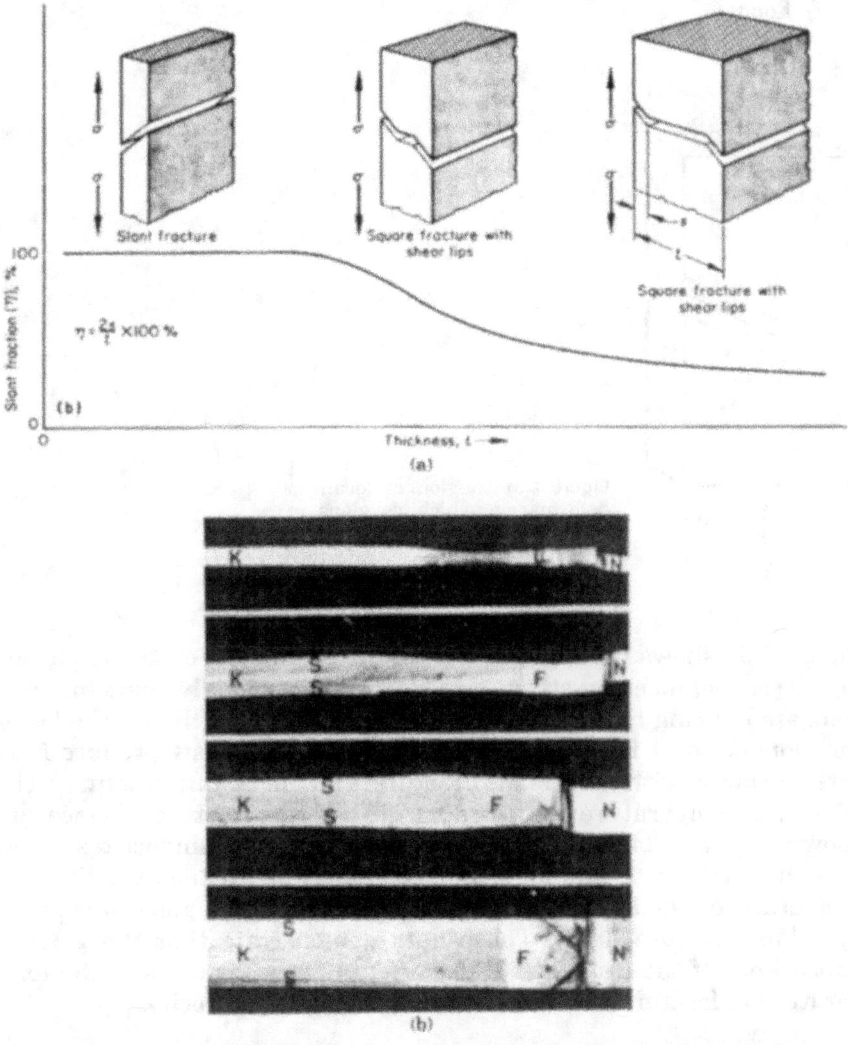

Figure 2.30 (*a*) Effect of the thickness of plates on the relative amount of slant and flat fracture. (*From* Metals Handbook.[7]) (*b*) Influence of plate thickness on fracture appearance. Slant fracture in thin sheet (top); square fracture in plate (bottom); transitional behavior (center). N = central starter notch; F = fatigue fracture area; K = final fracture area; S = shear lips. (*From Broek.*[16])

ined. Most materials become stronger but more brittle with increasing strain rate and lower temperature. The typical effect of temperature on the true-stress–true-strain curve is shown in Fig. 2.32. Note that the yield strength is increased with decreasing temperature, and that the elongation at fracture is reduced. The fracture surfaces show more flat fracture and less slant fracture as the temperature is reduced.

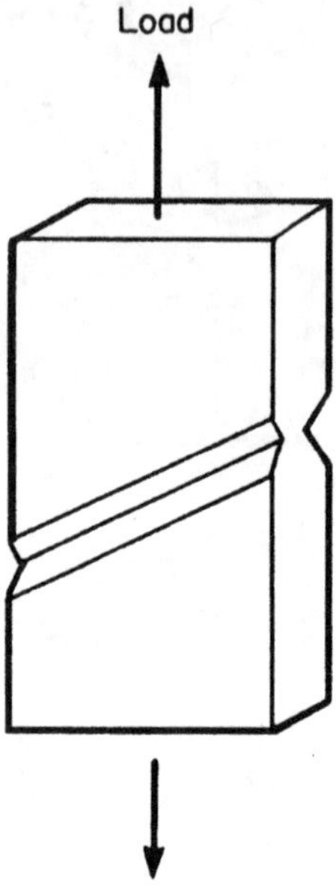

Figure 2.31 Region of localized deformation which develops in thin sheet.

Figure 2.33 shows the typical effect of the strain rate on the engineering-stress–engineering-strain curve. The more rapidly the sample is elongated during the test, the higher the yield strength and the lower the elongation at fracture. The higher-strain-rate tests produce fracture surfaces with less slant fracture and more flat fracture. The effects of temperature and strain rate on the fracture surface are shown in Fig. 2.34. The samples were notched and impact-tested, so that the strain rate was high. Each sample was broken at a different temperature. The lower the temperature, the less the amount of shear lip (slant fracture area). These data demonstrate that the general appearance of the fracture-surface orientation gives clues to the temperature of fracture and the fracture propagation velocity.

2.10 Crack Propagation

The rate of crack growth is usually separated into two categories. One is those that propagate "slowly" [for example, 20 ft/s (6.1 m/s)], and only propagate on application of an external load. This is called *stable crack growth.* The other category is "fast" growth [for example, 3000 ft/s (915 m/s)], and these cracks can proceed without external loading if there is sufficient internal elastic stressing (such as residual stresses from heat treating). This is referred to as *unstable crack growth.* (The propagation of the fracture in a tensile sample after necking is also unstable crack growth.) The type of fracture that occurs depends on

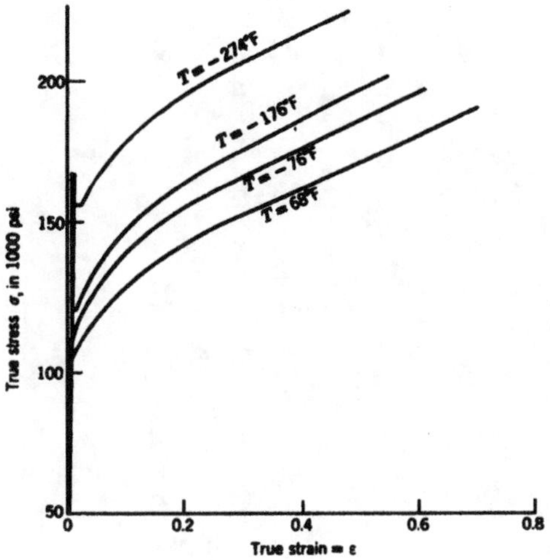

Figure 2.32 Effect of test temperature on the true-stress–true-strain curves in a tensile test of a pearlitic steel. Note that the yield strength increases with decreasing temperature and the ductility (elongation at fracture) decreases. (*Adapted from Zener and Holloman.*[17])

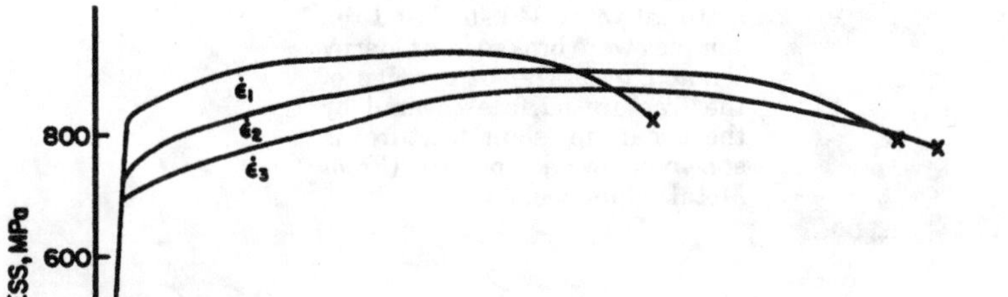

Figure 2.33 Effect of strain rate $\dot{\varepsilon}$ on the engineering-stress–engineering-strain curves in a tensile test of a 1040 steel. Note that as the strain rate increases, the yield and tensile strengths increase and the ductility (elongation at fracture) decreases. (*From Meyers and Chawla.*[18])

many factors, such as loading conditions, temperature, material, and geometry of the component. It is to be noted that the plane-strain condition is more conducive to unstable crack propagation.

Figure 2.30 shows that the relative amount of flat and slant fracture depends on the thickness of the plate. However, another factor that affects the type of fracture is the stress concentration, which changes as the crack propagates and less uncracked material remains to carry

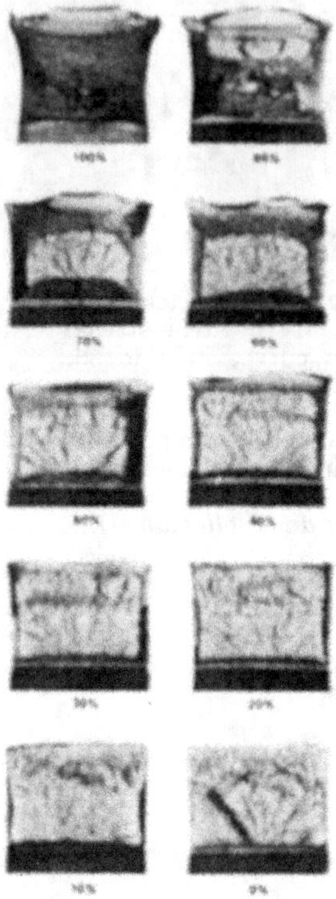

Figure 2.34 Fracture surfaces of a series of impact samples. The material was 4340 steel and the samples were broken in a Charpy V-notch test. The percentage of the fracture surface covered by the shear lip (slant fracture) is shown below each picture. (*From* Metals Handbook.[7])

the load. For example, in terms of round notches, r/d changes as the crack becomes deeper since d decreases (see Fig. 2.17). Thus a crack may begin propagating under plane-strain conditions and then change to plane stress (as for the tensile sample fracture described in Sec. 2.8).

An important complicating factor affecting the fracture-surface macroscopic features is the dynamics of crack propagation. This aspect is not well understood, and the common approach to interpreting the fracture-surface features is to rely on empirical observations and experience. Two features are cited here and described in more detail in Chap. 4. In many fractures of components of circular or square (or near-square) cross section, radial marks are observed which point back toward the origin of the fracture. Examples are presented in Fig. 2.35. Figure 2.35*a* shows the fracture-surface of a cylindrical tensile sample. The radial marks point back toward the center of the sample, where fracture initiated (see Sec. 4.2). Upon the development of a sufficiently acute notch, the plain-strain condition caused relatively rapid crack propagation, with the dynamics of crack movement leading to the crack moving from one inclined plane to another as it advanced, developing the radial marks. Figure 2.35*b* shows the fracture-surface of a tensile sample of the same steel but of a

Figure 2.35 Radial marks on fracture surfaces of tensile samples of 4340 steel having circular cross section. (*a*) Structure of tempered martensite with a hardness of Rockwell C 28. The inner fibrous zone is ridged circumferentially; the intermediate zone has coarse radial marks; the outer ring is the shear-lip zone. (*b*) Sample was fractured at −196°C (−321°F). The structure was tempered martensite; the hardness was Rockwell C 35. This fracture has no fibrous zone. The radial marks occupy the entire area inside the narrow shear-lip zone. (*Adapted from* Metals Handbook.[12])

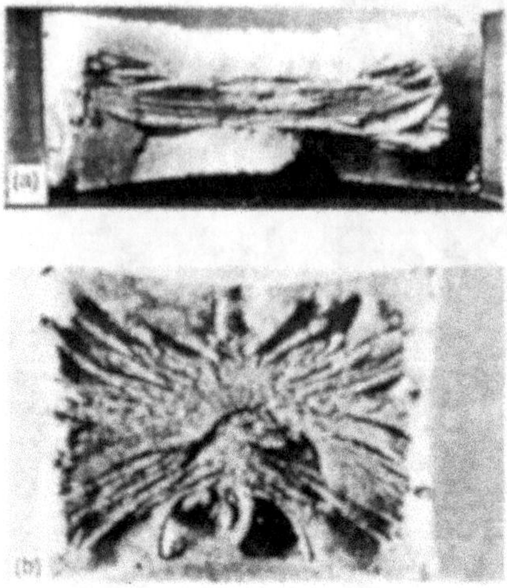

Figure 2.36 Fracture surfaces of a rectangular and a square tensile steel sample showing radial marks. (*From* Metals Handbook,[12] *as adapted from Nunes et al.*[19])

different hardness and tested at a lower temperature. The radial marks are finer and cover more of the fracture. The relationship between this macroscopic fracture-surface topography and the fracture process is discussed in detail in Chap. 4.

The fracture surfaces of tensile samples of square and rectangular cross section are shown in Fig. 2.36. In the square cross section the radial marks are evident, but in the rectangular cross section these are parallel to the long axis of the sample. As the thickness of such a sample decreases, the radial marks become symmetrical about the midline of the plate, forming *chevron* marks (also called *herringbone* marks), as shown in Fig. 2.37. One representation of the dynamics of crack propagation is given in Fig. 2.38. It is hypothesized that cracks form just ahead of the advancing crack front, and the intersection of these cracks with the front forms the chevron marks.

2.11 Meaning of Ductile and Brittle Fracture

The terms *ductile* and *brittle* are commonly taken as self-descriptive, but care should be taken in using them. They are used to indicate the general plasticity at fracture, relying on such quantities as the elongation at fracture (ductility) and the presence of necking to indicate whether the material is ductile or brittle. A difficulty here is to decide how much ductility is required to place the material in a ductile class; there is no common value to be used. More pertinent to our interests is that the terms ductile and brittle are used to distinguish ductile and brittle fracture, both the macroscopic and the microscopic aspects. As is discussed in Chap. 3, the void coalescence mechanism (on a microscopic scale) of crack formation involves plastic deformation by slip.

Figure 2.37 Fracture surface of a steel plate, showing well-developed chevron marks which point back toward the origin of fracture (at left), which is adjoined by a small fibrous zone. This is a high-velocity fracture. (*From* Metals Handbook.[12])

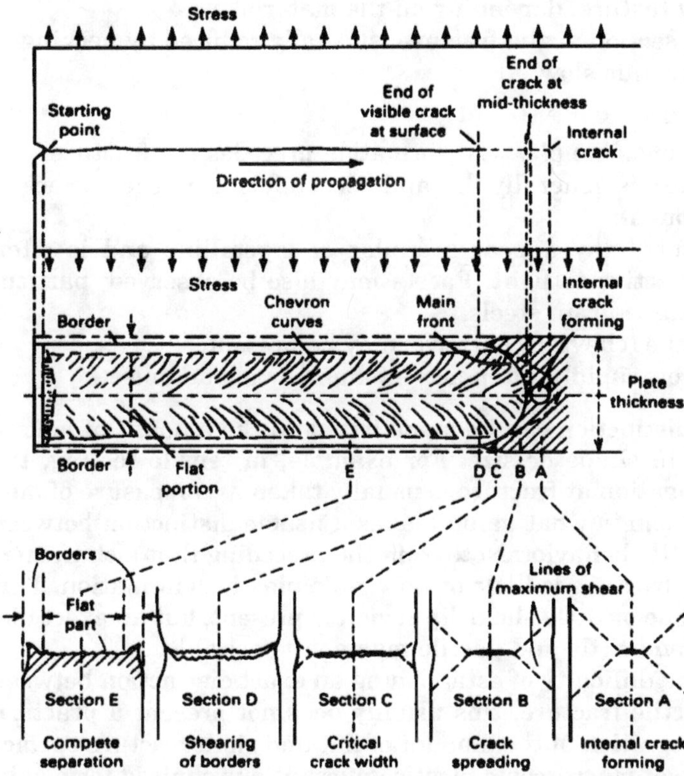

Figure 2.38 Formation of chevron marks on a fracture surface. The lower portion shows an analogy between chevron formation in a plate and cup-and-cone fracture. (*Adapted from Boyd.*[20])

Thus this mechanism is sometimes referred to as ductile. However, it is possible to have cracks form and fracture occur by this mechanism, yet the material shows no obvious or little macroscopic plastic deformation. On this basis, the fracture would be categorized as brittle.

In this book, the terms ductile and brittle are used to describe the macroscopic aspects of fracture and are not based on the microscopic mechanisms. The following characteristics of the two processes allow distinguishing them (adapted from Vander Voort[15]). They include macroscopic fracture-surface characteristics.

1. Ductile fracture
   - A relatively large amount of plastic deformation precedes the fracture.
   - Shear lips are usually observed at the fracture termination areas.
   - The fracture surface may appear to be fibrous or may have a matte or silky texture, depending on the material.
   - The cross section at the fracture is usually reduced by necking.
   - Crack growth is slow.
2. Brittle fracture
   - Little or no visible plastic deformation precedes the fracture.
   - The fracture is generally flat and perpendicular to the surface of the component.
   - The fracture may appear granular or crystalline and is often highly reflective to light. Facets may also be observed, particularly in coarse-grain steels.
   - Herringbone (chevron) patterns may be present.
   - Cracks grow rapidly, often accompanied by a loud noise.

A convenient distinction can be made based on the amount of macroscopic change in shape or size. For example, in tensile testing, the amount of elongation at fracture is usually taken as a measure of ductility. The problem is what value makes a usable distinction between ductile and brittle behavior. Note from the preceding that a characteristic for brittle fracture is little or no visible plastic deformation. This implies that none or little shear-lip region is present, but an exact distinction between brittle and ductile may not be possible.

In spite of the difficulty of establishing an exact distinction between brittle and ductile fracture, this usually does not present a practical problem. A recognition of the amount of flat and slant fracture surface and the degree of macroscopic plastic deformation required for a suitable working distinction usually can be discerned.

2.12 Fracture Mechanics and Failure

A major concern in the design of structural and machine components is the prevention of sudden catastrophic fracture. There are many documented cases in which the systems had functioned properly for many years, then suddenly broke. As experience increased, certain factors, such as low temperature and environment, appeared to exacerbate the problem. It was often found that the fracture originated from already existing cracks or surface (internal and external) discontinuities.

An example of such behavior is usually found in fatigue fractures (see Sec. 2.13). If a component is loaded elastically below the yield strength, and is not operating under conditions where creep (see Sec. 2.14) would occur, then it is expected that it will perform safely. However, if the load is cyclic (but not exceeding the yield strength), it is often found that small cracks will form, especially at surface discontinuities (such as machining notches) due to stress concentration at these locations (see Sec. 2.4). These cracks will expand with each cycle, and the remaining material must support the load. This decreasing support area eventually becomes insufficient to sustain the next increase in load, and fracture is expected. The important aspect of the fracture is that it usually occurs rapidly (catastrophically) and with little plastic deformation, that is, the final fracture is brittle. Thus there is something about the crack (notch) which eventually made the material break in this fashion. Two important characteristics of the notch are its size and its sharpness.

Thus loaded components may fracture suddenly and brittlely if cracks or notches have a certain size. Note that such behavior is not predictable from the results of a tensile test. Although a material which is relatively brittle (for example, low elongation at fracture in a tensile test) would be suspect, a relatively ductile material may fracture catastrophically. What is desired is a material property which can be measured and from which such behavior can be predicted. Fracture mechanics allows treatment of this problem. Here we briefly outline the ideas involved; more realistic treatments can be found in the references listed at the end of the chapter.

Consider a parallelepiped of the dimensions shown in Fig. 2.39*a*. If it is loaded elastically along its axis, the force-length curve is that shown in Fig. 2.39*b*, which is assumed (realistically) to be linear. This force-length curve can be converted to the engineering-stress–engineering-strain curve shown in Fig. 2.39*c*. The slope of the line in the elastic region is the elastic modulus E. The work to stretch the bar to a length l at force F is given by $\int F\,dl$. This work is the stored elastic energy U and is shown in Fig. 2.39*d* to be $\sigma^2/(2E)$ per volume of material.

Now consider a small region taken out of the center of the bar, as shown in Fig. 2.40. If the stress on the bar is still σ, then the elastic energy of this bar has been reduced since there is less material present. This reduction is shown in Fig. 2.40 to be $\sigma^2 a^2 l_0/(2E)$.

If the solid bar has a crack or notch in the center, as shown in Fig. 2.41, and the bar is still loaded to stress σ, the material in the region of the notch supports no load, so the presence of the notch lowers the energy of the bar. Assume that the volume which supports no load is

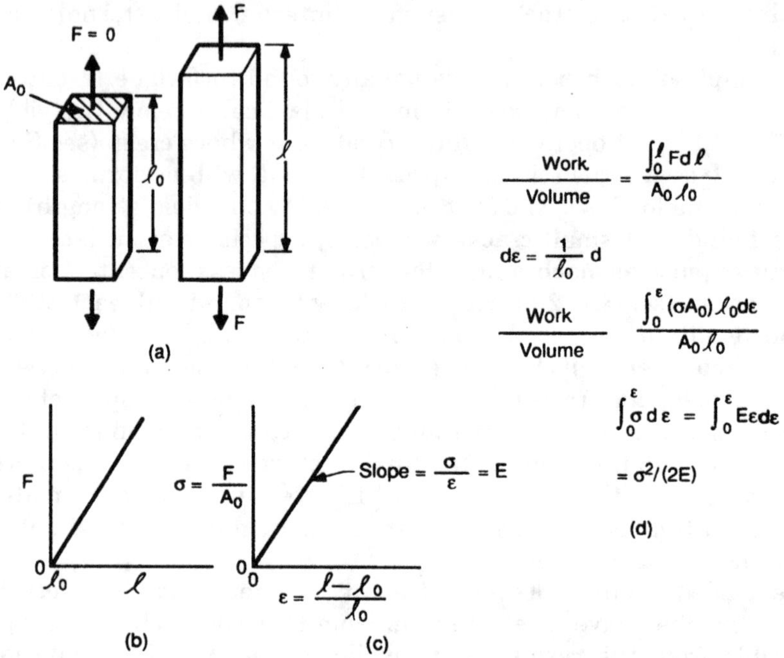

Figure 2.39 Derivation of the elastic energy in a cylinder when loaded axially to stress σ.

given by the elliptical shape shown. Then the reduction in energy is $\sigma^2 a^2 B/E$. Thus the elastic energy of the bar decreases with the area of the notch, assuming that the notch is small compared to the bar cross section. In Fig. 2.41, the crack extends through the bar thickness, so the energy reduction is proportional to a^2.

However, creating a crack forms a free surface which has an energy per surface area of γ. For the case in Fig. 2.41, the notch surface area is $2 \times 2aB$, and so the energy of the bar has been increased to $4aB\gamma$. This increase in energy is proportional to a. The total energy then is the sum of the reduction in elastic energy and the increase in surface energy. This is depicted graphically in Fig. 2.42.

It is seen that once the crack attains the size a^*, a further increase in its size is accompanied by a decrease in energy of the bar, which implies that a crack of this size will grow. This is the explanation why such catastrophic rapid crack propagation occurs.

The relationship of critical crack size to stress is obtained by differentiation of the relation for the total energy and equating it to zero (corresponding to the maximum in the total-energy curve; Fig. 2.42). This gives

$$\sigma\sqrt{\pi a} = \sqrt{2E\gamma} \qquad (2.1)$$

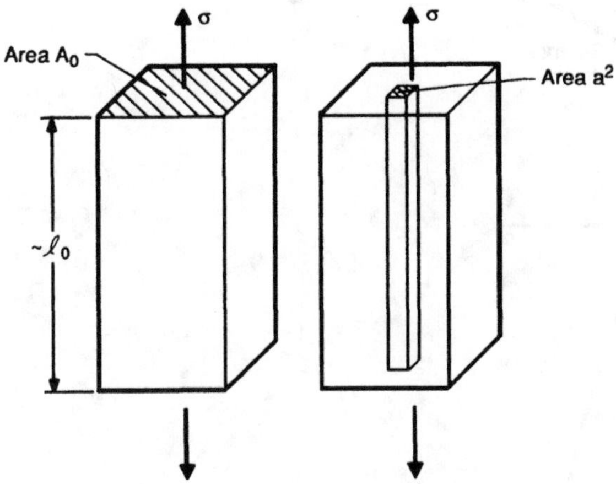

$$U = \frac{\sigma^2}{2E} A_0 \ell_0 \qquad U = \frac{\sigma^2}{2E} A_0 \ell_0 - \frac{\sigma^2}{2E} a^2 \ell_0$$

$$\Delta U = \frac{\sigma^2}{2E} a^2 \ell_0$$

Figure 2.40 Derivation of the reduction in energy upon removal of a small bar from the sample.

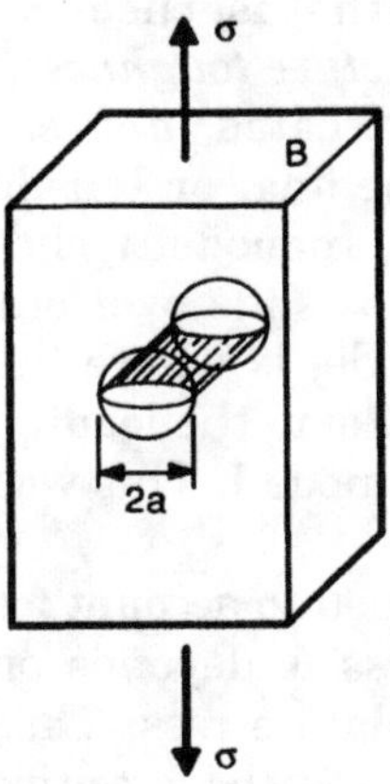

$$\Delta U_{elastic} \simeq \frac{\pi (2a)(a)B\sigma^2}{2E} \simeq \frac{\sigma^2 a^2 B}{E}$$

$$\text{Surface energy} = (2a)(B)2\gamma = 4aB\gamma$$

Figure 2.41 Derivation of the reduction in energy upon forming a small crack in a parallelepiped.

The right side contains only material constants, and hence we can write

$$\gamma \sqrt{\pi a} = K \tag{2.2}$$

A parallel treatment takes the approach of calculating (from elasticity theory) the stress field ahead of the crack tip and expressing the result in terms of a critical stress intensity factor K_I. It is similar to K in Eq.

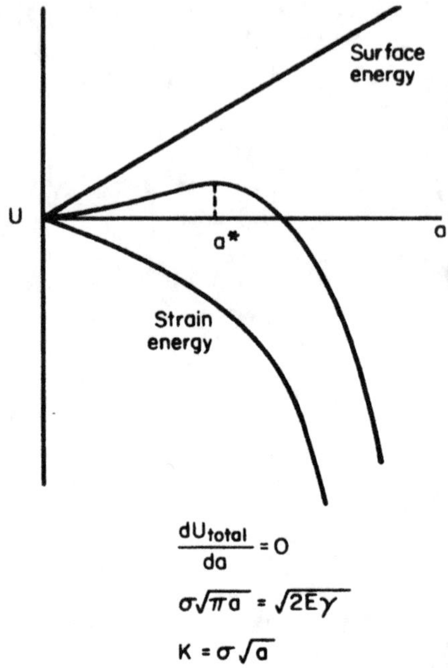

Figure 2.42 Variation of elastic strain energy (for a given stress) and surface energy with crack size. The maximum in the sum of these two curves defines the critical crack size a^*.

(2.2), but the exact form depends on the size and shape of the crack and plate. K_I is taken as a property of the material (that is, chemical composition and microstructure) and is called the *fracture toughness*.

There are three modes of crack opening that are treated in stress analysis. They are depicted in Fig. 2.43. Mode I opening, or tensile mode, is where the crack surfaces move directly apart. In mode II, sliding or in-plane shear occurs, where the crack surfaces slide over one another. In mode III, tearing or antiplane shear mode, is where the crack surfaces move relative to one another and parallel to the leading edge of the crack. By far most crack growth involves mode I. This was the type of crack opening discussed in Sec. 2.6.

The treatment of fracture toughness must be expanded to account for local plastic deformation. Also, the fracture toughness K depends on whether the loading conditions are plain strain or plain stress. Thus there are different types of K values. Also, the K value may be sensitive to the environment, and values are defined for crack propagation under stress corrosion cracking conditions, hydrogen embrittlement, and so on.

The toughness also depends on the material, its microstructure, and the temperature. These are illustrated in Fig. 2.44. In general, the stronger the material or the lower the temperature, the lower the toughness.

In failure analysis of a broken part, a fracture mechanics analysis of the failure will often reveal considerable information about the failure.

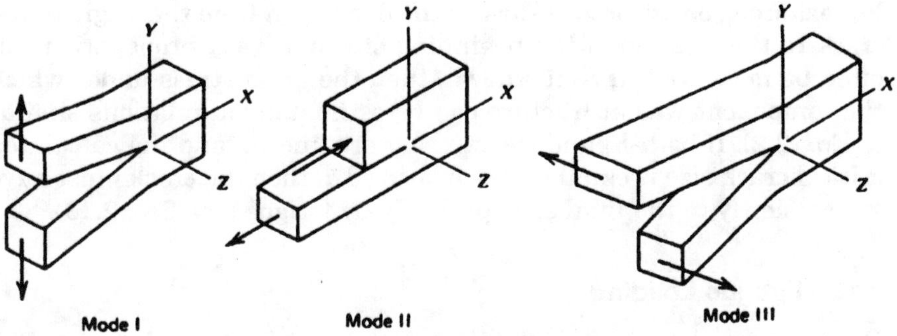

Figure 2.43 Illustration of the three modes of crack opening.

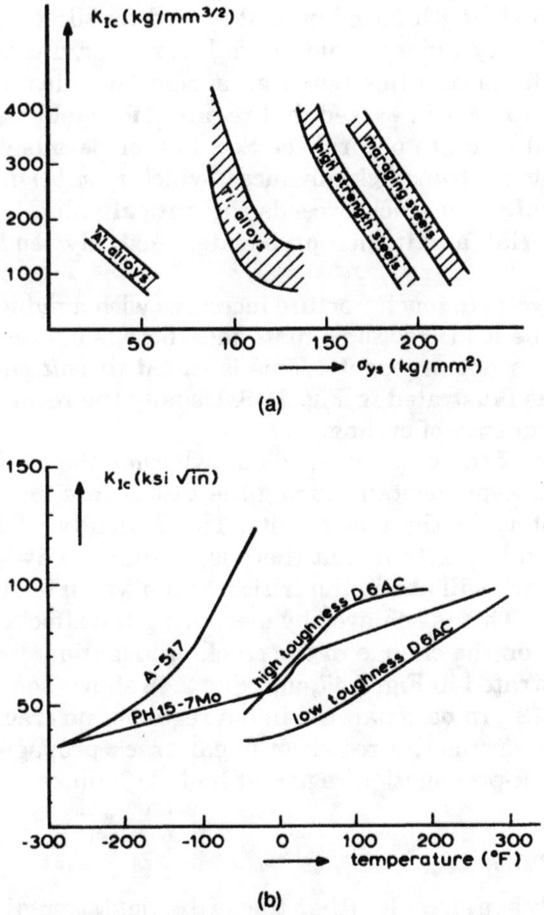

Figure 2.44 Fracture toughness (*a*) as a function of yield strength and (*b*) as a function of temperature. (*From Broek.*[16])

For example, on a fracture surface the distance a from the origin of the crack to the location of the beginning of rapid crack propagation can often be measured. If K_I is known, then the gross stress under which the component was at fracture can be estimated via relations similar to Eq. (2.2). If both K_I and the stress σ are known, and the calculated critical crack size is equal to the measured a, then the crack must have grown slowly to the final size, probably by fatigue (see Sec. 2.13).

2.13 Fatigue Loading

Consider a rod subjected to a time-dependent load, as shown in Fig. 2.45. The amplitude of the load at the maximum of the cycle is only sufficient to induce an elastic stress. That is, the cylinder is stressed at a value below the yield strength based on a standard tensile test, and thus it appears that the cylinder should not fail. However, due to stress concentration at discontinuities (such as surface notches or internal cracks), the local stress will exceed that required for yielding, and deformation and local fracture occurs (see Sec. 3.9 for details of mechanisms). With each cycle the crack advances, which is *subcritical* crack growth. Eventually the crack exceeds the critical value for the toughness of the material, and it then propagates unstably, leading to fracture.

The time or number of cycles to reach fracture increases with a reduction in the magnitude of the load. For some materials there is a stress below which the crack does not propagate. This is called the *fatigue limit* or *fatigue strength,* as illustrated in Fig. 2.46. Usually the results are not sensitive to the frequency of cycling.

The rate of propagation of the crack depends on whether the load, the stress, or the strain is kept constant during the test. These must be defined to properly interpret the test results. The difficulty with using data such as those in Fig. 2.46 is that there is no simple way to predict when a growing crack will attain the critical value which leads to rapid crack propagation. This is achieved by measuring the effect of the stress intensity factor on the change in the crack propagation distance. The concept is illustrated in Fig. 2.47, and Fig. 2.48 shows some data. The curve in Fig. 2.48 can be separated into a region of no crack propagation, where ΔK is very low, a region of linear crack propagation, and a very rapid crack-propagation region at high ΔK values.

2.14 Creep Deformation

If a cylinder is loaded axially at a stress less than that of the yield strength, then yielding and fracture are not expected. However, there are situations in which a slow deformation will occur. This means that the response of

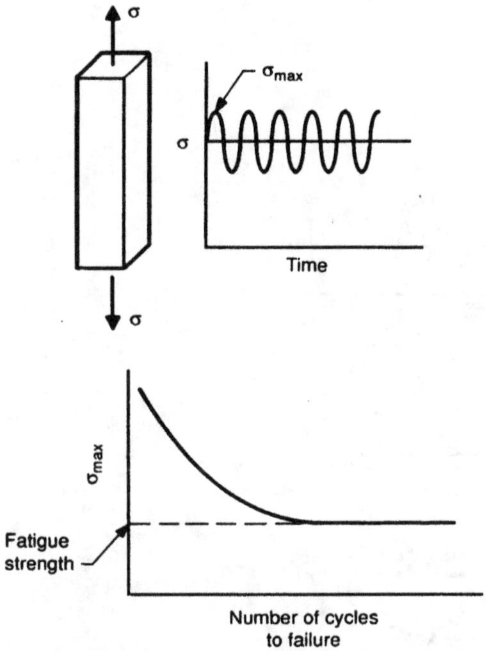

Figure 2.45 Depiction of a fatigue test.

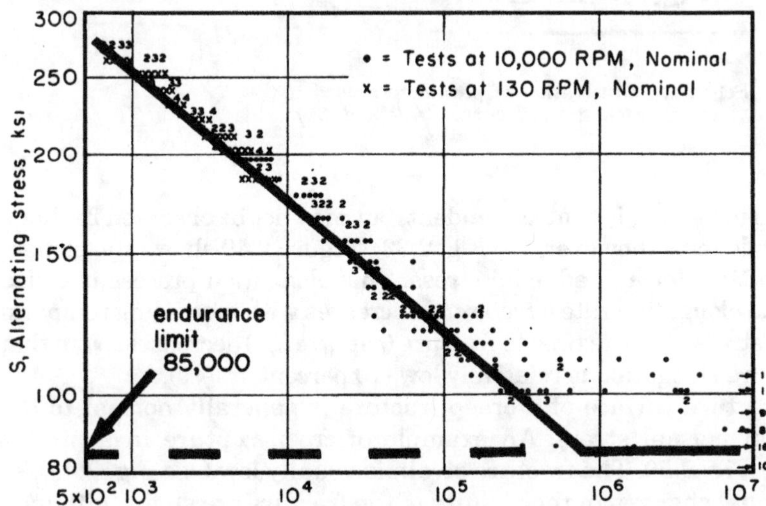

Figure 2.46 Typical fatigue curve. The 4340 steel was heat-treated to a tensile strength of 260,000 lb/in^2, yield strength 250,000 lb/in^2, hardness 53 R_c. The numbers next to the data points are the number of tests giving the same data point. The numbers next to the arrows indicate the number of specimens at the indicated stress which did not break at 10^7 cycles. The specimens were cylinders tested in a rotating-beam configuration. Note that the results are valid for two widely different rotational speeds. (*Adapted from Cummings et al.*[21])

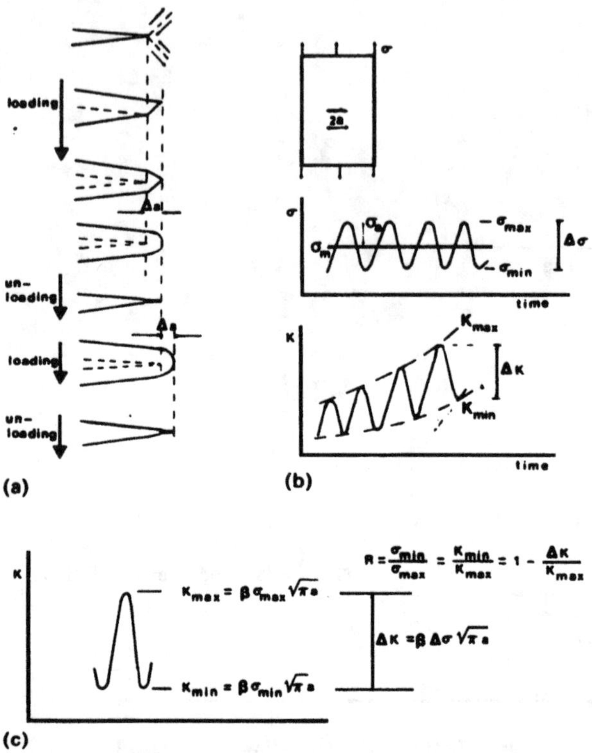

Figure 2.47 Schematic illustration of parameters used in obtaining a fatigue crack growth curve. (*Adapted from Broek.*[22])

materials to loading is time-dependent, but may not be observed in short-time tensile tests (however, see Fig. 2.33). Figure 2.49 shows the elongation with time for a fixed axial stress. This elongation process is called *creep*. The elongation rate (*creep rate*) increases with increasing temperature and stress, and the time to fracture (*rupture life*) decreases. Note that the fracture elongation is relatively low (10 percent or less).

The fracture surface of a creep fracture is generally normal to the maximum normal stress. An example of creep rupture in a pipe is shown in Fig. 2.50. The fracture strain is usually low (see Fig. 2.49), so dimensional changes in the vicinity of the fracture are slight, although local cracking can lead to apparent plastic deformation.

References

1. W. F. Smith, *Principles of Materials Science and Engineering,* 2d ed., McGraw-Hill, New York, 1986.
2. C. O. Smith, *The Science of Engineering Materials,* 2d ed., Prentice-Hall, Englewood Cliffs, N.J., 1977.

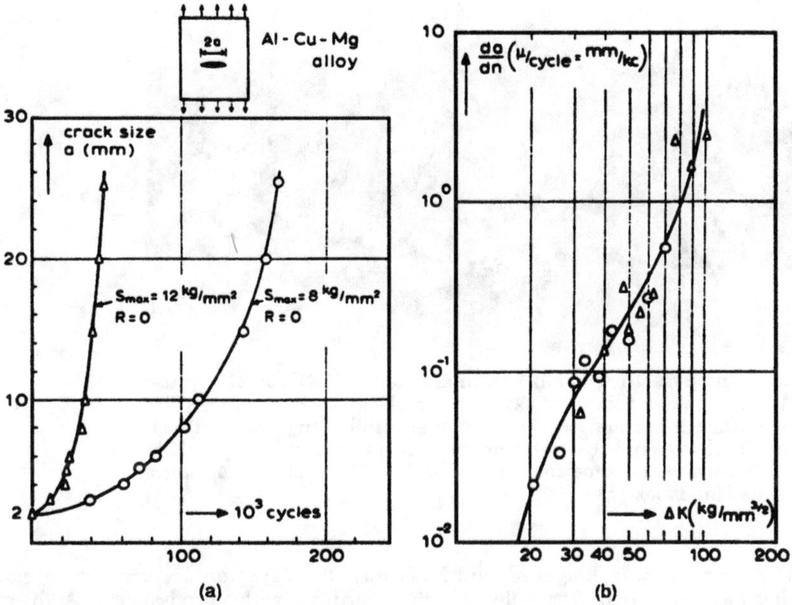

Figure 2.48 Fatigue crack propagation curves. (*a*) Crack-growth curves. (*b*) Crack-propagation rate. (*From Broek.*[16])

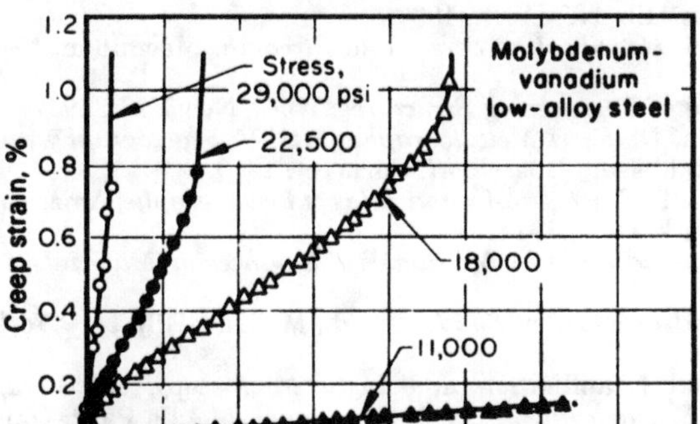

Figure 2.49 Typical creep curve. The sample was a Mo-V low-alloy steel under tension at 600°C (1110°F) at the four stress levels noted. (*From* Metals Handbook.[21])

3. S. Rabinowitz and N. Brown, "Microstrain Investigation of Polyethylene," *J. Polymer Science* A-2, vol. 5, pp. 143–156, 1967.
4. S. Matsuoka, "Nonlinear Viscoelastic Stress-Strain Relationships in Polymeric Solids," in W. Brostow and R. D. Corneliussen (eds.), *Failure of Plastics,* Hanser Publishers, New York, 1989, pp. 24–59.
5. D. J. Wulpi, *Understanding How Components Fail,* American Society for Metals, Metals Park, Ohio, 1985.

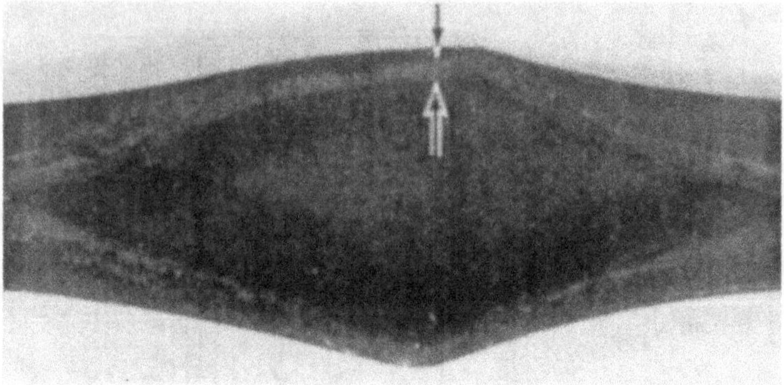

Figure 2.50 Creep fracture in a 321 stainless-steel (ASME SA-213, grade TP321H) superheater tube that failed by thick-lip stress rupture. The fracture surface is normal to the circumferential (tangential) stress. Note that the change in tube-wall thickness at the arrows is slight. This fracture appearance is sometimes called *fishmouth* rupture. (*Adapted from* Metals Handbook.[23])

6. R. P. Feynman, R. B. Leighton, and M. Sands, *The Feynman Lectures on Physics,* California Institute of Technology, 1964, reprinted with permission of Addison-Wesley, Reading, Mass.
7. *Metals Handbook,* 8th ed., vol. 10: *Failure Analysis and Prevention,* American Society for Metals, Metals Park, Ohio, 1975.
8. F. A. Avallone and T. Baumester III, *Mark's Standard Handbook for Mechanical Engineering,* 9th ed., McGraw-Hill, New York, 1987.
9. C. Lipson and R. C. Juvinall, *Handbook of Stress and Strength,* Macmillan, New York, 1963.
10. E. R. Parker, *Brittle Behavior of Engineering Structures,* Wiley, New York, 1957.
11. G. Henry and D. Horstmann, *De Ferri Metallographia,* vol. V: *Fractography and Microfractography,* Verlag Stahleisen, Düsseldorf, Germany, 1979.
12. *Metals Handbook,* 8th ed., vol. 9: *Fractography and Atlas of Fractographs,* American Society for Metals, Metals Park, Ohio, 1974.
13. R. W. Hertzberg, *Deformation and Fracture Mechanics of Engineering Materials,* 2d ed., Wiley, New York, 1983.
14. A. Nadai, *Theory of Flow and Fracture of Solids,* 2d ed., McGraw-Hill, New York, 1950.
15. G. F. Vander Voort, "Visual Examination and Light Microscopy," in *Metals Handbook,* 9th ed., vol. 12: *Fractography,* ASM International, Metals Park, Ohio, 1987.
16. D. Broek, *Elementary Engineering Fracture Mechanics,* Sijthoff and Noordhoff, The Netherlands, 1978.
17. C. Zener and J. H. Holloman, "Effect of Strain Rate upon Plastic Flow of Steel," *J. Appl. Phys.,* vol. 15, p. 22, 1944.
18. M. A. Meyers and K. K. Chawla, *Mechanical Metallurgy—Principles and Applications,* Prentice-Hall, Englewood Cliffs, N.J., 1984.
19. J. Nunes, F. L. Carr, and F. R. Larsen, "Macrofractographic Features," in R. F. Bunshah (ed.), *Techniques of Metals Research,* vol. 2, pt. 1, Wiley, New York, 1968, p. 379.
20. G. M. Boyd, "The Propagation of Fracture in Mild Steel Plates," *Engineering,* vol. 175, pp. 65 and 100, 1953.
21. H. M. Cummings, F. B. Stulen, and W. C. Schulte, "Relation of Inclusions to the Fatigue Properties of SAE 4340 Steel," *Trans. ASM,* vol. 49, p. 482, 1957.
22. D. Broek, *The Practical Use of Fracture Mechanics,* Kluwer, Boston, Mass., 1989.

23. *Metals Handbook,* 9th ed., vol. 11: *Failure Analysis and Prevention,* American Society for Metals, Metals Park, Ohio, 1986.

Bibliography

Campbell, J. E., W. W. Gerberich, and J. H. Underwood (eds.): *Application of Fracture Mechanics,* American Society for Metals, Metals Park, Ohio, 1982.
Ewalds, H. L., and R. J. H. Wanhill: *Fracture Mechanics,* Edward Arnold, Baltimore, Md., 1984.
Hertzberg, R. W.: *Deformation and Fracture Mechanics of Engineering Materials,* Wiley, New York, 1983.
Tetelman, A. S., and A. J. McEvily: *Fracture of Structural Materials,* Wiley, New York, 1967.

Chapter

3

Fracture Mechanisms and Microfractographic Features

"Fractography" is the name we have applied to this technique. Reticence in introducing a new term is counteracted by the advantage in substituting a word for a phrase as long as "the micrographic study of cleavage facets on fractured metal specimens," which is the definition of fractography. In addition, the word fractography has some etymological correctness in stemming from roots similar to those affording metallography, for "fracto-" is the combining form of the Latin "fractus," meaning fracture.

C. A. ZAPFFE AND M. CLOGG, JR.
Transactions of the American Society for Metals, *1945*

3.1 Introduction

In this chapter we review the common processes which occur on a relatively fine scale when a material is subjected to increasing loads, and which lead to material separation (fracture). Since at this scale the human eye cannot resolve individual features, the use of microscopes is required. These processes can be referred to as *fracture micromechanisms,* but we will use the term *fracture mechanisms.* This distinguishes the fine-scale processes from the coarser-scale processes, or fracture macromechanisms, which we call modes in Chap. 4.

Since the fracture mechanisms determine the fine topology of the fracture surface, its appearance gives clues, and frequently direct confirmation, of the fracture mechanism. Examination of this fine surface topography is called *microfractography.* Examination of the coarser fracture-surface topography, generally resolvable with the eye or low-magnification (such as 25×) optical instruments, is called macrofractography, which is reviewed in Chap. 4.

We begin this chapter by reviewing the common, basic mechanisms of the response of a metallic material to loading, namely, slip, twinning, and cleavage. We do not examine the underlying atomistic mechanisms, such as dislocation generation and movement, but instead restrict our comments to those features which most directly affect the fracture-surface topography. We then relate these basic mechanisms to the fracture mechanisms and to the resulting fracture-surface topography.

3.2 Slip and Cleavage

A common mechanism of plastic deformation is *slip,* which is the parallel movement of layers of material in a crystal past adjacent layers. This can be illustrated by the response of a single crystal to a shear stress, as depicted in Fig. 3.1. Here it is assumed that plastic deformation is initiated when the shear stress on the plane shown exceeds a critical value, *the critical resolved shear stress.* Continued application of this stress will cause separation and hence fracture.

A more realistic case is illustrated in Fig. 3.2. Here a cylindrical single crystal is loaded axially with an axial stress of F/A_0, where A_0 is the cross-sectional area of the cylinder. The plane on which slip occurs is oblique to the cylinder axis, with its normal at an angle ϕ to the cylinder direction (the loading axis). Slip occurs along only one direc-

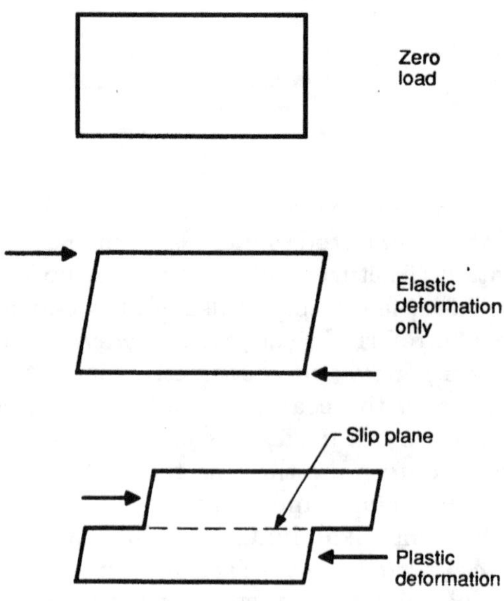

Figure 3.1 Illustration of plastic deformation of a single crystal by slip.

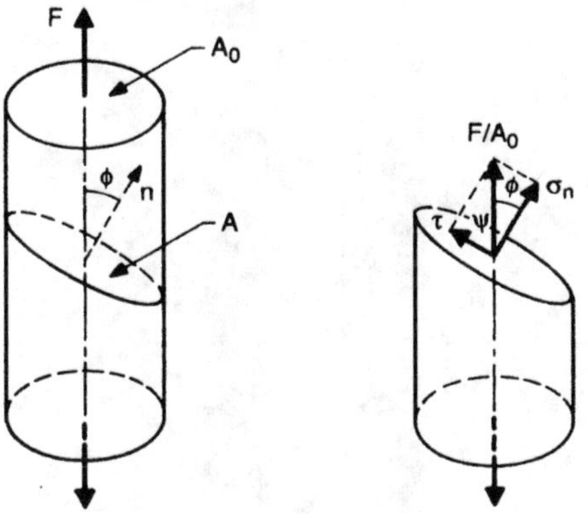

τ = resolved shear stress = $\frac{F}{A_0}$ cos ψ cos ϕ

σ = resolved normal stress = $\frac{F}{A_0}$ cos ϕ sin ψ

Figure 3.2 Illustration showing resolved shear stress τ [= (F/A_0) cos ψ cos ϕ] and resolved normal stress σ_n [= (F/A_0) cos ϕ sin ψ] for a cylindrical sample loaded axially.

tion in this plane, the slip direction, which makes an angle ψ with the loading axis. Thus the component of the axial load F resolved onto this plane (the shear force F_s) is F cos ψ. The area of the shear plane is given by A_0/cos ϕ. The resolved shear stress τ is (F/A_0) cos ψ cos ϕ. Upon subjecting the cylindrical single crystal to an increasing load, slip begins when the resolved shear stress on the slip plane reaches the critical resolved shear stress value. It is observed to occur at many locations, with elliptically shaped discs moving past each other, as depicted in Fig. 3.3*b*. If the crystal is relatively large (such as 1 cm in diameter) and the surface initially well polished, these offsets are visible to the unaided eye with sufficient deformation (Fig. 3.3*c*). Examination with a microscope shows that each coarse offset (slip bands) is made up of much finer steps (slip lines) (Fig. 3.4).

This effect can be seen on a polycrystalline sample. If the sample is metallographically polished, then loaded as shown in Fig. 3.5*a*, in each grain a slip system is activated and surface offset occurs when the critical resolved shear stress is exceeded. An example is given in Fig. 3.5*b*. Note that the slip direction changes at the grain boundaries, and hence the locations of these boundaries are revealed even though the sample is unetched.

The planes and directions on which slip occurs depend on the crystal structure of the material; usually these are the atomistically close-packed planes and directions. Thus in face-centered cubic crystals these are the {111}-type planes and the <110>-type directions. A slip plane and a slip direction in that plane constitute a slip system. In

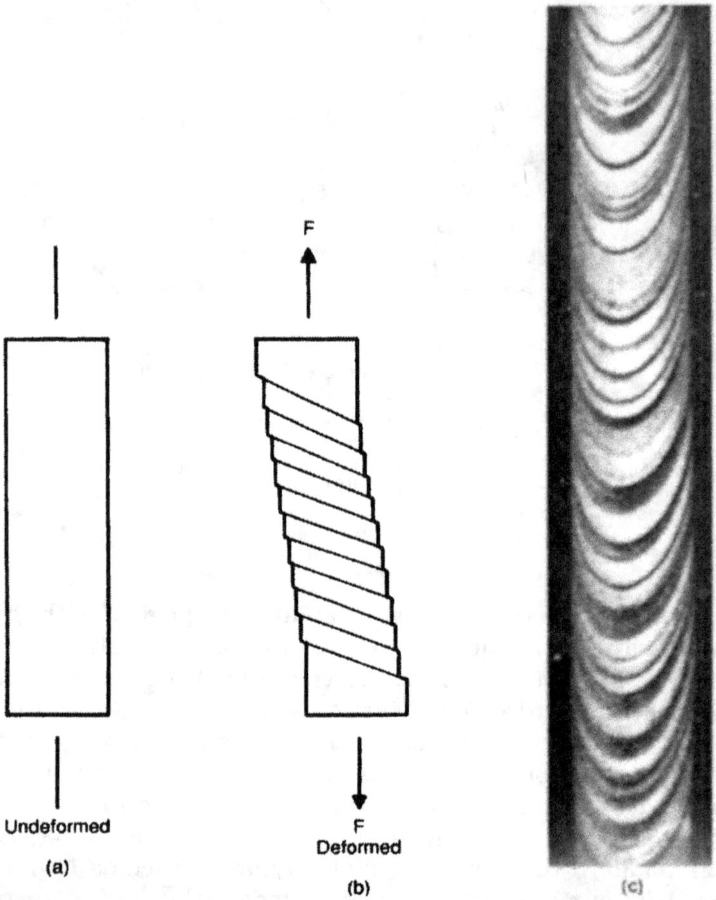

Figure 3.3 Illustration of plastic deformation of a single crystal. (*a*) Undeformed. (*b*) Deformed showing slip. (*c*) Deformed cylindrical single crystal of zinc, showing elliptical offsets due to slip (*Adapted from Boas and Schmid.*[1])

face-centered cubic metals there are four {111}-type planes, and in each {111}-type plane three <110>-type directions, so there are 12 slip systems. In the other close-packed structure, which is hexagonally close-packed, there is only one close-packed plane, but three identical slip directions, or only three slip systems.

When a crystal is loaded, the shear stress increases on all slip systems, but slip initiates only on the system which first attains the critical resolved shear stress for the material. Which system this is depends on the orientation of the crystal. It is possible to have an orientation such that two slip systems have identical resolved shear stresses, and that these two attain the critical resolved shear stress

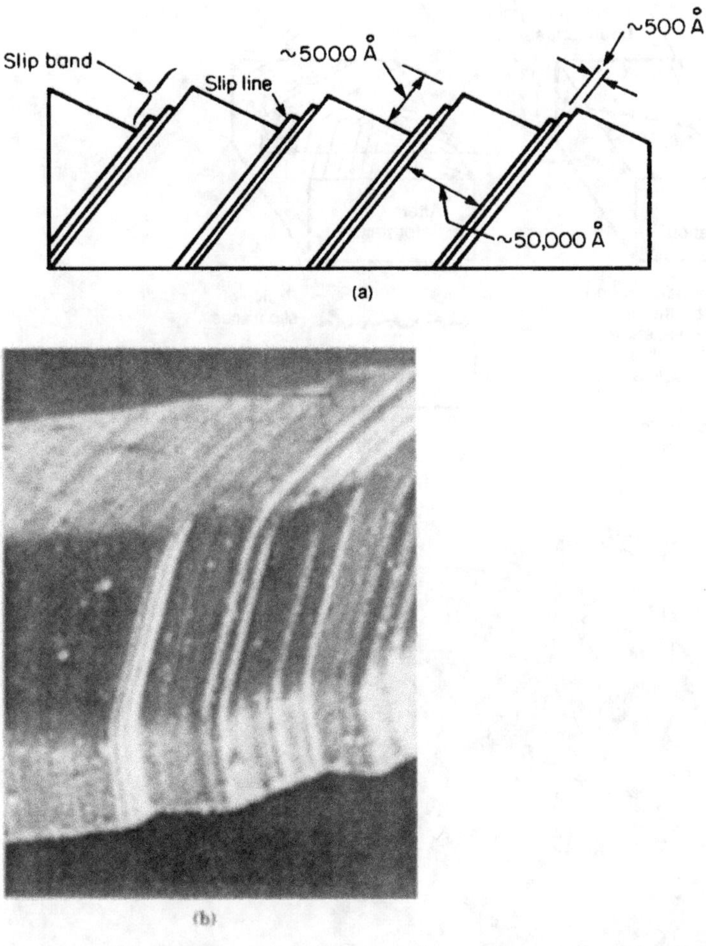

Figure 3.4 Illustration of fine offsets on a free surface produced by slip. (*a*) Appearance of the surface cross section at high magnification. (*b*) SEM of slip markings on the surface of a deformed single crystal of cobalt. (*Courtesy of Eric Lifshin, Research and Development Center, General Electric Company; from Guy.*[2])

first. This leads to simultaneous slip on these two slip systems, which is called *cross slip*.

In the hexagonal close-packed structure it is possible to have an orientation in which the resolved shear stress is zero. This is illustrated in Fig. 3.6. The basal plane, which is the slip plane, is normal to the loading force. Thus as the loading force is increased, it might be expected that slip will occur on another type of slip plane (not a close-packed plane). However, the critical resolved shear stress for this is very high. Another possibility is that the resolved *normal* stress on a

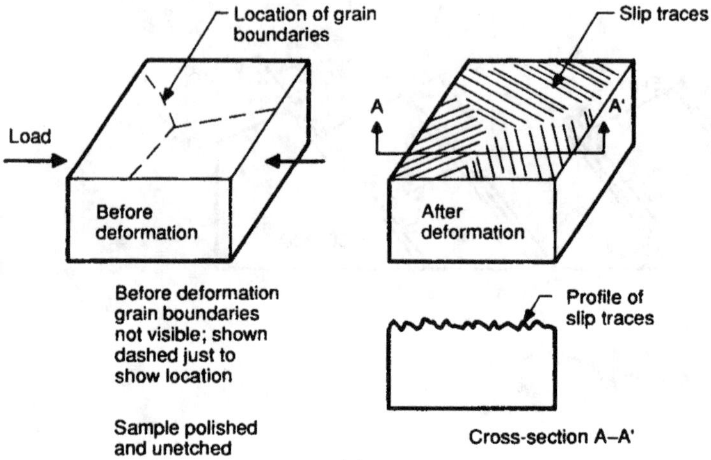

(a)

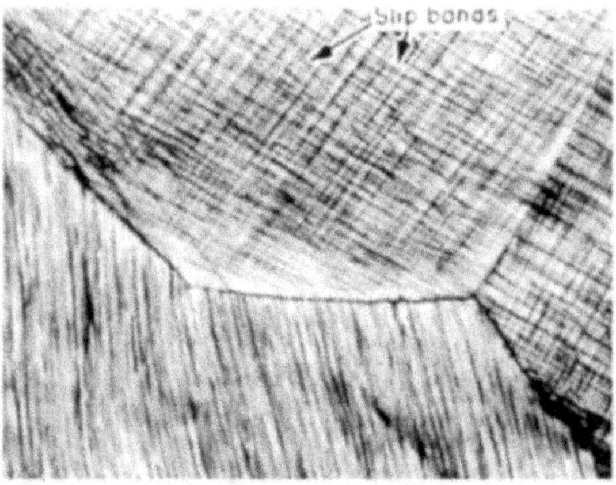

(b)

Figure 3.5 (*a*) Loading of a metallographically polished (but unetched) sample to produce slip markings on the polished surface. (*b*) Micrograph of such a sample of aluminum showing slip traces. (*Adapted from Verhoeven.*[3])

plane will exceed that which intrinsically bonds the planes together, and these planes will separate, or cleave. Note from Fig. 3.2 that a normal stress σ_n can be defined for a specified plane (specified by the angle ϕ), which is given by $(F/A_0) \sin \phi \cos \psi$. Thus cleavage will occur when the critical resolved normal stress for cleavage is exceeded.

If a crystal of the hexagonal close-packed metal zinc is loaded with the configuration shown in Fig. 3.6, slip cannot occur. Instead cleavage occurs on the basal plane, and the fracture surface will be relatively

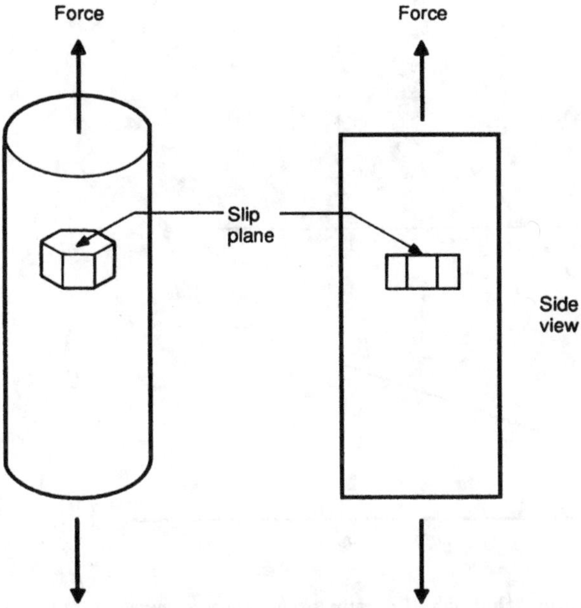

Figure 3.6 Illustration of the orientation of a single crystal of a hexagonal close-packed metal, with the basal (slip) plane normal to the loading axis, so that there is no resolved shear stress on it.

smooth, with a mirrorlike appearance. Such cleavage usually cannot occur with a face-centered cubic crystal. If the crystal is oriented so that a {111} slip plane is normal to the loading axis, there are still three other {111} planes at oblique angles to this axis, and hence these planes, each containing three <110> slip directions, have finite resolved shear stresses, and slip will occur on some of these.

Thus there is a competition between slip and cleavage. As the external load on a crystal increases, the shear stresses on the slip systems increase and the normal stresses on the cleavage planes increase. Which occurs first, slip or cleavage, depends on which stress attains the critical value first. This is depicted in Fig. 3.7. If the loading is such that τ and σ_n increase along line *A*, then slip occurs first; if along line *B*, then cleavage occurs. Clearly slip is more favored if τ_c is much lower than $\sigma_n{}^c$, which is the case for most close-packed metals and most body-centered cubic metals at sufficiently high temperature. The critical values of τ and σ_n are temperature-dependent, both increasing with decreasing temperature. However, in some cases, such as body-centered cubic metals, lower temperatures are more favorable for cleavage (Fig. 3.8). These critical values of τ and σ_n are also affected by the strain rate, so that a metal may exhibit cleavage if the loading which causes fracture is sufficiently rapid.

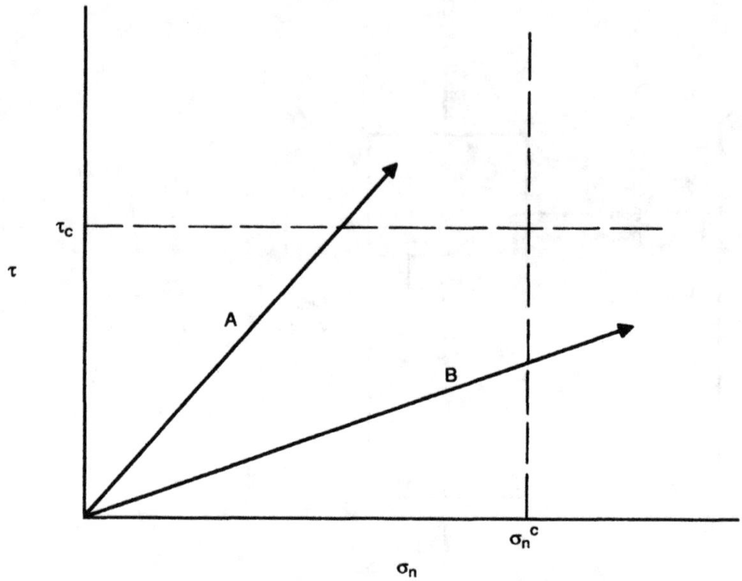

Figure 3.7 Schematic illustration that whether slip or cleavage occurs depends on the relations between resolved shear and resolved normal stress as the load increases, and values of critical resolved shear stress for slip τ_c and critical resolved normal stress for cleavage $\sigma_n{}^c$.

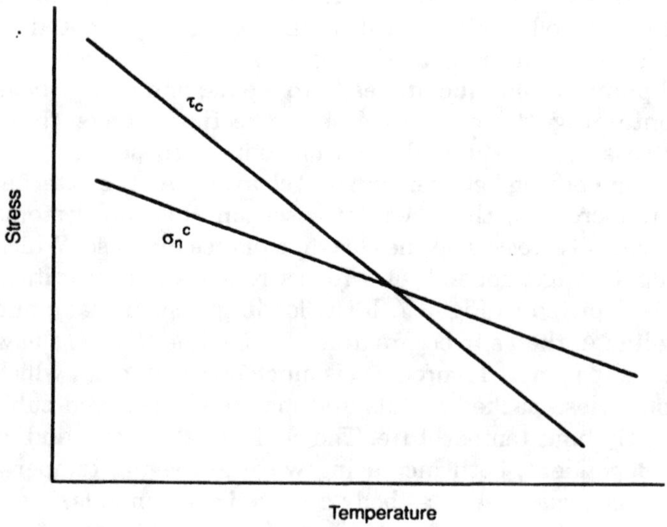

Figure 3.8 Temperature dependence of critical resolved shear stress and critical resolved normal stress for a body-centered cubic metal.

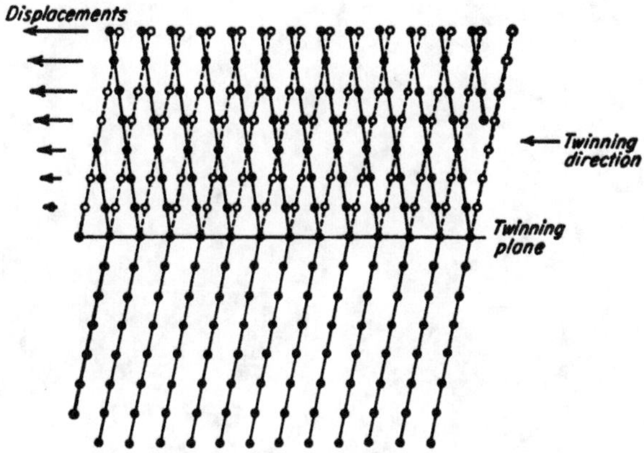

Figure 3.9 Schematic illustration of mechanical twinning. The twinned part lies above the twinning plane. Open circles—initial positions; closed circles—final positions, which are mirror images of the points below the twinning plane. (*From Birchenall.*[4])

3.3 Twinning

In some metals and alloys, crystals can respond to increasing loading by twinning. The atomistic rearrangement for the process is depicted schematically in Fig. 3.9. Note that the shift of the lattice along the twinning plane produces the same crystal structure, but as a mirror image across this plane. Hence it is said that the crystal on one side of this plane is a twin of the crystal on the other side.

Twinning is a rather complex process. It rarely occurs in face-centered cubic crystals, but does occur in both hexagonal close-packed and body-centered cubic crystals. Its occurrence may be favored by low temperature and high strain rates. Figure 3.10 shows a microstructure containing typical mechanical twins. The line separating the original crystal from the twinned material is the trace of the twinning plane. In fracture of body-centered cubic metals, twinning may play a role in crack nucleation. [Mechanical twins should not be confused with annealing twins. The latter are also crystallographic twins, but they form by other mechanisms. They are especially prominent in many face-centered cubic metals and appear as straight lines in the microstructure (Fig. 3.11).]

3.4 Cleavage Fracture Topography

From the preceding description of cleavage it would appear that the fracture surface of a single crystal would be atomistically smooth and hence featureless. Indeed, at low magnification such surfaces are mirrorlike,

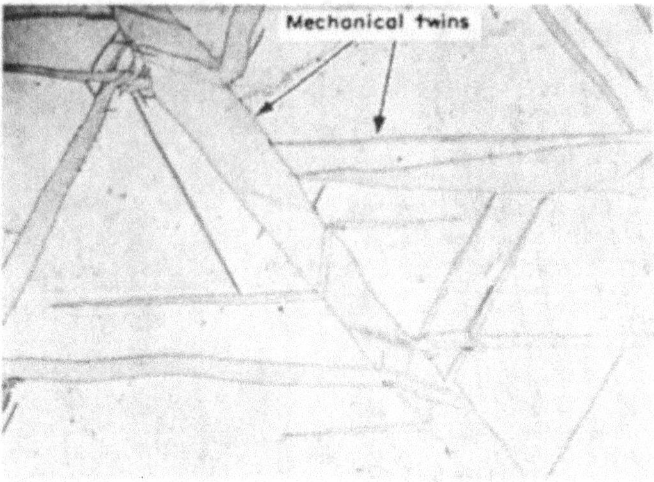

Figure 3.10 Microstructure of plastically deformed magnesium showing mechanical twins. (*Adapted from Roberts.*[5])

and polycrystalline samples which have fractured completely by cleavage may appear to the eye to consist of bright facets due to light reflection. However, even in single crystals of pure metals there are lattice defects which cause the propagating cleavage crack to deviate and hence generate topography. Also, the cleavage cracks may nucleate at more than one location, and thus the cracks propagate at different levels even when on the same type of parallel planes. Where these levels meet, fracture must occur by either slip or cleavage on another set of planes, so that the fracture surface has offsets on it.

The formation of the fractographic feature where the fronts of cleavage cracks moving on adjacent planes meet is illustrated in Fig. 3.12*a*. The connecting material fractures either by cleavage on another cleavage plane or by plastic deformation. The characteristics of the steps which form are revealed by cross-sectional views of the fracture surface. An example is presented in Fig. 3.13, where the steps connecting the parallel cleavage planes are apparent. Note that the straight-line profiles of these steps indicate that they fractured by cleavage. In some cases the connecting material may "tear" by plastic deformation (slip). In the example of Fig. 3.14, the connecting material does not show the straight lines associated with cleavage, but instead is bent.

In body-centered cubic and hexagonal close-packed structures, the presence of mechanical twins, or more commonly the formation of them during the deformation, allows the crack front to deviate by separation of the plane between the twin and the matrix. This is illustrated in Fig. 3.15. In many body-centered cubic metals (particularly

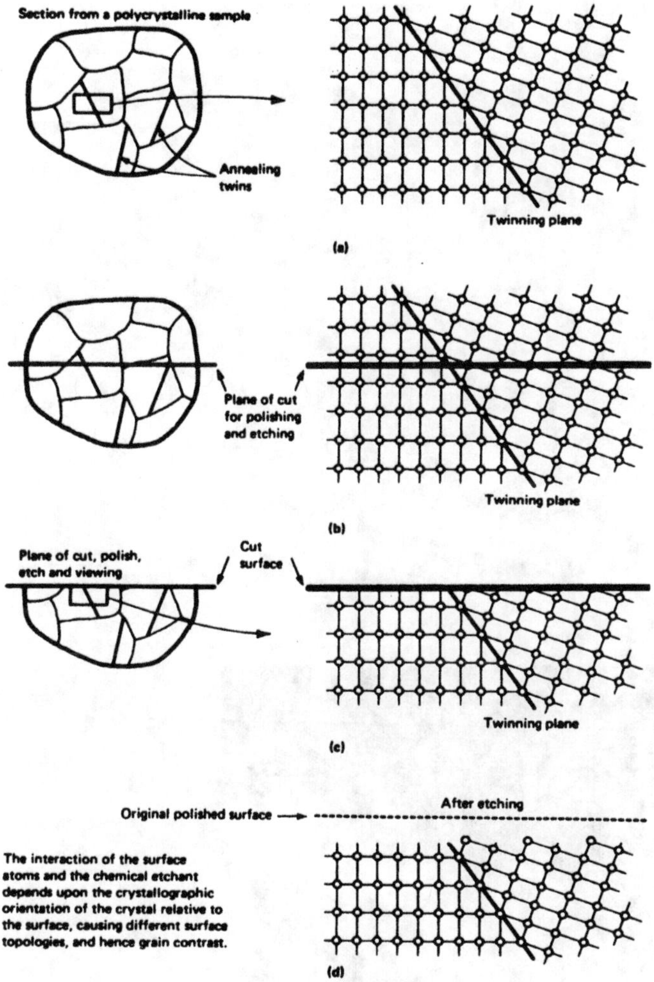

Figure 3.11 (*a*)–(*d*) Illustration of why annealing twins are revealed upon etching a polished surface. (*e*) SEM of Cu–5% Zn alloy in the annealed condition. (*From Brooks.*[6])

ferrite in steels), the protrusions or matching depressions shown in Fig. 3.16 are sometimes seen on cleavage fracture surfaces. These are called *tongues*. Cross-sectional views (Fig. 3.17), which allow measurements of the geometry of these regions, have shown that the tongue plane is a {112} type, and it is generally agreed, therefore, that these are created by fracture along the twin matrix interface. It appears that these twins form ahead of the advancing cleavage crack.

The fractures described appear to illustrate the main mechanism of cleavage crack formation. However, the topography is usually more

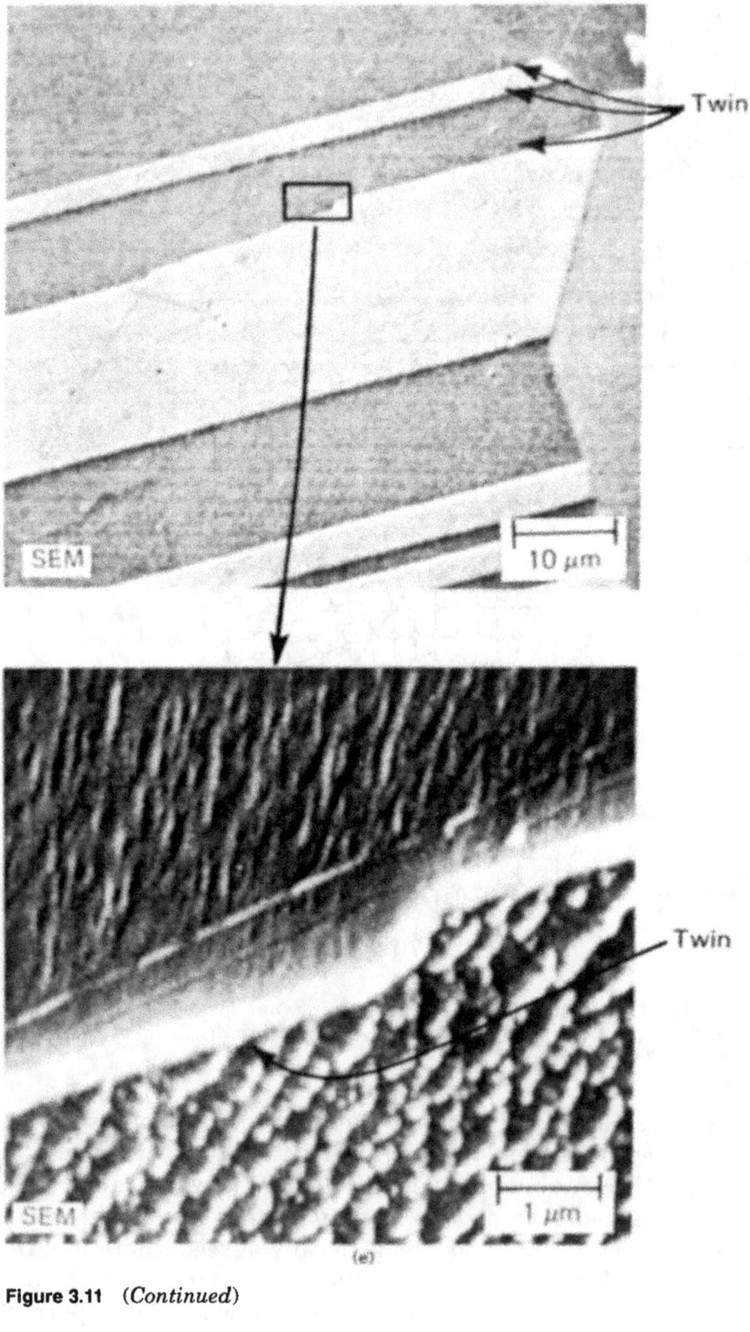

Figure 3.11 (*Continued*)

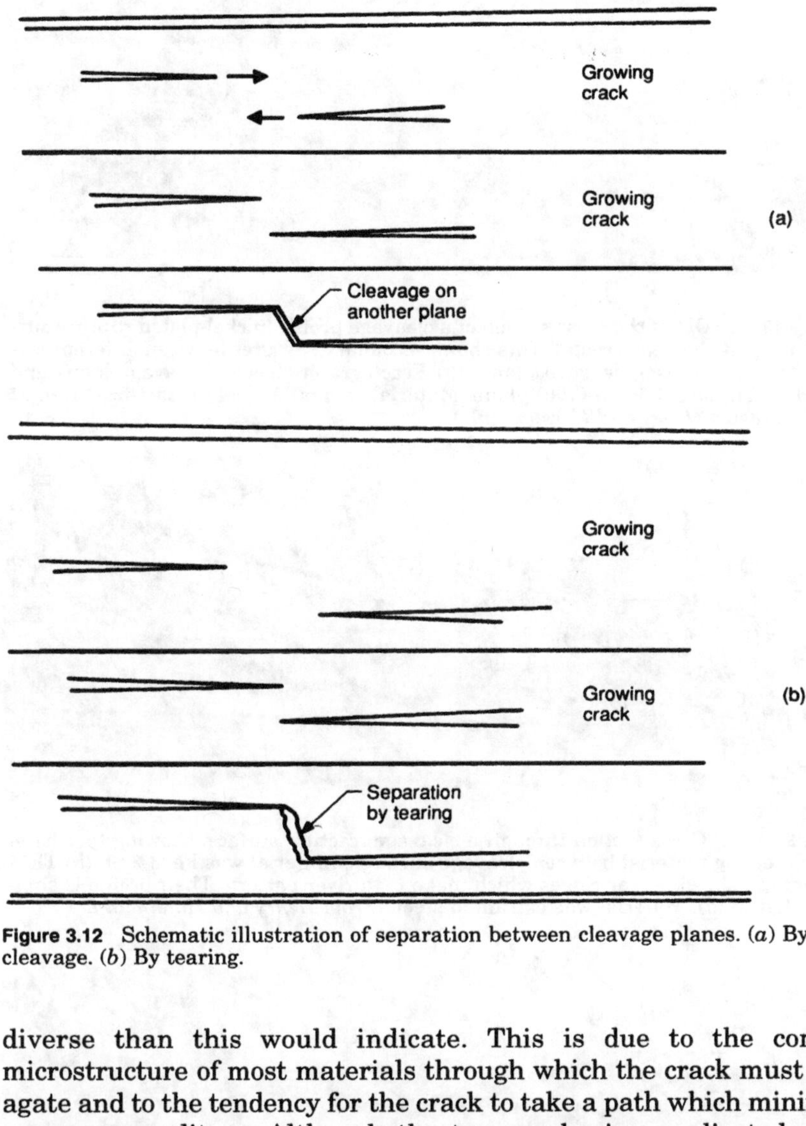

Figure 3.12 Schematic illustration of separation between cleavage planes. (*a*) By cleavage. (*b*) By tearing.

diverse than this would indicate. This is due to the complex microstructure of most materials through which the crack must propagate and to the tendency for the crack to take a path which minimizes energy expenditure. Although the topography is complicated, fortunately there are usually distinctive features associated with cleavage. One feature is a fan, shown in Fig. 3.18. In this case the advancing crack (from left to right) has encountered a grain boundary (marked *A–A*), whereupon locally a fresh crack was initiated at *B*. This cleavage crack then propagated in a fanlike fashion onto planes of slightly different elevation. The local crack propagation direction is noted on the fractograph by arrows.

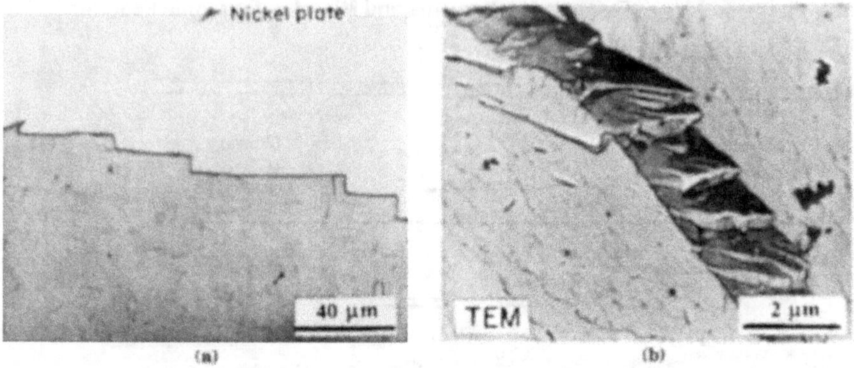

Figure 3.13 (*a*) OM of the cross section of a cleavage profile (nickel-plated rupture surface) in an Fe–Si single crystal. This shows secondary ruptures between different levels of fracture on the cleavage plane. (*b*) Fractograph showing cleavage steps and secondary rupture along a (100) plane. Material was mild steel, impact-fractured at −196°C. (*From Henry and Horstmann.*[7])

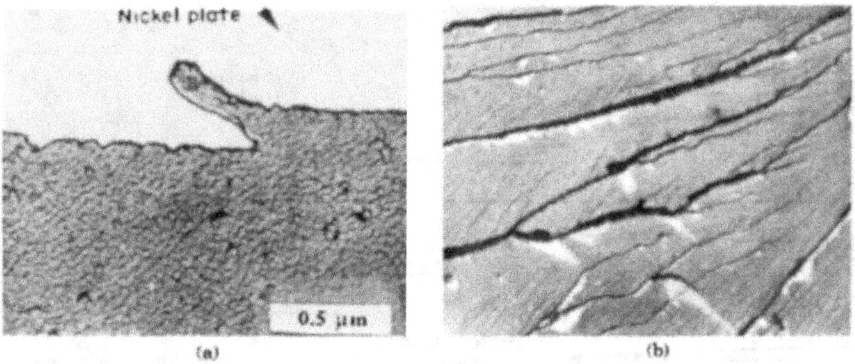

Figure 3.14 (*a*) Cross section through a cleavage fracture surface, showing tearing of the connecting material between cleavage surfaces. Material was Fe–3% Si. (*b*) TEM fractograph showing tear ridges which make up a river pattern. Their profile is similar to that in (*a*). Material was extramild steel. (*From Henry and Horstmann.*[7])

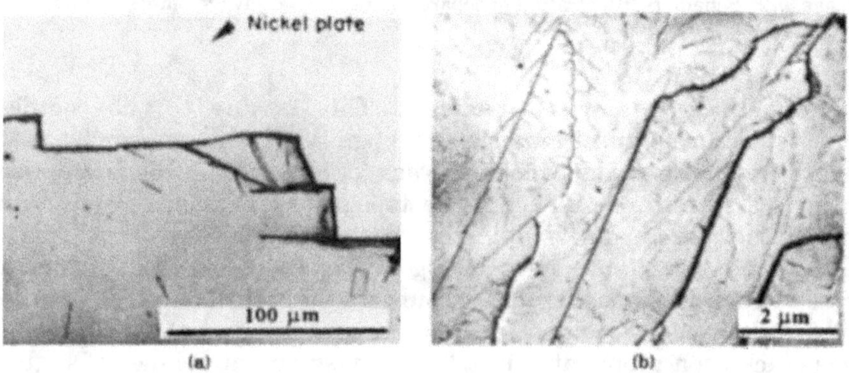

Figure 3.15 (*a*) Cross section through a cleavage fracture surface, showing interaction with mechanical twins. Material was an Fe–Si single crystal. (*b*) TEM fractograph showing preferential propagation along a [100] direction, and features due to fracture of twinned regions. Material was an Fe–25% Cr alloy. (*From Henry and Horstmann.*[7])

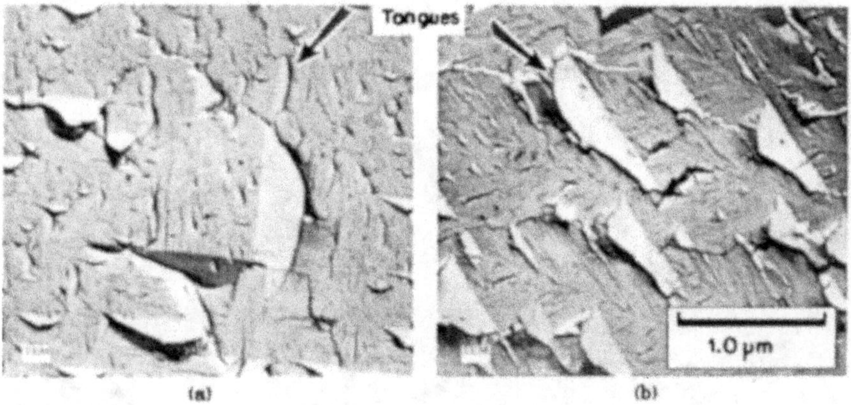

Figure 3.16 Fractographs showing the general appearance of "tongues." Samples were impact-fractured at −196°C. Note that the intersections of the cleavage plane and the tongues are oriented along one or the other of two perpendicular directions. The tongue plane is a (112) twin plane. (*a*) Extramild steel. (*b*) Fe–1.8% Si steel. (*From Henry and Horstmann.*[7])

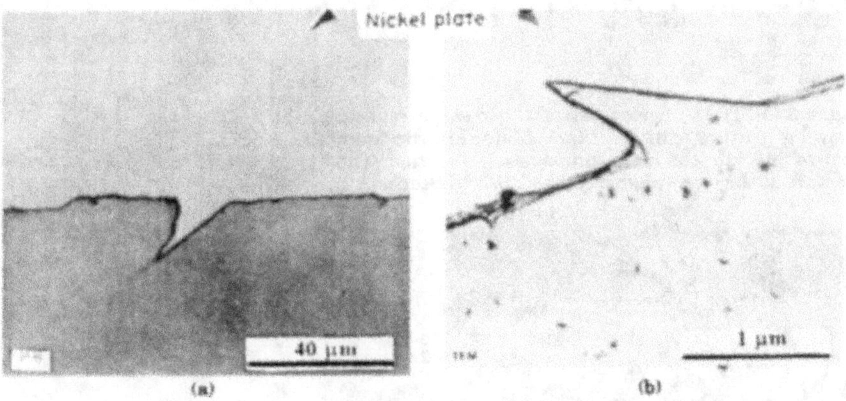

Figure 3.17 Cross section through fracture surface, showing separation along the interface between mechanical twins and the matrix. (*a*) OM of an Fe–Si single crystal, impact-fractured at −196°C. The fracture surface was nickel-plated (top). In the metal (bottom), a small twin which initiated the tongue is noticed. (*b*) TEM replica micrograph of fracture in Armco iron, showing profile of a tongue. (*From Henry and Horstmann.*[7])

In another example, shown in Fig. 3.19, the crack advanced from left to right. Along the line *A–A* it met a low-angle boundary. This causes the initiation of many cracks on the new cleavage plane, but at slightly different elevations. As the crack then propagated, these levels combined to form a crack advancing on fewer parallel planes. The reason is that there is less area associated with the fracture of the connecting material, and hence less expenditure of energy as the crack propagates. Such a mechanism generates a pattern similar to a river and its

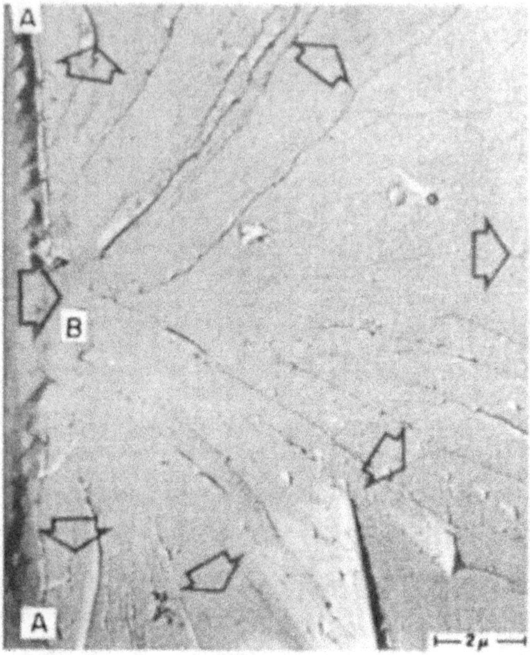

Figure 3.18 TEM micrograph of a two-stage replica, showing feature called a fan. Material was iron. Arrows indicate the crack propagation direction. The fracture surface was slightly etched. (*From Beachem.*[8])

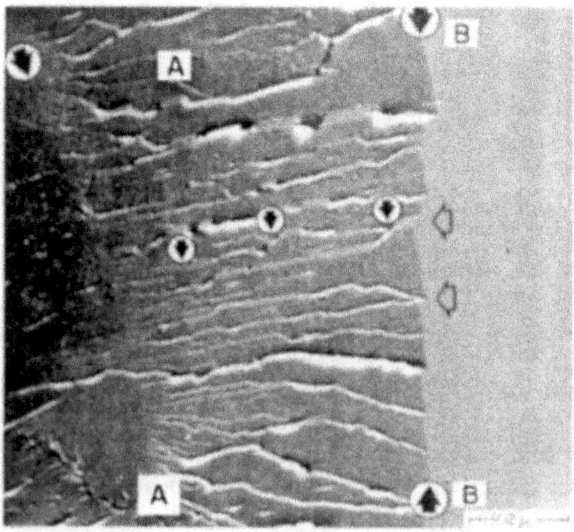

Figure 3.19 TEM micrograph of a two-stage replica, showing cleavage steps, river patterns, grain boundary (left), and mechanical twin interface (right) in cleaved iron. (*From Beachem.*[8])

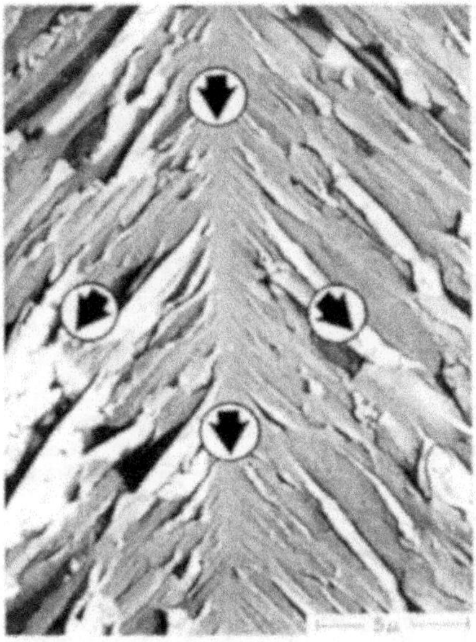

Figure 3.20 TEM micrograph of a two-stage replica, showing feature called a herringbone. Material was an Fe–Cr–Al alloy. (*From Beachem.*[8])

branches, and hence is called a river pattern. Note that in this sense the local crack propagation direction is "downstream." (Another example is shown in Fig. 3.15.) As pointed out, cleavage in iron and ferrite can also occur on the mechanical twin interface ({112}-type plane), and this is what happened at *B–B*. To the right is an extremely smooth fracture facet along such an interface.

The interaction of the fracture surface with twins sometimes creates the feature shown in Fig. 3.20, called a herringbone pattern. Cleavage on the basic {100} plane has changed on both sides of the midrib to cleavage on the {112} twin interface planes.

The processes described are shown schematically in Fig. 3.21. Examples of fracture surfaces containing these features are illustrated in Fig. 3.22. In some cases it is possible to see slip bands or lines on the cleavage facets. An example is given in Fig. 3.23. This apparently occurs after cleavage, when the connecting material is beginning to fracture by void coalescence. (See dimples, which are described in Sec. 3.5.)

3.5 Void Coalescence

If the metal begins to slip instead of cleavage occurring, then a different fracture mechanism occurs. A single crystal might deform due to cross slip, as shown in Fig. 3.24, with fracture occurring due to eventual "pinching off" of the final connecting material. A similar process can occur in polycrystalline material.

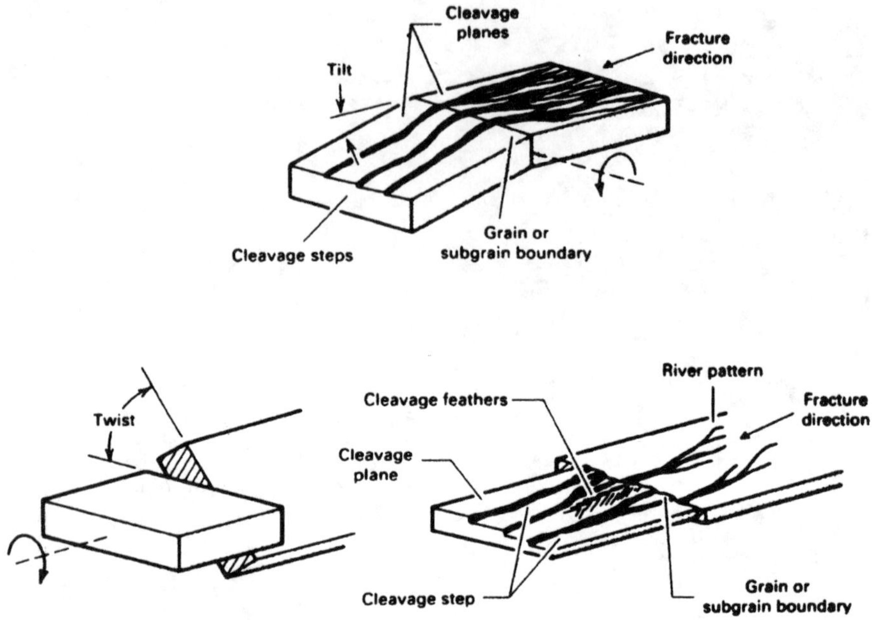

Figure 3.21 Cleavage surface formation, showing the effect of subgrains and grain boundaries. (*From Kerlins and Phillips.*[9])

However, a more common mechanism associated with slip is due to the separation of material internally, forming voids which then join to develop the fracture surface. This process is called void coalescence and is depicted schematically in Fig. 3.25. Note that once the void forms, the connecting material continues to deform by slip, allowing the voids to expand until they begin to connect. This fracture mechanism develops a fracture topography consisting of cusps on each fracture surface. Such a surface is called dimpled, and an example is given in Fig. 3.26.

In most engineering alloys the voids form at second-phase particles. This is shown schematically in Fig. 3.27, where the voids are initiated either by decohesion of the particle-matrix interface or by fracture of the particles. Examination of the microstructure of samples in the early stages of plastic deformation confirms this model, as shown in Fig. 3.28. An example of a fracture surface showing such particles is presented in Fig. 3.29. Some particles are in the dimples on the mating fracture surface, or they fell out during separation, so that in the fractograph some dimples are empty.

The shape of the dimples depends on the loading conditions. This is illustrated schematically in Fig. 3.30. Void coalescence by a stress normal to the overall plane of fracture creates dimples which are equiaxed. (However, this geometry will appear distorted if the surface

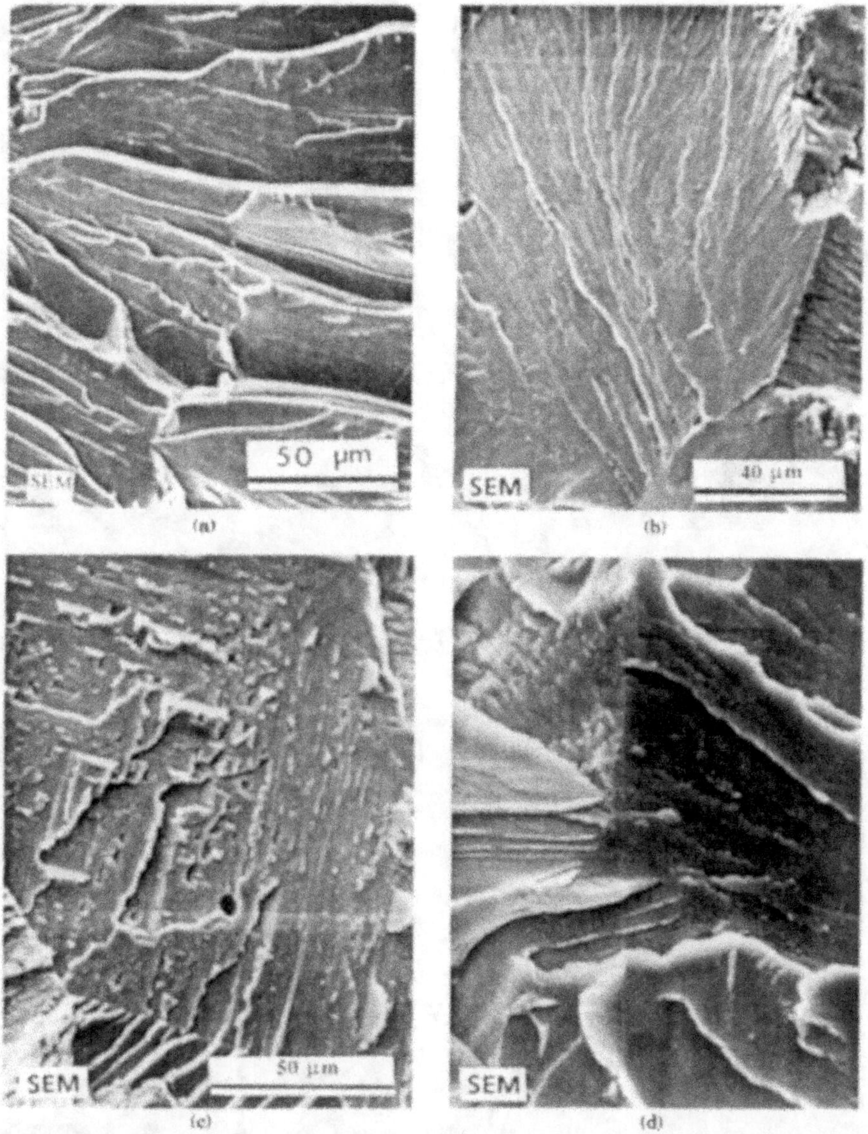

Figure 3.22 Examples of cleavage fracture surfaces. (*a*) Armco iron, impact-ruptured at −196°C. (*From Henry and Horstmann.*[7]) (*b*) Weld metal of 30% chromium steel. Owing to the varying crack propagation direction on the cleavage surface of this single grain, a featherlike structure has resulted, consisting of fine steps delineating the individual fracture paths which indicate the direction of propagation. The direction of crack advancement is from bottom to top. (*c*) Thanks to the presence of numerous twins, various crack propagation directions in a single plane can be deduced. The crack has divided in two along a straight front with a small section moving obliquely away from the main cleavage surface, subsequently returning to it by means of an irregular curved path. Where the crack has crossed into neighboring grains, many new subsidiary cracks form, as seen at the bottom of the picture. (*d*) 0.5% Mn steel pin. In the center of the picture the crack propagating from the right-hand side crosses a tilted grain boundary. As a result, many new crack surfaces are reinitiated in the grain on the left-hand side of the picture. In this grain a twin can be seen, running obliquely upward from the surface, this having the typically semicircular tonguelike shape. [(*b*)–(*d*) *From Engel and Klingele.*[10]]

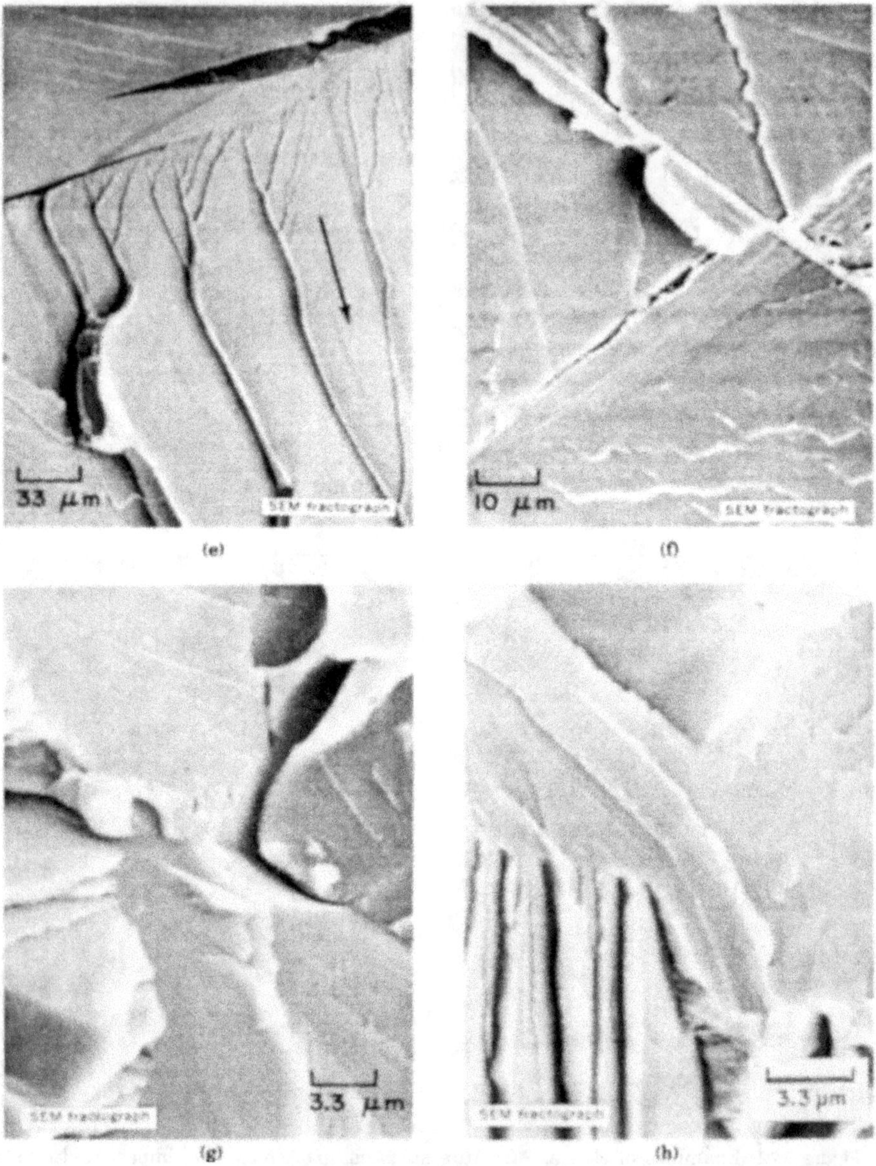

Figure 3.22 (*Continued*) Examples of cleavage fracture surfaces. (*e*) Fe, 0.01% C, 0.24% Mn, and 0.02% Si, heat treated at 950°C (1742°F) ½ h, air-cooled. The structure is ferrite. The fracture was generated by impact at −196°C (−321°F). Cleavage steps beginning at the twin at top form a sharply defined river pattern. Crack propagation was in direction of arrow. (*f*) Higher-magnification view of the specimen(s), showing the blocklike cleavage pattern that is characteristic of low-temperature impact of low-carbon iron. The well-defined tongue at the center of the fractograph is typical. Secondary cracks are visible as well. (*g*) 1021 steel heat treated at 900°C (1652°F) for ½h, air-cooled. Ferrite-pearlite structure. Hardness, 120 dph. The specimen was notched, and was broken by impact at −196°C (−321°F). The cleavage occurred on various {100} planes having different orientations. (*h*) Another SEM view of the fracture in (*g*), showing similar flat cleavage facets along crystallographic planes, but also very pronounced steps that form a gross example of a river pattern beginning at either a twist or a tilt boundary. [(*e*)–(*h*) *From* Metals Handbook.[11]]

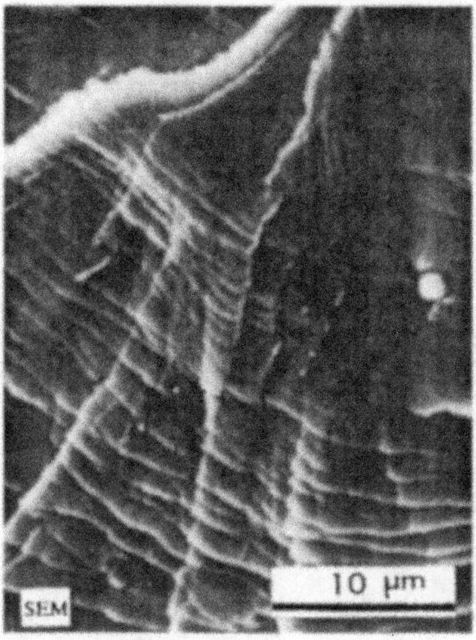

Figure 3.23 Fractograph of cleavage fracture surface showing slip traces. Slip lines reveal plastic deformation which occurs following the formation of ductile zones. (*From Henry and Horstmann.*[7])

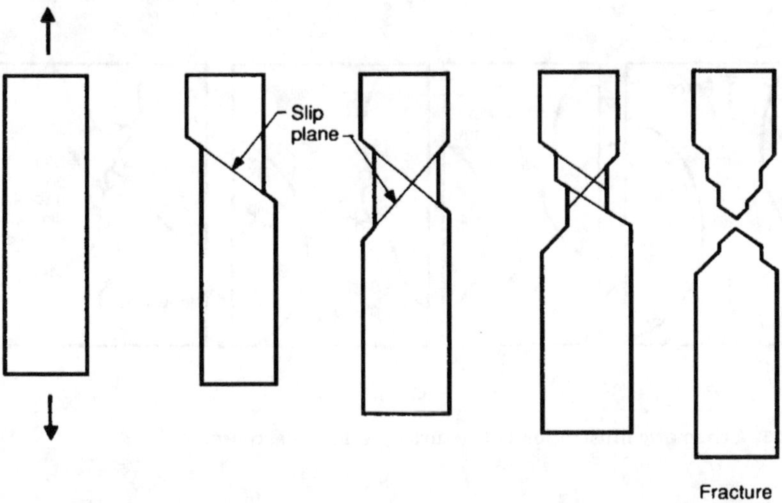

Figure 3.24 Schematic illustration of fracture of a single crystal by cross slip.

is examined at an angle.) A shear or tear loading (Fig. 3.30) will create elongated dimples. The two cases can be distinguished by examination of both fracture surfaces, knowing the original alignment of the two pieces. In the case of shear, the elongated dimples on the two surfaces will be in opposite directions. Examples of shear dimples are shown in Fig. 3.31 and of tear dimples in Fig. 3.32.

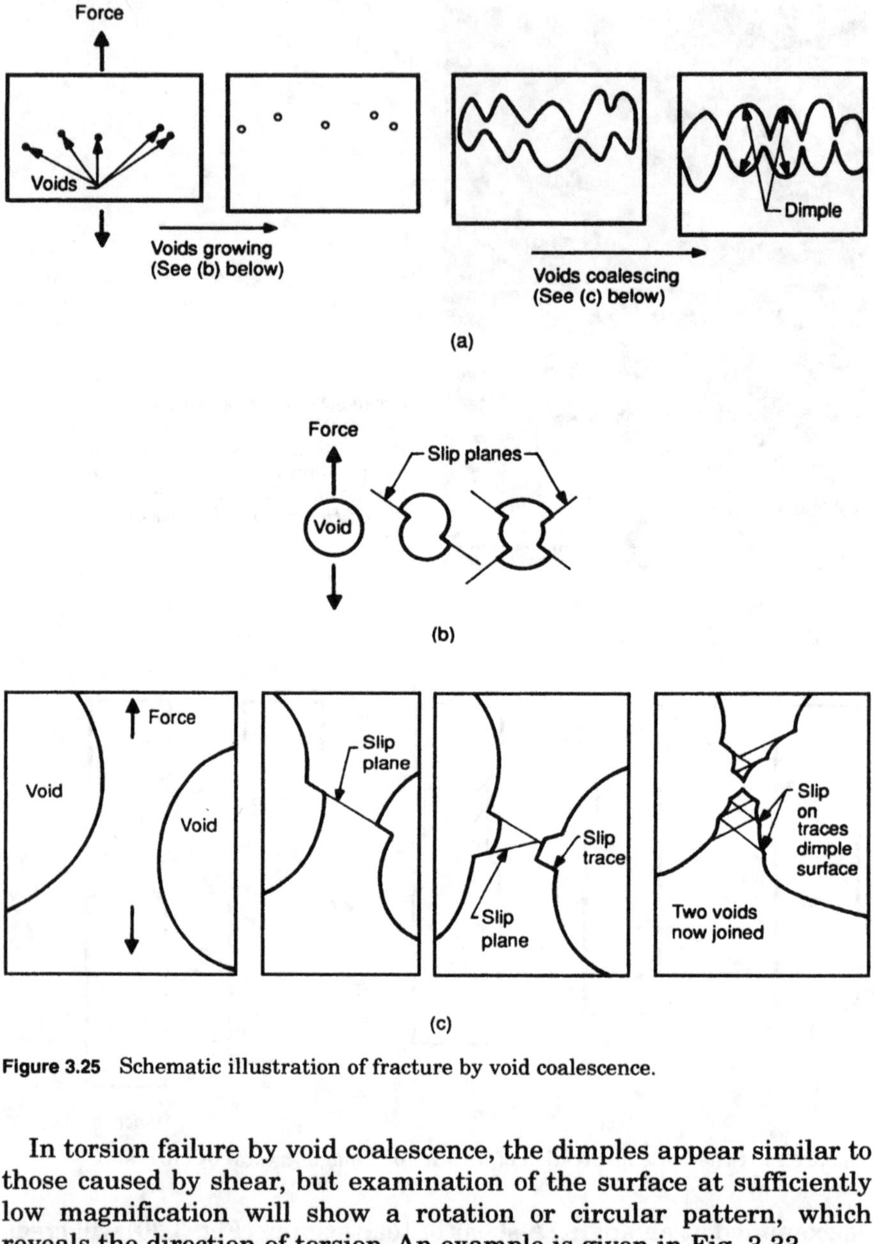

Figure 3.25 Schematic illustration of fracture by void coalescence.

In torsion failure by void coalescence, the dimples appear similar to those caused by shear, but examination of the surface at sufficiently low magnification will show a rotation or circular pattern, which reveals the direction of torsion. An example is given in Fig. 3.33.

The size and shape of the dimples are directly related to the size, shape, and dispersion of the second-phase particles. This is seen in Fig. 3.29, where the larger voids have the larger particles in them. Figure 3.34*a* shows the fracture surface of a steel with elongated MnS

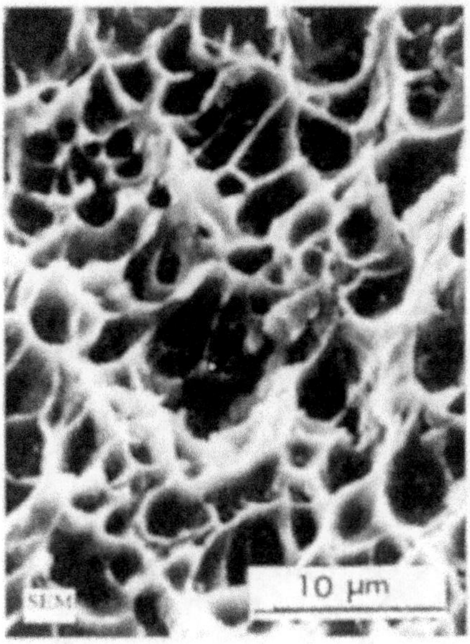

Figure 3.26 Fractograph showing dimples. Material was a steel containing 0.10%, C, 0.02% Si, 0.30% Mn, 0.032% P, 0.024% S, and 0.003% N. The sample was notched and broken in bending. (*From Henry and Horstmann.*[7])

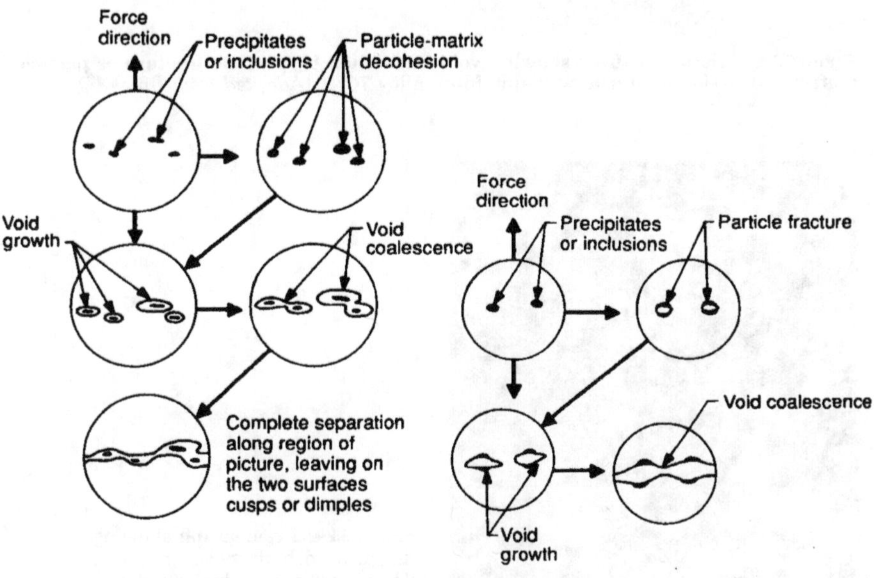

Figure 3.27 Schematic illustration of void nucleation at particles. (*a*) Initiation by particle-matrix interface decohesion. (*b*) Initiation by particle fracture.

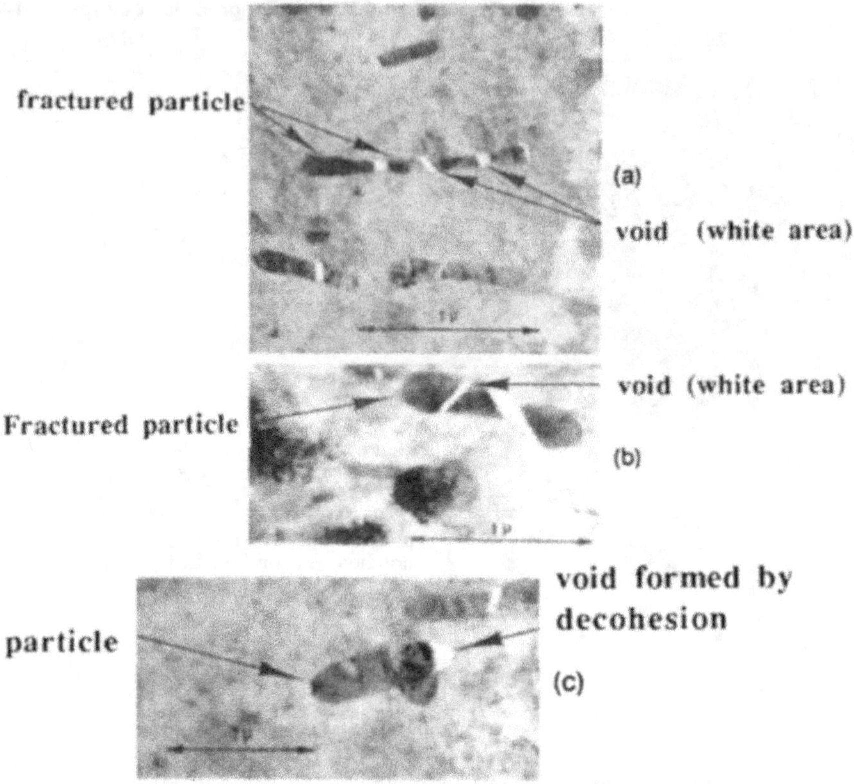

Figure 3.28 Microstructures showing void nucleation by particle fracture or particle-matrix decohesion. Material was aluminum alloy 7079. (*Adapted from Broek.*[12])

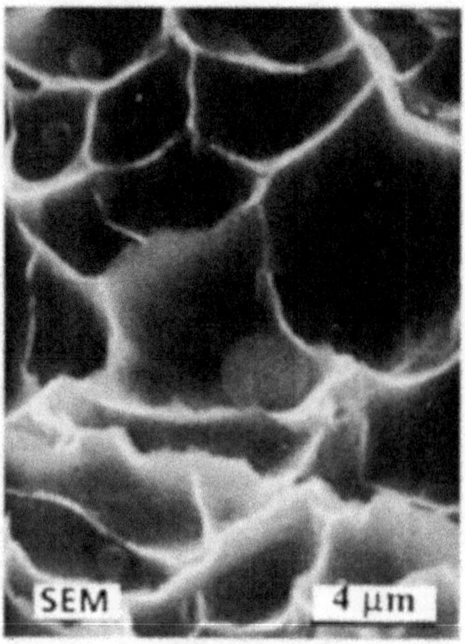

Figure 3.29 Fractograph showing dimples with particles in them. These silicate inclusions determined the form and size of the dimples in this weld-seam fracture of an unalloyed carbon steel. (*From Engel and Klingele.*[10])

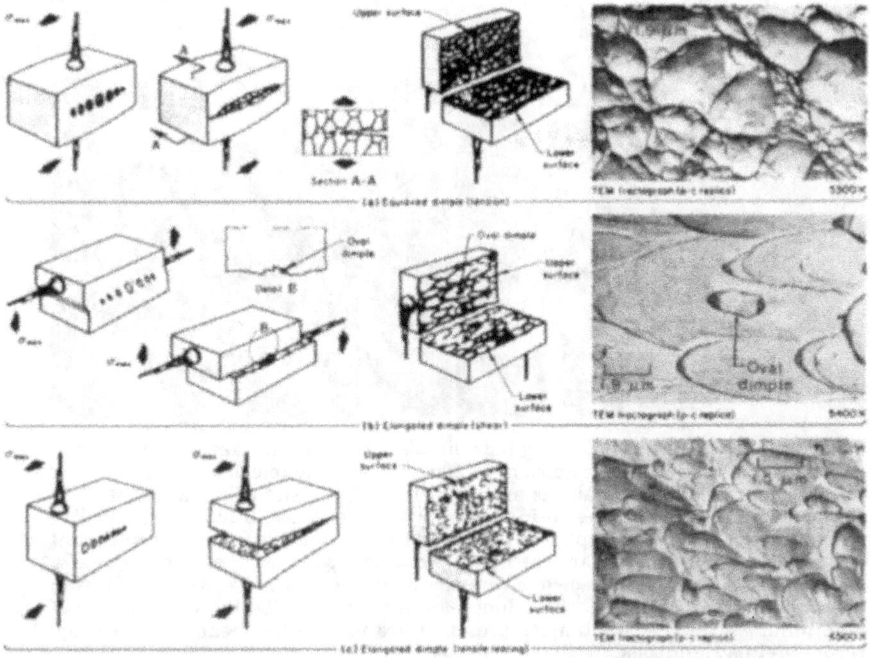

Figure 3.30 Schematic illustration of the influence of loading conditions on the geometry of the dimples found on the fracture surface. The fractographs show the shapes of the dimples formed. (*a*) In tension, equiaxed dimples are formed on both fracture surfaces. (*b*) In shear, elongated dimples point in opposite directions on mating fracture surfaces. (*c*) In tensile tearing, elongated dimples point toward fracture origin on mating fracture surfaces. (*From* Metals Handbook.[11])

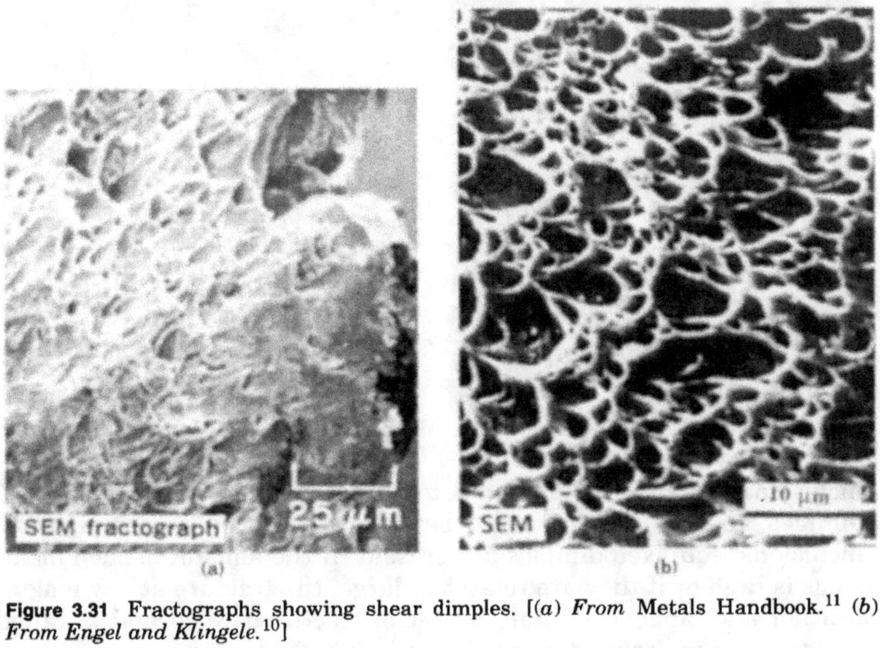

Figure 3.31 Fractographs showing shear dimples. [(*a*) *From* Metals Handbook.[11] (*b*) *From Engel and Klingele.*[10]]

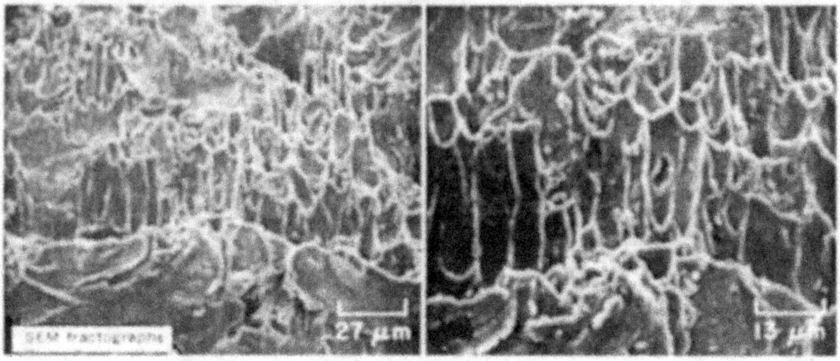

Figure 3.32 Fractographs showing tear dimples. These were taken from the fracture region at the root of the notch of a Charpy impact sample of a 1040 steel. Tear dimples resemble slim parabolas and point toward the site of crack initiation at the notch root, which is shown in the lower third of these fractographs. Unlike shear dimples, the tear dimples on the mating surface of this fracture would point in the same direction—toward the notch. Occasionally, when the overall shape of the fracture surface is unknown, a comparison of the mating fracture surfaces to observe the relative directions of dimple orientation may be the only method of determining whether the dimples present were formed by shear or by tearing. (*From* Metals Handbook.[11])

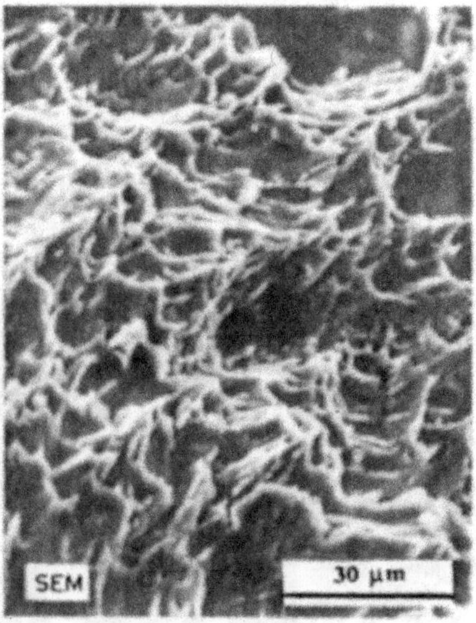

Figure 3.33 Fractograph showing torsional dimples. The wire sample was fractured by means of a monotonic torsional loading. The resulting arrangement of the dimples in a clockwise fashion indicates the direction of torsion. (*From Engel and Klingele.*[10])

inclusions. Note the correlation between the geometry of the elongated dimples and the inclusions. In between the regions containing the inclusions, equiaxed dimples are present. If the amount of such inclusions is high or if they are relatively large, the fracture at low magnification may appear similar to that of fractured wood and is called *woody fracture.* Examples are shown in Fig. 3.34*b* and *c*.

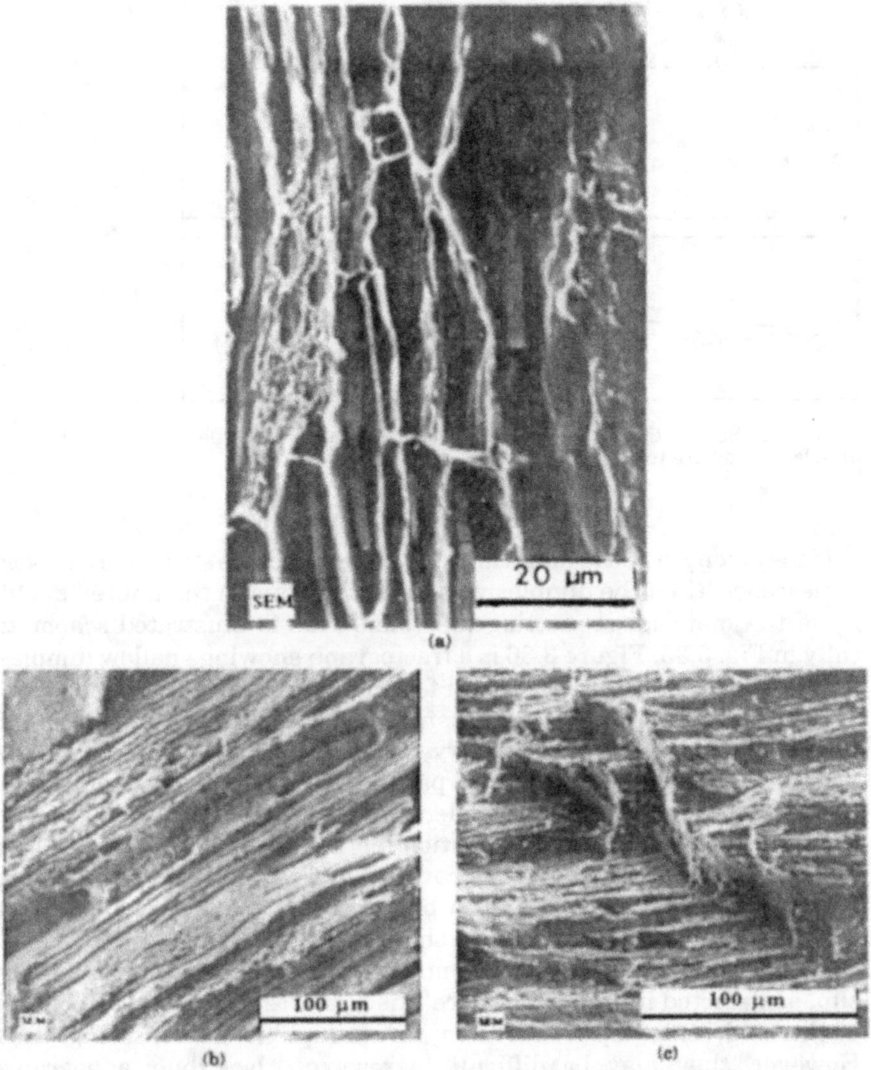

Figure 3.34 (*a*) Fracture showing elongated dimples due to elongated inclusions. Material is steel containing 0.12% C, 0.02% Si, 0.44% Mn, 0.059% P, and 0.068% S. Cracks are initiated along sulfide inclusion lines. (*b*) Fracture showing woody structure, 0.35% C, 0.62% Mn, 0.028% S structural steel, water quenched, and tempered at 550°C. Tensile rupture at 20°C (transverse direction). (*c*) Fracture showing woody structure, 0.44% C, 1.5% Mn, 0.30% S free cutting steel, water quenched and tempered at 550°C. Tensile rupture at 20°C (transverse direction). (*From Henry and Horstmann.*[7])

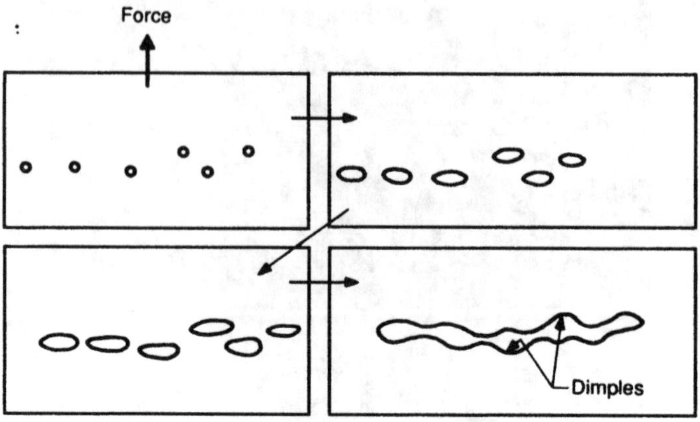

Figure 3.35 Schematic illustration of the formation of shallow (flat) dimples. (Compare to Fig. 3.25.)

If the strength of the material is high, but fracture still occurs by void coalescence, then the dimples will be shallow due to the limited ductility of the material connecting the voids. This is illustrated schematically in Fig. 3.35. Figure 3.36 is a fractograph showing shallow dimples.

If the grain boundaries contain features that make this region initiate voids, then appearances illustrated in Fig. 3.37*a* may be found. This might occur if the boundary contains particles, or if there is a region of particles adjacent and parallel to the boundary. (Note that this type of fracture is not considered to be intergranular in the sense that fracture occurs by separation of the boundary between two planes. This type of fracture is covered in Sec. 3.8.) The effect of the structure on and near the grain boundary on the topography of the intergranular fracture surface is shown in Fig. 3.37*b*.

The plastic deformation of the material between the voids occurs by slip, as depicted in Fig. 3.25. Since this slip extends to the free surface of the void, then slip traces form; an example is shown in Fig. 3.38. However, they may be difficult to resolve. Also, their appearance depends on the local crystallographic orientation of the region slipping relative to the free surface. In some cases, slip traces appear as wavy lines, probably associated with cross slip. This feature is called serpentine glide; an example is shown in Fig. 3.39. If the slip plane is quite oblique to the free surface of the cusp, the slip traces are spread out so that the surface appears very smooth, but with faint traces of the slip offset (Fig. 3.40). These features are called ripples, and the mechanism is called stretching. [The mechanism has also been called glide plane decohesion (because the slip planes eventually are highly offset; see Fig. 3.24), and it has been called ductile cleavage.]

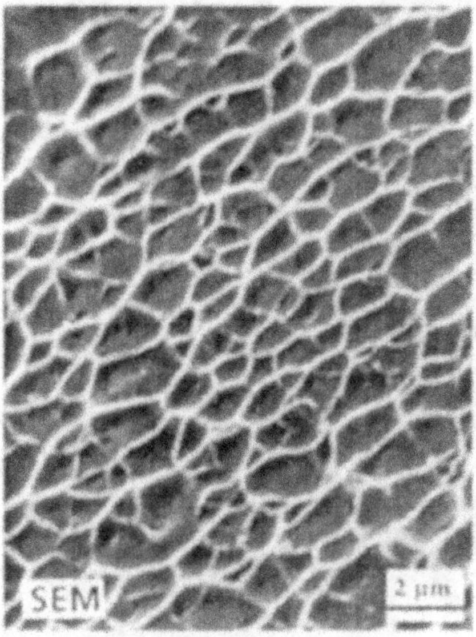

Figure 3.36 Fractograph showing shallow dimples. The sample was a maraging steel (0.02% C, 0.1% Si, 0.1% Mn, 18% Ni, 9% Co, 5% Mo, 0.9% Ti) whose formability had been exhausted due to severe cold working. The specimen failed under static loading with a minimum amount of plastic deformation. (*From Engel and Klingele.*[10])

Usually macroscopically ductile fractures, which show gross dimensional changes, occur microscopically by void coalescence. However, it is important to realize that macroscopically brittle fracture, with little dimensional change, can also occur by void coalescence. This is favored by a high-strength material and by a fine dispersion of particles. An

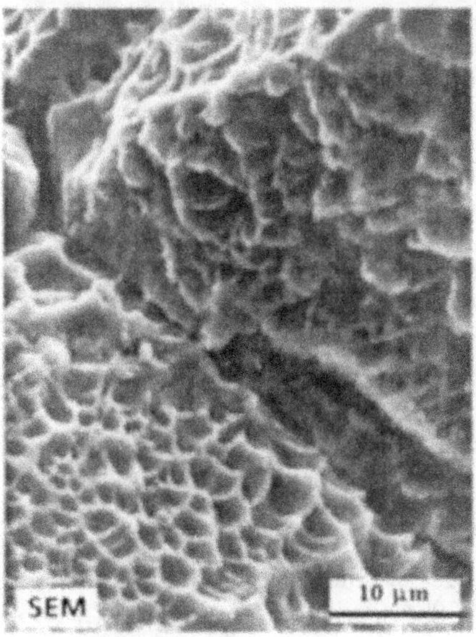

(a)

Figure 3.37 (*a*) Intergranular fracture surface showing dimples on the surface. (*From Engel and Klingele.*[10])

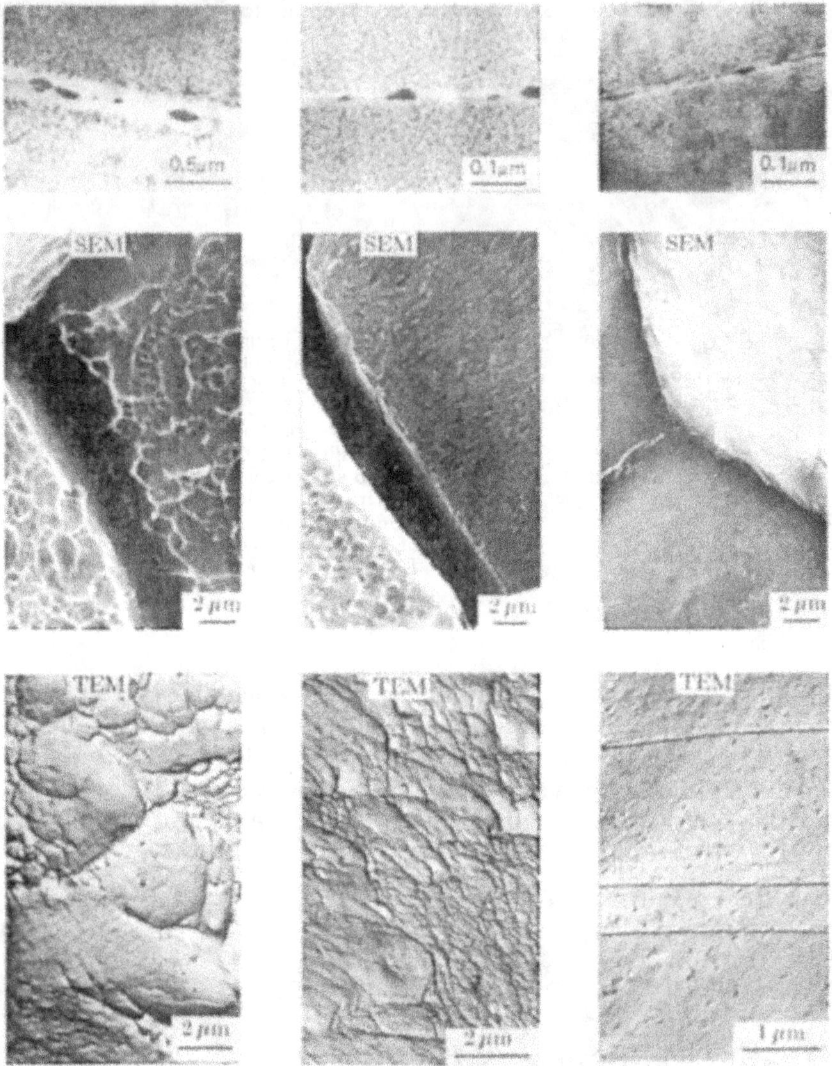

(b)

Figure 3.37 (*Continued*) (*b*) Effect of grain-boundary structure on fine topology of the intergranular fracture surface. (*From Lynch.*[13])

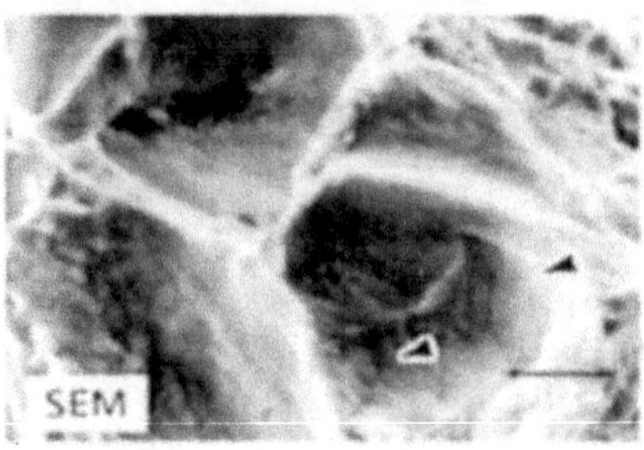

Figure 3.38 Fractograph showing slip traces (arrows) on the side of the dimples. (*From Gabriel.*[14])

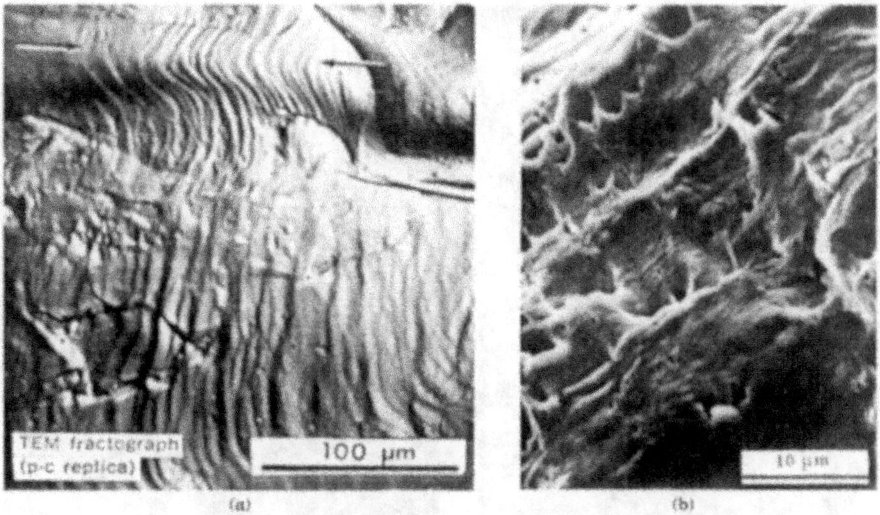

Figure 3.39 Fractographs showing examples of serpentine glide. (a) TEM replica of fracture surface of Armco iron. (*From* Metals Handbook.[11]) (b) Fracture in oxygen-free copper tube, produced by forcing open the crack in a precracked specimen. (*From Engel and Klingele.*[10])

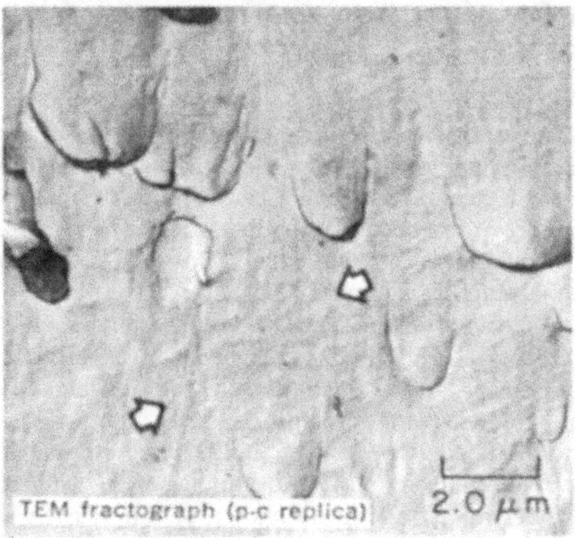

Figure 3.40 Two-stage carbon TEM replica showing ripples (at arrows). The sample was a 302 stainless steel. (*From* Metals Handbook.[11])

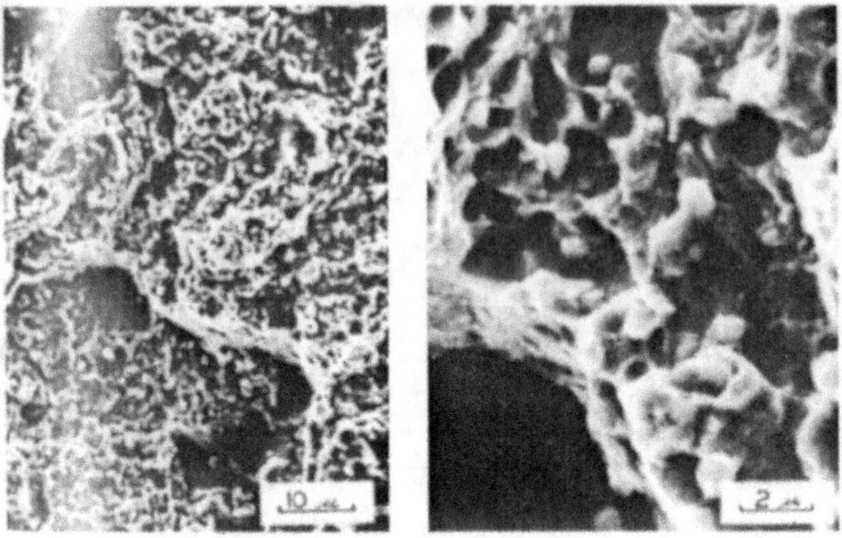

Figure 3.41 SEM fractographs showing shallow dimples in a material which failed in a brittle mode. (*From Syniuta and Corrow.*[15])

example of such a fracture is shown in Fig. 3.41. Note that the dimples are relatively flat.

3.6 Mixed Mechanism and Quasicleavage Fracture

In many materials, fracture may first occur by cleavage, then fracture of the connecting material by void coalescence. Examples are shown in Fig. 3.42. This is the case particularly in body-centered cubic materials (such as ferrite), when fracture occurs in the vicinity of the transition temperature (which depends on the strain rate). This is an example of a mixed mechanism of fracture.

In some materials, especially steels, a fracture surface forms which consists of regions showing ill-defined or relatively faint signs of cleavage (sometimes called *rosettes*) separated by regions showing extensive deformation. These latter regions are *tear ridges*. Examples of such surfaces are given in Fig. 3.43. They appear to develop by the formation of microcracks and then tearing (plastic deformation) of the connecting material, as shown schematically in Fig. 3.44. This mechanism is called *quasicleavage*. It is to be noted that this is actually cleavage, accompanied by tearing or dimples, where the fine microstructural details make the identification less apparent than described earlier for cleavage. In this sense, it is really a mixed-mechanism fracture mechanism.

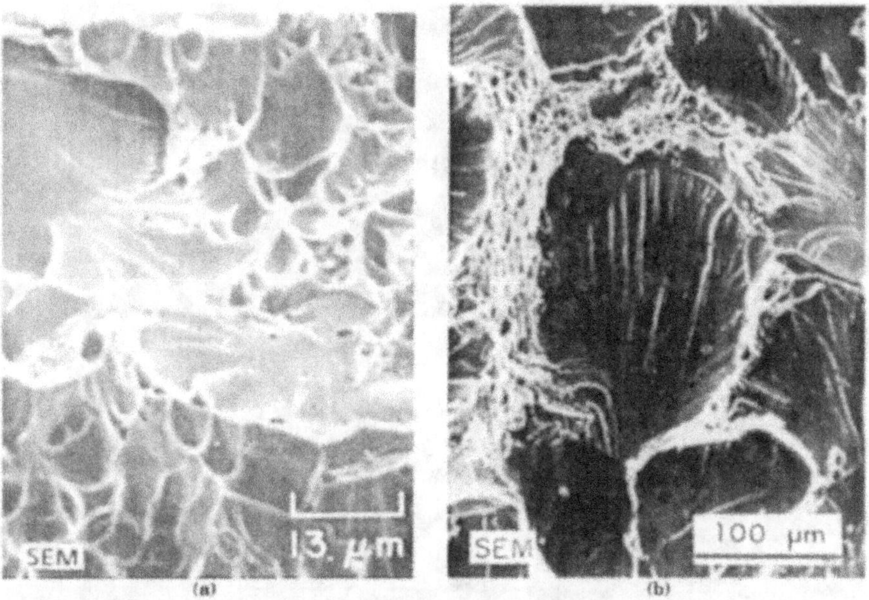

Figure 3.42 Examples of cleavage and void coalescence mixed mechanisms of fracture. (*a*) 1040 hot-rolled steel. (*From* Metals Handbook.[11]) (*b*) Tensile sample broken at 0°C. Steel contained 0.08% C, 0.14% Si, 0.34% Mn, 0.021%, P, 0.018% S, and 0.034% Al. (*From Henry and Horstmann.*[7])

3.7 Tearing Topography Surface

In some cases, a fractographic topography as shown in Fig. 3.45 is found. It consists of very fine "facets" connected by tear ridges and is referred to as *tearing topography surface* (TTS). This appearance seems to be caused by the particular complex microstructures involved in the fracture. The underlying mechanisms are probably those already described.

3.8 Intergranular Separation

When the grain boundaries are the weakest locations in the microstructure, then decohesion occurs on these surfaces before cleavage or slip. In the strict sense, this decohesion mechanism of fracture is called intergranular fracture. However, in complex microstructures, the fracture surface may appear intergranular, even though the fracture path may have been, for example, along particles on the boundary (by decohesion or separation along the particle-matrix interface), by void coalescence associated with particles on the boundary, or by void coalescence and tearing in a weak zone parallel and adjacent to the boundary. Even in these cases, the fracture topography is referred to as

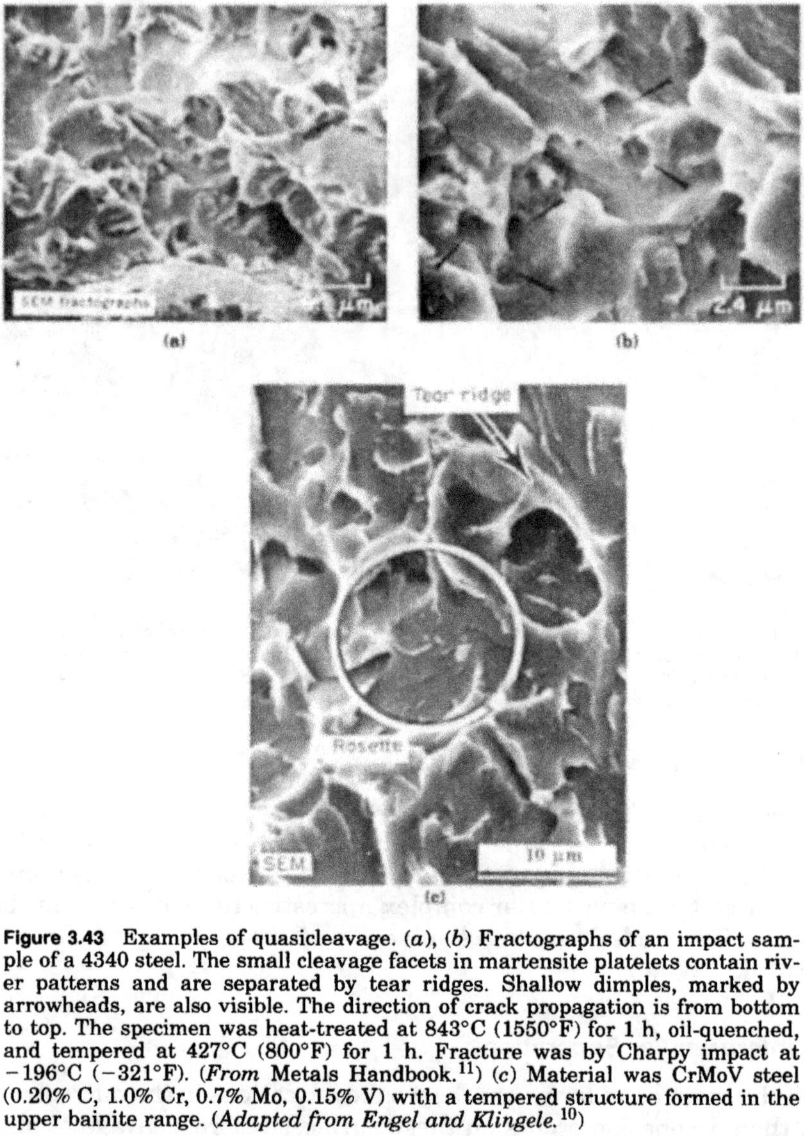

Figure 3.43 Examples of quasicleavage. (*a*), (*b*) Fractographs of an impact sample of a 4340 steel. The small cleavage facets in martensite platelets contain river patterns and are separated by tear ridges. Shallow dimples, marked by arrowheads, are also visible. The direction of crack propagation is from bottom to top. The specimen was heat-treated at 843°C (1550°F) for 1 h, oil-quenched, and tempered at 427°C (800°F) for 1 h. Fracture was by Charpy impact at −196°C (−321°F). (*From* Metals Handbook.[11]) (*c*) Material was CrMoV steel (0.20% C, 1.0% Cr, 0.7% Mo, 0.15% V) with a tempered structure formed in the upper bainite range. (*Adapted from Engel and Klingele.*[10])

intergranular fracture. In such cases the fracture surface will show topographical details, whereas in decohesion the surface will be smooth if the boundary contains no particles. The shape of the intergranular fracture topography is determined by the grain shape, as illustrated in Fig. 3.46. Typical intergranular fracture fractographs are shown in Fig. 3.47. In some cases the structure on or near the grain boundaries causes the fracture surface to be relatively rough on a fine scale, as shown in Fig. 3.37.

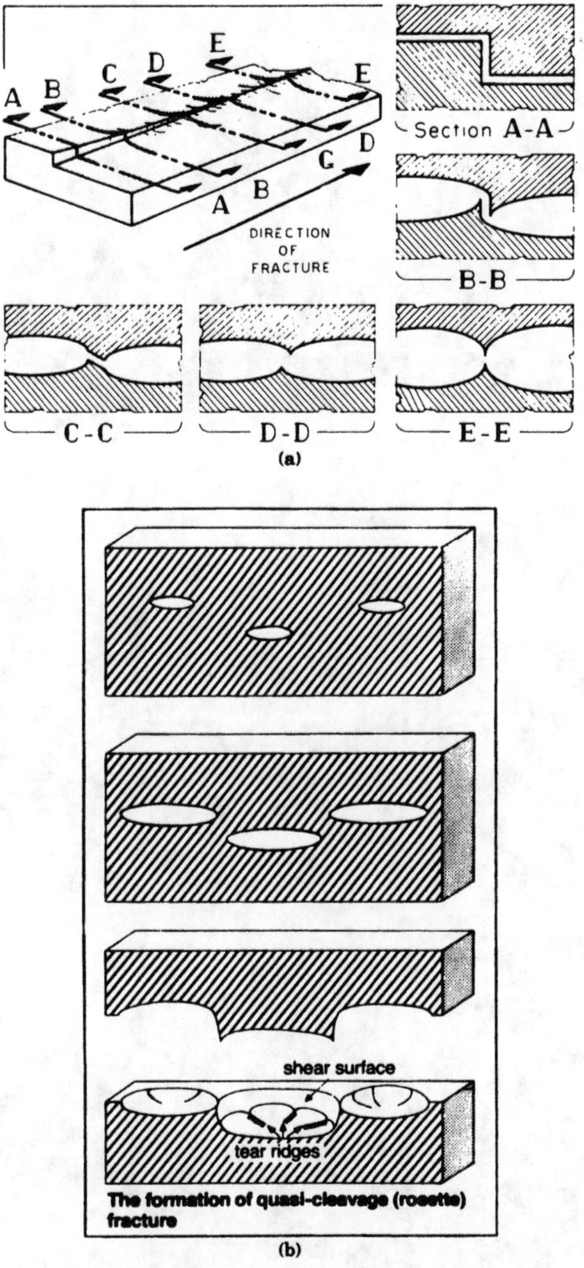

Figure 3.44 Schematic illustration of the formation of fracture by quasicleavage. (*a*) Model showing a cleavage step blending in with a tear ridge. At the top left is the lower surface of a fracture, showing a step at the lower left and a ridge at the upper right. At right and bottom are sections through the fractured member, showing profiles of both the upper and the lower fracture surfaces. (*From* Metals Handbook.[11]) (*b*) Model of the formation of quasicleavage rosettes. (*From Engel and Klingele.*[10])

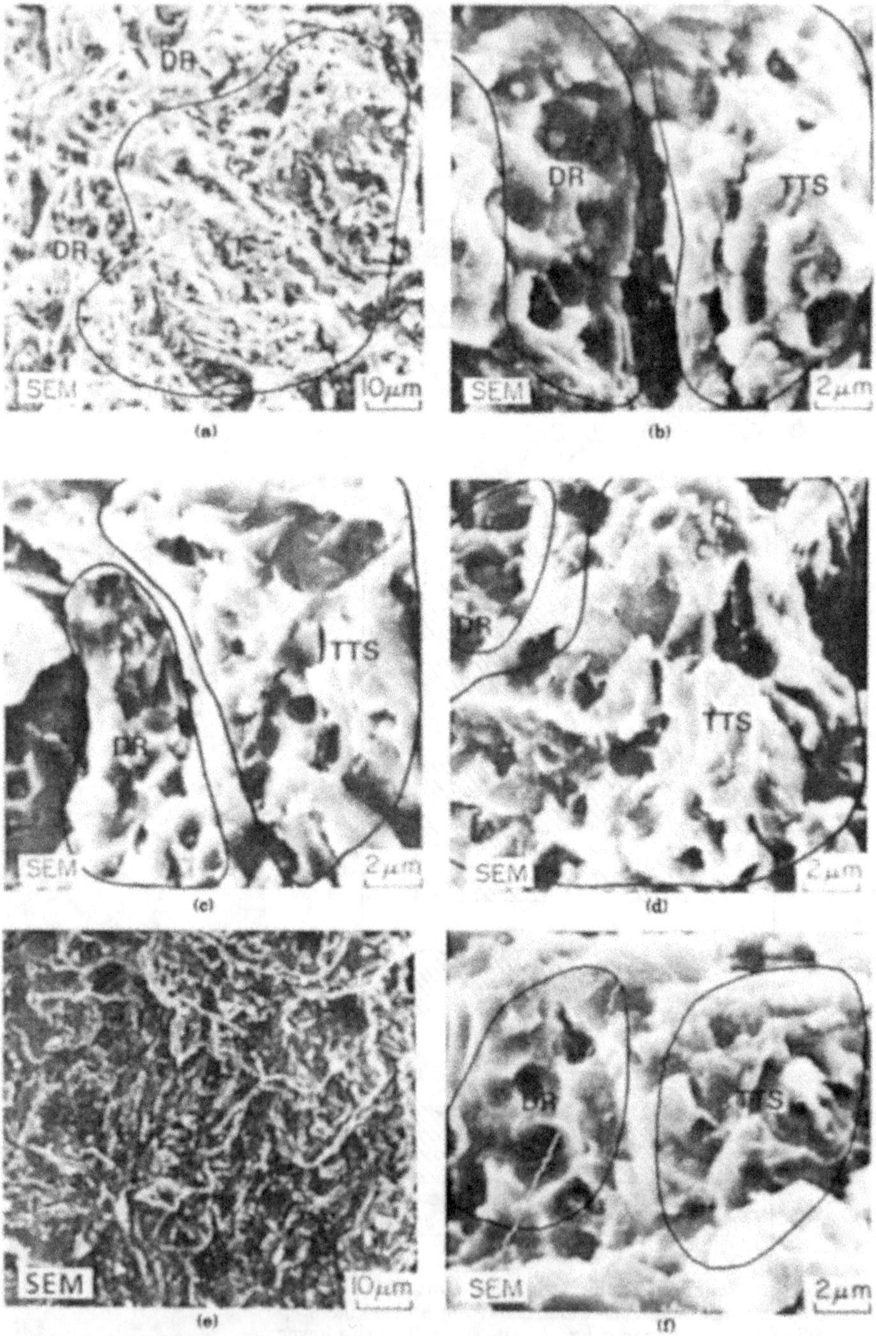

Figure 3.45 Examples of tearing topography surfaces (TTS). In these cases dimple rupture (DR) is also found. (*a*)–(*d*) HY-130 steel. (*a*) Areas of complex tearing (T) and dimple rupture (DR). (*b*) Detail from upper left corner of (*a*) showing particle-nucleated dimples (DR) and region of TTS. (*c*), (*d*) Additional examples of TTS. (*e*) TTS in an essentially 100% pearlitic eutectoid steel (similar to AISI 1080) where fracture propagates across pearlite colonies. (*f*) Fractograph showing dimple rupture (DR) and TTS fracture in a quenched-and-tempered (martensitic) HY-130 steel. (*From Thompson and Chesnutt.*[16])

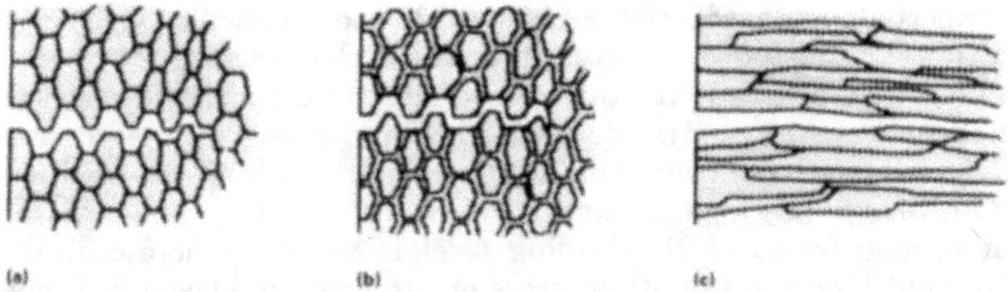

Figure 3.46 Schematic illustration of grain-boundary fracture. (*a*) Decohesion along grain boundaries of equiaxed grains. (*b*) Decohesion through a weak grain-boundary plane. (*c*) Decohesion along grain boundaries of elongated grains. (*From Kerlins and Phillips.*[9])

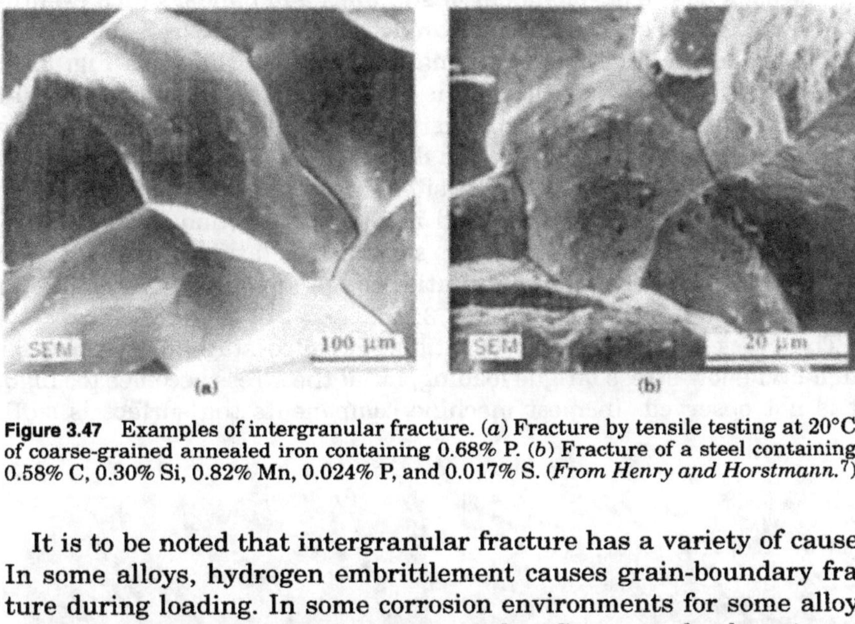

Figure 3.47 Examples of intergranular fracture. (*a*) Fracture by tensile testing at 20°C of coarse-grained annealed iron containing 0.68% P. (*b*) Fracture of a steel containing 0.58% C, 0.30% Si, 0.82% Mn, 0.024% P, and 0.017% S. (*From Henry and Horstmann.*[7])

It is to be noted that intergranular fracture has a variety of causes. In some alloys, hydrogen embrittlement causes grain-boundary fracture during loading. In some corrosion environments for some alloys, stress corrosion cracking is intergranular. Some steels show temper embrittlement fracture along the former austenite grain boundaries, which are the high-angle boundaries of the ferrite matrix. In some steels this is due to segregation of minor impurities (such as P or Bi) to these boundaries, and in some cases it is due to carbide networks on these boundaries.

3.9 Fatigue Fracture Topography

Consider a material which has a certain yield strength, obtained from testing a cylinder of this material in a simple tensile test. Thus if a cylindrical sample of this material is loaded to an axial stress two-thirds of the yield strength, then plastic deformation and fracture are not expected to occur. However, if the axial load is cycled between zero and this two-thirds value, after sufficient cycles fracture may occur. This is called a fatigue fracture. Since most structural and machine components operate under alternating loading, fatigue failures are relatively common, and being able to recognize them from the fracture-surface topography is important.

In fatigue fractures the loading level is usually macroscopically below that required to initiate gross or macroscopic plastic deformation. However, due to surface irregularities, microstructural features, and so on, the local microscopic stress exceeds the yield strength, in which case, upon repeated fatigue loading, plastic deformation occurs locally. On highly polished free surfaces the local plastic deformation can be seen as the development of slip lines and bands, which eventually reach a size such that they can act as the initiator of the fatigue crack. An example of such deformation bands is shown in Fig. 3.48. Generally this local deformation is associated with a shear stress at approximately 45° to the tensile stress created by the load. In some alloys the initial stages of plastic deformation create extrusions and intrusions at the surface at these sites (Fig. 3.49). These are shown on the free surface in Fig. 3.50*a* and in the cross-sectional view in Fig. 3.50*b*. Such surface discontinuities serve as the fatigue crack initiator. The fracture surface in this initiation stage appears cleavagelike, as shown by the fractographs in Fig. 3.51.

The crack initiation stage of fatigue is called stage I. It occurs for high-cycle–low-stress fatigue loading, but if the stress becomes too high, it is not observed. In most machine components the surface is suffi-

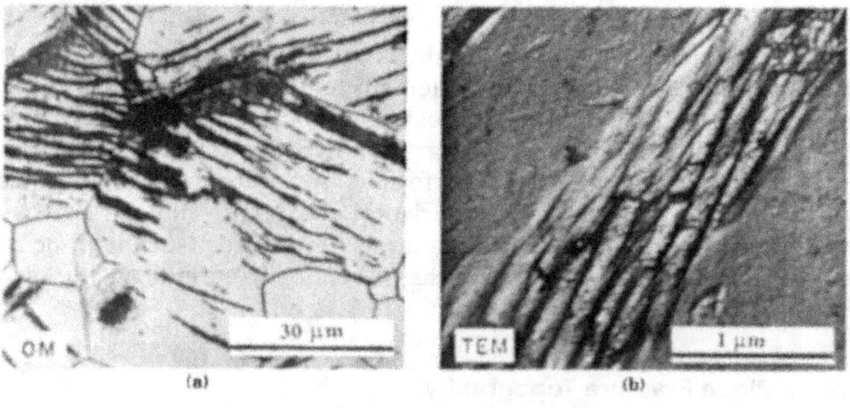

(a) (b)

Figure 3.48 Formation of deformation bands during fatigue loading. (*a*) OM of a mild steel specimen surface after 525,000 cycles of alternate bending under ±13 daN/mm^2. (*b*) TEM replica micrograph of the surface of a mild steel after 125,000 cycles of alternate bending under ±10 daN/mm^2. (*From Henry and Horstmann.*[7])

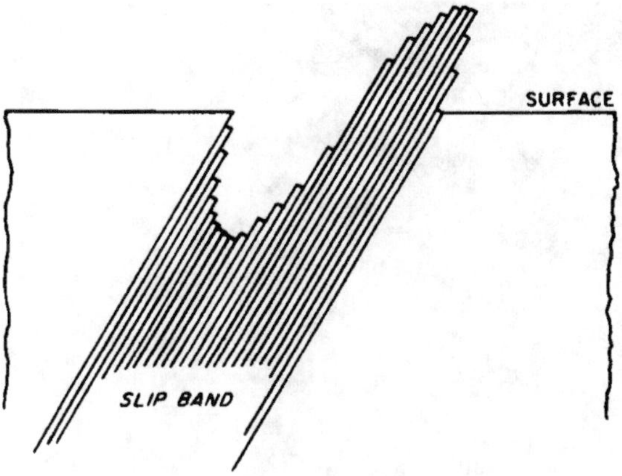

Figure 3.49 Schematic illustration of the formation of extrusions and intrusions in stage I of fatigue. (*From Grosskreutz.*[17])

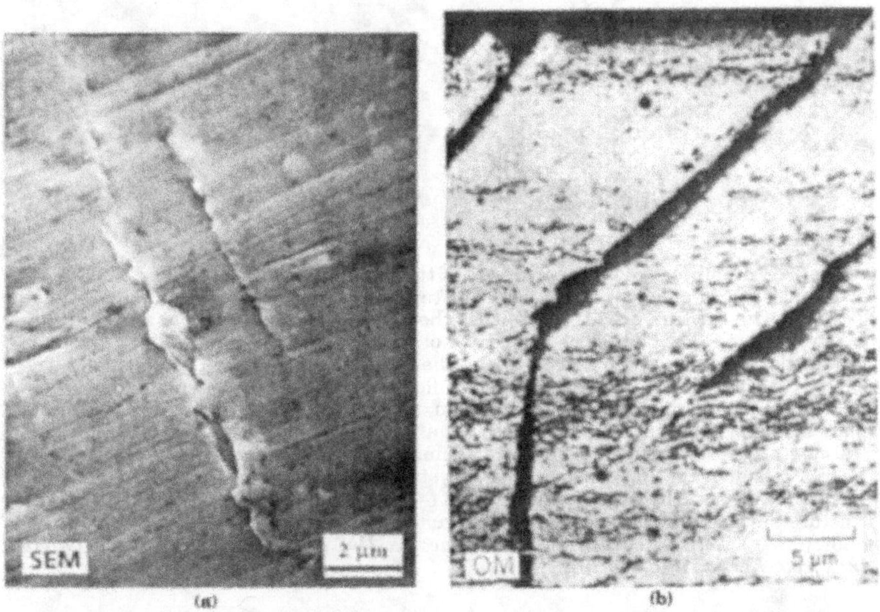

Figure 3.50 (*a*) Example of extrusions on the surface of a fatigue test sample. Material was an Al alloy containing 4% Cu, 1.2% Mg, 0.3% Mn. Test time was 10,000 cycles. (*From Engel and Klingele.*[10]) (*b*) Cross section of the surface of a fatigue sample showing stage I cracks at the surface and the beginning of stage II. Material was pure aluminum. (*Courtesy of Forsyth; from Henry and Horstmann.*[7])

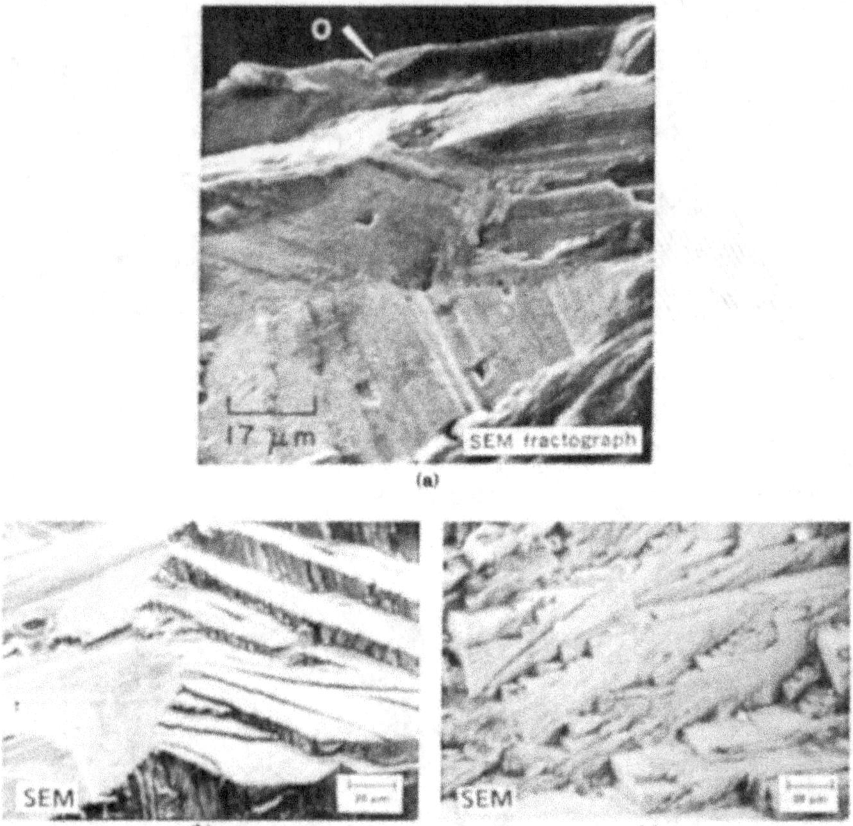

(a)

(b) (c)

Figure 3.51 Examples of the appearance of the fracture surface of stage I. (*a*) Stage I of fatigue fracture progress in a smooth (unnotched) specimen of aluminum alloy 2024-T3. The fracture origin is at 0, near the top of the fractograph, in a free surface of the specimen. The change of orientation of the fracture plane from grain to grain is well demonstrated. Within a given grain, the fracture is along a {111}-type plane. The ridge patterns within each grain are parallel to the direction of crack propagation. Each ridge is probably in a <110> direction, which is the intersection of two (111) slip planes. The presence of inclusions does not affect the fracture path. Stage I of fatigue-fracture progress extended over 3 or 4 grains; stage II began at bottom right. (*From* Metals Handbook.[11]) (*b*) Cleavagelike, crystallographically oriented stage I fatigue fracture in a cast Ni–14Cr–4.5Mo–1Ti–6A1–1.5Fe–2.0(Nb + Ta) alloy. (*c*) Stair-step fracture surface indicative of stage I fatigue fracture in cast ASTM F75 cobalt-base alloy. [(*b*), (*c*) *Courtesy of R. Abrams, Howmedica, Div. Pfizer Hospital Products Group Inc.; from Kerlins and Phillips.*[9]]

ciently rough that this stage of fatigue crack formation occurs at surface irregularities or defects. Stage I may not comprise a significant portion of the total crack propagation. For these reasons, recognition of stage I may require careful examination of the fracture surface in the vicinity of the crack initiation as deduced from macroscopic observations.

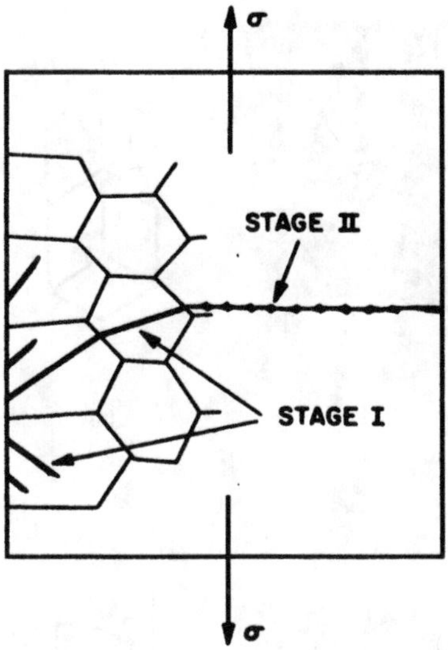

Figure 3.52 Schematic illustration of stage I and stage II fatigue crack formation. (*From Laird.*[18])

When the crack has progressed a small distance from the surface by the mechanism just described (stage I), the crack begins to propagate by a different mechanism, called stage II. The general crack path becomes perpendicular to the tensile stress, as illustrated in Fig. 3.52. The crack propagates only during the tensile stress part of the cycle. This is illustrated in Fig. 3.53. Here a notched sample was loaded in tension-compression, and the polished surface near the notch was observed in a scanning electron microscope. Note the opening of the crack during the tensile loading (1 to 2), the closing during compressive loading (3 to 4), then the reopening during the next tensile cycle (5 to 6). Also note the advancement of the crack tip from point 2 to point 6. It is further seen that slip bands have formed on the prepolished surface, showing that the crack opening is associated with slip.

The advancement of a fatigue crack can be seen in the micrographs in Fig. 3.54. Note that the crack advances about 140 μm from the location after 96,400 cycles to the location after 96,800 cycles, or a distance of about 0.3 μm per cycle (3000 Å, or about 1000 atom distances). Figure 3.55 shows schematically how the crack grows by slip.

As the fatigue crack propagates in stage II, it usually leaves on the fracture surface, behind the advancing crack front, regions of depression and elevation. An example is shown in Fig. 3.56. These *striations* are the most common characteristic of fatigue fractures seen on a microscopic scale. However, as shown later, fatigue cracks can propagate without forming striations; thus their absence does not mean the part did not fail by fatigue. Also their resolution will depend on

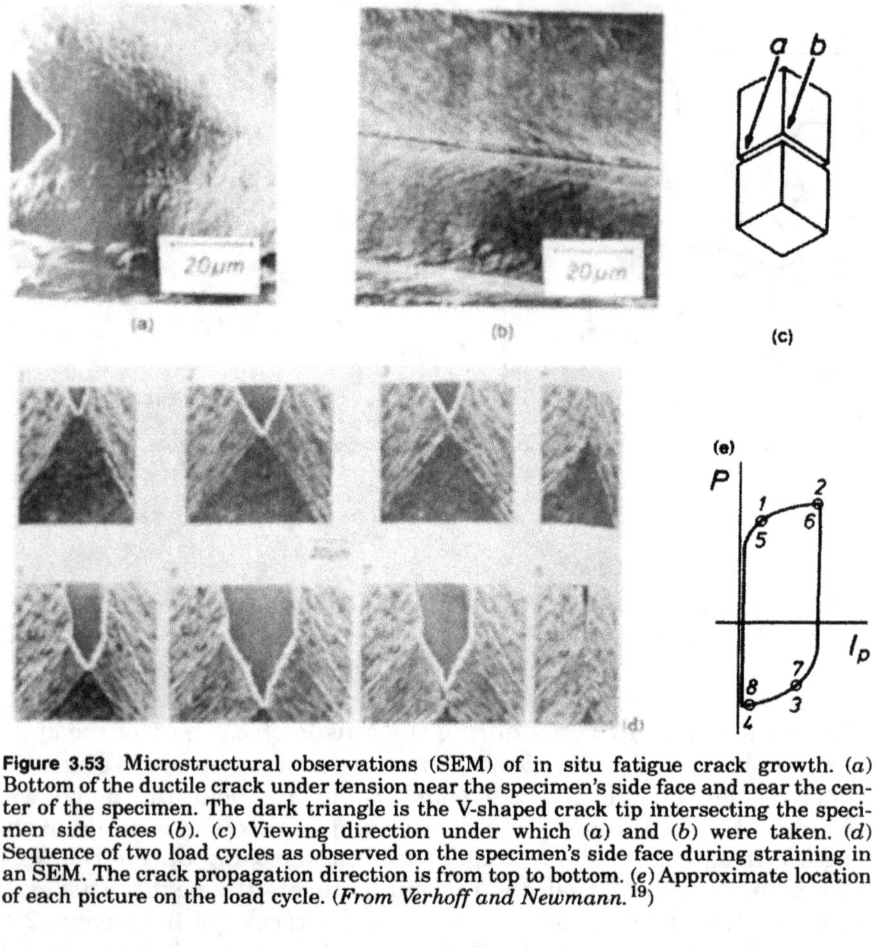

Figure 3.53 Microstructural observations (SEM) of in situ fatigue crack growth. (*a*) Bottom of the ductile crack under tension near the specimen's side face and near the center of the specimen. The dark triangle is the V-shaped crack tip intersecting the specimen side faces (*b*). (*c*) Viewing direction under which (*a*) and (*b*) were taken. (*d*) Sequence of two load cycles as observed on the specimen's side face during straining in an SEM. The crack propagation direction is from top to bottom. (*e*) Approximate location of each picture on the load cycle. (*From Verhoff and Newmann.*[19])

the type of damage to the surface when the crack closes (in the compression part of the loading cycle) and on the reaction of the fracture surface with the environment.

It has been clearly established that each striation is associated with the growth of the crack during one loading cycle. Although the fine-scale mechanism of the formation of the striations is not clear, Fig. 3.57 illustrates one way these striations could form. On a microscopic scale, the direction of crack propagation will depend on the microstructural features through which the crack is advancing, and this causes the striations to alter orientation locally, as depicted schematically in Fig. 3.58. This variation aspect of the orientation is seen in the fractograph of Fig. 3.59. The overall direction of crack movement, of course, will be that of the macroscopic crack propagation direction.

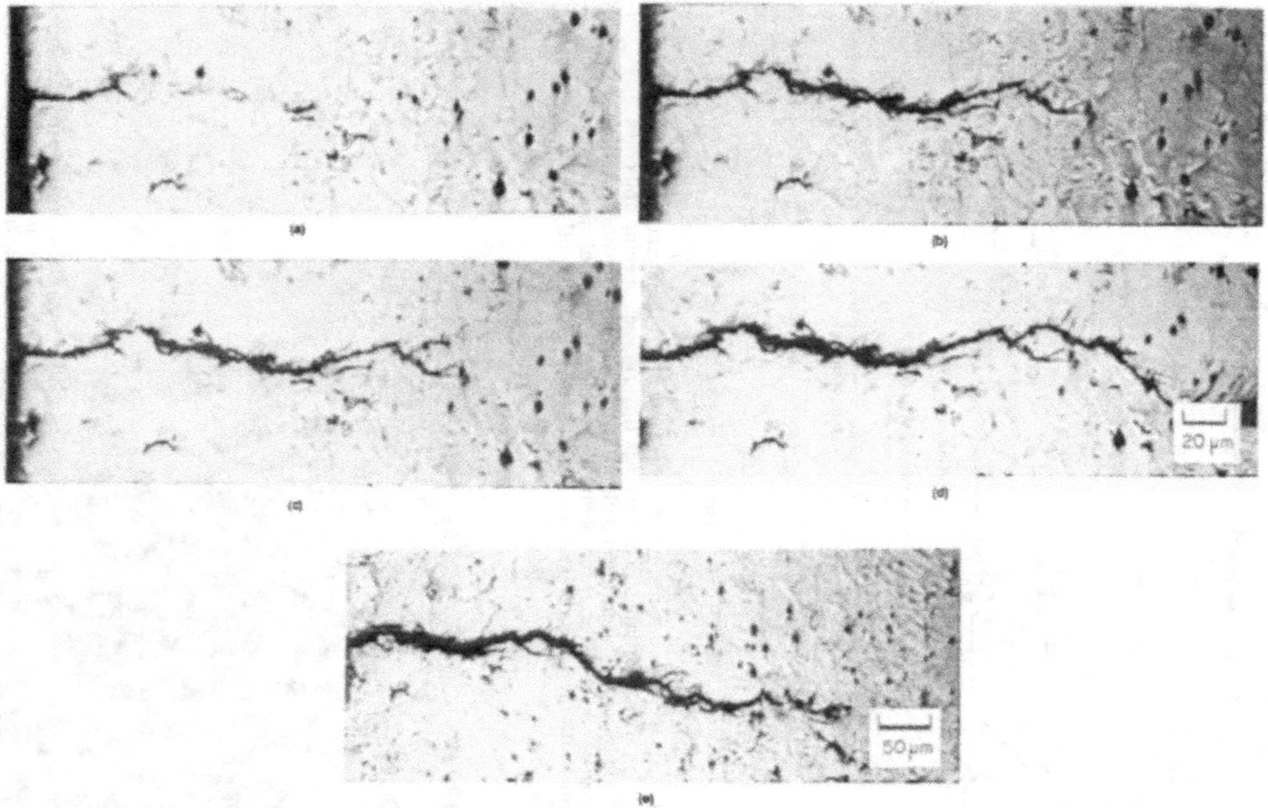

Figure 3.54 OMs of the growth of a fatigue crack on the surface of a fatigue sample. (*a*) 96,000 cycles, crack length 60 μm. (*b*) 96,300 cycles, crack length 235 μm. (*c*) 96,400 cycles, crack length 240 μm. (*d*) 96,800 cycles, crack length 380 μm. (*e*) 97,400 cycles, crack length 570 μm. (*From Haworth, Singh, and Mueller.*[20])

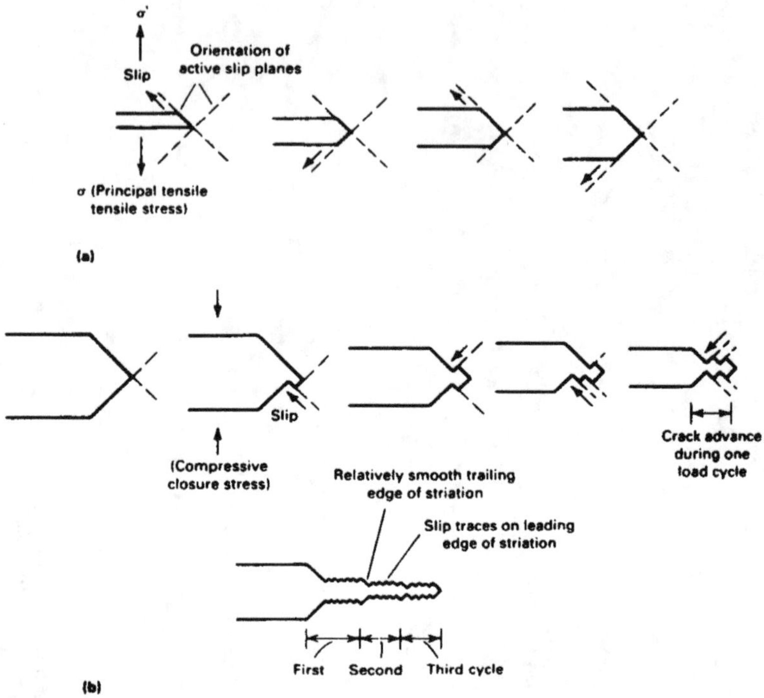

Figure 3.55 Schematic illustration of a mechanism of fatigue crack propagation in stage II by alternate slip at the crack tip. (*a*) Crack opening and crack tip blunting by slip on alternate slip planes with increasing tensile stress. (*b*) Crack closure and crack tip resharpening by partial slip reversal on alternate slip planes with increasing compressive stress. (*From Kerlins and Phillips.*[9])

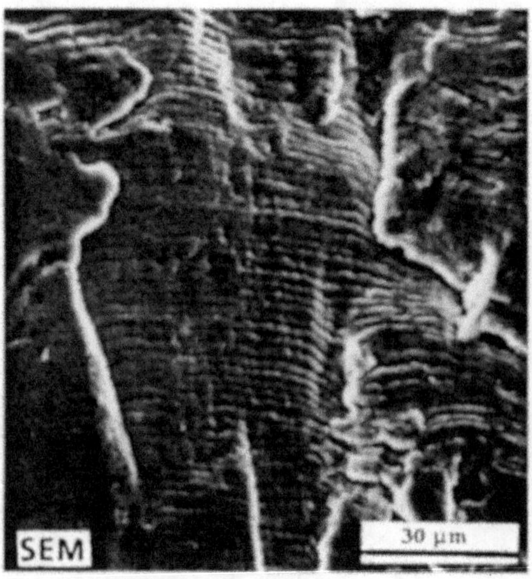

Figure 3.56 Fatigue fracture surface showing typical striations. Material was an 18:8 austenitic stainless-steel screw fastener. (*From Engel and Klingele.*[10])

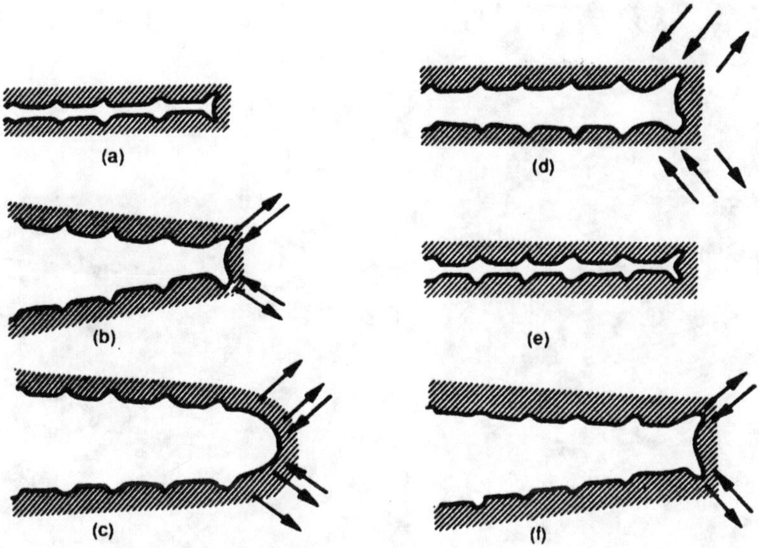

Figure 3.57 Possible mechanisms of the formation of fatigue striations. (*a*) Unstressed. (*b*) Small tensile stress. (*c*) Maximum tensile stress. (*d*) Small compressive stress. (*e*) Maximum compressive stress. (*f*) Small tensile stress. (*From Laird.*[18])

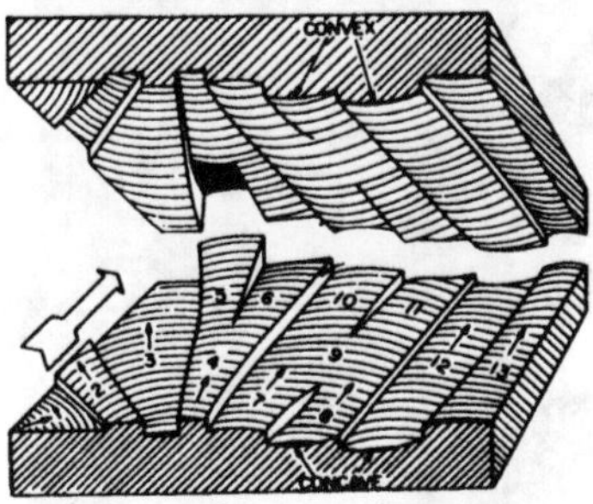

Figure 3.58 Local variation of the orientation of the fatigue striations due to microstructural features. (*From Beachem.*[8])

Several examples of fatigue fracture surfaces showing striations are given in Fig. 3.59. It is seen that the striation spacing is from approximately 1 to 10 μm, which is consistent with the distance the crack advances per cycle for the cases shown in Figs. 3.53 and 3.54. The correlation between the striation spacing and loading has been established from controlled experiments. Examples showing the effect of the loading on striation spacing are given in Fig. 3.60. Thus examination of the striation spacing gives clues to the loading spectrum. The higher the loading frequency and the lower the stress magnitude, the finer the striations. Also as the crack propagates, the remaining material supporting the load decreases, thereby increasing the stress, and thus

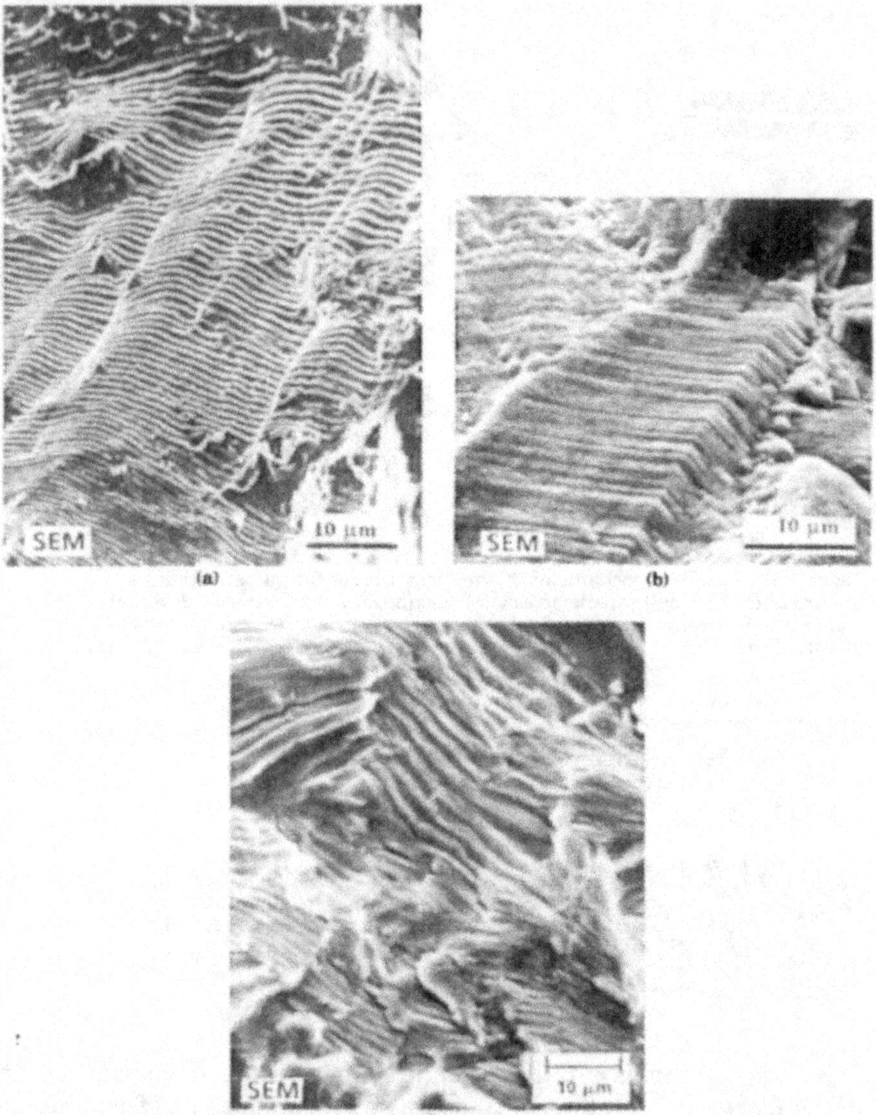

Figure 3.59 Fractographs of fatigue fractures showing typical striations. (*a*) Nickel alloy specimen (Nimonic) tested at 650°C under rotating-bending conditions. The propagation direction is from bottom to top. (*From Engel and Klingele.*[10]) (*b*) Crystallographically oriented fatigue striations on a fatigue fracture surface (stage II) from a laboratory tested nickel-based alloy (Inconel 718) tested at 600°C. (*From Engel and Klingele.*[10]) (*c*) Fatigue fracture in commercially pure titanium. (*Courtesy of O. E. M. Pohler, Institut Straumann AG; from Kerlins and Phillips.*[9])

the crack propagation distance will eventually show an increase (if the maximum load in the cycle is constant).

It is important to note that the fatigue striation spacing may be at or below the limit of resolution of an optical microscope. In addition, the roughness of the surface makes it difficult to focus due to the low depth of field. An example of this effect is shown in Fig. 3.61. Thus examination must usually be made with an electron microscope, either of the surface directly with an SEM or of a replica with an SEM or a TEM. Note that this requirement is reflected in the relatively high magnification (>2000×) used for most fatigue fractographs.

A common feature associated with fatigue striations is *secondary cracks*. Examples are shown in Fig. 3.62. The mechanism of their formation is not known.

Occasionally, fatigue cracks propagate by cleavage and thus without significant local plastic deformation. An example of the resulting fracture-surface topography is shown in Fig. 3.63. In such cases it is more difficult to correlate the fracture-surface features with the cyclic loading. Figure 3.64 shows a case where "striations" are found. However, these may be made prominent by the secondary cracking associated with each one. In rare cases the fatigue crack may propagate intergranularly, as illustrated in Fig. 3.65.

Another feature sometimes found on fatigue fracture surfaces is called *tire tracks*. Examples are given in Fig. 3.66. This feature is formed

(a)

Figure 3.60 Examples of fatigue fracture surfaces showing the effect of loading on the fatigue striation spectrum. (*a*) One large striation at 65 N/mm^2 cycle, three striations closer to each other (75 N/mm^2), and 20 striations very close to one another (60 N/mm^2). (*From Henry and Horstmann.*[7])

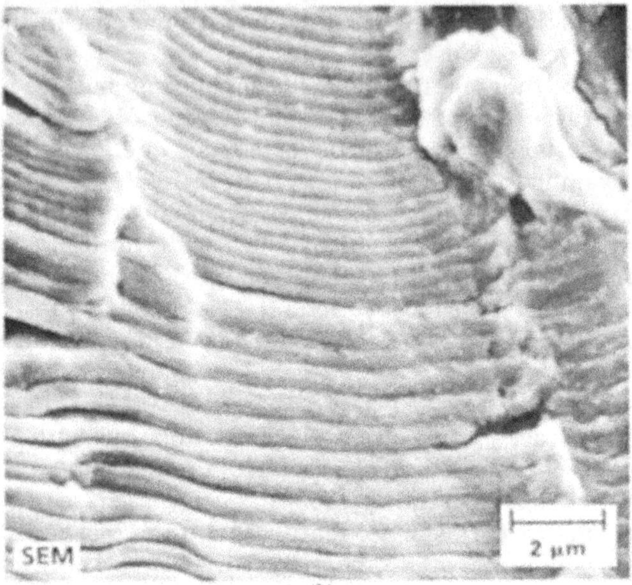

(b)

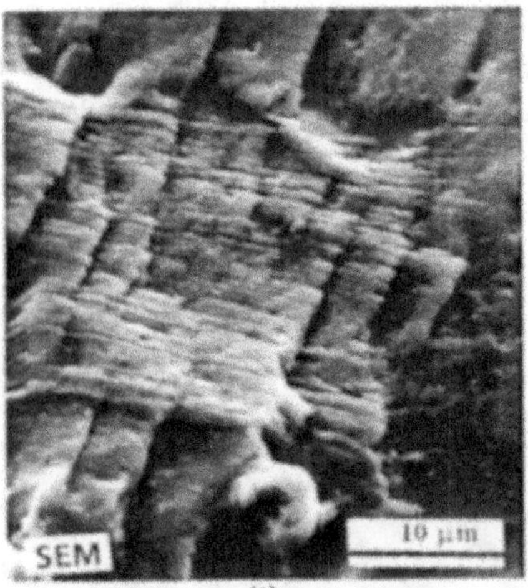

(c)

Figure 3.60 (*Continued*) Examples of fatigue fracture surfaces showing the effect of loading on the fatigue striation spectrum. (*b*) Spectrum-loaded fatigue fracture of a 7475-T7651 aluminum alloy test coupon showing an increase in the striation spacing due to higher alternating stress. (*From Kerlins and Phillips.*[9]) (*c*) Fatigue test of an aluminum alloy. In the center is a band of fatigue striations which originated from 45 cycles of higher stress amplitude. Above and below this band are fracture regions which originated from load cycles of lower amplitudes. (*From Engel and Klingele.*[10])

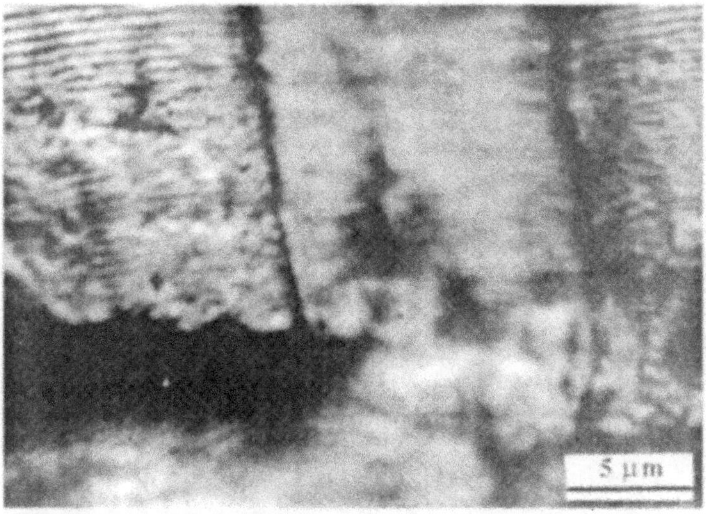

Figure 3.61 Fatigue striations as observed using an optical microscope. Material was an aluminum alloy tested in complete reversed bending, at a maximum stress of 172 MPa (25 ksi) at room temperature, to failure at 336 × 10^3 cycles. (*From Zapffe and Worden.*[21])

by the repeated impingement on one surface of a particle embedded in the other surface, as shown in Fig. 3.67.

Although in most cases fatigue cracks form striations, in many cases they do not, or they are extremely difficult to see or find. This is especially true in materials which have complex microstructures (such as tempered steels with a fine dispersion of carbides in ferrite). Figure 3.68 shows an example taken from a fatigue test. In Fig. 3.69 are examples of fracture surfaces in which the fatigue striations are either missing or ill-defined. Thus the absence of striations does not preclude fracture by fatigue loading.

The final stage of fatigue crack propagation is stage III, which is the final overload fracture. The crack propagates in stage II to the point where the remaining supporting material is not sufficient to support the load on the next cycle (or the next few cycles), and fracture occurs usually catastrophically and usually with little gross plastic deformation. The fracture-surface topography depends on the fracture mechanism, which is usually by void coalescence or cleavage. It is usually easy to distinguish stage III from the other stages.

It might be expected that there would be a relationship between the fine fracture topography and the "macroscopic" conchoidal markings (see Sec. 4.5), which are characteristic of fatigue fracture. Figure 3.70*a* shows an SEM fractograph at low magnification in which the conchoidal marks are seen. (They were also quite visible

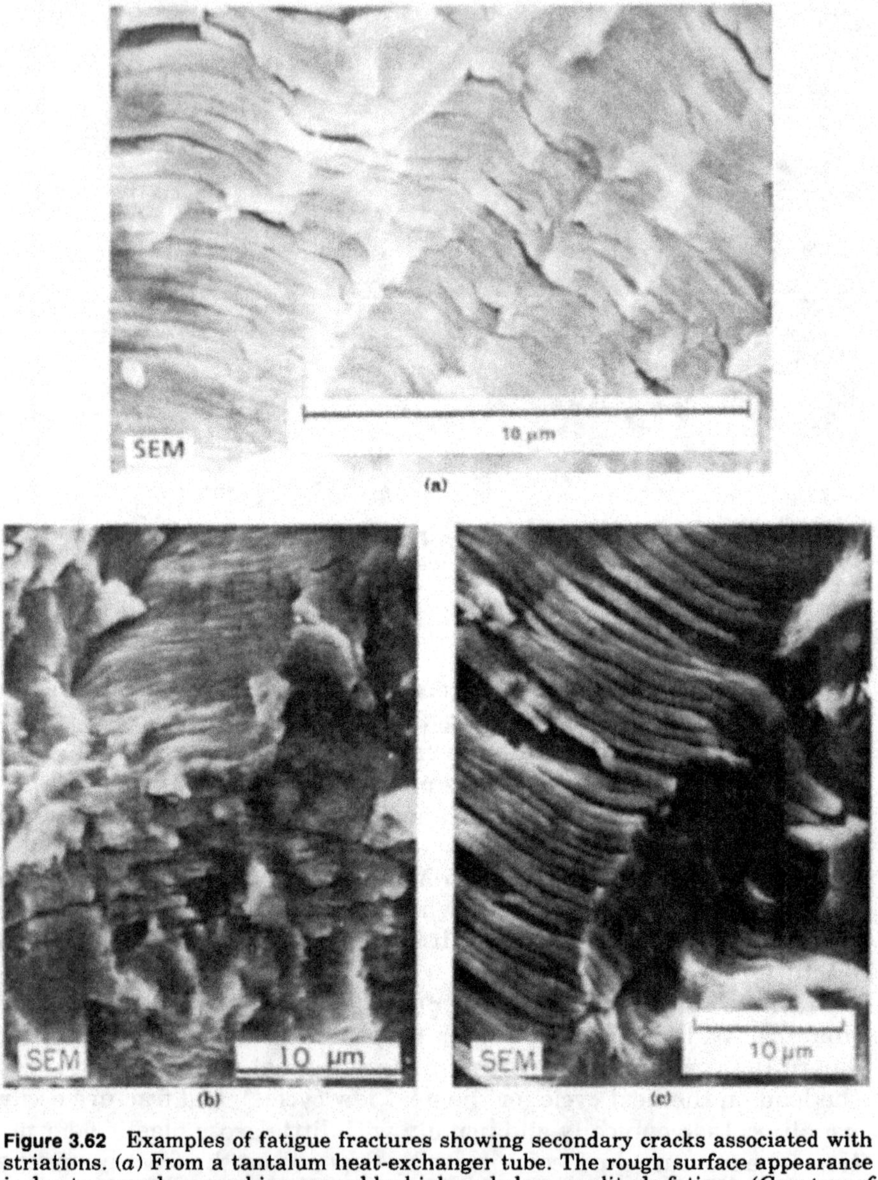

Figure 3.62 Examples of fatigue fractures showing secondary cracks associated with striations. (*a*) From a tantalum heat-exchanger tube. The rough surface appearance is due to secondary cracking caused by high-cycle low-amplitude fatigue. (*Courtesy of M. E. Blum, FMC Corporation; adapted from Kerlins and Phillips.*[9]) (*b*) From steel (0.38% C, 1.19% Si, 1.23% Mn, and 0.023% P) axle broken in service. (*From Henry and Horstmann.*[7]) (*c*) From steel containing 0.10% C, 0.27% Si, 0.43% Mn, 0.027% P, and 0.022% S. (*From Henry and Horstmann.*[7])

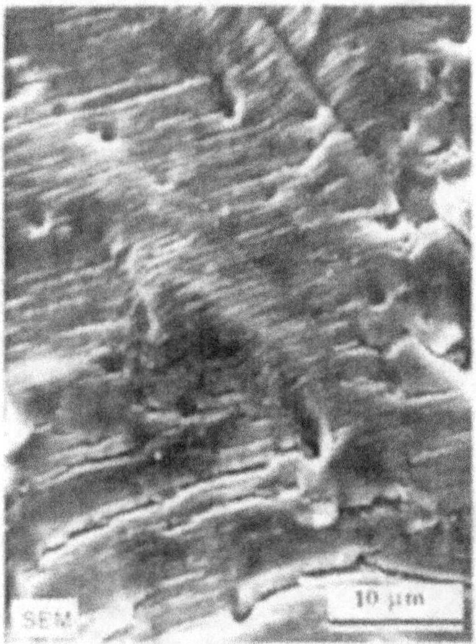

(d)

Figure 3.62 (*Continued*) Examples of fatigue fractures showing secondary cracks associated with striations. (*d*) From Inconel 718 tested at room temperature. The crack propagation is from bottom to top. (*From Engel and Klingele.*[10])

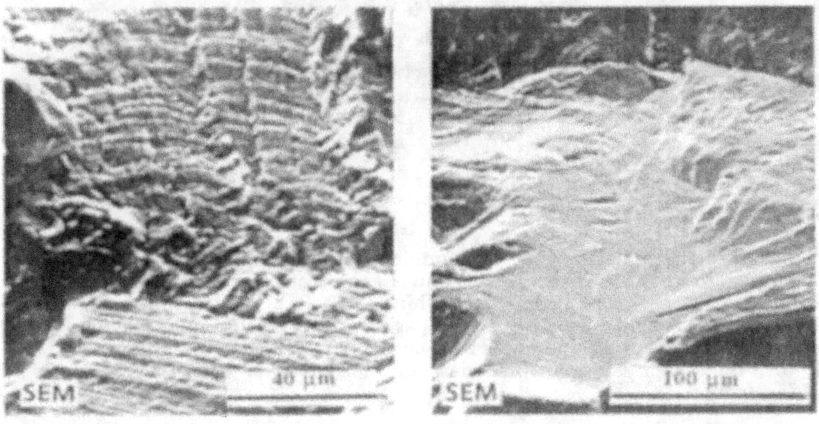

Figure 3.63 Fatigue fracture surface which fractured by brittle crack propagation on cleavage planes. This was produced by testing at room temperature a 1-mm-thick silicon sheet. Crack propagation is from bottom to top. (*From Engel and Klingele.*[10])

in light microscopy.) However, upon imaging at increasing magnification the region containing one of these "lines," it is seen that the cause of the markings is less clear. There is a slight difference in coarseness of the surface on one side of the line compared to the other side (see Fig. 3.70*e*), which is what gives rise to the markings. (Note that in this case only faint fatigue striations are observed.) In this

Figure 3.64 Striations on a fatigue fracture surface in which the crack propagated in a brittle manner. Material was a titanium casting alloy. (*From Engel and Klingele.*[10])

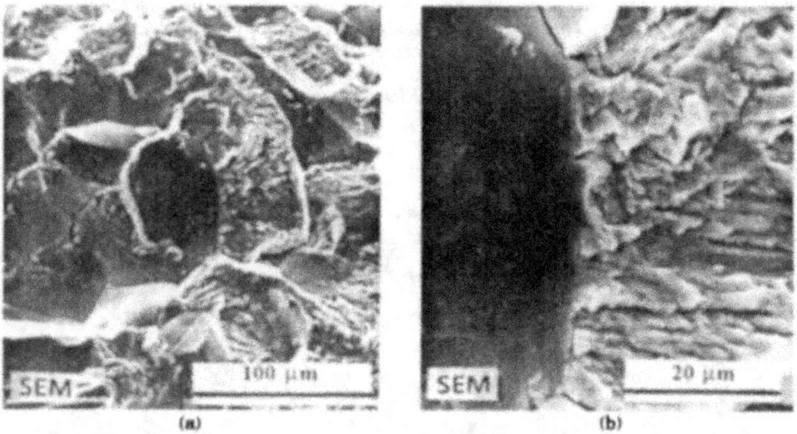

Figure 3.65 Fractographs showing an example where the fatigue crack propagated intergranularly. The fracture is from a gear tooth made from a case-hardening steel (0.16% C, 0.25% Si, 1.15% Mn, 0.95% Cr). The fracture in the core is on the right-hand side of (*a*) and propagated in a transgranular fashion, but in the 0.8-mm-deep case-hardened zone on the left, it propagated intergranularly. (*b*) Fractograph is a magnified view from the center of (*a*). (*From Engel and Klingele.*[10])

example the loading spectrum appears to have been such that the fracture surface topography changed locally, which gave rise to the conchoidal markings.

The point of this example is that it is often not obvious exactly what causes the conchoidal markings to be visible. This can be caused by the intrinsic effect of the loading on the fine fatigue surface topography, by variations in rubbing of the fatigue fracture surface during service, and by different degrees of corrosion of the fracture surface as the crack advances.

Figure 3.66 Examples of fatigue fracture surfaces showing tire tracks. (*a*) TEM replica of quenched and tempered 4140 steel. (*Courtesy of I. Le May, Metallurgical Services Ltd.; from Kerlins and Phillips.*[9]) (*b*) Alternate bending of a steel containing 0.09% C, 0.28% Si, 0.43% Mn, 0.025% P, and 0.024% S. (*From Henry and Horstmann.*[7]) (*c*) Threaded bar broken in service. Steel contained 0.61% C, 0.32% Si, 0.75% Mn, 0.027% P, and 0.028% S. (*From Henry and Horstmann.*[7])

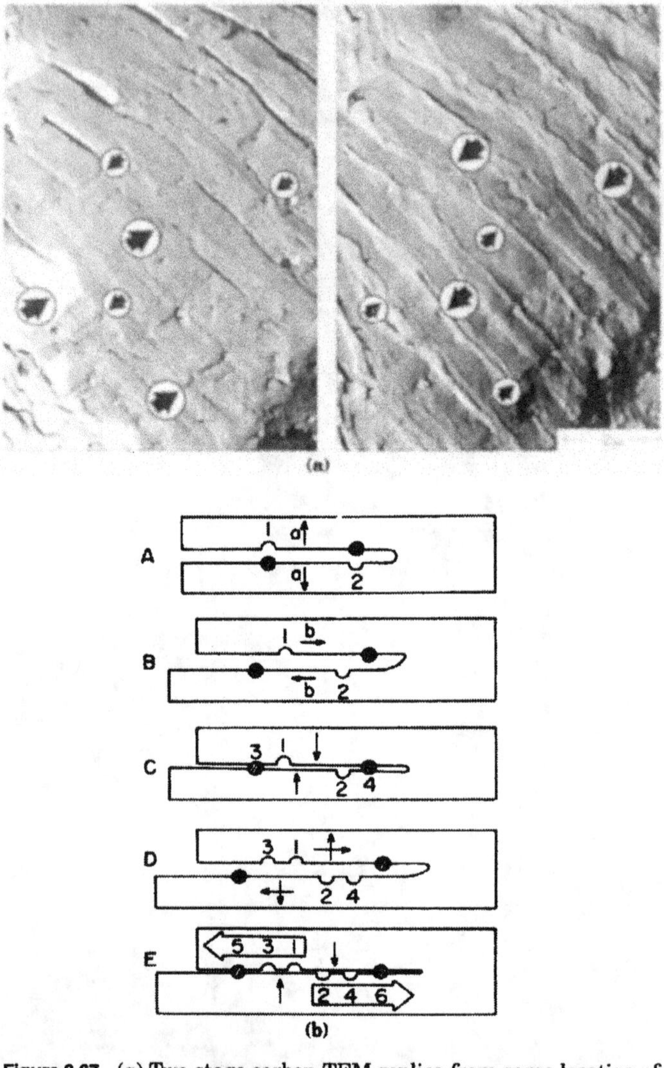

Figure 3.67 (*a*) Two-stage carbon TEM replica from same location of mating surfaces of a fatigue fracture, showing relationship between particles (large arrows) on one surface and tire tracks (small arrows) on the other 2024-T3 aluminum alloy. (*b*) Schematic illustration of how tire tracks form, showing how the two surfaces offset and then reclose to form a series of impressions. (*From Beachem.*[8])

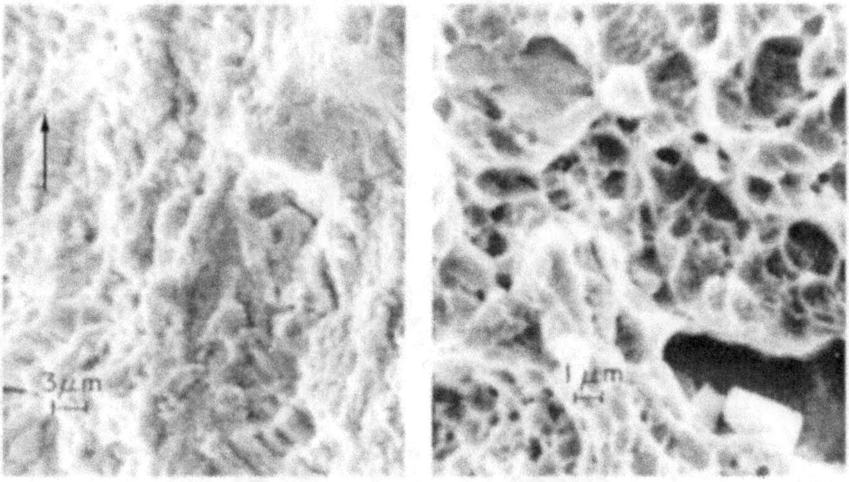

Figure 3.68 SEM fractographs taken from the fracture surface of a fatigue sample, showing that no striations are present. Arrow indicates crack propagation direction. This is the region of stage II propagation. The material was a 4140 steel tempered at 400°C. (*a*) $\Delta K = 21$ MN/m$^{3/2}$. (*b*) $\Delta K = 67$ MN/m$^{3/2}$. (*From Thielen and Fine.*[23])

3.10 High-Temperature Fracture Topography

In this section we examine the surface topography of fractures that occur at temperatures high enough that the mechanism may be basically different from those just described. Thus the term "high temperature" here does not necessarily mean elevated temperature, but refers to a temperature at or near the melting point, or to a temperature where creep-type processes control fracture. For example, these could occur in lead-based alloys at room temperature. From a practical standpoint, though, we are mainly concerned with fracture considerably above 25°C.

An important practical problem with fractography of high-temperature fractures is that after fracture the features of the fracture surface are usually affected by the environment. For example, the surface may be covered with a heavy oxide scale. This effect on the surface may make identification of the mechanism difficult. Examples of this difficulty are illustrated in Figs. 3.71 and 3.72. The clear dimpled morphology of the fracture surface, allowing identification of the mechanism as void coalescence, is obliterated upon exposure of the fresh fracture surface to air. It also may be difficult to separate the effect of the environment on the fracture process, per se, from the effect it has on the fresh fracture surface after fracture or during fracture.

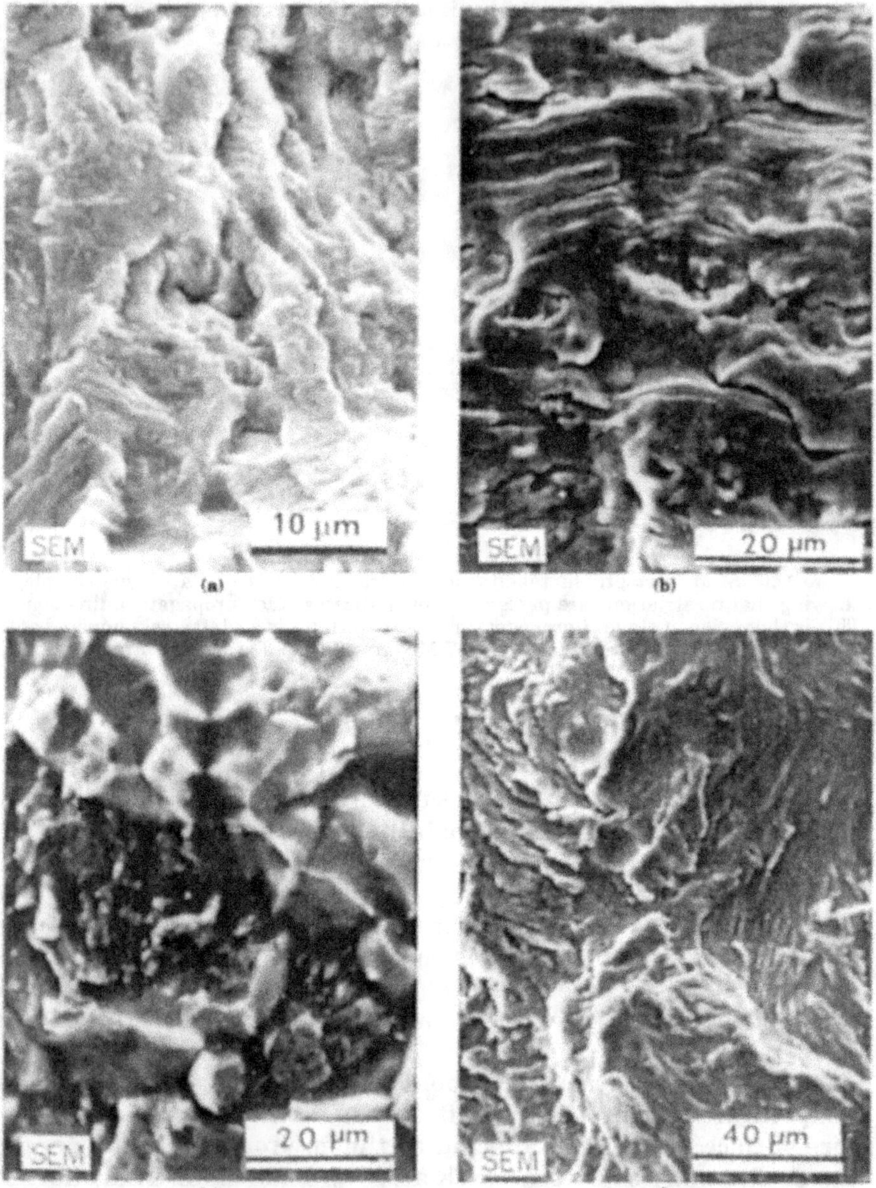

Figure 3.69 Examples of fatigue crack surfaces in which striations are absent or ill-defined. (*a*) Unalloyed carbon steel (0.45% C, 0.25% Si, 0.65% Mn). (*From Engel and Klingele.*[10]) (*b*) Drawn and galvanized annealed steel wire (0.46% C, 0.20% Si, 0.48% Mn, 0.027% P, 0.023% S). Rotative bending specimen. The striations are crushed and one mostly sees the secondary cracks. (*From Henry and Horstmann.*[7]) (*c*) Fatigue rupture under tension-compression of a steel containing 0.51% C, 0.32% Si, 0.65% Mn, 0.013% P, 0.015% S, 1.12% Cr, and 0.28% V. Fatigue rupture along grain boundaries. Striations can only be seen in several parts of the rupture. (*d*) Threaded steel bar (0.61% C, 0.32% Si, 0.75% Mn, 0.027% P, and 0.021% S) broken in service. Pearlite structure. Here one can observe characteristic steps and striations besides pearlite. (*From Henry and Horstmann.*[7])

As the temperature of fracture increases, some alloys show a change in mechanisms from void coalescence to intergranular separation, especially if a liquid forms at the boundaries. An example is shown in Fig. 3.73. It is interesting to note that the tensile elongation of the sample tested at 25°C was about 70 percent, while that tested at 1253°C, which fractured intergranularly (Fig. 3.73*b*), only exhibited 8 percent elongation.

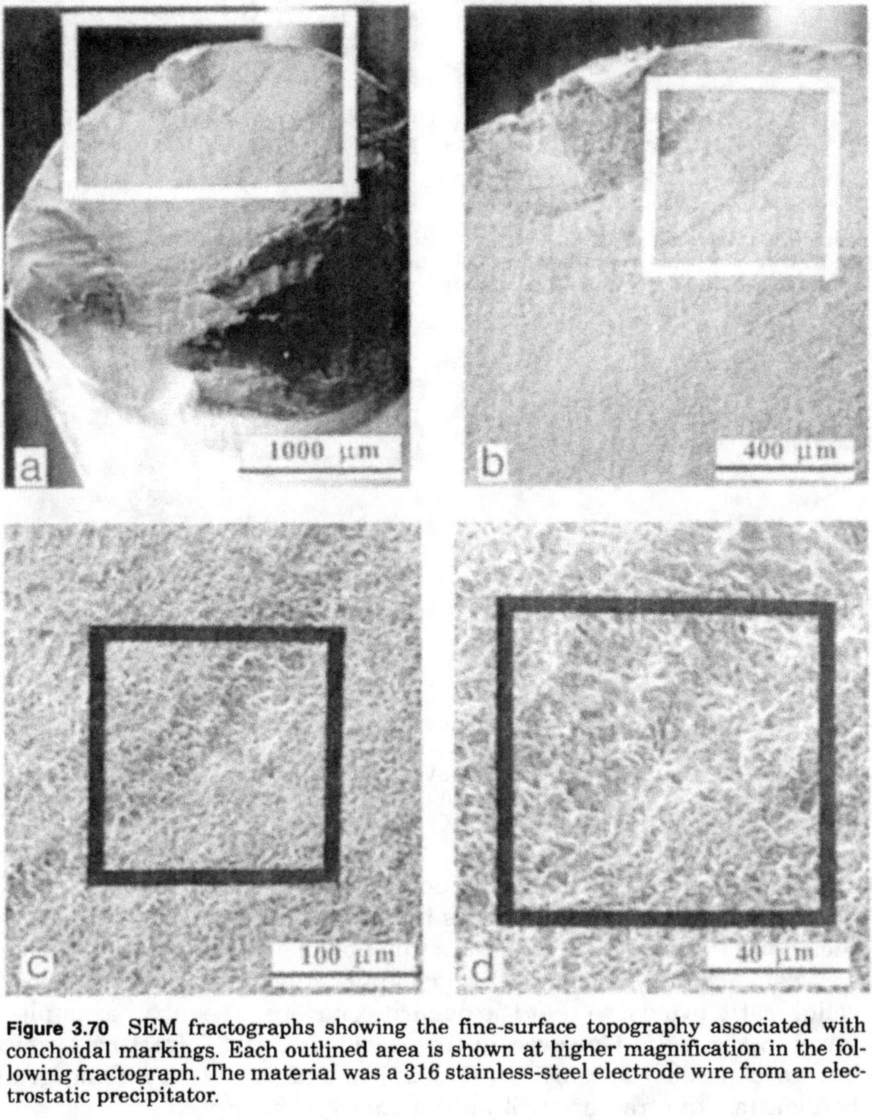

Figure 3.70 SEM fractographs showing the fine-surface topography associated with conchoidal markings. Each outlined area is shown at higher magnification in the following fractograph. The material was a 316 stainless-steel electrode wire from an electrostatic precipitator.

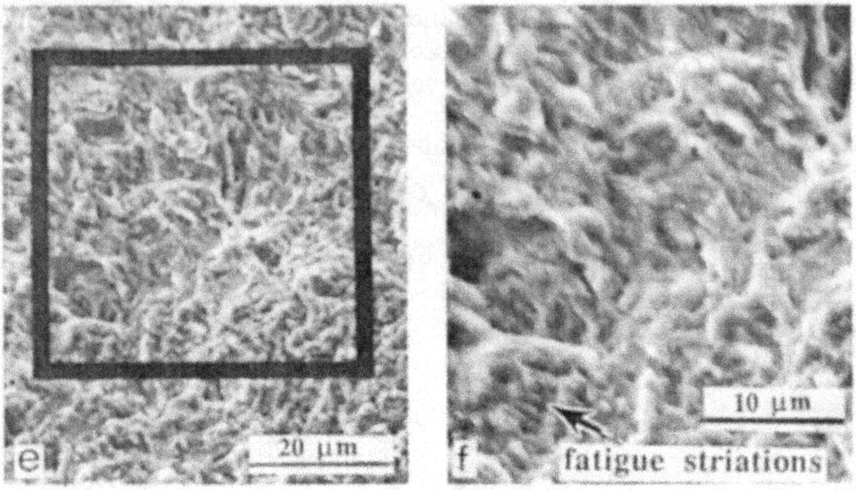

Figure 3.70 (*Continued*) SEM fractographs showing the fine-surface topography associated with conchoidal markings. Each outlined area is shown at higher magnification in the following fractograph. The material was a 316 stainless-steel electrode wire from an electrostatic precipitator.

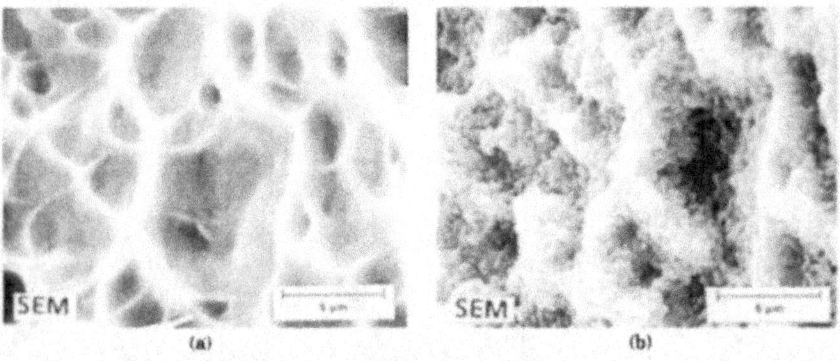

Figure 3.71 Fractographs illustrating the effect of oxidation on fracture-surface features. Material was an annealed Ti–6Al–6V–2Sn alloy. (*a*) Original fracture surface appearance. (*b*) Effect of a 15-min exposure in air at 800°C (1470°F). (*From Kerlins and Phillips.*[9])

If fracture is associated with local melting, then a characteristic dendritic topography is found, as illustrated in Fig. 3.74. In this case the actual fracture occurred at the junction of a few dendrite arms (see arrows). A similar topography is found when the fracture exposes solidification voids, so that the dendritic surface is seen; an example is given in Fig. 3.75. However, in this case this is not a fracture surface but an exposed void surface. If incipient melting occurs along grain boundaries, the fracture will appear intergranular.

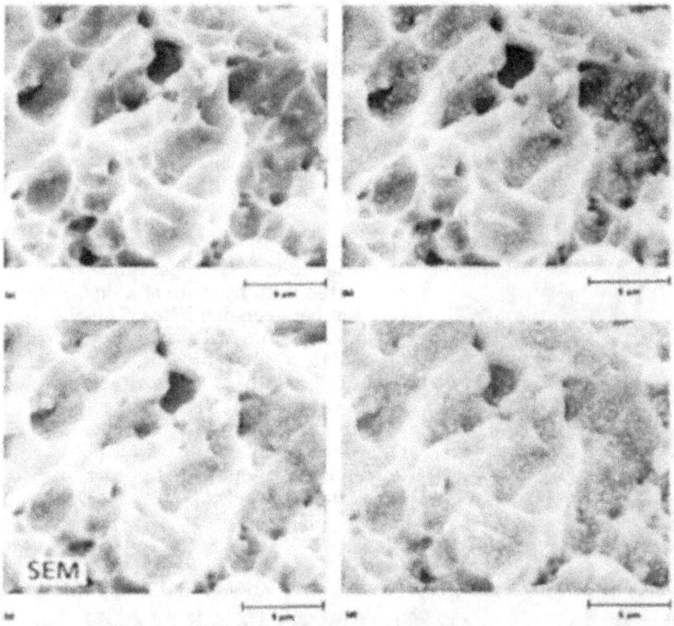

Figure 3.72 Fractographs illustrating the effect of oxidation on the fracture-surface features. Effect of a 700°C (1290°F) air exposure on an annealed Ti–6Al–2Sn–4Zr–6Mo alloy with an initial dimpled fracture surface. (*a*) As fractured. (*b*) Identical area after 3-min exposure. (*c*) After 10 min. (*d*) After 30 min. (*From Kerlins and Phillips.*[9])

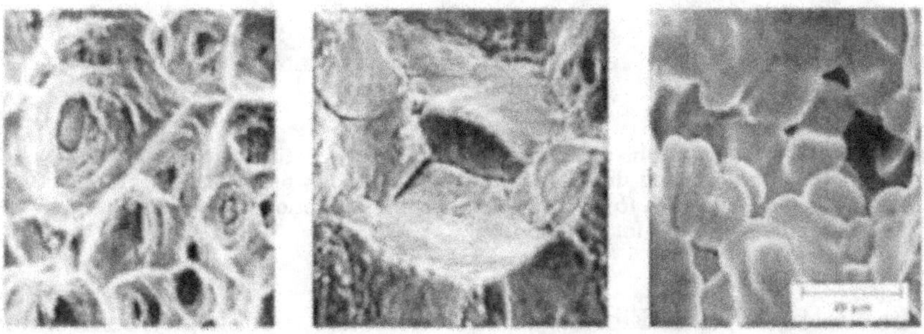

Figure 3.73 Effect of test temperature on fracture-surface topography. The sample was a Haynes 556 alloy and was tested at a strain rate of 1 s^{-1}. (*a*) Dimple rupture fracture at 1015°C (1860°F). At the bottom of many dimples are TaC inclusions. (*b*) Intergranular decohesion at 1523°C (2287°F). Secondary cracks are also visible. (*c*) Local eutectic melting of TaC + austenite at 1333°C (2431°F). (*Courtesy of J. J. Stephens, M. J. Cieslak, and R. J. Lujan, Sandia National Laboratories; from Kerlins and Phillips.*[9])

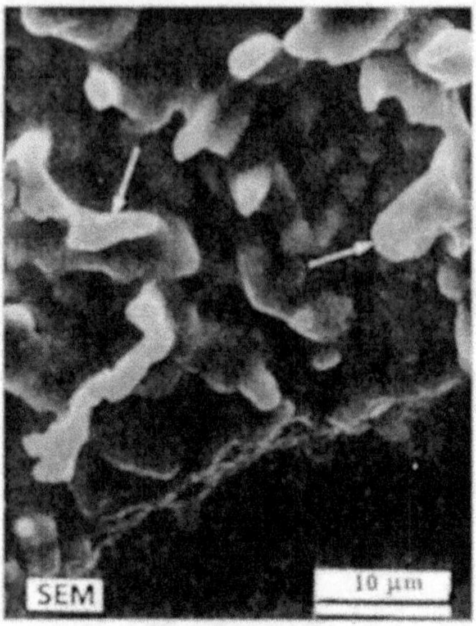

Figure 3.74 Fracture surface showing a dendritic appearance, associated with fracture of a liquid film between dendrites. This was a crack in a Waspaloy alloy immediately adjacent to an electron beam weld. The crack shows the characteristics of fracture in the pasty state, namely, the surface consisting of droplets and drawn-out peaks. (*From Engel and Klingele.*[10])

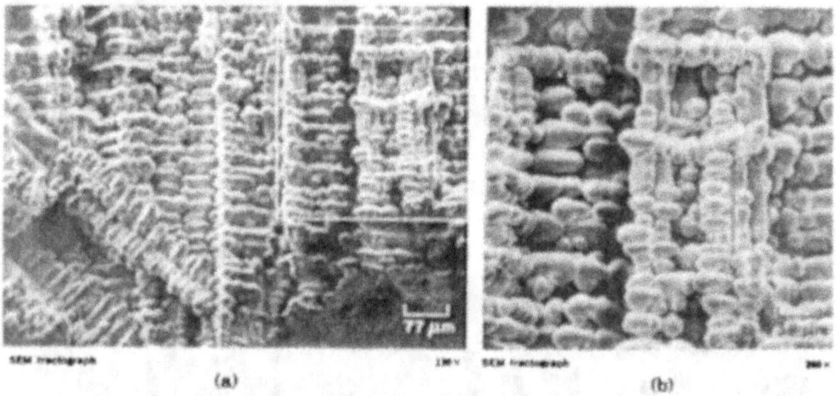

Figure 3.75 SEM fractographs of dendrites found on a fracture surface, but which are the surface of preexisting shrinkage cavities. The alloy was an 18% Ni, grade 300 maraging steel. Fractograph (*b*) shows at higher magnification the area outlined by the rectangle in (*a*). (*From* Metals Handbook.[11])

The fractograph in Fig. 3.76 illustrates the topography sometimes seen in fractures in which the alloy is partly liquid. The smooth regions were molten at the temperature at which the part fractured. The rougher regions are the location of actual separation which occurred by void coalescence.

At high temperature, after the formation of microcracks and microvoids, the remaining connecting material can undergo extensive plastic deformation before final separation occurs. This creates highly

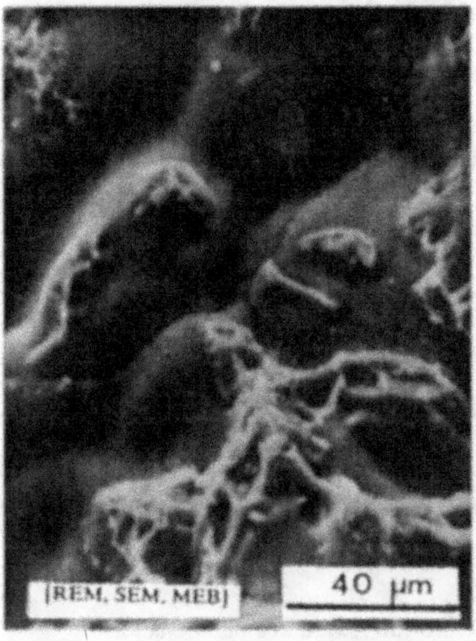

Figure 3.76 Fracture surface showing evidence of melting. This is from a crack in a brazed tube. The smooth regions are the intergranular zones that fractured at high temperature in the presence of liquid copper; the dimpled regions are where the material fractured during opening of the crack at room temperature to reveal the fracture surface. (*From Henry and Horstmann.*[7])

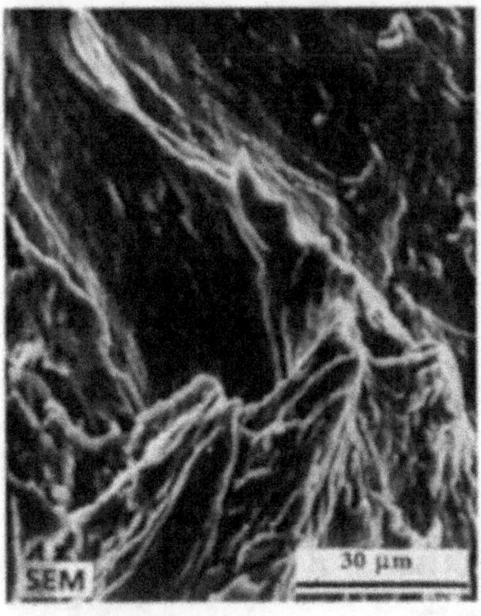

Figure 3.77 Fractograph showing extensive plastic deformation due to fracture at high temperature. (*From Engel and Klingele.*[10])

elongated features on the fracture surface, as illustrated in Fig. 3.77. In some cases, very elongated "necks" are seen, as shown in Fig. 3.78.

Creep fracture is a separate mechanism. Usually cracks are initiated at grain boundaries and at grain-boundary junctions, as shown schematically in Fig. 3.79. The former are called *creep voids* and the latter *wedge cracks*. Examples of these are shown in the microstructures in Fig. 3.80. Fractographs of creep voids and wedge cracks are

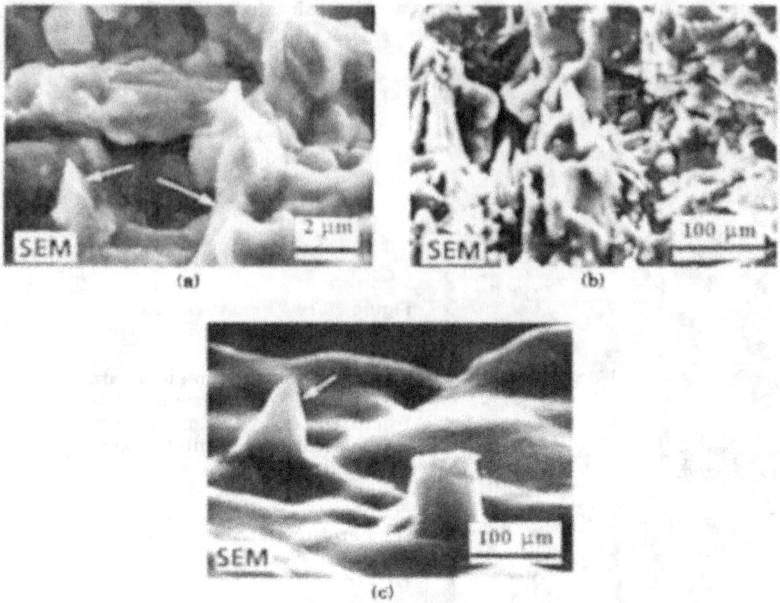

Figure 3.78 Fractographs showing regions exhibiting extensive plastic deformation (arrows). (*a*) Separation with the formation of drawn-out peaks in an electron-beam weld bead in a maraging steel (0.02% C, 0.1% Si, 0.1% Mn, 12% Co, 5% Mo, 18.5% Ni, 0.9% Ti). (*b*) Separation (hot tearing) produced in a nickel alloy (Al Si 12 Cu Mg) casting which involved sharp changes in cross section. The fracture surface shows rounded quench structures resembling shrinkage porosity, dimple formation and the characteristic, drawn-out peaks. (*c*) Intercrystalline separation, showing peaks formed at the softened grain boundaries in a steel forging. (*From Engel and Klingele.*[10])

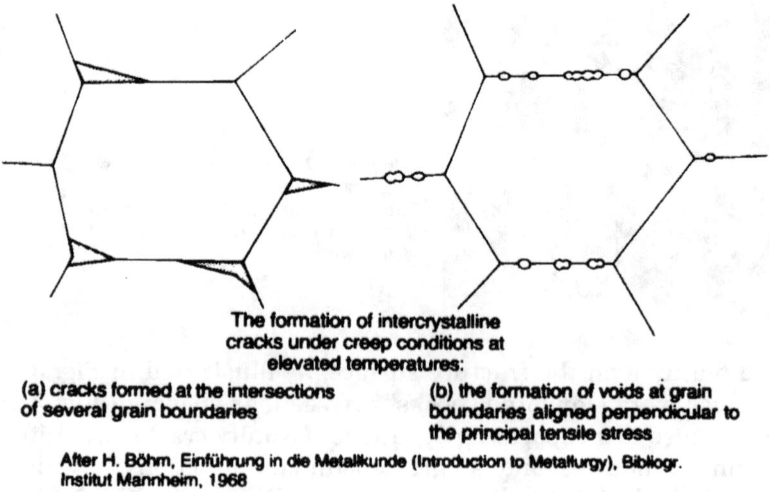

Figure 3.79 Schematic illustration of the formation of (*a*) wedge and (*b*) creep cavities. (*After Böhm*[24]; *from Engel and Klingele.*[10])

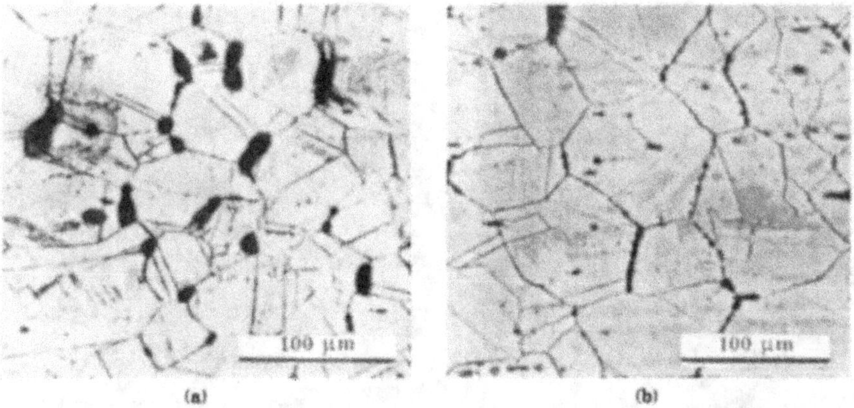

Figure 3.80 Microstructures from creep specimens showing creep cavities and wedge cracks. (*a*) Cracks initiated at triple boundaries. 17Cr–14Ni–Ti austenitic stainless steel creep-tested at 650°C (1202°F) under 80 N/mm^2 for 3383 h. (*b*) Beadlike cracks along grain boundaries. 17Cr–14Ni–Ti–B austenitic stainless steel creep-tested at 650°C (1202°F) under 130 N/mm^2 for 429 h. (*From Henry and Horstmann.*[7])

given in Fig. 3.81. As pointed out, interaction of the fracture surface with the environment during and after the fracture may make it difficult to identify such features.

3.11 Environmentally Assisted Fracture

The environment can have a potent effect on crack nucleation and propagation in alloys. In spite of extensive research on the environmental effect, the underlying mechanisms are not always understood since the processes have been found to be very complex. This is reflected in the wide variety of fracture-surface topographies found, showing that the type of fracture which occurs depends on the alloy and the environment, and therefore a single underlying fracture mechanism has not been found.

Thus the approach to recognizing that the environment may have influenced fracture cannot rely so clearly on the type of fracture topography. In addition, the "fresh" fracture surface after it forms may be altered by subsequent chemical interaction with the environment, which makes interpretation difficult. In this section we present some examples of typical fracture topographies found to illustrate the influence of the environment.

An important type of fracture mechanism is *stress corrosion cracking.* It has been found that components loaded to below the yield strength may suddenly fracture in certain corrosive environments. Figure 3.82 shows examples of such fracture surfaces, which illustrate

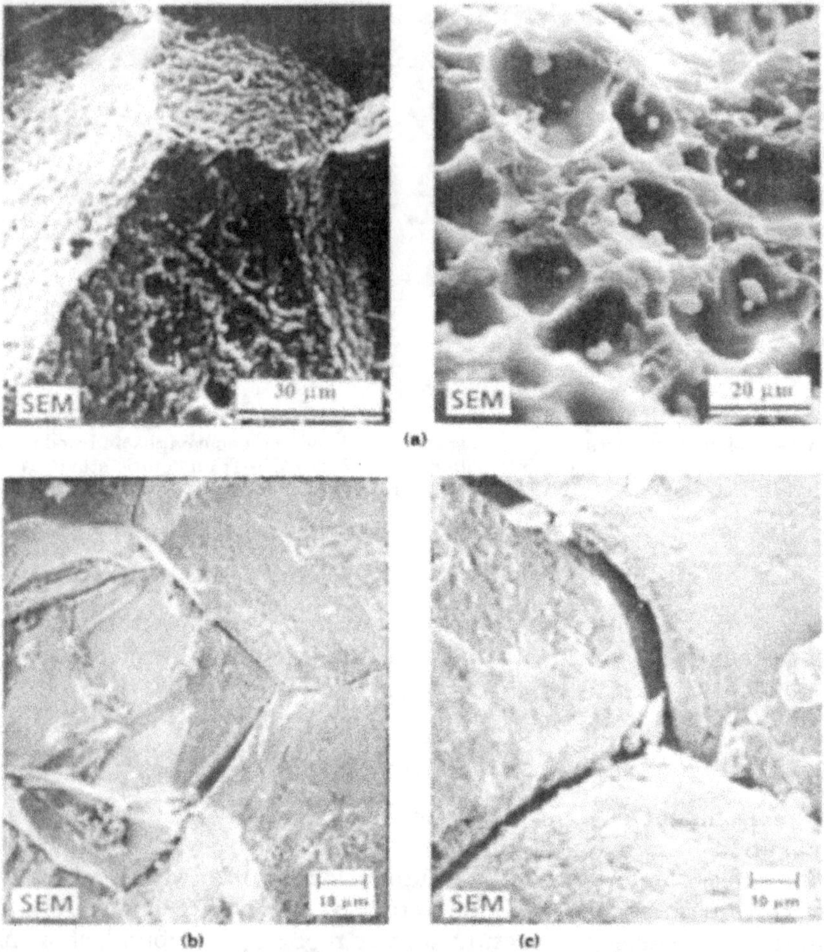

Figure 3.81 Fractographs showing creep cavities and wedge cracks. (*a*) Creep cavities (voids) which formed at 800°C (1472°F) at the grain boundaries of Nimonic 105. These were revealed by breaking the sample at −196°C (−321°F). (*b*) Wedge cracking in Inconel 625 and (*c*) wedge cracking in Incoloy 800. [(*a*) *From Engel and Klingele*[10]; (*b*), (*c*) *from Kerlins and Phillips.*[9]]

the diversity of the topography. Note that both intergranular and transgranular fractures are found. It may be difficult just from the fracture-surface appearance to conclude that the fracture was caused by corrosion assistance.

Hydrogen embrittlement is another environmentally assisted fracture mechanism commonly encountered in a number of alloys. The hydrogen can already be present in the material (such as by retention during casting), but the more common source is hydrogen picked up in

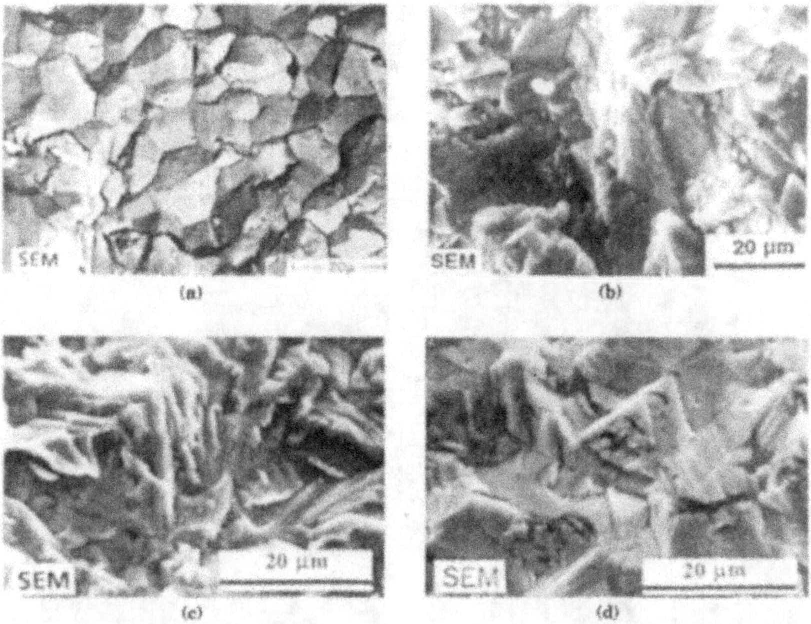

Figure 3.82 Fractographs illustrating a variety of stress corrosion cracking features. (*a*) Two-stage TEM replica of intergranular stress corrosion crack surface in AMS-6434 steel. (*From Beachem.*[8]) (*b*) Transcrystalline stress corrosion cracking in a ferritic welding metal (unalloyed, high-purity filler material) which initiated due to the presence of aqueous NaOH. This type of attack occurs only with high-purity unalloyed steels and under the influence of liquid NaOH, which formed as a result of the evaporation of water in a crevice formed due to bad manufacturing of a boiler. (*c*), (*d*) Stress corrosion cracking in an austenitic stainless steel (0.1% C, 1% Si, 2% Mn, 18% Cr, 10% Ni, 2% Mo, 0.6% Ti) caused by $MgCl_2$ solution at over 100°C. Noticeable features are intergranular separation (*c*), and featherlike structures and large cleavagelike areas (*d*). (*From Engel and Klingele.*[10])

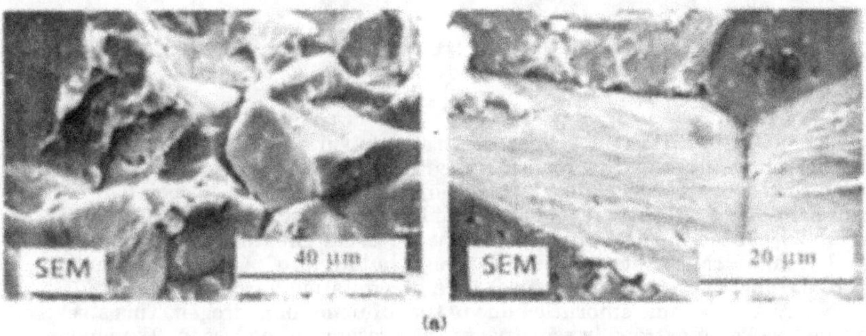

Figure 3.83 Fractographs illustrating a variety of hydrogen embrittlement fracture surfaces. (*a*) A gas-carburized steel screw (0.3% C, 0.25% Si, 0.75% Mn, 1% Cr) broke after being galvanized. Characteristic features of hydrogen-initiated fracture are the yawning grain boundaries, micropores, and hairline patterns (partly in the form of crows-feet markings). (*From Engel and Klingele.*[10])

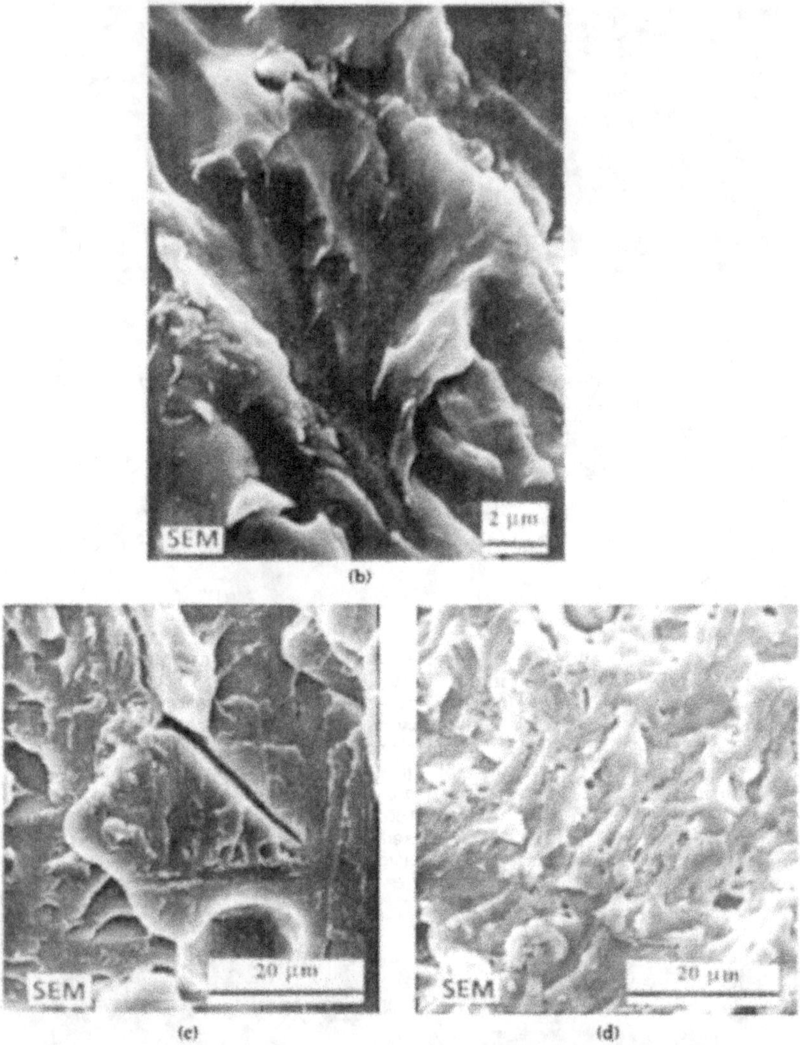

Figure 3.83 (*Continued*) Fractographs illustrating a variety of hydrogen embrittlement fracture surfaces. (*b*) A ferritic fine-grained steel (0.2% C, 0.5% Si, 1.5% Mn) was tested in the laboratory by straining it beyond the yield point in an aqueous solution of H_2S. Hydrogen, produced by the corrosion reaction $Fe + H_2S \rightarrow FeS + 2H$, caused the brittle fracture of the material. The fracture mechanism is mainly that of transcrystalline quasicleavage, but areas of pearlite can also be recognized (roughly in center of picture). (*c*) A precipitation hardening steel (17% Cr, 3% Ni, 2% Cu) became embrittled due to the influence of hydrogen. This steel is regarded as being extremely sensitive to the presence of hydrogen. The characteristic features indicating hydrogen embrittlement are: microquasicleavage facets, micropores, and ductile hairlines. (*d*) Flake cracking in unalloyed carbon steel. (*From Engel and Klingele.*[10])

service or during processing (such as plating). The hydrogen can be absorbed in sufficient quantities at quite low temperature, such as during chemical cleaning of the surface of the component or during plating.

The mechanism of the embrittlement is not clear. The type of crack which forms and propagates is sensitive to several factors, such as the alloy and the loading conditions. This may make it difficult to ascertain whether hydrogen embrittlement was involved from just the fracture surface appearance. Figure 3.83 shows examples of the fracture surfaces of hydrogen-embrittled alloys.

Stress corrosion cracking and hydrogen embrittlement are usually found in statically loaded situations. If the load varies, then corrosion-assisted fatigue is encountered. Typical fracture surfaces are shown in Fig. 3.84. Usually the surfaces show signs of corrosion. Fatigue striations may be seen, but not always. They may have been obliterated by the corrosive attack of the fracture surface during crack propagation, or after fracture but before the part could be removed from the environment.

3.12 Flutes

A fractographic feature which has been found in alloys of complex microstructure (such as Ti alloys) is called *flutes*. Examples are shown in Fig. 3.85. In some alloys their formation is associated with a variety of fracture modes, such as fatigue, overload, and stress corrosion cracking. They appear to form from planar slip as tear regions form between cleavage cracks. The elongated geometry is associated with the elongated grains which are present in the microstructure.

3.13 Wear

The removal of small particles from surfaces by contact with other surfaces or by interaction with the moving surrounding environment (such as high-velocity water) is called *wear.* The basic characteristic of the process is fracture of material from the surfaces, and thus involves the fracture mechanisms described. However, it usually is difficult to characterize the wear process in a simple fashion due to factors such as melting and oxidation at the interface of the mating surfaces due to frictional heating, localized solid-state welding of surface asperities followed by fracture, embedding of debris in the surfaces, and so on. Because it is such a complicated process, a treatment of the subject is not included in this book. However, publications listed in the Bibliography (Phol, Tucker, and Wulpi in particular) contain treatments which will serve as starting points for understanding wear failures. Also, the Glossary at the end of the book contains some of the common terms in wear failures.

3.14 Fracture in Ceramics

Ceramics, with their strong ionic bonding, are usually brittle, with very little plastic deformation occurring prior to fracture. Even the amorphous ceramics (e.g., glasses) are brittle. Brittle materials are susceptible to crack nucleation at internal and external flaws, and the fracture surface may show features that indicate the location of the fracture initiation.

Crystalline ceramics can fracture by intergranular separation (Fig. 3.86) and by cleavage. As in brittle metallic materials, the cleavage fracture topography may be complex. If the fracture surface is very smooth, it is referred to as a fracture mirror. An example that is more typical to that of the fracture surface of amorphous ceramics is shown in Fig. 3.87. A fractograph typical of both polycrystalline and glass ceramics is shown in Fig. 3.88*b*. To describe the appearance of the features on the surfaces, the terminology shown in Fig. 3.88*a* is used.

3.15 Fracture Mechanisms in Polymeric Materials

The mechanisms of plastic deformation and fracture of polymeric materials are fundamentally different from those in ceramics and metallic

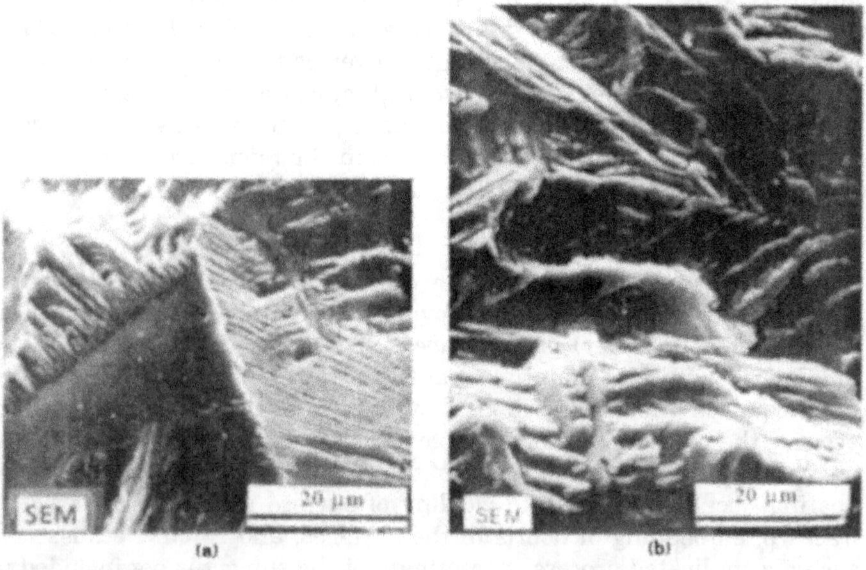

Figure 3.84 Fractographs of alloys which failed due to corrosion-assisted fatigue. (*a*) Corrosion fatigue in an austenitic stainless steel (17% Cr, 15% Ni, 4.3% Mo) caused by chloride-containing bleaching solution with a pH of 2. The fracture surface shows little evidence of plastic deformation and consists of smooth, crystallographically oriented surfaces. (*b*) Brass condenser tube (Cu–Zn–28Sn). The crystallographically oriented, featherlike structure on the fracture surface has resulted from transcrystalline fracture which has been accompanied by only a minimal amount of deformation. (*From Engel and Klingele.*[10])

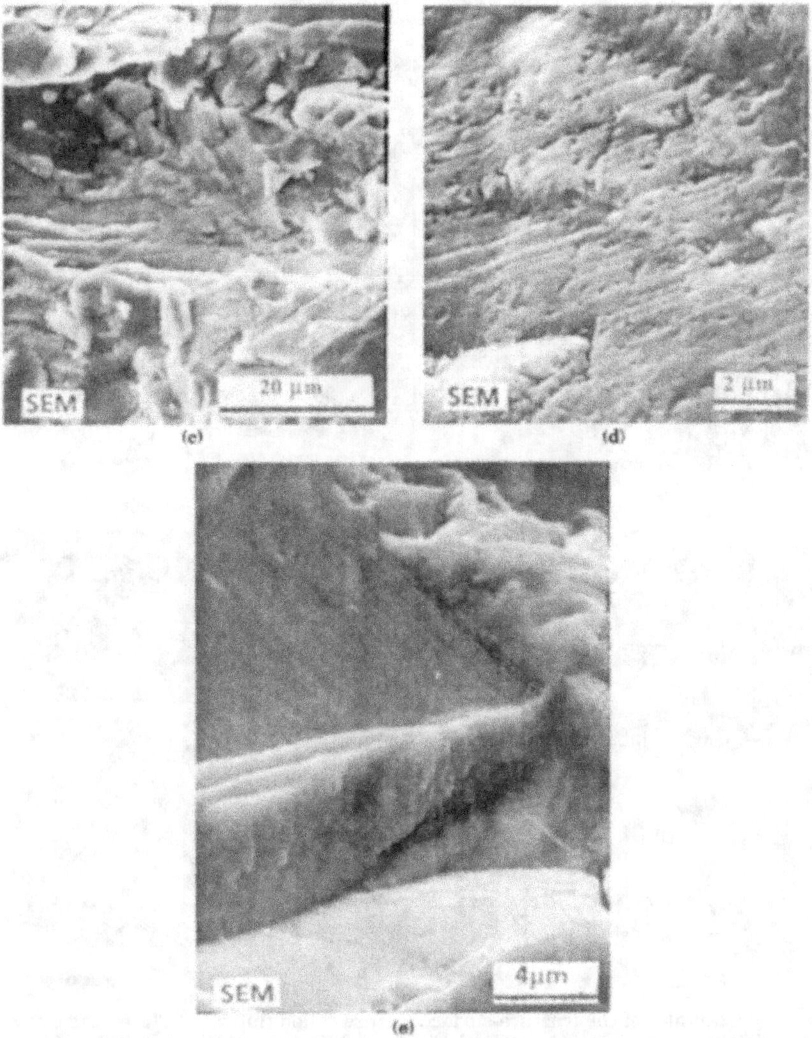

Figure 3.84 (*Continued*) Fractographs of alloys which failed due to corrosion-assisted fatigue. (*c*) Cu–Zn–28Sn. Fatigue striations on the smooth blocks are characteristic of this type of fracture. (*d*), (*e*) Cu–Zn–20Al. The smooth crack paths with fatigue striations are characteristic features of corrosion fatigue. [(*c*), (*d*) *From Engel and Klingle.*[10] (*e*) *From Kerlins and Phillips.*[9]]

materials. This is a consequence of the underlying molecular architecture. Polymers consist basically of long carbon chains, from 1000 to 100,000 atoms, with an extremely strong bond between the carbon atoms.[32] The structure is illustrated schematically in Fig. 3.89. The bond between chains is relatively weak. However, interchain strength depends on the complexity of the molecule, being affected by cross-linking and the specific molecules substituted for carbon atoms or attached to them.

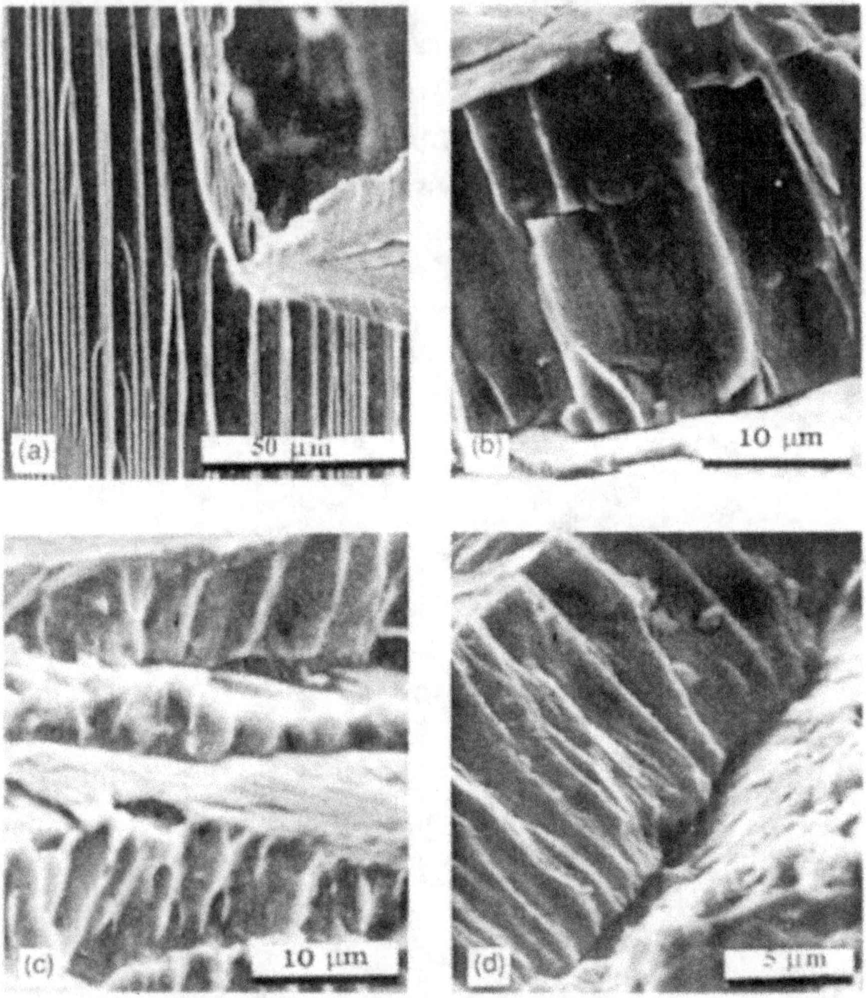

Figure 3.85 Examples of the fracture-surface feature called flutes. (*a*) Flutes and cleavage resulting from mechanical overload of a Ti–0.35O alloy. (*b*) Flutes and cleavage resulting from stress corrosion cracking of beta-annealed Ti–8Al–1Mo–1V alloy in methanol. (*c*) Flutes and cleavage in beta-annealed Ti–8Al–1Mo–1V resulting from sustained-load cracking in vacuum. (*d*) Flutes near the notch on the fracture surface of mill-annealed Ti–8Al–1Mo–1V alloy resulting from corrosion fatigue in salt water. (*Adapted from Meyn and Brooks.*[25])

In the liquid state, polymers consist of a spaghettilike, entangled mass of very long molecules. As the temperature decreases, the secondary bonding between polymer chains becomes strong enough to align molecules into a periodic arrangement of a crystal. For example, the arrangement of the molecules in linear polyethylene, the simplest polymer molecule, can be characterized by the arrangement in Fig. 3.90. In a cool-

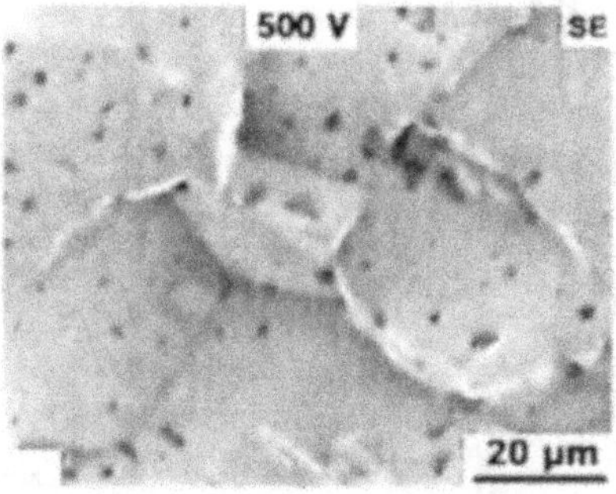

Figure 3.86 Fractograph of high-purity Al_2O_3 broken in bending, showing intergranular fracture. Note the pores in the ceramics from the sintering process. (*From Brooks and McGill.*[26] *Reprinted with permission of Elsevier Science.*)

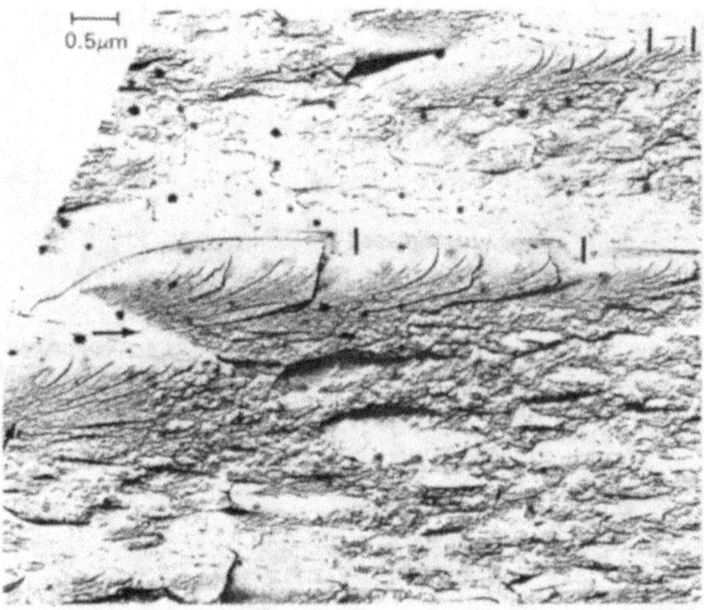

Figure 3.87 Fractograph of brittle fracture in a glass. Crack propagation is from left to right. Local cracks fan out from arrows in association with hackle formation and diminishing height of hackle (near vertical marks) associated with distinct turning of fracture steps across hankle ridges. (*From Rice.*[27] *Copyright © ASTM; reprinted with permission.*)

ing liquid, the crystals nucleate at various locations in the liquid, but due to the great length of the molecular chains, links exist between the crystals from molecules in different crystals. This is illustrated schematically in Fig. 3.91. The crystals form into plates, or lamellae, with the molecules folded, with configurations such as shown in Fig. 3.92. Thus the boundary between crystals consists of folds in which the atoms do not fit well into either crystal. In addition, molecules extend into neighboring crystals (Fig. 3.91). This leads to the polycrystalline arrangement

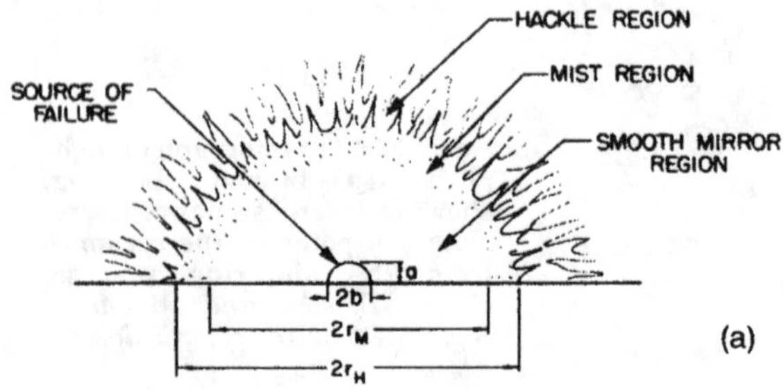

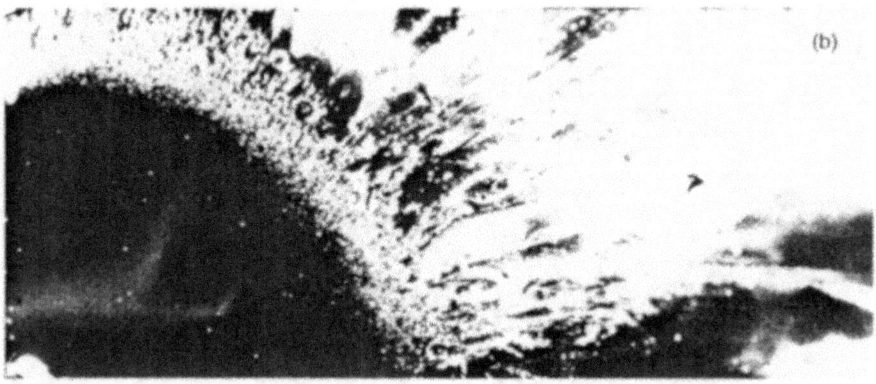

Figure 3.88 (*a*) Schematic diagram showing different fracture regions and approximate textural details in fracture of brittle ceramics. (*b*) Fractograph of brittle glassy ceramic revealing mirror, mist, and hackle regions. [(*a*) *From Mecholsky et al.*[28]; (*b*) *from Orr.*[29] *Copyright © ASTM; reprinted with permission.*]

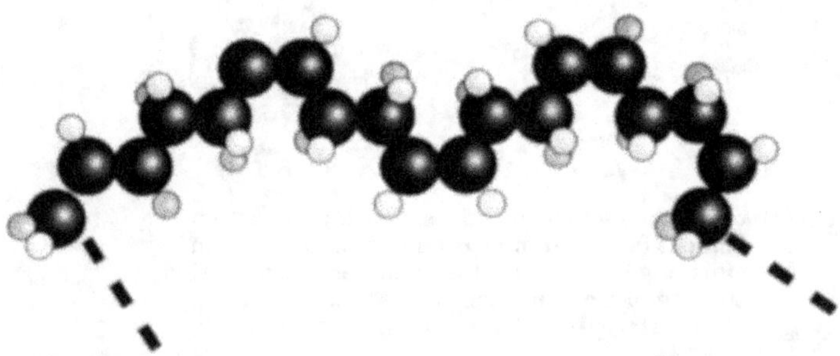

Figure 3.89 Schematic illustration of a simple polymer chain. (*Adapted from Cotterill.*[30] *Reprinted with permission of Cambridge University Press.*)

similar to that of metallic and ceramic materials, except that the molecules between the crystals are disorganized (Fig. 3.91), constituting an amorphous or noncrystalline region. The crystals, called spherulites, can be imaged by use of polarized light of thin sections, and an example is shown in Fig. 3.93.

The degree of crystallinity depends on the type of polymer. Linear chains consisting of simple —CH_2— links (mers) will have a relatively

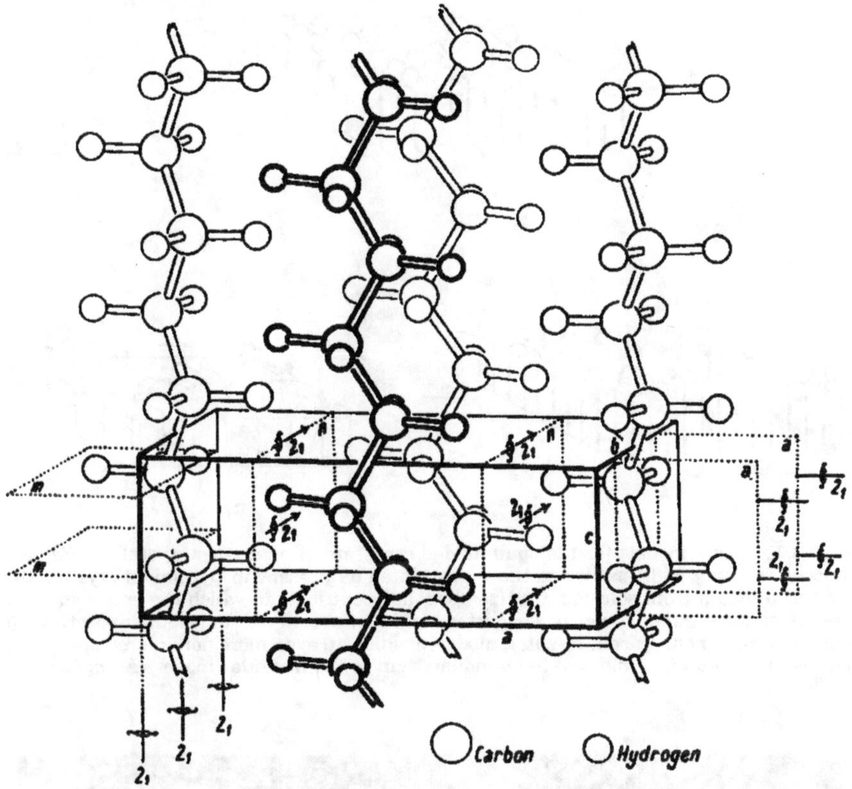

Figure 3.90 The crystal structure of polyethylene. (*From Bunn.*[31] *Copyright © 1961 by Oxford University Press. Reprinted with permission.*)

Figure 3.91 Schematic illustration of the connection of molecules from one crystal to another. This causes the crystals to be separated by an amorphous layer. (*From Flory.*[32] *Copyright © 1953 by Cornell University and copyright © 1981 by Paul J. Flory. Reprinted with permission of Cornell University Press.*)

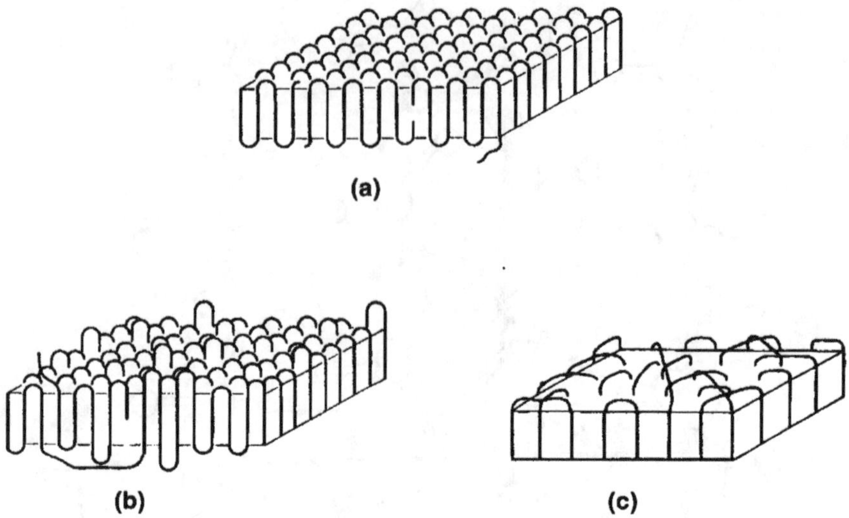

Figure 3.92 Some models for the chain-folded structure of a polymer crystal. (*a*) Regular, adjacent re-entry folds similar to those postulated as present in pyramidal crystals that have been grown from solution. (*b*) Irregular, adjacent folds in which the extent or thickness of the irregular layer is suggested to be proportional to the temperature. (*c*) Switchboard, or nonadjacent re-entry, model in which an even more nonordered amorphous layer is present on both sides of the lamallae than in the irregular model. (*From Geil.*[33])

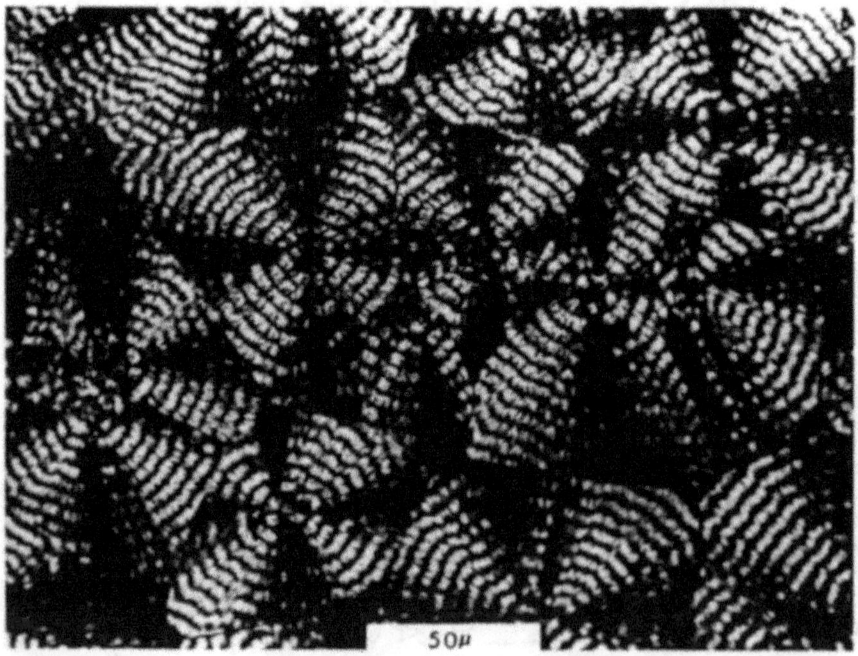

Figure 3.93 A photomicrograph using polarized light showing the crystals (spherulites) of polyethylene. (*From Clark.*[34])

Figure 3.94 Diagram illustrating a polymer molecule with side chains.

high degree of crystallinity. An example is high-density polyethylene. Simple polymers with chains containing side branches (Fig. 3.94) will have less crystallinity (e.g., low-density polyethylene). Polymers consisting of a network of intertwined complex molecules with side branches or with interconnecting molecules (Fig. 3.95) may have sufficient difficulty in aligning to form crystals upon cooling from the liquid that they may be entirely amorphous. The degree of crystallinity has considerable effect on the fracture process in polymers.

Plastic deformation in simple polymer crystals cannot occur by slip, as occurs in metallic crystals, but instead involves the unfolding or unraveling of the long chains to produce a change in orientation. This is illustrated in Fig. 3.96. Note that continued deformation produces a reorientation of the crystals. Fracture can occur by two mechanisms: breaking of the very strong bonds along the chain (e.g., —C—C—), or by separation of groups of molecules from each other. The latter involves breaking the weaker secondary bonds between molecules and is more likely. Deformation by the formation of long groups of molecules leads to the formation of thin extended threads, called fibrils, which are the last to fracture. This is shown in Fig. 3.97. Just before

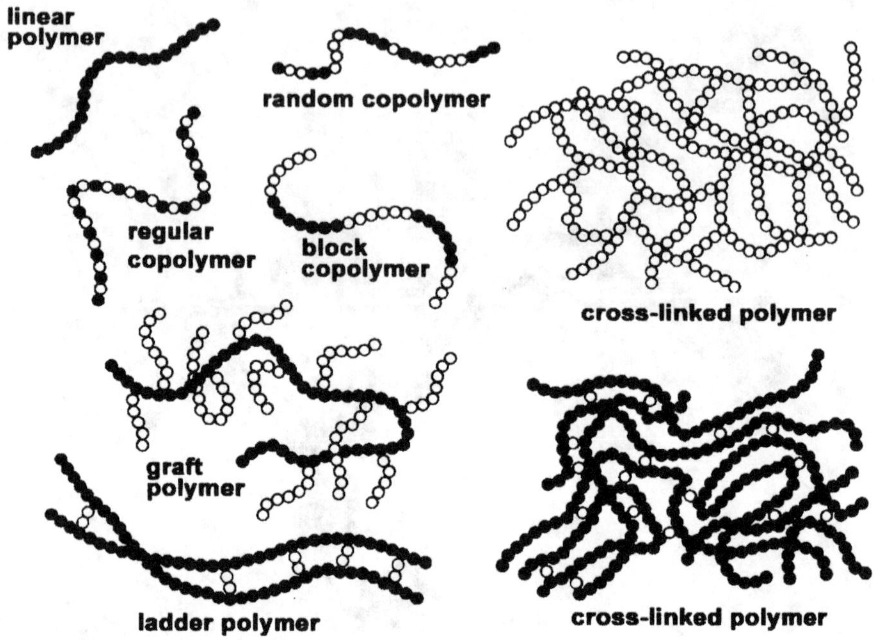

Figure 3.95 Schematic illustration of various types of cross-linking in polymers. The requirements for polymer formation are not difficult to satisfy: There are numerous chemical groups that possess an appropriate number of unsaturated bonds that are free to provide links with neighboring groups. Advances in synthetic techniques have made it possible to produce virtually any imaginable arrangement of the individual monomers. The dark circles denote either single atoms or groups of atoms. (*Adapted from Cotterill.*[30] *Reprinted with permission of Cambridge University Press.*)

the fibrils fracture, they are highly elastically elongated, and upon fracture there is considerable springback, causing the ends to curl (Fig. 3.97*b*).

If the temperature is not too low, crystalline polymers tend to be ductile (macroscopically) and show considerable plastic deformation before fracture (see Sec. 2.2). The crack can take various paths through the spherulite morphology. Fracture between the spherulites (intergranular) can occur, as shown in Fig. 3.98*b*, and through the spherulite (transcrystalline), as shown in Fig. 3.98*c*. However, usually the extensive plastic deformation makes it difficult to identify the spherulites from the fracture surface appearance. The molecules undergo extensive rearrangement and elongation (Fig. 3.96), and fracture can occur by local separation, forming voids, with the final separation of the connecting fibrils. (This is analogous to void coalesence in metals.) Figure 3.99 shows the elongated appearance of such a fracture surface. Note the curled appearance of the end of some of the fibrils.

Just as in metallic and ceramic materials, the fracture topography in polymers is sensitive to the microstucture. Second phases (e.g., rubber or

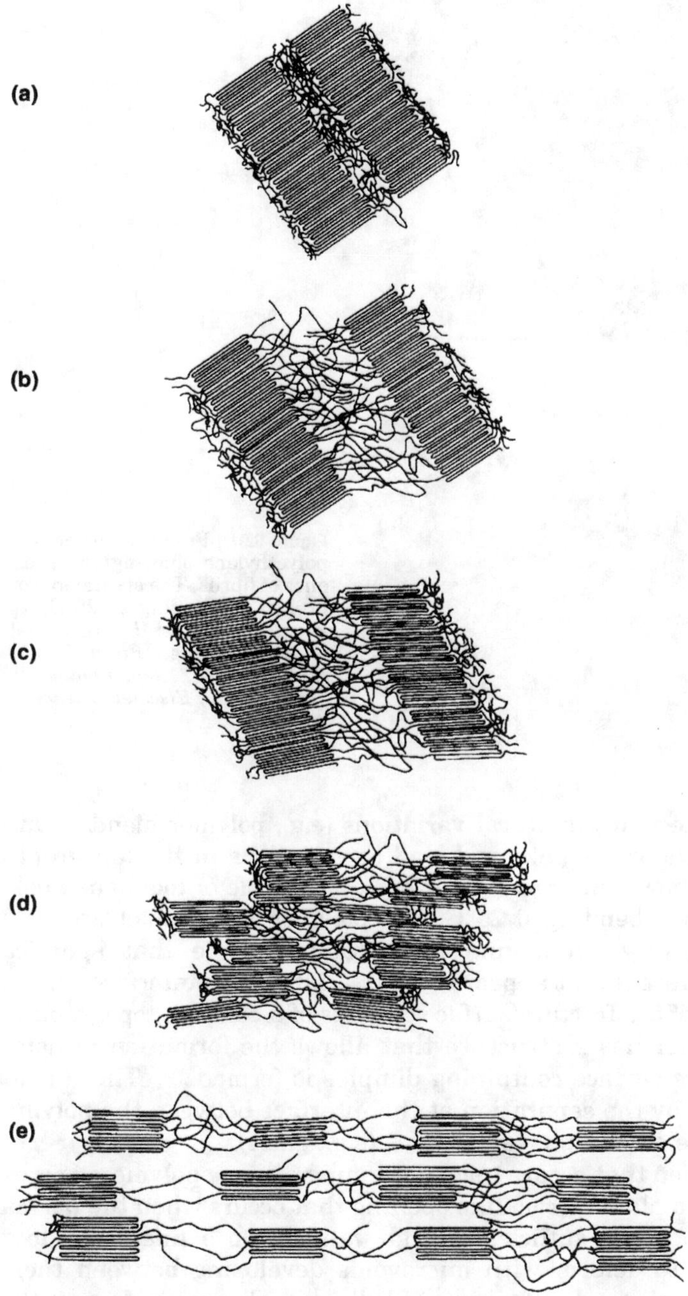

Figure 3.96 Schematic illustration of the unfolding of chains and their rearrangement to form differently oriented crystals. (*From Schultz.*[35] *Reprinted with permission of Pearson Education, Inc.*)

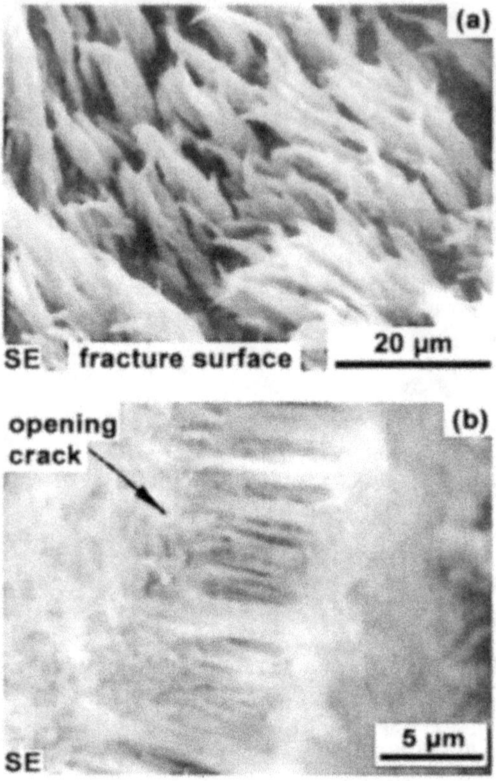

Figure 3.97 Fracture surfaces of polyethylene showing the presence of fibrils. The specimen was broken by bending at 25°C. (*a*) Fracture surface; (*b*) crack that is just opening. (*From Brooks and McGill.*[26] *Reprinted with permission of Elsevier Science.*)

glass) and molecular structural variations (e.g., polymer blends) cause deviations in the crack path and lead to variations in the topography. Figure 3.100 shows the fracture surface of a sample of toughened polystyrene broken in bending at 25°C. This polymer is a blend of two molecular species, polystyrene and polybutadiene. Note that spherical particles are present that appear to be associated with interface separation. The rest of the fracture surface consists of a complex topography.

If the polymer has a structure that allows the formation of holes, then a fracture surface containing dimples is formed.[30,35] These holes may originate by the separation at the interface between the polymer matrix and the second-phase particles (e.g., glass).

A phenomenon that occurs upon deforming of some polymers is crazing. A craze is a planar, cracklike opening that occurs when the normal stress has reached a sufficiently high value.[37] On a fine scale, local regions begin to deform with microvoids developing between them. These local regions develop into fibrils, as shown in Fig. 3.101. Although on a fine scale there is considerable plastic deformation, on a macroscopic scale the material fractures in a brittle manner because

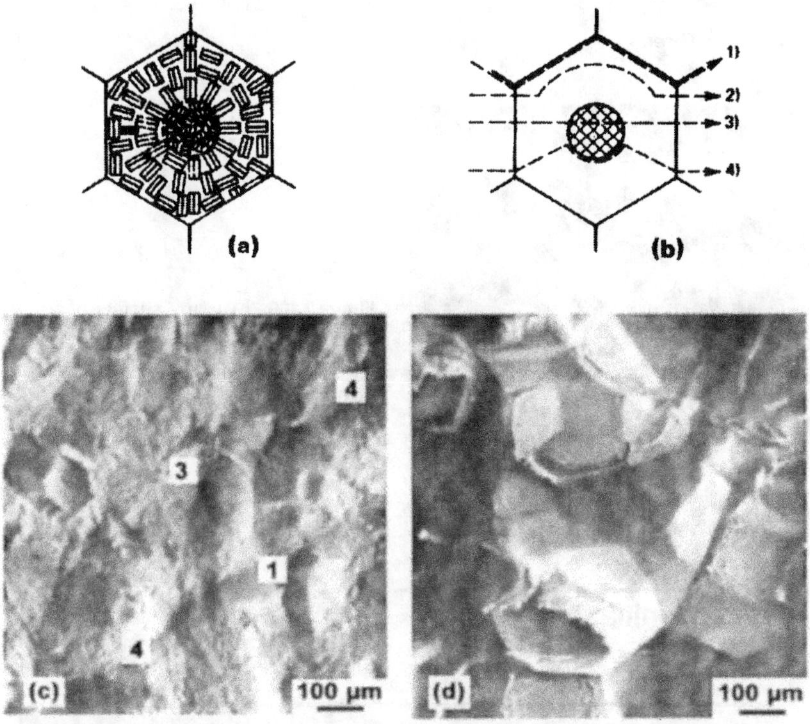

Figure 3.98 (*a*) Schematic of the orientation of crystal lamellae in spherulites. (*b*) Schematic of possible fracture path through spherulites in polymers. (*c*) Fractograph from fast fracture through polypropylene [paths 2–4 in (*b*)]. (*d*) Fractograph from slow fracture through polypropylene showing interspherulitic fracture [path 1 in (*b*)]. (*From Friedrich.*[36])

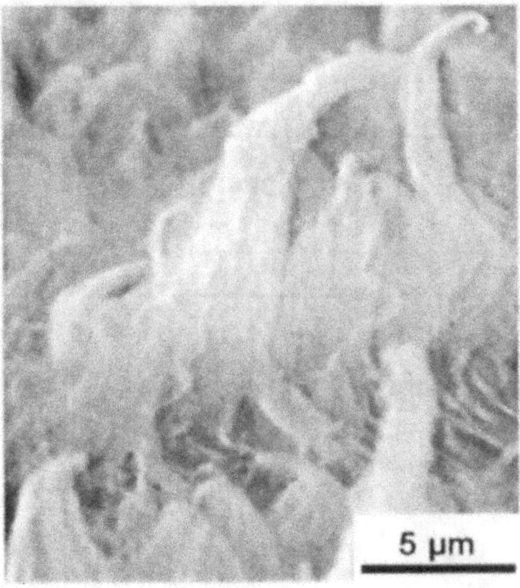

Figure 3.99 Fractograph of polyethylene showing broken fibrils. (*From Brooks and McGill.*[26] *Reprinted with permission of Elsevier Science.*)

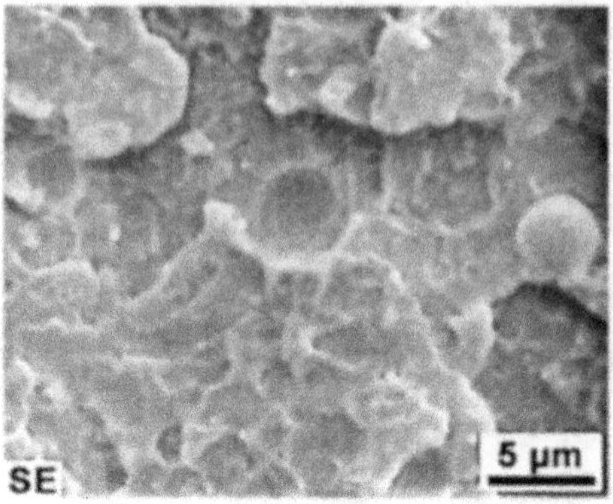

Figure 3.100 Fracture surface of a sample of toughened polystyrene broken in bending at 25°C. (*From Brooks and McGill.*[26] *Reprinted with permission of Elsevier Science.*)

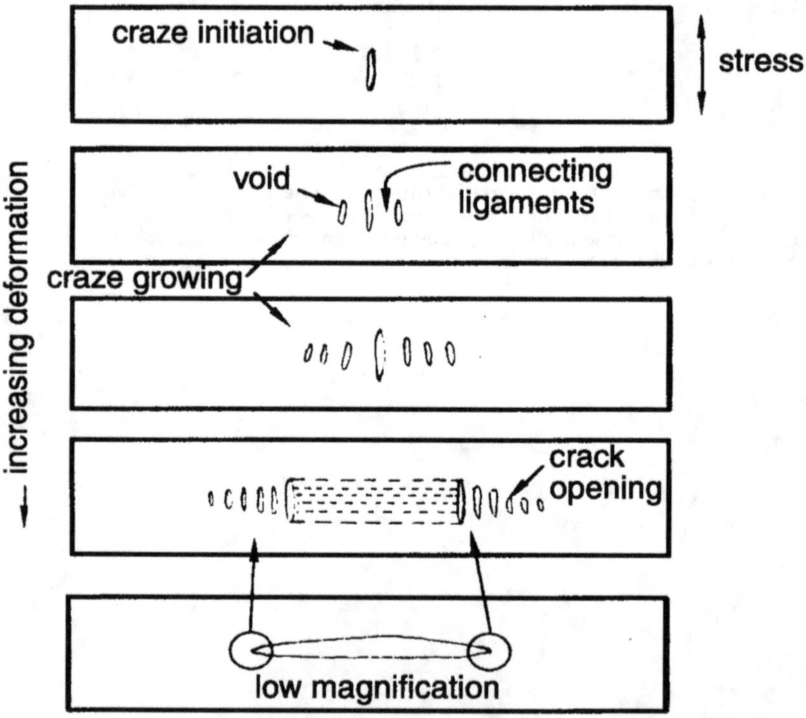

Figure 3.101 Schematic diagram of a craze in a polymer. (*From Brooks and McGill.*[26] *Reprinted with permission of Elsevier Science.*)

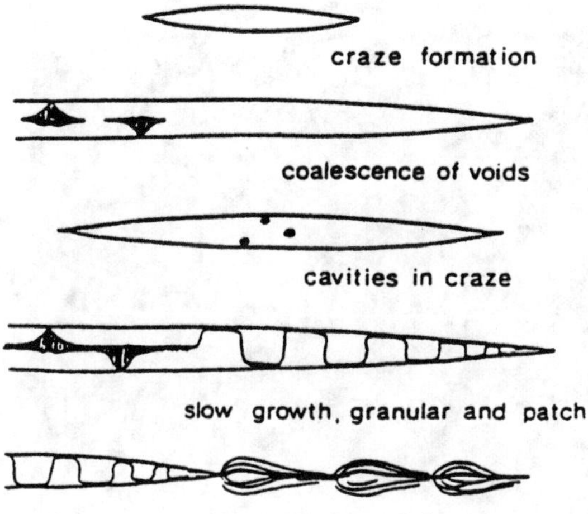

Figure 3.102 Schematic diagram of the fracture path through a craze in a polymer. (*From Hull.*[37])

of the high density of holes and because plastic deformation is restricted to these regions. Crazes frequently form first on the free surface, but also form ahead of an advancing crack.[34]

The fracture path may pass through the crazed regions or may pass around them, as illustrated in Fig. 3.102. A crack moving relatively slowly may pass through the midplane of the craze as the fibrils pull apart. A more rapidly moving crack may move along the craze-matrix interface and back and forth across the craze thickness.[37] This path gives rise to the fracture surface pattern (patchy) such as shown in Fig. 3.103*a*. The fracture surface at the matrix-craze interface consists of broken fibrils (Fig. 3.103*b*) and of elongated, broken fibrils where it passes through craze (Fig. 3.103*b*, top). Cracks can also show the pattern in Fig. 3.104*a*, called hackle markings, which consist of patch fracture (Fig. 3.104*b*). This morphology is caused by the formation of crazes ahead of the crack and the movement of the crack from one group of crazes to another. Figure 3.102 summarizes the fracture paths that a crack follows through crazes.[37]

The fracture surfaces of brittle polymers have macroscopic features similar to those found in the fracture of ceramics and brittle metallic materials (Sec. 3.4). Figure 3.105 shows the fracture surface of polymethyl methacrylate (PMMA) that was broken by bending at 25°C. The radial marks point back to the general origin of fracture initiation

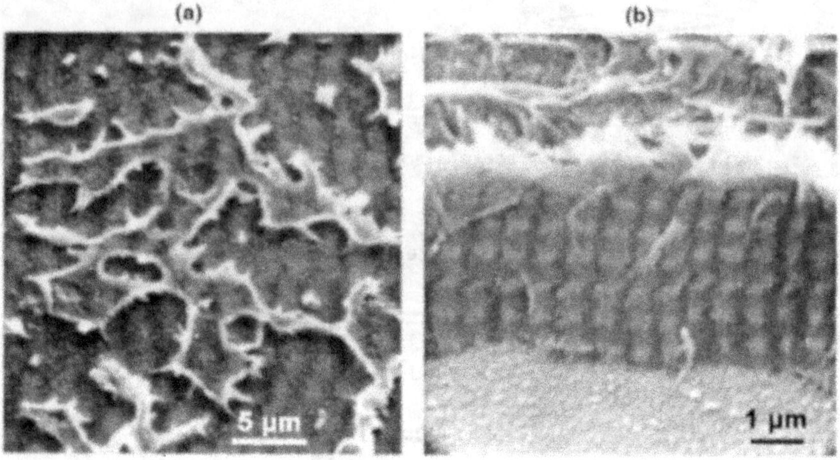

Figure 3.103 Fractographs of amorphous polymer typical of that of fracture through crazes, showing (*a*) a patchy appearance. High magnification shows that the crack has passed through [(*b*) bottom] the matrix-craze interface and [(*b*) top] the center of the crazes, leaving broken fibrils. (*From Hertzberg.*[38])

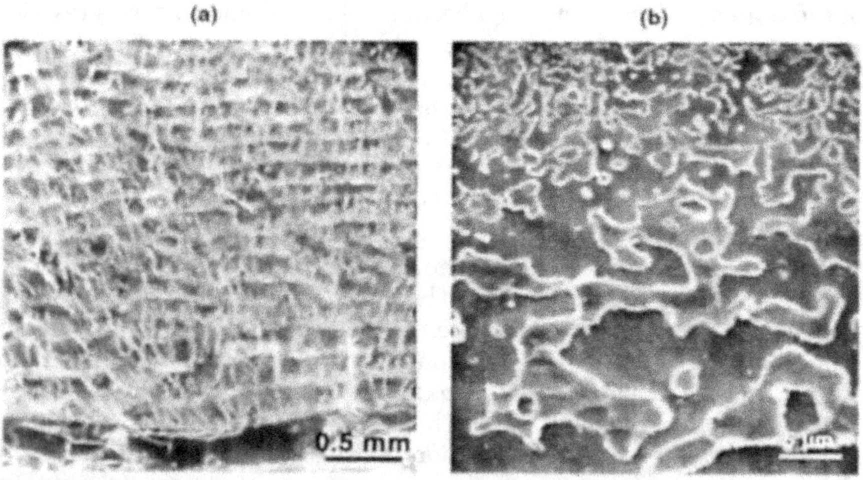

Figure 3.104 Fractographs of amorphous polymer showing hackle pattern (*a*), which consists of patchy fracture through crazes (*b*). (*From Hertzberg.*[38])

(the dark region). Higher magnification shows that the fracture surface has a fine, complex topography.

Fatigue fracture in polymers is a complicated process, affected by the polymer structure and the loading conditions. Although striations are commonly found on the fracture surface (Fig. 3.106), a complex morphology may be associated with this type of fracture in polymers.

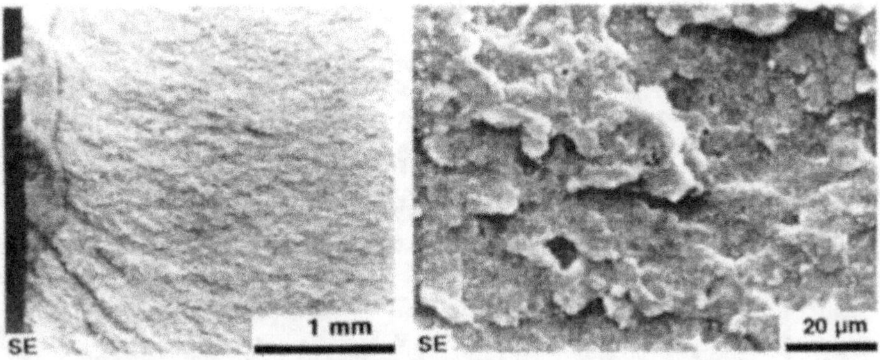

Figure 3.105 Fracture surface of sample of polymethyl methacrylate broken in bending at 15°C. (*From Brooks and McGill.*[26] *Reprinted with permission of Elsevier Science.*)

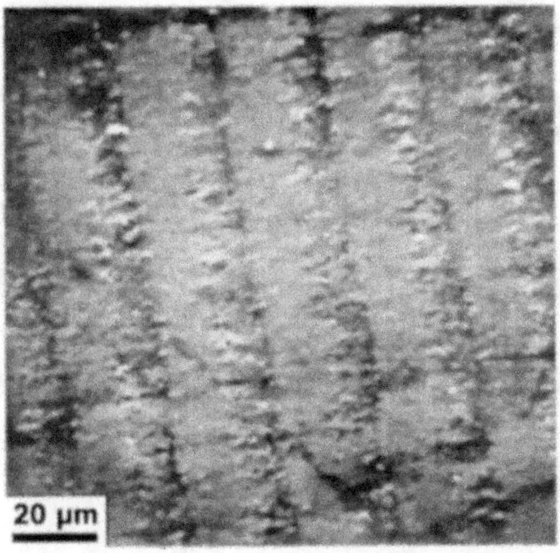

Figure 3.106 Fractograph of fatigue fracture in a polymer showing microscopic fatigue striation markings. (*From Rimnac et al.*[39] *Copyright © ASTM; reprinted with permission.*)

Hertzberg[38] has summarized the status of fatigue fracture and fractography of polymers, and the following is excerpted from his description: At relatively high stress intensity levels, the fatigue fracture surface of many polymers may exhibit fatigue striations developed during each cycle. An example is shown in Fig. 3.107*a*. However, striations may be present (Fig. 3.107*c*) that do not correlate with the crack growth cycle. In Fig. 3.108 the fracture surface shows what appears to be secondary cracks parallel to the fatigue crack growth direction, but these do not correlate with the loading frequency. Thus detection of striations on the fracture surface of polymers is an indication of fatigue fracture, and it may be difficult to correlate the spacing with the loading cycles.

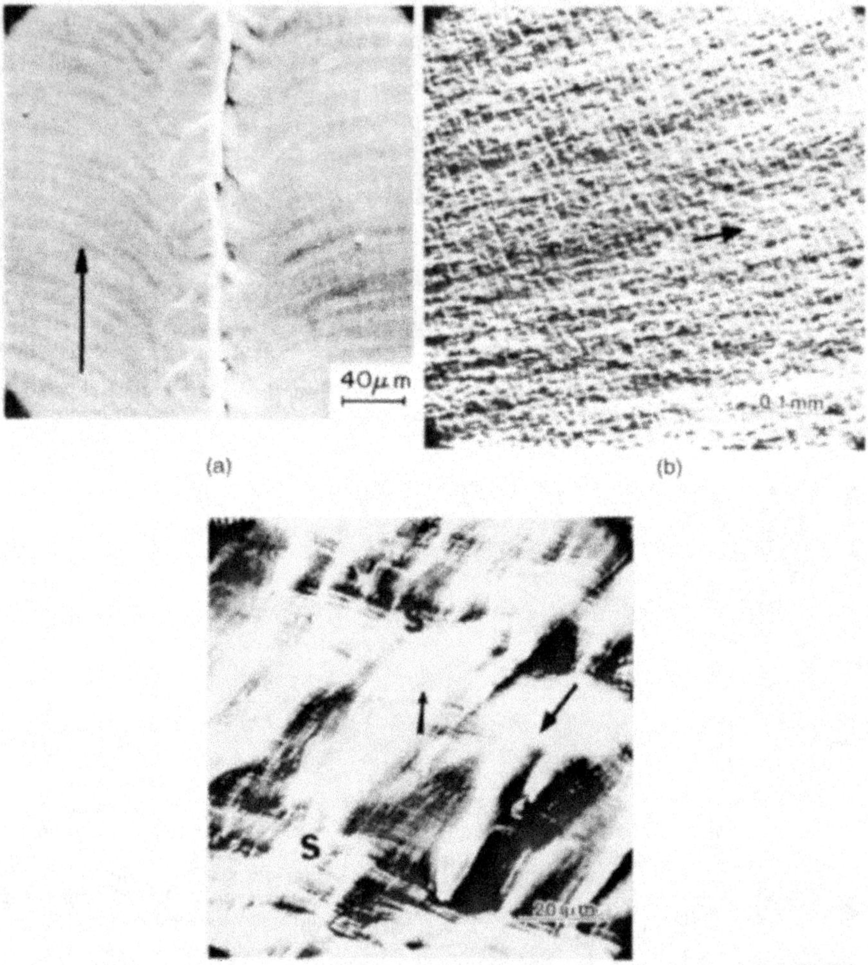

(a) (b)

(c)

Figure 3.107 Fractographs of fatigue fracture surfaces in engineering plastics: (*a*) polysulfone; (*b*) Nylon 66 + 2.7% water; (*c*) higher magnification of (*b*). Arrows show crack propagation direction. [(*a*), (*b*) *From Skibo et al., in Hertzberg*[38]; (*c*) *from Hertzberg.*[38]]

An example of clear fatigue markings that do not correlate with the loading frequency is shown in Fig. 3.109. In this case, the crack is stationary for several cycles, then jumps by advancement through crazes.

Figure 3.110 shows the morphology of the fatigue fracture surface of a semicrystalline polymer, where the crack has passed through a spherulite. However, if the polymer exhibits more viscoelastic behavior, the type of fracture surface marking shown in Fig. 3.111 is seen. There is a drawn appearance to the surface.

(a)

(b)

Figure 3.108 Fatigue fracture surface of impact-modified N66 polymer. Arrow shows crack growth direction. (*a*) Rumple markings in fatigue fracture region. (*b*) Rumple markings in fast fracture zone. (*From Hahn et al., in Hertzberg.*[38])

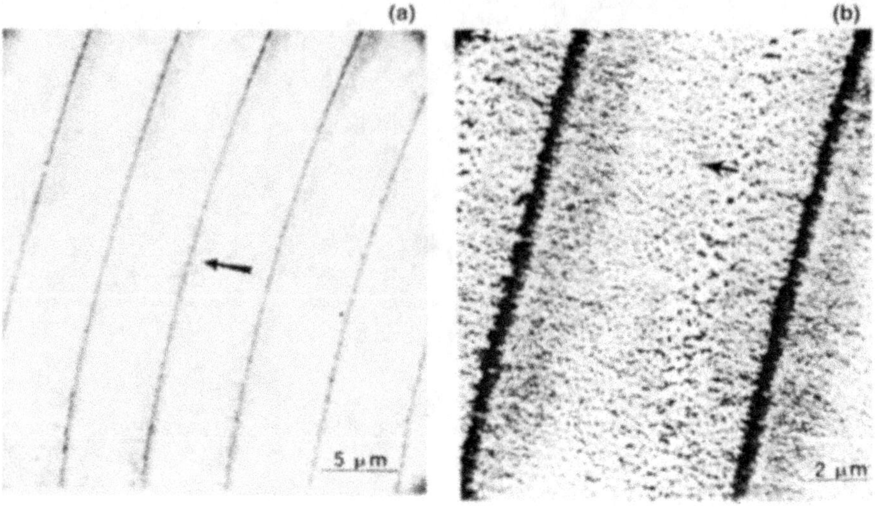

Figure 3.109 Discontinuous growth bands on a fatigue fracture surface of polyvinyl chloride. Arrows show crack propagation direction. Note the fine void gradient in (*b*). (*From Rimnac et al., in Hertzberg.*[38])

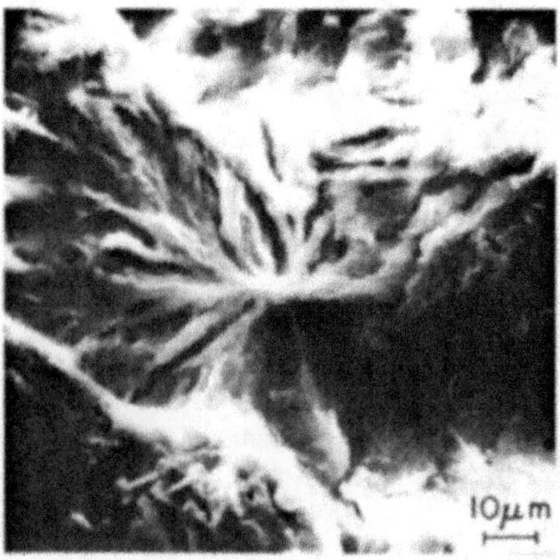

Figure 3.110 Fatigue fracture surface in semicystalline polyacetal homopolymer (Delrin). (*From Bretz et al., in Hertzberg.*[38])

3.16 Stereo Examination of Fracture Surfaces

Although electron micrographs of fracture surfaces (especially scanning electron micrographs) have a three-dimensional appearance, due to the great depth of field of the instruments, a much better feeling of the depth profile can be obtained by stereo imaging the surface (see App. 1A). Such pictures are easy to obtain. They require taking a pic-

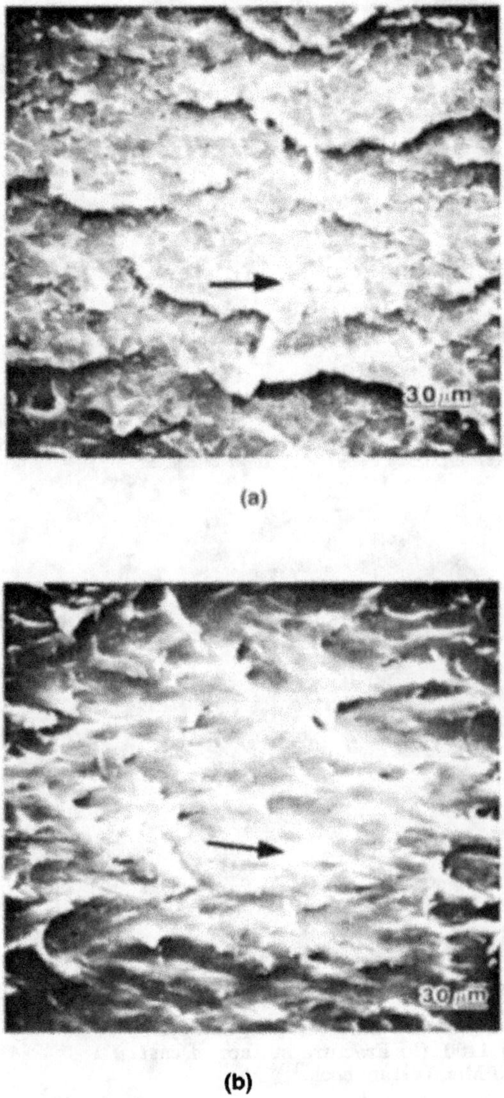

Figure 3.111 Fatigue fracture surface in semicrystalline polymer. Arrow indicates crack propagation direction. (*a*) Dry Nylon 66 and (*b*) Nylon 66 + 2% water. (*From Bretz et al. in Hertzberg.*[38])

ture of the region of interest, then tilting the specimen a few degrees and photographing the new image of the same area. The two pictures are then appropriately arranged and viewed with a stereo viewer to bring out the depth detail.

The added information obtained by stereo viewing cannot be overemphasized. To emphasize the advantage of stereo viewing, several stereo views of fracture surfaces, which should be examined using a stereo viewer, are presented here. Figure 3.112 shows optical stereo

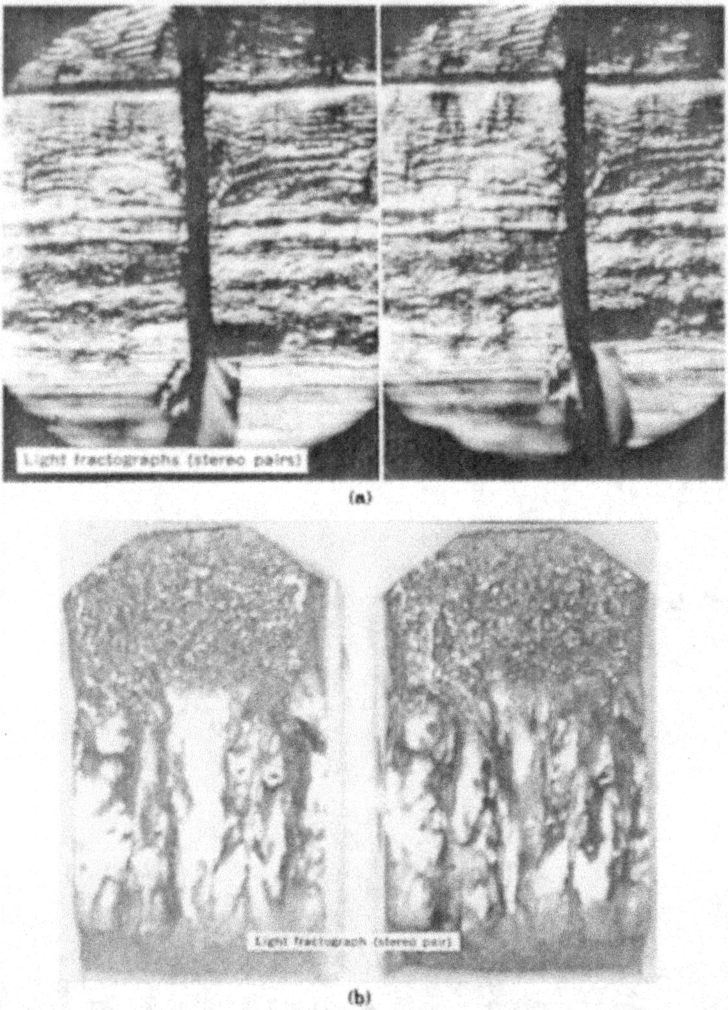

Figure 3.112 Optical stereo pairs of fracture surfaces. (*a*) Low-cycle fatigue fracture in aluminum alloy 1100. (*b*) Fracture surface of cast experimental low-carbon steel. (*From* Metals Handbook.[11])

pairs, illustrating that this method can be used for viewing with an optical microscope. However, the image suffers from low resolution, so most stereo pictures are obtained using an electron microscope. Figures 3.113 and 3.114 present several stereo pairs for a variety of fracture surface topographies.

Stereo pairs can be used to obtain quantitative depth profiles. The procedure is outlined in detail in App. 1A.

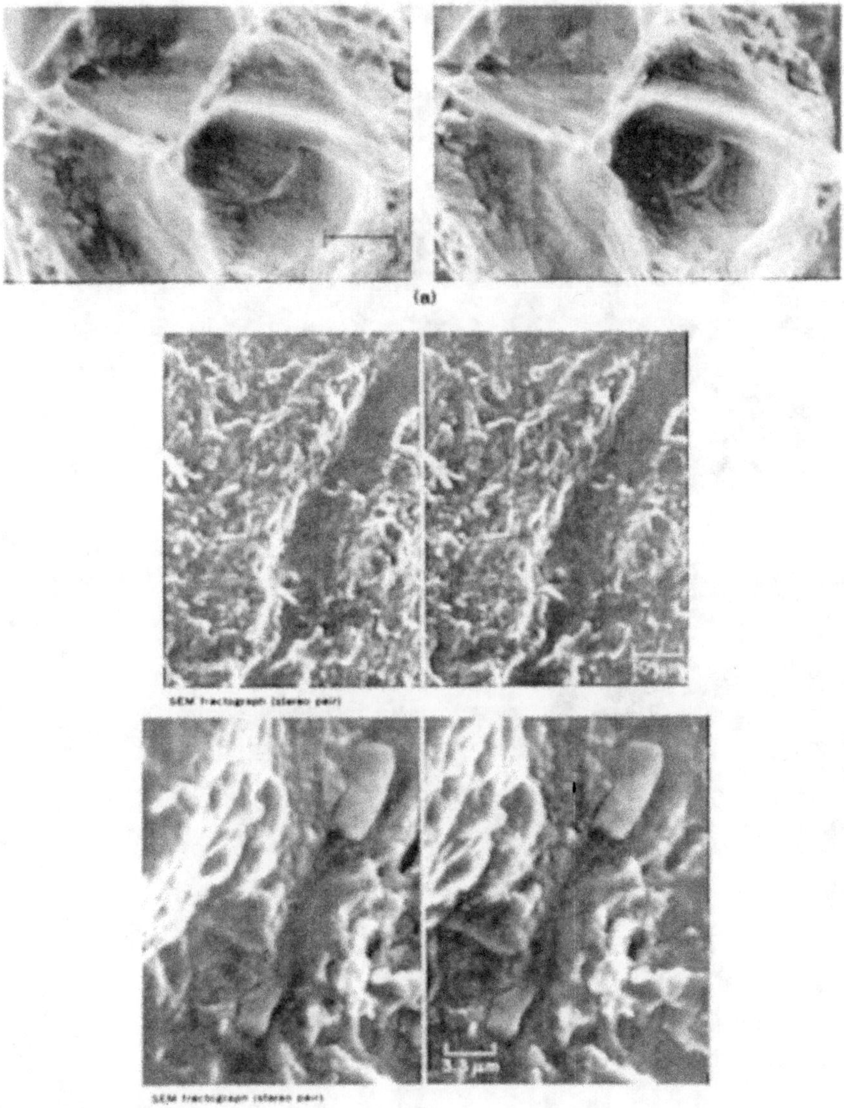

Figure 3.113 Stereo pairs of several types of fracture surface topographies. (*a*) Stereo pair showing deep dimples in the fracture surface of commercially pure titanium. (*Courtesy of M. Erickson-Natishan, University of Virginia; from Gabriel.*[14]) (*b*) Stringer and two stringer troughs on the fracture surface of 4340 steel heat-treated to a tensile strength of 1310–1450 MPa (190–210 ksi). Top—trough that earlier contained a stringer; bottom—trough that still contains two portions of a stringer. (*From* Metals Handbook.[11])

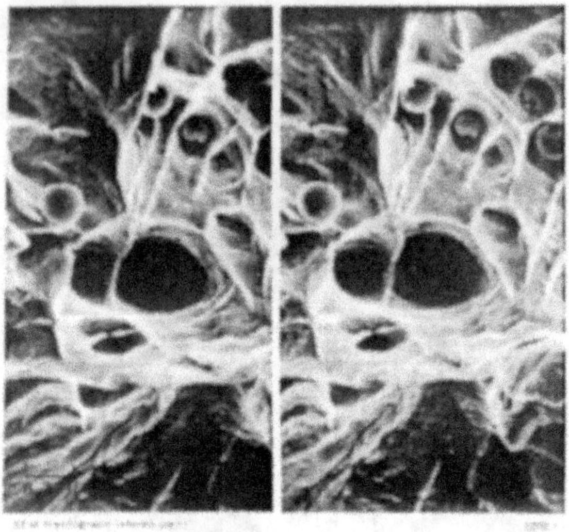

(c)

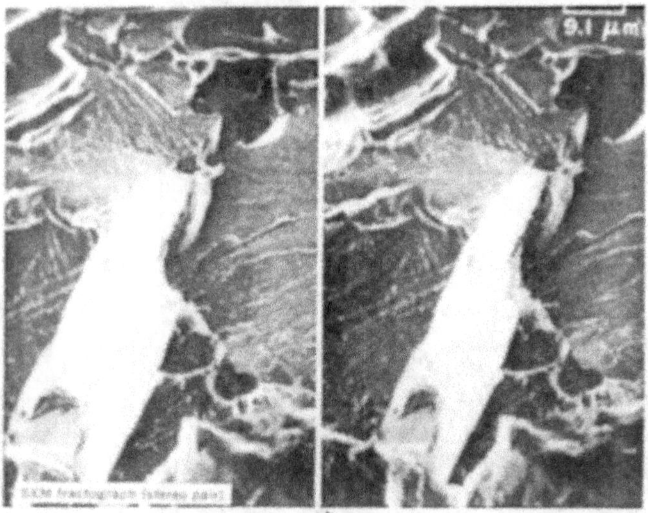

(d)

Figure 3.113 (*Continued*) Stereo pairs of several types of fracture surface topographies. (*c*) SEM fractographs (stereo pair) of the surface of a tensile-test fracture obtained at room temperature. The alloy was a low-carbon iron to which an appreciable amount of Fe_2O_3 had been added. (*d*) SEM fractograph (stereo pair) of the fracture surface of a Charpy impact test bar of high-purity iron. (*From* Metals Handbook.[11])

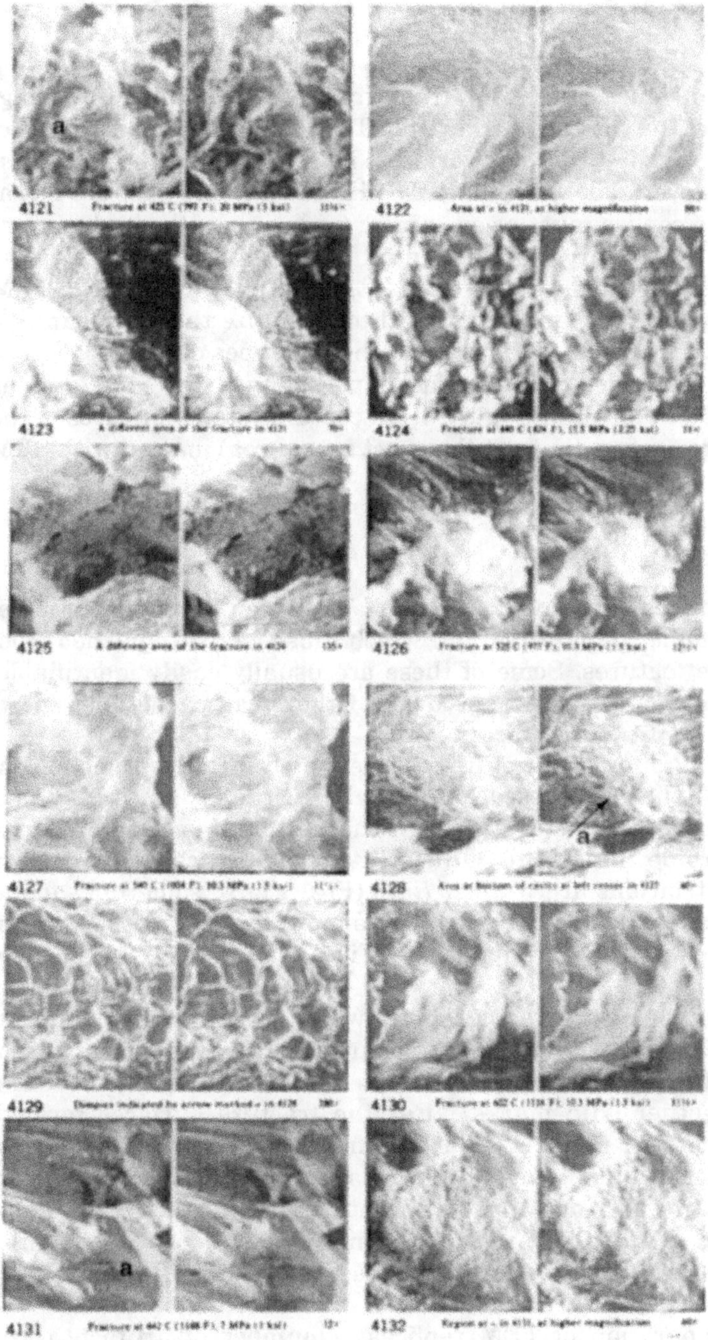

Figure 3.114 SEM stereo fractographs of several types of fractures. (*From* Metals Handbook.[11])

3.17 Comparison of SEM and TEM Fractographs

Prior to the advent of the SEM, fractographs were obtained by imaging replicas of the fracture surface in a TEM, and thus the earlier documentation of the fracture surfaces is in this form. Now most observations are made with the SEM. However, there still are situations where a replica must be examined; for example, the broken component may be too large to place in an SEM. The appearance of the same fracture surface as observed in SEM and from a replica in TEM is somewhat different, and it is useful to recognize the appearance of common features (such as dimples) in the two types of images. Figure 3.115 shows examples of the surface topography of the *same area* as imaged in SEM and TEM. Figure 3.116 contains stereo pictures of the same fracture surface (but not the exact same area) imaged with SEM and TEM.

3.18 Artifacts

There may be features on fracture surfaces which are artifacts introduced by sample preparation and which must be distinguished from true fracture features. Some of these are usually easily identifiable, such as debris and corrosion products. Others can cause false interpretation of the fracture process.

Figure 3.117 shows a fracture surface which includes regions containing parallel lines. These are similar to striations, but in this case they are caused by fracture through the parallel lamellae of ferrite and iron carbide in pearlite. Other features which can be mistaken for fatigue striations are *Wallner lines* (Fig. 3.118) (sometimes found on fracture surfaces of brittle materials and phases, but their origin is not well established), slip traces (Figs. 3.39 and 3.23), and tire tracks (Fig. 3.67).

Figure 3.119 shows a fractograph in which smooth regions appear. This is a true feature of the fractured surface; however, it is not associated with the fracture process but was formed by rubbing of the two surfaces together (rub marks) after the fracture surfaces formed. This is a common feature in fatigue fractures. Figure 3.120 illustrates a feature called *mud cracks.* This is a region of deposited corrosion products and not an actual feature of the fracture process. Examination of the fracture surface in between repeated cleaning (for example, with replication tape; see Sec. 1.4.2) will allow detection of such artifacts.

In the preparation of TEM replicas, a number of artifacts occur which have been recognized. One problem is tearing, stretching, or

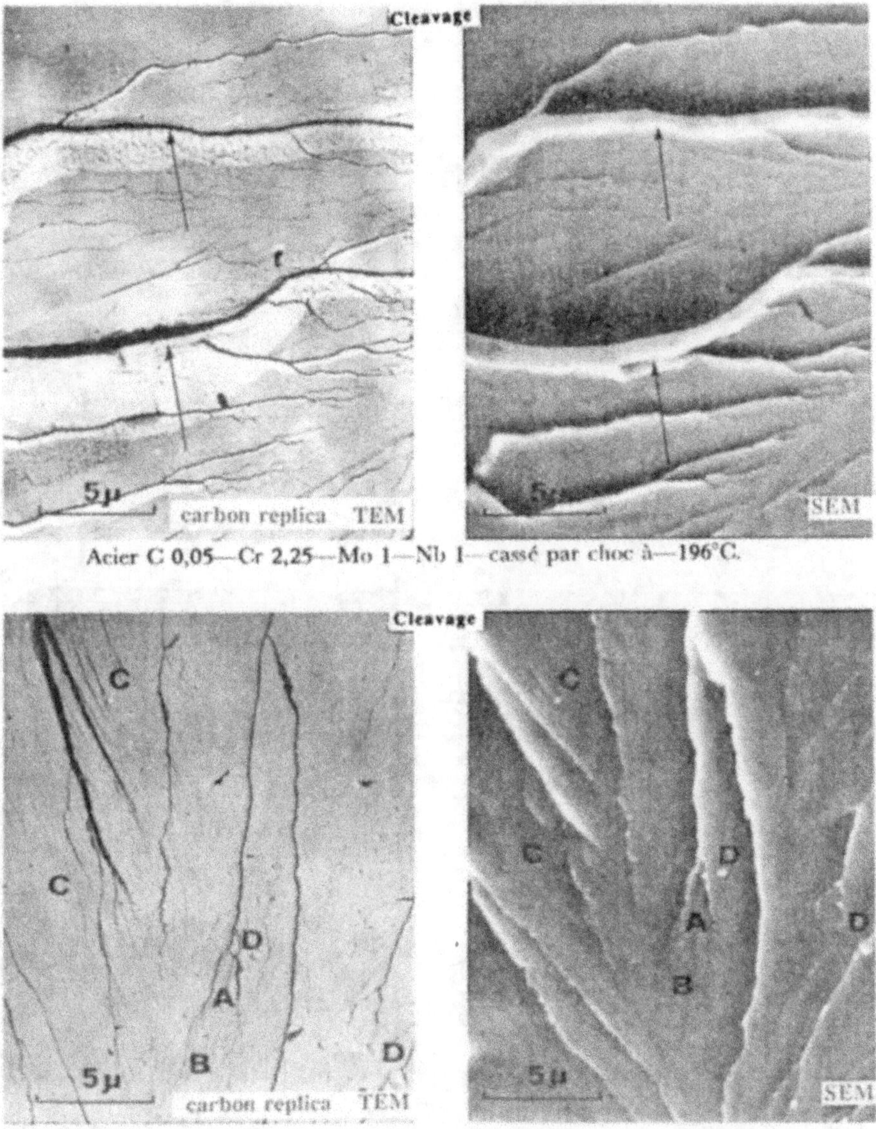

Figure 3.115 Fractographs comparing the exact same areas as observed on a carbon replica in TEM (left) and in SEM (right). (*From Maillard, Meny, and Champigny.*[41])

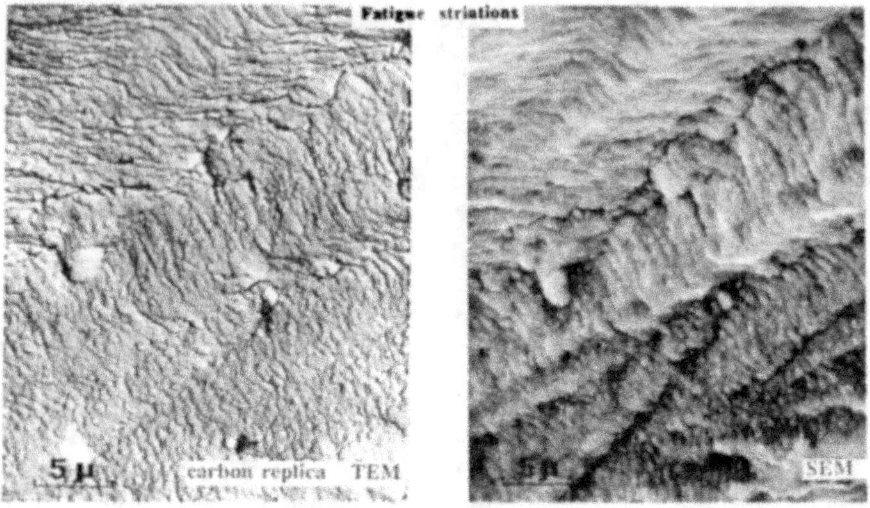

Acier C 0,03 – Cr 16 – Ni 12 – Mo 2 – Mn 1,2. Fatigue en traction compression à 580 C dans l'argon – nombre de cycles 25.535 – ±0,5 – fréquence 20 s.

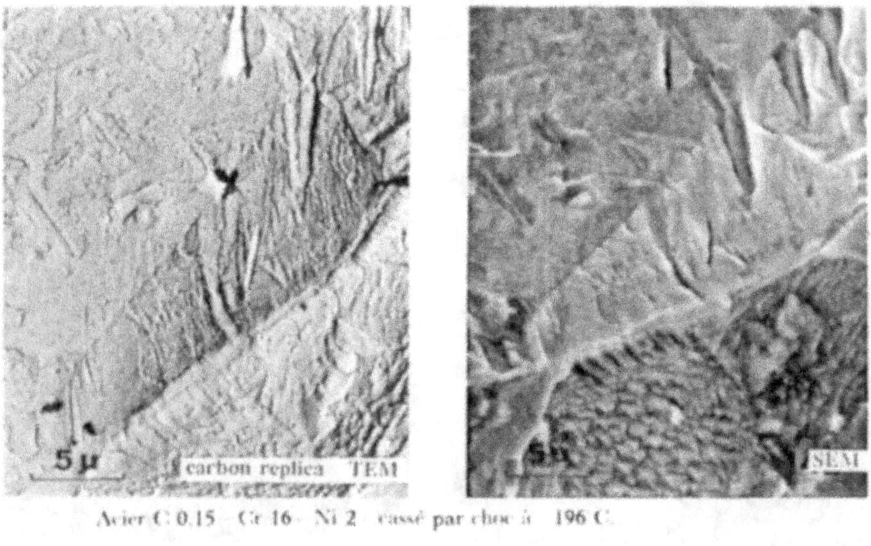

Acier C 0,15 – Cr 16 – Ni 2 – cassé par choc à – 196 C.

Figure 3.115 (*Continued*)

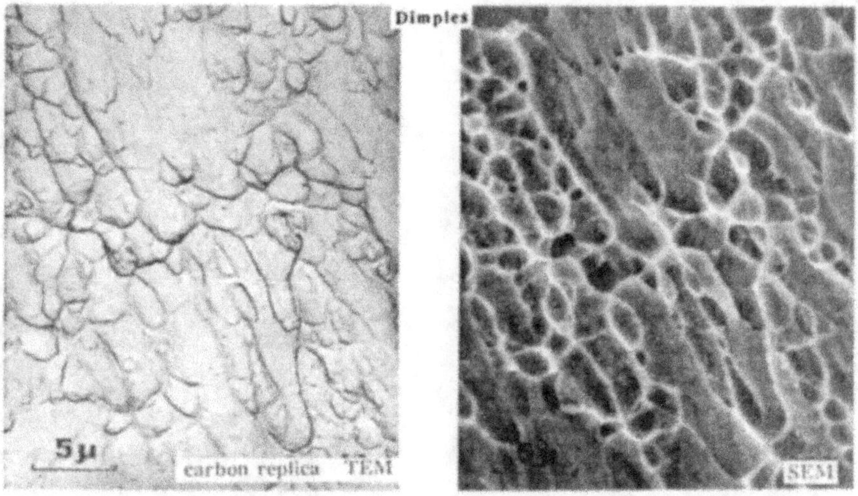

Acier C<0,08—Cr 17—Ni 12—traction à 20°C.

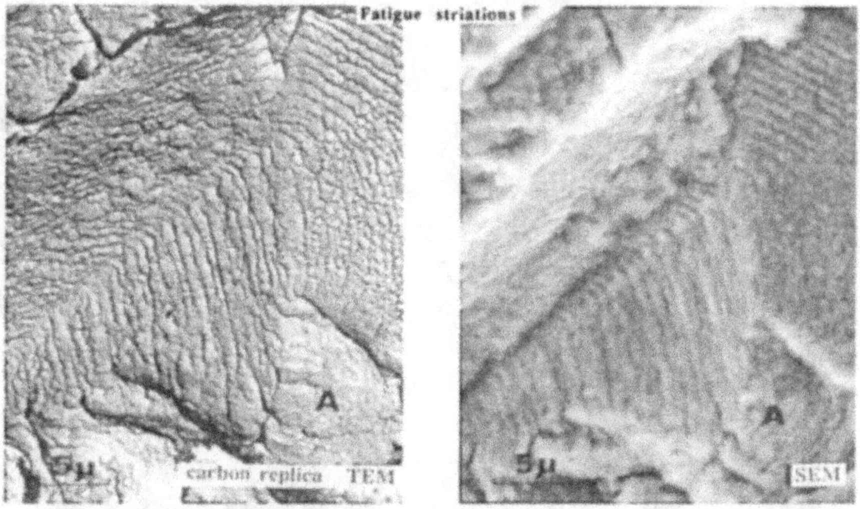

Acier C<0,03—Cr 16—Ni 12—Mo 2—Mn 1,2. Fatigue en traction—compression à 580°C dans l'argon—nombre de cycles 25.535—ε0,5—fréquence 20 s.

Figure 3.115 (*Continued*)

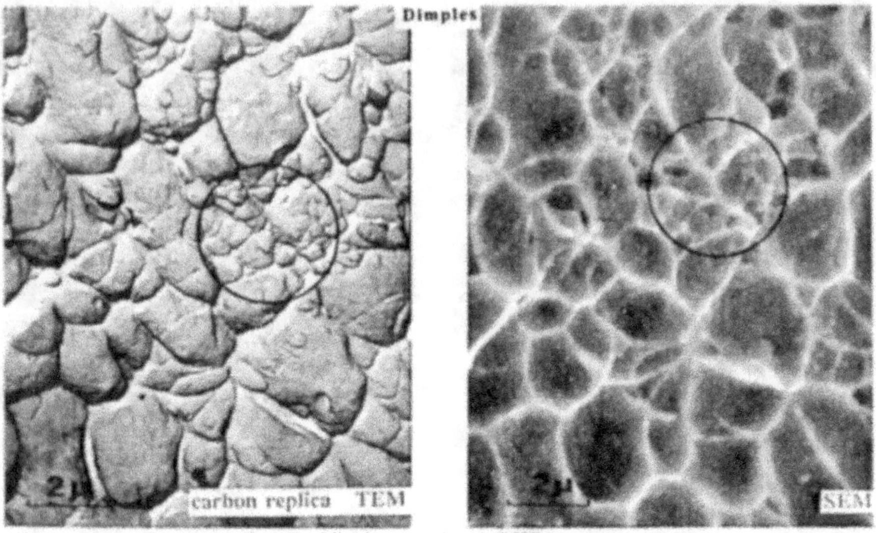

Acier C<0.08—Cr 17—Ni 12—traction à 20°C.

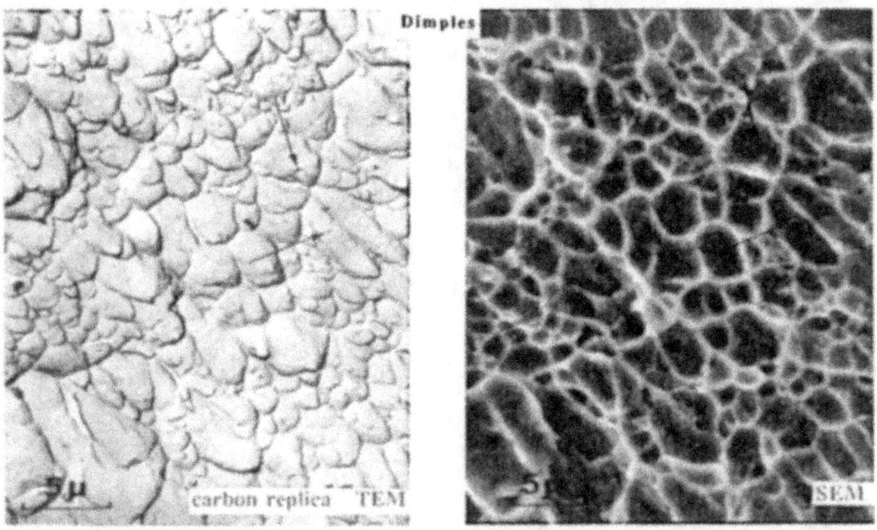

Acier C<0,08—Cr 17—Ni 12—traction à 20°C.

Figure 3.115 (*Continued*)

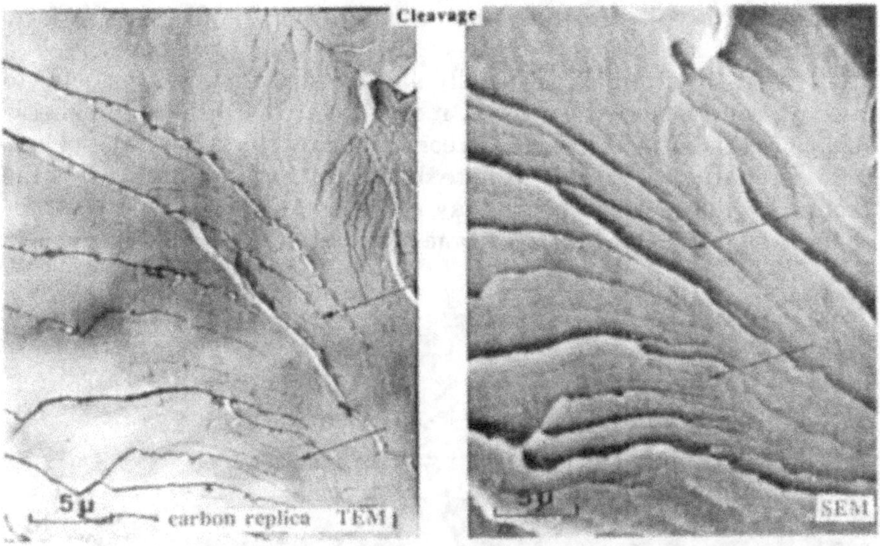

Fer pur cassé par choc à −196°C.

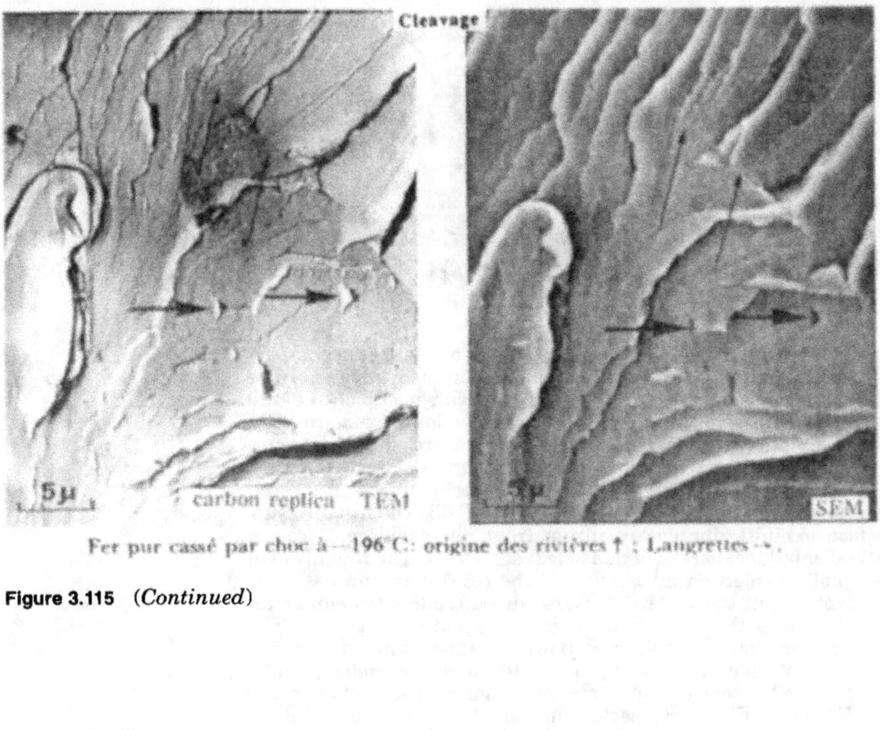

Fer pur cassé par choc à −196°C: origine des rivières ↑ ; Langrettes →.

Figure 3.115 (*Continued*)

scraping of the plastic replica on removal from the fracture surface. Examples of the effect of these on fractographs are shown in Fig. 3.121. Since the replica is usually shadowed with a heavy metal to improve contrast (see Sec. 1.4.2), it is possible to overheat the plastic replica in this step. Testing of the consistency of the appearance of successive replicas will usually isolate this effect. Excessive heating of the replica by the electron beam may cause the heavy metals used for shadowing (see Sec. 1.42) to become reticulated (patterned), as shown

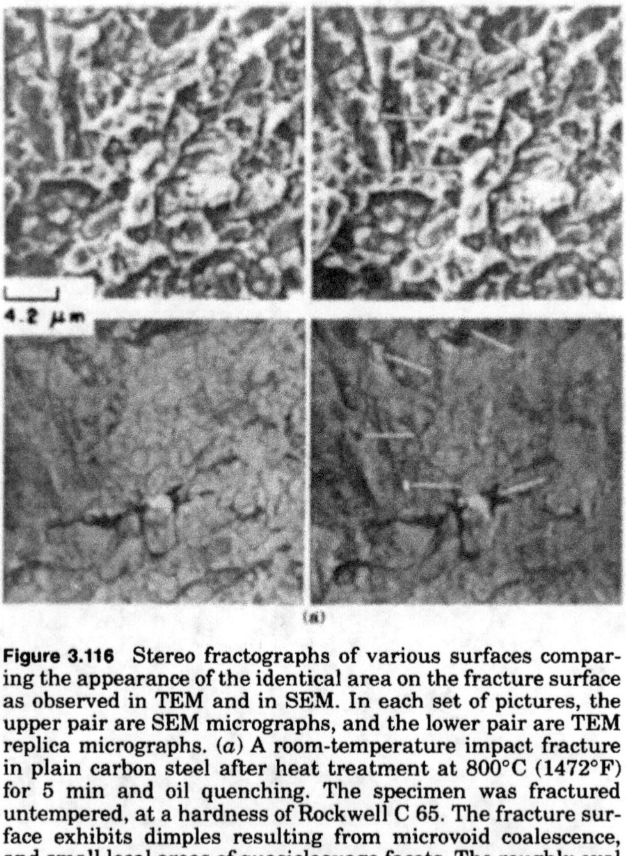

Figure 3.116 Stereo fractographs of various surfaces comparing the appearance of the identical area on the fracture surface as observed in TEM and in SEM. In each set of pictures, the upper pair are SEM micrographs, and the lower pair are TEM replica micrographs. *(a)* A room-temperature impact fracture in plain carbon steel after heat treatment at 800°C (1472°F) for 5 min and oil quenching. The specimen was fractured untempered, at a hardness of Rockwell C 65. The fracture surface exhibits dimples resulting from microvoid coalescence, and small local areas of quasicleavage facets. The roughly oval tip of a projection at arrow 1 in the SEM view can easily be identified in the TEM. A feature with a crudely triangular tip pointing to the left (arrow 2) is also readily recognized. The upper portion of this feature, however, appears much flatter in the TEM view than in the SEM, although local marks can be identified in each. Another site of conformance between the SEM and TEM stereofractographs is at arrow 3. Some details are lost at arrows 4 and 5 in the TEM pair, because of local tearing of the replica. (*From* Metals Handbook.[11])

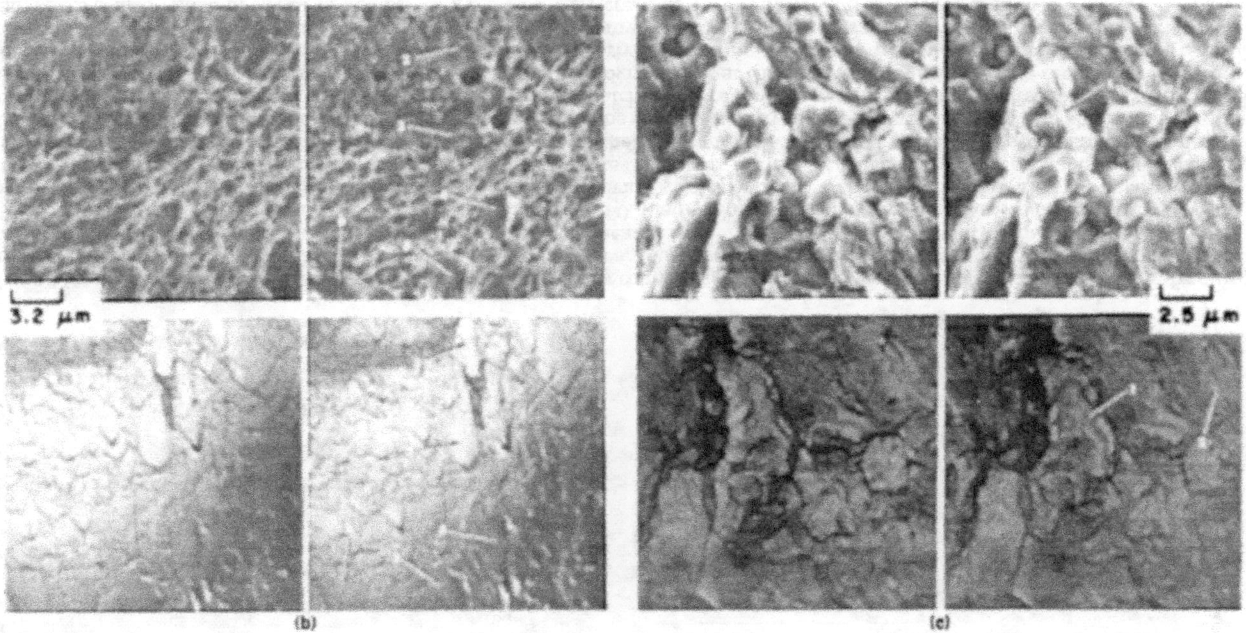

Figure 3.116 (*Continued*) Stereo fractographs of various surfaces comparing the appearance of the identical area on the fracture surface as observed in TEM and in SEM. In each set of pictures, the upper pair are SEM micrographs, and the lower pair are TEM replica micrographs. (*b*) A room-temperature impact fracture in low-alloy manganese steel (0.60 to 0.70% C, 1.0 to 1.1% Mn) that was oil quenched from 900°C (1652°F) before being broken. The surface shows mainly shear dimples, some of which (at upper right in each stereo pair) are large enough to display areas of stretching. (It should be noted that a TEM replica of such large dimples may sag or collapse and, as in this comparison, introduce a very different appearance to the dimples from that displayed in an SEM view.) Specific dimples may be identified in each pair of views, and a number of these have been indicated by arrows 1 to 6. Observe, however, that the sense of depth, particularly in the larger dimples in the SEM pair, is quite lacking in the TEM. (*c*) This is a remarkably complex area, containing many small and varied quasicleavage facets and exhibiting a number of spheroidal inclusions. The most prominent feature is the high peak at arrow 1, which is reproduced in very faithful and identical detail in both stereo pairs. Note that the two inclusions between the twin tips of the peak are shown in both pairs, but that similar inclusions slightly nearer the top of the SEM pair are missing in the TEM because of damage to the replica. The top center area of the SEM view appears to record a greater number of surface complexities than does the same area in the TEM view; at arrow 2, however, the TEM view displays very fine fracture marks that cannot be found in the same location in the SEM view. (*From* Metals Handbook.[11])

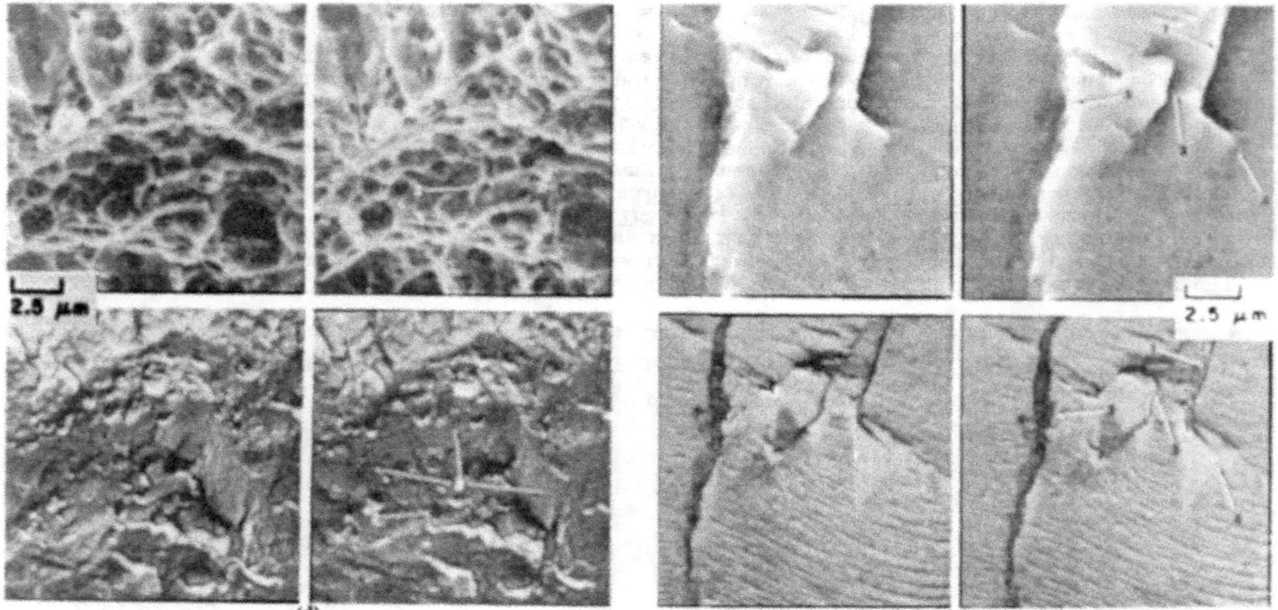

(d) (e)

Figure 3.116 (*Continued*) Stereo fractographs of various surfaces comparing the appearance of the identical area on the fracture surface as observed in TEM and in SEM. In each set of pictures, the upper pair are SEM micrographs, and the lower pair are TEM replica micrographs. (*d*) A notched specimen of 4340 steel heat treated ½ h at 816°C (1500°F) and oil quenched, tempered 1 h at 260°C (500°F) and air cooled to give a hardness of Rockwell C 50, then broken by tension overload at room temperature. The surface is dimpled, with at least one area (at arrow 1) showing some evidence of stretching. At first glance, the TEM pair appears to bear little resemblance to the SEM, but careful scrutiny in 3D of the sites at arrows 1 and 2 establishes that they are indeed identical. The major area of departure appears to lie in the regions marked 3 in the TEM pair. (*e*) A low-cycle fatigue-test fracture in aluminum alloy 7075-T6. The fracture surface shows well-defined fatigue striations that are clearly visible in both the SEM and the TEM pairs but with more detail and contrast in the TEM. Note the excellent match of the reproductions made by the two techniques. The vertical surface at arrow 1 is quite accurately registered in the TEM fractograph, but at the other side (at arrow 2) the replica has been torn. The detail in the crevices marked by arrows 1 and 3 is more clearly displayed in the TEM pair but could be matched in the SEM if the exposure were adjusted for this purpose. The sharp line at arrow 4 in the TEM fractograph bears some resemblance to a replica tear, but it is also visible in the SEM and actually is a secondary crack in the aluminum. (*From* Metals Handbook.[11])

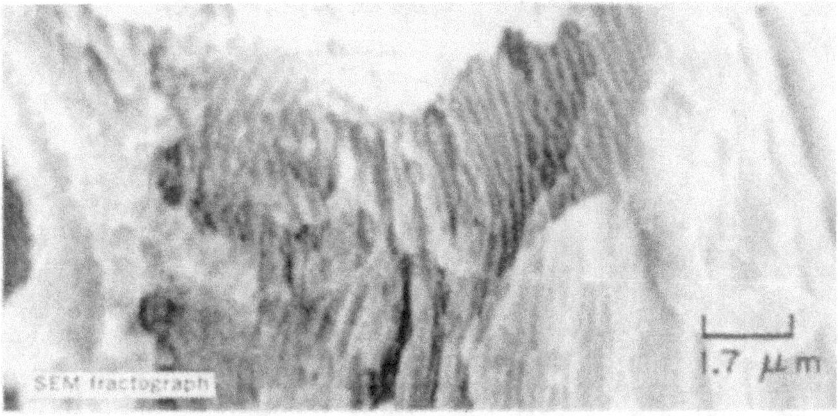

Figure 3.117 Fractograph showing what appear to be fatigue striations, but these are actually caused by fracture through the alternate lamellar layers of ferrite and iron carbide in pearlite. Note that in several local areas the pearlite lamellae have been separated by secondary cracks. (*From* Metals Handbook.[11])

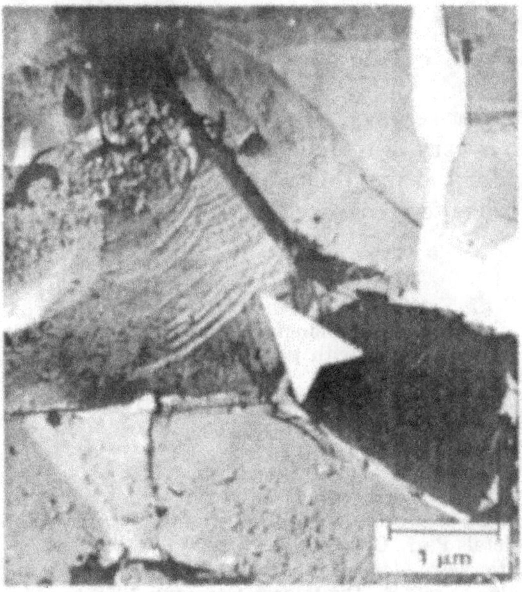

Figure 3.118 TEM replica of Wallner lines, a feature which sometimes occurs on the fracture surface of very brittle materials (such as glass). The specimen was WC–Co. Etched with 5% HCl. (*Courtesy of S. B. Luyckx, University of Witwatersrand; from Kerlins and Phillips.*[9])

in Fig. 3.122. Again, whether such an appearance is an artifact can be determined by the preparation of successive replicas. A problem in two-stage plastic replica preparation (see Sec. 1.42) is that the plastic may not be completely removed from the carbon film. The appearance of the residual plastic film in TEM is shown in Fig. 3.123. Even if the

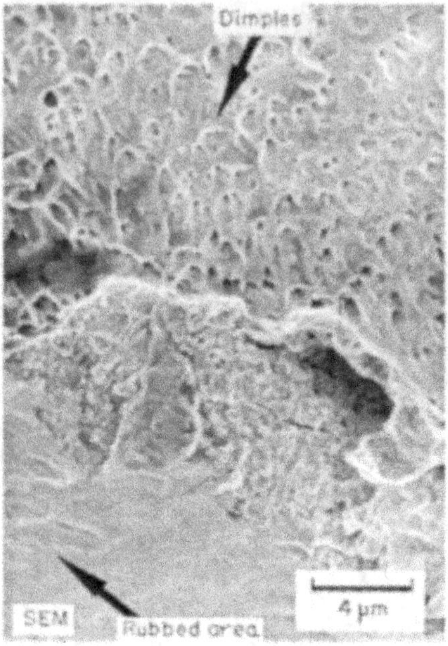

Figure 3.119 Fractograph showing smooth regions which are rub marks, due to contact of the two surfaces after fracture occurred.

replica contains no artifacts after removal from the surface to be replicated, the carbon film, which is very thin (such as 200 Å), may tear and will give an appearance similar to that illustrated in Fig. 3.124; however, this artifact is usually easy to recognize.

References

1. W. Boas and E. Schmid, "Über die Temperaturabhängigkeit der Kristallplastizität," *Z. Physik,* vol. 61, p. 767, 1930. (Reprinted with permission of Springer-Verlag, Heidelberg, Germany.)
2. A. Guy, *Introduction to Materials Science,* McGraw-Hill, New York, 1972.
3. J. D. Verhoeven, *Fundamentals of Physical Metallurgy,* Wiley, New York, 1975. (Reprinted with permission of John Wiley & Sons, Inc.)
4. C. E. Birchenall, *Physical Metallurgy,* McGraw-Hill, New York, 1959.
5. C. S. Roberts, *Magnesium and Its Alloys,* Wiley, New York, 1960.
6. C. R. Brooks, *Heat Treatment, Structure and Properties of Non-Ferrous Alloys,* American Society for Metals, Metals Park, Ohio, 1982.
7. G. Henry and D. Horstmann, *De Ferri Metallographia,* vol. V: *Fractography and Microfractography,* Verlag Stahleisen m.b.H., Düsseldorf, Germany, 1979.
8. C. D. Beachem, "Microscopic Fracture Processes," in H. Liebowitz (ed.), *Fracture,* vol. 1: *Microscopic and Macroscopic Fundamentals,* Academic Press, New York, 1968.
9. V. Kerlins and A. Phillips, "Modes of Fracture," in *Metals Handbook,* 9th ed., vol. 12: *Fractography,* American Society for Metals, Metals Park, Ohio, 1987, p. 12.
10. L. Engel and H. Klingele, *An Atlas of Metal Damage,* Carl Hanser Verlag, Munich, Germany, 1981.

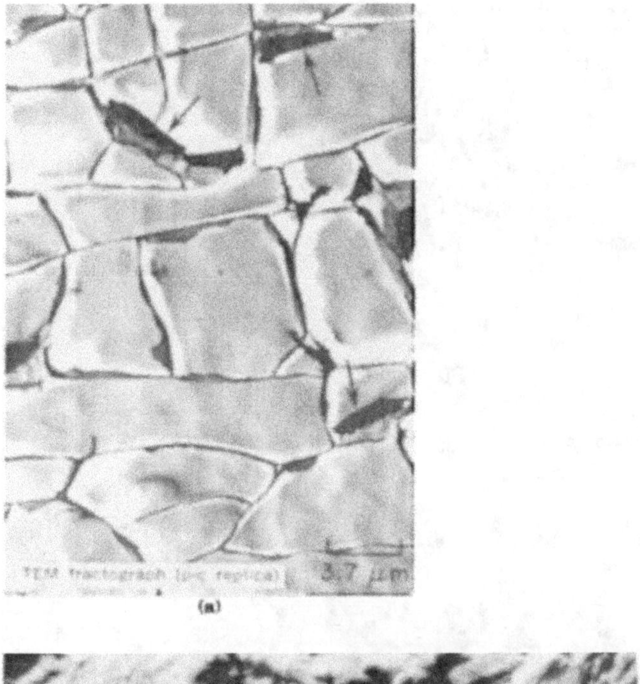

(a)

(b)

Figure 3.120 Fractographs showing mud cracks, a feature associated with corrosion products on the fracture surface. (*a*) Surface of an intergranular fracture in aluminum alloy 7079-T651 caused by stress-corrosion cracking in a $3^1/_2\%$ NaCl solution. (*From Meyn.*[42]) (*b*) Fracture surface of 316L austenitic stainless-steel surgical implant, broken in service in body fluid. (*From* Metals Handbook.[40])

(a)

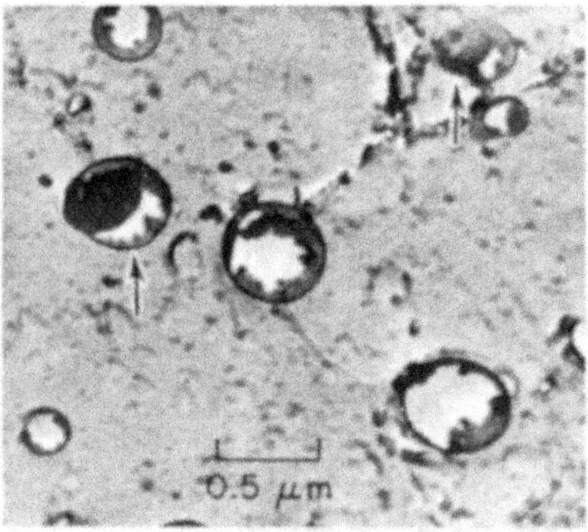

TEM fractograph (p-c replica) 20,000×

(b)

Figure 3.121 Examples of artifacts in TEM replicas. (*a*) Scraping artifact frequently encountered in the two-stage process. (*b*) The thin film of plastic between the bubble and the replicated surface is seen to be partly removed from the plastic replica at the bubbles indicated by arrows. (*From* Metals Handbook.[11,40])

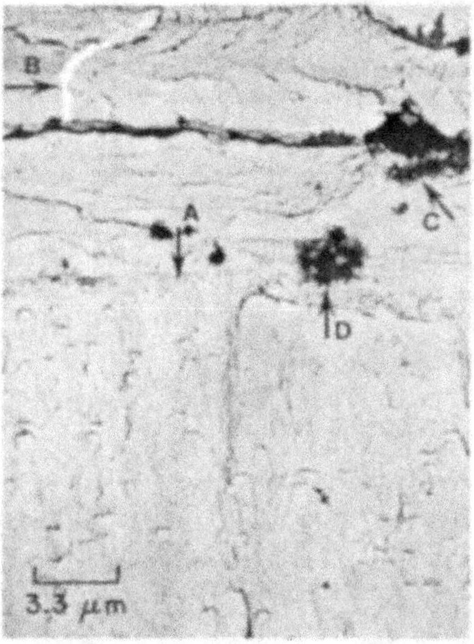

TEM fractograph (p-c replica) 3000×

(c)

Figure 3.121 (*Continued*) Examples of artifacts in TEM replicas. (*c*) The appearance of the bottom half of the fractograph (below arrow A) is the result of tears in the first-stage plastic replica, and is not at all representative of the fracture surface. A crack in the carbon second-stage replica is indicated by arrow B, a particle from the improperly cleaned fracture surface is indicated by arrow C, and residual plastic from incomplete dissolution of the first stage is indicated by arrow D. (*From* Metals Handbook.[11,40])

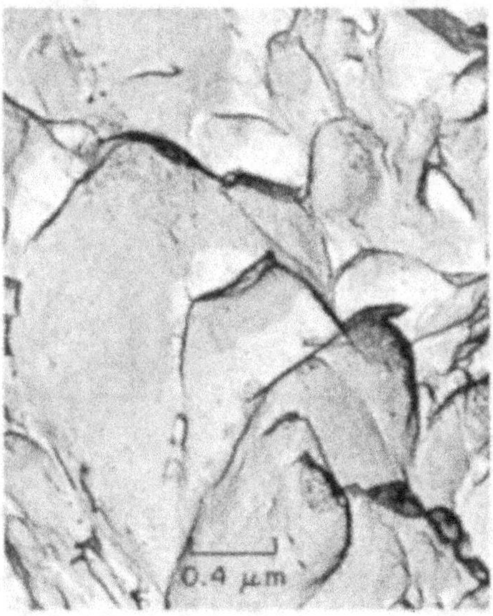

TEM fractograph (p-c replica) 22,500×

Figure 3.122 Example of reticulation of a replica caused by excessive heating in the electron beam of the TEM. This was a Pd-shadowed plastic-carbon replica of a fracture in nickel showing dimples. Reticulation was caused by melting of the shadowing metal in the microscope and the formation of globules. (*From* Metals Handbook.[40])

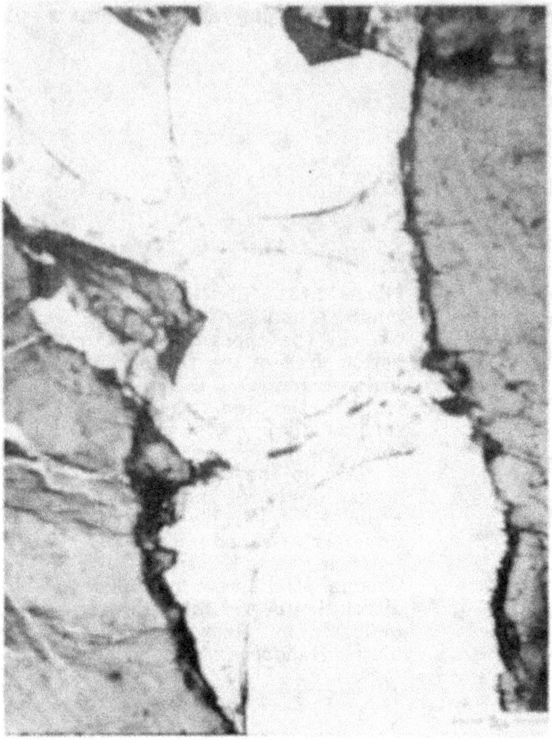

(a)

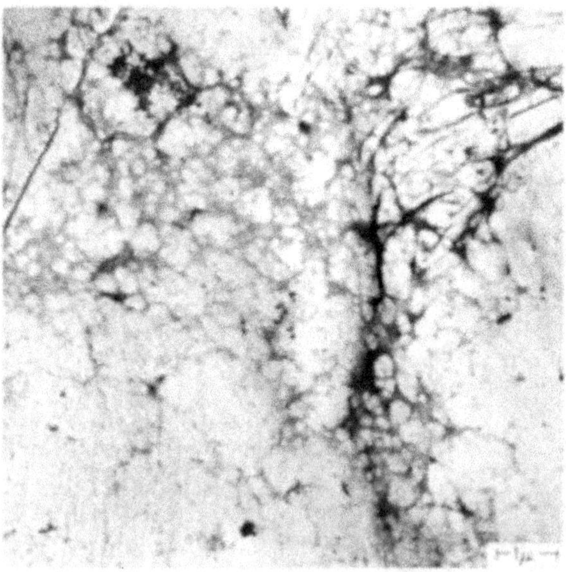

(b)

Figure 3.123 TEM replicas showing incompletely removed plastic from a carbon film. (*From Beachem.*[8])

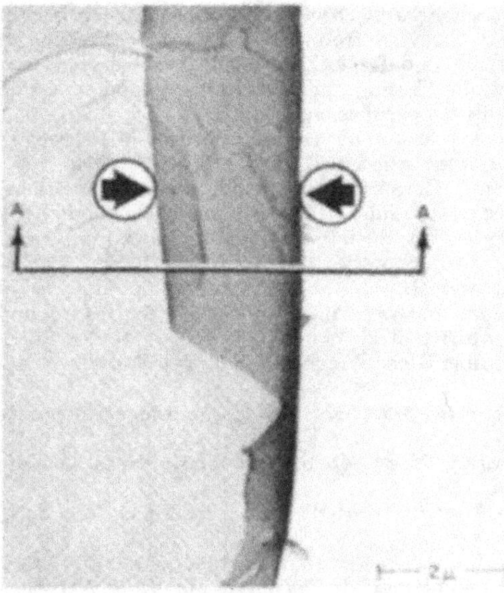

Figure 3.124 TEM replica showing the effect of curling of the carbon film. This was a direct carbon replica of a flat fracture surface (iron cleaved at dry-ice temperature). A region of gradually changing density is shown between the arrows. (*From Beachem.*[8])

11. *Metals Handbook,* 8th ed., vol. 9: *Fractography and Atlas of Fractographs,* American Society for Metals, Metals Park, Ohio, 1974.
12. D. Broek, "Some Contributions of Electron Fractography to the Theory of Fracture," *Int. Met. Rev.,* vol. 19, p. 135, 1974.
13. S. P. Lynch, "Ductile and Brittle Crack Growth: Fractography, Mechanisms and Criteria," *Mat. Forum,* vol. 11, p. 268, 1988.
14. B. L. Gabriel, "Scanning Electron Microscopy," in *Metals Handbook,* 9th ed., vol. 12: *Fractography,* American Society for Metals, Metals Park, Ohio, 1987, p. 166.
15. W. D. Syniuta and C. J. Corrow, "Scanning Electron Microscopic Studies of Fracture Mechanisms of SAE 52100 Bearing Steel," *Wear,* vol. 15, p. 171, 1970.
16. A. W. Thompson and J. C. Chesnutt, "Identification of a Fracture Mode: The Tearing Topography Surface," *Met. Trans.,* vol. 10A, p. 1193, 1979.
17. J. C. Grosskreutz, "Fatigue Mechanisms in the Sub-Creep Range," in S. S. Manson (ed.), *Metal Fatigue Damage—Mechanism, Detection, Avoidance, and Repair,* American Society for Testing and Materials, Philadelphia, Pa., 1971.
18. C. Laird, "The Influence of Metallurgical Structure on the Mechanisms of Fatigue Crack Propagation," in *Fatigue Crack Propagation,* American Society for Testing and Materials, Philadelphia, Pa., 1967.
19. H. Verhoff and P. Newmann, "In situ SEM Experiments Concerning the Mechanism of Ductile Crack Growth," *Acta Met.,* vol. 27, p. 915, 1979.
20. W. L. Haworth, V. K. Singh, and R. K. Mueller, "Holographic Detection of Fatigue-Induced Surface Deformation and Crack Growth in a High-Strength Aluminum Alloy," *Met. Trans.,* vol. 11A, p. 219, 1980.
21. C. A. Zapffe and C. O. Worden, "Fractographic Registrations of Fatigue," *Trans. ASM,* vol. 43, p. 958, 1951.
22. C. D. Beachem, "Microscopic Fatigue Fracture Surface Features in 2024-T3 Aluminum and the Influence of Crack Propagation Angle upon Their Formation," *Trans. ASM,* vol. 60, p. 324, 1967.
23. P. N. Thielen and M. E. Fine, "Fatigue Crack Propagation in 4140 Steel," *Met. Trans.,* vol. 6A, p. 2133, 1975.
24. H. Böhm, *Einführung in die Metallkunde,* Bibliogr. Institut, Mannheim, Germany, 1968.

25. D. A. Meyn and E. J. Brooks, "Microstructural Origin of Flutes and Their Use in Distinguishing Striationless Fatigue Cleavage from Stress-Corrosion Cracking in Titanium Alloys," in L. N. Gilbertson and R. D. Zipp (eds.), *Fractography and Materials Science,* American Society for Testing and Materials, Philadelphia, Pa., 1981. (Copyright © ASTM; reprinted with permission.)
26. C. R. Brooks and B. L. McGill, "The Application of Scanning Electron Microscopy to Fractography," *Microstructural Characterization,* vol. 33, pp. 195–243, 1994.
27. E. K. Beauchamp, cited in R. W. Rice, "Ceramic Fracture Features, Observations, Mechanisms and Uses," in J. J. Mecholsky and S. R. Powell (eds.), *Fractography of Ceramic and Metal Failures,* American Society for Testing and Materials, Philadelphia, Pa., 1984, pp. 5–103.
28. J. J. Mecholsky, S. W. Freiman, and R. W. Rice, "Fractographic Analysis of Ceramics," in *Fractography in Failure Analysis,* American Society for Testing and Materials, Philadelphia, Pa., 1978, pp. 363–379.
29. L. Orr, "Practical Analysis of Fracture in Glass Windows," *Mat. Res. Stand.,* vol. 12, pp. 21–23, 1972.
30. R. Cotterill, *The Cambridge Guide to the Materials World,* Cambridge University Press, New York, 1985.
31. C. W. Bunn, *Chemical Crystallography,* 2d ed., Oxford University Press, London, 1961.
32. P. J. Flory, *Principles of Polymer Chemistry,* Cornell University Press, New York, 1953.
33. P. H. Geil, *Polymer Single Crystals,* Interscience, New York, 1963.
34. E. S. Clark, "Structure of Crystalline Polymer," in *Polymeric Materials,* American Society for Metals, Metals Park, Ohio, 1975, pp. 1–54.
35. J. Schultz, *Polymer Materials Science,* Prentice-Hall, Englewood Cliffs, N.J., 1992.
36. K. Friedrich, in D. M. R. Raplin (ed.), *Fracture,* vol. 3, Pergamon Press, New York, 1977, pp. 1119–1126.
37. D. Hull, "Nucleation and Propagation Processes in Fracture," in *Polymeric Materials,* American Society for Metals, Metals Park, Ohio, 1975, pp. 487–550.
38. R. W. Hertzberg, "Fracture Surface Micromorphology in Engineering Solids," in J. E. Masters and J. J. Au (eds.), *Fractography of Modern Engineering Materials,* American Society for Testing and Materials, Philadelphia, Pa., 1987, pp. 5–36.
39. C. M. Rimnac, R. W. Hertzberg, and J. A. Manson, "Fatigue Fracture Surface Micromorphology in Poly (Vinyl Chloride)," in L. N. Gilbertson and R. D. Zipp (eds.), *Fractography and Materials Science,* American Society for Testing and Materials, Philadelphia, Pa., 1981, pp. 291–313.
40. *Metals Handbook,* 9th ed., vol. 12: *Fractography,* ASM International, Metals Park, Ohio, 1987.
41. A. Maillard, L. Meny, and M. Champigny, "Comparaison de Microfractographies Types Obtenues par Microscopie à Balayage et par Microscopie Conventionnelle," *Micron,* vol. 2, p. 290, 1971.
42. D. A. Meyn, "Fractographic Diagnosis of Stress Corrosion Cracking of Al–Zn–Mg Alloys," *Corrosion,* vol. 26, p. 427, 1970.

Bibliography

Among the books and articles that cover microfractography, the following have been found to be especially useful. From these, and from the references cited in this chapter, other references can be traced.

Bhattacharyya, E. S., V. E. Johnson, S. Agarwal, and M. A. H. Howes (eds.): *IITRI Fracture Handbook: Failure Analysis of Metallic Materials by Scanning Electron Microscopy,* IIT Research Institute, Chicago, Ill., 1979.

Brostow, W., and R. D. Corneliussen (eds.): *Failure of Plastics,* Hanser Publisher, Munich, Germany, 1989.

Engel, L., and H. Klingele: *An Atlas of Polymer Damage,* Carl Hanser Verlag, Munich, Germany, 1981.

Gilbertson, L. N., and R. D. Zipp (eds.): *Fractography and Materials Science,* STP 733, American Society for Testing and Materials, Philadelphia, Pa., 1981.

Hertzberg, R. W.: "Fracture Surface Micromorphology in Engineering Solids," in J. E. Masters and J. J. Au (eds.), *Fractography of Modern Engineering Materials: Composites and Metals,* STP 948, American Society for Testing and Materials, Philadelphia, Pa., 1987, pp. 5–36.

Jacoby, G.: "Application of Microfractography to the Study of Crack Propagation under Fatigue Stresses," AGARD Rep. 541, North Atlantic Treaty Organization, London, 1966.

Lange, G. A. (ed.): *Systematic Analysis of Technical Failures,* DGM Informationsgesellschaft-Verlag, Braunschweig, Germany, 1986.

Louthan, M. R., Jr., and T. A. Place (eds.): *Microscopy, Fractography and Failure Analysis,* Failure Analysis and Prevention Lab., Virginia Polytechic Institute, Blacksburg, Va., 1986.

Mecholsky, J. J., and S. R. Powell (eds.): *Fractography of Ceramic and Metal Failures,* American Society for Testing and Materials, Philadelphia, Pa., 1984.

Phol, M.: "Material Failure through Wear," in G. A. Lange (ed.), *Systematic Analysis of Technical Failures,* DGM Informationsgesellschaft-Verlag, Braunschweig, Germany, 1986.

Rice, R. C. (ed.): *Fatigue Design Handbook,* 2d ed., Society of Automotive Engineers, Warrendale, Pa., 1988.

Ryder, D. A.: "The Elements of Fractography," AGARD Rep. 155, North Atlantic Treaty Organization, London, 1971.

Tucker, R. C., Jr.: "Wear Failures," in *Metals Handbook,* 9th ed., vol. 11: *Failure Analysis and Prevention,* American Society for Metals, Metals Park, Ohio, 1986.

Whiteson, R. V., A. Phillips, and V. Kerlins: "Electron Fractographic Techniques," in R. F. Bunshah (ed.), *Techniques for the Direct Observation of Structure and Imperfections,* Interscience, New York, 1968, pp. 445–497.

Wulpi, D. J.: *How Components Fail,* American Society for Metals, Metals Park, Ohio, 1985.

Chapter

4

Fracture Modes and Macrofractographic Features

Persian SIGLOS, fifth century B.C., *showing impression of fractured bronze (?) punch (4×).*

C. S. SMITH

A History of Metallography,

University of Chicago Press, 1960

4.1 Introduction

When the fracture surface of a broken component is examined with the unaided eye or at low magnification (such as <20×), distinct topographical features are usually seen. Deduction of the mode of fracture based on this macroscopic examination relies on the correlation of characteristic features with known fracture conditions. Fortunately there exists a wealth of experimental observations of the fracture surfaces of test samples fractured under known loading conditions, and in most cases the topographical features are quite unique for each type of loading condition. The orientation of the fracture surface of a broken component often reveals unambiguously the loading condition which led to fracture, and this was examined in Chap. 2. The microscopic process whereby metallic materials fracture was called mechanism (or micromechanism), and the associated fracture-surface fine-scale topography was the subject of Chap. 3.

In Chap. 4 the types of fracture mode and the associated macroscopic fracture-surface topographical features are examined. The term *fracture mode* is used to indicate the type of external loading that causes fracture; this could be called fracture macromechanism. The simplest modes are described first (such as simple tensile loading), then more complex cases are examined. We conclude the chapter by illustrating the microfractography of test samples of known fracture modes which have specific macrofractographic appearances.

4.2 Tensile Overload

In the tensile overload mode it is assumed that the external load is uniaxial, and that it is monotonically increasing until the material begins to neck. For samples tested in a conventional tensile tester, the test is conducted at approximately a constant extension rate, so that after necking the load actually decreases until fracture (see Sec. 2.2).

The fracture-surface appearance depends on the type of material, its microstructure, and the testing conditions (such as test temperature and strain rate). Under tensile loading, the material may show extensive plastic deformation prior to fracture, in which case it is considered to be ductile (see Sec. 2.11). If little plastic deformation occurs, the fracture is called brittle. However, the amount of plastic deformation which separates these two categories is somewhat arbitrary, although less than 5 percent elongation at fracture in a tensile test is a convenient value to consider a material brittle (see Sec. 2.11).

First we consider a ductile overload mode. In this case there is extensive elongation at fracture, and the cross section will be decreased in the vicinity of the fracture. In a cylindrical test sample, this usually results in the fracture appearance shown in Fig. 4.1. There is a central rough region whose overall plane is normal to the loading axis. On the periphery is a surface inclined at 45° to this plane, called a *shear lip*. This type of fracture is called *cup-and-cone fracture*.

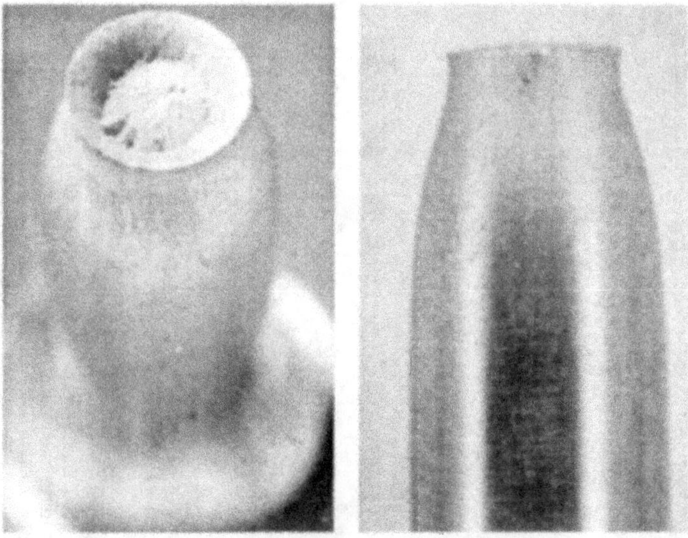

Figure 4.1 Broken tensile sample showing ductile fracture with the common cup-and-cone appearance. (*After Vander Voort.*[1])

The origin of such a macroscopic geometry was discussed in Sec. 2.8. It was pointed out there that in the tensile test, once necking begins, the triaxial stress which develops favors plastic deformation by shear in the center of the plane of the minimum cross section (at the neck location). Thus the examination of cross sections on the axial plane of such samples loaded to near fracture shows that fracture initiates in the center of the cylinder (Fig. 4.2). When the cracks develop sufficiently, the intense triaxiality at the periphery of the internal crack causes the development of large shear stresses at 45° in the outer ring of connecting material, and this region then fractures along a general plane at this angle, forming the shear-lip region. Tensile test samples of rectangular cross section which fracture in a ductile manner also show similar characteristics (Fig. 4.3).

The fracture surface of such necked samples usually looks similar to that in Fig. 4.4. In the center there is a *fibrous zone,* so called because its appearance is similar to that of a broken fiber structure. Emanating from the outer periphery of this zone are somewhat coarser markings, called *radial marks.* These terminate at the outer circumferential zone containing the shear lip. The extent of each zone depends on the material, its structure, and the test conditions. Examples are shown in Fig. 4.5. In general, as the strength increases, the amount of necking is less, and there is a reduction in the extent of the shear-lip zone. Also, the central fibrous zone covers less area and the radial marks are less coarse. These effects are illustrated in Fig. 4.6. Similar effects are seen

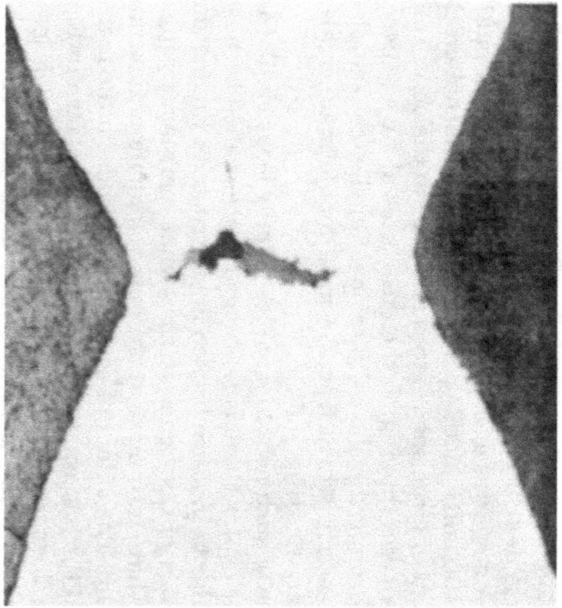

Figure 4.2 Cross section through a tensile sample just prior to fracture. Note that the fracture crack has initiated in the center of the sample. (*After Vander Voort.*[1])

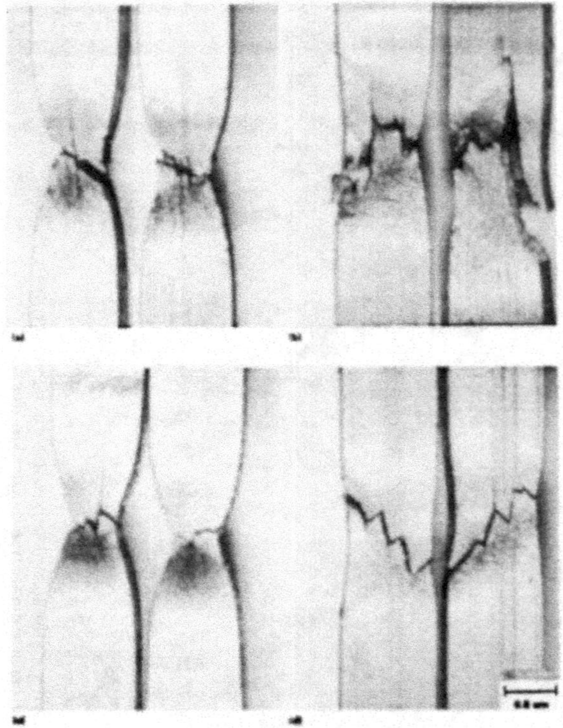

Figure 4.3 Broken rectangular steel tensile samples showing necking. (*a*) Longitudinal specimen finished rolled at 315°C (600°F). (*b*) Transverse specimen. (*c*) Longitudinal specimen finished rolled at 150°C (300°F). (*d*) Transverse specimen. (*From B. L. Bramfitt and A. R. Marder,* Met. Trans., *vol. 8A, p. 1262, 1977; after Vander Voort.*[1])

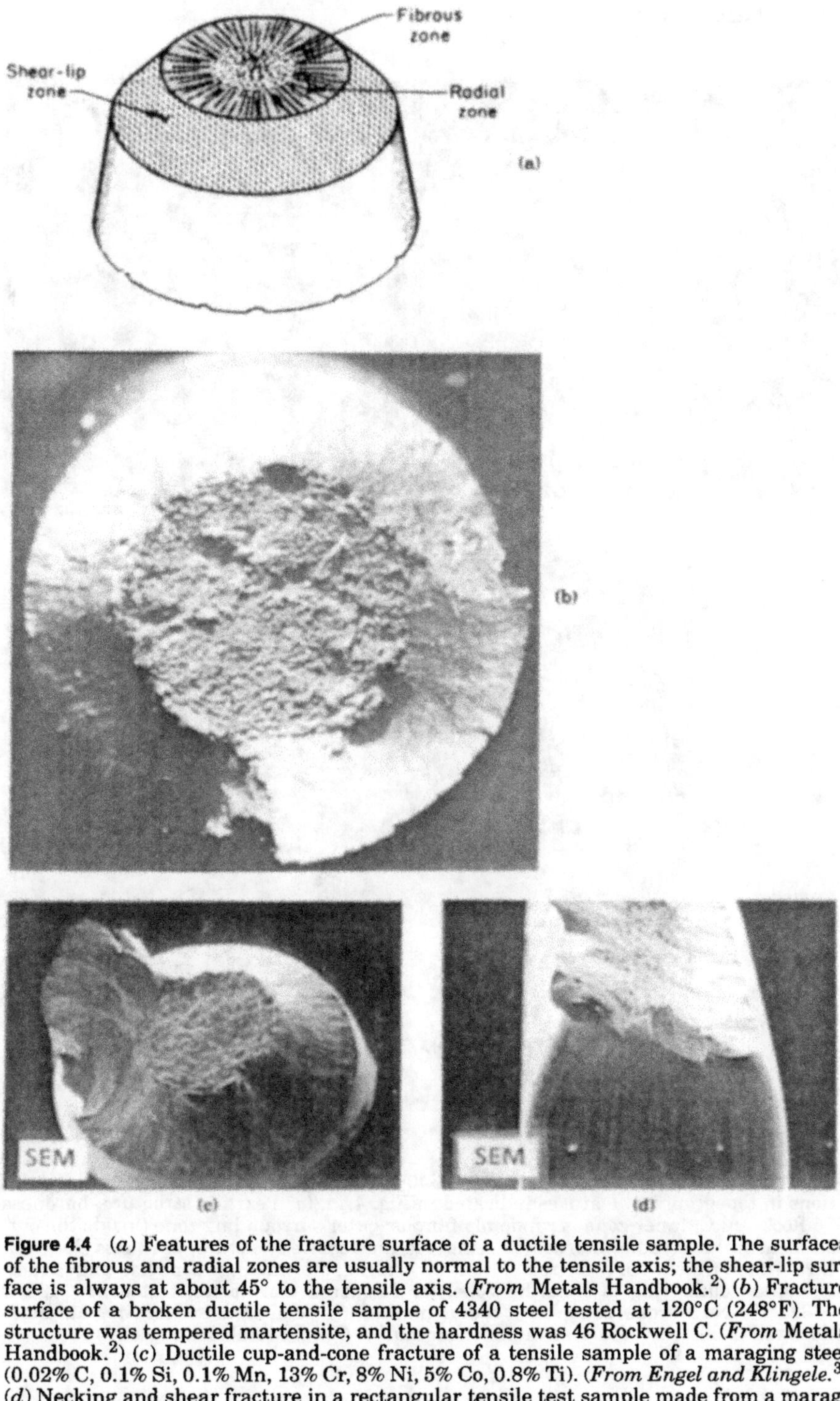

Figure 4.4 (*a*) Features of the fracture surface of a ductile tensile sample. The surfaces of the fibrous and radial zones are usually normal to the tensile axis; the shear-lip surface is always at about 45° to the tensile axis. (*From* Metals Handbook.[2]) (*b*) Fracture surface of a broken ductile tensile sample of 4340 steel tested at 120°C (248°F). The structure was tempered martensite, and the hardness was 46 Rockwell C. (*From* Metals Handbook.[2]) (*c*) Ductile cup-and-cone fracture of a tensile sample of a maraging steel (0.02% C, 0.1% Si, 0.1% Mn, 13% Cr, 8% Ni, 5% Co, 0.8% Ti). (*From Engel and Klingele.*[3]) (*d*) Necking and shear fracture in a rectangular tensile test sample made from a maraging NiCoMo-containing steel. (*From Engel and Klingele.*[3])

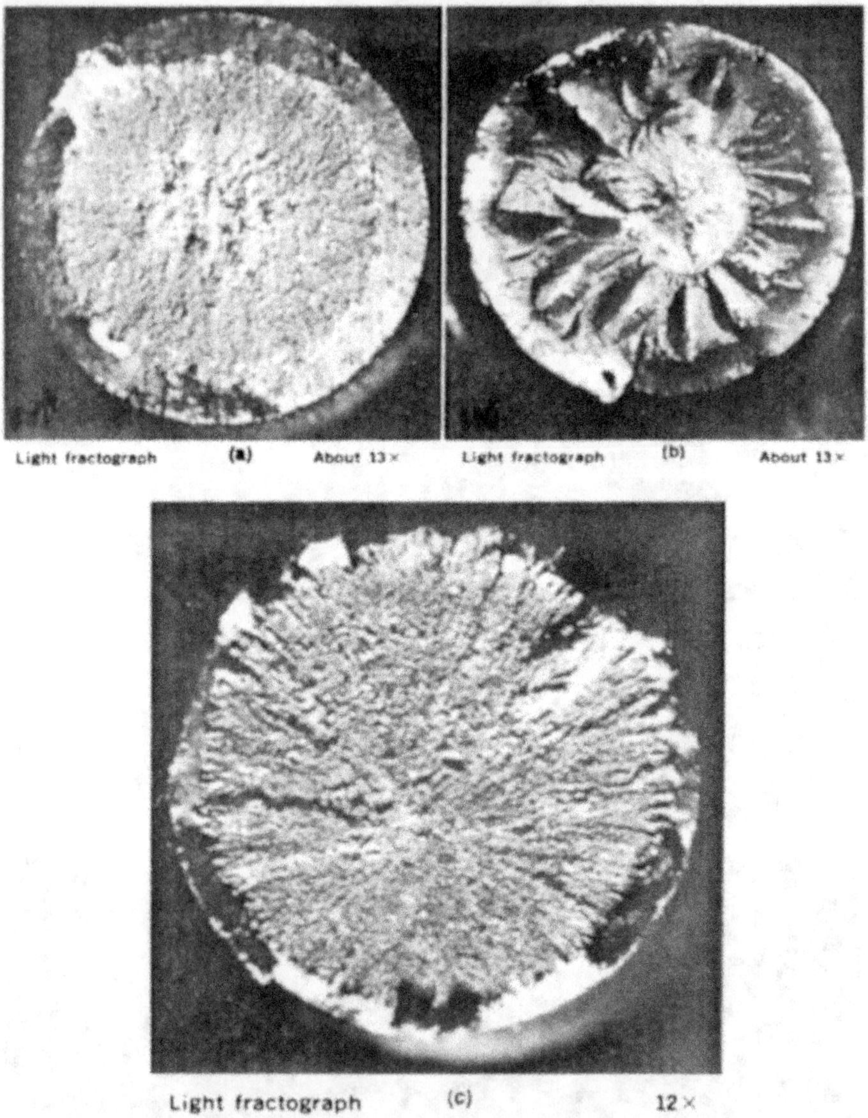

Figure 4.5 Fracture surfaces of broken 4340 steel tensile samples, showing the variations in topographical features indicated in Fig. 4.4*a*. (*a*) Pearlitic structure, hardness 15 Rockwell C. Inner zone is randomly fibrous; outer surrounding zone (inside the outer shear lip) has radial marks. (*b*) Tempered martensite structure, hardness Rockwell C 28. Inner, fibrous zone is circumferentially ridged; intermediate zone has coarse radial marks; outer ring is shear-lip zone. (*c*) Tempered martensite, tested at −196°C (−321°F). No fibrous zone, only radial marks. (*d*) Tempered martensite, hardness Rockwell C 28. Curved coarse radial shear marks. (*From* Metals Handbook.[2])

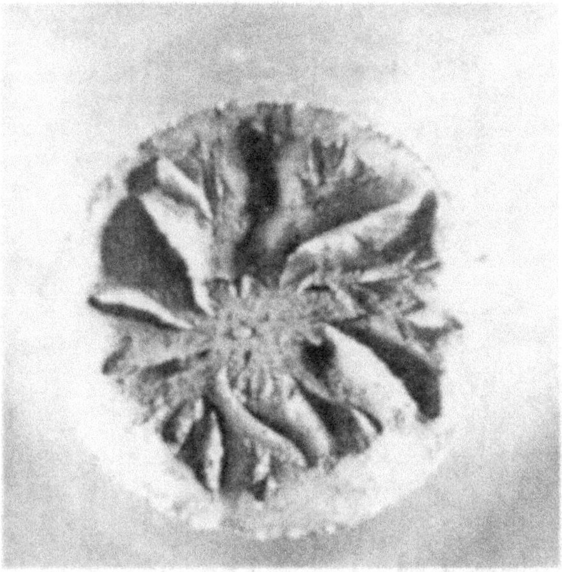

Light fractograph (d) 14×

Figure 4.5 *(Continued)*

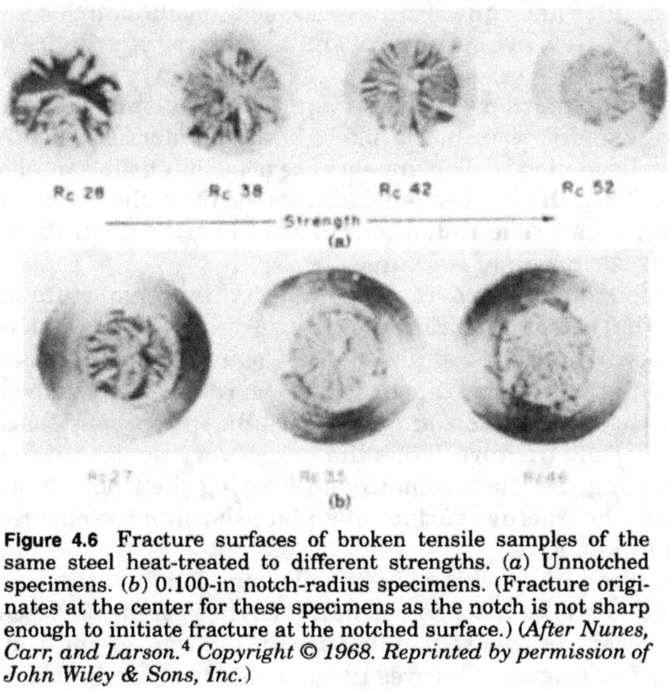

Figure 4.6 Fracture surfaces of broken tensile samples of the same steel heat-treated to different strengths. (*a*) Unnotched specimens. (*b*) 0.100-in notch-radius specimens. (Fracture originates at the center for these specimens as the notch is not sharp enough to initiate fracture at the notched surface.) (*After Nunes, Carr, and Larson.*[4] *Copyright © 1968. Reprinted by permission of John Wiley & Sons, Inc.*)

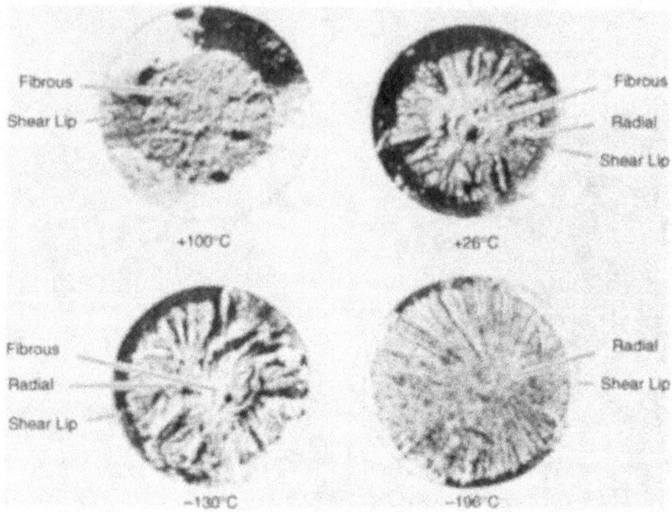

Figure 4.7 Fracture surfaces of tensile samples of the same material broken at different temperatures. (*From Nunes, Carr, and Larson.*[4] *Copyright © 1968. Reprinted by permission of John Wiley & Sons, Inc.*)

as the test temperature decreases, as shown in Fig. 4.7. The general features for rectangular and square cross-sectioned tensile samples are presented in Fig. 4.8; they are seen to be similar to those of cylindrical samples.

If the sample is brittle, there is little macroscopic plastic deformation and fracture usually occurs on a plane which is macroscopically normal to the tensile load axis. The appearance of such tensile samples is shown in Fig. 4.9. On the fracture surface there is little shear lip and the fibrous zone is small. The radial marks are finer, but their direction still indicates the origin of fracture.

The macroscopic fracture topography is affected by constraints on the sample. As the width-to-thickness ratio increases, the relative amount of the fracture surface containing the shear-lip zone increases also (Fig. 4.10). Of special importance is that the radial marks develop into a pattern called *chevron marks* (Fig. 4.8), which point back toward the origin of the fracture. Thus they are a very important feature in reconstructing the fracture path and locating the fracture origin. An example of the fracture surface of a plate showing the chevron pattern is given in Fig. 4.11.

Constraints induced by notching the tensile sample affect the fracture topography. Due to the intense triaxiality at the root of the notch, the cracks will initiate in this region and not in the center, and then progress inward. The fracture features usually observed are indicated

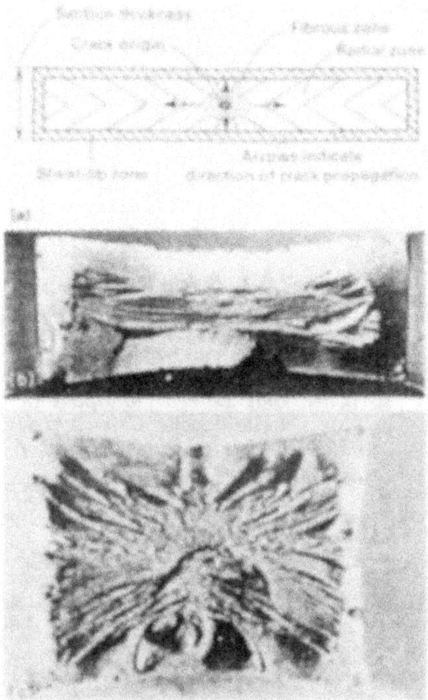

Figure 4.8 Fracture surfaces of broken rectangular tensile samples. (*a*) Schematic representation of tensile fracture features. (*b*) and (*c*) show that the features are affected by the width-to-length ratio; (*c*) shows a much smaller shear lip zone. (*From Nunes, Carr, and Larson*[4]; *adapted in* Metals Handbook.[5] *Copyright © 1968. Reprinted by permission of John Wiley & Sons, Inc.*)

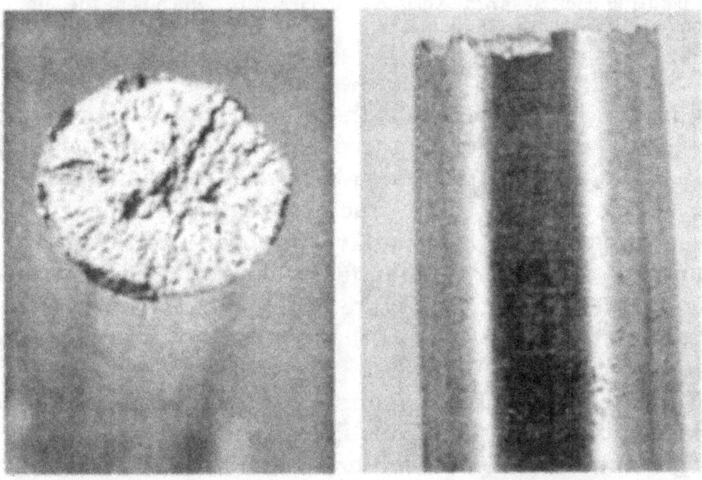

Figure 4.9 Broken tensile sample showing brittle fracture. (*After Vander Voort.*[1])

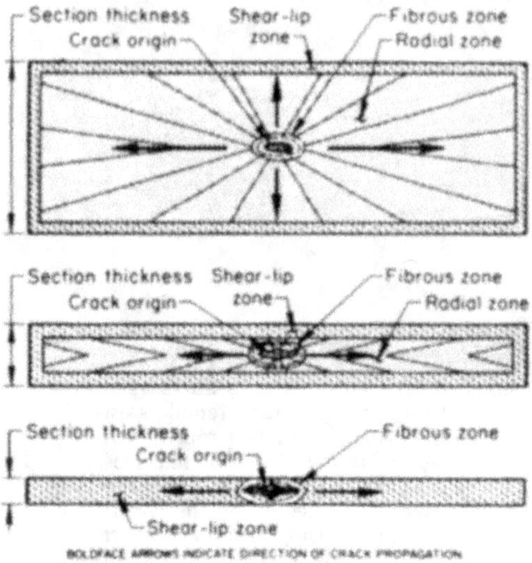

Figure 4.10 Schematic illustration of the fracture-surface features of plate samples of different thicknesses broken in tensile loading. (*From Nunes, Carr, and Larsen*[4]; *adapted in* Metals Handbook.[5] *Copyright © 1968. Reprinted by permission of John Wiley & Sons, Inc.*)

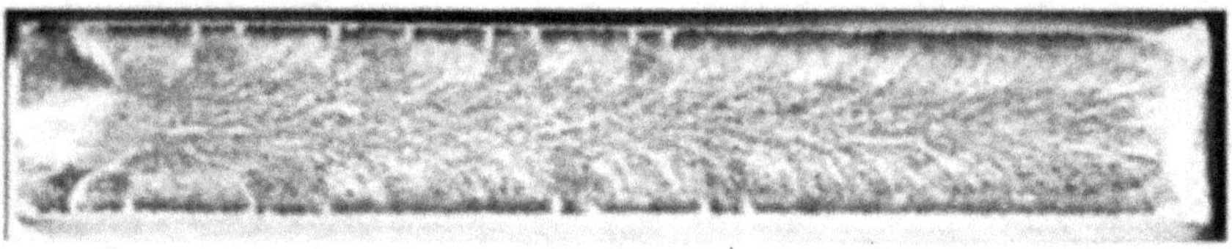

Figure 4.11 Fracture surface of a broken plate which shows radial marks in a chevron pattern, pointing back toward the origin of the fracture, which is at the left end of the specimen. (*From* Metals Handbook.[2])

schematically in Fig. 4.12*a*. Note that the fibrous zone is on the outside, compared to the case of an unnotched sample (Fig. 4.4), and there is no shear lip. The radial marks emanate from the periphery of the fibrous zone, in toward the final fracture region at the center. However, these radial marks are similar in appearance to those for unnotched samples (Fig. 4.4). Similar features are found in tensile tests of notched plates (Fig. 4.13). Figure 4.14 shows the effect of the test temperature on the appearance of the fracture surface of notched tensile samples.

Figure 4.15 illustrates two sets of bolts broken by tensile overload. The bolt on the right in Fig. 4.15*a* is ductile, showing necking and considerable elongation; the one on the left broke in a brittle manner with little elongation. In Fig. 4.15*b*, the bolt on the left has a cup-and-cone fracture surface, and the distance between the threads increases as the fracture surface is approached, showing increased axial elongation. The bolt on the right in Fig. 4.15*b* broke in a brittle manner with

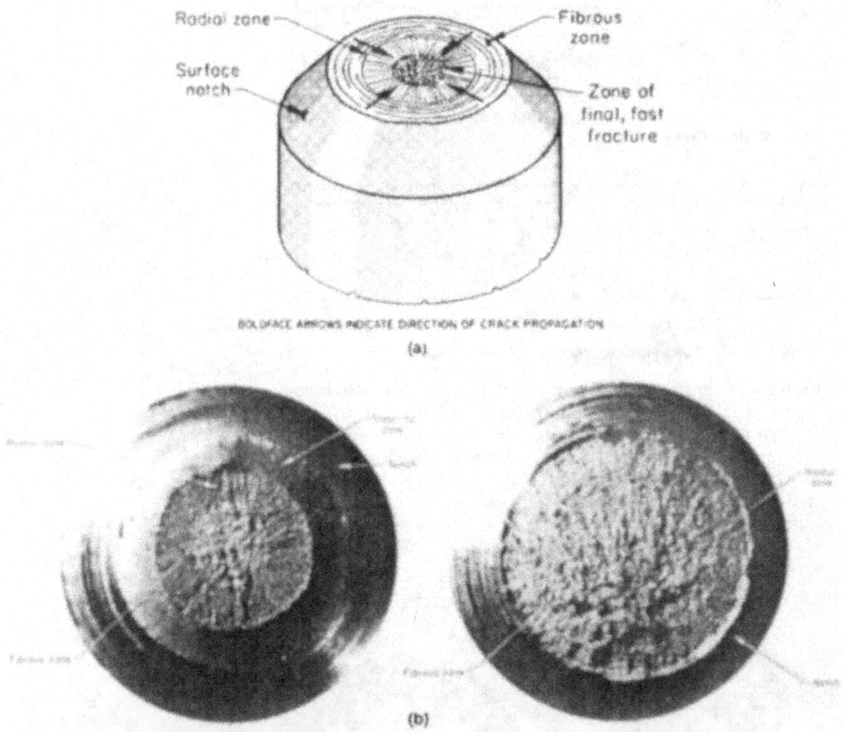

Figure 4.12 (*a*) Schematic illustration of the fracture surface zones of a notched tensile specimen. There is no shear lip, and the final fracture zone is in the center. (Compare to Fig. 4.4*a*.) (*b*) Fracture surfaces of notched tensile samples showing the effect of notch radius on the fracture appearance. The material is 4340 steel, tested at −40°C (−40°F). Left: notch-root radius 0.1 in, tensile strength 1544 MPa (224 ksi). Right: notch-root radius 0.01 in, tensile strength 1758 Mpa (255 ksi). (*From* Metals Handbook.[2])

no obvious macroscopic dimensional change, and the fracture plane is relatively smooth and normal to the tensile load axis.

4.3 Torsion Overload

The stress state associated with pure torsion was described in Chap. 2. In cylinders the maximum normal stress is at 45° to the axis, and the maximum shear stress is at 90°. Fracture can occur in a ductile or in a brittle manner. If the material fails in a ductile fashion, then obvious plastic deformation will be detected. For example, if the cylinder has axial marks on the surface (such as drawing lines), then these will be spiraled around the surface. A clear example is shown in Fig. 4.16. A case showing twisted splines which reveal torsion fracture is illustrated in Fig. 4.17. If this effect is not revealed by such surface markings, it sometimes can be revealed by surface etching, as shown in Fig. 4.18.

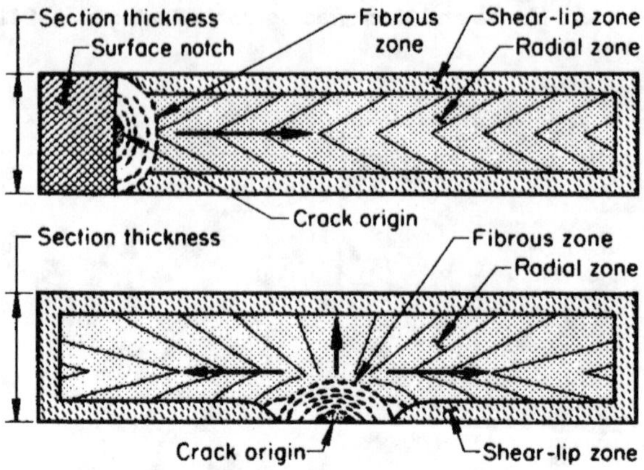

Figure 4.13 Schematic illustration of the fracture-surface features for notched rectangular tensile samples. (Compare to Fig. 4.10.) A fibrous zone is generated from the notch and extends for a short distance. A radial zone then supplants the fibrous zone, and a shear-lip zone is formed wherever the radial zone approaches the edge of the specimen. (*From Nunes, Carr, and Larson*[4]; *adapted in* Metals Handbook.[5] *Copyright © 1968. Reprinted by permission of John Wiley & Sons, Inc.*)

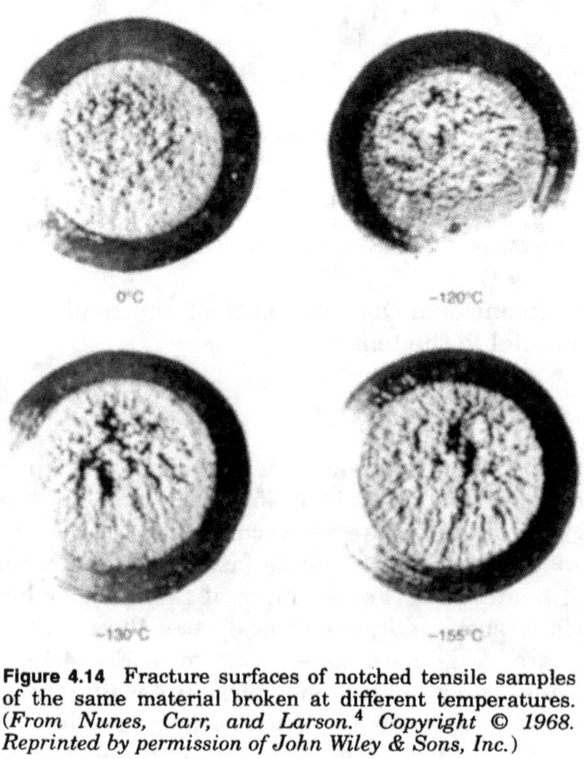

Figure 4.14 Fracture surfaces of notched tensile samples of the same material broken at different temperatures. (*From Nunes, Carr, and Larson.*[4] *Copyright © 1968. Reprinted by permission of John Wiley & Sons, Inc.*)

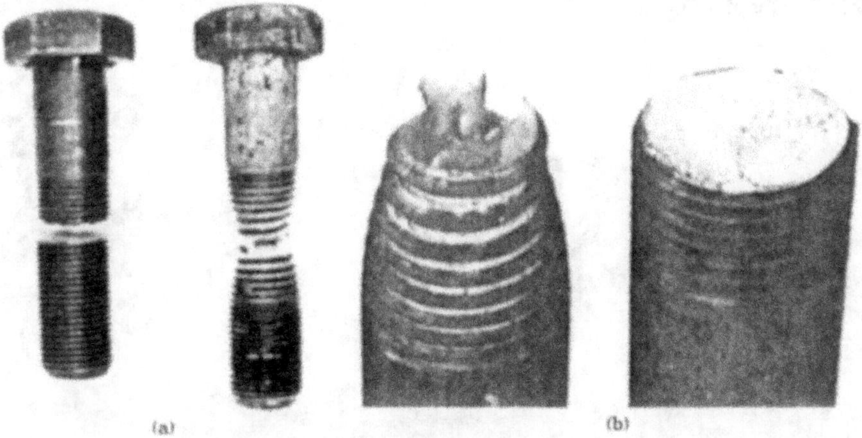

Figure 4.15 Steel bolts broken in tensile overload. (*a*) Brittle versus ductile fracture. The bolt on the left was water-quenched with a hardness of 47 Rockwell C and has no obvious deformation. The bolt on the right was annealed to a hardness of 95 Rockwell B (equivalent of 15 Rockwell C) and shows extensive permanent deformation. (*b*) Two identical steel bolts given different heat treatments. The annealed bolt on the left shows extensive necking and a cup-and-cone fracture. The austenitized then brine-quenched bolt on the right shows a relatively flat fracture surface, normal to the loading axis, and no obvious plastic deformation. (*From Wulpi.*[6])

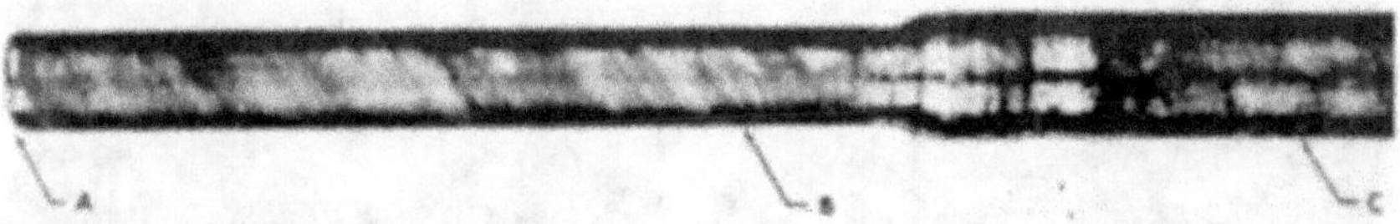

Figure 4.16 Surface of a shaft which broke in torsion overload. The fracture surface at point *A* is perpendicular to the page. Between *A* and *B* helical lines are seen, which are torsion deformation bands. (*From* Metals Handbook.[7])

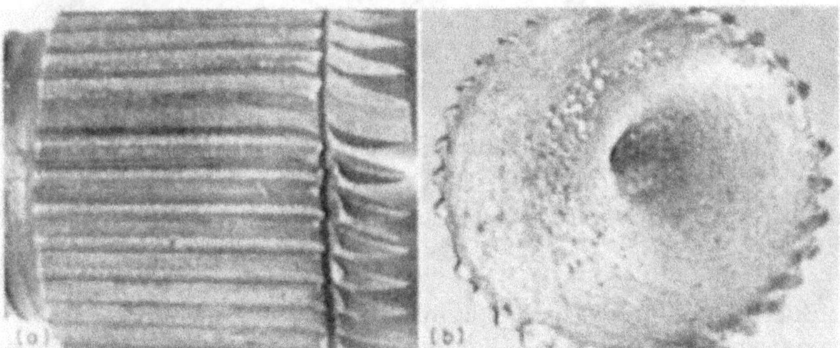

Figure 4.17 Splined shaft which broke in torsion overload. The 6118 steel shaft had a hardness of 23 Rockwell C. (*a*) The splines on the left were constrained in the holder, whereas the part on the right was free to twist, showing that the fracture was ductile. (*b*) The fracture surface shows a swirl pattern, characteristic of torsion, and is normal to the shaft axis. If there is combined bending, the region of final fast fracture will be off center. (*From* Metals Handbook.[2])

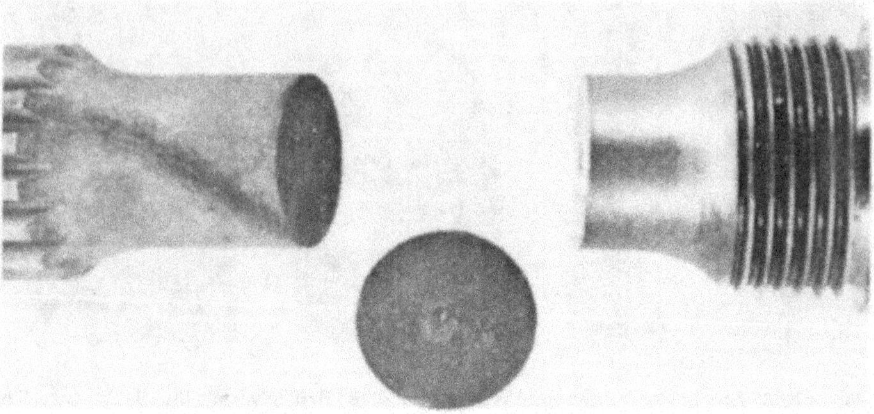

(a)

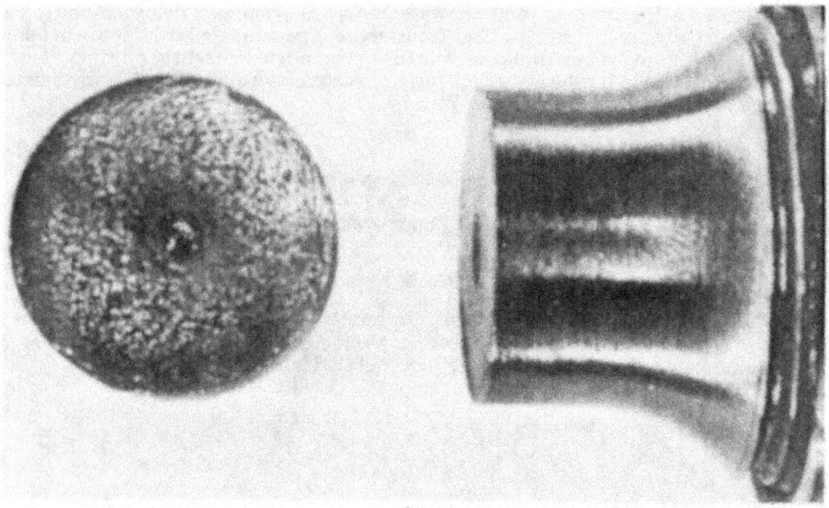

(b)

Figure 4.18 (*a*) Shaft that broke in torsion overload. (*b*) Higher magnification view of (*a*). 1035 steel drive shaft. Hardness of Rockwell C 34. Note that the fracture surface is flat and perpendicular to the shaft axis and that it shows a swirl pattern. The dark spiral on the shaft in (*a*) was revealed by etching, and shows that the shaft was subjected to plastic deformation by torsion. (*From* Metals Handbook.[2])

In ductile torsion overload fracture, the fracture surface is at 90° to the axis, as shown in Figs. 4.17 and 4.18. The surface appearance usually shows a "swirl" pattern (Fig. 4.17), but such a pattern can be made by the surfaces rubbing after fracture, and hence is not necessarily indicative of torsional overload fracture.

If the material is brittle, then it is expected that the fracture will be along planes sustaining the maximum normal stress, which is at 45° to the cylinder axis (see Fig. 2.12). Figure 4.19 shows a sketch of the fracture surface of a rod of chalk which has been broken in pure torsion. The fracture surface is a spiral at 45° to the axis. Figure 4.20 shows the fracture surface of a hardened steel rod which fractured in a brittle manner in torsion overload. The cracks on the surface of a shaft which fractured in a brittle manner are shown in Fig. 4.21; the cracks are at 45° to the axis and hence normal to the maximum normal stress.

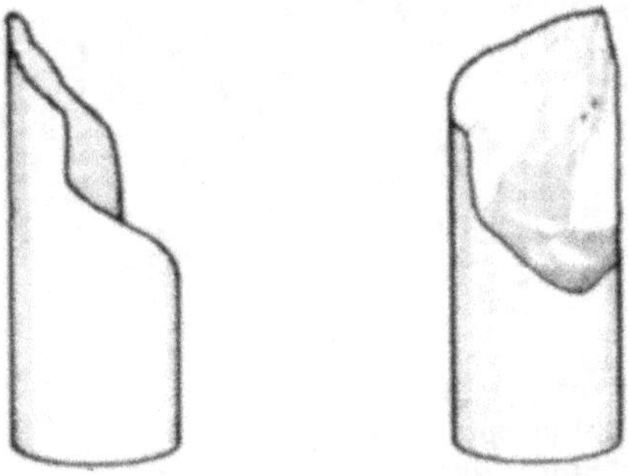

Figure 4.19 Fracture appearance of a piece of brittle chalk after breaking it in torsion overload. (*From* Metals Handbook.[2])

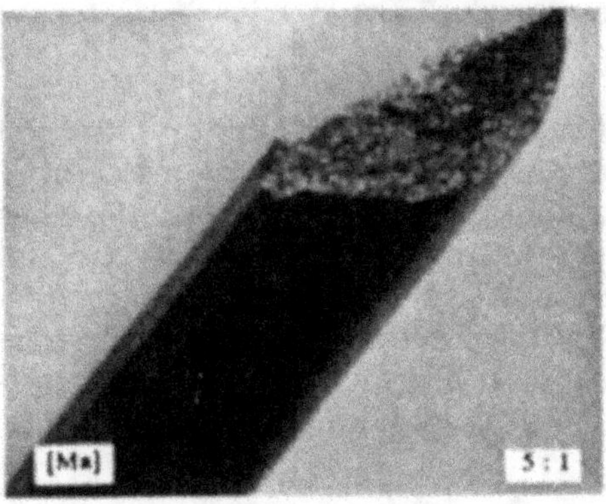

Figure 4.20 Shaft broken in torsion overload, showing brittle fracture. The steel contained 0.8% C, 0.15% Si, 0.60% Mn, 0.021% P, 0.032% S, 0.03% Ni, and 0.03% Cr and was quenched and tempered. (*From Henry and Horstmann.*[8])

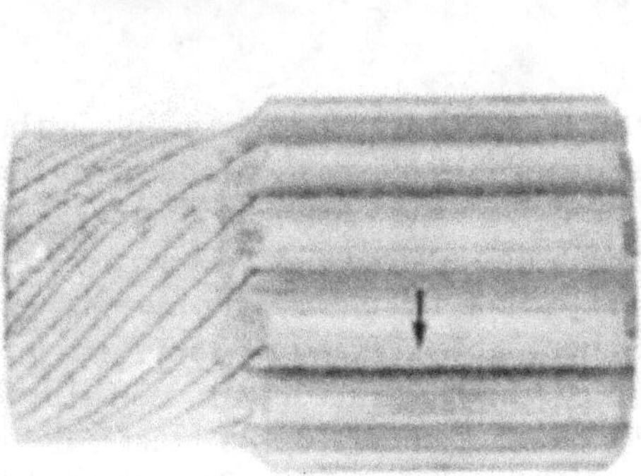

Figure 4.21 Case-hardened steel shaft which failed in torsion in a brittle manner. The arrow shows the direction that this end was twisted. The black lines at 45° are cracks, which show that the fracture is brittle (see Fig. 2.12). (*From Wulpi.*[6])

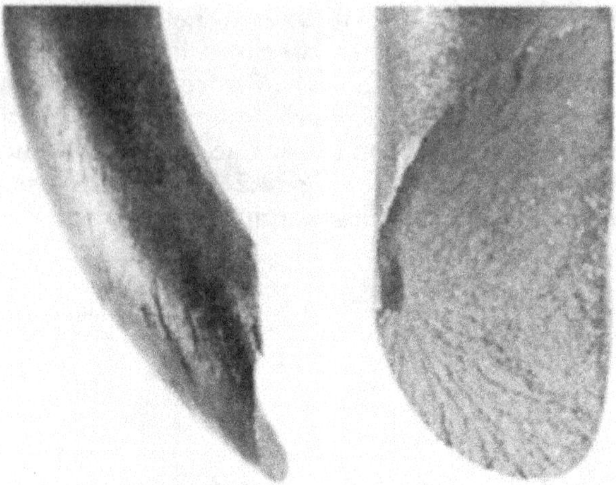

(a) (b)

Figure 4.22 Spiral spring which broke in a brittle manner. (*a*) Fractured 13-mm-diameter spring of AISI 10B62 steel wire with a hardness of 477 Brinell. Note the spiral gouges, which are screw marks that were generated during coiling. (*b*) Fracture surface shows a small fatigue crack which originated at a screw mark, but only grew slightly before torsion overload fracture occurred. (*From* Metals Handbook.[5])

In brittle fracture, the fracture-surface appearance is that associated with fast fracture. A very common type of brittle torsion fracture is found in spiral springs made of high-strength steel. An example which typifies the surface appearance is given in Fig. 4.22. In this case, initially a small fatigue crack had formed and served as the origin of the torsion overload fracture. Note that the fracture surface is helical and at 45° to the wire axis. The fracture surface shows only radial marks, characteristic of fast fracture, and they point back to the location of the original fatigue crack.

4.4 Bending Overload

In bending fracture, generally the fracture surface is similar to that produced by tensile overload. The difference arises due to the fact that one side of the component is in tension and the other in compression. Thus the crack forms on the tension side and propagates across until separation occurs. Figure 4.23 shows a broken bending sample that was ductile. The sample was notched, which introduced stress triaxiality, but the loading rate was low. Thus the fracture proceeded from the tension side to the opposite side with considerable plastic deformation and the formation of shear lips on the sides at 90° to the notch.

Figure 4.23 Ductile bending overload fracture. This was a notched bending specimen made of a steel containing 0.10% C, 0.02% Si, 0.30% Mn, 0.032% P, 0.024% S, and 0.003% N. (*From Henry and Horstmann.*[8])

Figure 4.24 shows the surface of a brittle sample broken in bending. No gross plastic deformation occurred, and a secondary crack formed. Note that there are no clear distinguishing surface features to help locate the origin of the fracture. This is more clearly seen on the fracture surface of Fig. 4.25. However, in some cases radial marks do form and point back to the origin of the fracture. An example is shown in Fig. 4.26, the case of a broken crane hook. (In this case the fracture began below the surface.)

Shafts frequently break by overload bending, and a typical fracture appearance is shown in Fig. 4.27. The direction of bending is obvious

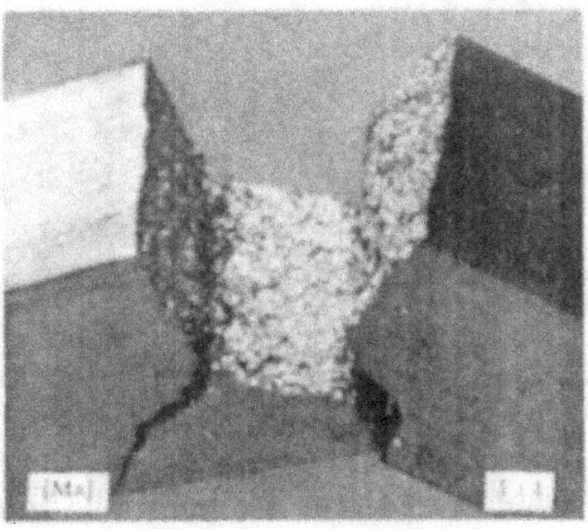

Figure 4.24 Brittle bending overload fracture. This steel sample contained 0.09% C, 0.02% Si, 0.34% Mn, 0.042% P, 0.028% S, and 0.010% N and was in the aged state. The brittle cleavage rupture first propagated in the zone under tension perpendicular to the principal stress direction, then branched out. (*From Henry and Horstmann.*[8])

Figure 4.25 Brittle bending overload fracture of a sample notched on both sides. The crack propagated from notch to notch. (Material as in Fig. 4.24.) (*From Henry and Horstmann.*[8])

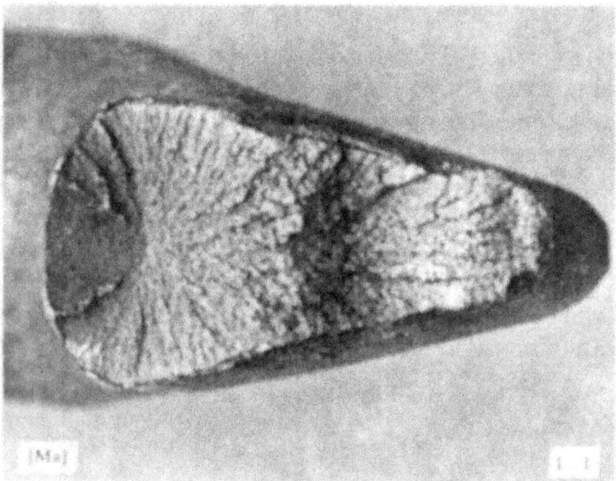

Figure 4.26 Fracture surface of a steel crane hook broken by bending overload. The steel contained 0.24% C, 0.40% Si, 0.39% Mn, 0.095% P, and 0.024% S. (*From Henry and Horstmann.*[8])

from the slant of the fracture. The location of the origin of the fracture is deduced from the clear chevron marks on the periphery of the fracture surface (Fig. 4.28). In this case the distinct difference in appearance of the periphery and the center of the fracture surface is caused by the different microstructural characteristics in the two locations. The

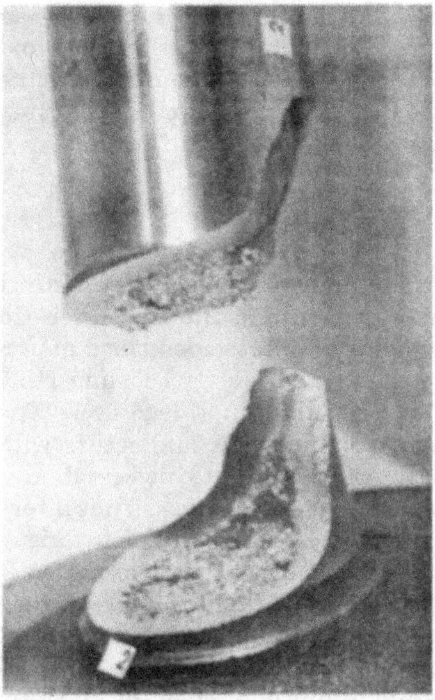

Figure 4.27 Steel axle shaft broken in a laboratory by a single bending impact overload. The steel was modified SAE 1050 containing 0.50% C, 0.95% Mn, 0.25% Si, 0.01% S, and 0.01% P. The hot-rolled and upset shaft had an induction-hardened case (60 Rockwell C) and a pearlite–primary ferrite core (20 Rockwell C). (*From* Metals Handbook.[5])

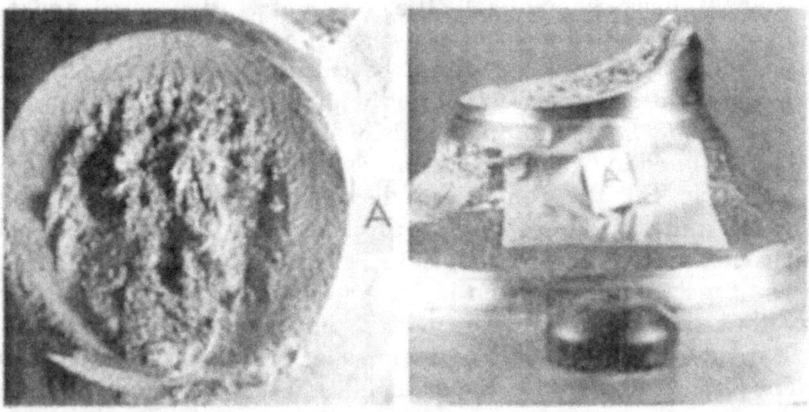

Figure 4.28 Fracture of an axle shaft broken in a laboratory by a bending overload. Note that the chevron marks (in the induction-hardened surface) and radial marks (in the softer core) point to the origin of the fracture. The material was modified AISI 1050 steel, induction-hardened and tempered to 60 Rockwell C at the bearing surface. (*Courtesy of Z. Flanders, Packer Engineering Associates; from* Metals Handbook.[5])

outer layer was induction-hardened, and hence here the crack ran through a brittle material. In the center the material was tougher, and coarse radial marks were formed. Note that these also point toward the origin of the fracture, and they merge into finer radial marks, which in turn merge with the chevron marks in the vicinity of the origin.

4.5 Fatigue Fracture

A fatigue fracture surface usually displays macroscopically two distinct regions: the fatigue crack propagation region and the final overload region. The latter has characteristics similar to those described in previous sections, since this surface was created when the supporting material was too slight to sustain the next (and final) load cycle. The final fracture is nearly always a catastrophic event and occurs with almost no plastic deformation. The lack of gross plastic deformation is also a characteristic of the propagation of the fatigue crack. Thus a very distinguishing characteristic of a component which broke due to fatigue loading is that the parts exhibit almost no gross plastic deformation.

The main interest in this section is the fractographic features which fatigue fractures exhibit upon visual examination or at low (such as 20×) magnification. We first present the appearance of test samples broken under a loading cycle of constant amplitude. Then the effects of loading conditions and specimen geometry are examined. Finally, a few examples of the fracture surface of machine components which fractured in fatigue follow.

Figure 4.29 shows schematically the growth of a fatigue crack and its relation to the number of loading cycles. It is emphasized that the stress generated by the load would be calculated to be below that required for yielding. However, due to the intense stress concentration at the root of the crack, local plastic deformation will occur upon each cycle. As discussed in Chap. 3, frequently striations are associated with each load cycle. But the propagation distance per cycle is quite small and the associated fatigue striations are too small to be resolved, and hence are not macrofractographic features. Figure 4.30 shows fracture surfaces of test samples broken in fatigue loading. It is important to realize that these samples were tested continuously until fracture, using a cyclic load of constant frequency and of a constant maximum load. The distinction between the area of fatigue crack growth and that of final overload separation is not necessarily very clear.

However, in most (but not all) fatigue fractures a surface feature is found which is usually an indicator that fracture was by fatigue. This is illustrated in Fig. 4.31. The markings are called by several names based on their appearance: *oyster-shell marks, tide marks, ripple marks, beach*

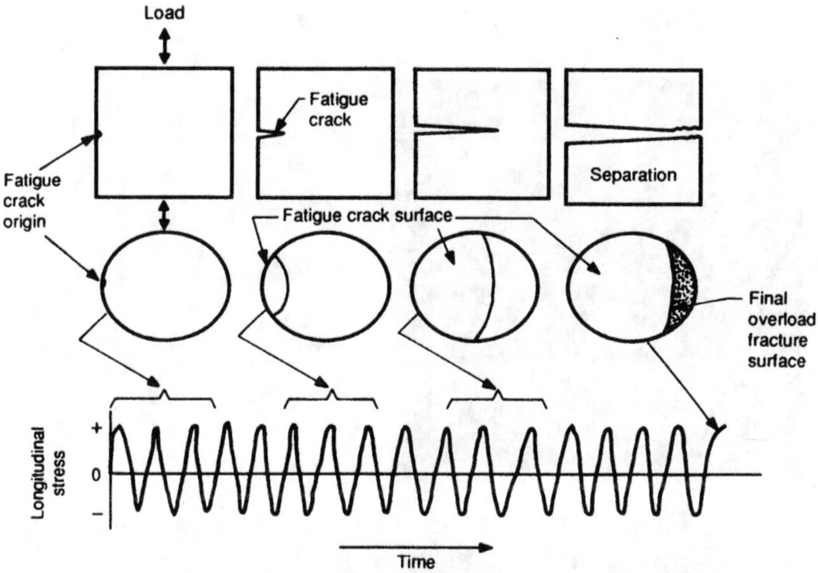

Figure 4.29 Schematic illustration of the location of a fatigue crack and number of cycles.

marks, stop marks, arrest marks, clamshell marks, or *conchoidal marks.* We will use the last term for them. (This term is also used in mineralogy to describe the fracture appearance of shells.) These markings are associated with surface-topography changes that cause the surface appearance to vary locally. They are caused by such factors as variation in the crack growth velocity or surface oxidation. We now examine how these markings develop, assuming that such factors allow them to form, and we see what information they convey about the loading conditions.

To illustrate how these markings form, consider a cylindrical test bar loaded cyclically in tension-compression-tension, as shown in Fig. 4.32. Let the crack form only on one side and progress across toward the other side. After a certain advancement, let the loading cease and the surface oxidize. Now resume the loading, with the crack advancing, and then let the loading cease again. The new surface now oxidizes, and the previous surface oxidizes further. Thus the initial fracture surface has a thicker oxide layer on it. This process continues until fracture occurs, as shown in Fig. 4.32. The final overload fracture surface will show features associated with overload fracture, and the part of the surface along which the fatigue crack grew will show conchoidal marks because of the different thicknesses of oxide. Thus in this example the appearance of these marks relies on the crack not moving continuously until fracture, but growing intermittently. Other mechanisms also cause

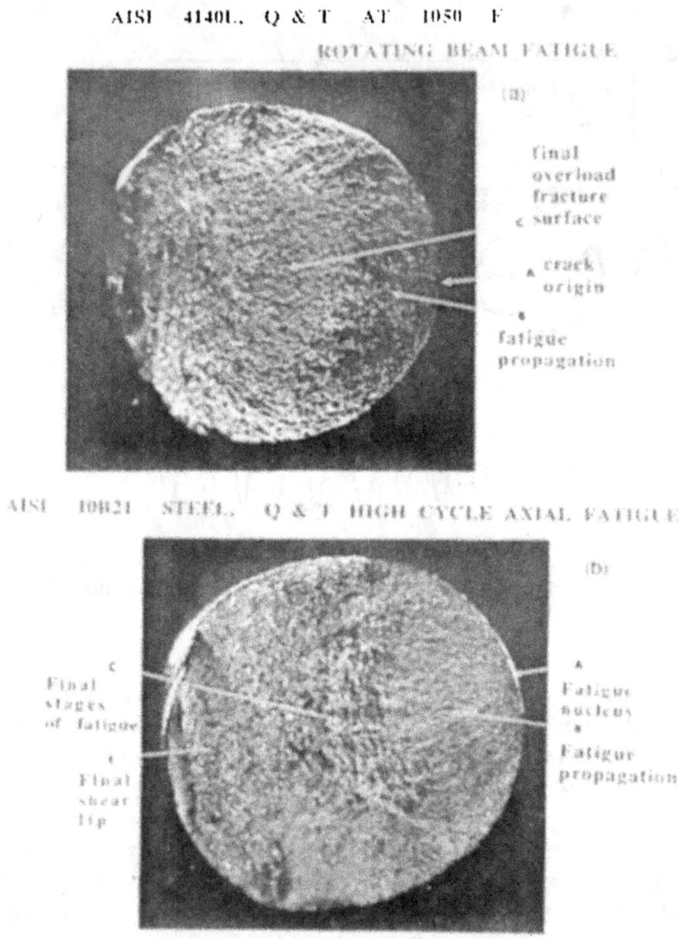

Figure 4.30 Fracture surfaces of samples tested in fatigue with a constant maximum load and frequency. (*a*) AISI 4140L steel, quenched and tempered at 1050°F (566°C), rotating beam fatigue. (*b*) AISI 10B21 steel, quenched and tempered, high-cycle axial fatigue. (*From Bhattacharyya et al.*[9])

these conchoidal marks to form, such as rubbing of the mating surfaces in different fashions if the load amplitude varies. It is important to note that fatigue fracture propagation without any arrests will probably not produce conchoidal marks. This was the case in Fig. 4.30, where the fatigue specimens were tested uninterruptedly to fracture.

The example in Fig. 4.32 can be used to illustrate the effect of the load level (stress level) on the fractographic features. If the load is relatively

Figure 4.31 Fracture surface of a sample which broke in fatigue loading, illustrating the conchoidal marks and the final overload region typical of fatigue fractures. This was an axle of 8640 steel with a hardness of approximately 30 Rockwell C. The loading was nonrotational. The arrow indicates the location of the origin of crack initiation, which was at a discontinuity in a cut thread. (*From* Metals Handbook.[2])

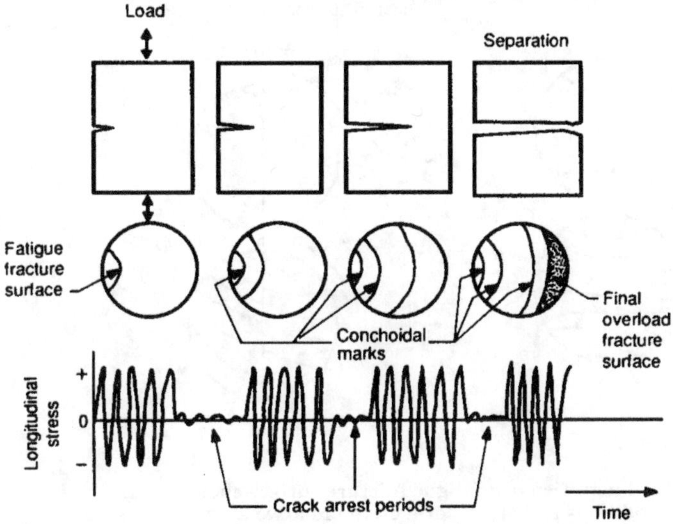

Figure 4.32 Schematic illustration of the formation of conchoidal marks in a fatigue test.

low, then it takes many cycles for the fatigue crack to propagate to a location where the remaining material will fracture catastrophically on the next (and final) load application. Thus the relative amount of the fracture surface covered by the conchoidal marks will be large. If the stress level is high, then the crack will not propagate far before final fracture occurs. This is illustrated in Fig. 4.33*a*. A similar situation exists if the sample is loaded in alternating unidirectional bending, as shown in Fig. 4.33*b*. If the sample is loaded in reversed bending, then the crack will probably initiate simultaneously on opposite sides, as

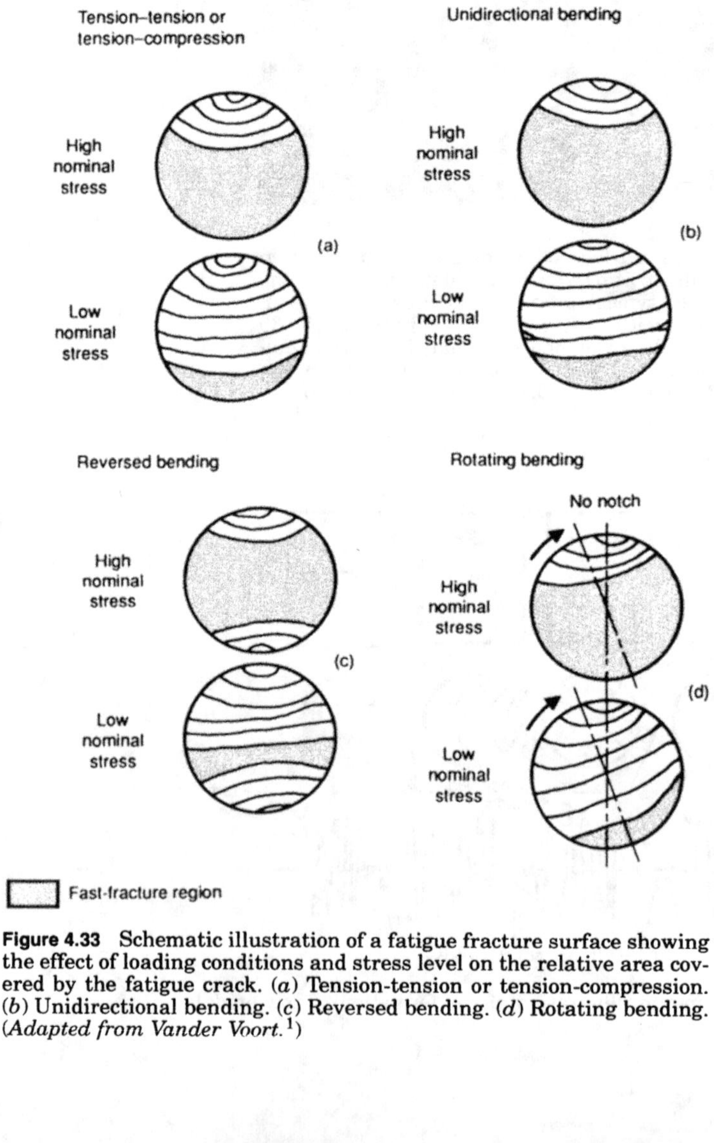

Figure 4.33 Schematic illustration of a fatigue fracture surface showing the effect of loading conditions and stress level on the relative area covered by the fatigue crack. (*a*) Tension-tension or tension-compression. (*b*) Unidirectional bending. (*c*) Reversed bending. (*d*) Rotating bending. (*Adapted from Vander Voort.*[1])

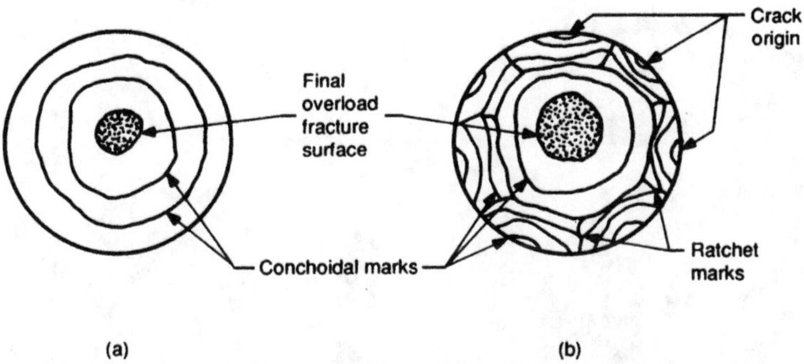

Figure 4.34 Schematic illustration of the fracture-surface appearance of a fatigue fracture which occurred in rotating bending. The fatigue crack originated simultaneously (*a*) at different locations on the same circumferential plane, (*b*) at different locations but not on the same circumferential plane, forming ratchet marks.

shown in Fig. 4.33*c*. Again, a low nominal stress creates a fracture surface with a large amount of it covered with conchoidal marks.

In simple rotating bending, every point on the surface is placed in longitudinal tension sometime during a rotation. If only one fatigue crack initiates, then the appearance of the fracture surface will be similar to that shown schematically in Fig. 4.33*d*. However, the fatigue crack may initiate at many locations simultaneously and advance from the periphery. This is shown schematically in Fig. 4.34*a*. If the fatigue cracks originate at several locations but at different locations axially, then when the various advancing fatigue cracks meet at the different levels, accommodation occurs to join these into one fatigue crack. This leaves offsets on the surface called *ratchet marks,* as illustrated in Fig. 4.34*b*. Figure 4.35 shows a fatigue fracture surface with many faint ratchet marks along the periphery. The fracture surface in Fig. 4.36 depicts an extreme case, where the individual fatigue crack growth regions did not join before fracture occurred.

The fatigue fracture surface appearance is affected by the stress state, which can be illustrated by considering the effect of a notch. The more acute the notch is, the more likely that a fatigue crack will initiate at more than one location on the periphery. However, the subsequent growth of the crack is not greatly affected. This is due to the fact that once the crack forms (enters stage II, see Chap. 3), the root of the fatigue crack is sharp, and this will be true whether the initial crack formed at a notch or not. However, if the nominal stress is relatively low, then it is more likely that the crack will move along the periphery faster than it moves across the diameter, giving a different appearance to the fracture surface. These features are illustrated schematically in Fig. 4.37, and Fig. 4.38 shows examples of this effect in terms of stress concentration at the surface.

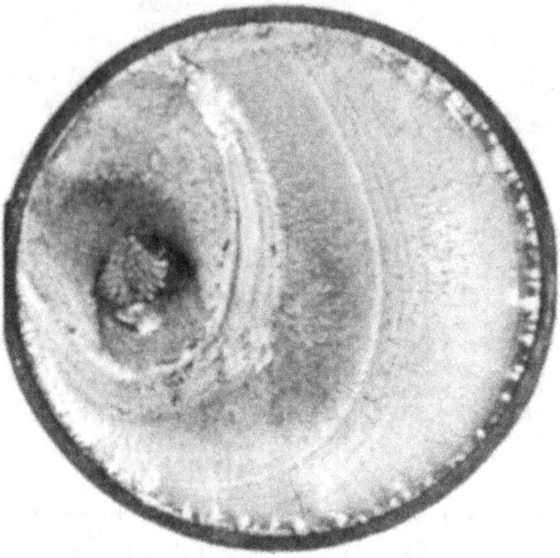

Figure 4.35 Fracture surface of a fatigue failure showing fine ratchet marks on the periphery, which formed along a sharp snap-ring groove. The eccentric pattern of oval conchoidal marks indicates that the load on the shaft was not balanced. This was a 1050 steel shaft with a hardness of about 35 Rockwell C. It was subjected to rotating bending. (*From Wulpi.*[6])

Figure 4.36 Fatigue fracture in which the fatigue cracks formed at six locations around the periphery, but at distinctly different levels. The locations were near the runouts of six grooves. Grinding damage in the fillet was the cause of the fracture. The component was a 4817 steel shaft, carburized and hardened to a surface hardness of 60 Rockwell C. It was loaded in rotating bending. (*From Wulpi.*[6])

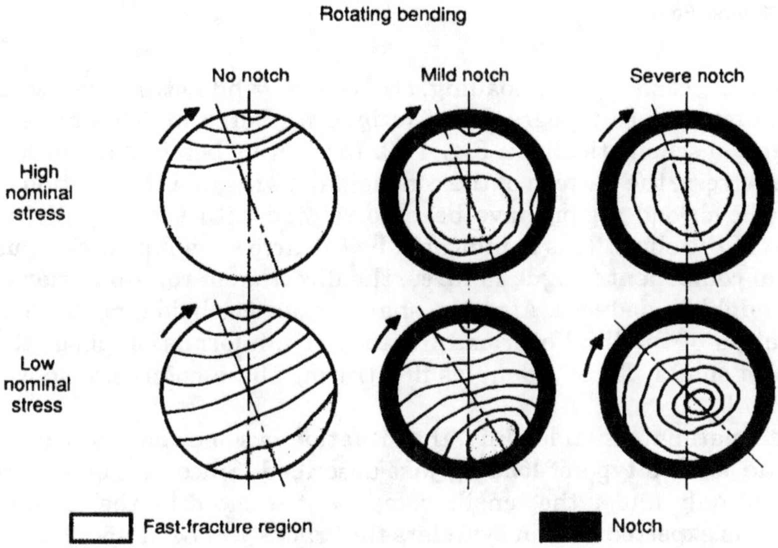

Figure 4.37 Schematic illustration of the effect of notch severity and stress level on the fatigue fracture surface appearance in rotating bending of cylinders. (*Adapted from Vander Voort.*[1])

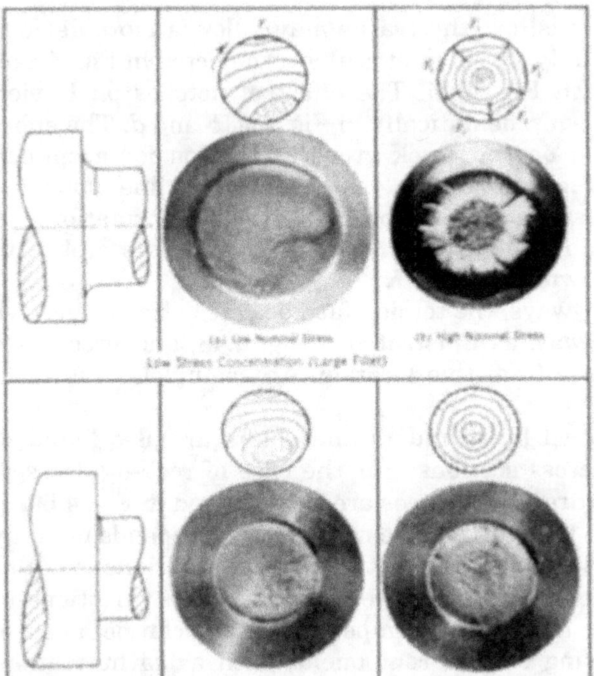

Figure 4.38 Fatigue fracture surfaces illustrating the effect of stress concentration on the fracture-surface appearance. A rotating shaft was subjected to bending in a fixed plane. (*From Peterson.*[10])

The effects of the type of loading, stress level, and notch severity on the fracture-surface topography of fatigue fractures in cylinders are summarized schematically in Fig. 4.39. In rotating bending under low nominal stress, the pattern of the conchoidal marks and the location of the final overload region have been correlated with the direction of rotation of the shaft. This is indicated by the arrows in Fig. 4.39. Thus in actual component fatigue failures, the direction of rotation prior to failure might be deduced. Also, the shape of the conchoidal marks and the location of the final overload fracture give information about the loading of shafts. Three examples illustrating these points are shown in Fig. 4.40.

In alternating torsion loading, the situation is somewhat more complex than for the type of loading just described. Since fatigue cracks propagate only under the tensile component induced by the load, in torsion it is expected that in cylinders the cracks will be at 45°. This is shown schematically in Fig. 4.41*a*, and an example of such a fatigue failure is given in Fig. 4.42. If the cracks originate at several circumferential locations, then the type of fractures shown in Figs. 4.43 and 4.44 may occur.

If the nominal stress is low, the cracks may follow a more devious path, growing both axially and diametrically, as depicted in Fig. 4.41*c*. An example is given in Fig. 4.45. The effect of notches on torsion fatigue fracture is shown schematically in Fig. 4.41*b* and *d*. The effect of stress concentration on the crack growth direction for a splined shaft is illustrated in Fig. 4.46. In Fig. 4.46*b* the contour lines show the maximum tensile stress distribution prior to the crack formation and growth. Examples of the fracture surface of splined shafts broken in torsion fatigue are shown in Fig. 4.47.

In cylinders with keyways, the torsion fatigue crack often originates at the root of the keyway, as illustrated n Fig. 4.48, and progresses along the circumference from this location. An example is shown in Fig. 4.49.

The features described for cylinders and shafts are also found in components of other cross sections. For the case of rectangular sections, the fatigue fracture appearances are summarized in Fig. 4.50. It is important to realize that the fatigue fracture path depends not only on the geometry of the component and the loading conditions, but also on the structural inhomogeneity of the material. Such variations in structure are common in machine components. They can be a consequence of the processing of that component, such as carburization, induction hardening, and peening of a surface. They also can be due to the anisotropy of mechanical properties caused by microstructural features such as elongated inclusions.

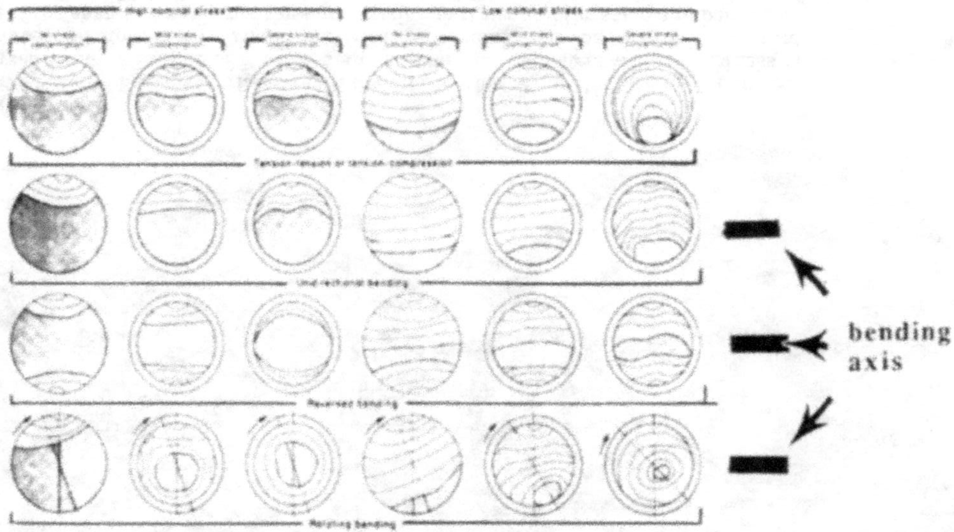

Figure 4.39 Schematic illustration of the effects of the type of loading, stress level, and notch severity on the fatigue fracture surface appearance of cylinders. (*After Vander Voort.*[1])

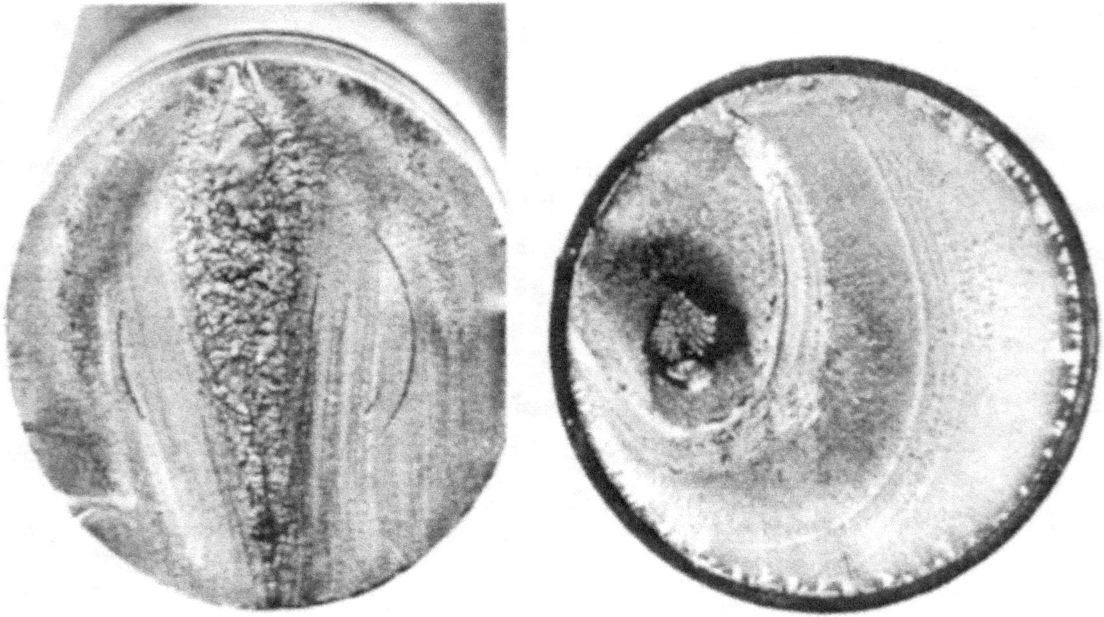

(a) (b)

Figure 4.40 Fatigue fracture surfaces revealing information about the loading conditions. (*a*) Reversed bending fatigue of a 1.6-in-diameter shaft of 1046 steel with a hardness of approximately 30 Rockwell C. The symmetrical conchoidal mark pattern indicates that each side of the shaft was subject to the same maximum stress and the same number of load applications. (*From Wulpi.*[6]) (*b*) Rotating bending of a 1050 steel shaft with a hardness of about 35 Rockwell C. The numerous ratchet marks indicate that fatigue cracks initiated at many locations along a sharp snap-ring groove. The eccentric pattern of oval conchoidal marks indicates that the load was not balanced. (*From Wulpi.*[6])

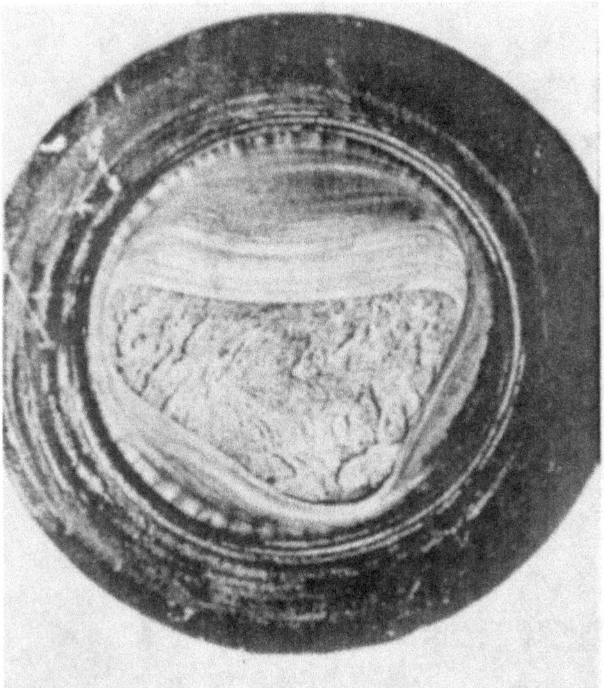

Light fractograph 0.65×

(c)

Figure 4.40 (*Continued*) Fatigue fracture surfaces revealing information about the loading conditions. (*c*) 4-in-diameter axle of 8640 steel with a hardness of about 30 Rockwell C. The shaft was contained within a shrink-fitted collar, which is the outer circular area. This nonrotating axle was subjected to bending stresses in three directions, which produced this unusual pattern of conchoidal marks. (*From* Metals Handbook.[2])

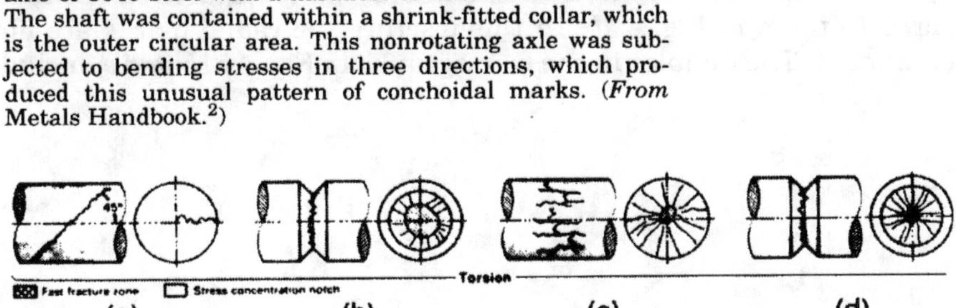

Figure 4.41 Schematic illustration of the effects of stress level and notch severity on the fatigue fracture of cylinders in reversed torsion loading. (*After Vander Voort.*[1])

4.6 Correlation of Micro- and Macrofractographic Features

In this section examples of the fracture surfaces of test samples are used to compare the topographic features seen macroscopically and the corresponding microscopic features. These examples also illustrate the sensitivity of the macroscopic fracture features and the fracture mechanism to the microstructure of the material.

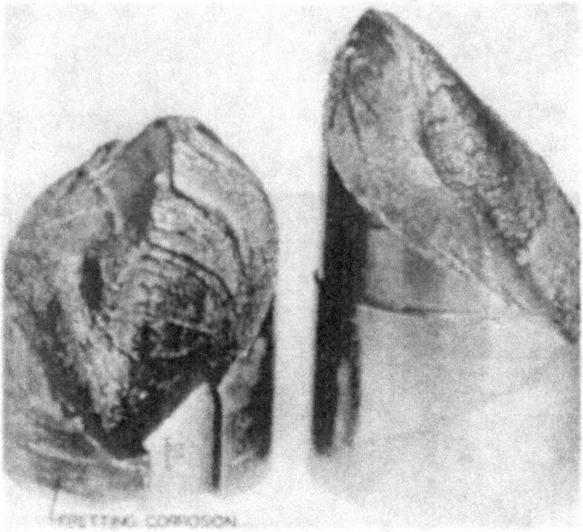

Figure 4.42 Fatigue fracture of a shaft which broke due to reversed torsion loading. (*From Peterson.*[10])

Figure 4.51*a* shows the fracture surface of a tensile sample of a 10B21 steel. The mode of fracture is thus tensile overload, and the high ductility [56 percent reduction in area (RA)] and the large shear-lip zone show that this was a ductile fracture. The features can be compared to those in Fig. 4.4*b*. In this fracture the radial marks are not prominent. The dimples in the fractographs in Fig. 4.51*b* and *c*, includ-

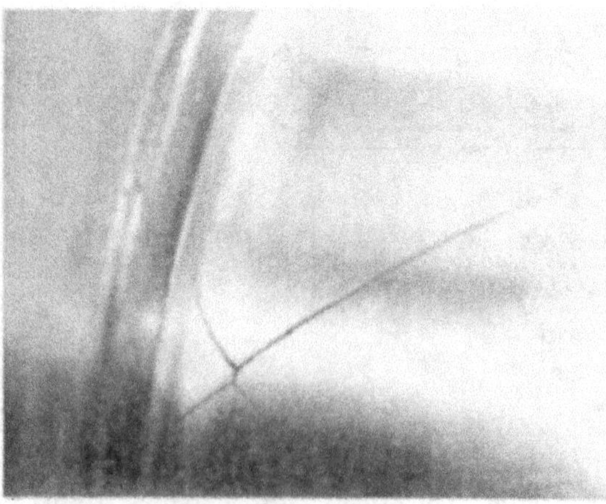

Figure 4.43 Fatigue crack geometry in a shaft fractured by reversed torsion loading. This was a 1045 steel crankshaft. (*After Wulpi.*[6])

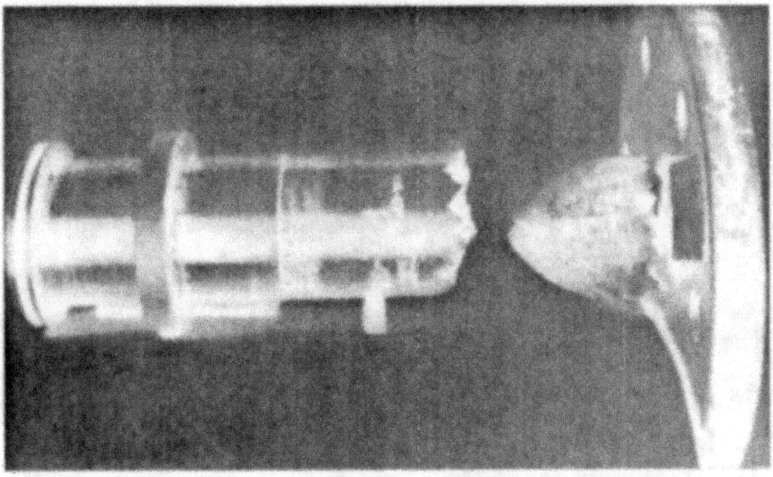

Figure 4.44 Fatigue fracture in a steel shaft broken by reversed torsion loading. (*From Peterson.*[10])

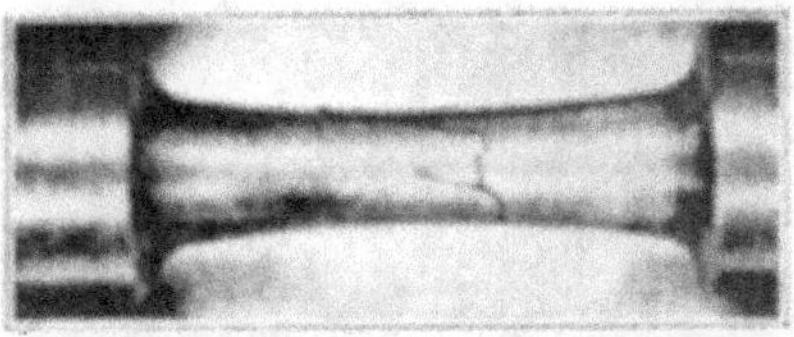

Figure 4.45 Fatigue crack geometry showing axial and radial crack growth due to reversed torsion loading. The material was ductile steel (Dolan). (*From Peterson.*[10])

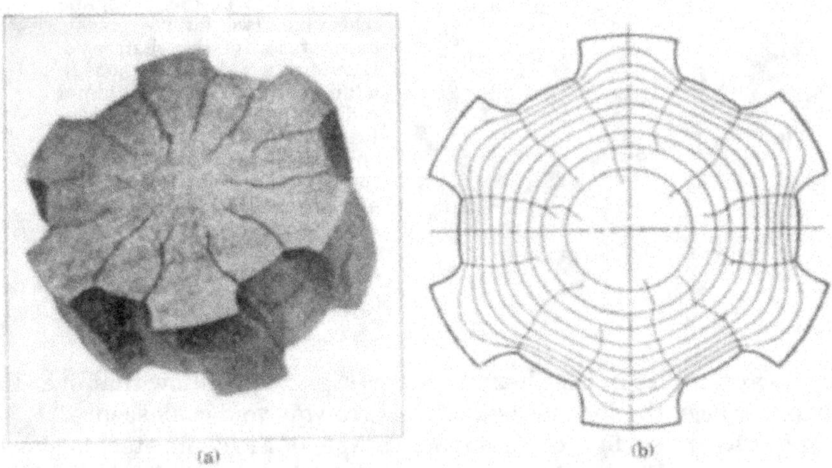

Figure 4.46 (*a*) Fatigue cracks in a spline due to reversed torsion loading (Oschatz). (*b*) Maximum tensile stress distribution. (*From Peterson.*[10])

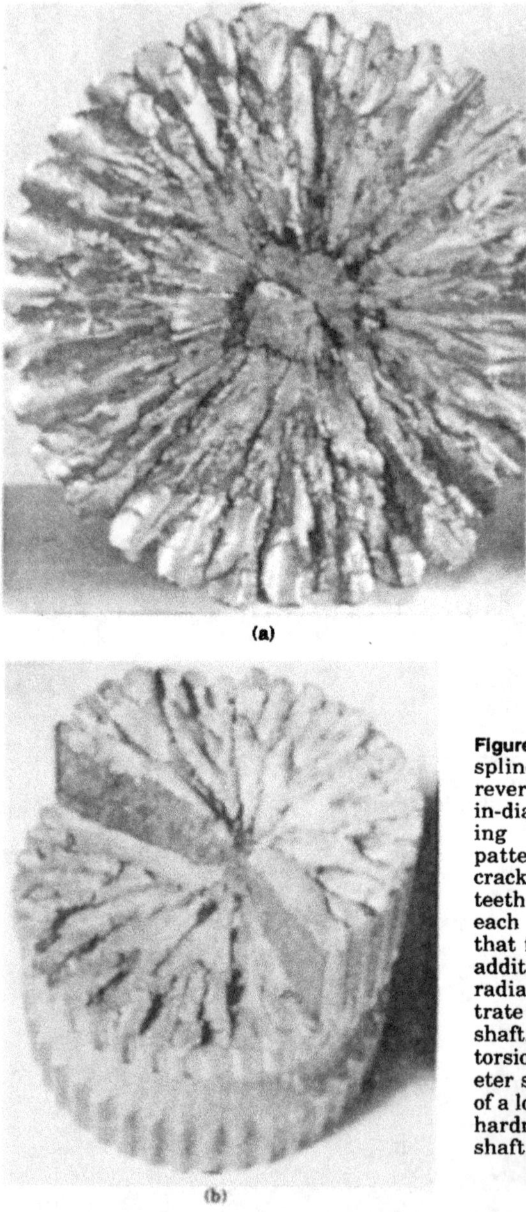

Figure 4.47 Fracture surface of spline shafts broken due to reversed torsion loading. (*a*) 6¾-in-diameter spline shaft showing characteristic "starry" pattern of multiple fatigue cracks. Each of the 32 spline teeth has two fatigue cracks, each at 45° to the shaft axis, that form a V-shaped region. In addition there are longitudinal radial fatigue cracks that penetrate nearly to the center of the shaft. (*b*) Fracture in reversed torsion loading of a 1½-in-diameter spline. The shaft was made of a low-carbon alloy steel with a hardness of 24 Rockwell C in the shaft area. (*From Wulpi.*[6])

ing shear dimples in the shear-lip zone (Fig. 4.51*d*), show that over the entire surface, the mechanism of fracture was void coalescence.

The fracture surface of an impact sample of the same steel is shown in Fig. 4.52*a*. Fine radial marks point back toward the root of the machined notch where fracture initiated. Due to the nature of the test,

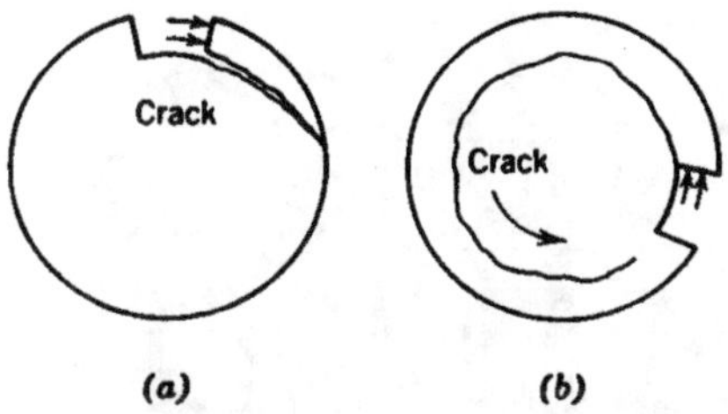

Figure 4.48 Schematic illustration of the formation of fatigue cracks from the root of a keyway of shafts loaded in reversed torsion. (*From Peterson.*[10])

Figure 4.49 Fracture surface due to torsion fatigue loading, as illustrated in Fig. 4.48. The component was a 3³/₈-in-diameter keyed tapered shaft of 1030 steel. The fatigue crack originated in corner *A* on the keyway from pressure of the key aligning a large hub to the shaft. The fatigue crack progressed completely around the shaft, under the keyway, and started around again, all in a clockwise direction, before final separation occurred, and nearly broke off the "wing" at left. The only thing necessary to prevent this type of fracture is to make sure that the large nut which forces the hub to fit tightly against the tapered surface does not come loose. If the nut is loose, the key and keyway, rather than the frictional fit of the conical joint, carry the torsional force. (*From Wulpi.*[6])

the mode of fracture is bending overload, and the material fractured in a ductile manner, as evidenced by the large amount of shear lip. The fractographs in Fig. 4.52*b* and *c* show that everywhere the mechanism of fracture was by void coalescence.

Figure 4.53*a* shows the fracture surface of the same 10B21 steel broken in a high-cycle axial fatigue test. In this test the loading frequency was constant and the loading uninterrupted, and thus there are no conchoidal marks. As expected from the loading condition, the fatigue crack initiated at only one location on the surface. Because of the lack of conchoidal marks, the location of the initiation of the frac-

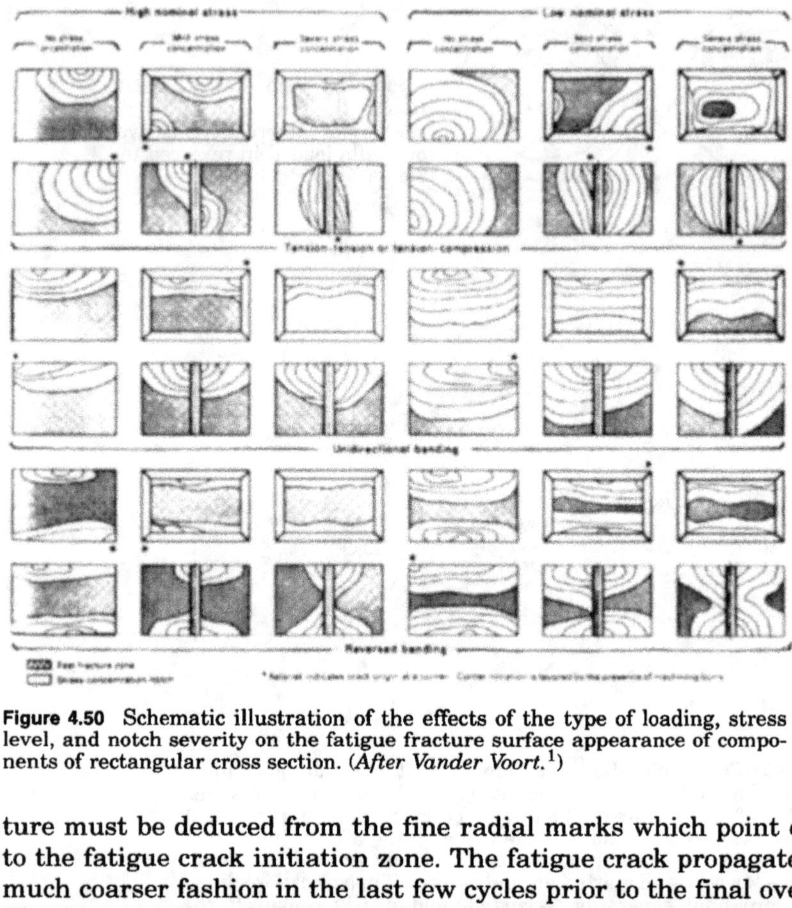

Figure 4.50 Schematic illustration of the effects of the type of loading, stress level, and notch severity on the fatigue fracture surface appearance of components of rectangular cross section. (*After Vander Voort.*[1])

ture must be deduced from the fine radial marks which point clearly to the fatigue crack initiation zone. The fatigue crack propagated in a much coarser fashion in the last few cycles prior to the final overload. The fractograph in Fig. 4.53*b* purports to show fatigue striations, but they are not prominent, even though this is a fatigue test sample. However, faint striations can be found on the surface (Fig. 4.53*c* to *e*). The final overload fracture region shows a dimple structure (Fig. 4.53*f*), characteristic of void coalescence.

The next example is a 4140L steel which has been quenched and tempered at 1050°F (566°C). Figure 4.54*a* shows the fracture surface of a tensile sample. The ductility was high (54 percent RA). A small fibrous zone in the center, coarse radial marks, and a shear-lip zone are present. The radial marks are much coarser than those for the 10B21 steel (Fig. 4.51*a*). All three regions show fine dimples (Fig. 4.54*b* to *d*), and these are much finer (and somewhat less well defined) than those for the tensile sample of the 10B21 steel (Fig. 4.51*b*). The difference in the morphology of the radial marks and of the dimples is related to the greater amount of carbides in the 0.4 percent carbon 4140 steel and perhaps to a finer carbide distribution.

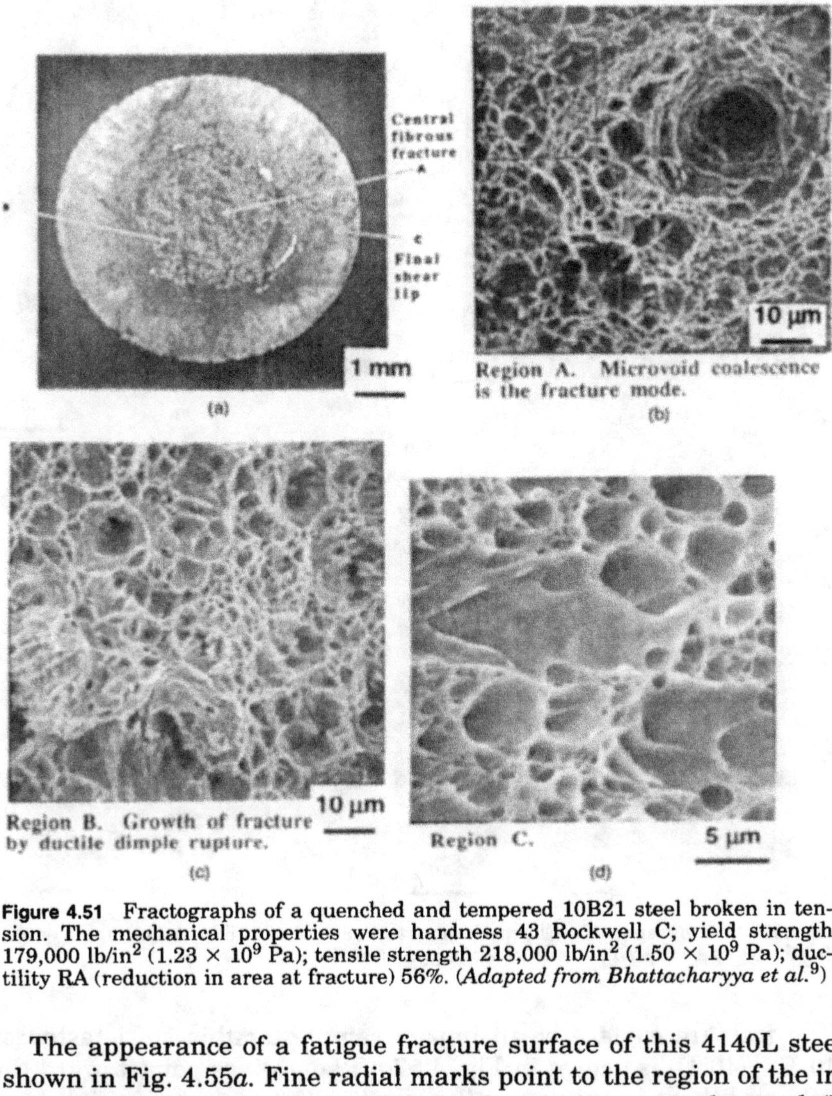

Figure 4.51 Fractographs of a quenched and tempered 10B21 steel broken in tension. The mechanical properties were hardness 43 Rockwell C; yield strength 179,000 lb/in^2 (1.23×10^9 Pa); tensile strength 218,000 lb/in^2 (1.50×10^9 Pa); ductility RA (reduction in area at fracture) 56%. (*Adapted from Bhattacharyya et al.*[9])

The appearance of a fatigue fracture surface of this 4140L steel is shown in Fig. 4.55*a*. Fine radial marks point to the region of the initiation of the fatigue crack. Ill-defined striations are observed (Fig. 4.55*b* and *c*), best revealed by the secondary cracking. The final overload region shows fine dimples (Fig. 4.55*d*).

We now look at several examples of a 4340 steel which has been quenched and tempered at 400°F (204°C). Figure 4.56*a* shows a tensile sample fracture surface. A small fibrous zone at the center, a region of fine radial marks, and a large shear-lip zone are present. The fine dimple structure of the fracture surface is shown in Fig. 4.56*b* to *d*. The morphology of this sample is similar to that of the 10B21 steel (Fig. 4.51).

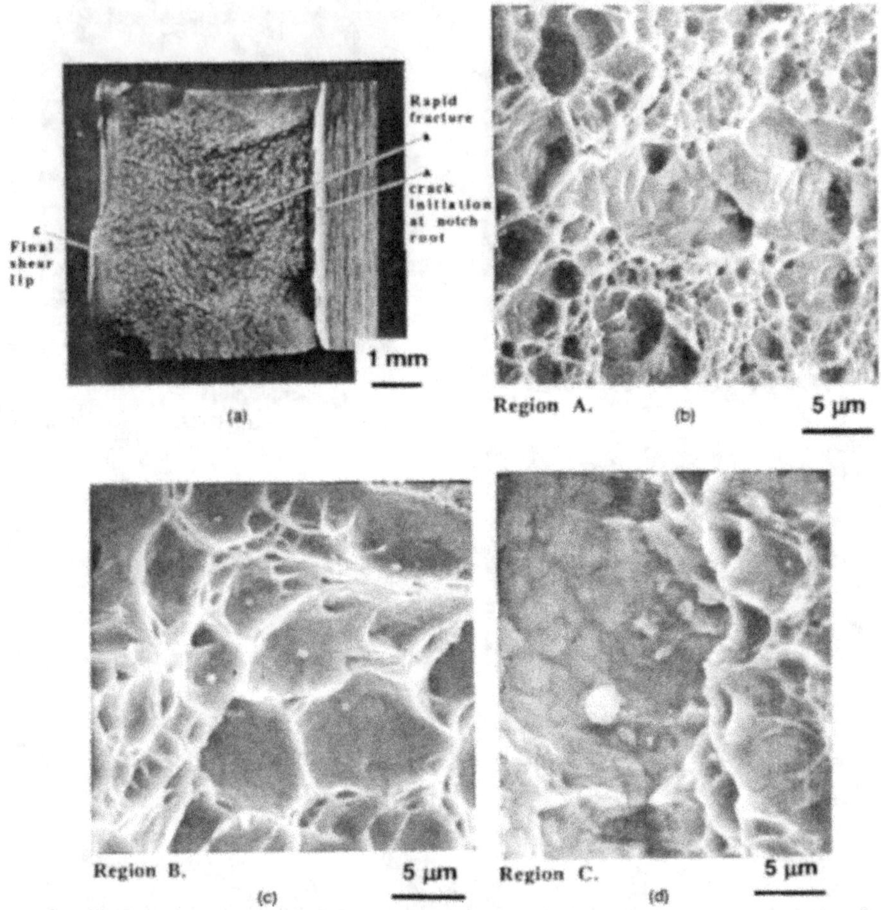

Figure 4.52 Fractographs of a Charpy impact test sample of 10B21 steel. Impact energy 45 ft · lb (61 J). (*Adapted from Bhattacharyya et al.*[9])

The fracture surface of an impact sample for this steel tested at −40°F (−40°C) is shown in Fig. 4.57*a*. Radial lines point toward the notch, and there is only a small shear lip. The region of initial crack growth contains dimples (Fig. 4.57*b*), but the rest of the fracture surface (except for the small shear lip) contains a mixture of dimples and cleavage, or quasicleavage (Fig. 4.57*c* and *d*). Figure 4.58*a* shows the impact fracture surface obtained upon fracture at high temperature, 70°F (21°C). Again, the radial marks point clearly toward the origin of the fracture (the notch), and there is more shear-lip region than for the lower test temperature. For this higher test temperature the entire surface fractured by void coalescence, as evidenced by the dimple morphology (Fig. 4.58*b* and *c*).

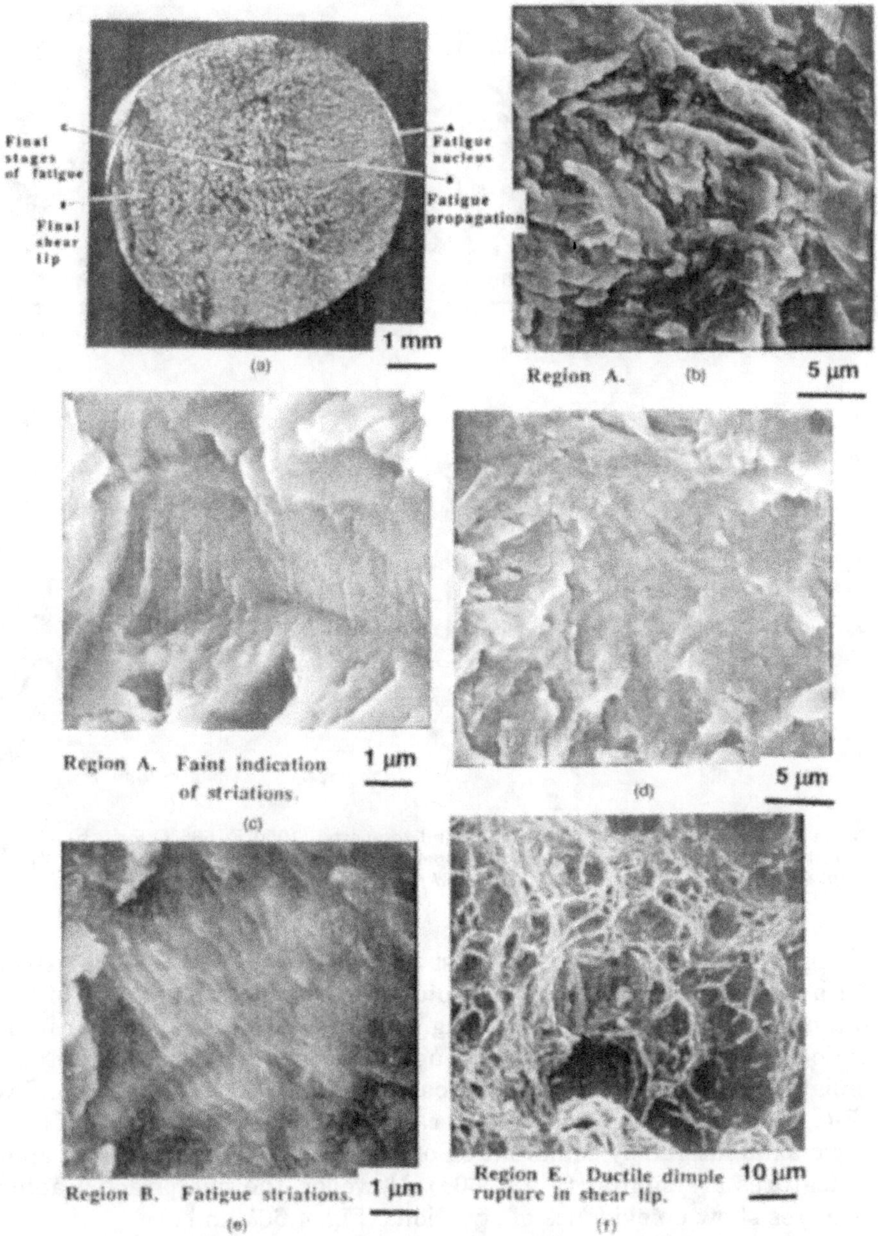

Figure 4.53 Fractographs of a fatigue test sample (high-cycle axial fatigue test). Stress 43,300 to 168,000 lb/in^2 (2.98×10^8 to 1.16×10^9 Pa); frequency 2600 cycles/min; number of cycles to failure 221,000. (*Adapted from Bhattacharyya et al.*[9])

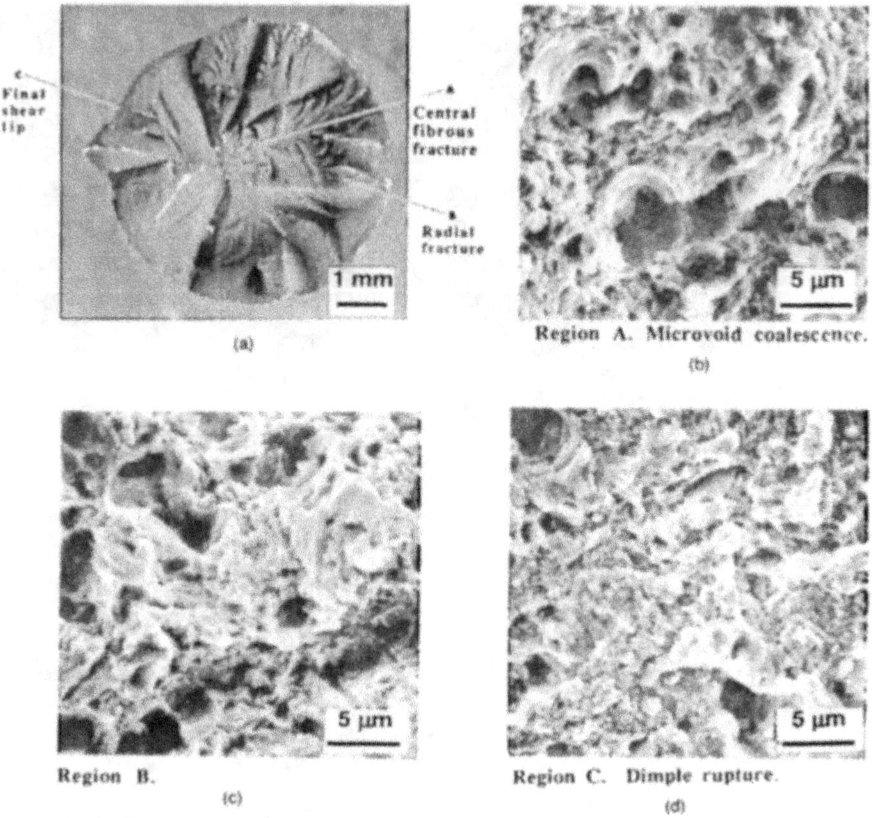

Figure 4.54 Fractographs of a quenched and tempered (1050°F; 566°C for 2 h) 4140L steel tensile sample. The mechanical properties were tensile strength 144,800 lb/in^2 (9.98×10^8 Pa); ductility RA 54%. (*Adapted from Bhattacharyya et al.*[9])

Figure 4.59*a* shows a fatigue test fracture surface for this 4340 steel for a high-cycle axial test. The radial marks point to the origin of the fatigue fracture. Fractography (Fig. 4.59*b* to *e*) shows little clear indication of fatigue striations, although some are revealed by the secondary cracking. The final overload fracture region shows dimples (Fig. 4.59*f*). These characteristics can be compared to the case of low-cycle axial fatigue by examination of Fig. 4.60. The macrofractographic features are similar (Fig. 4.60*a*). However, the microfractographic features show no evidence of striations (Fig. 4.60*b* and *c*).

Figure 4.61*a* shows the fracture surface of a fatigue sample of this material for tempering at 1050°F (566°C), and hence of lower strength. The fractographs (Fig. 4.61*b* to *d*) do not clearly reveal striations, but they appear to be faintly resolved in places (Fig. 4.61*b*).

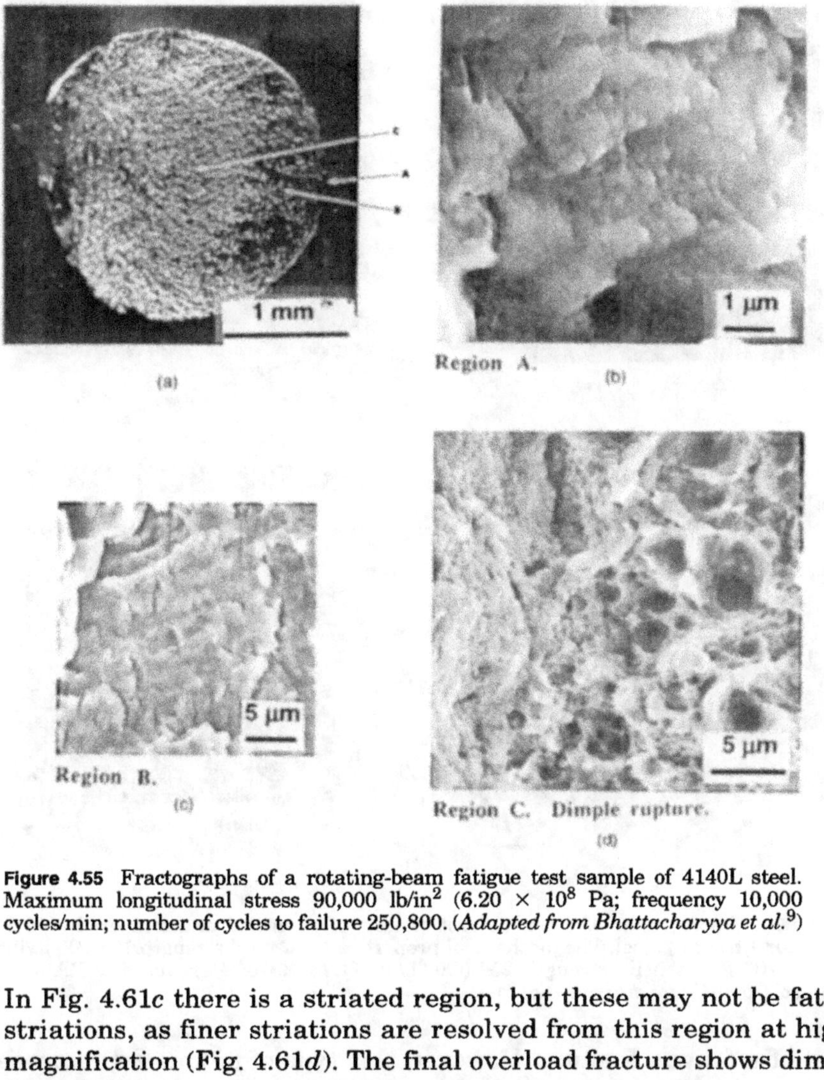

Figure 4.55 Fractographs of a rotating-beam fatigue test sample of 4140L steel. Maximum longitudinal stress 90,000 lb/in^2 (6.20 $\times$ 10^8 Pa; frequency 10,000 cycles/min; number of cycles to failure 250,800. (*Adapted from Bhattacharyya et al.*[9])

In Fig. 4.61*c* there is a striated region, but these may not be fatigue striations, as finer striations are resolved from this region at higher magnification (Fig. 4.61*d*). The final overload fracture shows dimples (Fig. 4.61*e*).

The fracture surface of the tensile sample shown in Fig. 4.56 was for the steel quenched and tempered, with a microstructure of fine carbides in ferrite. Figure 4.62 shows the fracture-surface features for this steel in the normalized condition, which has a heterogeneous microstructure of primary ferrite and pearlite. This is much lower strength than in the quenched and tempered condition. The central fibrous region shows only faint radial marks (Fig. 4.62*a*). The fractographs in Fig. 4.62*b* and *c* show striations, which could be taken as

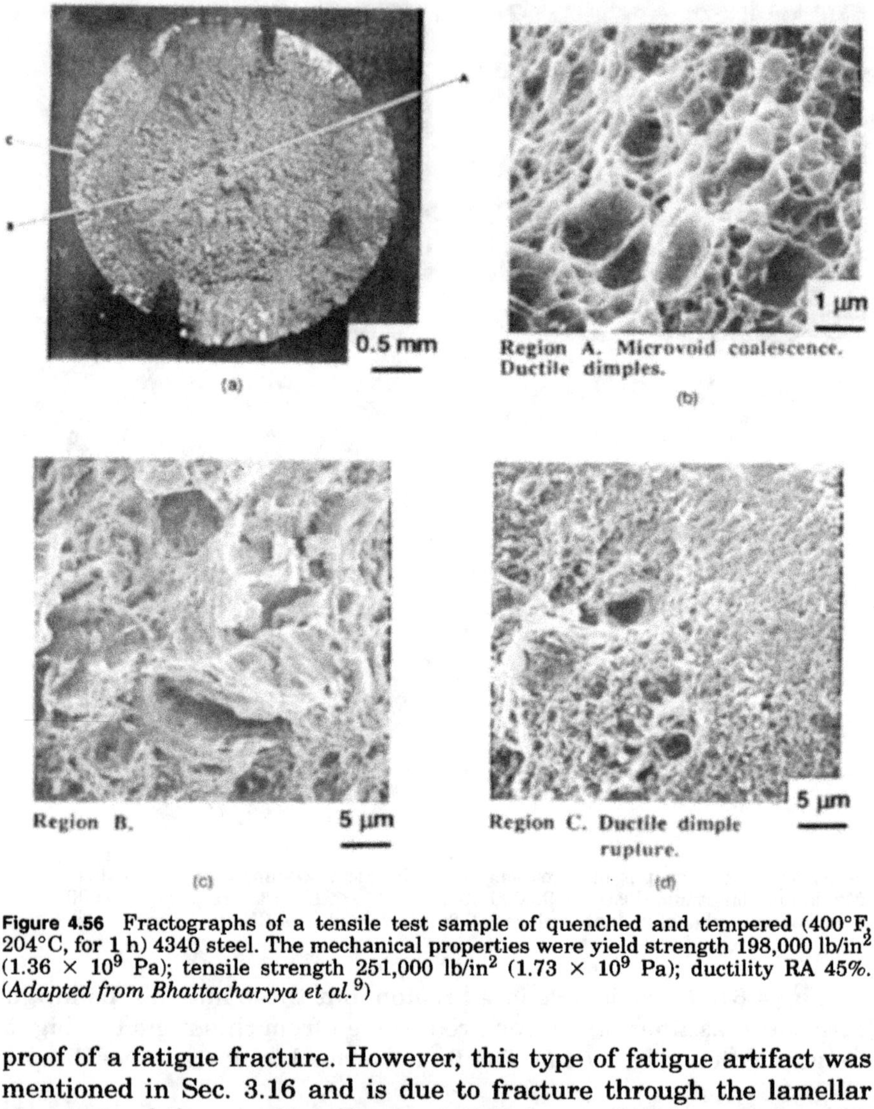

Figure 4.56 Fractographs of a tensile test sample of quenched and tempered (400°F, 204°C, for 1 h) 4340 steel. The mechanical properties were yield strength 198,000 lb/in^2 (1.36 × 10^9 Pa); tensile strength 251,000 lb/in^2 (1.73 × 10^9 Pa); ductility RA 45%. (*Adapted from Bhattacharyya et al.*[9])

proof of a fatigue fracture. However, this type of fatigue artifact was mentioned in Sec. 3.16 and is due to fracture through the lamellar structure of the pearlite. Figure 4.62*d* shows dimples, as does the region of the shear-lip zone (Fig. 4.62*e*).

The fracture surface of an impact test sample of this normalized steel is shown in Fig. 4.63. Faint radial marks point to the origin of fracture (Fig. 4.63*a*). The crack initiation site at the root of the notch consists of dimples (Fig. 4.63*b*), and the striated regions are where fracture occurred in the lamellar pearlite. However, most of the fracture propagated by a mixed mechanism of cleavage and void coalescence, as revealed in Fig. 4.63*c*.

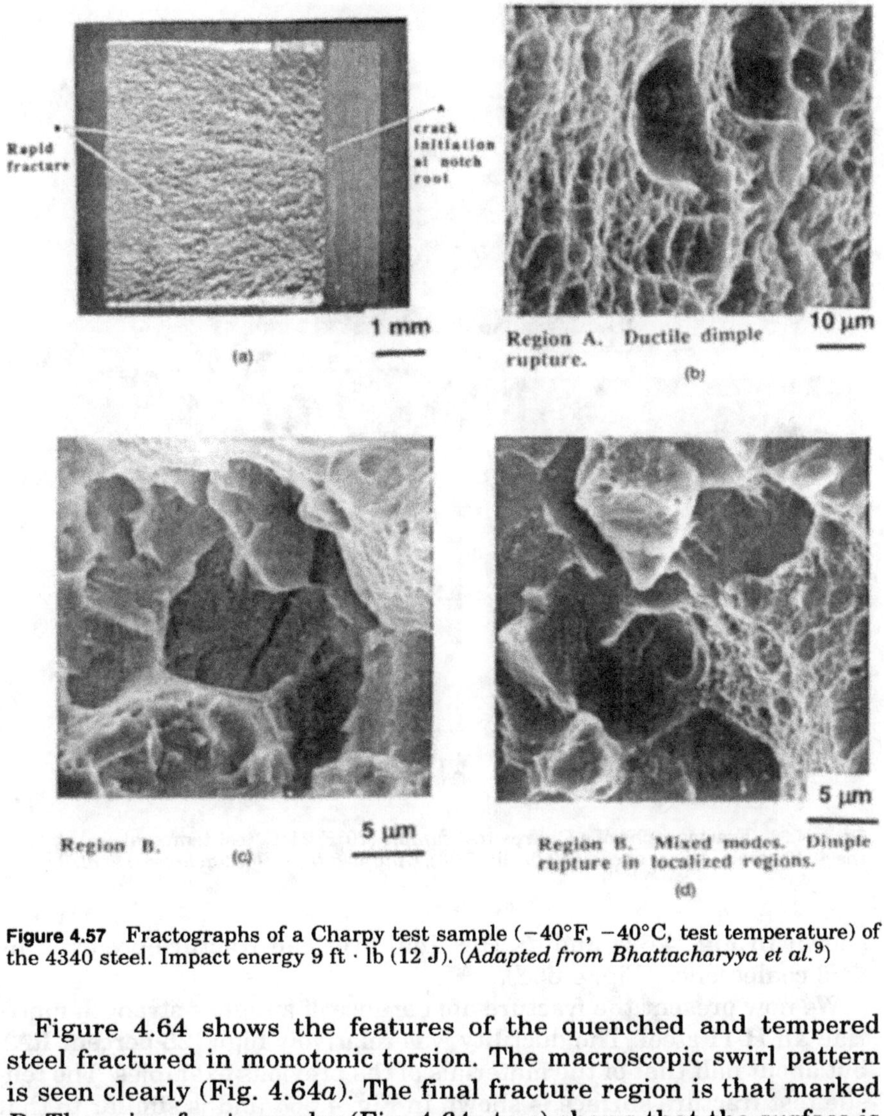

Figure 4.57 Fractographs of a Charpy test sample (−40°F, −40°C, test temperature) of the 4340 steel. Impact energy 9 ft · lb (12 J). (*Adapted from Bhattacharyya et al.*[9])

Figure 4.64 shows the features of the quenched and tempered steel fractured in monotonic torsion. The macroscopic swirl pattern is seen clearly (Fig. 4.64*a*). The final fracture region is that marked *D*. The microfractography (Fig. 4.64*c* to *e*) shows that the surface is covered with shear dimples whose orientation reflects the torsion loading.

For comparison of the microfractographic features of a fatigue fracture in this normalized steel, the fractographs in Fig. 4.65 are presented. Macroscopically, fine radial marks point to the fatigue fracture initiation site (Fig. 4.65*a*). Faint striations are resolved in places on the fatigue crack propagation region (Fig. 4.65*b* and *c*). Compare their appearance to that of fracture through the pearlite

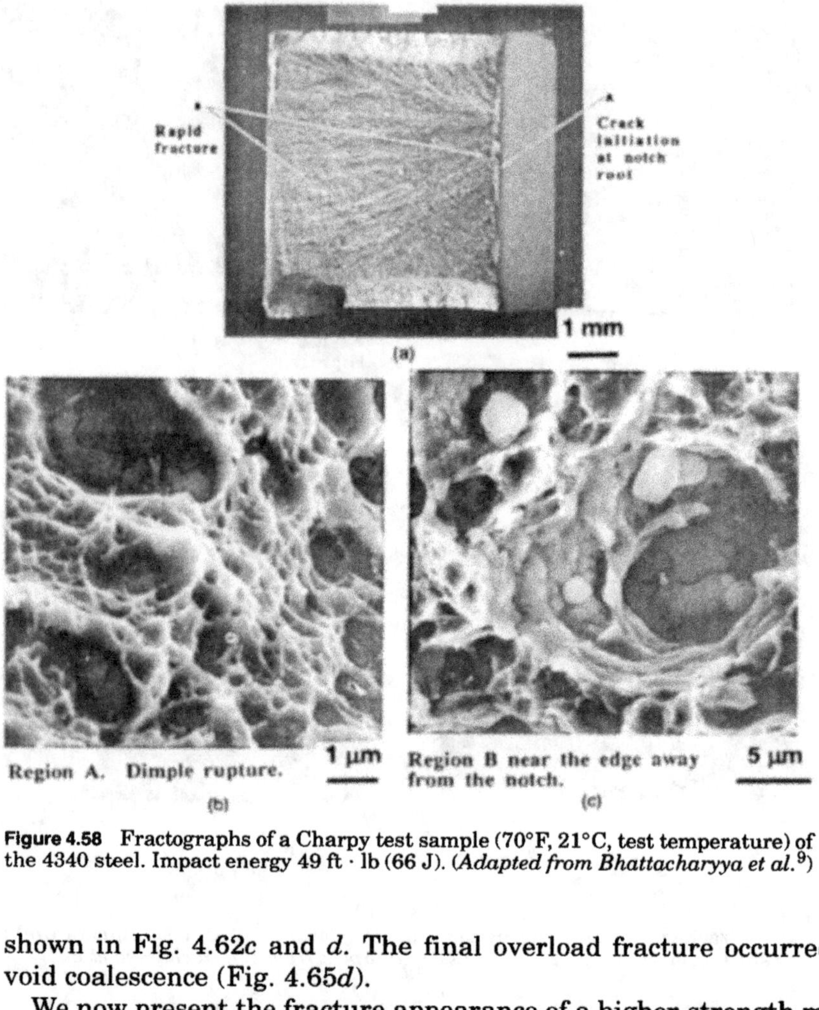

Figure 4.58 Fractographs of a Charpy test sample (70°F, 21°C, test temperature) of the 4340 steel. Impact energy 49 ft · lb (66 J). (*Adapted from Bhattacharyya et al.*[9])

shown in Fig. 4.62*c* and *d*. The final overload fracture occurred by void coalescence (Fig. 4.65*d*).

We now present the fracture appearance of a higher-strength material, an H-11 steel. The ductility was relatively high (28 percent RA), but about half that of the materials of the previous examples. The tensile test fracture surface is shown in Fig. 4.66*a* and is similar to that of the sample of Fig. 4.56*a*. The fracture mechanism was by void coalescence (Fig. 4.66*b* and *c*), but the dimples are less defined than those shown in Fig. 4.56. The impact fracture surface in Fig. 4.67*a* reveals well-defined radial marks. The initial stage of fracture propagation occurred by void coalescence (Fig. 4.67*b*), but the later stages by quasicleavage (Fig. 4.67*c*). The microstructures of both this high-strength H-11 steel and the 4340L steel consisted of fine carbides in ferrite. Both steels had about the same amount of carbides as the carbon content

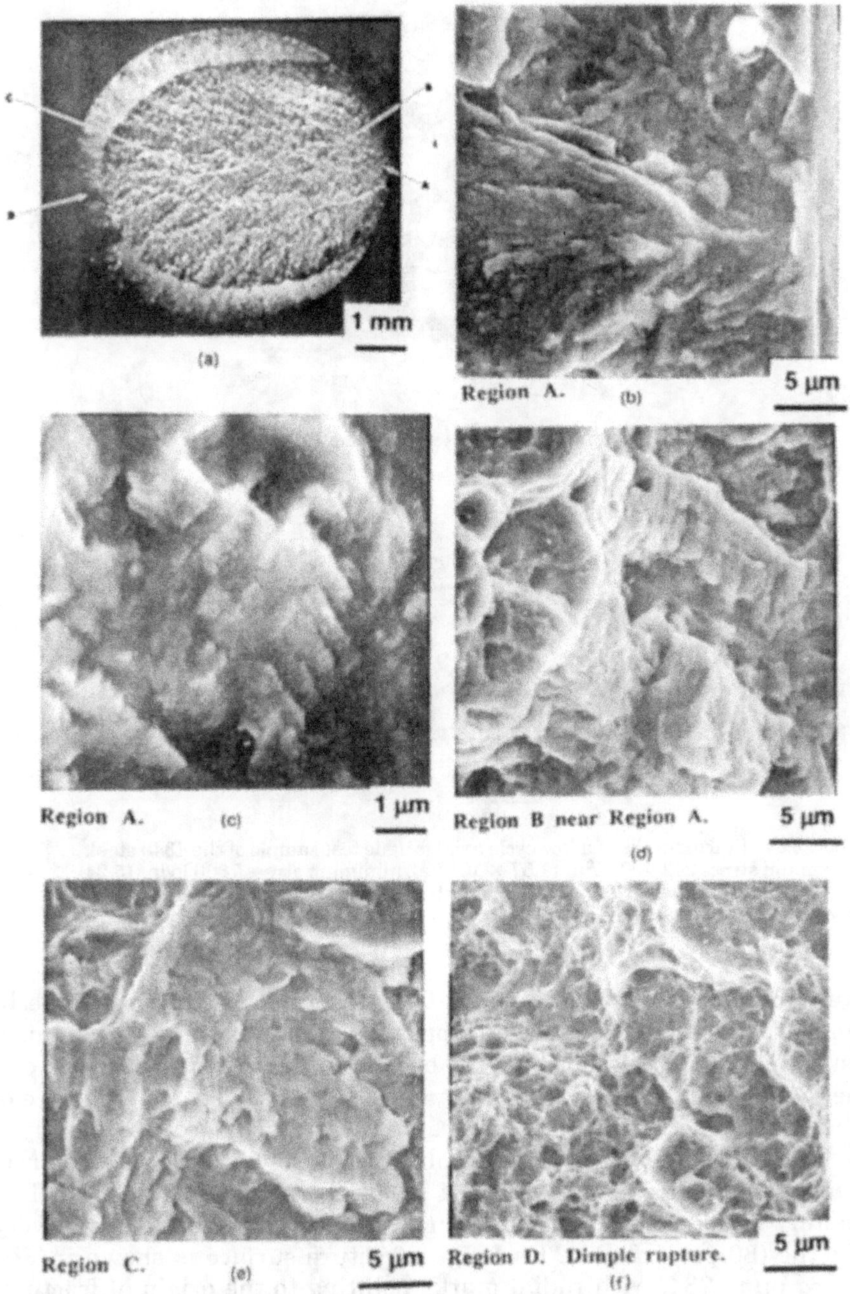

Figure 4.59 Fractographs of a high-cycle axial fatigue test sample of the 4340 steel. Maximum stress 96,000 lb/in^2 (6.61×10^8 Pa); minimum stress 0; test frequency 3600 cycles/min; number of cycles to failure 148,000. (*Adapted from Bhattacharyya et al.*[9])

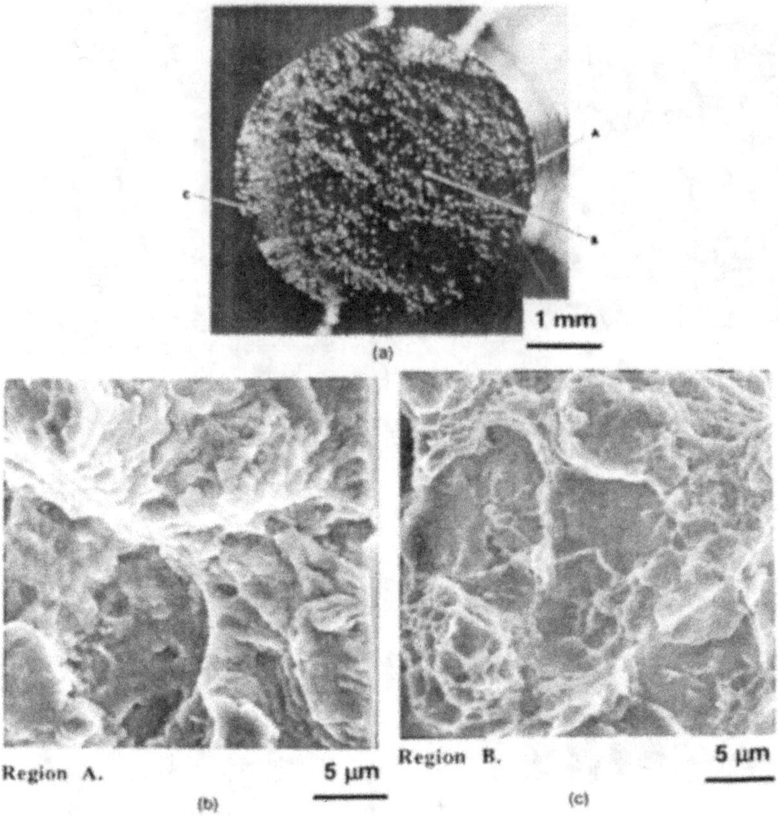

Figure 4.60 Fractographs of a low-cycle axial fatigue test sample of the 4340 steel. Maximum stress 242,400 lb/in^2 (1.67 × 10^9 Pa); minimum stress 7600 lb/in^2 (5.24 × 10^7 Pa); test frequency 6 cycles/min; number of cycles to failure 1317. (*Adapted from Bhattacharyya et al.*[9])

was about 0.4 percent for each. The different fracture micromechanisms, and the difference in appearance for the same mechanism (such as void coalescence), must be due to the difference in the dispersion of the carbides, which is an example of the strong influence of the microstructure on the fracture process and surface morphology.

The previous examples all involved relatively complex microstructures. The next example is a 304 austenitic stainless steel, which is mainly single-phase (except for inclusions). This material was very ductile (80 percent RA). Its fatigue fracture surface is shown in Fig. 4.68*a* on p. 284, with radial marks pointing to the origin of fracture. Even though this test was a continuous constant-frequency, constant-maximum-load test, there are faint conchoidal marks (see arrows).

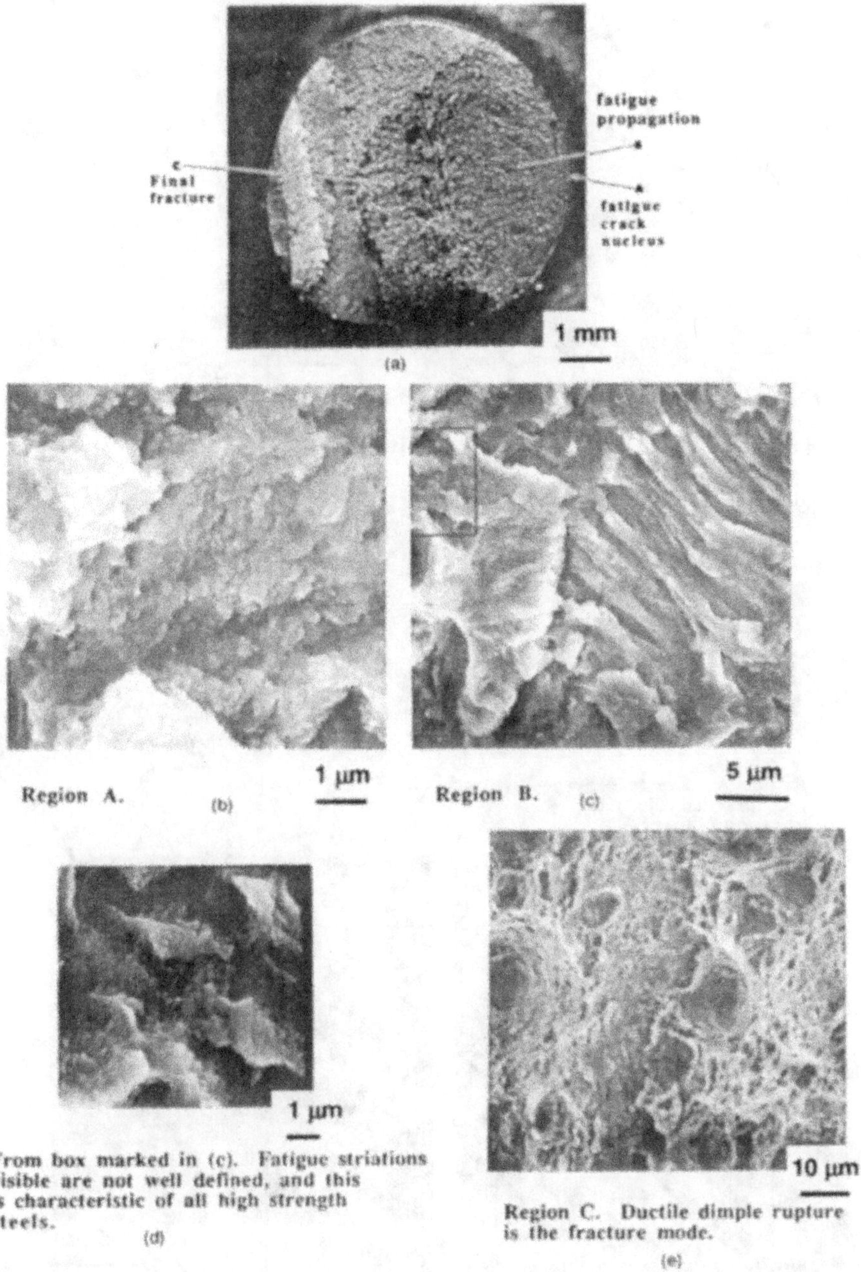

Figure 4.61 Fractographs of a high-cycle axial fatigue test sample of the 4340 steel, tempered at 1050°F (566°C). Maximum stress 89,000 lb/in^2 (6.13×10^8 Pa); minimum stress 2500 lb/in^2 (1.72×10^7 Pa); frequency 3600 cycles/min; number of cycles to failure 217,000. (*Adapted from Bhattacharyya et al.*[9])

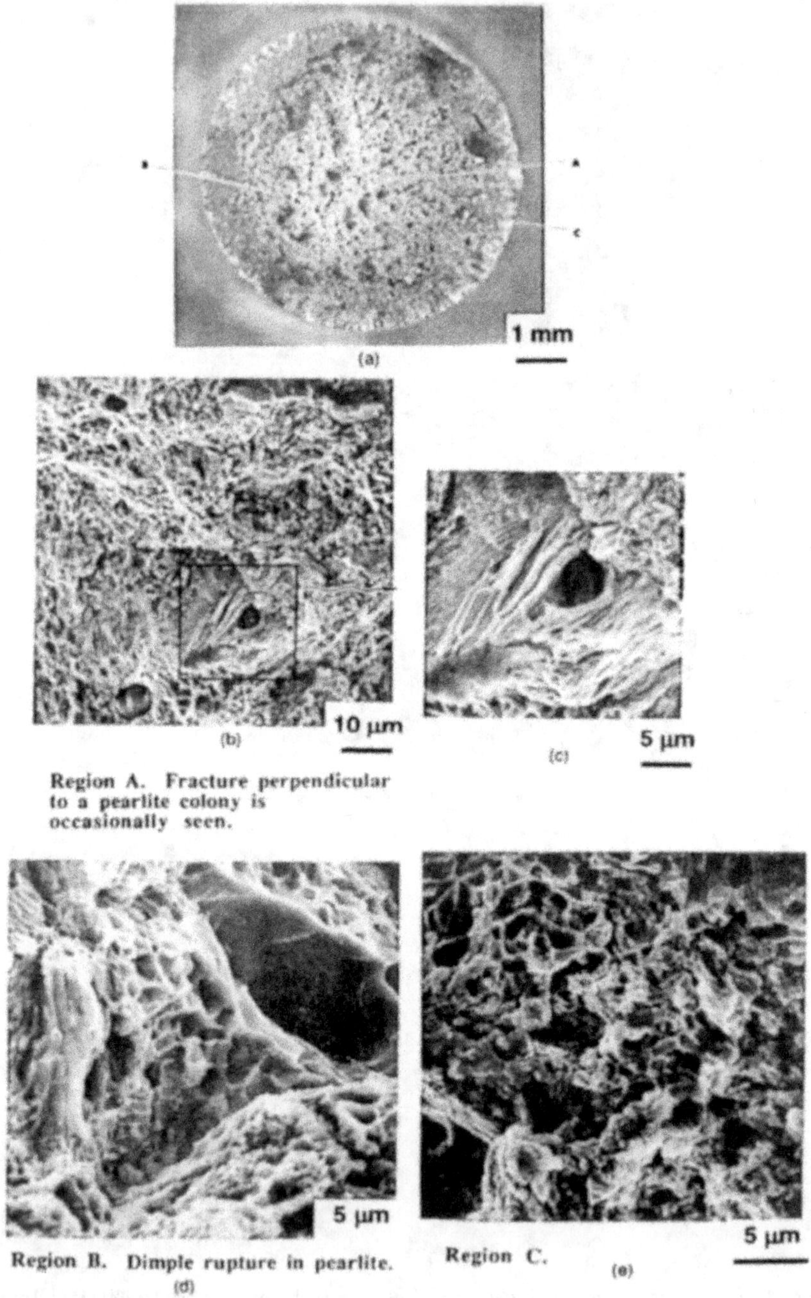

Figure 4.62 Fractographs of a 4340 steel tensile test sample in the normalized condition. The mechanical properties were yield strength 57,500 lb/in^2 (3.96×10^8 Pa); tensile strength 101,500 lb/in^2 (6.99×10^8 Pa); ductility RA 45%. (*Adapted from Bhattacharyya et al.*[9])

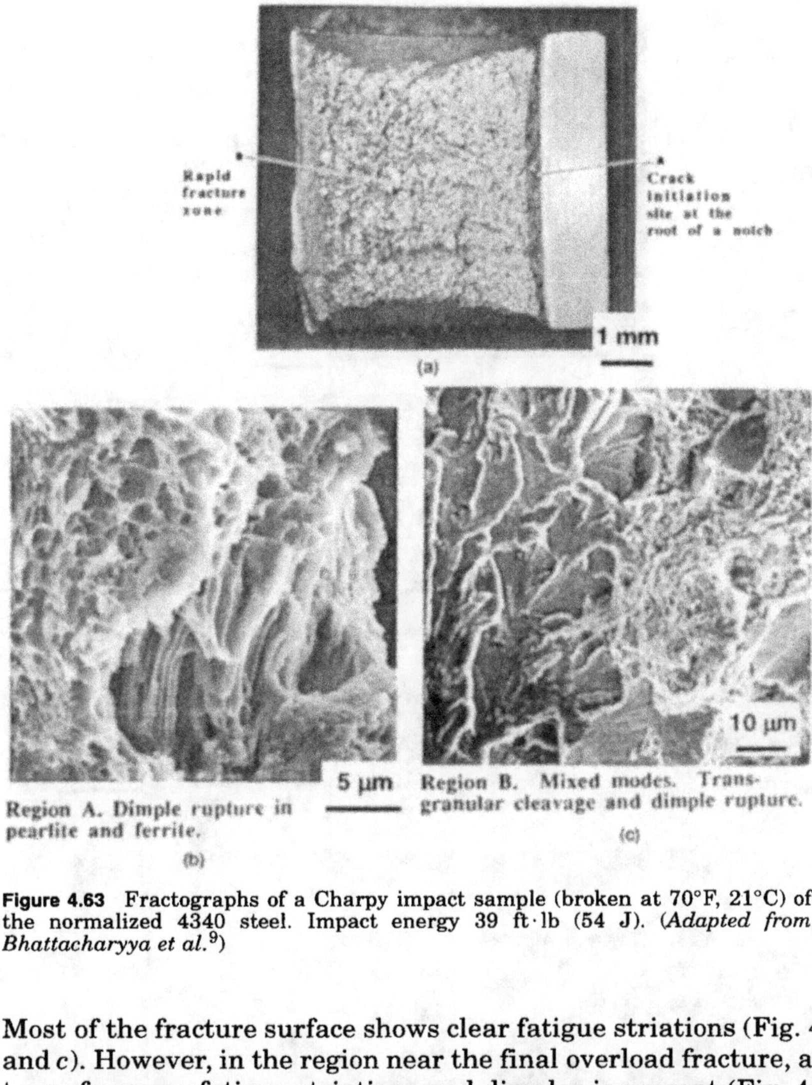

Figure 4.63 Fractographs of a Charpy impact sample (broken at 70°F, 21°C) of the normalized 4340 steel. Impact energy 39 ft·lb (54 J). (*Adapted from Bhattacharyya et al.*[9])

Most of the fracture surface shows clear fatigue striations (Fig. 4.68*b* and *c*). However, in the region near the final overload fracture, a mixture of coarser fatigue striations and dimples is present (Fig. 4.68*c*). The final overload fracture occurred by void coalescence (Fig. 4.68*d*).

Figure 4.69*a* on p. 285 shows the fracture surface of a tensile sample of a brittle material (1008 steel). Note the relatively shiny areas which are present. These are due to reflections from the relatively flat and coarse cleavage planes illustrated in Fig. 4.69*b*. A similar shiny appearance can be caused by intergranular fracture in coarse-grained material if the grain-boundary facets are smooth. An example is shown in Fig. 4.70 on p. 286.

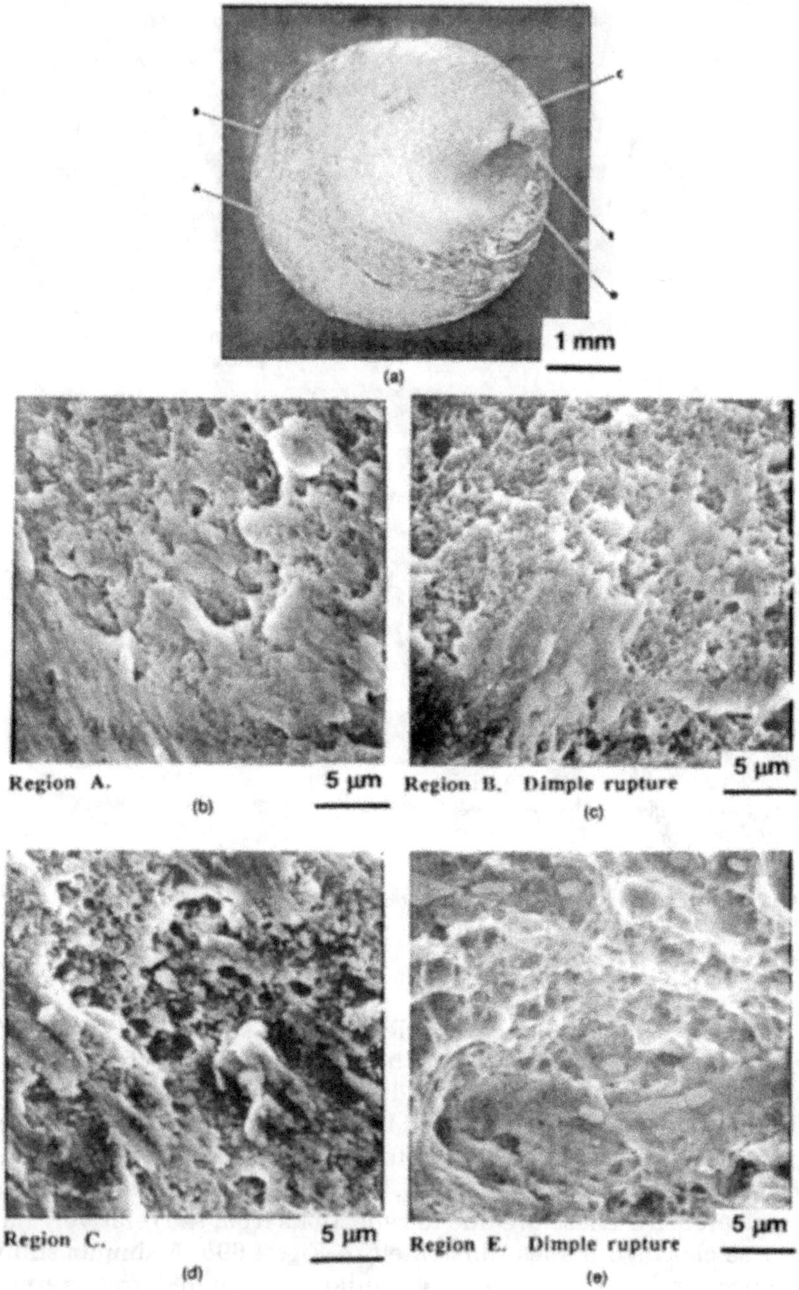

Figure 4.64 Fractographs of a torsion test of a 4340 steel, quenched and tempered (800°F, 426°C, for 1 h). Torque to failure 52 ft · lb (70 J); arc 720°. (*Adapted from Bhattacharyya et al.*[9])

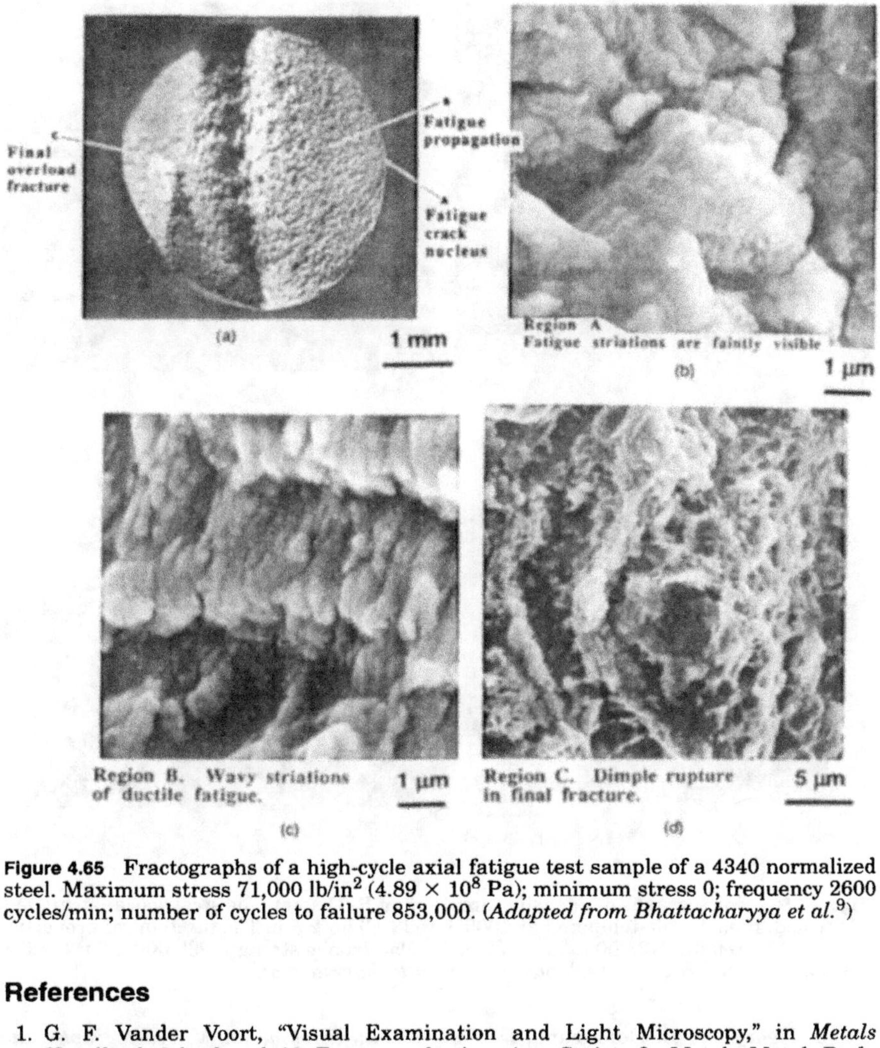

Figure 4.65 Fractographs of a high-cycle axial fatigue test sample of a 4340 normalized steel. Maximum stress 71,000 lb/in^2 (4.89×10^8 Pa); minimum stress 0; frequency 2600 cycles/min; number of cycles to failure 853,000. (*Adapted from Bhattacharyya et al.*[9])

References

1. G. F. Vander Voort, "Visual Examination and Light Microscopy," in *Metals Handbook,* 9th ed., vol. 12: *Fractography,* American Society for Metals, Metals Park, Ohio, 1987, p. 91.
2. *Metals Handbook,* 8th ed., vol. 9: *Fractography and Atlas of Fractographs,* American Society for Metals, Metals Park, Ohio, 1974.
3. L. Engel and H. Klingele, *An Atlas of Metal Damage,* Carl Hanser Verlag, Munich, Germany, 1981.
4. J. Nunes, F. L. Carr, and F. R. Larson, "Macrofractographic Techniques," in R. F. Bunshah (ed.), *Techniques for the Direct Observation of Structure and Imperfections,* pt. 1, Wiley, New York, 1968, p. 379.
5. *Metals Handbook,* 9th ed., vol. 12: *Fractography,* American Society for Metals, Metals Park, Ohio, 1987.
6. D. J. Wulpi, *Understanding How Components Fail,* American Society for Metals, Metals Park, Ohio, 1985.
7. *Metals Handbook,* 8th ed., vol. 10: *Failure Analysis and Prevention,* American Society for Metals, Metals Park, Ohio, 1975.

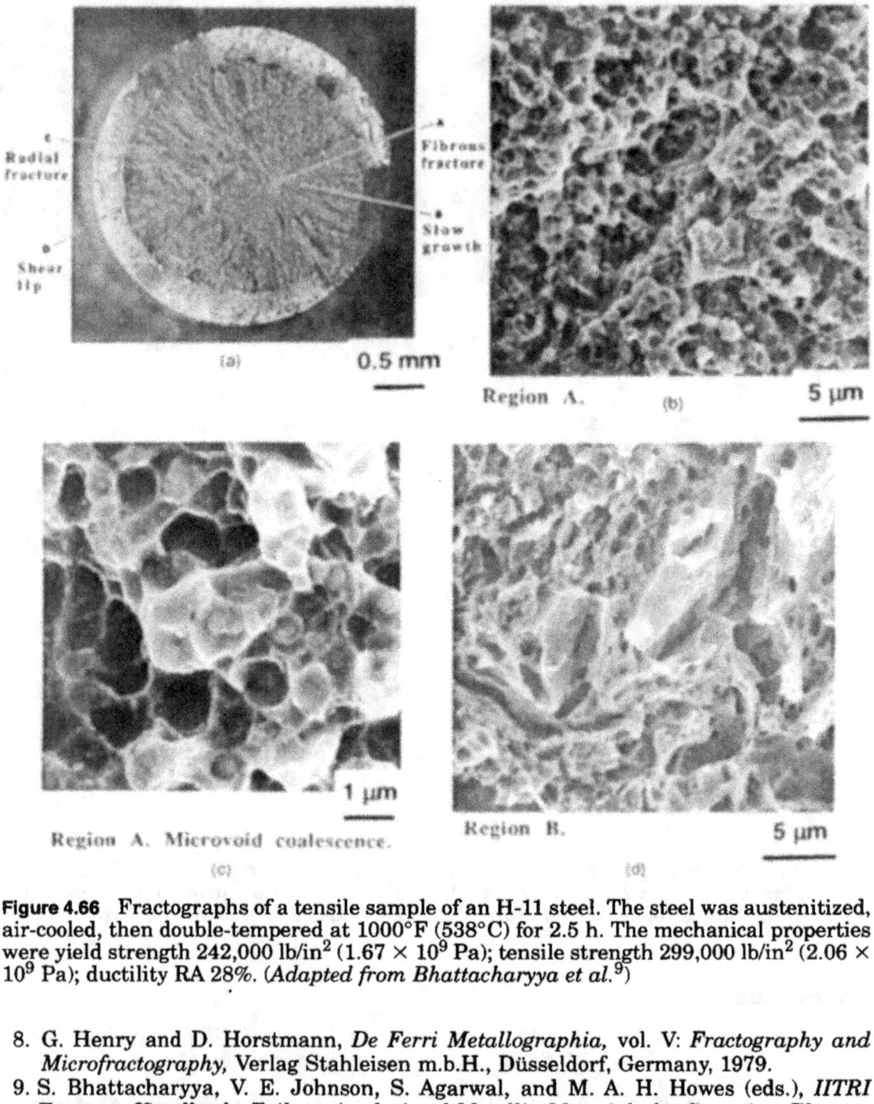

Figure 4.66 Fractographs of a tensile sample of an H-11 steel. The steel was austenitized, air-cooled, then double-tempered at 1000°F (538°C) for 2.5 h. The mechanical properties were yield strength 242,000 lb/in^2 (1.67×10^9 Pa); tensile strength 299,000 lb/in^2 (2.06×10^9 Pa); ductility RA 28%. (*Adapted from Bhattacharyya et al.*[9])

8. G. Henry and D. Horstmann, *De Ferri Metallographia,* vol. V: *Fractography and Microfractography,* Verlag Stahleisen m.b.H., Düsseldorf, Germany, 1979.
9. S. Bhattacharyya, V. E. Johnson, S. Agarwal, and M. A. H. Howes (eds.), *IITRI Fracture Handbook: Failure Analysis of Metallic Materials by Scanning Electron Microscopy,* IIT Research Institute, Chicago, Ill., 1979.
10. R. E. Peterson, "Interpretation of Service Failures," in M. Hetenyi (ed.), *Handbook of Experimental Stress Analysis,* Wiley, New York, 1950, p. 593.

Bibliography

Cottell, G. A.: "Fatigue Failures, with Special Reference to Fracture Characteristics," in F. R. Hutchings and P. M. Unterweiser, *Failure Analysis: The British Engine Technical Reports,* American Society for Metals, Metals Park, Ohio, 1981.

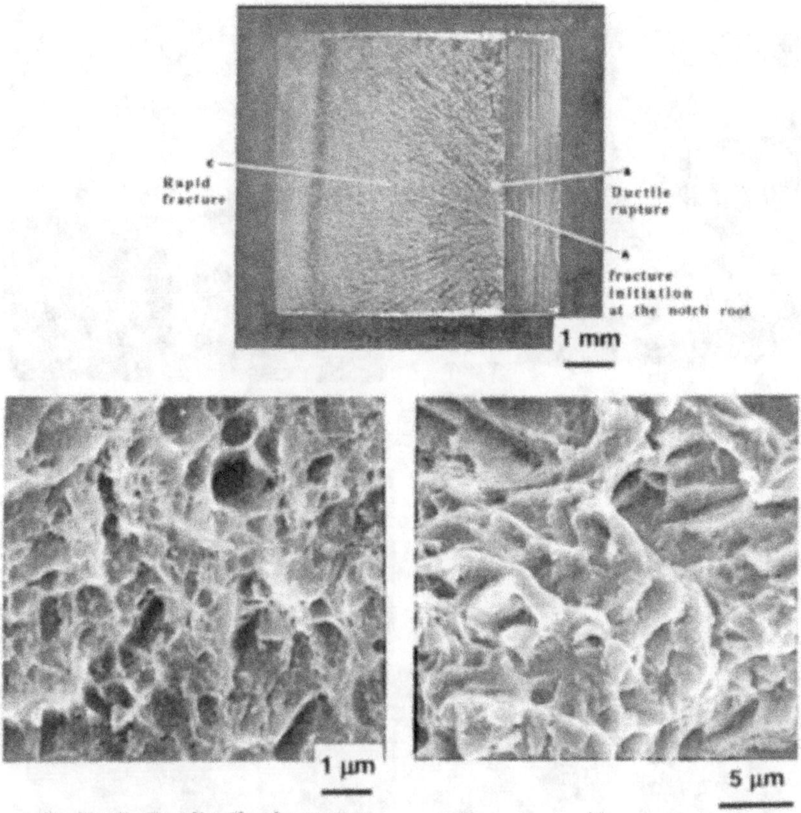

Region B. Ductile dimple rupture in local areas.

(b)

(c)

Figure 4.67 Fractographs of a Charpy impact sample of the H-11 steel. Impact energy 5 ft · lb (7 J). (*Adapted from Bhattacharyya et al.*[9])

Lang, G. A (ed.): *Systematic Analysis of Technical Failures,* DGM Informationsgesellschaft Verlag, Braunschweig, Germany, 1986.

Metals Handbook, 9th ed., vol. 11: *Failure Analysis and Prevention,* American Society for Metals, Metals Park, Ohio, 1986

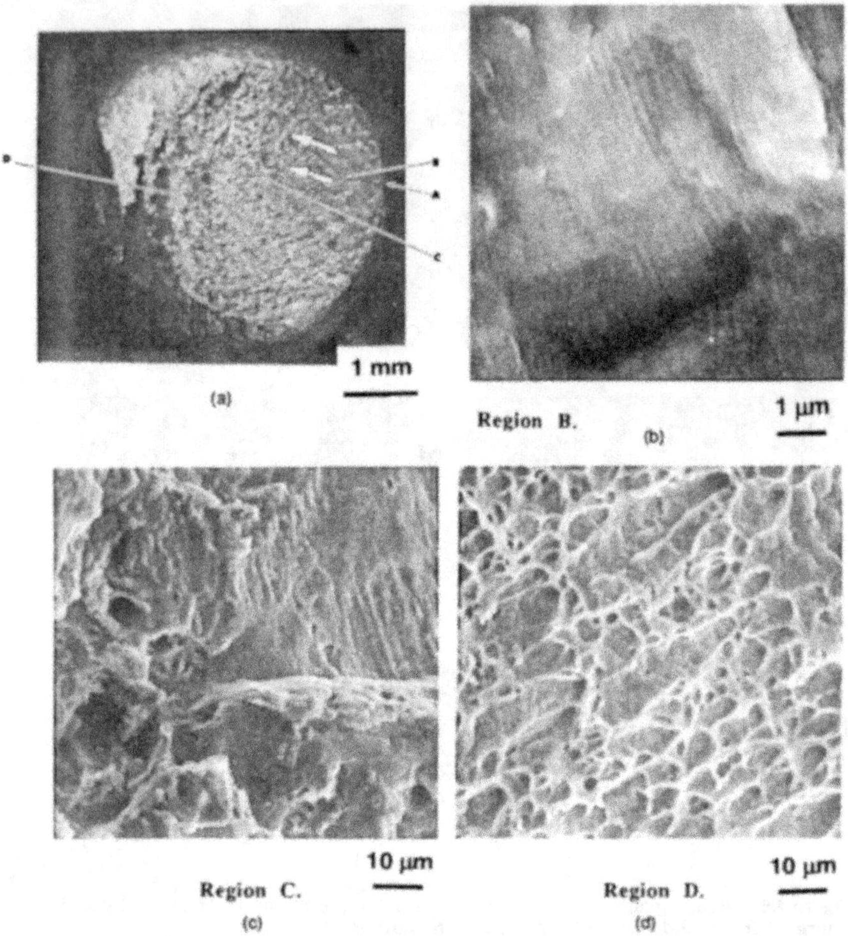

Figure 4.68 Fractographs of a high-cycle axial fatigue test sample of a 304 austenitic stainless steel. Maximum stress 89,000 lb/in^2 (6.13×10^8 Pa); minimum stress 4000 lb/in^2 (2.76×10^7 Pa); frequency 3600 cycles/min; number of cycles to failure 945,000. (*Adapted from Bhattacharyya et al.*[9])

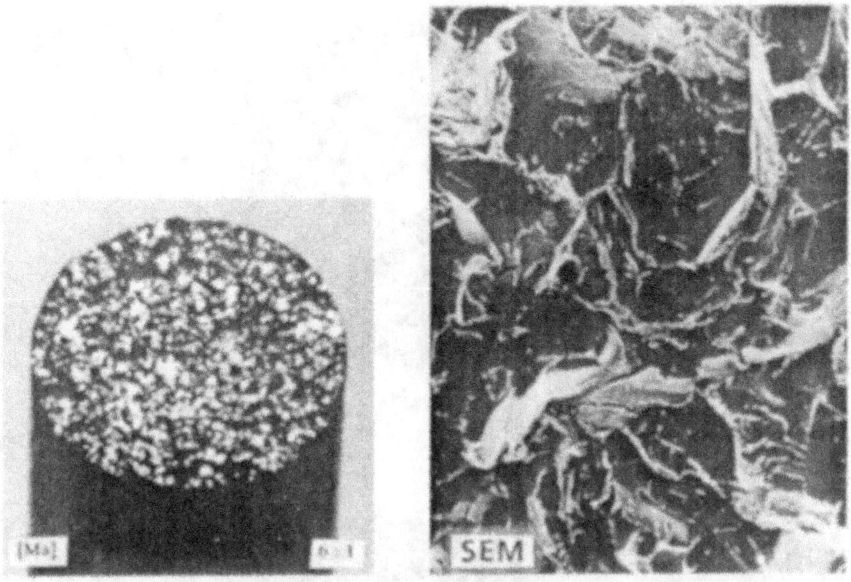

(a) (b)

Figure 4.69 (*a*) Tensile fracture at −196°C; coarse-grained 0.08% C steel. (*b*) High magnification showing cleavage facets. (*From Henry and Horstmann.*[8])

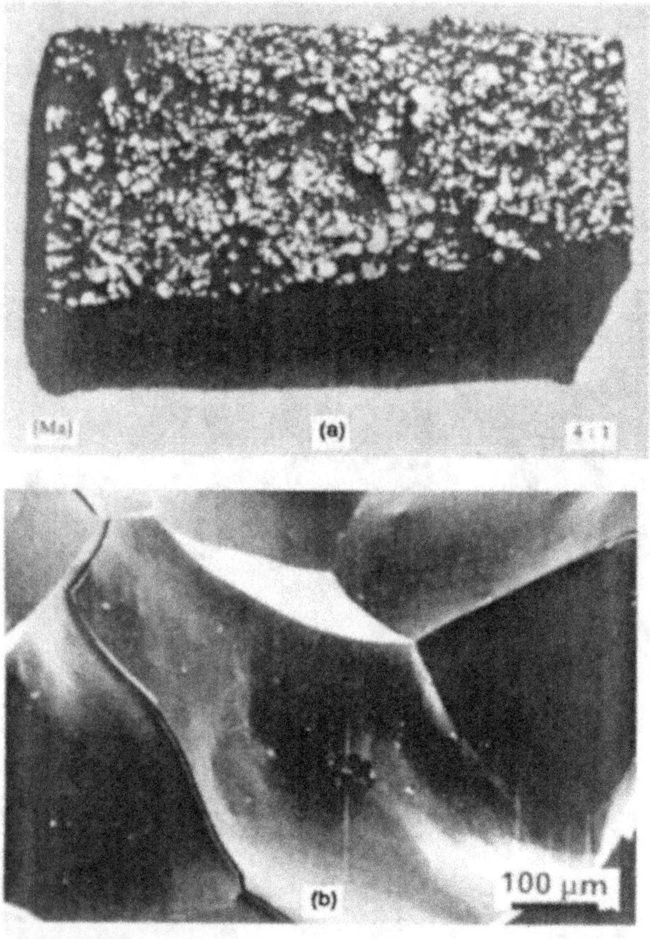

Figure 4.70 Tensile fracture surface of a sample of large-grained iron containing 0.68% P, showing intergranular fracture. (*From Henry and Horstmann.*[8])

Chapter

5

Failure Analysis of Composites

Metals, ceramics, glasses and polymers can be combined in materials that have unique properties on their own. Nature uses this principle in wood and bone; man applies it in a new family of supermaterials.

ANTHONY KELLY

"The Nature of Composite Materials," in Materials, *W.H. Freeman and Co., San Francisco, 1967*

5.1 Introduction

A common method of strengthening engineering materials is the development of a structure of one phase dispersed in another. For many metallic alloys, this is produced by specific heat treatments. An example is the formation of fine carbide particles in ferrite by tempering a hardened (martensite) steel, giving a usable combination of strength and toughness. Such a dispersion of one phase in another is an example of a composite material.

Composites are heterogeneous structural materials consisting of two or more distinct phases, which are bonded chemically or mechanically. They can be produced by appropriate thermal and mechanical treatment, such as the tempering of steels cited above. This type of composite is referred to as an in situ composite, and it usually has a fine structure. However, many composites are manufactured by mixing the individual components (phases) in a manner to bond them. An example is the embedding of glass fibers in epoxy to produce fiberglass. The structure of this type of composite is relatively coarse.

The properties of composites depend not only on the intrinsic properties of the dispersed component and those of the matrix, but also on the geometry (size and shape) of the dispersed component. In addition, the properties depend on the characteristics of the interface between the dispersed component and the matrix. Thus mechanical failure of the composites is associated not only with fracture of the dispersed component and the matrix, but also with separation of the two at their interface. This latter characteristic depends on the bonding between the two components. Also important is the surface geometry (e.g., roughness) since this affects the degree of mechanical locking of the components.

The majority of materials used in industrial, domestic, civil, and technological applications are composites. They can be either manufactured (e.g., fiberglass, reinforced concrete) or natural (e.g., bamboo, wood). Table 5.1 shows a broad classification of composite materials. The following variables have to be considered in the classification of composites:

1. Type of majority component (e.g., metal, mineral, organic, ceramic, glass, carbon)
2. Type of minority components (e.g., metal, mineral, organic, ceramic, glass, carbon)
3. Number of minority components (e.g., two-component concrete)
4. Amount of majority component
5. Amount of minority components
6. Form of minority components (e.g., fiber, particle, laminae)
7. Dispersion of minority components (e.g., bulk, agglomerates, surface enriched)

TABLE 5.1 Broad Classification of Composite Materials

Class	Examples
Natural composite materials	Wood Bone Bamboo Muscle and other tissue
Microcomposite materials	Metallic alloys, e.g., steel Toughened thermoplastics Sheet molding compounds Reinforced thermoplastics
Macrocomposites (engineered products)	Galvanized steel Reinforced concrete beams Helicopter blades Skis

SOURCE: From Hull.[1]

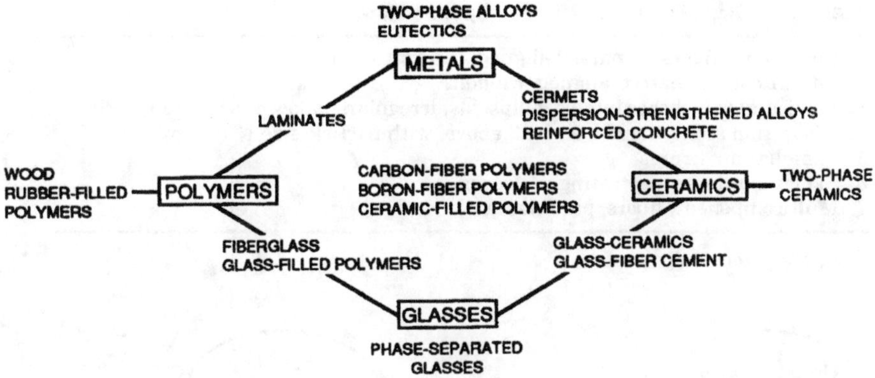

Figure 5.1 The classes of engineering materials and the composites that can be formed within a class and between classes. (*From Beaumont et al.*[2])

Figure 5.1 shows one classification scheme for composites made within and between classes of materials, i.e., metals, ceramics, polymers, etc. Consideration of the size, shape, and distribution of the two components in a composite leads to the classification scheme in Table 5.2. The performance of a composite material is dependent on the four major factors shown in Fig. 5.2. Note from this figure that in addition to the choice of individual components (in terms of their synthesis, composition), their mechanical properties, and the processing technique used to produce the composite, another factor is of crucial importance in determining the ultimate properties of the composite. This factor is interfacial properties. Considerable effort has been expended in understanding and engineering the properties of the various interfaces between the components of a composite. Optimization of interfacial properties can mean the difference between the success or failure of the use of a specific composite for a specific application.

The tensile stress-strain behavior of fiber-reinforced composites can illustrate the effect on the properties of the two constituents. For this illustration,

σ_{cu} = ultimate tensile strength (UTS) of composite

σ_{fu} = ultimate tensile strength of fiber

σ_{mu} = ultimate tensile strength of matrix

V_m = volume fraction of matrix

V_f = volume fraction of fiber = $1 - V_m$

Note that in the case of brittle materials such as glasses and ceramics, the ultimate tensile strength is the same as the tensile strength in

TABLE 5.2 Classification of Microcomposite Materials

1. Continuous fibers in matrix: aligned, random
2. Short fibers in matrix: aligned, random
3. Particulates (spheres, plates, ellipsoids, irregular, hollow or solid) in matrix
4. Dispersion strengthened, as for 3 above, with particle size $< 10^{-8}$ m
5. Lamellar structures
6. Skeletal or interpenetrating networks
7. Multicomponent, fibers, particles, etc.

SOURCE: From Hull.[1]

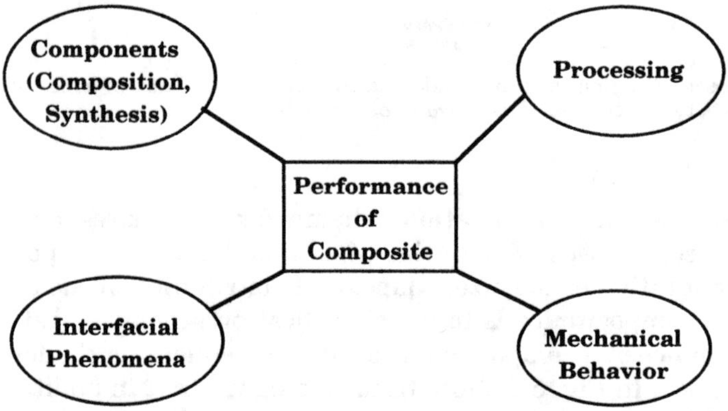

Figure 5.2 Performance of a composite is linked to some important factors: composition of components, their mechanical behavior, processing, and the characteristics of the interface between matrix and reinforcement. (*From Chawla.*[3])

that the engineering stress-strain curve for such materials does not display a maximum; rather fracture stress obviates a maximum in the curve.

If we first consider the impractical case where the failure strains of the fiber and the matrix are equal, the behavior of the composite can be represented by

$$\sigma_{cu} = \sigma_{fu} V_f + \sigma_{mu} V_m$$

However, in most metal matrix composites and some polymer matrix composites, the fiber failure strain will be less than that of the matrix. This is shown in the schematic representation of Fig. 5.3*a* for the variation of composite strength as a function of fiber volume fraction. Note that if the volume fraction of fibers is less than a minimum value V_{min}, the composite will be weaker than the matrix material (per equal volume). In order for the composite to be stronger than the matrix, a certain critical volume fraction of fiber (shown as V_{crit}) is needed. It can be

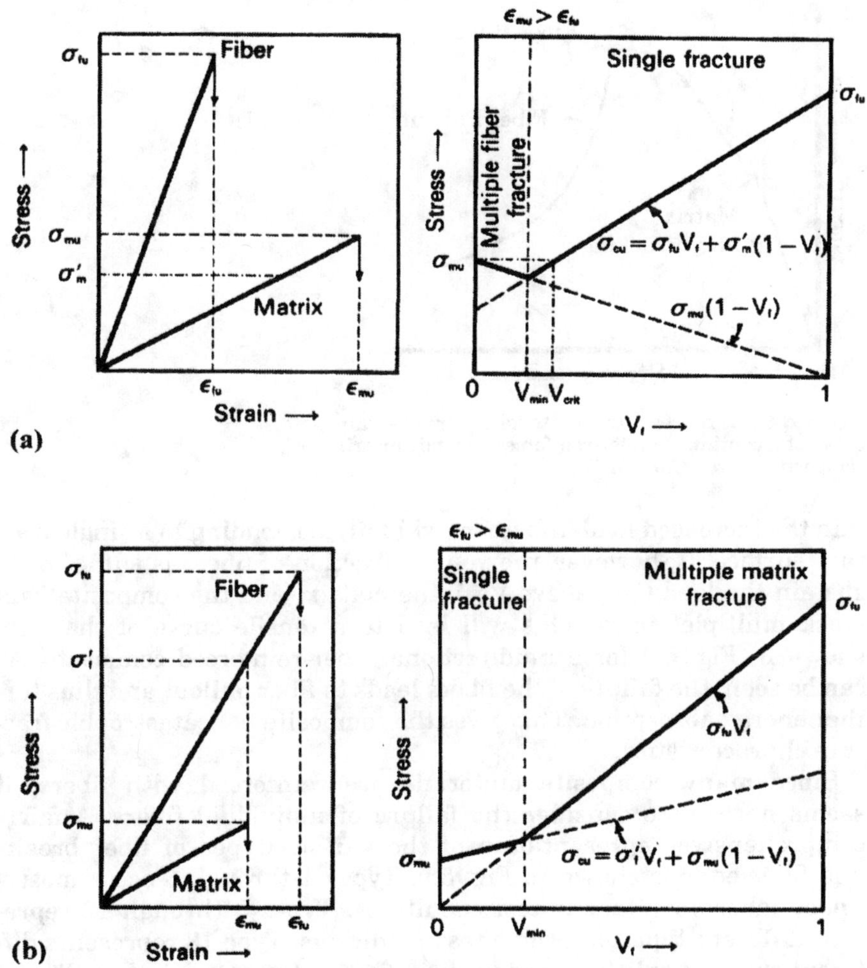

Figure 5.3 Stress-strain curves and strength of composite versus fiber volume fraction showing conditions for single and multiple cracking, in a unidirectionally aligned fiber composite: (*a*) matrix strain to failure greater than that of the fiber and (*b*) matrix strain to failure less than that of the fiber. (*From Chawla.*[3])

seen that at volume fractions below $V_{\min}$, the fracture surface will show multiple fiber fractures whereas above $V_{\min}$, a single fracture surface will be seen.

In most ceramic matrix composites, however, the fiber failure strain is higher than that of the matrix, leading to the schematic representation of Fig. 5.3*b* for the variation of composite strength with fiber volume fraction. Since the fiber failure strain is larger than that of the matrix, the matrix will fail first. Thereupon, if the fibers cannot sus-

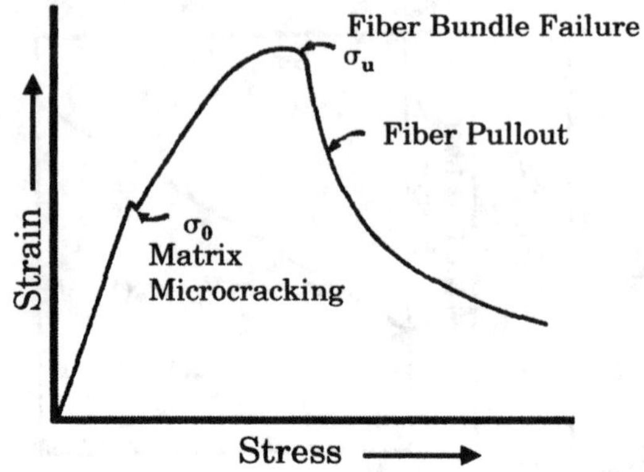

Figure 5.4 A representative tensile stress-strain curve of a continuous fiber-reinforced ceramic matrix composite. (*From Chawla.*[3])

tain the increased load, the fibers will fail, too, leading to a single fracture surface. If, however, the volume fraction of fibers is sufficient to sustain the load (i.e., above V_{crit}), the matrix (and the composite) will crack multiple times. This will lead to a tensile curve of the form shown in Fig. 5.4 for a unidirectional fiber-reinforced composite. As can be seen, the failure of the fibers leads to fiber pullout and thus further energy absorption. This gives the composite noncatastrophic fracture characteristics.

Since many composite materials are reinforced with fibers, it seems natural to consider the failure of individual fibers. A fairly comprehensive representation of the various types of fiber breaks and fiber ends is shown in Fig. 5.5. Types 1 through 6 occur mostly under laboratory tensile test conditions. Types 8 through 12 represent different types of fatigue test conditions. Type 13 represents different types of splitting and peeling. Type 14 represents rounding of the fiber due to wear following breakage in use. Type 15 represents blunt cutting or crushing while type 16 represents a sharp cut. Type 17 represents melting at the tip while ends of natural fibers are represented by type 18. It is also not uncommon for fibers to develop kink bands on the compression side of bent fibers (Fig. 5.6). Upon repeated bending, the kink bands undergo the sequence of events depicted in Fig. 5.7. Examples of flexural fatigue fracture of fibers are shown in the photographs of Fig. 5.8.

Figure 5.9 schematically depicts the three damage mechanisms operative in fatigue loading of unidirectional fiber composites, namely, fiber breakage and interfacial debonding, matrix cracking, and interfacial shear failure. In terms of the fatigue life, these mechanisms lead to the kind of behavior shown schematically in Fig. 5.10.

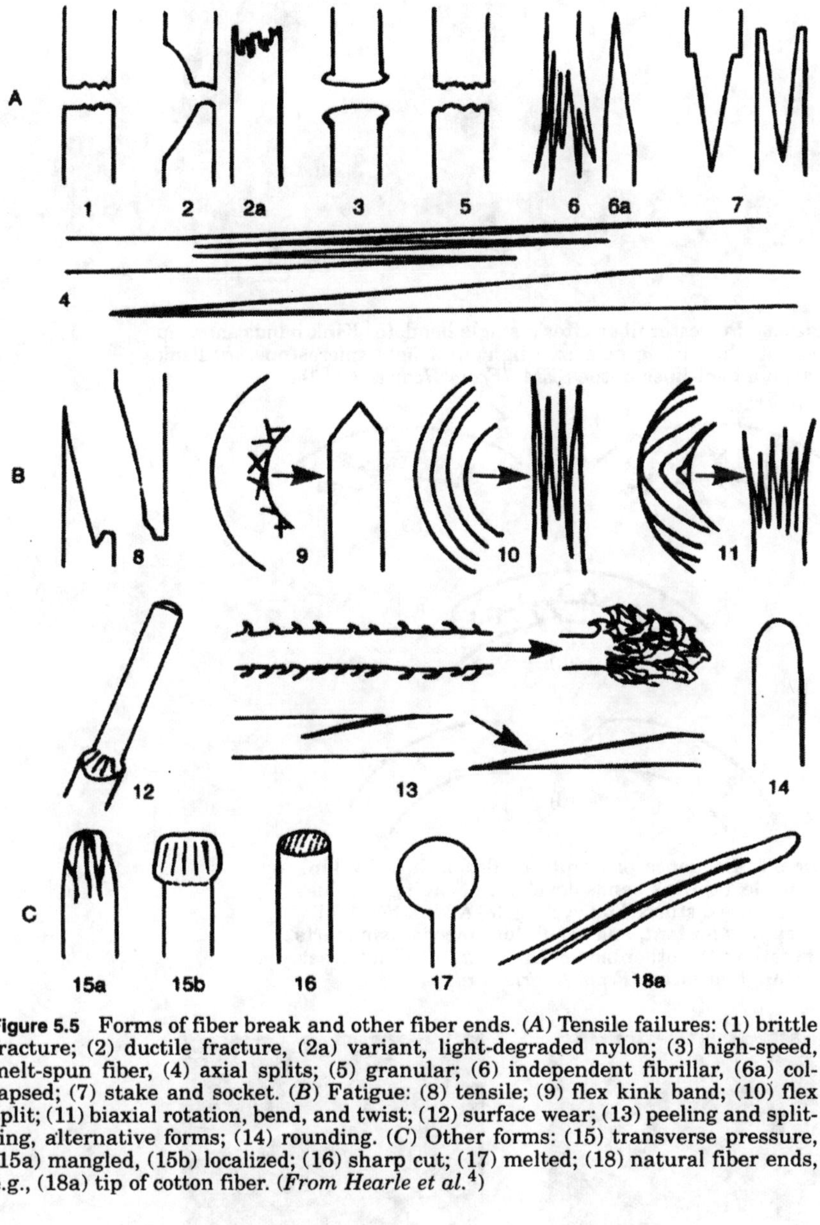

Figure 5.5 Forms of fiber break and other fiber ends. (*A*) Tensile failures: (1) brittle fracture; (2) ductile fracture, (2a) variant, light-degraded nylon; (3) high-speed, melt-spun fiber, (4) axial splits; (5) granular; (6) independent fibrillar, (6a) collapsed; (7) stake and socket. (*B*) Fatigue: (8) tensile; (9) flex kink band; (10) flex split; (11) biaxial rotation, bend, and twist; (12) surface wear; (13) peeling and splitting, alternative forms; (14) rounding. (*C*) Other forms: (15) transverse pressure, (15a) mangled, (15b) localized; (16) sharp cut; (17) melted; (18) natural fiber ends, e.g., (18a) tip of cotton fiber. (*From Hearle et al.*[4])

5.2 Polymer Matrix Composites

5.2.1 Continuous versus discontinuous fibers

A discussion of failure mechanisms in fiber-reinforced composites must recognize an essential difference between short- and long-fiber-

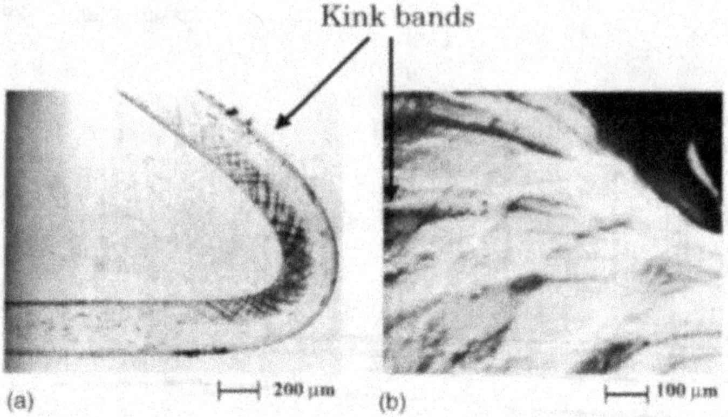

Figure 5.6 Polyester fiber after a single bend. (*a*) Kink bands show up in a bent fiber, using polarized light in a light microscope. (*b*) Kink bands in a bent fiber in the SEM. (*From Hearle et al.*[4])

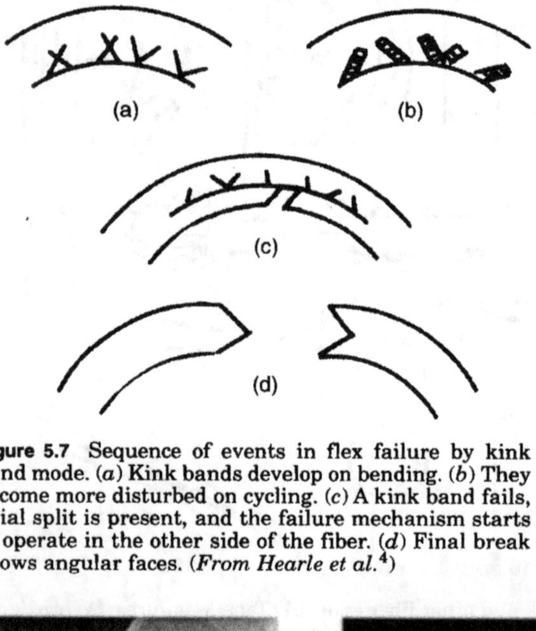

Figure 5.7 Sequence of events in flex failure by kink band mode. (*a*) Kink bands develop on bending. (*b*) They become more disturbed on cycling. (*c*) A kink band fails, axial split is present, and the failure mechanism starts to operate in the other side of the fiber. (*d*) Final break shows angular faces. (*From Hearle et al.*[4])

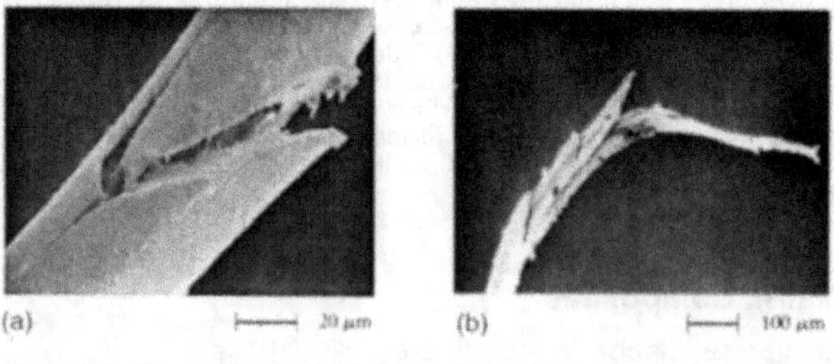

Figure 5.8 Flex fatigue of nylon 6 at 20°C, 65% relative humidity. (*a*) Example of kink band failure. (*b*) Kink bands with more pronounced splitting. (*From Hearle at al.*[4])

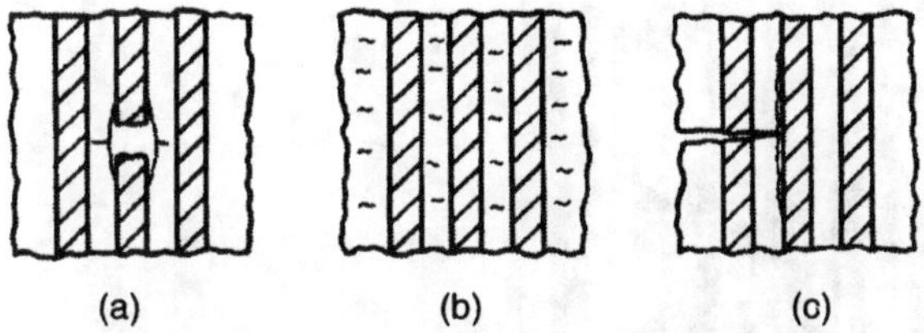

Figure 5.9 Fatigue damage mechanisms in unidirectional composites under loading parallel to fibers: (*a*) fiber breakage, interfacial debonding; (*b*) matrix cracking; (*c*) interfacial shear failure. (*From Talreja.*[5])

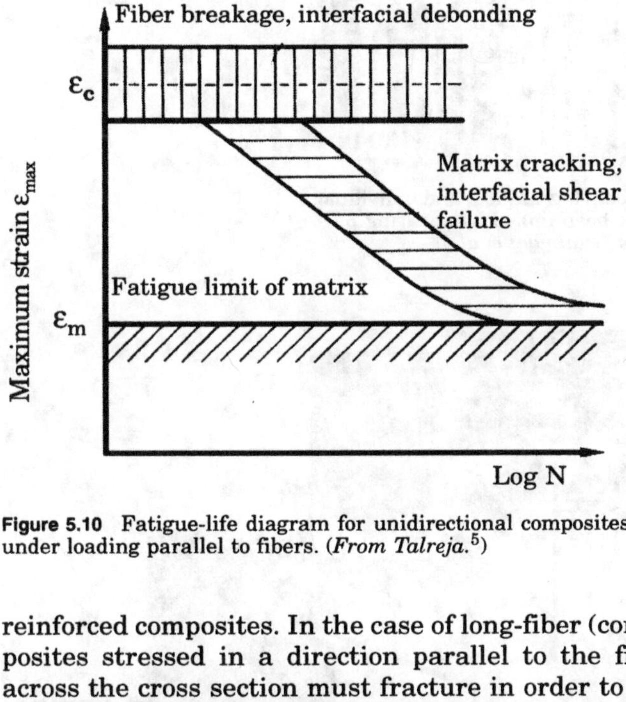

Figure 5.10 Fatigue-life diagram for unidirectional composites under loading parallel to fibers. (*From Talreja.*[5])

reinforced composites. In the case of long-fiber (continuous fiber) composites stressed in a direction parallel to the fiber axis, all fibers across the cross section must fracture in order to cause failure of the composite (Fig. 5.11). Conversely, in the case of short-fiber (discontinuous fiber) composites, the fracture path may evade all the fibers, traversing exclusively through the matrix as shown in Fig. 5.12. In both cases, the energy-absorbing mechanisms that contribute to the failure process are

- Debonding at the crack tip, followed by crack deflection
- Debonding in the crack wake, followed by crack deflection
- Fiber-matrix debonding (frictional sliding)
- Fiber pullout

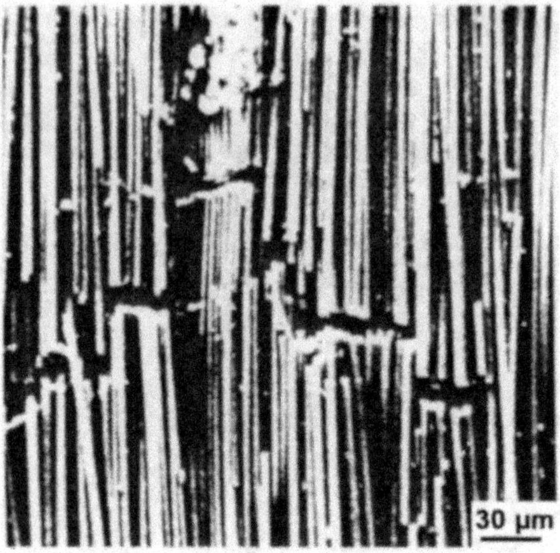

Figure 5.11 Groups of fibers have failed, and individual subbundle failures have been linked by shearing parallel to the fibers. (*From Beaumont et al.*[2])

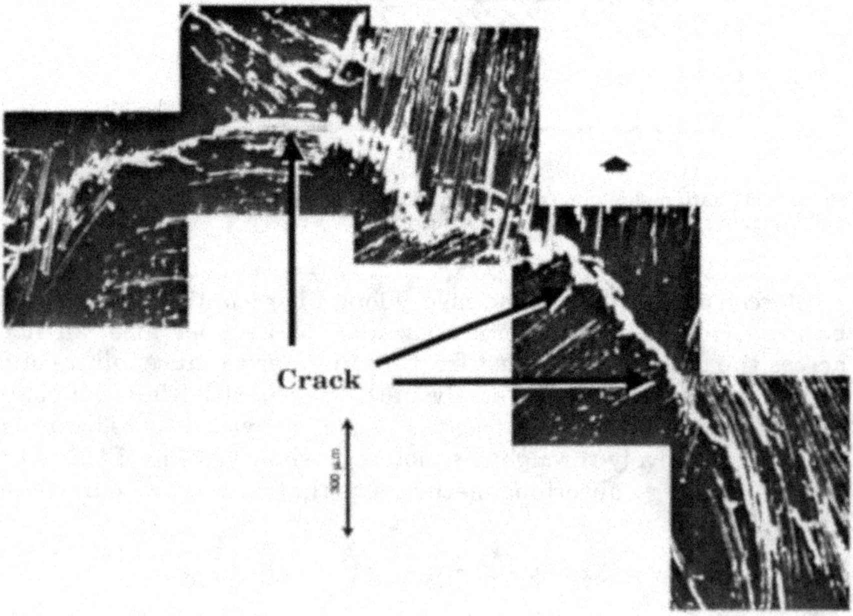

Figure 5.12 Crack path in short glass-fiber sheet molding compound after 15,000 cycles in tension-null fatigue loading. Crack path is one of "fiber avoidance." (*From Beaumont et al.*[2])

- Fiber fracture
- Crazing and matrix fracture

5.2.2 Influence of fiber brittleness and compressive strength

While glass and carbon fibers generally exhibit brittle fracture (Fig. 5.13*a* and *c*), some fibers such as Kevlar 49 exhibit significant necking and fibrillation prior to fracture, as shown in Fig. 5.13*b*. In the case of bending of Kevlar 49 fibers, high surface tensile and compressive stresses are set up on the outer and inner radii of the bend, respectively. Since Kevlar 49 has a low compressive strength, the material on the inner radius undergoes yielding (permanent deformation) by the formation of deformation bands (Fig. 5.14) prior to tensile fracture.

5.2.3 Influence of degree of adhesion

The overall fracture process and appearance of the fracture surface are affected by the degree of adhesion between the fiber and the matrix. This is illustrated by the example of injection-molded short fibers of

(a) (b)

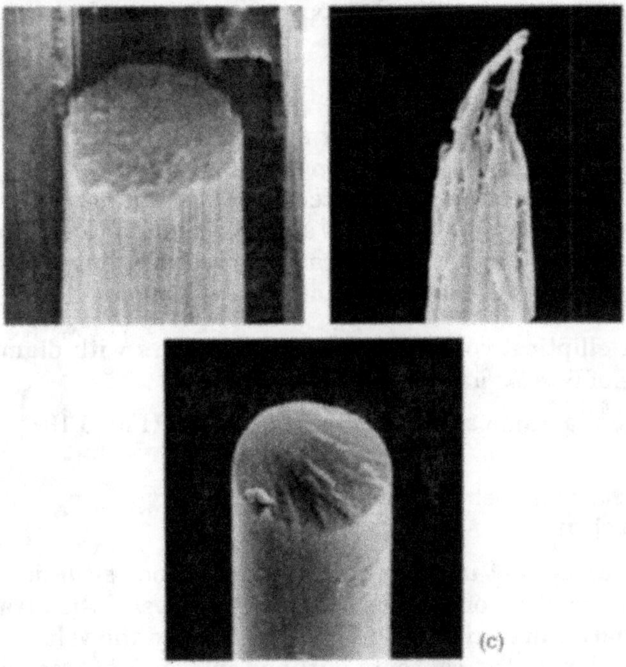

(c)

Figure 5.13 Scanning electron micrographs of fibers fractured in tension: (*a*) carbon, (*b*) Kevlar 49, (*c*) glass. (*From Hull.*[1])

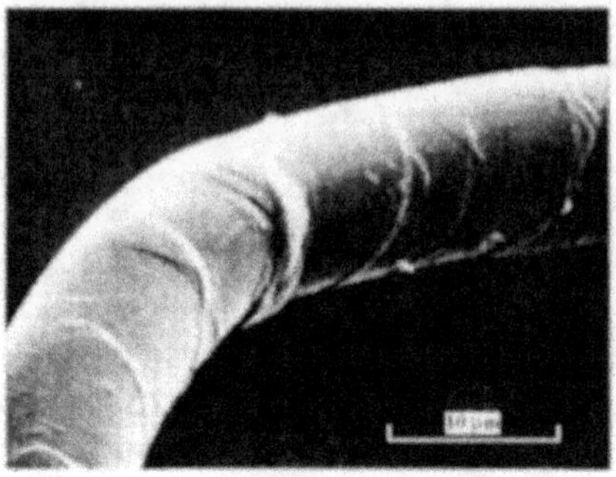

(a)

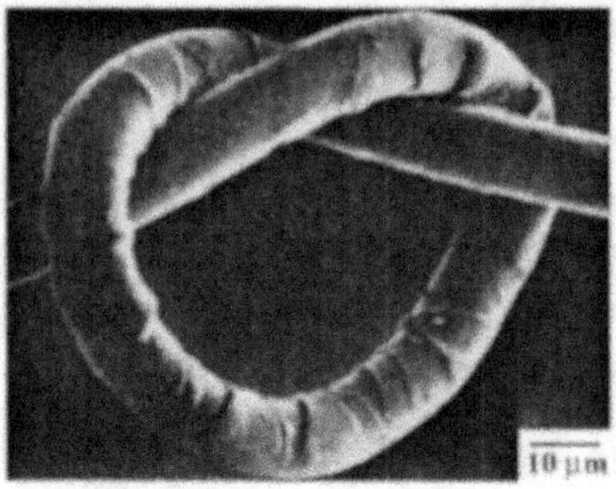

(b)

Figure 5.14 (*a*) Scanning electron micrograph of a Kevlar 49 fiber showing deformation bands on the compression side of a sharp bend. (*b*) Knotted Kevlar fiber showing buckling marks on compressive side. The tensile side is smooth. (*From Chawla.*[6])

glass in a polypropylene matrix. When the fibers are not treated with a coupling agent, the fracture surface shows extensive fiber pullout with smooth fiber surfaces as shown in Fig. 5.15. However, when the fibers are treated with a chemical coupling agent, the fracture surface shows a lesser degree of fiber pullout. In addition, the fiber surfaces show plastic adhering to them, as shown in Fig. 5.16.

5.2.4 Influence of voids

Interlaminar shear strength and other mechanical properties of composite materials are affected by the type, content, and distribution of voids in the composites. The void content and distribution in turn is dependent on the fiber volume fraction and distribution, resin properties, and processing conditions (temperature, pressure, time). In fiber-reinforced composites, there are two main types of voids:

1. Spherical or elliptical voids along individual fibers with diameters in the range of 5 to 20 μm (Fig. 5.17)
2. Voids between laminae and in resin-rich pockets (Fig. 5.18)

5.2.5 Longitudinal tensile strength of unidirectional laminae

For the failure process of unidirectional laminae, four situations can be envisaged, depending on the relative magnitudes of the strain to failure of the matrix material ε_m and the fibers ε_f and the volume fraction of fibers V_f. Using the concepts outlined in Fig. 5.3, consider the behavior of unidirectional lamina shown schematically in Fig. 5.19*a*.

(a) (b)

Figure 5.15 SEM micrographs of fractured specimens of injection-molded glass-filled polypropylene. Fibers without chemical coupling agent showing smooth fiber surfaces and holes left in the plastic matrix. (*From Hull.*[1])

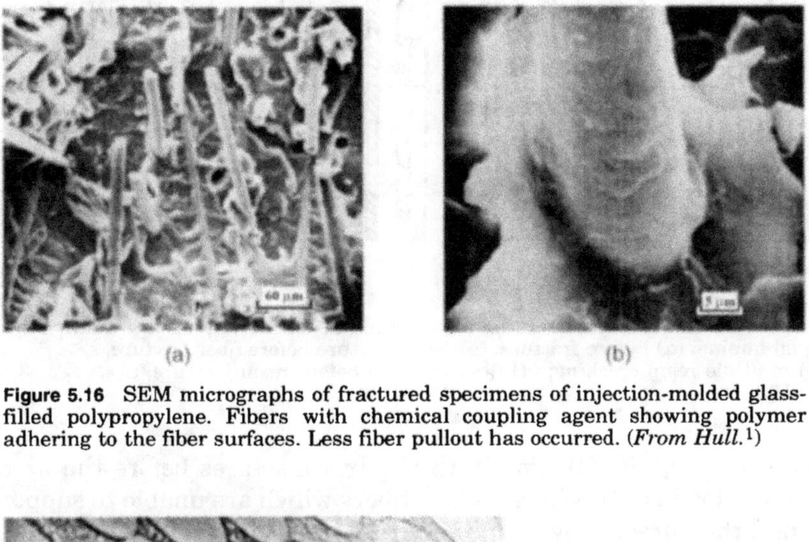

(a) (b)

Figure 5.16 SEM micrographs of fractured specimens of injection-molded glass-filled polypropylene. Fibers with chemical coupling agent showing polymer adhering to the fiber surfaces. Less fiber pullout has occurred. (*From Hull.*[1])

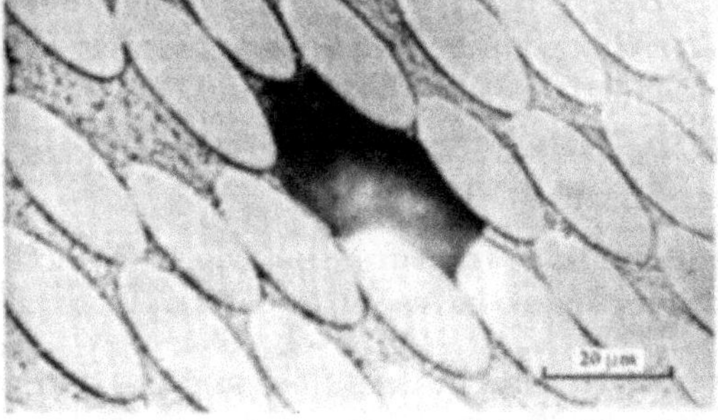

Figure 5.17 Void between fibers in a glass fiber–polyester resin lamina. (*From Hull.*[1])

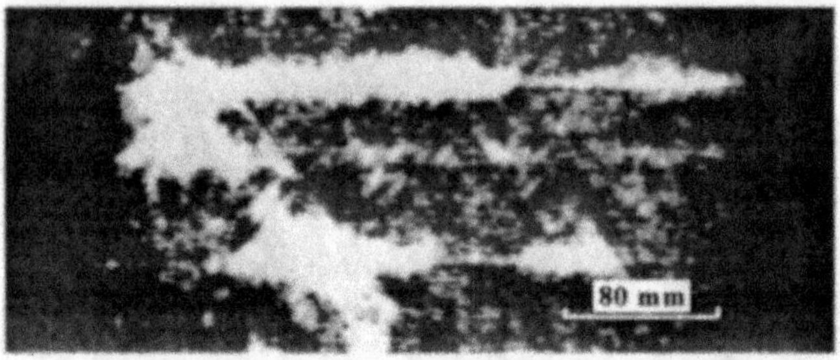

Figure 5.18 C-scan photograph of a multi-ply carbon fiber–epoxy resin laminate. Light regions are due to voids and delaminations in the material. (*From Hull.*[1])

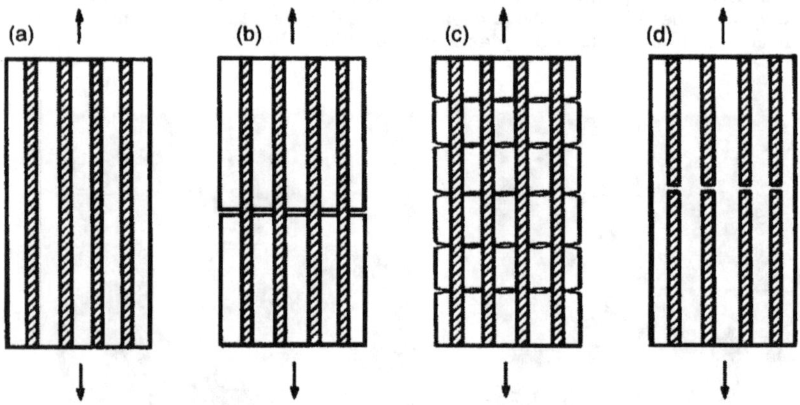

Figure 5.19 Schematic representation of longitudinal tensile failure of a unidirectional lamina. (*a*) Before fracture; (*b*) resin fracture before fiber fracture, $\varepsilon_f > \varepsilon_m$; (*c*) multiple resin cracking; (*d*) fiber fracture before resin fracture, $\varepsilon_m > \varepsilon_f$. (*From Hull.*[1])

1. When $\varepsilon_f > \varepsilon_m$, if V_f is small, the matrix fractures before the fiber, and the load is transferred to the fibers which are unable to support it and thus break (Fig. 5.19*b*).
2. When $\varepsilon_f > \varepsilon_m$, if V_f is large, the matrix takes a small proportion of load. Hence upon its fracture there is not much load transferred to the fibers, and thus multiple matrix cracking occurs first (Fig. 5.19*c*).
3. When $\varepsilon_m > \varepsilon_f$, if V_f is small, the fibers fracture but the extra load is insufficient to cause matrix fracture (Fig. 5.19*d*).
4. When $\varepsilon_m > \varepsilon_f$, if V_f is large, upon fiber fracture, the load transferred to the matrix is large, and hence matrix fracture results.

Three types of interactions between the broken fiber and the matrix control the manner in which the crack propagates:

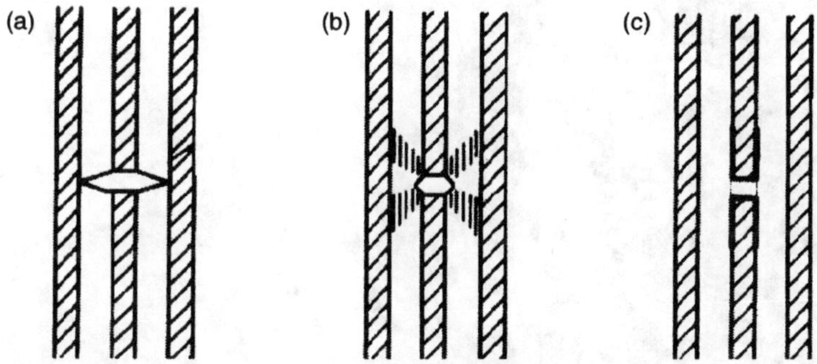

Figure 5.20 Failure processes around a fiber fracture: (*a*) brittle cracking of matrix, (*b*) shear yielding of matrix, and (*c*) interface cracking. (*From Hull.*[1])

1. The crack in the fiber can propagate in the matrix in a brittle manner (Fig. 5.20*a*).
2. The matrix can yield leading to crack blunting and spreading of the yield zone along the fiber (Fig. 5.20*b*).
3. Shear failure in the fiber or matrix at the interface can lead to shrinkage of the unloaded fiber (Fig. 5.20*c*).

The relative amount of interface failure and the magnitude of the frictional forces determine the overall appearance of the fracture surface. Figure 5.21 shows the relatively smooth fracture surface of a strongly bonded carbon fiber–epoxy resin system. In contrast the fracture surface of glass fiber–polyester resin (Fig. 5.22) shows extensive debonding and fiber pullout. Figure 5.23 shows similar extensive fiber pullout and fiber necking and kink band formation in a Kevlar 49–epoxy resin composite. Note that fiber pullout is more effective than debonding as an energy absorber.

5.2.6 Transverse tensile strength of unidirectional laminae

In general, the transverse strength of unidirectional laminae is poor compared to the longitudinal tensile strength. In fact, the presence of fibers has a negative reinforcing effect on the transverse tensile strength of the parent resin. The transverse tensile strength is affected by the properties of the fiber and the matrix, the interface bond strength, the presence and distribution of voids, and interactions between these factors. In the case of matrices that display linear stress-strain responses, stress magnification occurs due to the presence of voids. Conversely, for matrices that display pronounced nonlinear stress-strain responses, strain magnification between fibers

Figure 5.21 SEM micrograph of fracture surface of a carbon fiber–epoxy resin lamina tested in longitudinal tension. Fracture surface is relatively smooth and consists of a network of blocky outcrops of fibers and resin at different levels. (*From Hull.*[1])

Figure 5.22 SEM micrograph of fracture surface of a glass fiber–polyester resin lamina tested in longitudinal tension illustrating brushlike appearance associated with extensive fiber pullout. (*From Hull.*[1])

occurs. When stable crack growth is the operative mechanism, as depicted in Fig. 5.24*a,* in areas of high fiber density, debonding cracks nucleate ahead of the main crack (e.g., location *X*). When crack growth becomes unstable, as depicted in Fig. 5.24*b,* the matrix (resin bridges between fibers) plastically deforms ahead of the main crack, e.g., at location *Y*. Hence the debonding cracks are joined by tearing

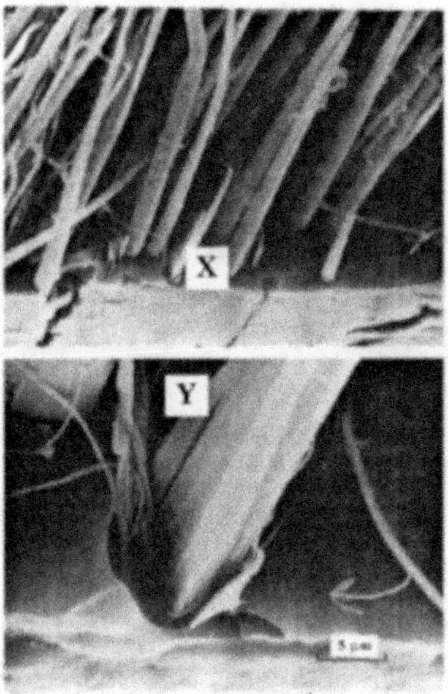

(a)

(b)

Figure 5.23 SEM micrograph of fracture surface of a Kevlar 49–epoxy resin lamina tested in longitudinal tension. (*a*) Fibrous fracture showing necking of individual fibers. (*b*) High magnification view of fiber X in (*a*) showing fiber fibrillation and kink band formation within fibrils at Y. (*From Hull.*[1])

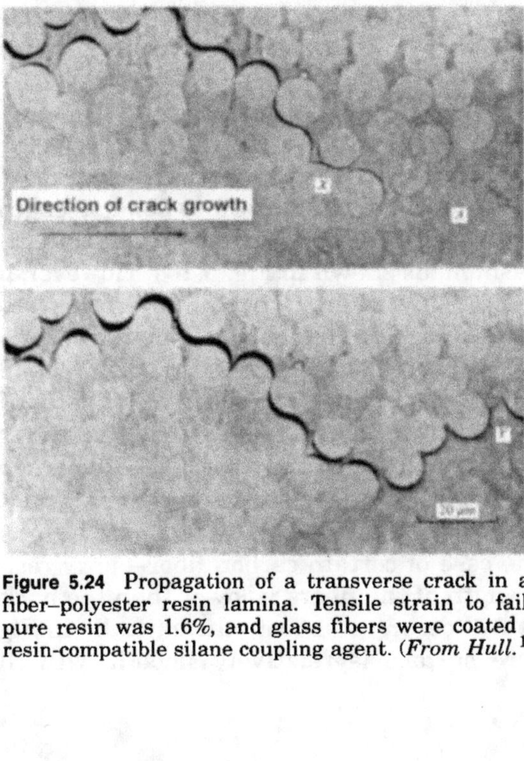

(a)

(b)

Figure 5.24 Propagation of a transverse crack in a glass fiber–polyester resin lamina. Tensile strain to failure of pure resin was 1.6%, and glass fibers were coated with a resin-compatible silane coupling agent. (*From Hull.*[1])

Figure 5.25 SEM micrograph of a transverse tensile fracture in a unidirectional carbon fiber–epoxy resin lamina. (*From Hull.*[1])

or shear cracks. Note that this sequence of events has led to the deflection of the crack around resin-rich regions, e.g., location *A* in Fig. 5.24*a*. The fracture surface morphology of a carbon fiber–epoxy resin composite (Fig. 5.25) shows clean fiber surfaces and brittle matrix fracture.

5.2.7 Longitudinal compressive strength of unidirectional laminae

The longitudinal compressive strength of unidirectional laminae is dependent on fiber and resin properties, the interface bond strength, and the void content. At low volume fraction of fibers, the fibers can buckle in an out-of-phase fashion, as shown in Fig. 5.26*a*. However, in the more realistic case of high V_f, the fibers buckle in an in-phase mode, as shown in Fig. 5.26*b*. In strongly bonded materials, the matrix fracture is dominated by shear. The fibers fracture in a bending mode. If the fibers are brittle (e.g., glass or carbon), the fibers will fracture in a brittle manner on the tensile load side, as shown schematically in Fig. 5.27*b* and in the photograph of Fig. 5.28*a* and *b*. If the fibers are more pliant (e.g., Kevlar 49), unfractured kink zones are formed, as shown schematically in Fig. 5.27*c* and in the photographs of Fig. 5.29. In the case of certain carbon fiber–epoxy resin composites, if the shear strength of the fibers is lower than the buckling strength, shear failure will occur, as shown schematically in Fig. 5.30. The resulting fracture surface is relatively smooth with ill-defined fiber ends.

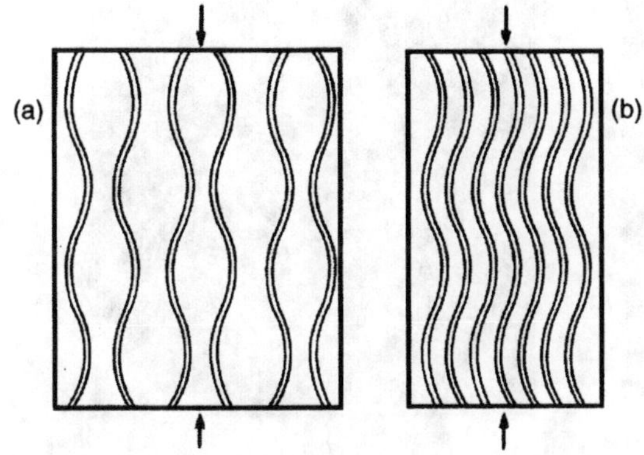

Figure 5.26 Schematic representation of (*a*) out-of-phase and (*b*) in-phase modes of buckling in unidirectional laminae in longitudinal compression. (*From Hull.*[1])

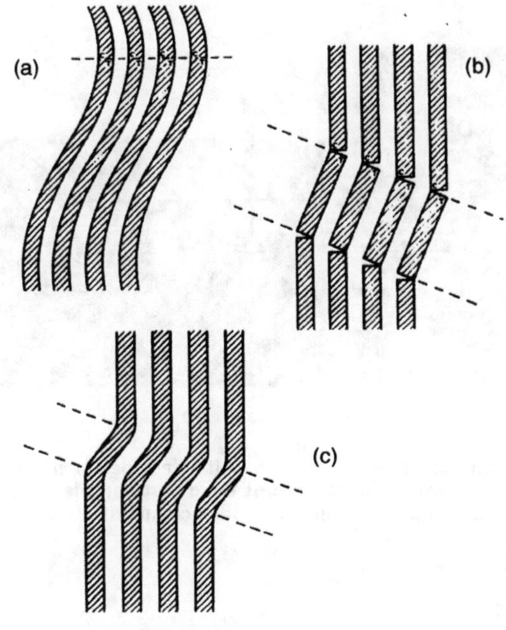

Figure 5.27 (*a*) Tensile and compressive stresses in a fiber due to in-phase buckling leading to a kink zone; (*b*) two planes of fracture formed with brittle carbon fibers; (*c*) unfractured kink zone formed with Kevlar 49 fibers. (*From Hull.*[1])

5.2.8 Transverse compressive strength of unidirectional laminae

As shown schematically in Fig. 5.31, the arrangement of fibers in the maximum shear stress surface is dependent on fiber orientation. Shear on plane *ABCD* can occur without fiber fracture while shear on plane *BEDF* will lead to fiber fracture.

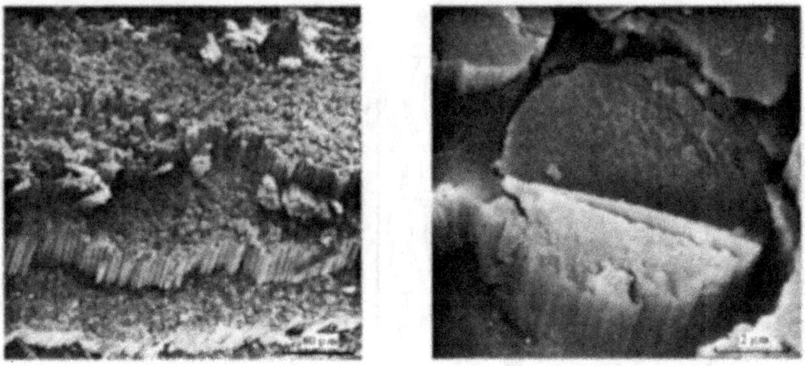

(a) (b)

Figure 5.28 SEM micrographs of the fracture surface of a carbon fiber–epoxy resin lamina after a fiber buckling mode failure due to a longitudinal compressive stress. (*a*) Low-magnification view showing smooth fracture surface. (*b*) High-magnification view showing tensile and compressive fracture in a single fiber. (*From Hull.*[1])

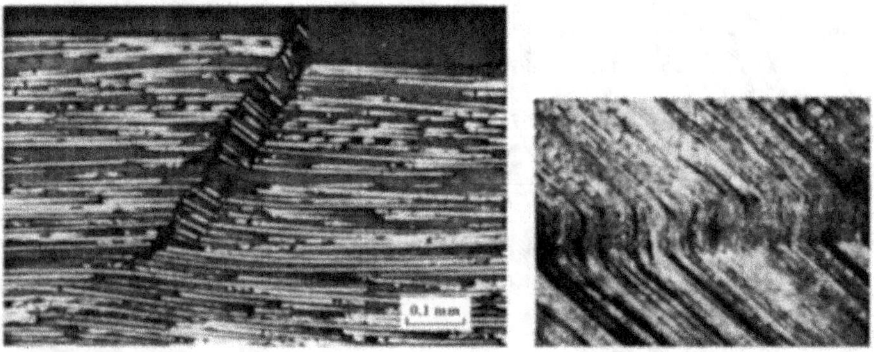

(a) (b)

Figure 5.29 (*a*) Polished section of a carbon fiber–epoxy resin pultruded specimen showing a kink band formed in compressive zone of a four-point bend test sample. (*From Hull.*[1]) (*b*) Fracture of uniaxial Technora-epoxy composite under compression at 0°C. (*From Hearle et al.*[4])

5.2.9 In-plane shear strength of unidirectional laminae

Figure 5.32 shows an edge view of an intralaminar shear crack with shear failure of the resin. Shear zones have grown ahead of the crack, and some local interface debonding has occurred. As the fracture surface of a carbon fiber–epoxy resin composite fractured under shear loading shown in Fig. 5.33 demonstrates, the resin displays significant plastic deformation (compared to Fig. 5.25).

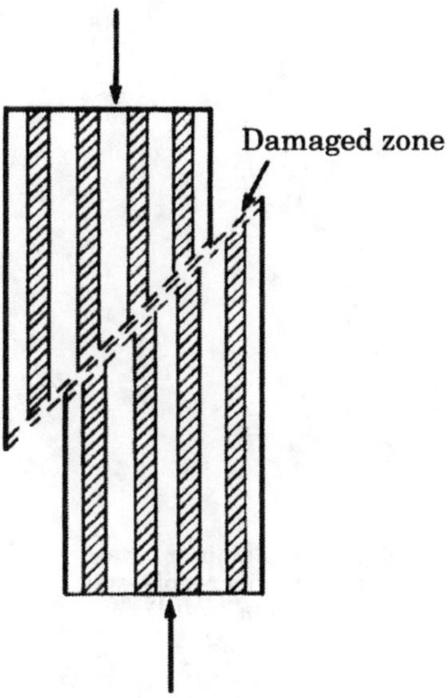

Figure 5.30 Schematic view of shear failure due to longitudinal compressive stresses. (*From Hull.*[1])

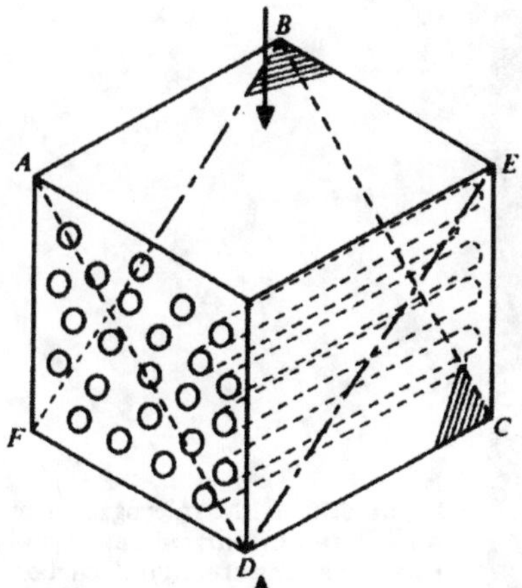

Figure 5.31 Transverse compression of a unidirectional lamina. Shear on *ABCD* can occur without fiber fracture; shear on *BEDF* must involve fiber fracture. (*From Hull.*[1])

5.2.10 Effect of environmental conditions on the deterioration of properties

The thermal and chemical environment under which composites perform greatly affects their properties. The effects of temperature are threefold:

1. Internal stresses result from differential expansion between fibers and resin.

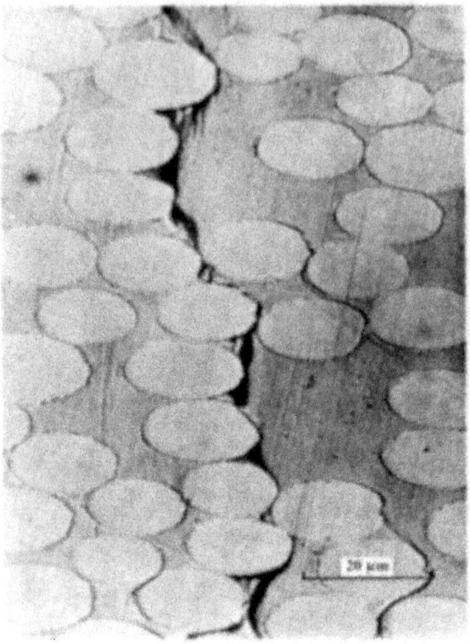

Figure 5.32 Intralaminar shear crack in a glass fiber–polyester resin lamina. (*From Hull.*[1])

Figure 5.33 SEM micrograph of an intralaminar shear fracture in a unidirectional carbon fiber–epoxy resin lamina. (*From Hull.*[1])

2. The properties of the constituents (particularly the resin) change with temperature.
3. Creep resistance is a strong function of temperature.

The chemical environment's effect is best illustrated by the deleterious effect of moisture on the properties of the resin. In addition, water

TABLE 5.3 Aspects of Property Deterioration in Polymer-Based Composite Materials

Component	Reversible changes	Irreversible changes
Resin	Water swelling Temperature flexibilizing Physical ordering of local molecular regions	1. Chemical breakdown by hydrolysis 2. Chemical breakdown by UV radiation 3. Chemical breakdown by thermal activation 4. Chemical breakdown by stress-induced effects associated with swelling and applied stress 5. Physical ordering of local molecular regions 6. Chemical composition changes by leaching 7. Precipitation and swelling phenomena to produce voiding and cracks 8. Nonuniform deswelling to produce surface cracks and crazes 9. Chemical effect of thermoplastic polymer content on long-term stability
Interface	Flexibilizing interface	1. Chemical breakdown as above 1, 2, 3, 4 2. Debonding due to internally generated stresses associated with shrinkage and swelling to the applied stress 3. Leaching of interface
Fiber		1. Loss of strength due to corrosion 2. Leaching of fiber 3. Chemical breakdown by UV radiation

SOURCE: From Hull.[1]

can leach soluble oxides such as K_2O and Na_2O from glass fibers, leading to the formation of surface pits. In acidic environments, glass fiber composite materials are subject to a stress corrosion effect due to a hydrogen ion exchange process such as

$$\text{—Si—O—Na} + H^+ \rightarrow \text{—Si—O—H} + Na^+$$

This process leads to shrinkage of the surface layers of the glass, which in turn leads to weakening of the glass by the creation of large tensile stresses at the surface. Table 5.3 summarizes reversible and irreversible changes in the properties of polymer-based composites due to environmental and thermal effects.

5.2.11 Effect of fiber-matrix interface strength on fracture mechanism

Figure 5.34 shows a montage of fracture surfaces from a glass fiber–polypropylene composite under fatigue and fast fracture conditions.

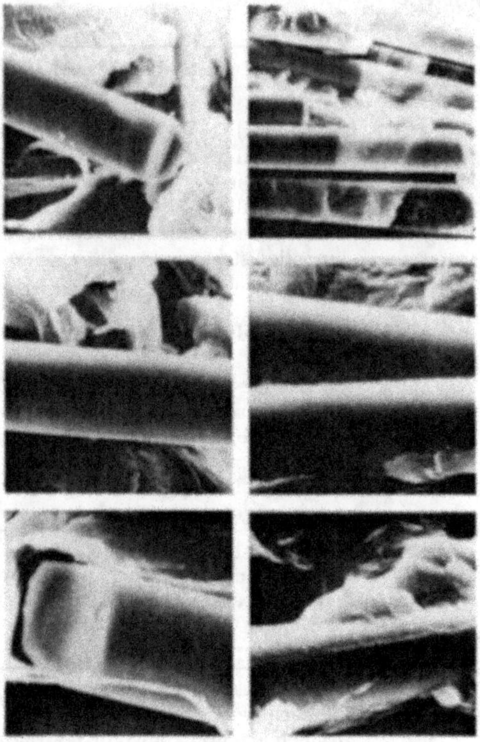

(a) (b) (c) (d) (e) (f)

Figure 5.34 Fracture surface morphology in glass fiber–reinforced polypropylene in fatigue [(*a*), (*c*), (*e*)] and fast fracture [(*b*), (*d*), (*f*)] as a function of low, intermediate, and high bond strength. (*From Hertzberg.*[7])

Figure 5.34*a, c,* and *e* corresponds to fatigue fractures for low, medium, and high interfacial strength levels. As can be seen, the crack path always follows the fiber-matrix boundary.

Figure 5.34*b, d,* and *f* corresponds to fast fractures for low, medium, and high interfacial strength levels. In *b* and *d,* the crack continues to follow the fiber-matrix interface; however, in the case of *f,* the length of the fiber shows remnant polypropylene adhesion.

5.2.12 Graphite fiber-epoxy composite fractography

In view of the importance of graphite fiber-epoxy composites in aerospace applications, the fractographic features of fatigue-tested specimens of such composites are described here. In the overload region of the fracture surface, the fracture surface consists of graphite fibers

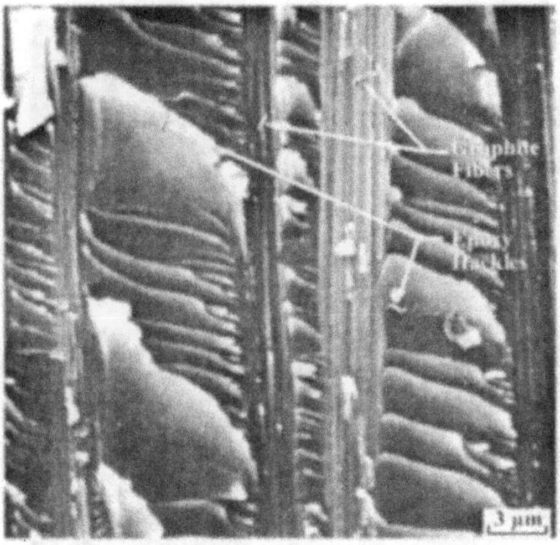

Figure 5.35 Hackles (fractured epoxy) and graphite topographic features on an overload fracture surface of graphite-epoxy composite. (*From Morris and Hetter.*[8])

and an overlapping platelet topography in the resin. The latter has been termed *hackles* or *lacerations,* as shown in Fig. 5.35. Note that in Chap. 3 similar terminology was used in the discussion of fractographic features in polymers and ceramics. The overlap of these hackles can be used to determine the fracture directions of various exposed plies. These overlaps in turn can be traced back to determine the location of fracture origin. In the fatigue fracture region, striations can be seen on both the fibers and the epoxy (Fig. 5.36). Those on the fibers are oriented normal to the fiber axis while the epoxy striations were more nearly parallel to the fiber axes. The epoxy striations were also more numerous and closer spaced than the fiber striations. There were some interesting striation configurations, as shown in Fig. 5.37:

1. Figure 5.37*a:* Striations in the epoxy propagate from adjacent fibers in opposite directions toward each other. The striations were concave to the nearest fibers and propagated such that one set of striations overlapped at the fracture surface created by the other set of striations.

2. Figure 5.37*b:* Epoxy striations initiated at the common sides of each fiber and propagated in a common direction toward the adjacent fiber. The striations propagated away from the top of each fiber and terminated near the bottom of the adjacent fiber.

3. Figure 5.37*c:* Fracture propagation is by a mixture of epoxy striations and hackles. The epoxy appeared to fracture initially by fatigue (striations) and finally by overload (hackles).

4. Figure 5.37*d:* Striations are evident on hackles.

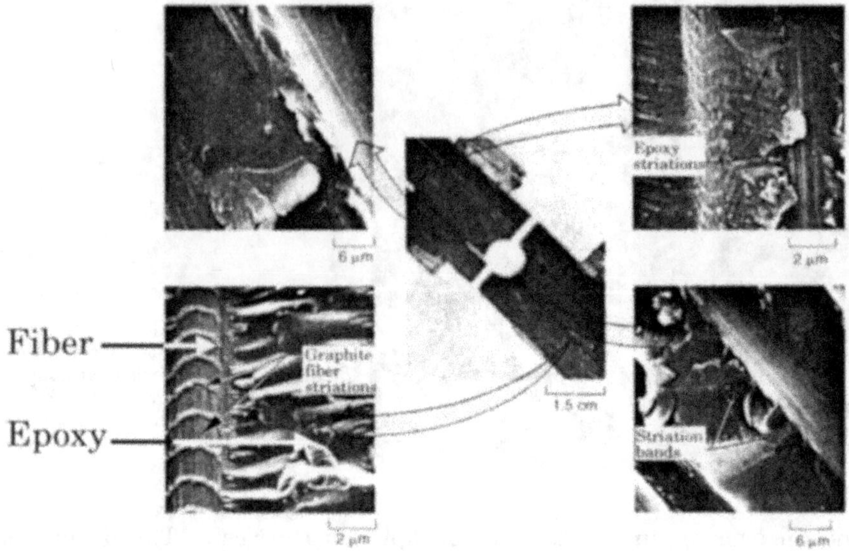

Figure 5.36 Variations in fatigue striation morphology. (*From Morris and Hetter.*[8])

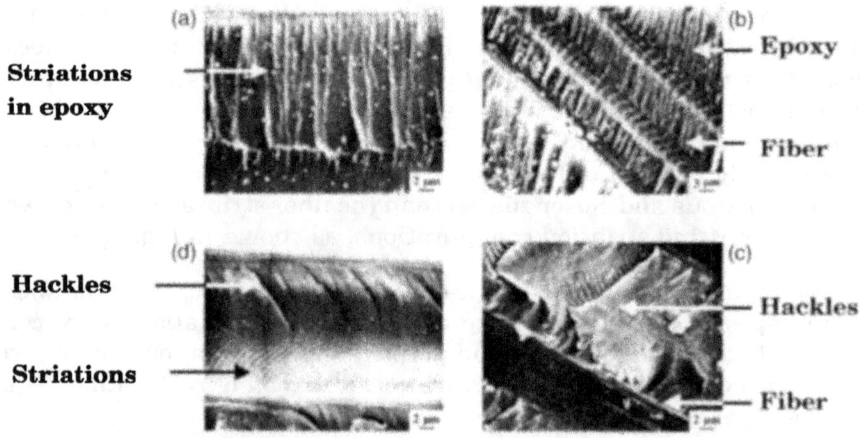

Figure 5.37 Various fatigue striation configurations. (*From Morris and Hetter.*[8])

The micromechanisms of fracture of graphite fiber-epoxy composites in mode I, mode II, and mixed-mode loading conditions (Chap. 2) can be affected by the toughening of the epoxy resin. Such toughening can be achieved (1) by progressively lower degrees of cross-linking in the resin compared to a baseline, relatively brittle resin and (2) addition of an elastomeric material to the resin. Figure 5.38 (from Ref. 9) shows the crack front under in

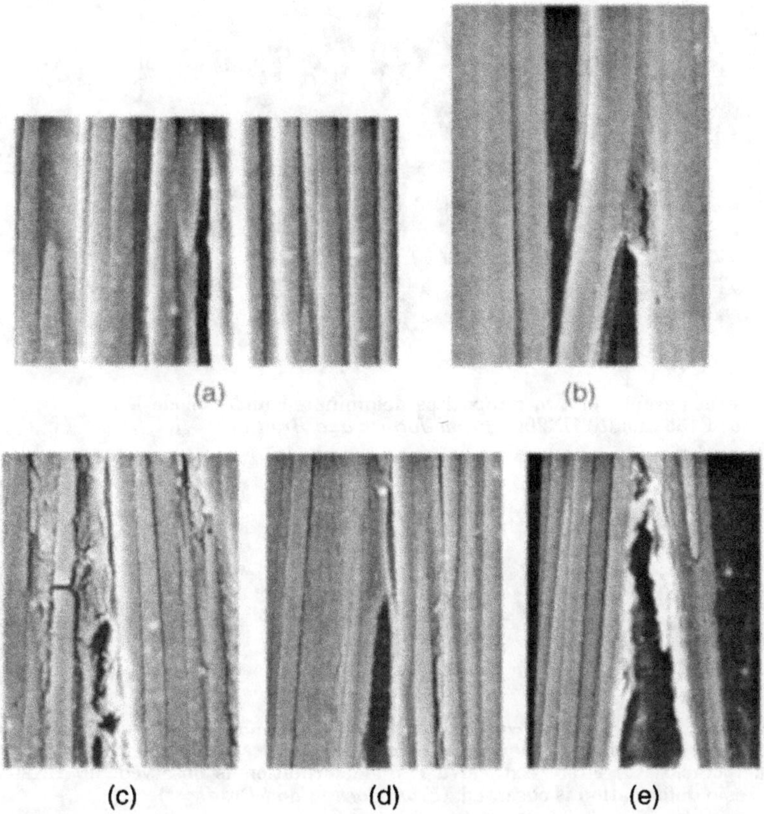

Figure 5.38 Delamination of composites observed in situ in the SEM. (*a*) AS4/3502 shows debonding with very little resin damage. (*b*) In T6T145/F155, debonding as well as microcracking is visible. (*c*) HX205 system: Extensive microcracking that coalesces into macroscopic crack growth is observed. (*d*) T6T145/F185 system: Large microcrack zone ahead of crack tip, with a significant extent above and below plane of crack, is observed. (*e*) T6T145/F185 system: Resin tearing as well as microcracking is evident. (*From Jordan and Bradley.*[9])

situ (SEM) delamination conditions for the four systems studied. Note that as the matrix resin is progressively toughened (*a* through *d*), the degree of debonding decreases while the degree of microcracking and tearing of the resin increases corresponding to an increase in the damage zone ahead of the crack tip. Under mode I loading conditions, the degree of resin deformation increases with the toughness of the resin (Fig. 5.39). Under mode II loading conditions, the fracture surfaces again show increasing resin plastic deformation with resin toughness, with zipperlike artifacts formed by the coalescence of sigmoidal microcracks (Fig. 5.40). Similar results are obtained under mixed-mode loading conditions with

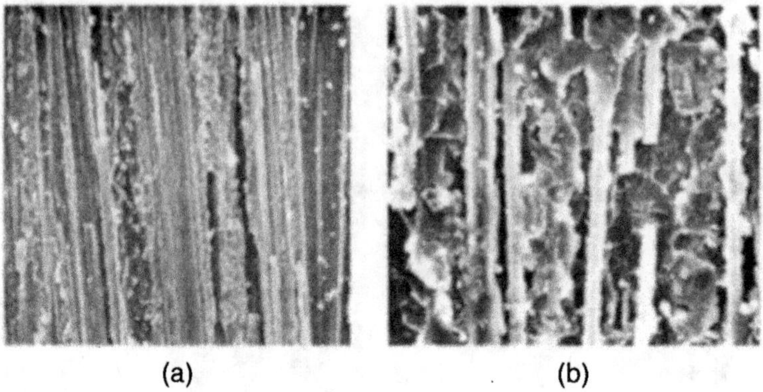

Figure 5.39 Fractographs of two composites delaminated under mode I conditions: (*a*) F155 and (*b*) HX205. (*From Jordan and Bradley.*[9])

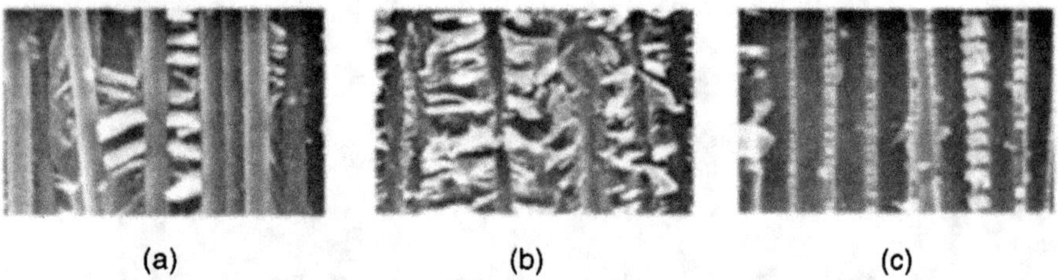

Figure 5.40 Fractographs of three composites delaminated under mode II conditions. (*a*) AS4/3502: Zipperlike artifacts appear to be formed by coalescence of a series of sigmoid-shaped microcracks. (*b*) F155: Extensive resin deformation is observed. (*c*) HX205: Extensive resin deformation is observed. (*From Jordan and Bradley.*[9])

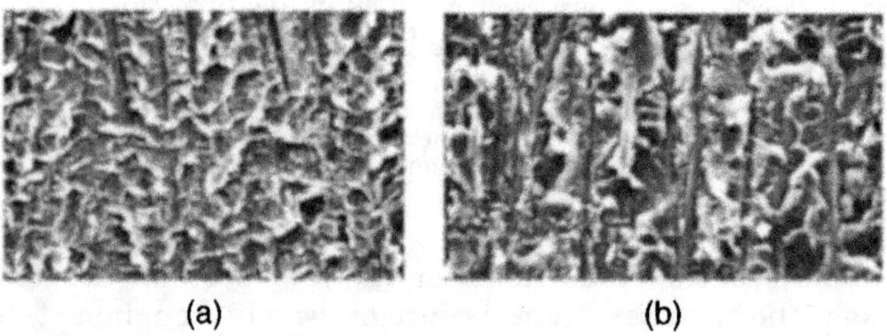

Figure 5.41 Fractographs of two composites delaminated in mixed-mode conditions: (*a*) F155 and (*b*) HX205. (*From Jordan and Bradley.*[9])

leaflike artifacts (Fig. 5.41). Similar trends in the correlation of fracture properties with resin toughness and loading mode are shown in summarized form in Figs. 5.42 and 5.43. The physical model proposed for mode I interlaminar fracture in thermoplastic matrix composite laminates is shown in Fig. 5.44.

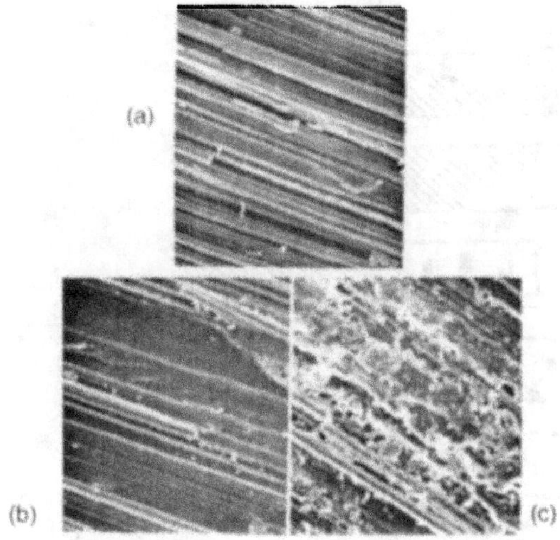

Figure 5.42 Fractography of mode I delamination in composites of AS-4 graphite fibers in three different epoxy matrices. Sizing refers to a surface treatment on the fibers. (*a*) System A: unsized fiber in a brittle resin showing smooth corrugated surface and fiber debonding. (*b*) System B: sized fiber in a resin of moderate toughness showing smooth corrugated surface and fiber debonding. (*c*) System C: sized fiber in a resin of high toughness showing some resin deformation with improved resin adhesion. (*From Hibbs et al.*[10])

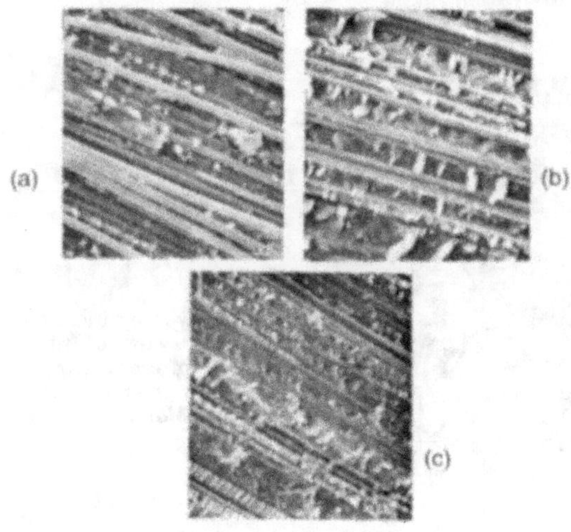

Figure 5.43 Fractography of mode II delamination in composites of AS-4 graphite fibers in three different epoxy matrices. Sizing refers to a surface treatment on the fibers. All the systems tested exhibit a rougher or more scalloped surface than seen in the fractographs of mode I failure in Fig. 5.42. (*a*) System A: unsized fiber in a brittle resin showing smooth corrugated surface and fiber debonding. (*b*) System B: sized fiber in a resin of moderate toughness showing smooth corrugated surface and fiber debonding. (*c*) System C: sized fiber in a resin of high toughness showing some resin deformation with improved resin adhesion. (*From Hibbs et al.*[10])

5.2.13 Short-fiber-reinforced thermoplastics

In short-fiber-reinforced thermoplastics, the fracture path in general avoids the fiber agglomerates (Fig. 5.45). In such cases, three distinct modes of local crack growth can be identified:

1. *Well-bonded, brittle matrix with brittle glass fibers.* The mechanism is local cracking and crack coalescence, depicted schematically in Fig. 5.46*a* and shown in the fractograph in Fig. 5.46*b*.

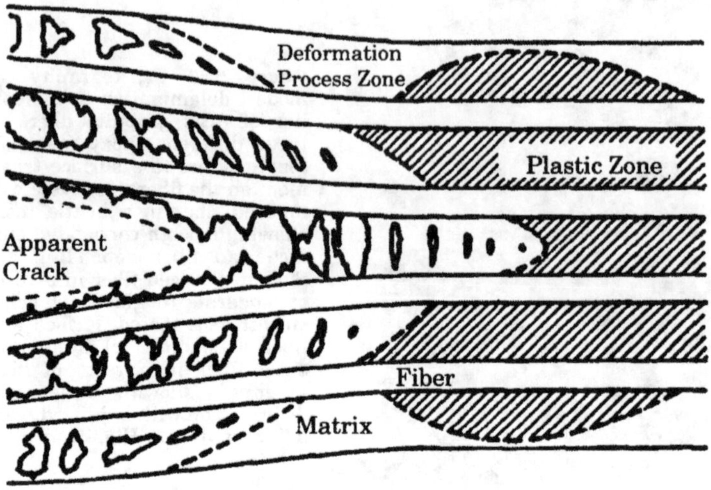

Figure 5.44 A physical model of the mode I interlaminar fracture in thermoplastic matrix composite laminates. (*From Su.*[11])

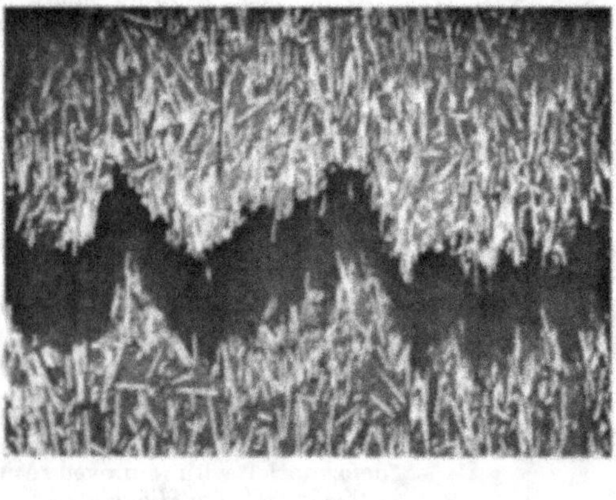

Figure 5.45 Low-magnification light microscope view of fiber avoidance mode of crack growth in carbon fiber–reinforced polycarbonate. (*From Mandell et al.*[12])

2. *Poorly bonded, brittle matrix with carbon fibers.* The mechanism is frequent fiber debonding and pullout, depicted schematically in Fig. 5.47*a* and shown in the fractograph in Fig. 5.47*b*.
3. *Ductile matrix systems.* The mechanism is local yielding ahead of the crack followed by coalescence into a main yielded zone, which eventually tears to form a crack. This is depicted schematically in Fig. 5.48*a* and shown in the fractograph in Fig. 5.48*b*.

The effect of increasing matrix strength and modulus (PPS, PEEK, and PAI in increasing order) on the fracture mechanism and resulting fractographic features in short carbon-fiber-reinforced composites

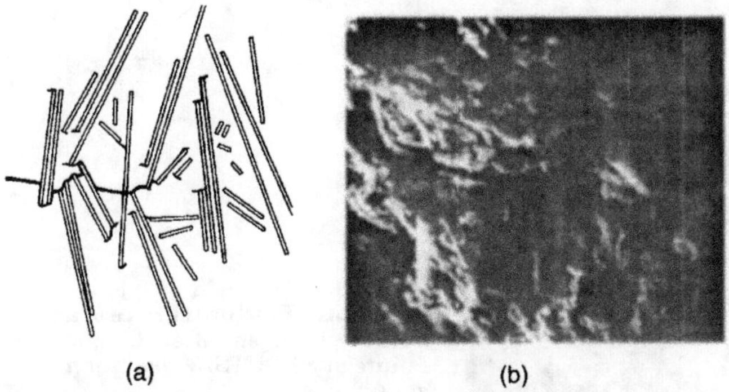

Figure 5.46 (*a*) Schematic of local mode of crack growth in well-bonded, brittle matrix, glass fiber materials. (*b*) Fracture surface of glass-reinforced polyphenylsulfide. (*From Mandell et al.*[12])

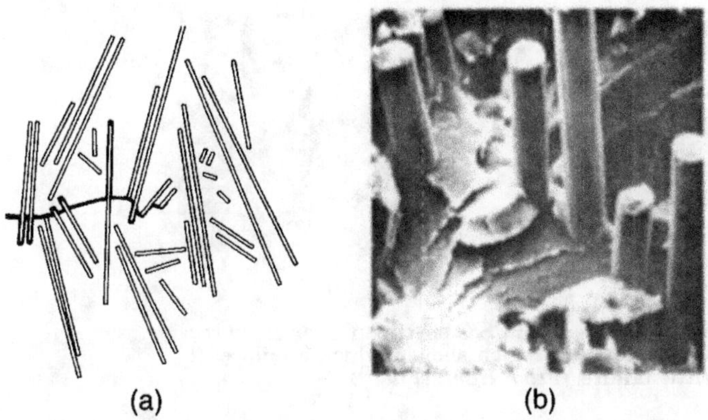

Figure 5.47 (*a*) Schematic of local mode of crack growth in poorly bonded, brittle matrix materials. (*b*) Fracture surface of carbon-reinforced poly(amide-imide). (*From Mandell et al.*[12])

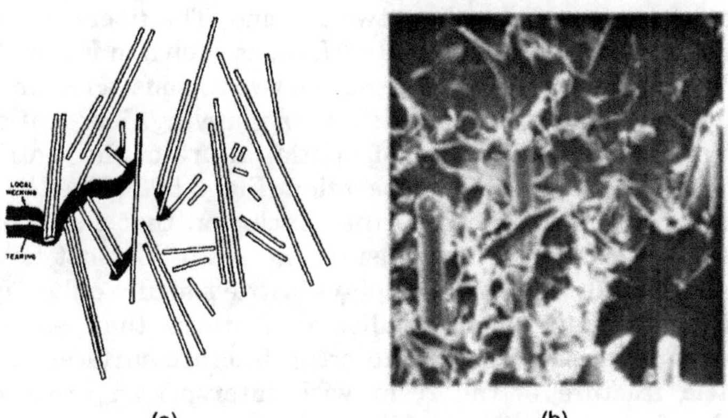

Figure 5.48 (*a*) Schematic of local mode of crack growth in ductile matrix material. (*b*) Fracture surface of carbon-reinforced poly(amide-imide). (*From Mandell et al.*[12])

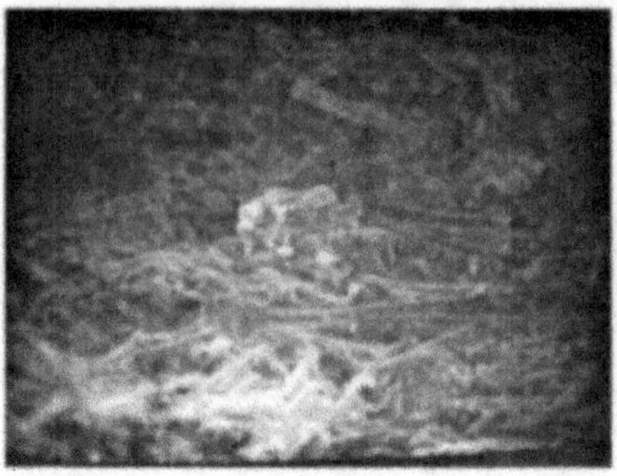

Figure 5.49 Fracture surface of a PPS/C specimen after fatigue failure at 80% UTS. (*From Hahn et al.*[13])

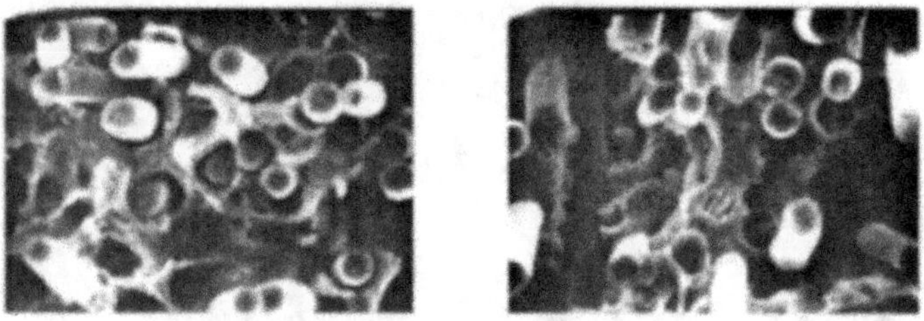

(a) (b)

Figure 5.50 Fracture surfaces of a PPS/C specimen after creep rupture at 40% UTS and 80°C: (*a*) region showing ductile failure; (*b*) region showing brittle failure. (*From Hahn et al.*[13])

under different loading conditions (static tension, tension-tension fatigue, and creep) can be summarized as follows:

1. In keeping with two regions of injection-molded composites (skin and core), the fracture surfaces show two regions. The fibers in the core tended to be parallel to the fracture surface, as shown in Fig. 5.49.

2. PPS/C: The creep fracture surface showed two regions: a region of ductile failure of the matrix (resin) with accompanying fiber-matrix separation (Fig. 5.50*a*) and a region of brittle failure of the matrix (resin) with little or no fiber-matrix separation (Fig. 5.50*b*). The clean fiber surfaces in both regions lead to the conclusion that the fiber-matrix interface is weak. In static tension, the resin does not have time to deform plastically and hence displays brittle fracture only (Fig. 5.51). The fatigue fracture surfaces display features that can be regarded as a combination of static and creep fracture surfaces, i.e., primarily brittle fracture of the resin with interspersed areas of debonding and resin yielding (Fig. 5.52).

3. PEEK/C: The creep fracture surfaces are similar to those of PPS/C. However, both the fatigue and static fracture surfaces reveal

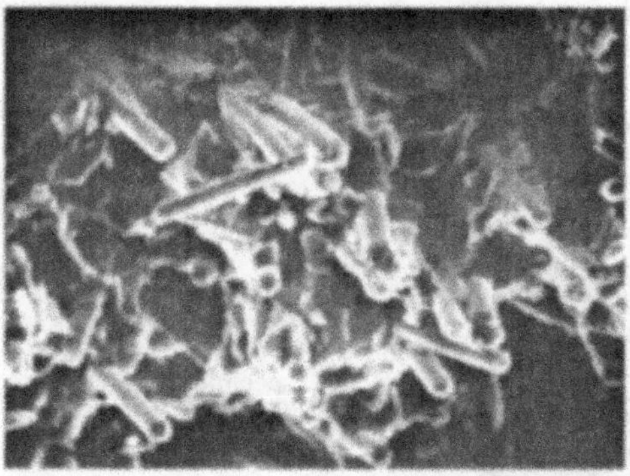

Figure 5.51 A static fracture surface of a PPS/C specimen. (*From Hahn et al.*[13])

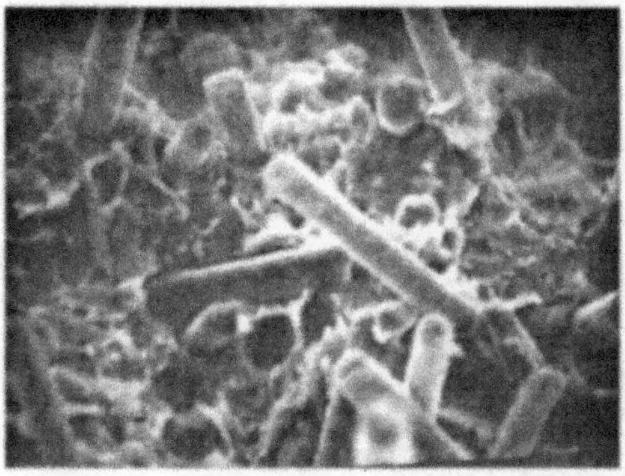

Figure 5.52 A fatigue fracture (70% UTS) surface of a PPS/C specimen. (*From Hahn et al.*[13])

excellent fiber-matrix bonding, resulting in very limited fiber pullout but with ends coated with resin corresponding to resin failure rather than interface failure (Fig. 5.53).

4. PAI/C: While this system exhibits good interfacial bonding, both static and fatigue fracture surfaces display a "crushed" resin appearance with fiber breakage (Fig. 5.54).

The mechanisms of crack propagation in a short fiber composite loaded in mode I are illustrated by the schematic representation of Fig. 5.55. Each pair of side view (profile) and fracture surface photographs of Fig. 5.56 shows each of these mechanisms in the case of glass-fiber-reinforced PET composites.

5.3 Ceramic Matrix Composites (CMCs)

5.3.1 Toughening mechanisms

Due to the inherent low ductility and toughness of the ceramic matrix, such composites are toughened by a variety of techniques, summarized

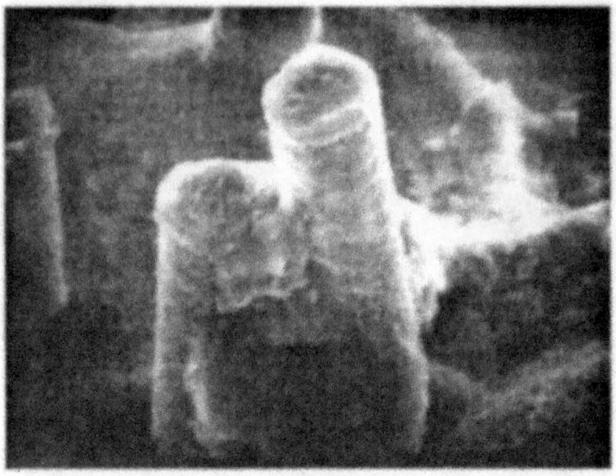

Figure 5.53 A fracture surface of a PEEK/C specimen after fatigue failure at 50% UTS. (*From Hahn et al.*[13])

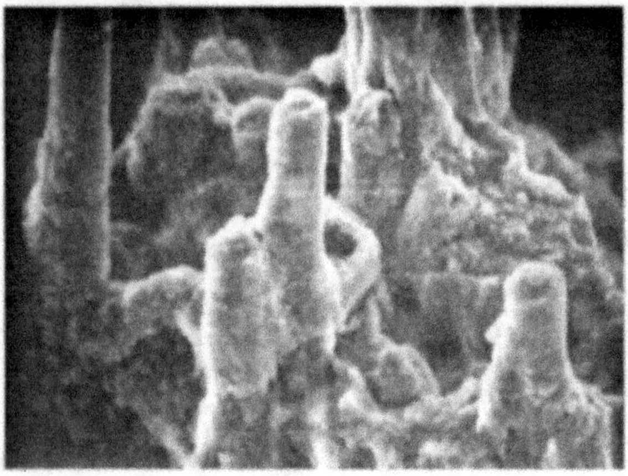

Figure 5.54 A fracture surface of a PAI/C specimen after fatigue failure at 40% UTS. (*From Hahn et al.*[13])

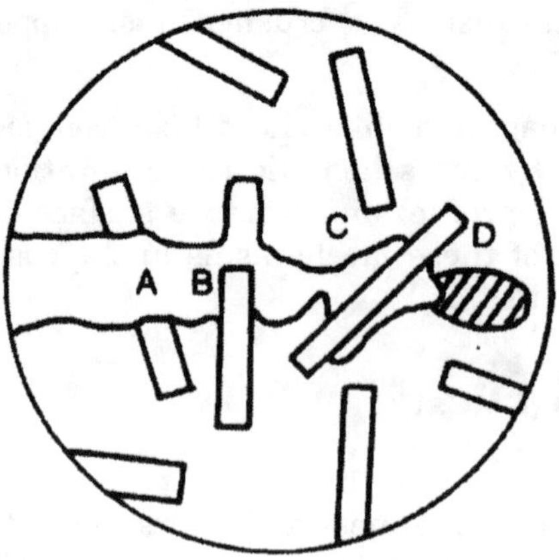

Figure 5.55 Mechanisms of failure at an advancing crack: (*A*) fiber fracture by overload; (*B*) fiber pullout; (*C*) delamination; (*D*) matrix flow. (*From Hearle et al.*[4])

(1) (2)
(3) (4)
(5) (6)

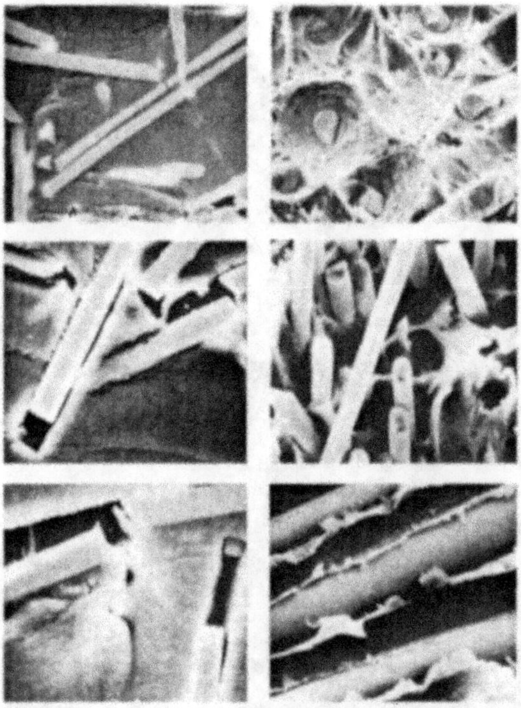

(a)

Figure 5.56 (*a*) Tensile fracture at an advancing crack tip in a compact tension specimen of a short glass fiber–PET composite. (1), (3), (5) Polished surface profiles; (2), (4), (6) fracture surfaces. (*From Hearle et al.*[4])

in Table 5.4. Figure 5.57 schematically shows some of the toughening mechanisms corresponding to the respective energy-absorbing processes.

5.3.2 Interface engineering in CMCs

Of the toughening techniques shown in Table 5.4, the coating and treatment of fibers to improve composite toughness deserve special attention. The type of interface between the fiber and the matrix affects the debonding characteristics and fiber pullout. Figure 5.58 shows a dramatic example of the efficacy of this technique. Figure 5.58*a* shows the flat, brittle failure characteristic of Nextel 480 (mullite) fiber–glass matrix composite. The fracture surface of a BN-coated Nextel fiber–glass composite is shown in Fig. 5.58*b* and *c* and displays the characteristics of toughened composites, namely, extensive fiber pullout and multiple matrix cracking. Figure 5.59 also shows fiber-

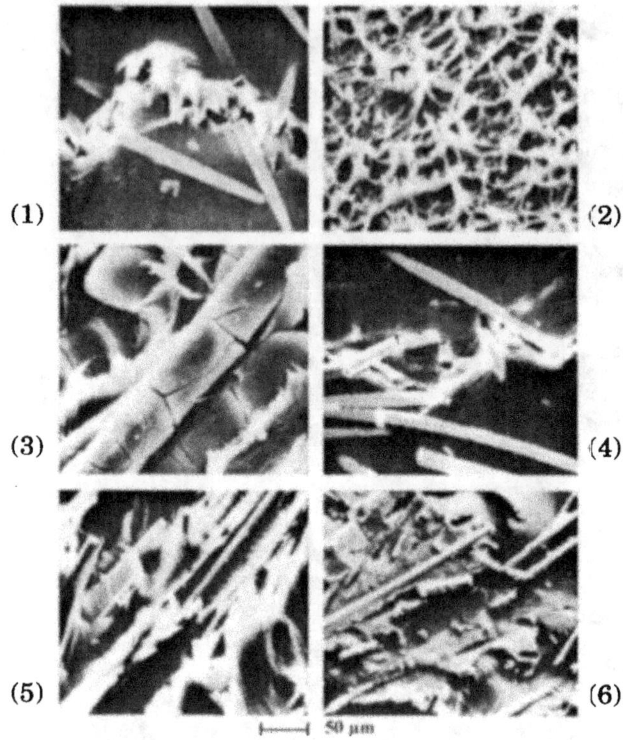

(b)

Figure 5.56 (*Continued*) (*b*) Fracture at crack tip in a compact tension specimen of a short glass fiber–PET composite. (1), (2) Plastic deformation in matrix; (3), (4) multiple fiber cracking due to an additional corrosive environment; (5) fracture surface, tested at +60°C; (6) fracture surface tested at −60°C. (*From Hearle et al.*[4])

matrix debonding and fiber pullout in a composite of SnO_2-coated Saphikon fibers in a glass matrix.

5.4 Metal Matrix Composites (MMCs)

5.4.1 Metal matrix composite systems and interfacial bonding

Tables 5.5, 5.6, and 5.7 show details of the most common fiber-reinforced MMC systems in use. Table 5.8 shows some important in situ composite systems. Some advantages of MMCs are

- High specific strength
- High specific modulus

TABLE 5.4 Ceramic Matrix Composite Toughening Mechanisms

Mechanism	Requirement
1. Compressive prestressing of the matrix	α_f and α_m will result in an axial compressive prestressing of the matrix after fabrication. *Note:* α is coefficient of thermal expansion.
2. Crack impeding	Fracture toughness of the second phase (fibers or particles) is greater than that of matrix locally. Crack either is arrested or bows out (line tension effect).
3. Crack deflection	Weak fiber-matrix interfaces deflect the propagating crack away from the principal direction.
4. Fiber (or whisker) pullout	Failure along fiber-matrix interface leads to fiber pullout on further straining.
5. Phase transformation toughening	The crack-tip stress field in the matrix can cause the second-phase particles (fibers) at the crack tip to undergo a phase transformation, causing expansion ($\Delta V>0$). The volume expansion can squeeze the crack shut.

SOURCE: From Chawla.[3]

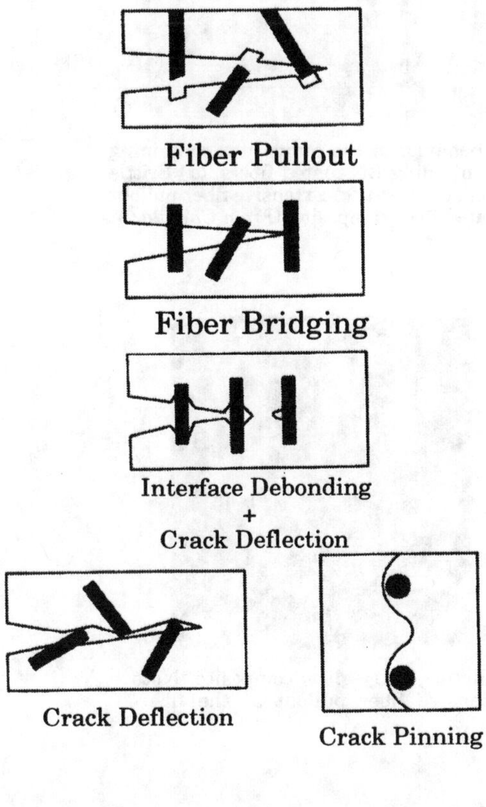

Figure 5.57 Schematic of some toughness or energy-dissipating mechanisms that can be brought to play in CMCs. (*From Chawla.*[14])

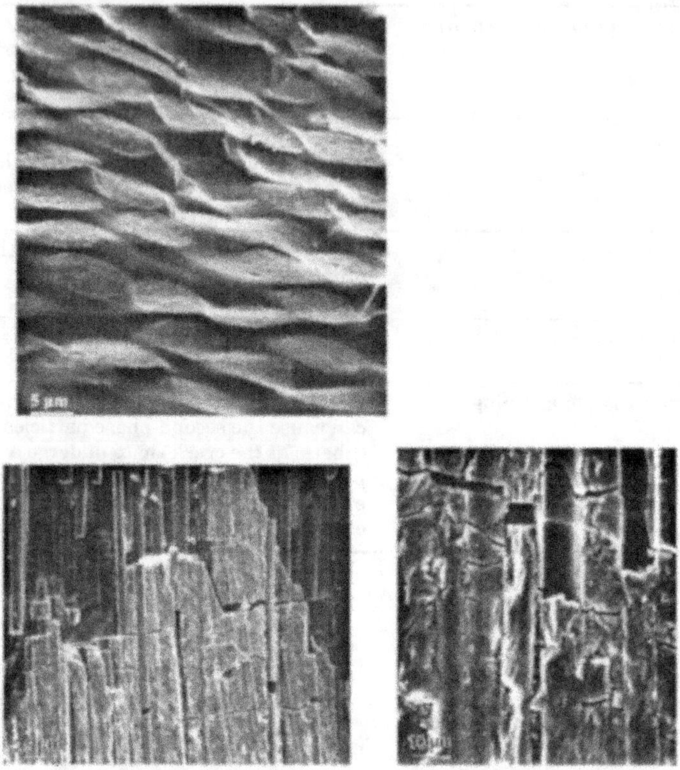

(a) (b) (c)

Figure 5.58 Difference in the fracture behavior of the composites containing uncoated Nextel 480 fiber and those containing BN-coated fibers. (*a*) Brittle planar fracture in the uncoated fiber composite. (*b*), (*c*) Extensive fiber pullout and multiple matrix cracking in the coated fiber composite. (*From Chawla.*[3])

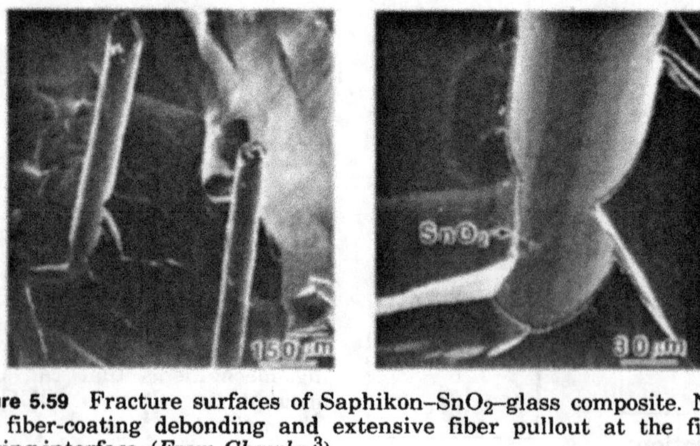

(a) (b)

Figure 5.59 Fracture surfaces of Saphikon–SnO_2–glass composite. Note the fiber-coating debonding and extensive fiber pullout at the fiber-coating interface. (*From Chawla.*[3])

TABLE 5.5 Metal Matrix Composite Systems with Metallic Fibers

Fibers	Matrix	Fabrication method	Field of application
Stainless steel	Al	Powder metallurgy, lamination	Aircraft industry
Be ribbons	Al, Ti	Al or Ti clad Be rods are inserted in drilled Al or Ti preforms; mechanical deformation of the preform	Shafts for high-speed rotating engines, e.g., rotor shafts in helicopters and central rods in aircraft
Ta	Mg	Infiltration technique	Aircraft industry
Mo	Ti or Ti alloy	Powder metallurgy, fiber alignment by extrusion, rolling, etc.	Supersonic aircraft rocket propulsion
Be or Ti clad Be	Ti alloy, pure Ti	Explosive welding of interposed layers of Ti and Be fibers	Aircraft construction
Be	Ti Ti-6Al-4V Ti-6Al-6V-2Sn Ti-5Al-2.5Sn	Hot extrusion of mixed precursor of Ti and Be; the latter can be present as fiber preforms	Aircraft industry
Ni_3Al	Ni (2–10% Al)	Powder metallurgy; the fiberlike phase is formed in situ during the mechanical working	Oxidation-resistant, high strength at high and low temperatures
Ni-Cr-Al-Y	Ni alloy	Powder metallurgy	Sealing elements in turbines and compressors
W/1%ThO_2	Superalloy	Investment casting	Jet engines
W	W-Ni-Fe alloy	Liquid-phase sintering; W fibers are recrystallized to avoid dissolution	
Stainless steel	Ni alloys	Electroforming	Rocket engines
Mo, Ti, Nb	Ni superalloys	Powder metallurgy	
W	Cu	Melt impregnation	Electrical machinery
Nb filaments	Ni, Cu, Ag	Nb filaments embedded in a Cu, Ni, or Ag matrix are passed through a molten bath of Sn to form Nb_3Sn	Superconductors

SOURCE: From Chawla.[6]

TABLE 5.6 Metal Matrix Composite Systems with Nonoxide Ceramic Fibers

Fibers	Matrix	Fabrication method	Field of application
SiC coated B	Al	Powder metallurgy; the composite article is clad with a sheet of Ti by diffusion bonding	Turbine blades
C (graphite, amorphous C)	Ni/Co Aluminide	Coating C fibers with Ni or Co; mixing with Ni-Co-Al powder, hot pressing	
C coated with boride of Ti, Zr, Hf	Al or Al alloys, Mg, Pb, Sn, Cu, Zn	Melt impregnation	
C	Al alloy containing carbide-forming metal, e.g., Ti, Zr	Melt impregnation	
SiC with W core	Al-Cu alloy	Coating filaments with Cu, passing Cu-coated filaments through an Al melt	
C	Mg or Mg alloy	Melt impregnation; molten Mg matrix contains small amounts of magnesium nitride to enhance wetting fibers	Turbine fan blades, pressure vessels, armor plates
SiC	Be or alloys with Ca, W, Mo, Fe, Co, Ni, Cr, Si, Cu, Mg, and Zr	Vacuum impregnation with molten Be or plasma spraying fibers with Be and consolidation by metallurigcal process	Aerospace and nuclear industries
B + stainless steel; Borsic + Mo fibers	Al, Ti	Impregnation; spraying etc.; combination of high-strength ductile and brittle fibers	Aerospace industry
SiC	Ti or alloy Ti-3 Al-2.5V	Hot pressing of interposed layers of fibers and matrix sheets; SiC fibers are previously coated with Zr diffusion barrier layer	Compressor blades, airfoil surfaces
Carbides of Nb, Ta, and W	Ni-Co and Fe-Cr alloys	Unidirectional solidification	Aircraft industry

SiC containing 0.01–20% free carbon	Cr-based alloys	Powder metallurgy; free carbon reacts with Cr to form carbides, thus improving bonding	High-strength, heat-resistant material, e.g., vanes and blades for turbines, rocket nozzles
SiC containing 0.01–80% free carbon	Co or Co-based alloys	Powder metallurgy or melt impregnation; carbide formation between fibers and Co matrix	High-strength, heat-resistant material, e.g., vanes and blades for turbines, rocket nozzles
SiC containing 0.01–20% free carbon	Mo-based alloys	Powder metallurgy	High-strength, heat-resistant material, e.g., vanes and blades for turbines, rocket nozzles
C coated with carbides	Ni or Ni alloys	Melt impregnation	Aeronautical industry
B	Cu-Ti-Sn alloy	Liquid-phase sintering	Cutting tools
C	Bronze	Various processes	Bearing materials
C	Cu alloy	Powder metallurgy; fibers are mixed with a slurry of Cu powder and 2% of a carbide-forming metal powder (Ti or Cr)	High-strength, electrically conductive materials
C coated with Ti boride	Al, Cu, Sn, Pb, Ag, Zn, and Mg	The matrix contains alloying elements of Ti and B to prevent deterioration of the TiB coating of fibers	Aeronautical industry
C coated with Ni	Metals with melting point lower than that of Ni	Melt impregnation	
C coated with SiO_2 + SiC	Al, Mg, Ti, and Ni	Melt impregnation, powder metallurgy	
Monocarbides of Ta, Ti, and W	Al, Al-Si alloy, Ag or Ag alloys, and Cu or Cu alloys	Melt impregnation	Abrasion-resistant materials
β-SiC	Ag or Ag alloys		Electrical conductors, contacts
C	Si	Powder metallurgy	Abrasive materials
SiC_w	Al		
C coated with TiB	Mg, Pb, Zn, Cu, Al	Melt impregnation	

SOURCE: From Chawla.[6]

TABLE 5.7 Metal Matrix Composite Systems with Oxide Ceramic Fibers

Fibers	Matrix	Fabrication method	Field of application
Al_2O_3, SiC, Al oxynitride	Al-Cu alloy	Mixing minute filaments in molten matrix; after solidification, filaments penetrate through grain boundary regions	
Al_2O_3	Al-Li alloy	Infiltration with a molten 1–8% Li alloy; reaction occurs between Al_2O_3 fiber and the Li	
Al_2O_3-SiO_2, Y Al_2O_3	Al, Al-Zn alloy	Melt impregnation, powder metallurgy, etc.	Aeronautical industry
Ni-coated glass ceramic fibers	Al	Powder metallurgy	Sliding parts
Al_2O_3 continuous and polycrystalline	Mg or Mg alloy	Melt impregnation of aligned fibers	Turbine blades, shafts
C, B, glass ceramic, and metal fibers	Mg alloy	Powder metallurgy; composite contains two different Mg alloy phases	Multiple applications
Metal-coated glass fibers	Fe, Be, Ti, Al, and Sn	Chopped metal-coated glass fibers are mixed with glass filaments and metal powder; hot pressing	Dimensionally stable machine parts, e.g., friction elements
Glass fibers	Pb	Metal impregnation	Battery plates, bearing materials, acoustic insulation

SOURCE: From Chawla.[6]

TABLE 5.8 Some Important In Situ Composite Systems

System	Carbide, vol. %	T_E, °C
Co-NbC	12	1365
Co-TiC	16	1360
Co-TaC	10	1402
Ni-HfC	15–28	1260
Ni-NbC	11	1330
Ni-TiC	7.5	1307

SOURCE: From Chawla.[6]

- Excellent toughness
- Good environmental resistance

5.4.2 Interfacial bonding

At the interface of ceramic-reinforced metal matrix composites, the following discontinuities in properties need to be contended with:

1. *Bonding:* ionic or covalent in the reinforcement versus metallic bonding in the matrix
2. *Crystallographic:* structure and lattice parameter difference between matrix and reinforcement
3. *Moduli:* difference in elastic modulus between matrix and reinforcement
4. *Chemical potential:* thermodynamic nonequilibrium between matrix and reinforcement
5. *Coefficient of thermal expansion (CTE)*: difference between matrix and reinforcement in CTE, to be expected

Two interfacial bonding mechanisms operate at the interfaces of MMCs. These are mechanical and chemical bonding:

- *Mechanical bonding.* It is obvious that the mechanical keying effect between a fiber's rough surface and the matrix will lead to increased interfacial strength. The compressive radial stress increases with the roughness amplitude and decreases with the fiber radius. Superconducting composites using Nb–Ti alloy filaments in a copper matrix utilize mechanical locking for their strength.
- *Chemical bonding.* Since most MMCs are thermodynamically nonequilibrium systems, exposure of these to high temperatures will lead to the formation of reaction layers at the fiber-matrix interface.

Figure 5.60 Photomicrograph showing the extensive cracking (arrow) of the interfacial M_6C layer in a Mar-M200 matrix composite tested at 250°C. (*From Chung and VanDrunen.*[15])

The thickness and nature of the reaction layer can lead to advantageous or deleterious effects on the properties of the MMC.

Two processes are often used to improve or modify the interfacial properties in MMCs. These are the following:

1. *Fiber surface treatment to improve wettability and adhesion or prevent adverse chemical interactions between the fiber and matrix.* Examples are SiC and B_4C coatings on boron fibers. Figure 5.60 shows an example of a cracking in the M_6C reaction zone of a MAR–M200 alloy reinforced with tungsten fibers. Application of a hafnium nitride barrier layer on the surface of the tungsten fibers eliminated the formation of this reaction zone. In fractographs, the treated surface of the fibers can have a variety of appearances. An illustration of this effect is shown in Fig. 5.61 where the fracture surface appearance of Ti- and B-coated graphite fiber-aluminum composites spans the range of stripped coating to an adherent "snakeskin." The heat treatment of such coated composites can change the fracture surface morphology significantly. This is shown in the fractographs of a Ti- and B-coated graphite fiber-aluminum shown in Fig. 5.62. In this figure, it is seen that a short (4.5-h) heat treatment at 571°C leads to a fracture surface where the fiber surfaces are smooth (Fig. 5.62*a*). Conversely, the fibers show extensive transverse cracking and complete lack of pullout (Fig. 5.62*b*) when the composite is treated for 24 h at the same temperature. An illustration of the deleterious effect of fiber coating on matrix properties is shown in Fig. 5.63. In this case, the graphite fibers are coated with nickel followed by melt infiltration in liquid aluminum. During

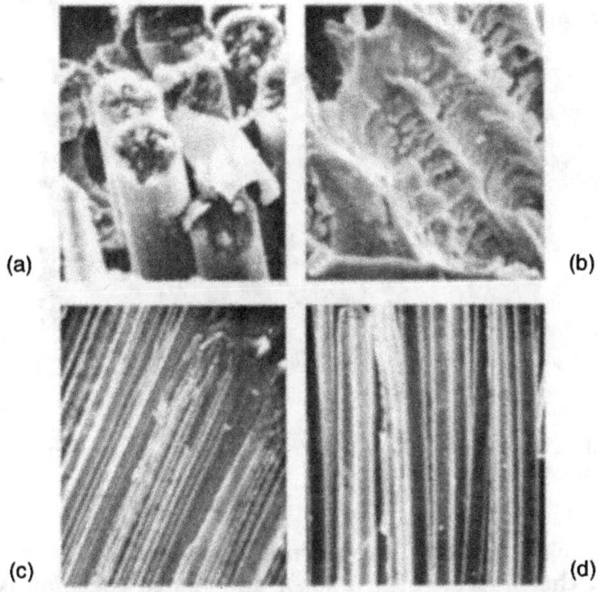

Figure 5.61 SEM micrographs of fracture surfaces in graphite-aluminum composites showing (*a*) coating separation after longitudinal tensile fracture, (*b*) coating breakup after local severe matrix deformation during longitudinal tensile fracture, (*c*) coating cracks after transverse tensile fracture, and (*d*) undisturbed vapor deposited "snakeskin" after transverse tensile fracture. (*From Harrigan and Flowers.*[16])

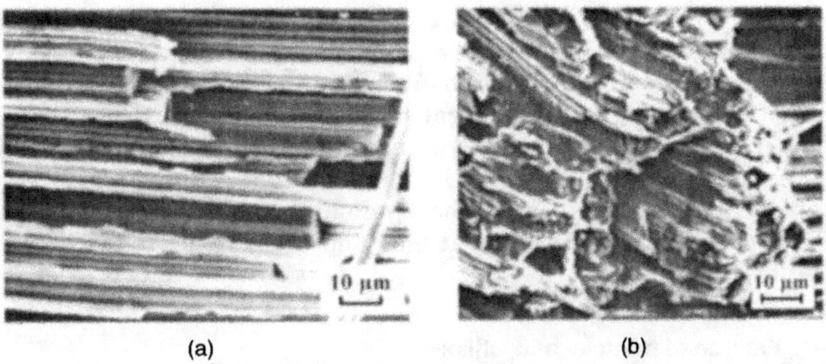

Figure 5.62 Fracture surfaces of T 50/201 (S) composite heat-treated near solidus temperatures for (*a*) 4.5 h and (*b*) 24 h. (*From Amateau and Dull.*[17])

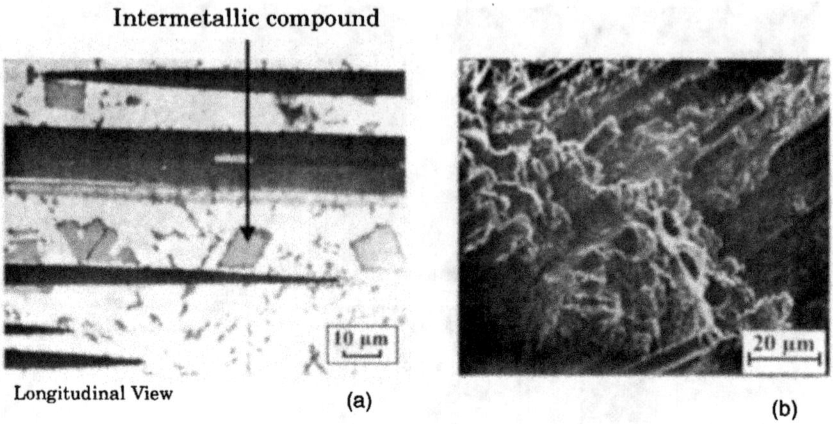

Figure 5.63 (*a*) Longitudinal section of HM 1000/201 (P) composite. (*b*) Fracture surface of HM 1000/201 (P) composite. (*From Amateau and Dull.*[17])

this process, brittle nickel aluminide intermetallic compounds are formed (see arrow) in the matrix (Fig. 5.63*a*), and hence the matrix shows brittle fracture (Fig. 5.63*b*).

2. *Modification of the matrix alloy composition to improve or modify wetting and bonding between the fiber and the matrix.* Examples are alloying of aluminum with lithium and silicon for MMCs reinforced with Al_2O_3 and SiC fibers, respectively.

5.4.3 Change in failure mode of MMCs as a function of temperature

The failure mode in MMCs can change significantly with temperature. As an example, Fig. 5.64 (cross-sectional view) shows extensive transverse cracking of tungsten fibers in tungsten fiber-reinforced MAR-M200 under room-temperature tensile loading. From this it can be inferred that the brittle fracture of fibers preceded matrix failure. Conversely, upon tensile loading at 700°C, Fig. 5.65 shows extensive fiber necking and fiber pullout at the fracture surface. Thus it can be concluded that at elevated temperatures, the matrix failed before the fibers.

5.4.4 Fatigue of particle- and whisker-reinforced composites

The fatigue properties of MMCs reinforced by particles or whiskers can be superior to those of the unreinforced matrix metal. Examples are aluminum alloys reinforced by SiC or Al_2O_3 particles. In these cases, the crack tip may either cut through the particles or go

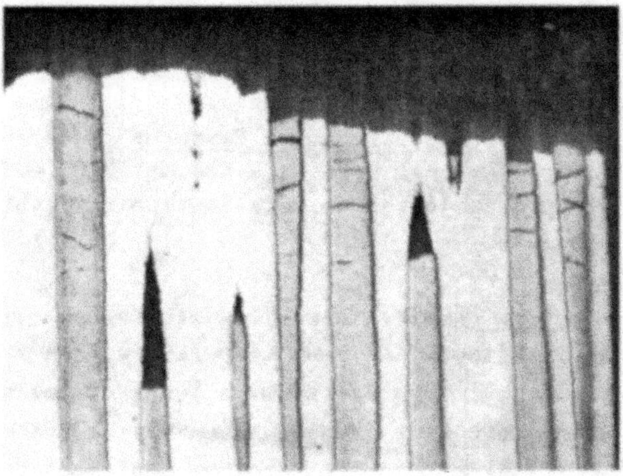

Figure 5.64 Photomicrograph showing the extensive transverse cracking in tungsten fibers of a Mar-M200 matrix composite tested at room temperature. (*From Chung and VanDrunen.*[15])

Figure 5.65 SEM photomicrograph showing tungsten fibers in the fracture surface of a Mar-M200 matrix composite tested at 700°C. (*From Chung and VanDrunen.*[15])

around them, as shown schematically in Fig. 5.66. In the case of monolithic materials, the fracture progresses via the damage zone ahead of a single crack tip. By comparison, the damage zone in composites is a diffuse zone in which many damage mechanisms are operating. Some of these mechanisms can be matrix cracking, interfacial debonding, delamination, fiber fracture, void growth, and multidirectional cracking. Figure 5.67 schematically compares the

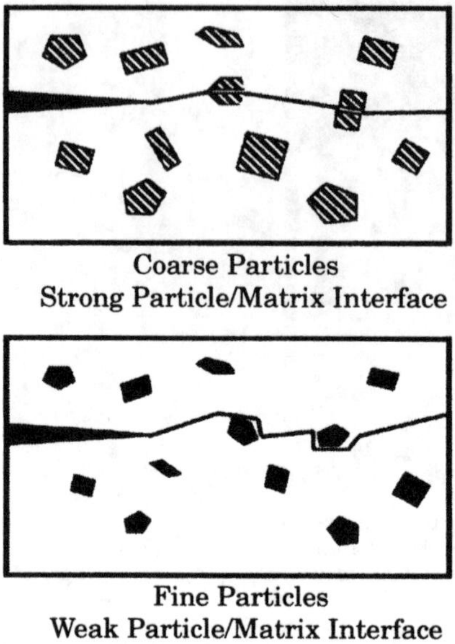

Figure 5.66 Some possible reinforcement and crack tip interactions. (*From Chawla.*[14])

extent of damage accumulation as a function of progress of fatigue between a homogeneous material and an advanced laminated composite. In this figure, *damage ratio* represents the damage level at a given point in time normalized to the damage level at final failure. Likewise, the *cycle ratio* represents the number of cycles at a given point in time normalized to the total cycles to failure. Note that unlike in homogeneous materials, the damage level in composites decelerates with cycling following an initial acceleration. The multiplicity of damage mechanisms responsible for such behavior is illustrated in the fractograph of Fig. 5.68. This fractograph of a B(W)/Al 6061 laminated composite displays ductile fracture in the aluminum, brittle fracture in the boron fiber, fiber pullout, and sheet delamination.

5.4.5 Fatigue of fiber-reinforced composites

Two types of responses to fatigue cracks can be found at the fiber-matrix interfaces based on the differences in properties between the fiber, the matrix, and the interface. Should the interface be weak compared to the matrix's yield strength, the stresses at the tip of the matrix crack can cause debonding of the fiber from the matrix, as shown schematically in Fig. 5.69*a* and in the illustrative fractograph of Fig. 5.70*a*. If the interface is strong, the fiber will in all

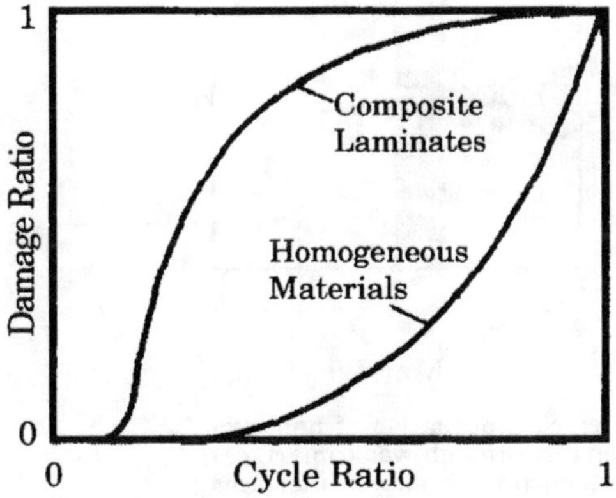

Figure 5.67 Comparison of damage accumulation as a function of fatigue cycles in a laminated composite made by appropriately stacking differently oriented plies and a monolithic, homogeneous material under constant stress amplitude fatigue. (*From Chawla.*[14])

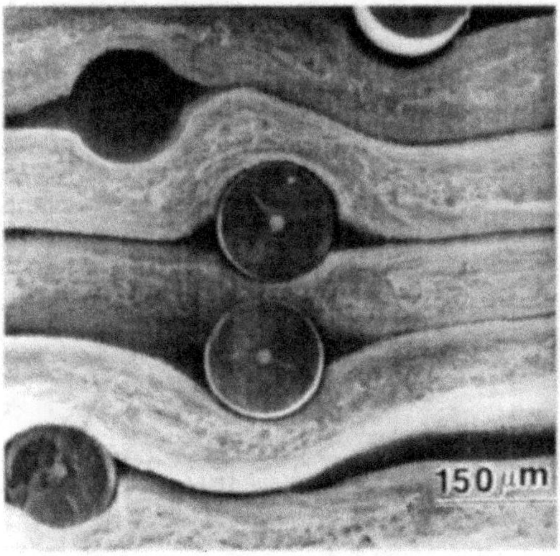

Figure 5.68 Fracture surface of B(W)/A1 6061 composite made by diffusion bonding. Note the ductile fracture in aluminum, brittle fracture in boron, fiber pullout (missing fiber in top left-hand corner), and sheet delamination. (*From Chawla.*[14])

probability crack in a brittle manner, as shown schematically in Figure 5.69*b* and in the illustrative fractograph of Fig. 5.70*b*. In the case of titanium-matrix composites, an intermediate-strength interface leads to a limited amount of fiber pullout, as shown schematically in Fig. 5.71.

In the case of materials with discontinuities (notched specimens), the damage growth/crack growth is different due to three factors:

- The rate of stiffness degradation of the material increases.
- Fiber failure versus debonding may occur (dependent on stress intensity).

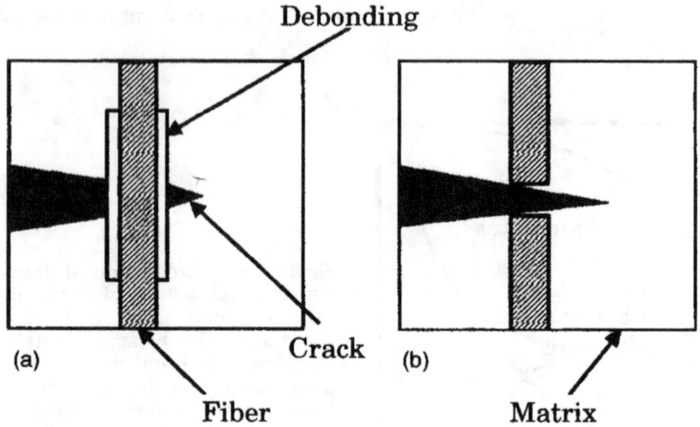

Figure 5.69 The effect of interface on crack propagation of unidirectional composites. (*a*) The crack propagates through weak interfaces by the debonding process; (*b*) propagation through strong interfaces achieved by fiber failure. (*From Rodopoulos.*[18])

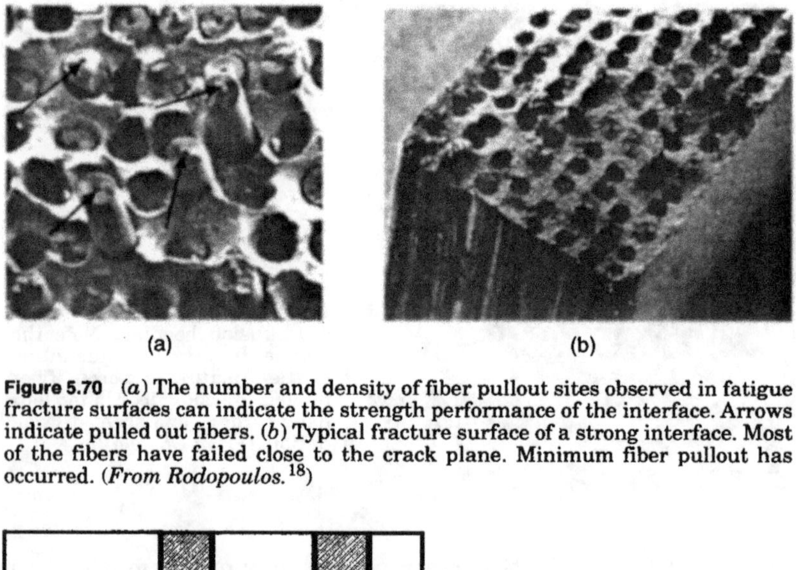

Figure 5.70 (*a*) The number and density of fiber pullout sites observed in fatigue fracture surfaces can indicate the strength performance of the interface. Arrows indicate pulled out fibers. (*b*) Typical fracture surface of a strong interface. Most of the fibers have failed close to the crack plane. Minimum fiber pullout has occurred. (*From Rodopoulos.*[18])

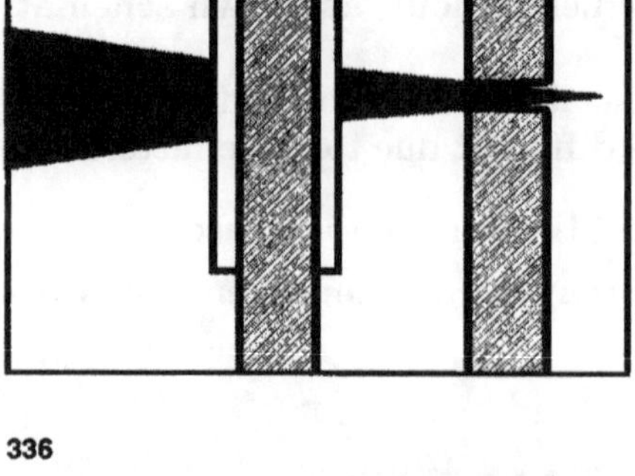

Figure 5.71 Crack propagation through an intermediate-strength interface. (*From Rodopoulos.*[18])

- The crack arrest capability of the material to the propagation of secondary cracks is reduced.

In the case of SCS-6/Ti-6-4 material, matrix cracks in mode I loaded specimens can grow in three distinct ways. This behavior is represented schematically in Fig. 5.72. The three ways can be described as follows:

1. In the case of high applied stress or stress concentration compared to the UTS of the notched material, a catastrophic rate of crack growth is observed, as shown by curve *A* in Fig. 5.72. In this case, the fibers fail ahead of the crack tip, as shown in the schematic representation of Fig. 5.73.
2. In the case of a medium level of applied stress or stress concentration, the sliding of fibers against the interface produces a crack closure

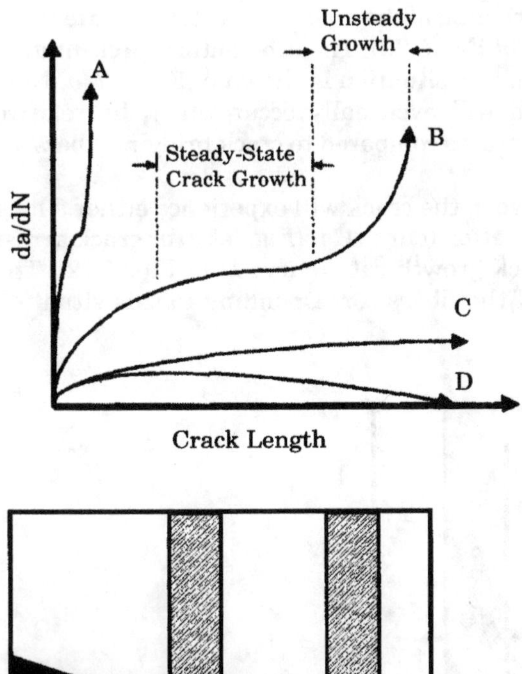

Figure 5.72 Different crack propagation modes with nominal loading conditions. Curve *A:* High nominal applied stress, catastrophic mode I crack propagation rate. Curve *B:* At slightly lower nominal applied stress, the fibers produce crack closure stress, leading to steady-state crack propagation rate followed by rapid increase. Curves *C* and *D:* At still lower nominal applied stress, the crack propagation rate decreases continuously, leading to either a long steady-state growth rate (curve *C*) or crack arrest (curve *D*). (*From Rodopoulos.*[18])

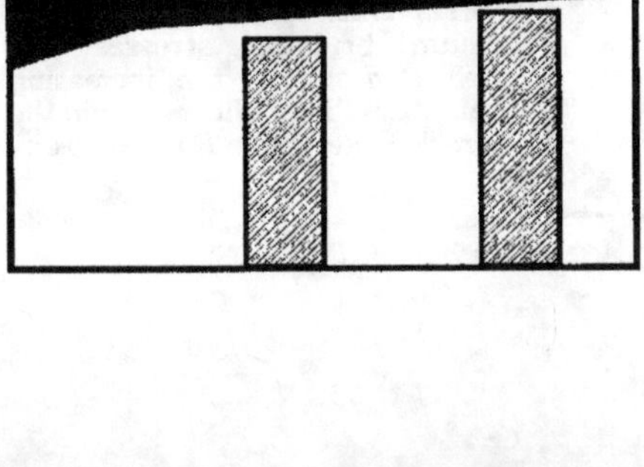

Figure 5.73 A schematic representation of mode A (Fig. 5.72) crack propagation. (*From Rodopoulos.*[18])

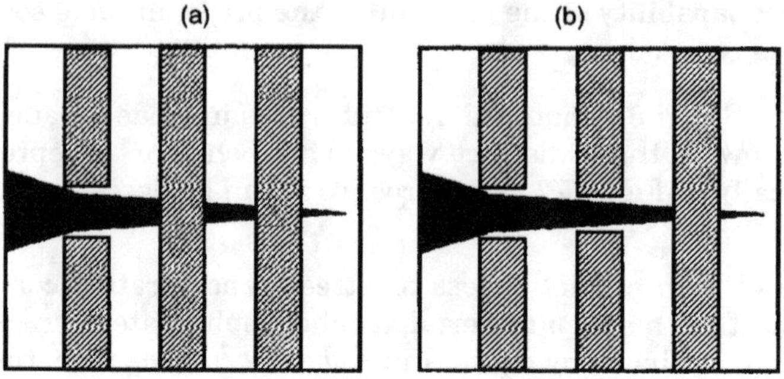

Figure 5.74 During steady-state crack growth (*a*) the number of intact fibers bridging the crack is larger than or equal to the number of broken fibers. As the crack advances, the number of broken fibers relative to those unbroken increases, signifying rapid crack propagation (unsteady crack growth) (*b*). (*From Rodopoulos.*[18])

stress. This produces fiber bridging of the crack and a steady-state crack growth rate region (curve *B* of Fig. 5.72). The schematic representation of the fibers corresponding to this situation is shown in Fig. 5.74*a*. Note that unsteady crack growth will eventually occur when the relative number of cracked fibers increases compared to crack bridging fibers, as shown in Fig. 5.74*b*.

3. At even lower stress levels, the crack will experience either a long period of steady-state propagation (curve *C* in Fig. 5.72) or crack arrest with a minimum in the crack growth rate (curve *D* in Fig. 5.72). The schematic representation of the fibers corresponding to this situation is shown in Fig. 5.75.

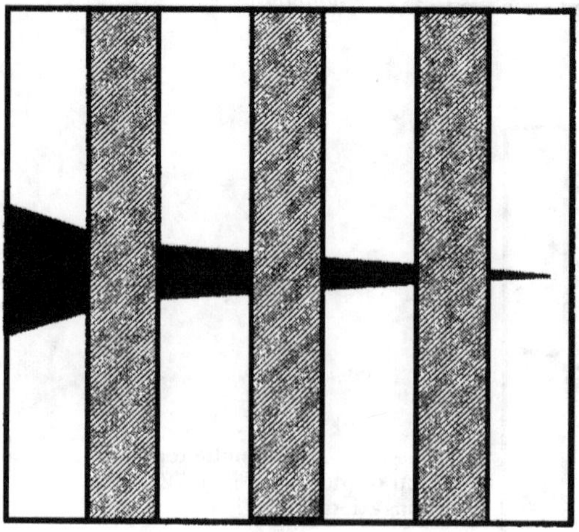

Figure 5.75 Mode C or D (Fig. 5.72) crack propagation. Maximum bridging stresses are expected due to the increasing number of intact fibers within the crack wake. (*From Rodopoulos.*[18])

5.4.6 Fracture behavior in MMCs as a function of fiber properties and loading conditions

The fracture behavior of MMCs depends on the properties of the reinforcing fibers and the type of loading imposed on the composite. An illustration of this can be found in the work of Ruzauskas and Hay.[19] In this work, the fracture behavior was studied as a function of the following variables:

- *Fiber type:* stainless steel (SS, ductile) and tungsten (W, brittle) fiber reinforcements in aluminum alloy matrices
- *Loading condition:* longitudinal tension and compression, transverse tension and compression, through-thickness compression

Their results can be summarized as follows:

1. *Longitudinal tension.* There were four stages of deformation, i.e., two-stage yielding, work hardening, and fracture. The SS-reinforced material showed dimpled matrix rupture along with cup and cone fracture of the ductile fibers (Fig. 5.76*a*). The W-reinforced material showed dimpled rupture of the matrix along with predominantly brittle fracture of the fibers with limited necking (Fig. 5.76*b*).
2. *Transverse tension.* Multistage deformation was absent, and fracture preceded yielding in both materials. Both materials exhibited dimpled rupture of the matrix. The SS-reinforced

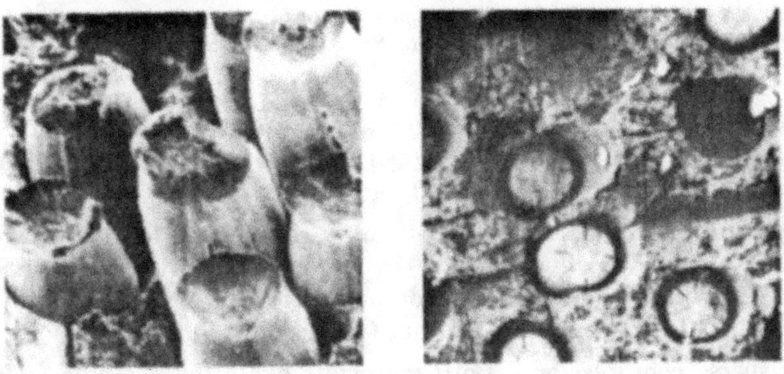

(a) (b)

Figure 5.76 (*a*) Fracture surface of an Al-SS specimen tested in longitudinal tension. (*b*) Fracture surface of Al-W specimen tested in longitudinal tension. (*From Ruzauskas and Hay.*[19])

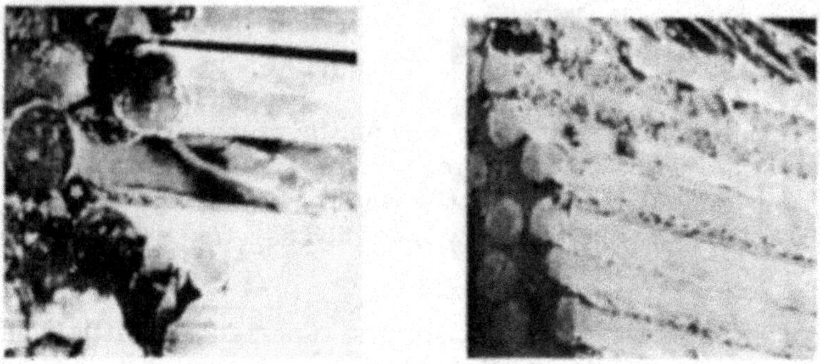

(a) (b)

Figure 5.77 (*a*) Fracture surface of an Al-SS specimen tested in transverse tension. (*b*) Fracture surface of Al-W specimen tested in transverse tension. (*From Ruzauskas and Hay.*[19])

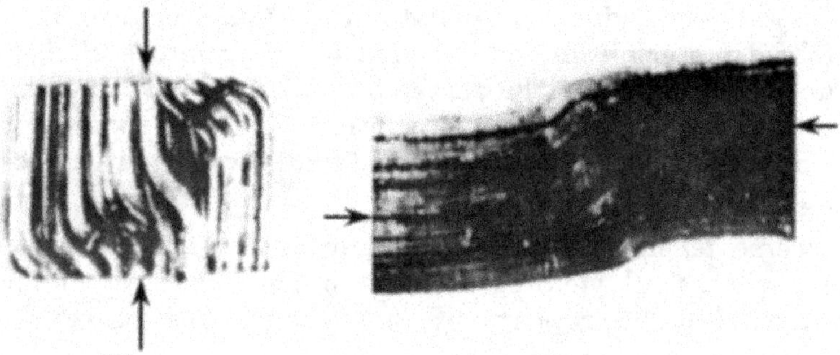

Figure 5.78 (*a*) Al-SS specimen deformed in longitudinal compression. (*b*) Al-W specimen deformed in longitudinal compression. (*From Ruzauskas and Hay.*[19])

material exhibited delamination at the fiber-matrix interface (Fig. 5.77*a*) while the W-reinforced composite exhibited longitudinal fiber splitting (Fig. 5.77*b*).

3. *Longitudinal compression.* Both materials exhibited two-stage yielding prior to shear failure by in-phase buckling. The SS-reinforced material exhibited delaminations (Fig. 5.78*a*) while the W-reinforced material did not (Figure 5.78*b*).
4. *Transverse compression.* Both materials exhibited a single yield point. The SS-reinforced material failed by shear on the planes of maximum shear stress along with fiber-matrix delamination (Fig. 5.79*a* and *b*). The W-reinforced material exhibited delamination in the matrix (Fig. 5.79*c*).

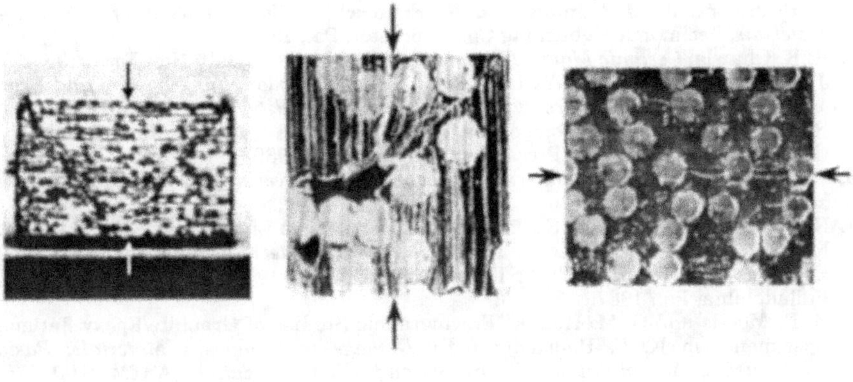

(a) (b) (c)

Figure 5.79 (*a*) Al-SS specimen deformed in transverse compression; (*b*) SEM micrograph of Al-SS specimen deformed in transverse compression; (*c*) Al-W specimen deformed in transverse compression. (*From Ruzauskas and Hay.*[19])

Figure 5.80 Al-W specimen deformed in through-thickness compression. (*From Ruzauskas and Hay.*[19])

5. *Through-thickness compression.* Both materials behaved in a manner identical to that under transverse compression. However, the W-reinforced material failed on the planes of maximum shear stress along with longitudinal fiber splitting (Fig. 5.80).

References

1. D. Hull, *An Introduction to Composite Materials,* Cambridge University Press, New York, 1987.

2. P. Beaumont, J. M. Schultz, and K. Friedrich, *Failure Analysis of Composite Materials,* Technomic Publishing Co., Lancaster, Pa., 1989.
3. K. K. Chawla, *Ceramic Matrix Composites,* Chapman and Hall, New York, 1993.
4. J. W. S. Hearle, B. Lomas, W. D. Cooke, and I. J. Duerdon, *Fibre Failure and Wear of Materials: An Atlas of Fracture, Fatigue and Durability,* Wiley, New York, 1989.
5. R. Talreja, "Fatigue of Composites," in A. Kelly (ed.), *Concise Encyclopedia of Composite Materials,* MIT Press, Cambridge, Mass., 1989.
6. K. K. Chawla, *Composite Materials, Science and Engineering,* Springer-Verlag, New York, 1987.
7. R. W. Hertzberg, "Fracture Surface Micromorphology in Engineering Solids," in J. E. Masters and J. J. Au (eds.), *Fractography of Modern Engineering Materials: Composites and Metals,* ASTM STP 948, American Society for Testing and Materials, Philadelphia, Pa., 1987.
8. G. E. Morris and C. M. Hetter, "Fractographic Studies of Graphite/Epoxy Fatigue Specimens" in K. L. Reifsnider (ed.), *Damage in Composite Materials: Basic Mechanisms, Accumulation, Tolerance, and Characterization,* ASTM STP 775, American Society for Testing and Materials, Philadelphia, Pa., 1982.
9. W. M. Jordan and W. L. Bradley, "Micromechanisms of Fracture in Toughened Graphite/Epoxy Laminates," in N. J. Johnston (ed.), *Toughened Composites,* ASTM STP 937, American Society for Testing and Materials, Philadelphia, Pa., 1987.
10. M. F. Hibbs, M. K. Tse, and W. L. Bradley, "Interlaminar Fracture Toughness and Real-Time Fracture Mechanism of Some Toughened Graphite/Epoxy Composites," in N. J. Johnston (ed.), *Toughened Composites,* ASTM STP 937, American Society for Testing and Materials, Philadelphia, Pa., 1987.
11. K. B. Su, "Delamination Resistance of Stitched Thermoplastic Matrix Composite Laminates," in G. M. Newaz (ed.), *Advances in Thermoplastic Matrix Composite Materials,* ASTM STP 1044, American Society for Testing and Materials, Philadelphia, Pa., 1989.
12. J. F. Mandell, D. D. Huang, and F. J. McGarry, "Crack Propagation Modes in Injection Molded Fiber Reinforced Thermoplastics," in B. A. Sanders (ed.), *Short Fiber Reinforced Composite Materials,* ASTM STP 772, American Society for Testing and Materials, Philadelphia, Pa., 1982.
13. H. T. Hahn, K. L. Jerina, and P. Burrett, "Fiber Orientation and Fracture Morphology in Short Fiber-Reinforced Thermoplastics," in G. M. Newaz (ed.), *Advances in Thermoplastic Matrix Composite Materials,* ASTM STP 1044, American Society for Testing and Materials, Philadelphia, Pa., 1989.
14. K. K. Chawla, *Composite Materials, Science and Engineering,* 2d ed., Springer-Verlag, New York, 1998.
15. N. Chung and G. VanDrunen, "Tensile Behaviour of Tungsten-Superalloy Composites," in J. A. Cornie and F. W. Crossman (eds.), *Failure Modes in Composites IV,* TMS-AIME, Warrendale, Pa., 1979.
16. W. C. Harrigan, Jr., and R. H. Flowers, "Graphite-Metal Composites: Titanium-Boron Vapor Deposit Method of Manufacture," in J. A. Cornie and F. W. Crossman (eds.), *Failure Modes in Composites IV,* TMS-AIME, Warrendale, Pa., 1979.
17. M. F. Amateau and D. L. Dull, "The Effect of Processing on the Transverse Strength of Graphite-Aluminum Composites," in J. A. Cornie and F. W. Crossman (eds.), *Failure Modes in Composites IV,* TMS-AIME, Warrendale, Pa., 1979.
18. C. A. Rodopoulos, "Fatigue Damage Tolerant Design in Unidirectional Metal Matrix Composites," in E. E. Gdoutos, K. Pilakoutas, and C. A. Rodopoulos (eds.), *Failure Analysis of Industrial Composite Materials,* McGraw-Hill, New York, 2000.
19. E. J. Ruzauskas and D. R. Hay, "Fracture Modes in Metal-Matrix Composites," in C. J. Hilado (ed.), *Carbon Composite and Metal Composite Systems,* Technomic Publishing Company, Westport, Conn., 1974.

Chapter

6

Electronic Failure Analysis

6.1 Introduction

Failure analysis of electronic systems presents certain unique challenges due to the small size of their components and the complexity of their construction. Failures of electronic components may involve conventional mechanical fracture, the mechanisms and modes of which are described in previous chapters. However, there are unique electrical phenomena involved in their failure which are not encountered in exclusively mechanical systems that must also be considered, e.g., arcing and electromigration. In addition, the small size of many of the electronic components produces microstructural effects that may play a role in failure. For example, the fine wires for connectors may have a grain size comparable to that of the dimensions of the wire, thereby making the wires susceptible to low ductility and creep fracture. Also inclusions become important since properties such as strength are greatly affected by their size in structures that have comparable dimensions.

The tools described in Chap. 1 are applicable and adequate for many failure analyses of electronic structures and components. However, in some cases, the electrical nature of the components and their fine architecture may require other tools, and these are described in Sec. 6.4.

A comment about terminology is in order. Electronic systems and their components are described by a variety of terms, acronyms, and abbreviations. In this chapter, when such terms are first mentioned, they are defined; subsequent use may involve the shorter notation. To

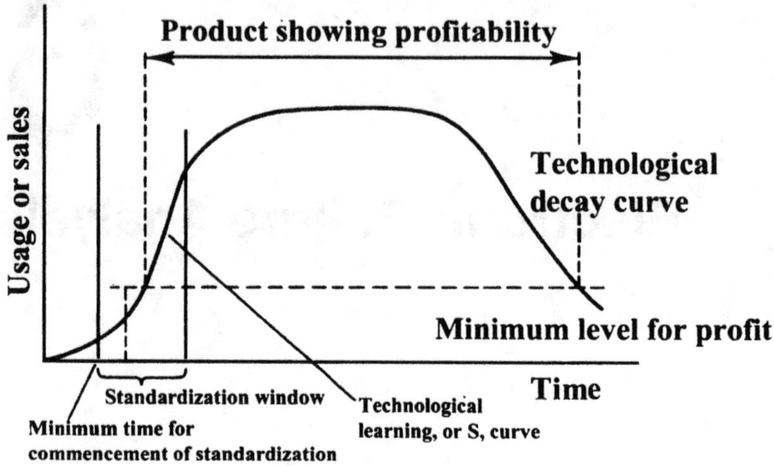

Figure 6.1 Schematic representation of the life cycle of a technology product. (*From Amerasekera and Najm.*[1])

assist in keeping the terminology straight, App. G lists acronyms and their definitions.

6.1.1 Correlation of component life and failure rates

A typical technology product is expected to undergo a product life cycle depicted schematically in Fig. 6.1. Correspondingly, the failure rate of such products has three stages. Early life failures (infant mortalities) can be ascribed to defects introduced into the component during the manufacturing or assembly process. This period is characterized by a high initial rate, followed by a progressively decreasing rate as defective units are eliminated. During the useful life of the hardware, random failures occur at infrequent intervals (intrinsic failure rate). This stage is followed by a rapid rise in the failure rate during the wear-out phase when the device deteriorates from physical and environmental effects. The general trends mentioned above can be represented schematically as the "bathtub curve" shown Fig. 6.2.

6.2 Failure Mechanisms

6.2.1 Terminology in electronics failure analysis

The terminology used in the failure analysis of electronic devices and materials is somewhat different from that used in other kinds of fail-

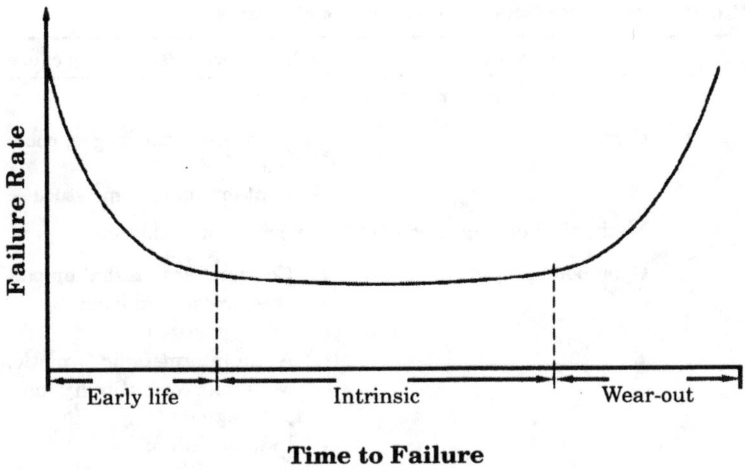

Figure 6.2 Schematic representation of the failure rate of a device as a function of time. (*From Marsh et al.*[2])

ure analyses. In this context, *failure mode* refers to the nature of malfunction of the device, which results in the observed failure characteristic (e.g., open circuit, short circuit, leakage). The nature of the actual fault that causes the device to malfunction is called the *failure defect* (e.g., microcracks and growths). Phenomena such as electromigration and corrosion that produce failure defects are called the *failure mechanisms.* The nature of the actions (e.g., high current density, stress concentrations) that cause failure mechanisms to occur are referred to as *failure causes.* When a device operates but not to the desired specification (e.g., a timing problem), it is referred to as a *specification failure,* while a device in which certain or all functions are inoperative is referred to as a *hard failure.* Using this terminology, Table 6.1 shows examples of the correlation between modes, defects, and mechanisms for some typical failures. The table is by no means exhaustive. Table 6.2 shows the distribution of failure mechanisms within integrated-circuit (IC) packages.

Failure defects can be classified in a variety of ways. Figure 6.3 shows one classification scheme that is useful when the purpose of the analysis is to assign the failure causes. An alternate classification scheme, shown in Fig. 6.4, is useful when the purpose is to determine requirements for diagnostic tools for an investigation. Note that the figure implies that the use of several complementary techniques may be necessary to get the information needed to arrive at the causes of a failure.

TABLE 6.1 Matching of Failure Modes and Failure Mechanisms

Failure mode	Failure defect	Failure mechanism/failure cause
Open circuit	Missing interconnects	Mask errors
	Corrosion	(i) Moisture (internal gas, poor seal) (ii) Contaminants + moisture
	Mechanical damage/scratch	Poor processing/handling
	Open bond	(i) Contaminant present on pad (ii) Overpressure in bonding (iii) Package stress (iv) Al/Au intermetallic formation (v) Materials compatibility (purple plague) (vi) Fatigue failure
	Open metallization	(i) EM in metallization (ii) Processing defects (step coverage, lithography) (iii) Corrosion (iv) Wrong passivation composition
	Metallization microcracks	Stress migration in as-deposited metallization
	Local disruption of interconnects	(i) Electrical overstress (EOS) (ii) Electrostatic discharge (ESD)
	Die cracking	Thermal/mechanical stress
Short circuit	Surface contamination	(i) Direct contaminant S/C between tracks (ii) Electrochemical metal migration/dendrites
	Pinholes in isolation layer	Direct S/C
	Hillocks	(i) Direct S/C via EM (ii) Whisker growth in double-layer metal
	Local disruption of interconnects	(i) Electrical overstress (ii) Electrostatic discharge
	Metal spikes	(i) Enhanced diffusion along dislocations (ii) Poor contact alloying (iii) Overstress giving spiking (iv) Electrothermal overstress

TABLE 6.1 Matching of Failure Modes and Failure Mechanisms (*Continued*)

Failure mode	Failure defect	Failure mechanism/failure cause
Short circuit (*continued*)	Metal shorts	(i) Poor bond placement (ii) Metallization definition defects (iii) Bond wire to substrate short
	"Grow-back" in PROMs	(i) Oxide reduction (ii) Stress/field-induced whisker growth
	Surface growths	EM of silicon
	Package lead short	(i) Whisker growth—pure tin plating (ii) Whisker growth—environmental corrosion (iii) Whisker growth—frit seal composition
Leakage	Surface contamination	(i) Surface leakage (ii) Ionic contamination at oxide-silicon interface
	Poor junction delineation	Processing faults
	Hot spots	(i) Metallization defects causing high power dissipation (ii) Crystallographic defects causing current concentration
	Metal spikes	Metal diffusion along crystal defects
	Oxide breakdown	(i) Voltage overstress (ii) Oxide defects (intrinsic)
High-temperature failure	Defective die attach	Thermal overstress
Low-temperature failure	Reduced depletion region width	Poor alignment/definition of diffusions

SOURCE: From Richards and Footner.[3]

6.2.2 Major failure mechanisms

The four major *failure mechanisms* to which most failures of electronic components can be attributed are distortion, fatigue, wear, and corrosion. The details of each are summarized in Table 6.3. Of these, the dominant mechanism can be broadly categorized as environmental effects, which can be specifically attributed to temperature extremes, vibration, moisture, sand and dust exposure, salt exposure, shock, and radiation damage. In addition, the manufacturing defects shown in Table 6.4 can be introduced during the manufacturing process.

TABLE 6.2 Major Failure Mechanisms within IC Package

Item	Percent of reports
Electromigration	25
Internal shorts	14
Internal opens	12
Parametric variations	10
Surface inversion	8
Material oxidation	8
Dielectric breakdowns	6
Intermetallic compounds	4
Parasitic action	4
Junction leakage	4
Dendrites	4
Threshold shifts	4

SOURCE: From Dicken.[4]

Failure mechanisms in discrete or IC semiconductor devices can be grouped into four major categories:

1. Event-dependent electrical stress failures
2. Materials-related intrinsic failure mechanisms
3. Extrinsic failure mechanisms related to interconnects, passivation, and packaging
4. Ionic contamination-induced inversion

6.2.3 Event-dependent electrical stress failures

Electrical overstress (EOS) and electrostatic discharge (ESD). There are two main types of electrical stress failures that account for more than 50 percent of all semiconductor field failures: (1) voltage-induced, caused by dielectric breakdown or oxide punch-through, and (2) current-induced, caused by metallization burnout or fusion by joule heating due to the discharge current. Damage caused by overvoltage or overcurrent conditions when the duration of stress is longer than 1 μs is known as an electrical overstress failure while that of shorter duration is known as an electrostatic discharge failure. Voltage and current transients resulting from power supply switching, relay operation, etc., can cause EOS. It can also be caused by the improper application of an IC in a system. In EOS, a runaway high-temperature condition resulting from high currents can cause either a short circuit by the melting of the silicon at 1688°C or an open circuit due to metallization meltdown (Fig. 6.5). ESD is caused by the rapid (a few hundred picoseconds

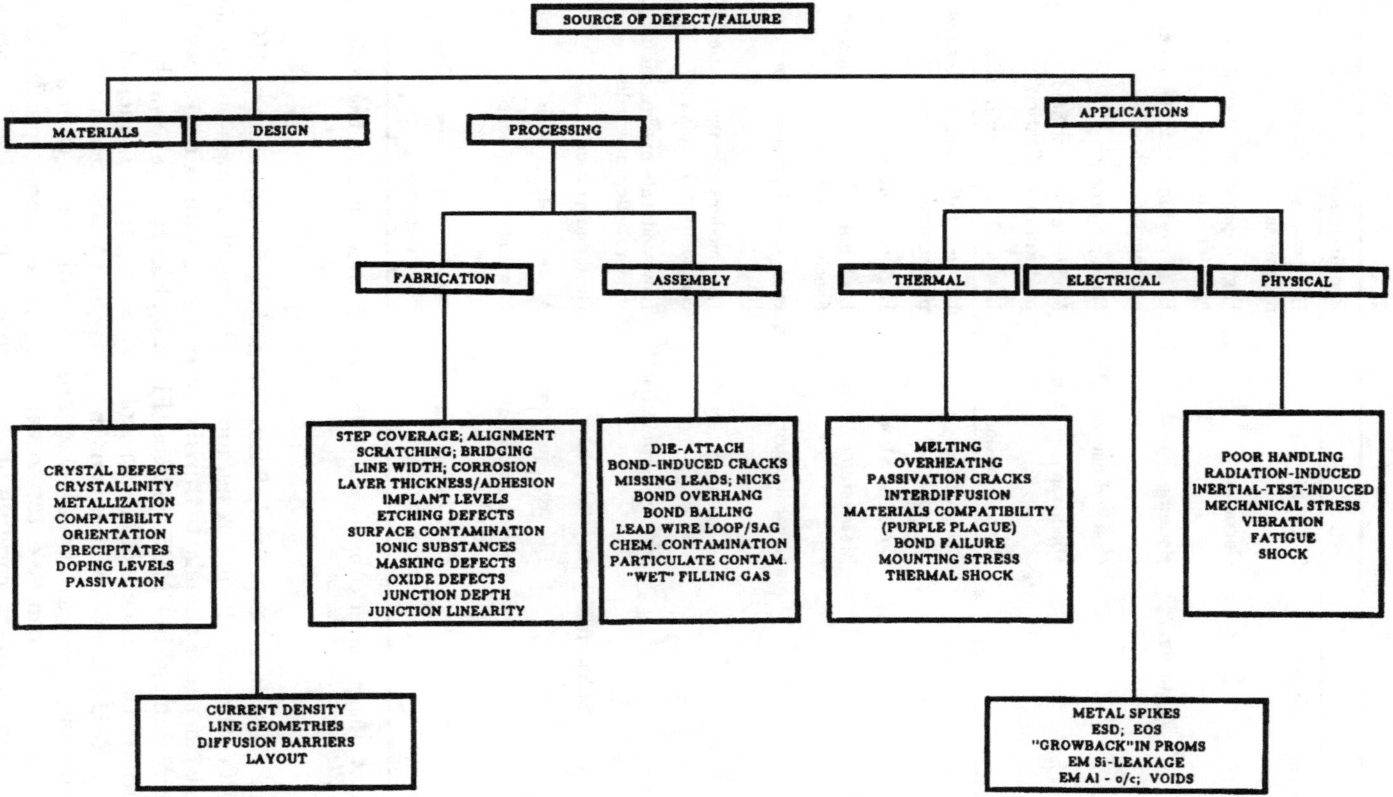

Figure 6.3 Classification of electronic failures by cause of failure. (*From Richards and Footner.*[3])

		Defect	Requirements of microscopes
Package	External	Mechanical	Long working distance; low magnification
		Corrosion	High depth of field; good sample manipulation
	Internal	Bonding	Long working distance; medium magnification
		Cracked package	Good penetration
		Die-attach	Moderate depth of field; low-medium magnification
		Contamination	Low-medium magnification
Die	Surface	Cracked die	Low magnification; some sample manipulation
		Contamination	High magnification
		Metal step coverage	High magnification; moderate depth of field
		Layer integrity	High magnification; colors useful
		Morphology/corrosion	Low-moderate magnification; compositional analysis
		Morphology/electrical	Low-moderate magnification; moderate depth of field
	Subsurface	Crystal defects	Good penetration
		Electrical junctions	Moderate depth of field
		Metal-semiconductor interface	Low-high magnification
		Ionic contamination/ charge related	Sometimes requires selective layer removal

Figure 6.4 Classification of package and die failures for selection of diagnostic tool. (*From Richards and Footner.*[3])

to 1 ms) discharge to ground of accumulated electrostatic charge (100 V to 20 kV) through the low-resistance IC. The resulting high-voltage discharge can cause damage to thin dielectrics (e.g., gate oxides) as well as thermal damage in CMOS systems (Fig. 6.6). Figure 6.7 shows the ESD sensitivity rating scale of all types of devices. Figure 6.8 schematically depicts ESD damage mechanisms in bipolar and MOS structures. Since ESD damage sites are small, they are generally found with an SEM, while EOS sites can be detected with an optical microscope. Table 6.5 shows a guide to interpreting physical damage due to electrical stress.

TABLE 6.3 Mechanical Mechanisms of Failure of Electronic and Mechanical Components

Distortion	Fatigue and fracture	Wear	Corrosion
Buckling	Ductile fracture	Abrasive wear	Corrosion fatigue
Yielding	Brittle fracture	Adhesive wear	Stress corrosion
Creep	Fatigue fracture	Subsurface-origin fatigue	Galvanic corrosion
Creep buckling	High-cycle fatigue	Surface-origin fatigue (pitting)	Crevice corrosion
Warping	Low-cycle fatigue	Subcase-origin fatigue (spalling)	Pitting corrosion
Permanent plastic deformation	Residual stress fracture	Cavitation	Biological corrosion
Temporary plastic deformation	Embrittlement fracture	Fretting wear	Chemical attack
Thermal relaxation	Thermal fatigue fracture	Scoring	Fretting corrosion
Brinnelling	Torsional fatigue, fretting fracture		

SOURCE: From Dylis.[5]

TABLE 6.4 Manufacturing Defects Uncovered during Manufacturing Process

Manufacturing defect	Percentage
Open circuit	34
Solder bridges	15
Missing parts	15
Misoriented parts	9
Marginal joints	9
Balls, voids	7
Bad parts	7
Wrong parts	7

SOURCE: From Dylis.[5]

Figure 6.9 shows the damage caused to a die bond due to power cycling. Filamentary "tentacles" of fused polysilicon formed on a power MOSFET die caused by ESD are shown in Fig. 6.10. Components are generally degraded and sometimes completely fail under the influence of static electricity. It is pointed out that the training of manufacturing personnel is crucial in the minimization of electrical stress damage

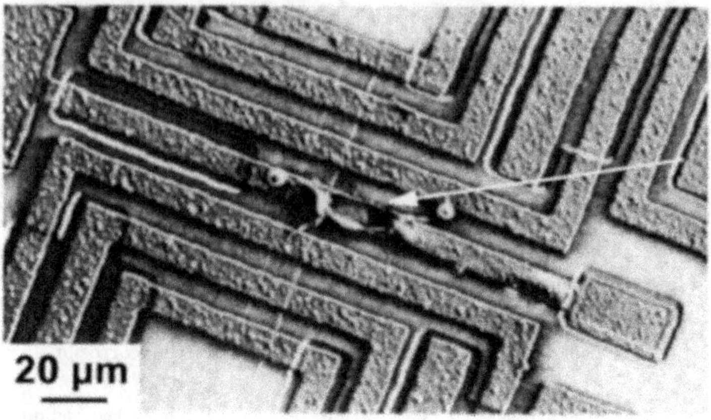

Figure 6.5 Failure due to melting of metallization (arrow) caused by electrical overstresses in a 16K D-RAM. (*From Amerasekera and Najm.*[1])

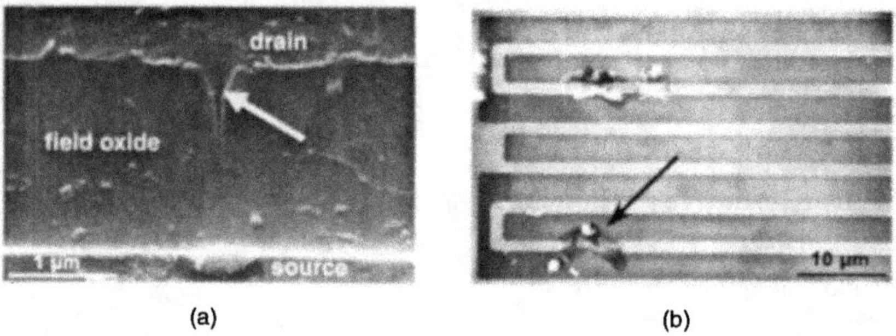

Figure 6.6 Examples of (*a*) ESD damage in an nMOS transistor (arrow) and (*b*) drain-source short circuit (arrow) due to ESD stress in a CMOS IC. (*From Amerasekera and Najm.*[1])

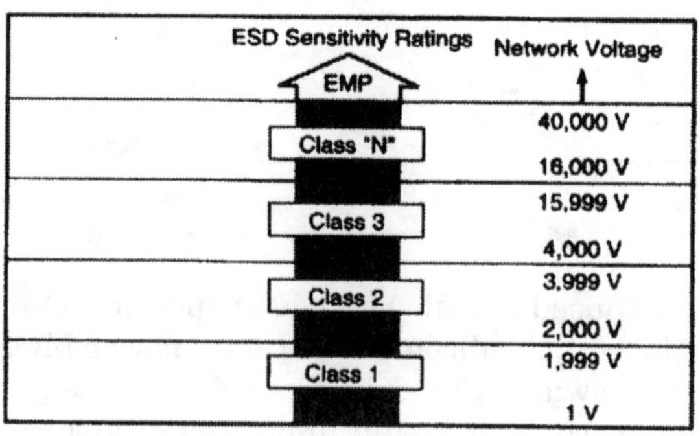

Figure 6.7 ESD sensitivity rating scale, which applies to all device types. (*From Lee.*[6])

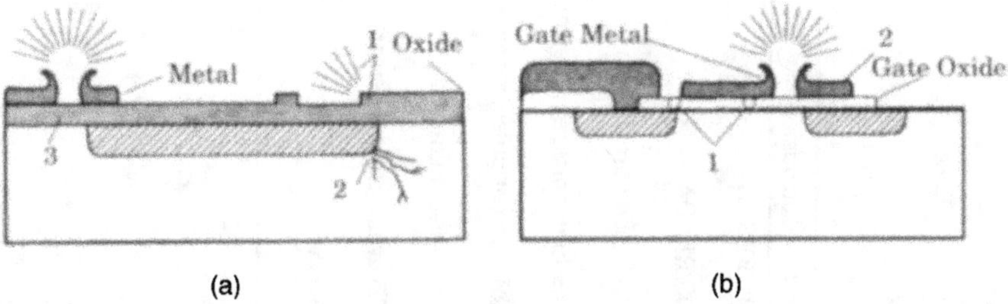

Figure 6.8 Schematic cross section showing ESD damage in (*a*) bipolar structures (1, metal transport across junction; 2, silicon melting in bulk; 3, fusing of metal or poly) and (*b*) MOS structures (1, gate oxide rupture; 2, gate metal fusing). (*From Lee.*[6])

since they are responsible not only for the majority of charge accumulation but also for the majority of discharges. Table 6.6 shows the electrostatic voltages developed by human activity and compares them to breakdown voltages for various types of electronic devices. It is also pointed out that electrostatic discharge can lead to degradation or failure of a component or device at any stage in its life. In addition, some electronic equipment spends a majority of its life cycle in a nonoperational mode during which deterioration can occur due to entrapment of moisture, chlorine, or other chemicals in addition to aging effects such as drying of lubricants, creep of metallic components, and embrittlement of polymers. The three ways to achieve ESD protection are

1. Minimization or elimination of charge buildup
2. Draining of accumulated charge in conductors
3. Neutralization of accumulated charge in insulators

Latchup. Latchup is a phenomenon that can be triggered in silicon controlled rectifiers (SCRs) when a low resistance path is formed between the supply pad and ground (Fig. 6.11), leading to overcurrents and subsequently catastrophic or noncatastrophic failures. Another concern in MOS processes is the integrity of the gate oxide, referred to as *gate oxide integrity* (GOI). An extrinsic source for current leakage could be ionic contamination by potassium, sodium, and chlorine. In addition, charge movement through the field oxide separating active diffusion regions is a concern. In memory devices such as dynamic random access memories (DRAMs) and static random access memories (SRAMs), it is critical that no charge leaks through the gate oxides where a critical charge is stored. One mechanism for such leakage is the entrapment of electrons by interstitials at the Si–SiO_2 interface (Fig. 6.12). Another mechanism is the injection of hot carriers (usually

TABLE 6.5 Guide to Interpretation of Physical Damage in Electrically Damaged Devices

Physical evidence	Most probable cause	Conclusion
1. Reconstructed metallization over entire die	FBSOA* exceeded, device running too hot, or excessive cycling	EOS/FBSOA
2. Reconstructed metallization in characteristic area of emitter	RBSOA* exceeded with any type of base drive, current-limited breakdown BVceo(sus) operation	EOS/RBSOA
3. Melt-through hole in characteristic area of emitter	Es/b operation or repetitive high-voltage transient	EOS, second breakdown
4. Melt-through holes in unusual area of emitter or base with edge arcovers	Nonrepetitive transient in application	EOS, high-voltage transient
5. Bonding wire fused, die usually electrically OK	Forward-biased junction(s)	EOS, current pulse, reverse
6. Bonding wire fused, melt-through hole in die, metal alloyed	Massive overstress	EOS, cause unknown
7. Tiny hole in oxide or silvery track visible in oxide between junctions	High voltage, low energy	ESD event, test transient
8. Shorted junction or gate—no other physical damage is obvious	ESD event with damage located in silicon bulk	ESD event

*FBSOA, forward bias safe operating area; RBSOA, reverse bias safe operating area.
SOURCE: From Lee.[7]

Figure 6.9 Die bond cracked (arrow) due to excessive power cycling. (*From Lee.*[7])

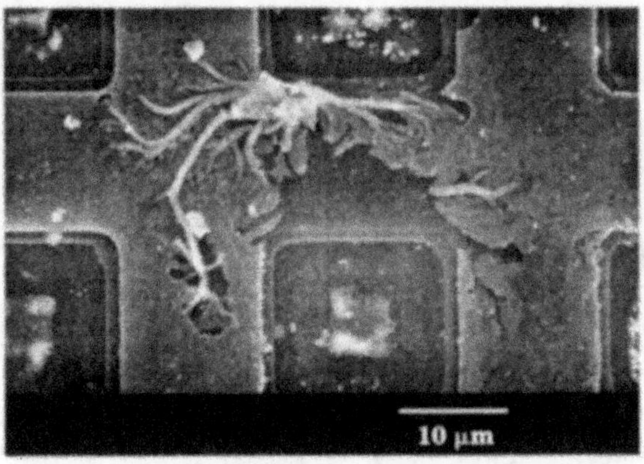

Figure 6.10 "Filamentary" fused polysilicon in a MOSFET device observed after removal of metallization and insulating oxide. (*From Johnson and Pote.*[8])

electrons) into the dielectric oxide due to the high lateral fields in certain submicrometer devices (Fig. 6.13). In the case of submicrometer devices, the use of plasma in various processing steps can lead to increased oxide leakage current due to damage of the oxide.

6.2.4 Materials-related intrinsic failure mechanisms

Crystalline defects such as interstitials (point), dislocation (line), stacking faults (area), and precipitates (volume) can all be deleterious

TABLE 6.6 Human Activity and Electrostatic Voltage and the Susceptibility of Various Devices to ES Damage

Electrostatic Voltage from Human Activity		
Source of static charge	Humidity 10–20%, kV	Humidity 65–95%
Walking across carpet	35	1.5 kV
Walking across vinyl flooring	12	250 V
Standing up from chair	1–1.5	—
Worker by the work table	6	100 V
Lifting poly bag from the work table	20	1.2 kV
Sitting at padded vinyl chair	18	1.5 kV
Wrapping in vinyl (instruction manuals)	7	600 V

Device Susceptibility to ESD	
Device type	ESD vulnerability range (V)
VMOS	30–1800
CMOS	250–3000
Power MOSFET	100–200
GaAs FET	100–300
Junction FET	140–10,000
Op-amp	190–2500
SCR	680–1000
ECL	500–1500
Bipolar transistors	380–7000
MOSFET	100–200
CMOS RAM	250–700
Schottky diode	300–2500
Schottky TTL	1000–2500
Film resistors (thick and thin film)	300–3000

SOURCE: From Viswanadham and Singh.[9]

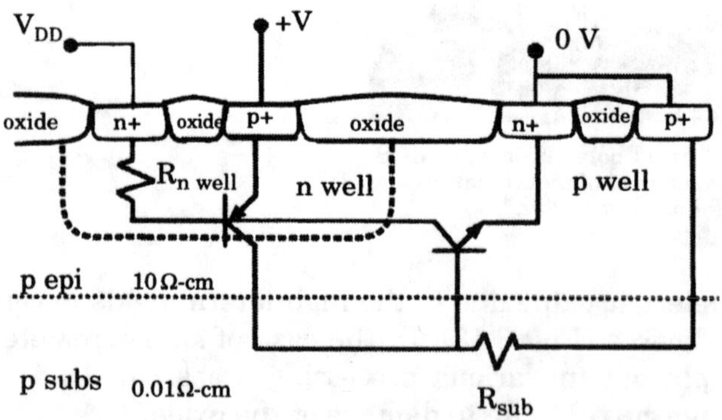

Figure 6.11 Schematic cross section showing latchup phenomenon in a *pnpn* device [also called a silicon controlled rectifier (SCR)]. Latchup occurs when upon triggering the SCR, a low-resistance path (R_{sub}) is formed between the supply pad (+V) and ground (0 V). It can be triggered by a voltage or current transient in a power supply or input/output pad. The low-resistance path remains after the transient has been removed. (*From Amerasekera and Najm.*[1])

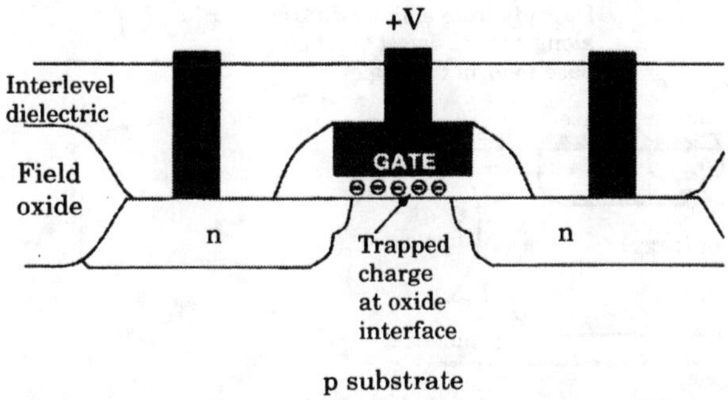

Figure 6.12 Schematic representation of leakage due to slow trapping of electrons in an nMOS transistor. (*From Amerasekera and Najm.*[1])

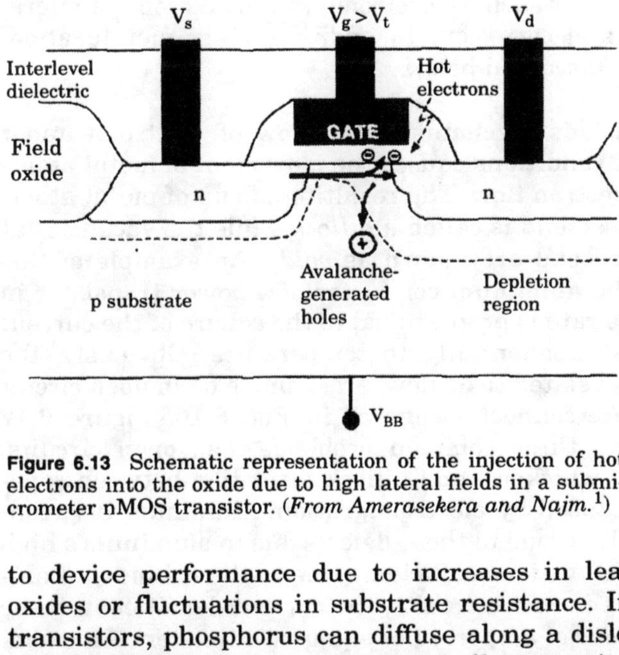

Figure 6.13 Schematic representation of the injection of hot electrons into the oxide due to high lateral fields in a submicrometer nMOS transistor. (*From Amerasekera and Najm.*[1])

to device performance due to increases in leakage current across oxides or fluctuations in substrate resistance. In the case of bipolar transistors, phosphorus can diffuse along a dislocation from the epitaxial collector region, causing a collector-emitter short (Fig. 6.14), known as a *pipe*. In addition, external radiation (γ rays, x-rays, cosmic rays) or internal radiation (α particles, β radiation) can interact with semiconductor materials to produce electron-hole pairs, which can interfere with circuit behavior (radiation damage).

6.2.5 Extrinsic failure mechanisms

Metallization. The performance of electronic circuits can be negatively affected by increases in interconnect resistance, which can be caused

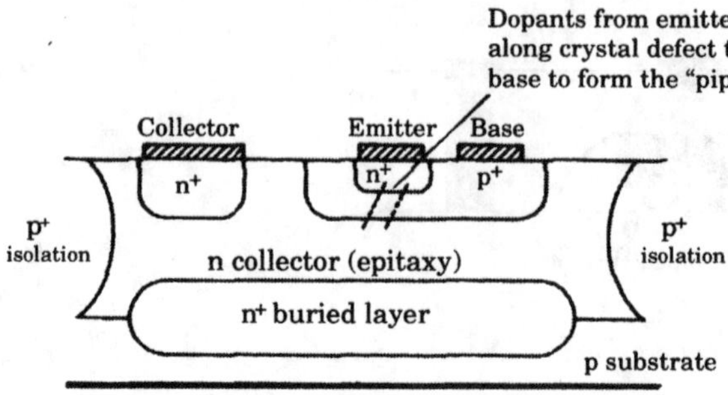

Figure 6.14 Schematic representation of enhanced phosphorous diffusion along a crystal defect. (*From Amerasekera and Najm.*[1])

by metallization problems such as electromigration and contact migration, via migration and corrosion. In addition, other metallization-related problems are discussed below.

Electromigration. In this mechanism, the flow of electrons under high current density conditions causes the movement of metal atoms in the direction of electron flow. The resulting pileup of metal atoms at the positively biased end is called a *hillock* while the vacancies at the negatively biased end are known as *voids*. An example of this effect is shown in the aluminum conductor of a power transistor in Fig. 6.15. The failure rate is proportional to the square of the current density and relates exponentially to temperature. Obviously, the problem is primarily related to dc flow. An example of an open circuit in an aluminum interconnect is shown in Fig. 6.16. Figure 6.17 schematically depicts three common problems, i.e., open circuits, short circuits due to whiskers, and hillock formation between separate wiring levels, caused by electromigration. Aluminum interconnections are especially prone to these defects due to aluminum's high grain boundary self-diffusion coefficient. Due to the thinness of aluminum interconnects in microcircuits, the current densities in these are often high (millions of amperes per square centimeter). An example of grain boundary preferential electromigration is shown in Fig. 6.18. Since the metal grain boundaries are responsible for electromigration, this problem can be alleviated by increasing the grain size of the metal lines to be comparable to the interconnection width and thickness, i.e., bamboo structures, as shown in Fig. 6.19. Other methods of alleviating this problem include the alloying of the aluminum with copper and limiting of the designed maximum current density. A recent advance in interconnects has been made by the use of layers of aluminum/silicon between titanium or tungsten layers. The use of refractory metals to alleviate the electromigration problem is shown

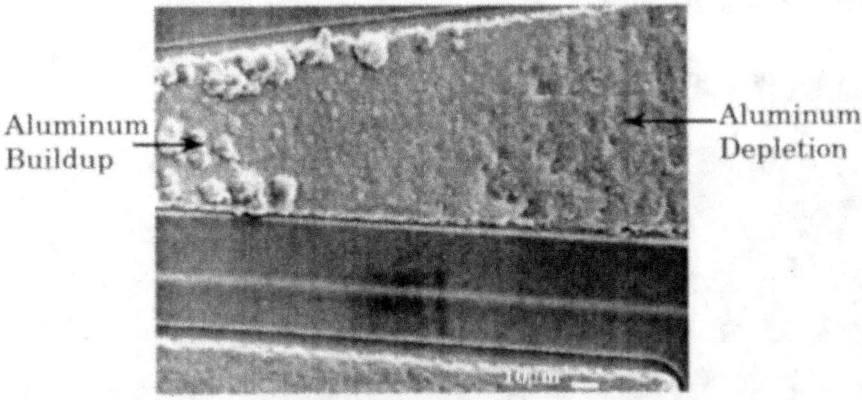

Figure 6.15 Electromigration of Al in a power transistor. Note Al buildup at left end and depletion at right end (*From Haythornthwaite.*[10])

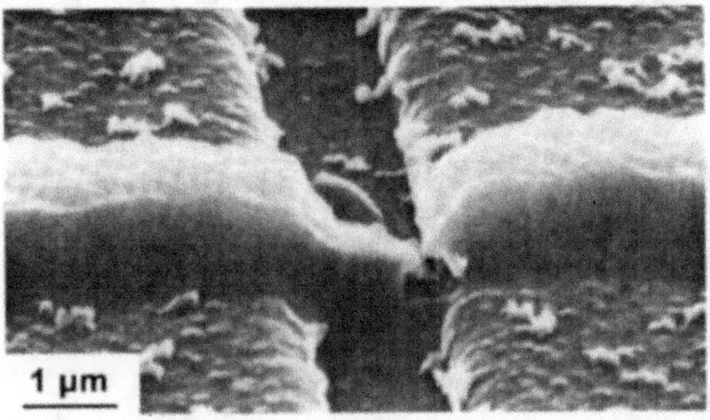

Figure 6.16 Example of open Al interconnect caused by electromigration. (*From Amerasekera and Najm.*[1])

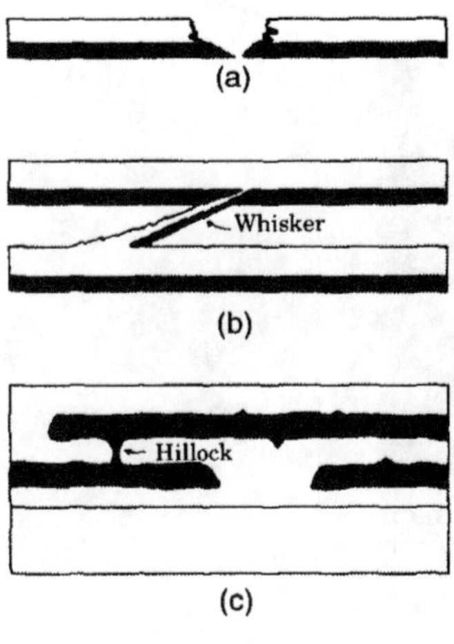

Figure 6.17 Electromigration-related failure modes: (*a*) broken line due to removal of metal atoms; (*b*) undesired short circuit due to a whisker; (*c*) short circuit due to a hillock. (*From Bakoglu.*[11])

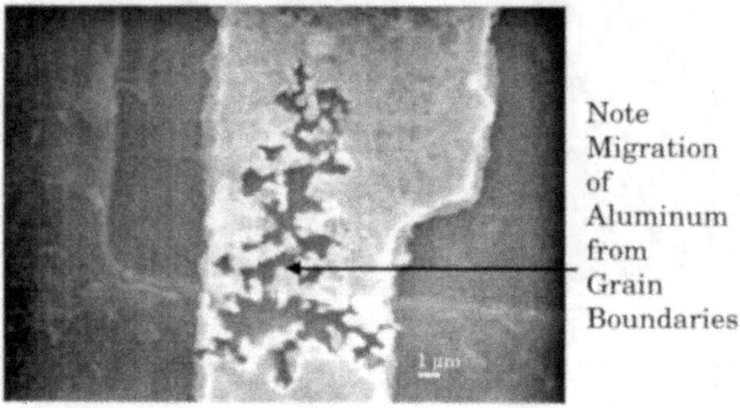

Figure 6.18 Preferential grain boundary electromigration of Al in an IC Al track. (*From Haythornthwaite.*[10])

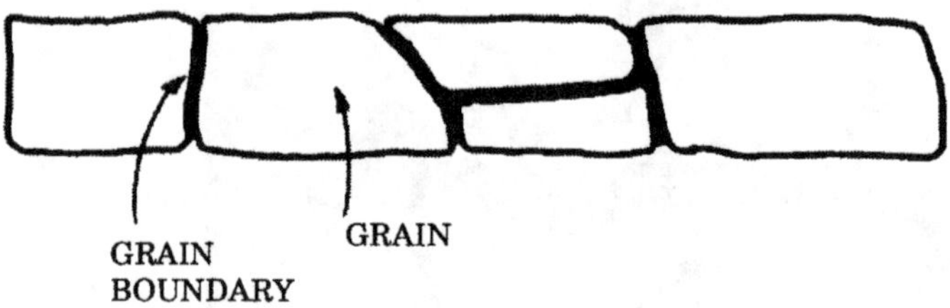

Figure 6.19 Schematic representation of bamboo structure in a metal wire. (*From Bakoglu.*[11])

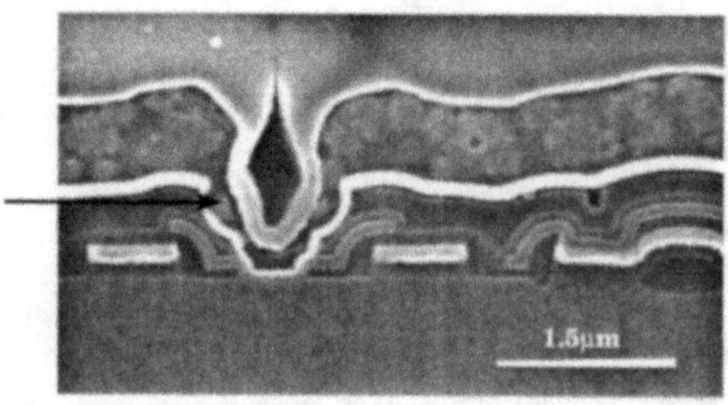

Figure 6.20 Metal continuity (arrow) in small contacts maintained by deposition of Ti/W layer (bright). (*From Haythornthwaite.*[10])

in Fig. 6.20. In this case, the upper aluminum layer would be subject to severe electromigration at the 45° slope where it goes through several intermediate layers to contact the silicon at the bottom. However, the bright layer of titanium-tungsten prevents such electromigration-caused open circuits from occurring.

Contact migration and alloy spikes. Under the action of high current density or high temperature, the aluminum and silicon in contact can interdiffuse. If silicon diffuses into the aluminum rails (contacts), voids can be created due to the Kirkendall effect, leading to an increase in the contact resistance, in the extreme causing open circuits. If the aluminum diffuses into the silicon region, spikes form which lead to junction leakage (Fig. 6.21). In this regard it is worth mentioning that (100) substrates are more prone to spiking than (111) substrates. Methods used to alleviate or obviate this problem include the use of aluminum alloys (as opposed to pure aluminum) or the use of barrier metals (titanium, tungsten, or titanium nitride), which prevents interdiffusion altogether (Fig. 6.22). A third approach involves the use of a triple structure of good contact/good diffusion barrier/good conductor, as shown in Fig. 6.23. Good contact material may be titanium disilicide ($TiSi_2$) or platinum silicide (PtSi) while the diffusion barrier may be titanium-tungsten (alloy) or separate titanium and tungsten layers.

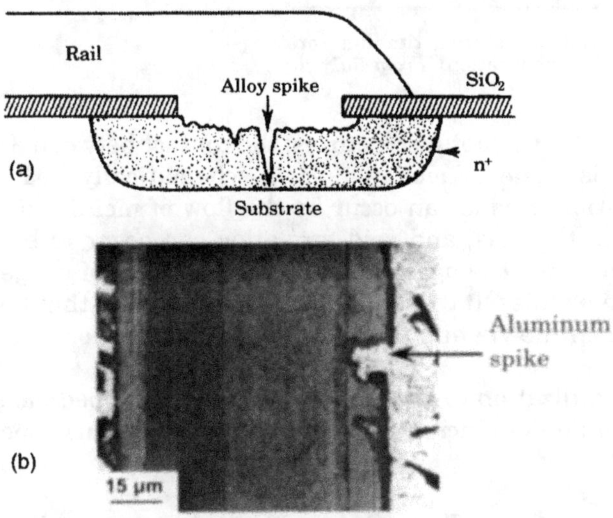

Figure 6.21 (*a*) Schematic representation of spike formation; (*b*) cross section of PIN diode showing Al spike (arrow) penetration. (*From Amerasekera and Najm.*[1])

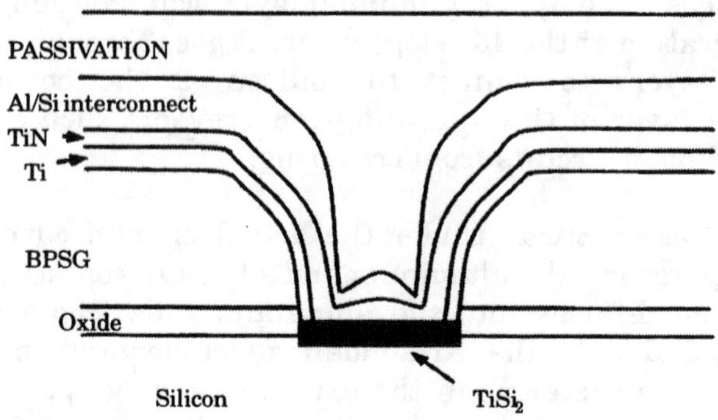

Figure 6.22 Schematic representation of a scheme used to prevent interdiffusion. (*From Amerasekera and Najm.*[1])

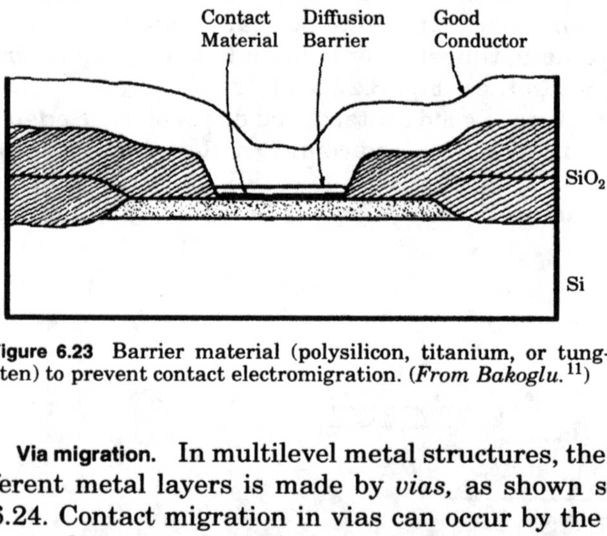

Figure 6.23 Barrier material (polysilicon, titanium, or tungsten) to prevent contact electromigration. (*From Bakoglu.*[11])

Via migration. In multilevel metal structures, the contact between different metal layers is made by *vias,* as shown schematically in Fig. 6.24. Contact migration in vias can occur by the flow of metal atoms toward or away from the vias, and void formation can occur in both directions of current flow. It is aggravated by the presence of oxygen residuals and silicon nodules. It can be alleviated by placing a thin layer of titanium between the via and the aluminum metallization.

Corrosion. The metallization in the die can react with trapped moisture or ionic contaminants to increase circuit resistance or cause open circuits.

Microcracks and step coverage. Topography effects during metallization such as those depicted in Fig. 6.25 can cause problems. Since smaller metallization thicknesses lead to higher current densities,

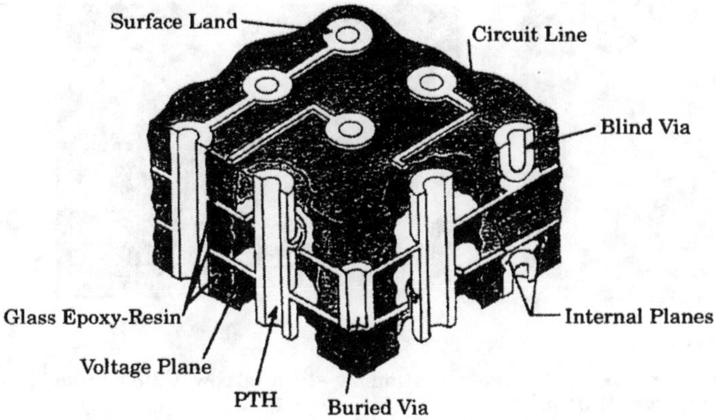

Figure 6.24 Schematic representation of three types of vias (blind, buried, and through) in PCBs. (*From Viswanadham and Singh.*[9])

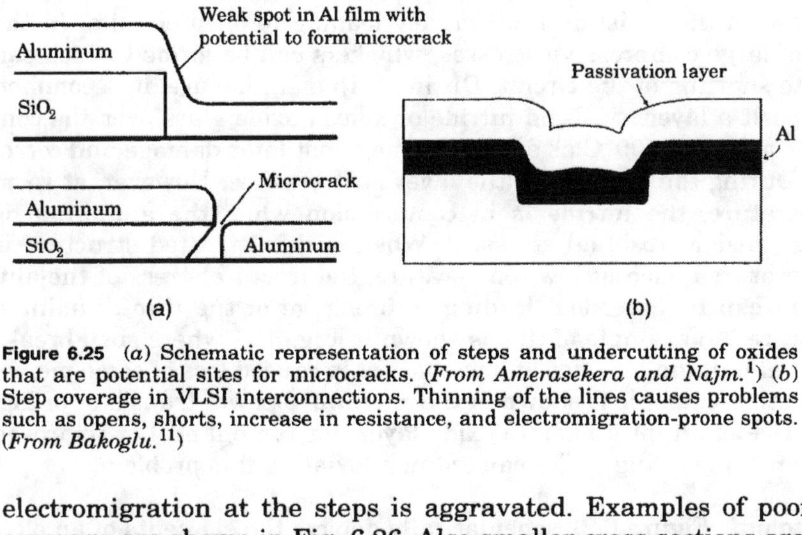

Figure 6.25 (*a*) Schematic representation of steps and undercutting of oxides that are potential sites for microcracks. (*From Amerasekera and Najm.*[1]) (*b*) Step coverage in VLSI interconnections. Thinning of the lines causes problems such as opens, shorts, increase in resistance, and electromigration-prone spots. (*From Bakoglu.*[11])

electromigration at the steps is aggravated. Examples of poor step coverage are shown in Fig. 6.26. Also smaller cross sections are more susceptible to microcracking and thus open circuits. Several techniques may be used to alleviate this problem. The first involves rotation of the wafer above the aluminum source to even out the deposition along the step. A second technique is to use planarization to reduce surface roughness. Sometimes, conductive *studs* are used to fill the via openings. A recent advance has been the introduction of CVD aluminum deposition technology.

Stress-induced migration. Thin metal layers are often subjected to large mechanical stresses during the manufacturing process. This in turn

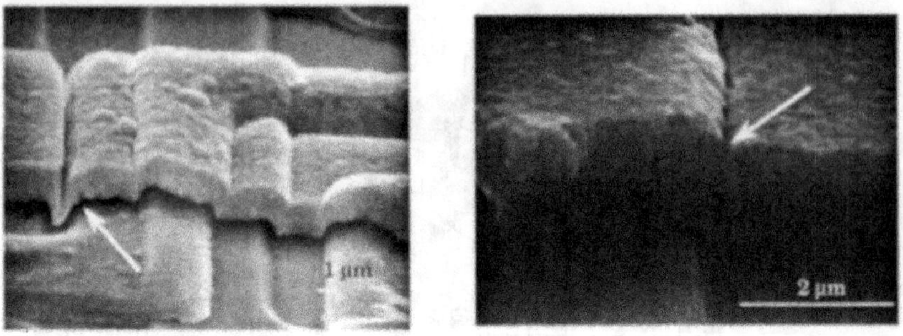

(a) (b)

Figure 6.26 Examples of inadequate metallization at steps (arrows). [(*a*) *From Haythornthwaite*[10]; (*b*) *from Martin.*[12]]

leads to plastic deformation of the layer. In the case of large tensile stresses, thinning of the metal layer (plastic deformation) can cause an increase in the resistance, ultimately leading to open circuits. In the case of large compressive stresses, whiskers can be formed which can lead to shorting of the circuit. During chip manufacture, it is common to deposit a layer of silicon nitride or silicon oxide glass over the conductor at 350 to 400°C as a protection against later damage and corrosion. During the deposition, the layer is stress-free. However, at room temperature, the nitride is in compression while the aluminum is under tension (residual stresses). When such a stressed structure is temperature-cycled at low temperature, the fracture stress of the aluminum can be exceeded, leading to breakage of the thin aluminum structure. An example of this is shown in Fig. 6.27 where such breaks in the aluminum bit lines of a memory chip caused loss of some memory. Another example is shown in the microwave power transistor of Fig. 6.28. Use of an intermediate oxide layer (for compliance) or improved temperature cooling cycles can aid in alleviating this problem.

Packaging. Figure 6.29 schematically depicts the elements of an electronic package. The electronic chip (die) is mounted on a lead frame. Electrical contact between the die and the lead frame is provided by thin bond wires (0.025- to 0.05-mm diameter). The lead frame provides the electrical contact with the macroscopic circuit outside the encapsulated package. The entire system is encapsulated in a hermetic ceramic package or a less expensive molded plastic package formed by transfer molding. Figure 6.30 schematically depicts the details of a wire bond. There are two points where the wire is metallurgically bonded. One is a ball bond between the wire and the die pad, and the other is a wedge bond between the wire and the lead frame.

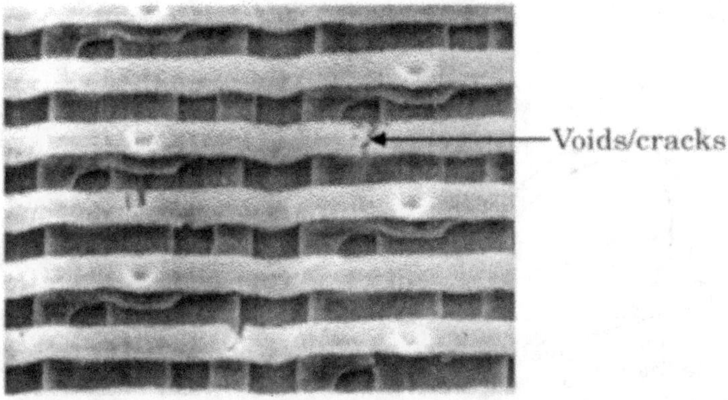

Figure 6.27 Stress voids (cracks) in Al tracks of memory circuit (after removal of glass coating). (*From Haythornthwaite.*[10])

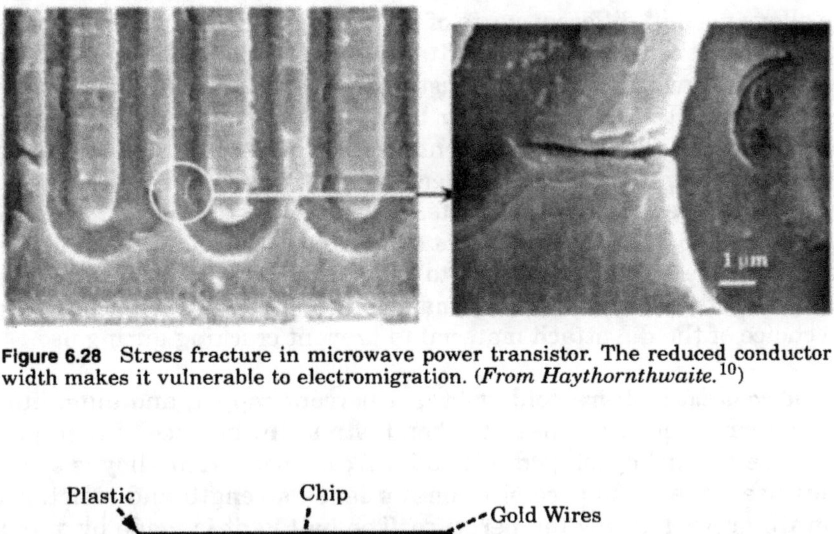

Figure 6.28 Stress fracture in microwave power transistor. The reduced conductor width makes it vulnerable to electromigration. (*From Haythornthwaite.*[10])

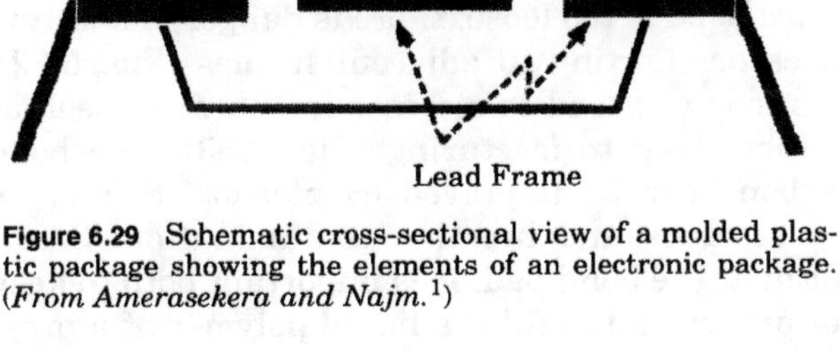

Figure 6.29 Schematic cross-sectional view of a molded plastic package showing the elements of an electronic package. (*From Amerasekera and Najm.*[1])

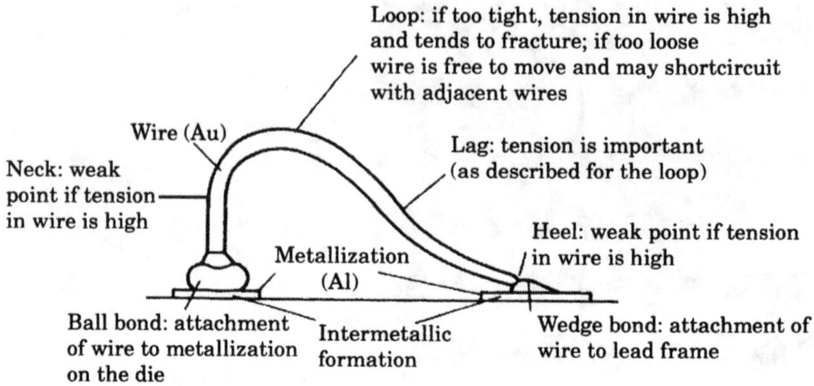

Figure 6.30 Schematic representation of a typical wire bond as used in IC applications. (*From Amerasekera and Najm.*[1])

Die attach failures. Table 6.7 lists the chemical composition of the various alloys used in different parts of the package. In the case of hermetic packages, the lead frame is attached to the die using gold-silicon eutectic bonding. In the case of plastic molded packages the attachment is by either soft solder or polymer bonding (silver-filled epoxy or polyimide adhesive). The die attachment process can result in the formation of voids between the die and the lead frame. Subsequently, these voids can result in cracking and hot spots. An example of a die attach crack is shown in Fig. 6.31. The entrapment of moisture and ionic contamination transfer to the interface can lead to die attachment problems. It is necessary that the coefficient of thermal expansion be matched properly in the choice of the die attach material to prevent cracking during use.

Bonding failures. Pure gold, gold + 1 percent copper, and aluminum + 1 percent silicon are used for bond wires. In the case of high-performance IC packages, gold + 1 to 2 percent beryllium alloy is sometimes used due to its precipitation-hardened strength and resistance to grain growth at low temperature. The ball joint is made by applying a combination of heat, pressure, and ultrasonic vibration while the wedge joint is generally made by applying only ultrasonic vibration. If the loop of a wire bond is too loose, there is danger of shorting between adjacent wires or shorting to adjacent frames (Fig. 6.32). Conversely, too tight a loop can lead to stresses in the wire, heel, or neck, which can, in turn, lead to fracturing (Fig. 6.33). The bond integrity of the wire bond can be improved by plasma cleaning to remove carbonaceous residues and bromine contamination. Sometimes whiskers can form at the bond pad. It is important both to keep the length of the wire proper and to fill the liquid polymer at a proper rate so as to prevent shorting between wires by wire sweeping. In

TABLE 6.7 Materials Used in Various Portions of an IC Package

Package part	Package type	Alloy used	Chemical composition (wt%)
Lead frame	All	Alloy 42	41Ni-0.02C-0.4Mn-0.15Si-58.43Fe
Eutectic bonding	Ceramic package	Gold-silicon	94Au-6Si
Soft solders	Plastic package		97Pb-1.5Ag-1Sn
			60Pb-40Sn
Polymer seal	Ceramic package		Silver-filled epoxy
			Polyimide adhesive

SOURCE: From Amerasekera and Najm.[1]

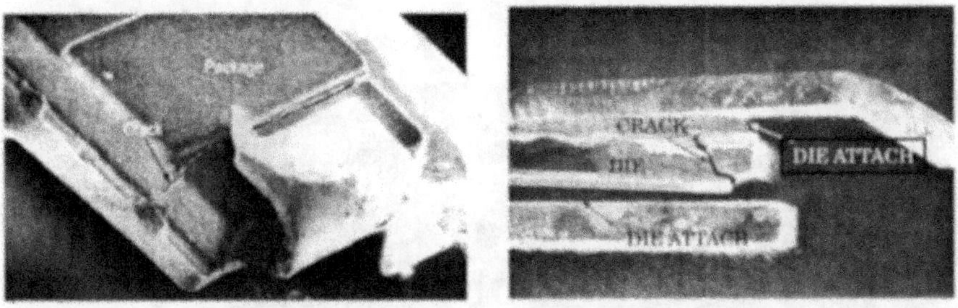

(a) (b)

Figure 6.31 (*a*) SEM view of crack in plastic package; (*b*) decapsulated device showing crack in die. (*From Martin.*[12])

Figure 6.32 Bond wires shorted (see arrow) to adjacent lead frame. (*From Pabbisetty et al.*[13])

addition, contamination from chlorine can cause corrosive damage, as shown in Fig. 6.34.

Intermetallics. A common kind of bond failure is caused by the formation of intermetallic compounds. The interdiffusion between the gold of the bond wires and the aluminum of the bond pad causes such

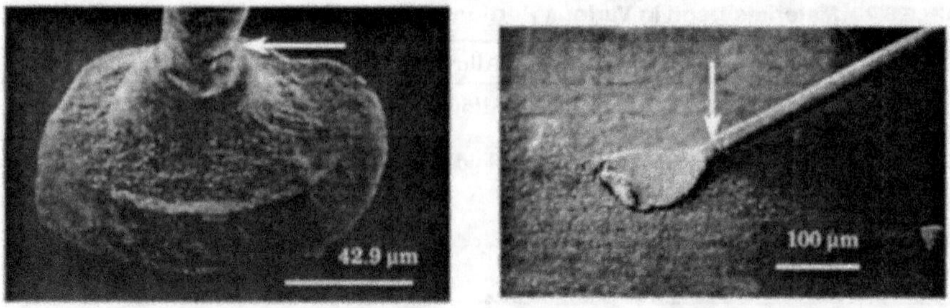

(a) (b)

Figure 6.33 Cracking at (*a*) ball bond neck and (*b*) stitch bond. (*From Pabbisetty et al.*[13])

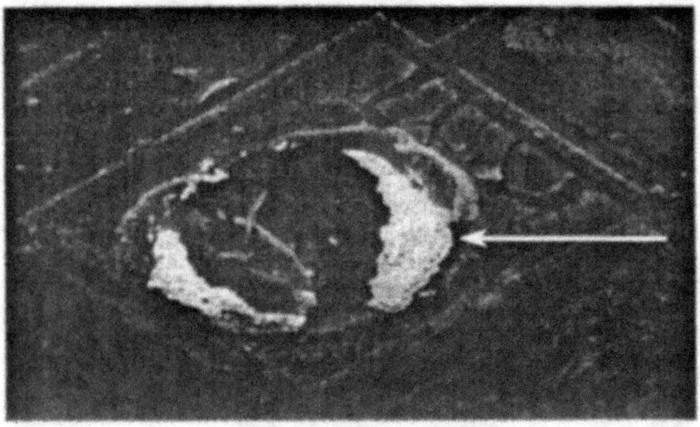

Figure 6.34 Corrosion of bond pad due to presence of chlorine. (*From Pabbisetty et al.*[13])

intermetallic compounds to form. Some of these (e.g., Au_2Al) are necessary for obtaining a proper bond and thus electrical contact; however, they can also cause contact embrittlement. Some compounds have colorful names, such as *purple plague* and *white plague*. In addition, contact quality can be impaired due to differing diffusion rates for aluminum and gold, leading to the formation of Kirkendall voids at the base of the wires. An example of this effect is shown in Fig. 6.35. The arrow in the photograph indicates the area where an open circuit has been created by the depletion of the thin aluminum pad due to the formation of the granular intermetallic compound. One possible cure for this is to use aluminum wire and gold thin film. In the example shown in Fig. 6.36, a hermetically sealed, gold-plated ceramic package is joined to an aluminum wire. Note that Kirkendall voids have formed in the gold film surrounding the aluminum wire

Figure 6.35 Formation of Au-Al intermetallics between Au wire and Al bond pad (square). Note depletion of Al surrounding the intermetallic. (*From Haythornthwaite.*[10])

Figure 6.36 Kirkendall voids under an Al wire bond. Open circuit in the underlying Au film is avoided by the continuous Ni film between Al and Au. (*From Haythornthwaite.*[10])

(arrows). However, circuit continuity is maintained by the introduction of a conductive nickel layer under the gold; this layer does not participate in the intermetallic formation.

Coefficient of thermal expansion (CTE) mismatch. The difference in CTE between the device and the chip carrier can lead to mechanical stresses which can vary as the temperature is cycled. Figure 6.37 shows a failure in a gold wire bond after 200 thermal cycles between 0 and 125°C. When a ball is newly formed after melting of a portion of the bond wire, the metal in the wire adjacent to the ball cools slowly, leading to the formation of large grains. Thermal cycling of such a wire bond can lead to cleavage-type failure, as depicted in Fig. 6.38.

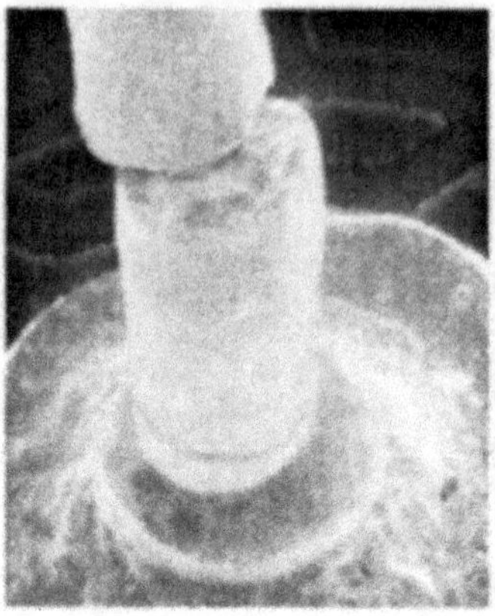

Figure 6.37 Wire bond failure on a gold wire after 200 thermal cycles. (*From Woychik and Senger.*[14])

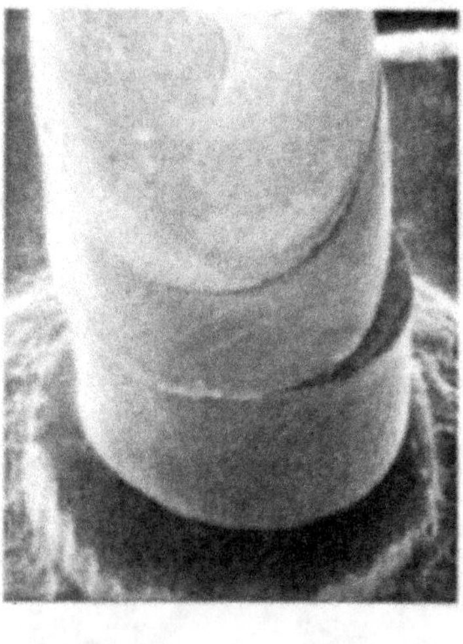

Figure 6.38 Cleavage-type failure in a gold wire bond after 10 thermal cycles. (*From Woychik and Senger.*[14])

Delamination and popcorn effect. In addition to moisture absorption in the polymeric encapsulant, moisture can enter the package through microcracks. During subsequent use and thermal cycling, absorbed moisture can cause delamination at various interfaces between the die, lead frame, and encapsulant, as depicted schematically in Fig. 6.39. In extreme cases, the die can crack with an audible sound ("popcorn").

6.2.6 Ionic contamination-induced inversion mechanism

A fourth mechanism, called *ionic contamination-induced inversion,* warrants a detailed discussion due the frequency of its occurrence. This occurs when the glass surface (dielectric) of a semiconductor die

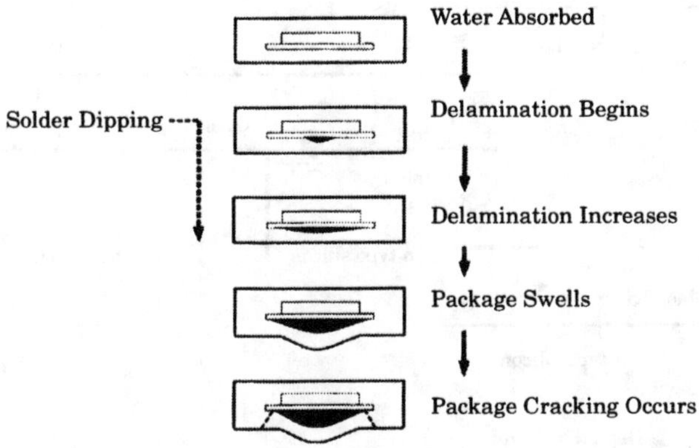

Figure 6.39 Schematic representation of popcorn: trapped moisture expands during subsequent solder dipping. (*From Amerasekera and Najm.*[1])

contains a contaminant such as sodium. Electrical biasing in certain devices can cause this effect through localized high concentration of contaminant in spite of the nominal concentration being low (less than 0.1 ppm). In the case of a silicon *p-n* junction, the passivation layer used is often silica (SiO_2), as shown in Fig. 6.40*a*, where the randomly distributed sodium ions in the silica layer are represented by the plus signs. When such a junction is reverse-biased, depletion layers form at the junction with the *p*-side layer being biased negatively with respect to the *n*-side layer. Hence the positively charged sodium ions will migrate (diffuse) to the interface between the dielectric and the *p*-type silicon, thereby causing a localized high concentration of sodium ions (Fig. 6.40*b*). This in turn will attract minority negative charge carriers in the *p*-type silicon to a localized area under the dielectric, which will begin to function as a local *n*-type silicon area (Fig. 6.40*c*). Such an area is called an *inversion region* and causes significant degradation of the junction's performance under both forward and reverse bias conditions.

A MOSFET device works by the intentional creation of an inversion layer, as shown in Fig. 6.41*a*. In this case, the application of a positive bias to the metallization (gate) causes an inversion region of negatively charged minority carriers to form at the interface between the dielectric and the *p*-type silicon, thereby forming an *n*-type conduction channel (the transistor is then "on"), as shown in Fig. 6.41*b*. If, however, the dielectric is contaminated with positively charged sodium ions, the positive biasing of the metallization leads to the repulsion of the sodium ions which concentrate in a layer in the dielectric adjacent to the *p*-type silicon. If the concentration of

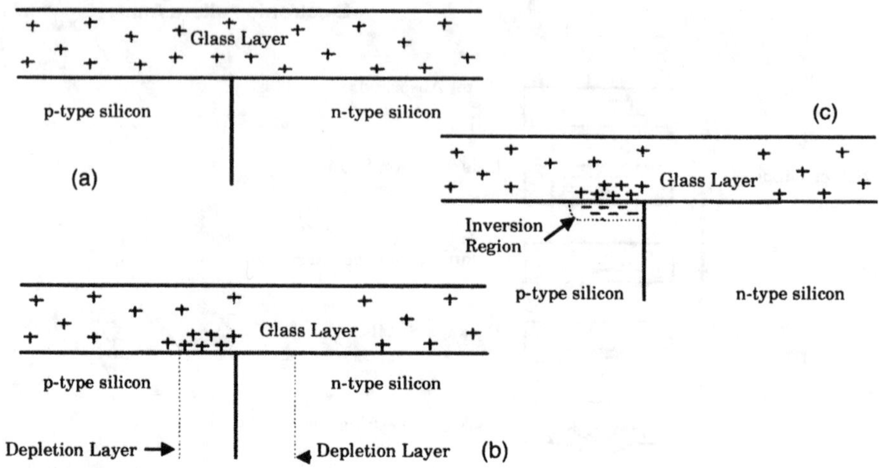

Figure 6.40 Schematic montage illustrating ionic contamination-induced inversion in a cross-sectional view of a *p-n* junction. (*a*) The plus symbols in the glass layer on the surface represent randomly distributed sodium ions in the glass layer. (*b*) With the junction reverse-biased, a depletion layer forms across the junction. The depletion layer in the *p*-type silicon will be biased negatively with respect to the bias on the depletion layer on the *n*-type silicon. This will cause the sodium ions to diffuse toward the *p*-type silicon and away from the *n*-type silicon. (*c*) After the reverse bias is removed, there is an accumulation of sodium ions in the glass over the *p*-type material. This attracts negative minority carriers in the surface of the *p*-type material. If sufficient minority carriers are attracted, the materials in this area may be inverted and act as *n*-type material with no bias on the gate metallization. (Thus, the FET is anomalously *on* with no gate bias.) (*From Erickson.*[15])

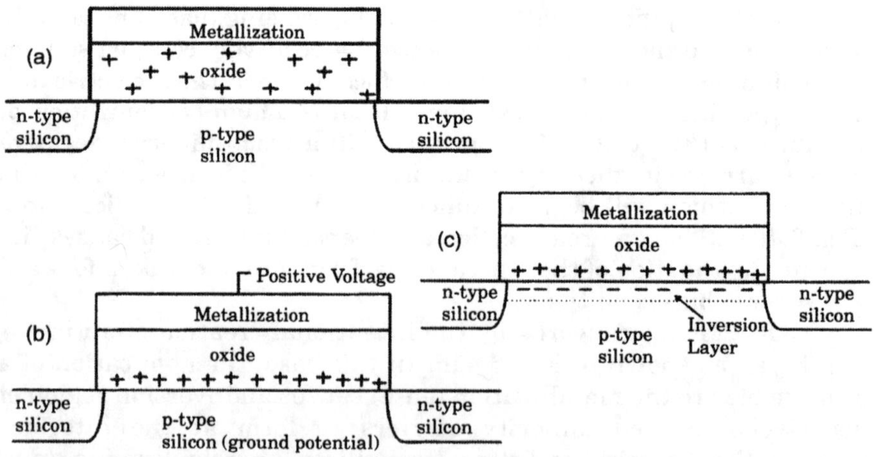

Figure 6.41 Schematic montage illustrating ionic contamination-induced inversion in a cross-sectional view of a portion of a MOS gate region on a silicon MOSFET device. (*a*) The plus symbols in the oxide layer represent randomly distributed sodium ions in the oxide layer. (*b*) With a positive bias on the metallization with respect to the *p*-type silicon, the sodium ions are repelled from the metallization and diffuse toward the silicon. (*c*) With a positive bias on the metallization with respect to the *p*-type silicon, the sodium ions are repelled from the metallization and diffuse toward the silicon. (*From Erickson.*[15])

such sodium ions is high enough, negatively charged minority carriers will form an inversion layer in the *p*-type silicon, even in the absence of any bias in the metallization; the transistor will be rendered "on," as shown in Fig. 6.41*c*. In both of these cases, the migration of sodium ions is by diffusion, and hence it is obvious that the problem can be cured faster at higher temperatures.

6.3 Failure Analysis Process

6.3.1 General introduction

The first indication of faulty electronic components usually is the inability to perform electrically in an expected manner. The observed phenomena in an electrical test are an open, a short, an undefined malfunction, or a shift in some parameter such as output. Table 6.8 shows the distribution of such phenomena in IC packages at the system level.

The sequence of steps used in the analysis of electronic devices, components, or systems is dependent on the type of item to be analyzed. This is illustrated in Fig. 6.42. In the case of systems and assemblies, electrical testing or circuit analysis techniques (respectively) may be required to isolate problems related to either design or defective parts. Once the specific root cause of failure of the specific part or process has been isolated, sophisticated failure analysis techniques can be used to determine solution(s) to the problem. This may involve interactions between manufacturer, user, and several test laboratories. The findings of such an exercise may then be extended to design for failure prevention and elimination.

6.3.2 Stages in failure analysis

There are usually seven stages in the failure analysis (FA) of most failed devices:

1. Examination prior to package opening

TABLE 6.8 Typical Observations of IC Failures at System Level

System failures	Percent of reports
Opens	30
Shorts	26
Malfunctions	25
Parameter shifts	19

SOURCE: From Dicken.[4]

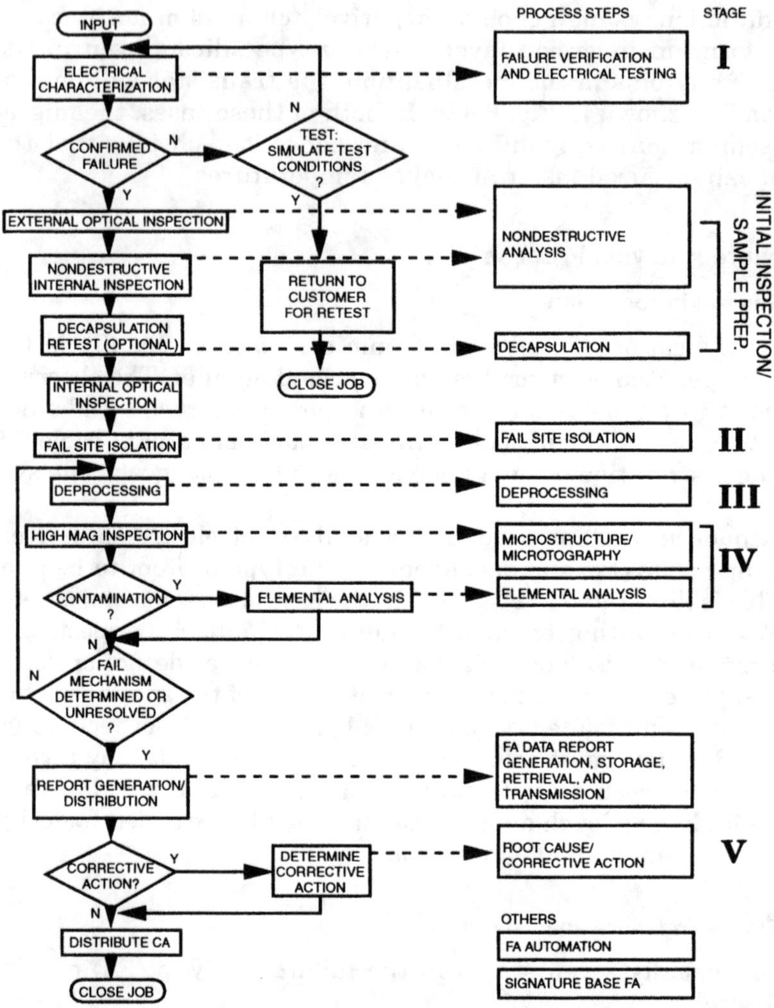

Figure 6.42 A typical failure analysis flowchart with key processes identified. (*From Pabbisetty.*[16])

2. Package opening and encapsulation removal
3. Internal examination
4. Selective layer removal
5. Location of failure site
6. Identification of failure cause, mechanism, and assignment
7. Simulation testing and final examination

In the case of components requiring destructive analysis, these seven stages are detailed in Fig. 6.43. Note the extra steps involved in cases entailing destructive analysis (Fig. 6.43) compared to the general case shown in Fig. 6.42. The nature and types of the various tests referred to in this figure are shown in greater detail in Table 6.9. Another possible sequence of steps in the performance of failure analysis of electronic components is shown in Table 6.10.

6.3.3 Sequence of FA steps for different products

Microprocessors

1. Use electrical and functional tests to eliminate functional units such as logic unit from the FA.
2. Use product knowledge to isolate the fault location, using functional tests.
3. Identify fault nodes.
4. Confirm the hypothesis, using electrical tests with microprobing, etc.

Logic products

1. Use electrical and functional tests to isolate periphery failures.
2. Isolate the fault location to the functional unit level.
3. Identify fault nodes, using failure signature and hot spot detection.
4. Confirm the hypothesis, using electrical tests with microprobing, etc.

Mixed signal products

1. Use electrical and functional tests to determine if the failure mode is analog or digital.
2. Isolate the fault location to the functional unit level.
3. Identify fault nodes, using failure signature and hot spot detection.
4. Confirm the hypothesis, using electrical tests with microprobing, etc.

Memory products

1. Use electrical and functional tests to isolate general fault location.
2. Decapsulate and confirm the hypothesis, using microprobing, liquid crystal, and optical and electron microscopy.

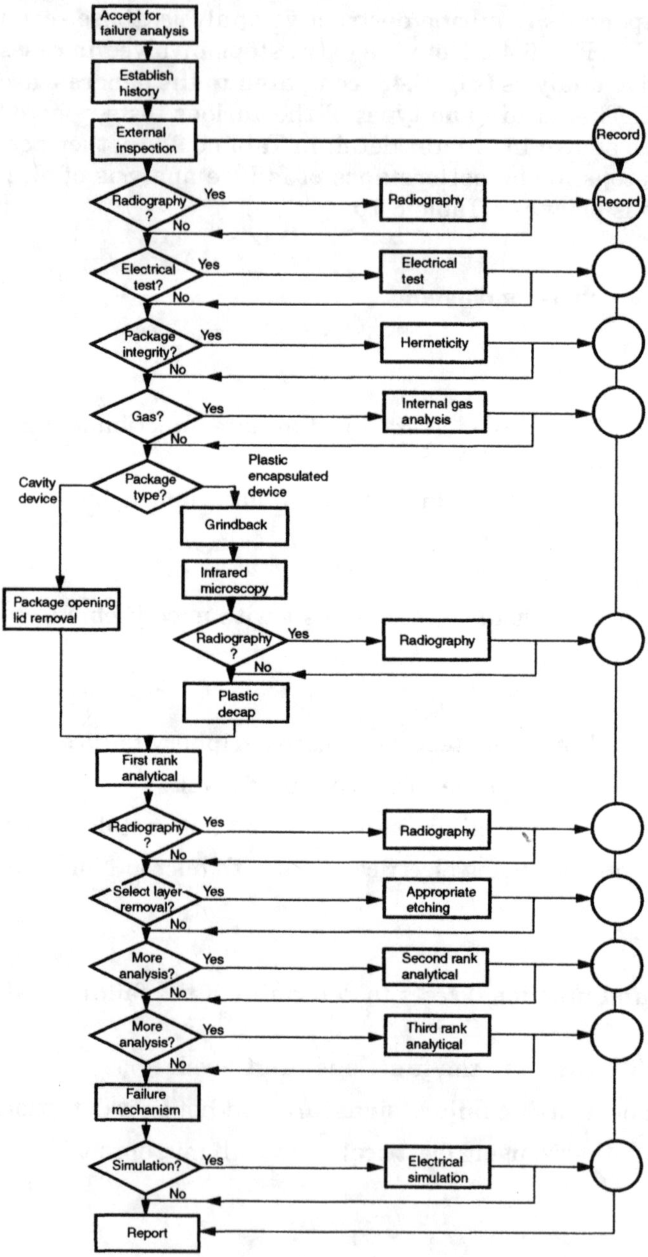

Figure 6.43 Failure analysis flow diagram for cases requiring destructive analysis. (*From Richards and Footner.*[3])

TABLE 6.9 Sequential Order of Failure Analysis Investigation/Procedure

No.	Technique	Comments
1	Visual/optical inspection	Packaging or gross defects
2	Radiography	Internal geometry and physical structure/defects
3	Simple electrical tests	(If appropriate)
4	Package tests	Hermeticity
5	Depackaging	Mechanical and/or chemical opening; gas analysis on cavity devices
6	First-rank analytical techniques	(Mainly imaging techniques)
	Optical/interferometric microscopy	Nondestructive
	SLM	Nondestructive
	SEM	Nondestructive
	SEM-EDX	Destructive?
	Metallography—sectioning	Destructive?
	Radiography	Destructive?
7	Second-rank analytical techniques	(Imaging, compositional, and constitutional techniques)
	IR microscopy	Partially destructive
	SEM-VC, SEM-CL	Nondestructive
	SEM-EBIC	Nondestructive
	SLM-OBIC	Nondestructive
	AES	Destructive?
	XPS	Destructive?
	SAM	Nondestructive
	X-ray diffraction/topography	Nondestructive
	Gas analysis—moisture content	Destructive?
	Trace analysis—chemistry	Destructive?
8	Third-rank analytical techniques	
	SIMS	Destructive
	RBS	Nondestructive?
	Thermal imaging	Nondestructive
	Liquid-crystal display/location	Nondestructive
	Laser scanning techniques	Nondestructive?
	Fluorocarbon vapor bubbles for location	Nondestructive
	Scanning electron acoustic microscopy (SEAM)	Nondestructible?
	Scanning laser acoustic microscopy (SLAM)	Nondestructive
9	Comparison of failure mechanisms with literature/library	
10	Simulation tests	(If appropriate)

SOURCE: From Richards and Footner.[3]

TABLE 6.10 Summary of Methods and Ranking of Device Failure Analysis Techniques

No.	Order of investigation/procedure
1	Visual/optical inspection
2	Radiography
3	Simple electrical tests
4	Package tests
5	Depackaging/decapsulation
6	First-rank analytical techniques (mainly imaging techniques) Optical/interferometric microscopy SLM SEM SEM-EDX, SEM-VC, SEM-CL Metallography—sectioning Radiography
7	Second-rank analytical techniques (imaging, compositional, and constitutional techniques) AES XPS SAM X-ray diffraction Gas analysis—moisture content Trace analysis—chemistry
8	Third-rank analytical techniques SIMS RBS Thermal imaging (IR microscopy) Liquid-crystal display/location Laser scanning techniques (LIMA)
9	Comparison of failure mechanisms with literature/library
10	Simulation tests

SOURCE: From Amerasekera and Najm.[1]

6.4 Tools and Techniques for Electronic Failure Analysis

The tools and techniques described in Chap. 1 are used extensively in the failure analyses of electronic systems and components. However, the unique characteristics of these systems (e.g., fine size, complex architecture, electrical phenomena) require additional methods for failure analysis. Some of these are described in detail in this section.

6.4.1 Techniques used in electronic failure analysis

These may be classified as follows:

- Photography and optical microscopy
- X-ray/radiographic component inspection
- Infrared thermography
- Acoustic microimaging
- Metallography
- Chemical characterization
- Electronic and electrical characterization
- Scanning electron microscopy (SEM), energy-dispersive x-ray analysis (EDXA), and wavelength-dispersive spectrometry (WDS)
- Miscellaneous techniques

Since the size range of most electronic components and the materials in them (layers) are generally very fine (and progressively getting smaller), microscopic techniques are necessary to image them. Many of these techniques are covered in detail in Chap. 1. Table 6.11 and Fig. 6.44 compare features, resolution, and depth of penetration of various microscopic techniques including some techniques that are unique in their application to electronic FA.

6.4.2 Photography and optical microscopy

The topic of optical microscopy and photography is covered in Chap. 1. The application of photodocumentation in the failure analysis of electronic components will be briefly discussed in this section. Visual inspection with the unaided eye and stereo viewing may comprise the incoming inspection of the as-received component. Photodocumentation should be used not only in this step but also in the subsequent progressive steps of nondestructive evaluation, destructive testing, disassembly, and simulation testing. For each of these steps, the following techniques may be used in the photodocumentation process: film, video, still video, and digital image processing. The use of reflected light optical microscopy in the detection of stress-induced delamination is illustrated in Fig. 6.45. In this figure, roughly circular stress patterns are observed in the passivation layer of the device wherever it contacts the channels in the oxide layer. In some places, the stresses are more extensive than desirable (e.g., region A) while at others, peeling is observed (e.g., region B). Nomarski differential interference contrast imaging can be used to detect slight variations in the thickness of thermally grown oxides on regions of different doping. This technique can also locate very shallow asperities, e.g., etch pits.

TABLE 6.11 A Comparison of Some Features of Various Microscopical Techniques

Technique	Lateral resolution, μm	Restriction on sample size	Specimen preparation	Destructive/ nondestructive	Depth of focus, μm	Sub-surface info.	Real-time imaging	Cost factor
OM (good)	0.5	X	None	ND	0.01	X		36,656
SEM-SE	0.005	X	None	ND	10.0	X		50–300
SEM-VC	0.1	X	Slight	ND		X		100
SEM-EBIC	0.5	X	Medium	ND				70
SEM-BSE	0.1	X	None	ND				65
SLM-imaging	0.3	X	Slight	ND	0.1	X		80
SLM-OBIC	0.5–1.0	X	Slight	ND				85
SAM-low frequency	20.0	X	None	ND	0.1			120
SAM-high frequency	1.0	X	None	ND	0.001	X	X	120
SLAM	10.0		None	ND	0.1		X	150
SEAM	1.0–5.0	X	Medium					80
IRM	0.5	X	Slight	ND?	0.05			40
PDM	10.0	X	None	ND	0.01		X	30
JDM	1.0	X	None	ND	0.01		X	50
STM	0.001		None	ND	0.001	X	X	100
AFM	0.001		None	ND	0.001	X	X	150
TEM	0.002		Much	D	1.0			150–400

SOURCE: From Richards and Footner.[3]

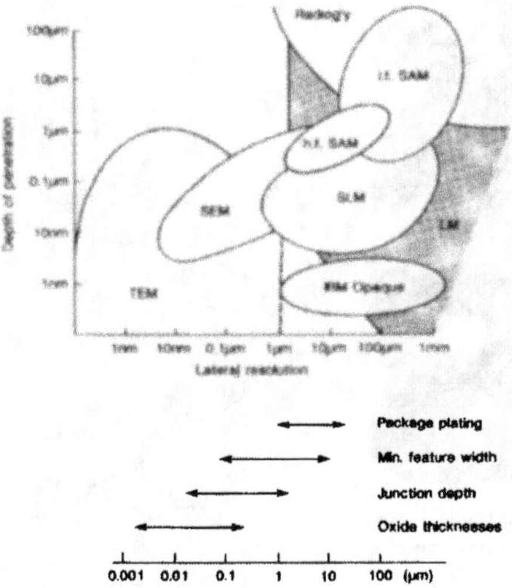

Figure 6.44 A schematic comparison of some of the various microscopies available to the failure analyst. (*From Richards and Footner.*[3])

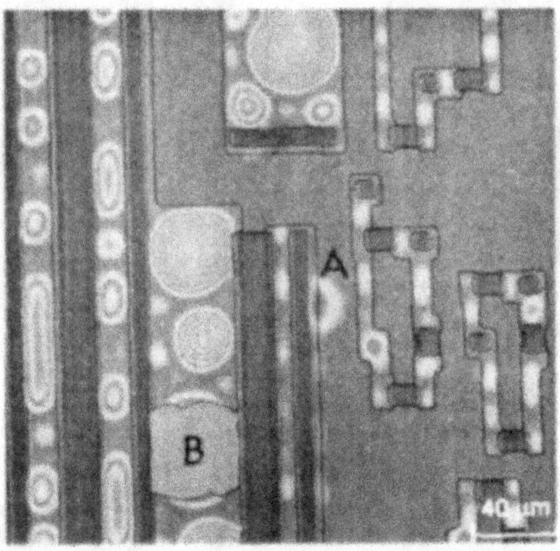

Figure 6.45 Stress-induced delamination of passivation detected by bright-field-reflected light microscopy (at A and B). (*From Richards and Footner.*[3])

6.4.3 X-ray/radiographic component inspection (microradiography)

The use of x-ray or radiographic inspection (RI) in electronic failure analysis can be for nondestructive evaluation, for monitoring component degradation, or as a geometric guide prior to disassembly/dissection of an assembly or component. Figure 6.46 illustrates the use of microradiography in the detection of wire displacement in a plastic encapsulated device, while Fig. 6.47 shows damaged emitter bond wires in transistors.

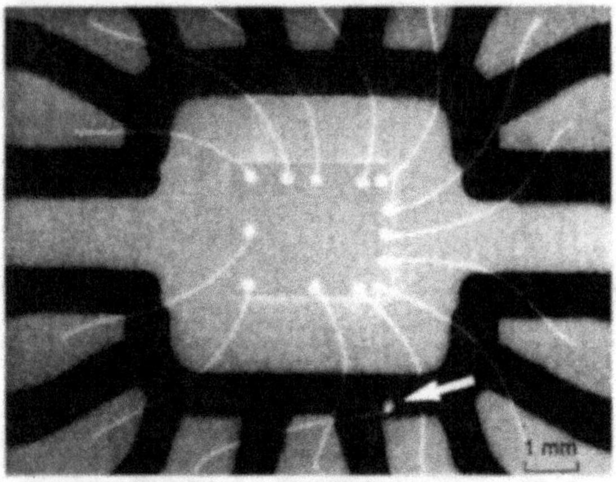

Figure 6.46 Microradiograph revealing gold wire displacement in a plastic encapsulated device (see arrow). (*From Richards and Footner.*[3])

Figure 6.47 Radiograph of missing emitter bond wire at arrow. (*From Ludwig.*[17])

6.4.4 Infrared thermography or infrared microscopy

Infrared (IR) thermal imaging instruments can be used to advantage in obtaining thermal maps (thermograms) of electronic packages, components, and assemblies since the flow of electricity through these produces heat. Such surface temperature (radiative) thermograms may be qualitative or semiquantitative. In the latter case, either gray scale or color scales can be used to indicate approximate temperatures. IR thermography has the advantages of being fast, noncontact, nonintrusive, and real-time. A useful extension of the semiquantitative thermograms is the comparative (benchmark) use of *standard thermal*

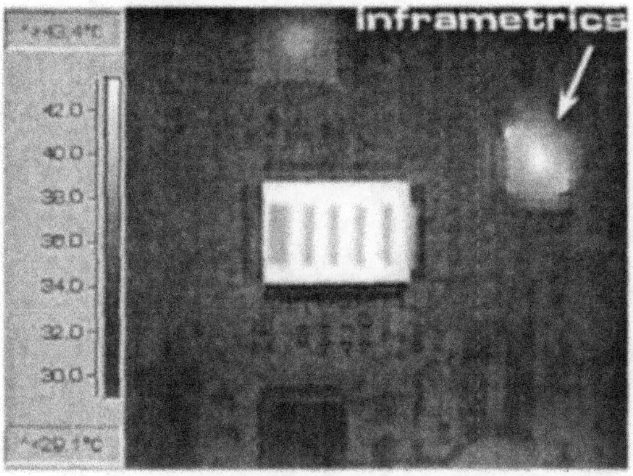

Figure 6.48 Thermogram of an overheated device (at arrow) on a hybrid electronic assembly. (*From Kaplan.*[18])

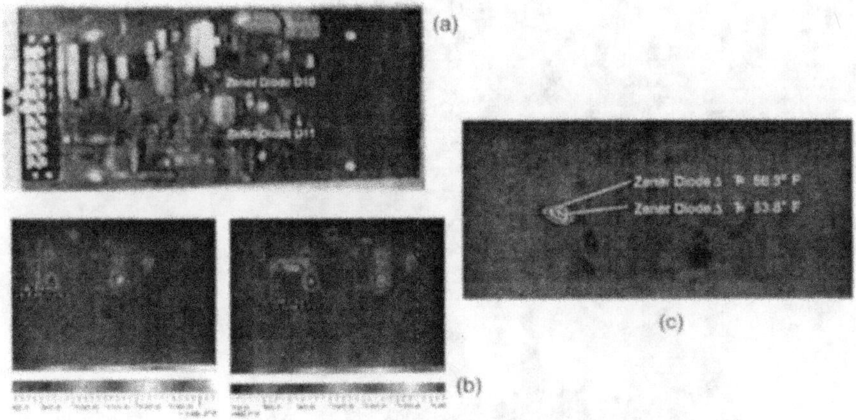

Figure 6.49 Differential thermography. (*a*) Photo of card; (*b*) thermograms of good and bad cards; (*c*) difference thermogram. (*From Kaplan.*[18])

profiles (STPs) of components, circuits, and modules using *differential infrared thermography* (DIT). Figure 6.48 shows a thermogram of an overheated device on hybrid electronic assembly while Fig. 6.49 illustrates the use of DIT. Figure 6.49*a* shows key components on a circuit card in a battery inverter system that had failed while Fig. 6.49*b* shows thermograms of a defective (right) and a nondefective (left) card. The DIT shown in Fig. 6.49*c* indicates the location of the two zener diodes responsible for the problem.

Infrared microscopy can also be used to detect intermetallic compounds formed between gold wire and aluminum metallization, as shown in Fig. 6.50. These intermetallics affect the bonding, which in

Figure 6.50 Micrograph showing intermetallics in footprint of gold ball bond on aluminum metallization. (*From Richards and Footner.*[3])

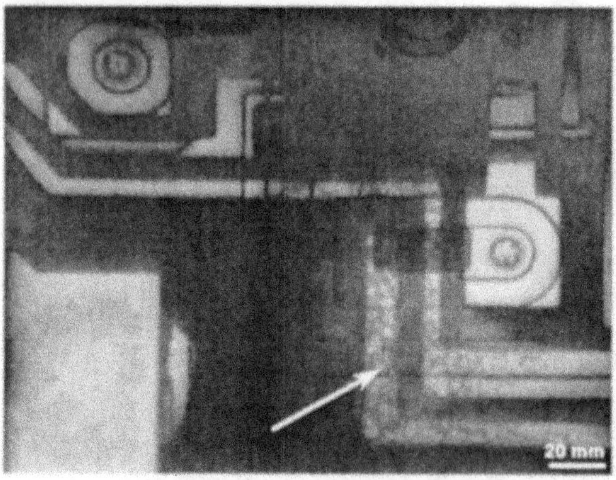

Figure 6.51 Subsurface corrosion (at arrow) of aluminum metallization imaged by IR micrography. (*From Richards and Footner.*[3])

turn affects the electrical continuity in the device. The use of IR microscopy in the detection of subsurface intergranular corrosion grain boundary attack is illustrated in Fig. 6.51, and damage due to electrostatic discharge is illustrated in Fig. 6.52.

The fabrication of ICs depends on good contact alloying during the contact sintering process. Should a high-temperature excursion occur during the process, a liquid phase is formed, leading to overalloying by the transfer of aluminum to areas depleted of aluminum by the migration of excess silicon. This results in high leakage currents or short circuits in the device. Figure 6.53 illustrates the use of IR micrography in

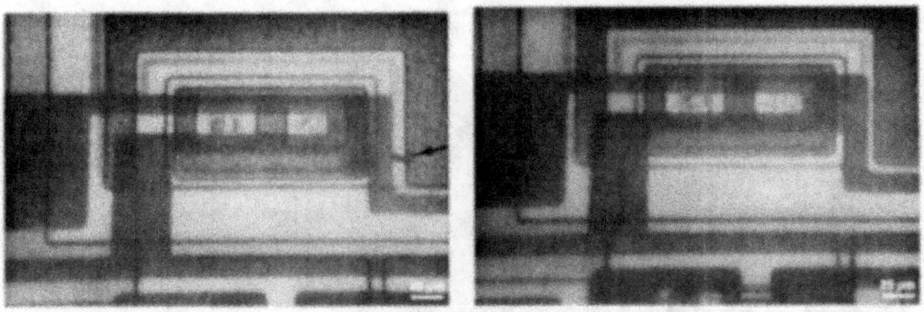

(a) (b)

Figure 6.52 (*a*) IR micrograph showing ESD damage (arrow) in a device; (*b*) corresponding undamaged device. (*From Richards and Footner.*[3])

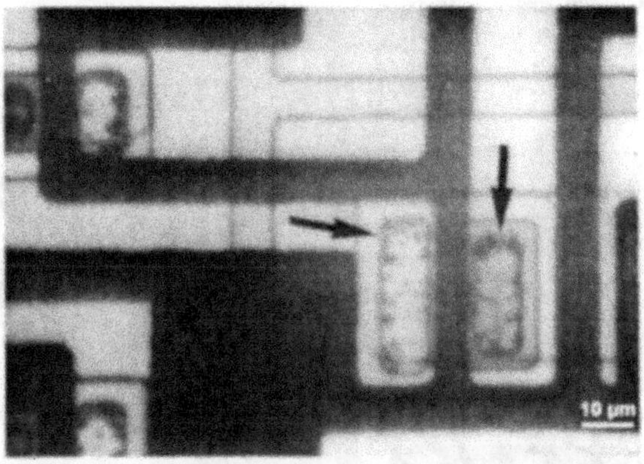

Figure 6.53 IR micrograph showing excessive alloying as discrete pits. (*From Richards and Footner.*[3])

the detection of such an overalloyed area. Intergranular precipitation of silicon in excess of the solid solubility limit can also be detected using IR micrography, as illustrated in Fig. 6.54.

6.4.5 Acoustic microimaging and scanning acoustic microscopy

Ultrasound travels through air and vacuum at rates slower than that through solid material. In *acoustic microimaging* (AMI), this property is used to nondestructively detect air gaps, cracks, voids, porosity, etc., in electronic components. Normally ultrasound in the frequency range of 10 to 200 MHz is used in either a transmission mode [scanning laser acoustic microscopy (SLAM)] or a depth-resolved reflection mode [C-mode scanning acoustic microscopy (C-SAM)] to obtain images of internal discontinuities in components. Figure 6.55 illustrates the use of both

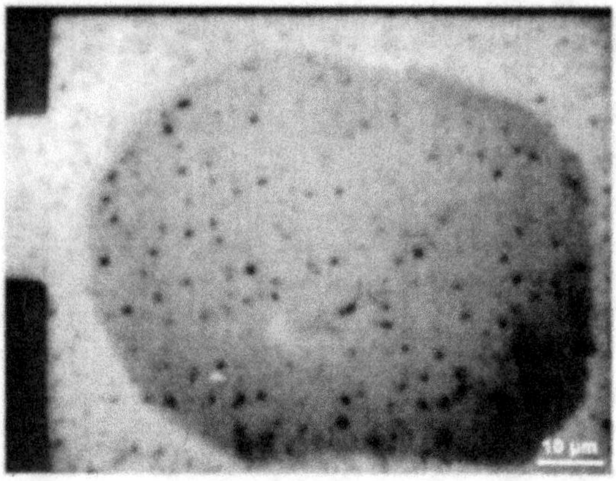

Figure 6.54 IR micrograph showing discrete silicon precipitates in footprint of gold ball bond. (*From Richards and Footner.*[3])

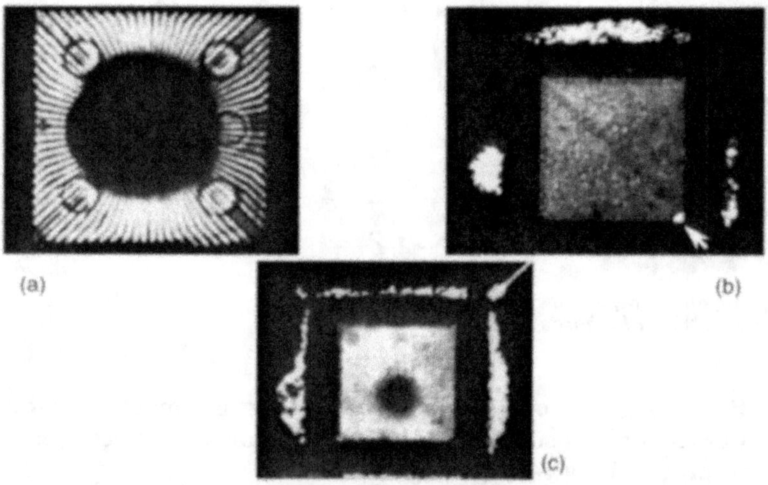

Figure 6.55 (*a*) Through-thickness SLAM image of PLCC with large popcorn crack (dark circle). (*b*) C-SAM image at die-to-mold compound interface (delamination at arrow). (*c*) C-SAM image at die attach interface; only the dark spot is bonded. (*From Kessler and Semmens.*[19])

of these techniques in detecting discontinuities in a plastic leaded chip carrier (PLCC). The scanning acoustic microscopy technique is used to study package-related failures such as cracking, delamination, and die attach problems. A SAM image is fundamentally a map of the sample showing regions of different acoustic impedance. Figure 6.56 shows a SAM image of a package; the delaminated area at the corner is marked,

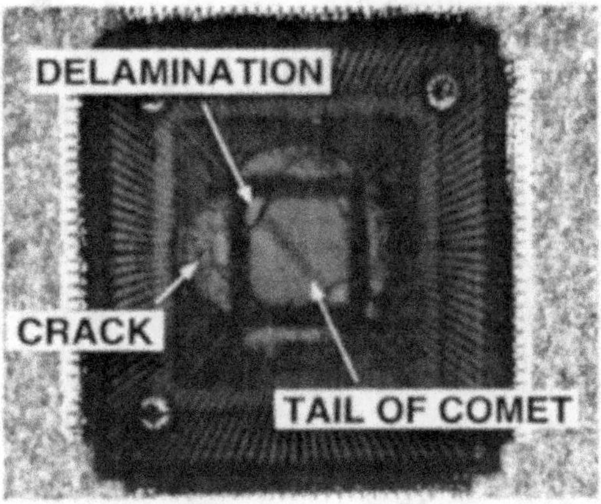

Figure 6.56 Scanning acoustic microscope image of a plastic package showing delamination and cracks. (*From Amerasekera and Najm.*[1])

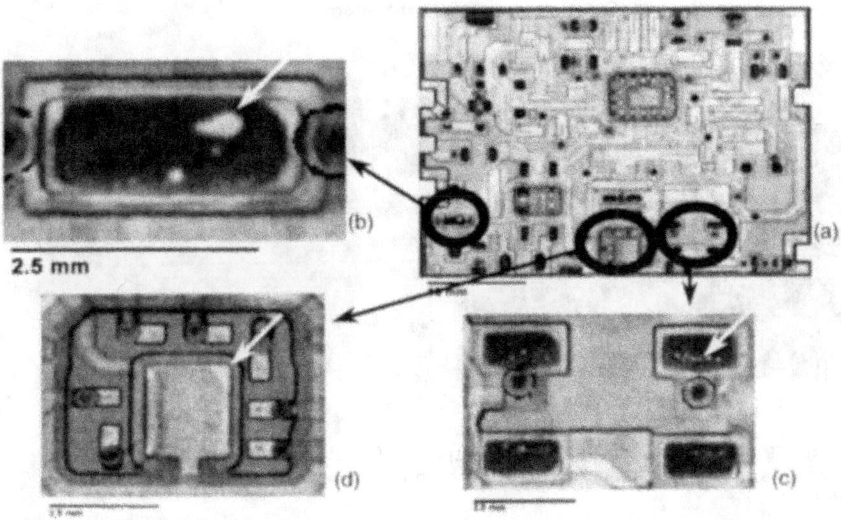

Figure 6.57 Montage of acoustic images of hybrid circuit devices. (*a*) Overall view; (*b*), (*c*) voids; (*d*) delamination (at arrows). (*From Adams.*[20])

and the low-density resin shows up as a "tail of comet." Figure 6.57 illustrates the detection of several types of defects in a hybrid circuit device using high-resolution reflection mode scanning acoustic microscopy.

6.4.6 Metallography

The utility of careful metallography in the failure analysis of any type of material is covered in detail in Chap. 1. With respect to its use in failure analysis of electronic materials and components, special care in

Figure 6.58 Nonuniformity of intermetallic penetration through aluminum metallization shown in light micrograph. (*From Richards and Footner.*[3])

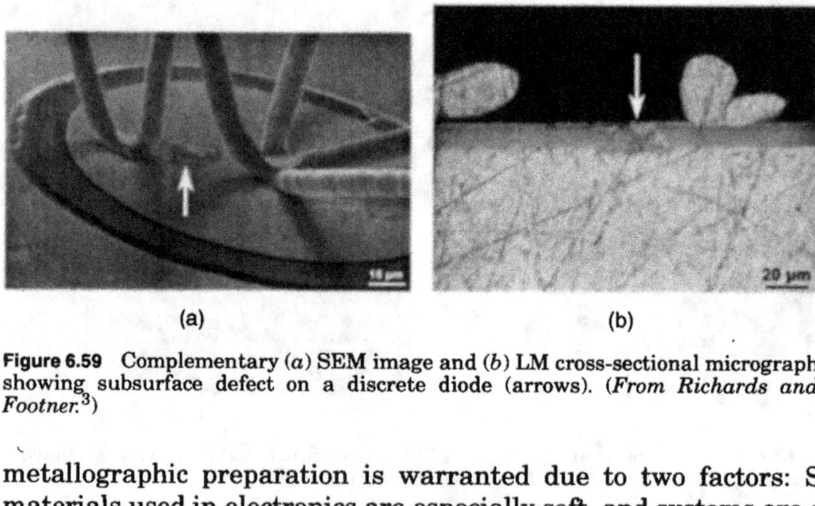

(a) (b)

Figure 6.59 Complementary (*a*) SEM image and (*b*) LM cross-sectional micrograph showing subsurface defect on a discrete diode (arrows). (*From Richards and Footner.*[3])

metallographic preparation is warranted due to two factors: Some materials used in electronics are especially soft, and systems are often designed with materials of widely differing metallographic characteristics in extreme proximity. The metallographic section of Fig. 6.58 shows nonuniformity of intermetallic formation and penetration through the aluminum metallization while Fig. 6.59 shows subsurface damage in a polished section of a discrete diode.

6.4.7 Chemical characterization

Table 6.12 shows the variety of chemical characterization techniques that may be used by the failure analyst.

TABLE 6.12 Some Techniques for Chemical Characterization

Category	Technique	Acronym
Atomic spectroscopy	Flame atomic absorption	FAA
	Graphite furnace atomic absorption	GFAA
	Atomic emission	AE
	Inductively coupled plasma—emission	ICP-E
	Inductively coupled plasma—mass spectrometry	ICP-MS
	Atomic fluorescence	AF
Fourier transform infrared microspectroscopy		
Thermal analysis	Thermomechanical analysis	TMA
	Differential scanning calorimetry	DSC
	Dynamic mechanical analysis	DMA
	High-pressure liquid chromatography	HPLC
	X-ray fluorescence	XRF
	Pyrolysis-gas chromatography	PGC
	Pyrolysis-gas chromatography-mass spectroscopy	PGC-MS
	Thermogravimetric analysis	TGA
	Infrared spectroscopy	
	Solvent extraction	

6.4.8 Electronic and electrical characterization

Electrical testing for failure analysis must of necessity be approached differently from standard electrical characterization. This is predicated on several factors such as these:

1. The device being tested is generally not a normal device.
2. In addition to testing for expected behavior, the test must encompass unexpected behavior (it should not do what it is not supposed to do).
3. Often the test must be able to detect intermittent behavior.

While the details of electrical and electronic testing are beyond the scope of this chapter, the following general sequence of steps is recommended for failure analysis electrical testing.

- Curve tracer testing
- Standard and modified electrical characterization
- Electrical characterization over a temperature range

- Testing under conditions reported at the time of failure
- Evaluation of electrical test data
- Additional diagnostic electrical characterization

6.4.9 TEM, SEM, EDXA, and WDS

As mentioned in Chap. 1, electron microscopy has distinct advantages in the failure analysis of electronic devices and materials. Depending on the type of information required, any combination of SEM, TEM, EDS, and WDS may be used to advantage in these types of failure analysis. The basics of electron microscopy are covered in Chap. 1. In this section, certain specialized techniques with specific application to electronic failure analysis are discussed.

The use of EDS and secondary electron imaging in the analysis of leakages in electron beam–welded hybrid packages is shown in the montage of Fig. 6.60.

While transmission electron microscopy (TEM) is used infrequently in electronic failure analysis, Fig. 6.61 shows an iron disilicide rod extending from the emitter to the base, causing device failure. Figure 6.62 shows a TEM image of $CuAl_2$ precipitate-decorated grain boundaries in a heavily annealed interconnect stripe.

The most common modes of imaging in a SEM are secondary electron imaging and backscatter electron imaging. The benefit of high depth of

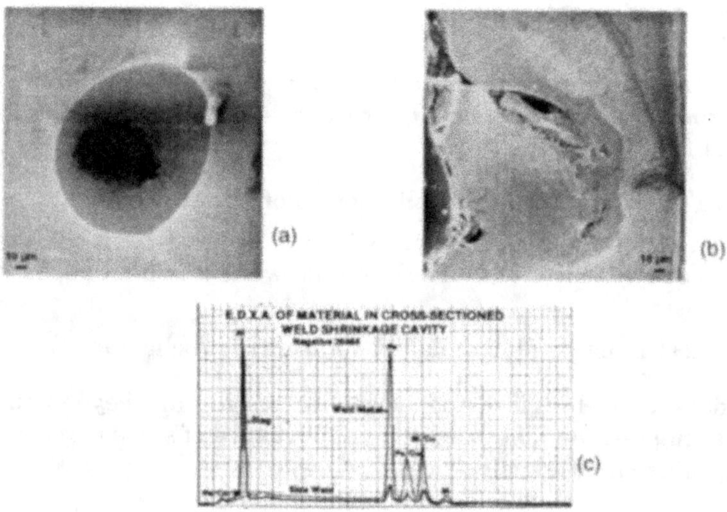

Figure 6.60 Use of EDS in failure analysis of leaks in electron beam–welded hybrid packages. Shrinkage cavity at a weld pass. (*a*) Plan view, (*b*) section through cavity, and (*c*) EDS confirming "slag" chemistry. (*From Kashar.*[21])

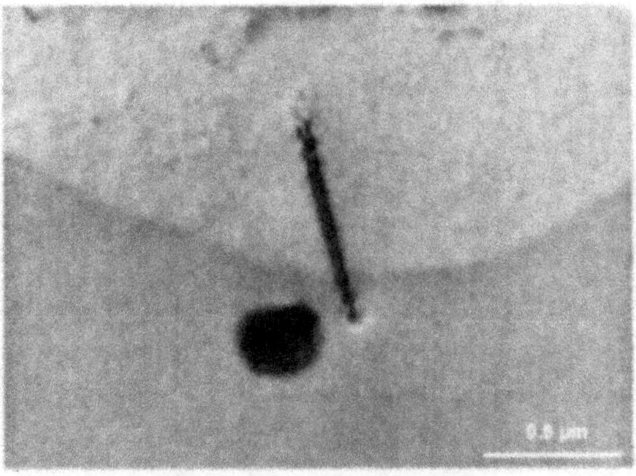

Figure 6.61 TEM image of iron disilicide rod and silicon precipitate in a bipolar transistor. (*From Richards and Footner.*[3])

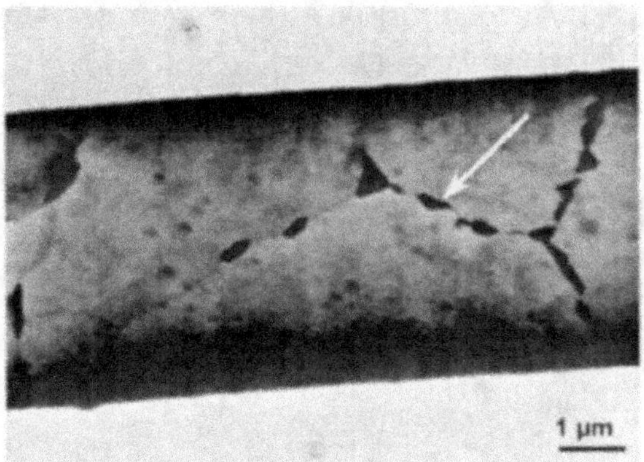

Figure 6.62 TEM image of $CuAl_2$ precipitates (arrow) decorating grain boundaries in a heavily annealed interconnect stripe. (*From Rose et al.*[22])

field in conjunction with high spatial resolution obtained in an SEM is demonstrated in the image shown in Fig. 6.63*a* and *b*. Both the nonuniformity of aluminum metallization thickness across a step and side voids in the metallization are imaged in Fig. 6.63*a* while a hidden "hot spot" between bond wires in a discrete diode is shown in Fig. 6.63*b*. The information depth in a backscatter image is usually greater than in the secondary electron mode, thus providing for a nondestructive way to image subsurface defects. The complementary secondary electron (*a*) and backscatter electron (*b*) images of Fig. 6.64 illustrate this. Some of the aluminum has electromigrated, leaving behind the voids seen in

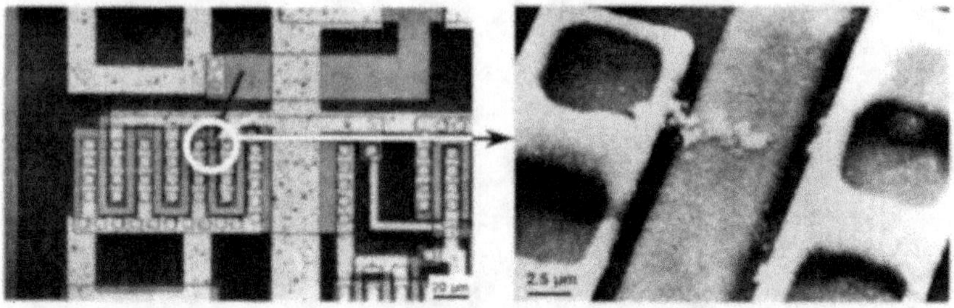

Figure 6.63 Surface contamination imaged using an SEM; could be confused as ESD damage. (*From Richards and Footner.*[3])

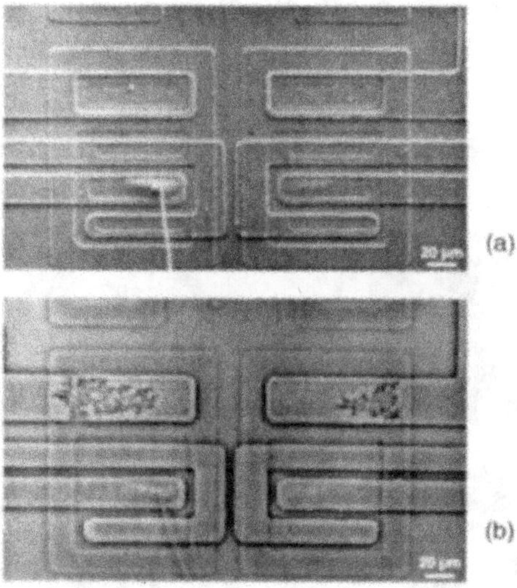

Figure 6.64 Use of BSE image (*b*) to show subsurface voids in metallization not evident in SE image (*a*). (*From Richards and Footner.*[3])

Fig. 6.64*b* and leading to the formation of the aluminum whisker shown protruding through the cracked passivation layer. In backscatter electron images, the contrast due to atomic number differences can be used to advantage in illustrating interdiffusion in heterogeneous layered structures, as shown in Fig. 6.65. An ohmic short between two metal lines caused by an embedded stainless-steel particle is shown in Fig. 6.66. Likewise, Fig. 6.67 shows an open site on a thermally cycled line caused by the presence of a tungsten particle.

In addition to these operational modes, the following special modes of operation of SEMs are particularly applicable to the failure analysis of electronic devices.

Specimen absorbed current mode. In an SEM, if the current reaching the specimen stage is amplified instead of grounded, the resultant

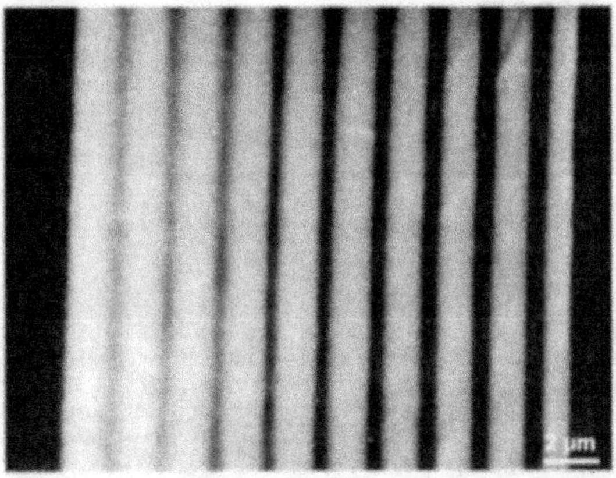

Figure 6.65 BSE image showing loss of sharp interfaces in alternating layers of CdTe and HgTe due to formation of Cd-Hg-Te by interdiffusion. (*From Richards and Footner.*[3])

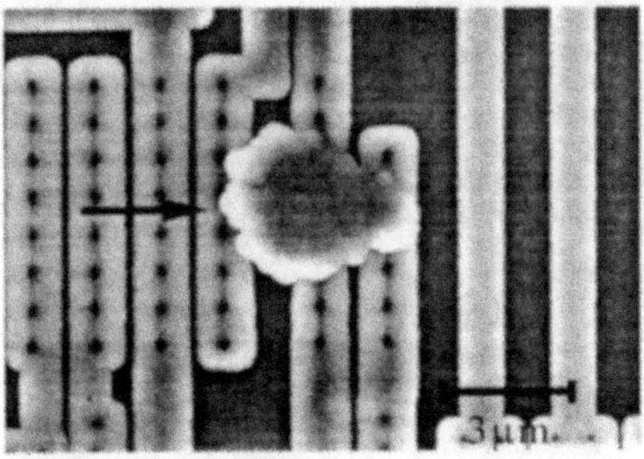

Figure 6.66 SEM image showing an embedded particle that produces an ohmic short between two adjacent metal lines. (*From Tangyunyong et al.*[23])

image can indicate a conduction path through a specimen or can be used for mapping differences in conductivity in a specimen. An alternate application is to isolate a portion of a complex printed wiring board by imaging only the desired portion in this mode while isolating the remaining portions from stage ground (Fig. 6.68).

Electron beam–induced current (EBIC) mode. This is a special case of specimen absorbed current mode used specifically in semiconductors. Generally, this imaging mode is conducted with large spot sizes, high beam voltages, and very slow scan rates. The atoms in the depletion

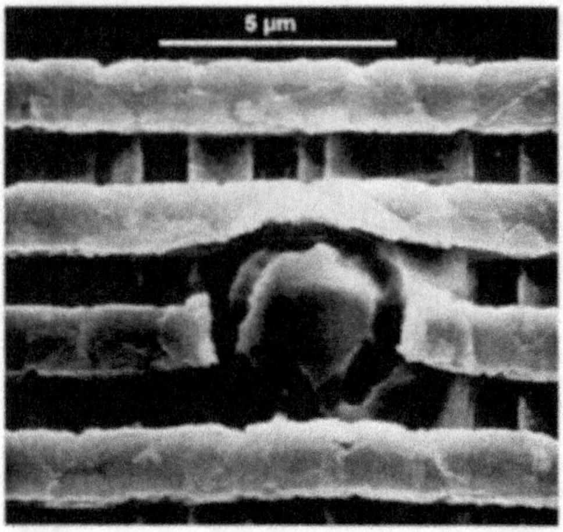

Figure 6.67 SEM image of line which opened during thermal cycling due to tungsten particle contamination. (*From Rose et al.*[22])

Figure 6.68 Specimen current image of a circuit. Only the black wire is connected to the stage ground. (*From Devaney.*[24])

layer of *p-n* junctions can be ionized under the influence of a slowly traveling electron beam. This ionization current sweeps across the junction under the influence of the depletion layer's field, causing a current flow. This current flow can be monitored via the specimen absorbed current amplifier, thereby imaging localized defects, e.g., as shown in Fig. 6.69. While EBIC is generally used on fabricated

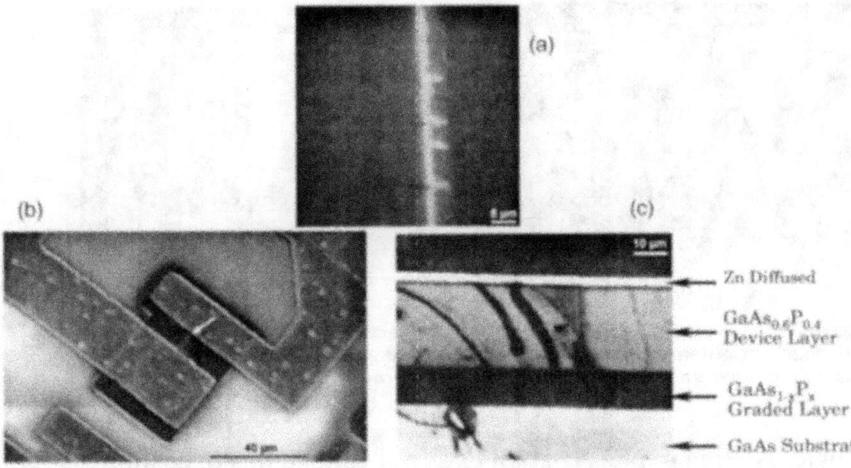

Figure 6.69 (*a*) EBIC image of Si *p-n* junction superimposed on SE image to reveal crystal damage (see spikes). (*From Richards and Footner.*[3]) (*b*) Superposition of emissive image over an EBIC image. Arrow denotes leakage site across the junction (EBIC image). (*From Devaney.*[24]) (*c*) CL image showing contrast variations due to doping. (*From Newbury et al.*[25])

TABLE 6.13 Different Modes of Charge Collection Microscopy Used for Characterization of Semiconductors

Bias source and type	Measured quantity	Mode
External constant voltage	Current	Beta conductivity
External constant current	Voltage	Beta conductivity
Internal (*p-n* or Schottky)	Open-circuit voltage	Electrovoltaic
Internal (*p-n* or Schottky)	Short-circuit current	Charge collection

SOURCE: From Newbury et al.[25]

devices, a modification of this mode called *cathodoluminescence* (CL) based on the recombination luminescence can be used for partially fabricated and starting materials. An example of changes in doping levels being detected by CL is shown in Fig. 6.69*c*. The current induced by the incident beam can be used with different combinations of bias source and measured quantity to gain different kinds of information about semiconductor materials, as summarized in Table 6.13.

Voltage contrast mode. If a metal surface is positively biased, secondary electrons cannot escape from such a surface and hence appear dark in a secondary electron image. This effect is used in the voltage contrast mode to locate electrical opens in metal lines and vias.

(a) (b)

Figure 6.70 (*a*) Low-magnification secondary electron image showing an electrical open (bright area at arrow); (*b*) higher-magnification secondary electron image obtained under reverse bias. (*From Rice and Chen.*[26])

Obviously this implies that an external voltage has to be impressed on the sample. Generally this imaging mode requires a secondary electron detector in very good condition, and the accelerating voltage may need to be finely adjusted. An example is shown in Fig. 6.70. Also note that this mode generally supplies only qualitative information since the contrast does not scale linearly with imposed voltages.

Low-voltage mode. Many electronic devices and components are made of a combination of materials some of which are electrically conductive while others are not. A typical example would be a "Winchester" type of hard disk and its drive head. A hard disk generally is made by spin-coating a magnetic paint onto a smooth aluminum substrate. The paint consists of needlelike gamma iron oxide particles with a few percent of hard alumina dispersed in an organic binder resin, which is cured after spin coating. A lubricating coating is then applied to the surface. When the disk is stationary, the head rests on the disk. When the disk spins at 3600 r/min, the head lifts off by aerodynamic thrust, subsequently landing back on the disk when the head stops spinning. The lubricating coating serves to reduce friction during head takeoff and landing. Furthermore, controlled porosity in the magnetic coating provides reservoirs for the storage of lubricating material. Typically the magnetic layer is thinner than 0.5 μm while the head flies over the disk at heights lower than 0.3 μm. It is obvious that microcontamination will seriously impair the performance of such drives (Fig. 6.71). The easiest way to image such microcontamination is by SEM; however, the insulating nature of some of the materials complicates such imaging. As shown in Fig. 6.72, the total electron emission [(secondary electrons + backscatter electrons)/(incident electrons)] is lower than unity at accelerating voltages between 20 and 30 keV. With decreasing accelerating voltage, the electron emission increases, reaching a maximum at about

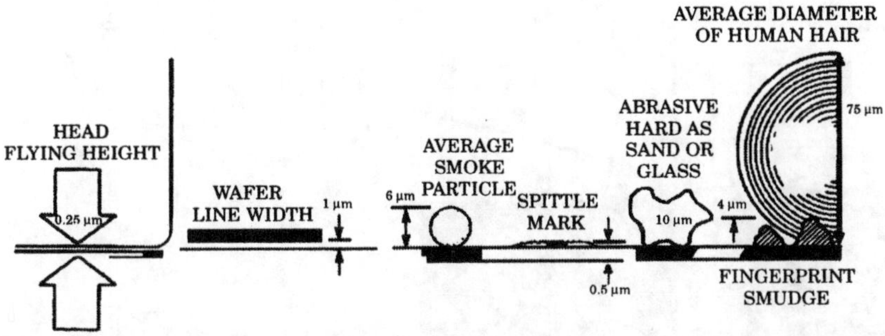

Figure 6.71 Approximate sizes of common contaminants compared with the critical dimension in disk drives and semiconductor devices. (*From Brar and Narayan.*[27])

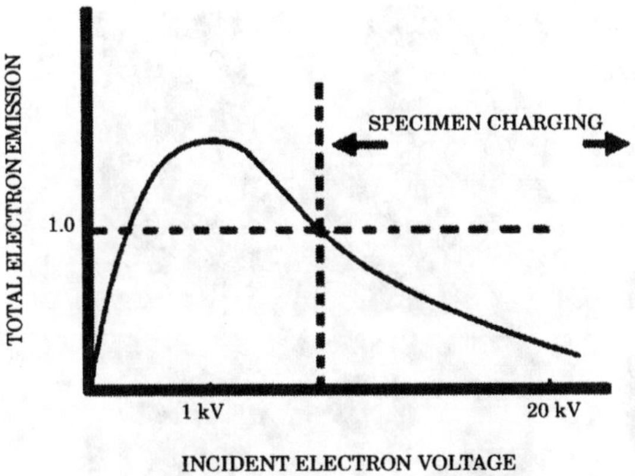

Figure 6.72 Total electron emission (secondary and backscattered) as a function of incident electron voltage. (*From Brar and Narayan.*[27])

1 to 1.5 keV, and then it drops with further decreases in accelerating voltage. It is noteworthy that when the total electron emission is less than unity, negative charge builds up in the specimen, thereby causing distortion of the incident beam and secondary electron image. This is the case during normal operation of an SEM at accelerating voltages of 20 to 30 keV. As mentioned above, there are nonconducting portions of hard disk drives, e.g., ceramic slider materials and organic resins. By operating the SEM at accelerating voltages of 1 to 5 keV, the problem of charging of these insulating portions can be alleviated. In addition to this, the low-voltage operation of SEMs obviates other problems such as polymerization of organic films and improves the topographic contrast. However, it must be recognized that the technique of low-voltage SEM has the problems of low brightness, chromatic aberration, and

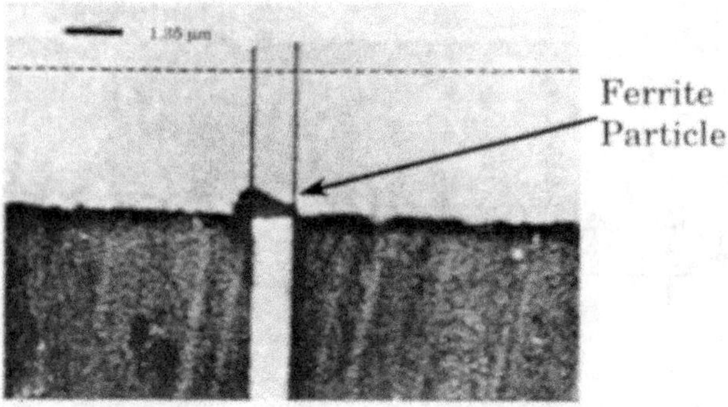

Figure 6.73 Ferrite core of a failed composite head showing ferrite contamination across the gap. (*From Brar and Narayan.*[27])

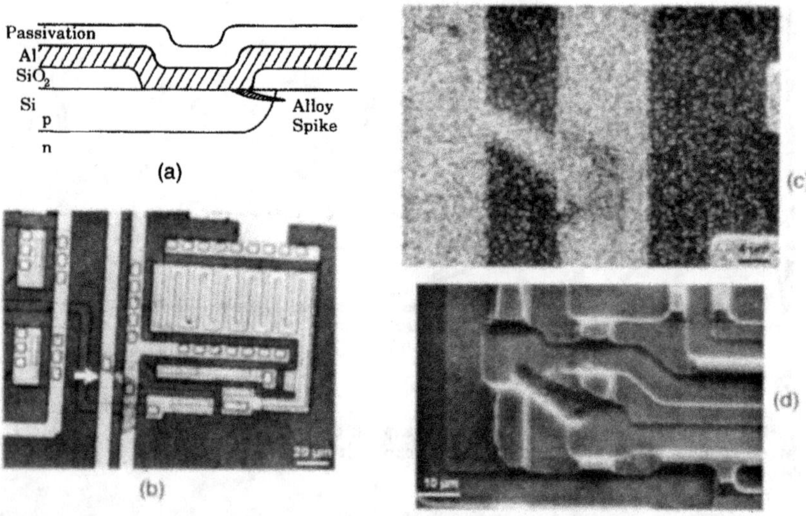

Figure 6.74 (*a*) Schematic representation of the growth of filamentary alloy spike initiated by ESD, (*b*) complementary light micrograph, (*c*) EDX Al-map, (*d*) SE image. (*From Richards and Footner.*[3])

sensitivity to stray fields. An example of the detection of a ferrite contamination causing failure is shown in Fig. 6.73.

An example of the use of EDS in the detection of a short and the cause of its occurrence is illustrated in the montage of Fig. 6.74.

The advent of advanced WDS systems has enabled the application of this technique to the identification of micron- and submicron-sized particles. In the example shown in Fig. 6.75, WDS and EDS linescans over a 1-μm particle on a silicon wafer are compared. The lower energy resolution of the EDS system used does not permit unequivocal elemental identification, i.e., silicon and tungsten appear under a single

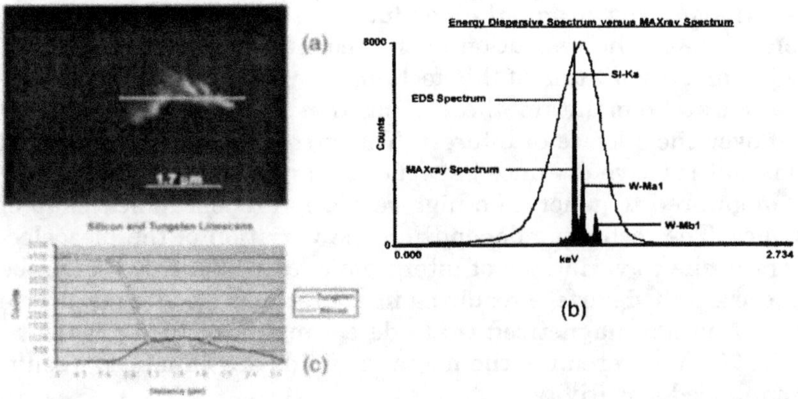

Figure 6.75 Identification of a particle on a silicon wafer. (*a*) Secondary electron image of a 1-μm particle on a silicon wafer showing the location of a 100-point linescan. (*b*) Comparison of EDS and advanced WDS energy spectra. The superior energy resolution of the WDS system permits the resolution of the W and Si peaks, while they are unresolved in the EDS spectrum. (*c*) Tungsten and silicon linescans showing increased tungsten X-ray counts as the electron beam of the linescan traverses the particle. (*By permission of ThermoNoran.*[28])

peak (Fig. 6.75*b*). However, the use of the advanced WDS system permits the three underlying peaks to be resolved as belonging to SiKα_1, WMα_1, and WMβ_1. The linescan (count rate plot) shown in Fig. 6.75*c* is further proof that the particle is tungsten.

The use of SEM in semiconductor analysis has four major drawbacks. First, subsurface defects are not easily detected due to the limited depth of penetration of the primary electron beam. Second, the passivation layers (dielectrics) in devices cause charging problems. Third, the SEM is not ideally suited for the detection of crystallographic or atomic number contrast. Fourth, in some devices the use of SEM renders the device inoperative and is thus considered a destructive technique.

6.4.10 Miscellaneous techniques

Several specialized techniques are used in the failure analysis of electronic materials. While a detailed discussion of each of these techniques is beyond the scope of this book, some of these techniques are described in brief.

1. The term *scanning probe microscopy* (SPM) is used to include the techniques of *scanning tunneling microscopy* (STM), *atomic force microscopy* (AFM), and *magnetic force microscopy* (MFM). In the case of STM, an accurate topographical description of the scanned surface is generated using the strong dependence of the tunneling current

between the probe tip and the conductive surface on the distance between the two. The resolution of the technique (depth) is approximately 0.1 nm. A variation of this technique is to use piezoelectric drives to accurately maintain the tip position constant as the tip is scanned over the surface of interest. The current needed to maintain this constant relative distance between the probe tip and the surface can be monitored to generate a high-resolution topographical map of the surface. The technique depends on the variation of the piezoelectric current due to variations of interatomic forces between the probe tip and the scanned surface resulting in the term AFM. Magnetic force microscopy uses a magnetized tip to detect magnetic field gradients. When the IC is powered up, the magnetic fields are enhanced, resulting in enhanced sensitivity.

2. In *low-energy electron microscopy* (LEEM), the strong interactions between the slow electrons from a low-energy electron beam and the material of the surface being studied results in electrons being back-diffracted by the first few atom layers. Some of these back-diffracted electrons can be used to form a high-resolution image of the surface. This technique requires the use of ultra-high vacuum systems.

3. Scanning near field light microscopy (SNFLM) uses an optical antenna that is small compared to the wavelength of the light used. When such an antenna is scanned across a surface of interest, the light emitted by the antenna is modulated by the near field interactions between the antenna and the surface. Thus an image with a resolution between 20 and 50 nm laterally and 1 nm vertically can be obtained.

4. *Photon emission microscopy* is a technique used for the nondestructive location of leakage currents, ESD/EOS damage, latchup, hot carriers, etc. In the damaged regions there is enhanced photon emission, which can be detected by using photon detectors.

5. Liquid-crystal analysis is an inexpensive technique used to identify locations of failures that result in leakage currents in the milliampere range. The chip is placed on a temperature-controlled chuck and observed under polarized light. The device is then powered up. At locations where there are leakage currents, the device heats up. At these locations, the liquid crystal becomes opaque, and a dark spot is thus observed. Obviously, the technique depends on the selection of an appropriate combination of chuck temperature and liquid-crystal phase transition temperature.

6.4.11 Choice of a microscopy technique

From the previous sections it is seen that the choice of a particular microscopy technique or a suite of complementary techniques is very much dependent on the specifics of the failure analysis to be conducted.

Not only are technical issues such as detectability, spatial resolution, and depth sensitivity important, so are business issues such as cost, availability, and speed of analysis. Table 6.14 shows a partial listing of several microscopy techniques that may be applicable to the detection of certain device defects. It is also pointed out that the choice of technique is often driven by the paradox of ever-decreasing feature size and ever increasing device size and complexity.

6.5 Failure Analysis of Electronic Packages

6.5.1 Packaging fundamentals

The term *packaging* of electronic components implies several levels of packaging, which can be summarized as

1. Zeroth-level packaging including semiconductors, attachment materials, and substrate materials.
2. First-level packaging of chips onto single-chip modules (SCMs) or multichip modules (MCMs) involving wire interconnects, tape interconnects, case materials, lid seals, and leads.
3. Second-level packaging of SCMs, MCMs, connectors, etc., onto printed-circuit boards (PCBs) involving reinforcement fiber materials, resins, laminates, flexible printed board materials, and conformal coatings.
4. Third-level packaging of PCBs, cables, power supplies, ancillary systems, etc., onto a frame or box (e.g., motherboard). From a materials standpoint, this involves backpanel materials, connector materials, cables, and flex circuit materials.

The ultimate objective of packaging is to ensure that all the components of the package operate in concert. Figure 6.76 schematically depicts the four levels of packaging.

At the first level, the chip or IC device is connected to its carrier by wire bonding, flip-chip attachment, tape-automated bonding and z-direction adhesives. In wire bonding, electrical connection between the pads on the chip and the external carriers is made by bonding fine gold or aluminum wires to both pads. In the case of flip-chip technology, the chip is inverted and bonding between chip and the carrier pads is achieved with either solder or gold bumps. Since the advent of organic (polyimide) substrates, the metal used for bonding has had to be modified due to their low tolerance for high temperatures. In the case of tape automated bonding, flat metal fingers provide connection between pads on the chip and its carrier. Figure 6.77 schematically depicts the various parts of an IC package using a dual-in-line package as an example.

TABLE 6.14 Microscopic Techniques Appropriate for Study of a Range of Device Defects

Defect	Package type	Microscopy techniques	Comments/examples
		External Defects	
Package damage	H/P	LM	Mechanical damage, chip-outs, microcracks
Corrosion contamination	H/P	LM, SEM, EDX, AES, XPS	Interlead dendrites/whiskers on plating
		Package Defects	
Hermetic seal	H	LM, SEM, LF-SAM, radiography	Choice of SAM or radiography determined by x-ray opacity of seal
Cracked plastic	P	LM, IRM, SAM	LM requires microsectioning
Plastic/die delamination	P	IRM	"Nondestructive"
Die attach	H/P	Radiography, SAM	Radiography will detect only voids, not delaminations; SAM detects both
		Internal Defects (Plastic Packages Decapsulated or Prepared for IRM)	
Wire bond defects	H	Radiography	Not possible for Al wires
Wire bond defects	H	LM, SEM, IRM	SEM has better depth of focus and field of view
			IRM for underside of Au ball bonds
Corrosion contamination	H/P	LM, SEM, EDX, micro-FTIR	Color via LM, morphology via SEM, elemental composition via EDX, FTIR good for organic contaminants
ESD	H/P	IRM, LM, SEM, SLM	IRM can damage under metal cracks
			SLM and SEM permit OBIC and EBIC for defects and location of low-level damage

EOS	H/P	IRM, LM, SEM, SLM	As above
EM	H/P	LM, SEM	For fine microvoids the extra resolution of SEM may be required
Layer delamination	H/P	SAM	Resolution determined by frequency
Alignment	H/P	LM, SEM, IRM	LM allows layer discrimination by virtue of layer color
Layer definition defects	H/P	LM, SEM, IRM	SEM-OBIC is excellent for study of junction alignment
			IRM allows viewing of junctions/layers under metal
Metal/Si contacts	H/P	IRM, SEM	For SEM examination the metal layer must be stripped first; IRM can damage directly from underside
Cracks in die	H/P	HF-SAM	Although a rare type of defect, HF-SAM is an excellent tool for studying such cracks
Microcracks in metallization	H/P	SEM+VC	Defect location via voltage contrast, examination via high-resolution SEM
Layer thickness	H/P	LM, SEM	Sample prepared by microsectioning or cleaving the die, depending on the need for precise area location
Oxide breakdown	H/P	EMMI, SEM	EMMI is the only technique for precise defect location, confirmed by etching and SEM
Silicon defects	H/P	LM, SEM, TEM	Defects require selective etching prior to LM, SEM examination

SOURCE: From Richards and Footner.[3]

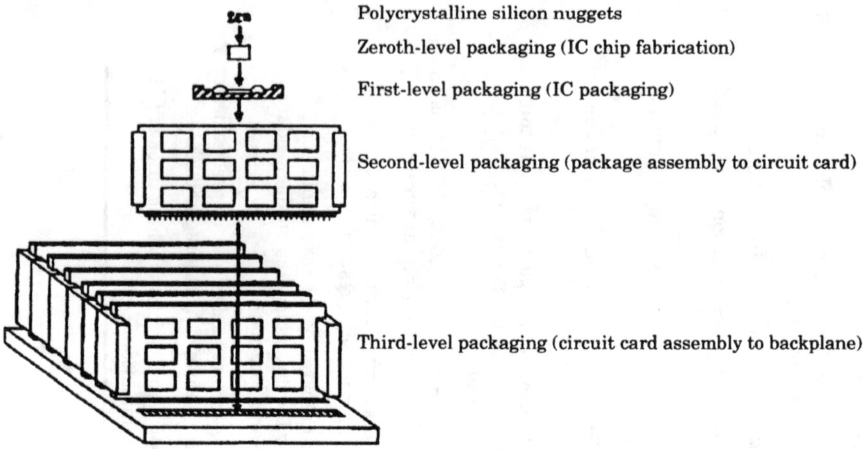

Figure 6.76 Hierarchy of electronic packaging. (*From Cohn and Shih.*[29])

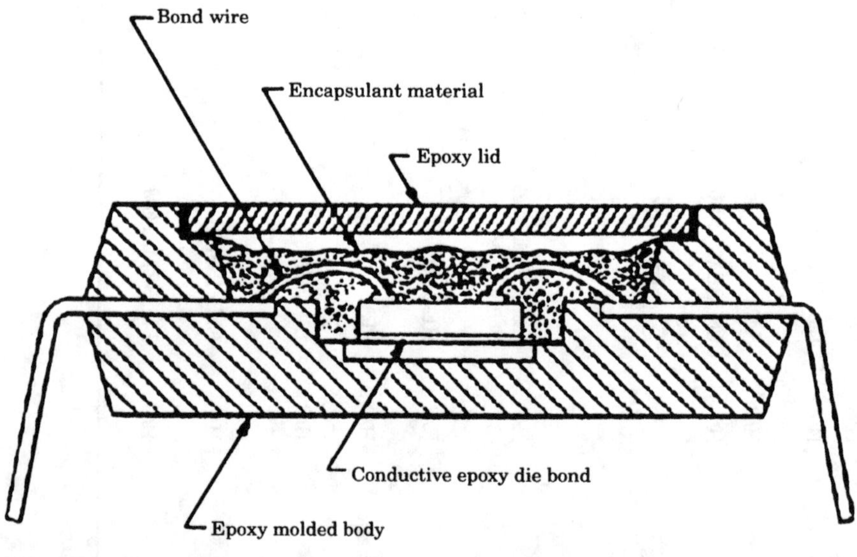

Figure 6.77 Cross-sectional view of a completed premolded plastic dual in-line package including the IC chip. (*From Cohn and Shih.*[29])

Figure 6.78 shows a montage of various options for chip attachment. Two other packaging technologies that deserve mention are ball grid array (BGA) packaging and chip scale packaging (CSP).

At the second level, the IC package (SCM or MCM) and discrete components are mounted on rigid or flexible "boards" to form an assembly known as a printed-circuit board (PCB) or a printed wiring board

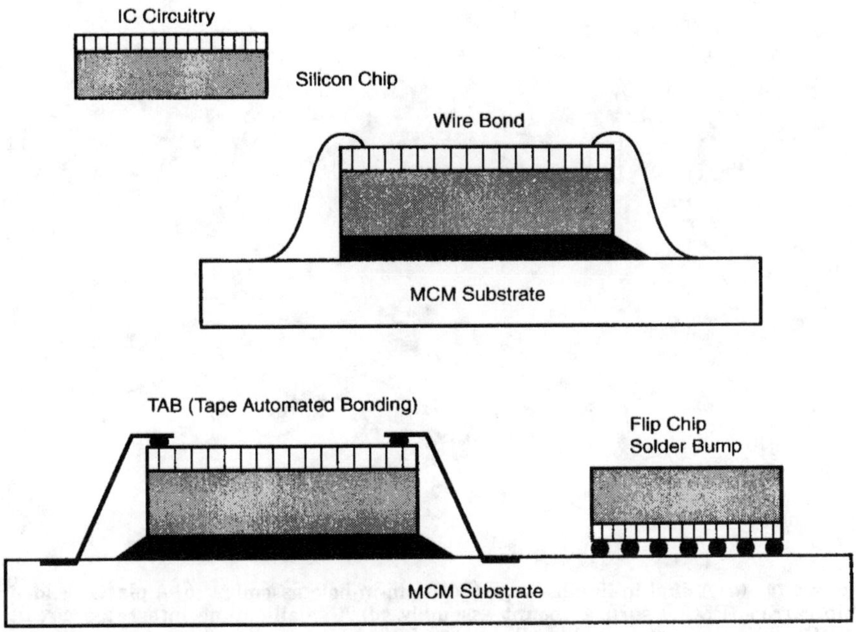

Figure 6.78 Chip attachment options. (*From Lumpp.*[30])

(PWB). There are two types of technologies for mounting components or chip packages on PCBs, and their use depends on the lead arrangement and mounting type. These are pin-in-hole (PIH) technology, in which the leads from components are inserted into the plated-through-holes (PTHs) of the cards and soldered in place; and surface mount technology (SMT), where the leads from components are soldered onto the pads or lands of the cards. These are shown schematically in Fig. 6.79.

At the third level, the boards are interconnected, and power supplies and other hardware (electrical and otherwise) are connected through backpanels, connectors, and cables.

6.5.2 Failures in zeroth-level packaging

At the zeroth level, semiconductor materials used to make IC chips can have atomic or crystallographic level defects associated with them that can adversely affect their performance. Such defects can be line defects such as slip-induced dislocations (Fig. 6.80), misfit dislocations (Fig. 6.81), and oxide-nitride edge dislocations. Areal defects, particularly stacking faults, can be generated in the semiconductor material during epitaxial growth (Fig. 6.82), oxidation (Fig. 6.83), and diffusion (Fig. 6.84) in an oxidizing atmosphere and ion implantation annealing. In addition to providing easy diffusion paths (pipes), such defects can

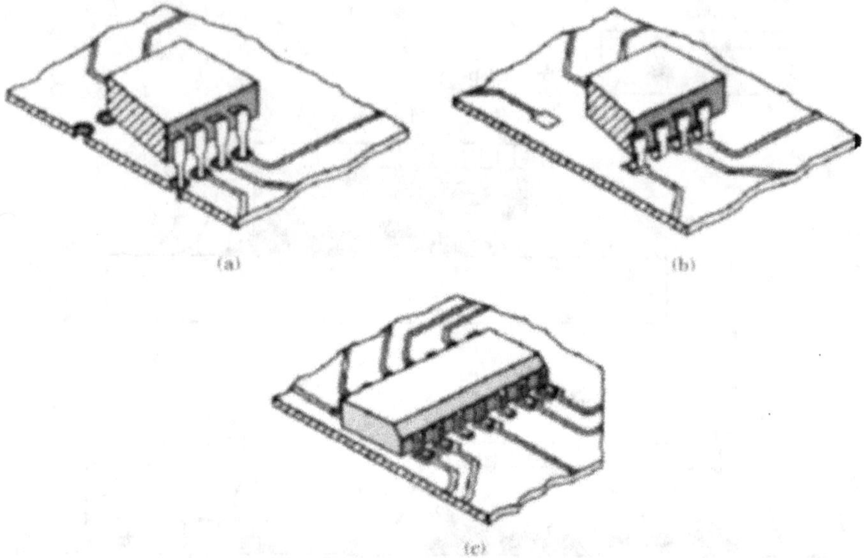

Figure 6.79 (*a*) A dual in-line package (DIP) pin-in-hole assembly. (*b*) A plastic-leaded chip carrier (PLCC) surface mount assembly. (*c*) A small-outline integrated circuit (SOIC) surface mount assembly. (*From Woychik and Senger.*[14])

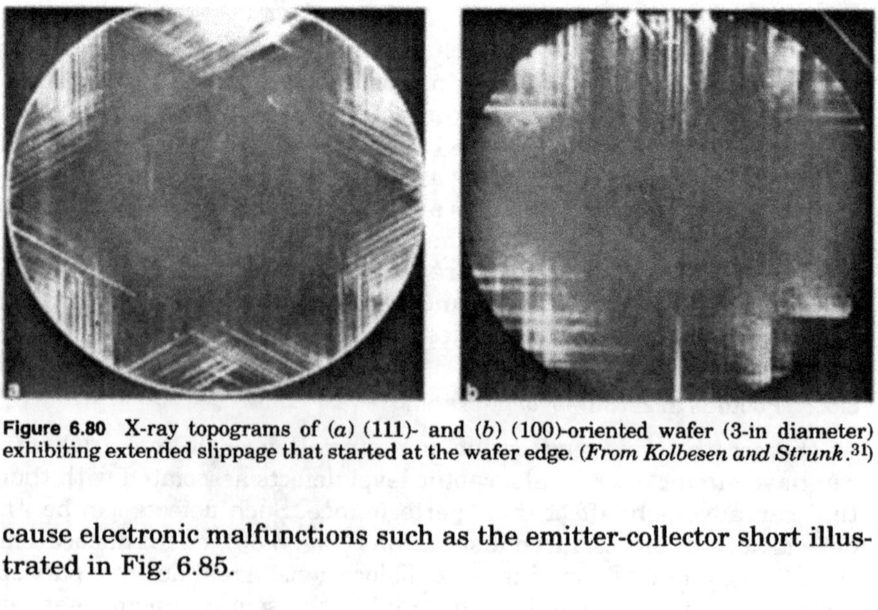

Figure 6.80 X-ray topograms of (*a*) (111)- and (*b*) (100)-oriented wafer (3-in diameter) exhibiting extended slippage that started at the wafer edge. (*From Kolbesen and Strunk.*[31])

cause electronic malfunctions such as the emitter-collector short illustrated in Fig. 6.85.

6.5.3 Failures in first-level packaging

Some aspects of the failure of wire bonds are covered in Sec. 6.2.5. In IC packages, the interconnection between the chip and the lead frame

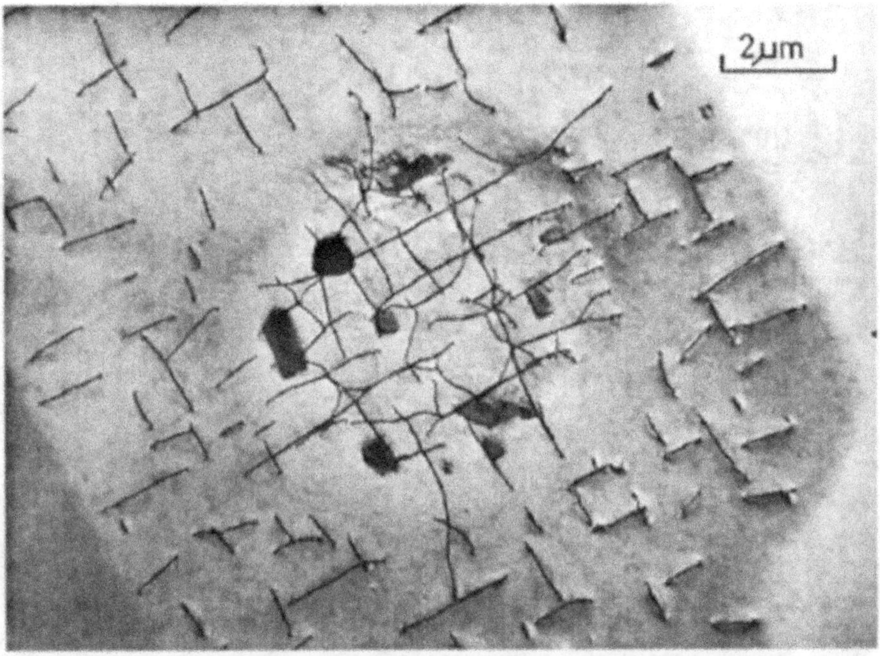

Figure 6.81 Example of misfit dislocations in a (100)-oriented wafer. The higher density in the circular inner contact area is due to a higher misfit (phosphorus concentration $>10^{20}/cm^3$) than in the outside (arsenic concentration $\sim 10^{20}/cm^3$) (HVEM). (*From Kolbesen and Strunk.*[31])

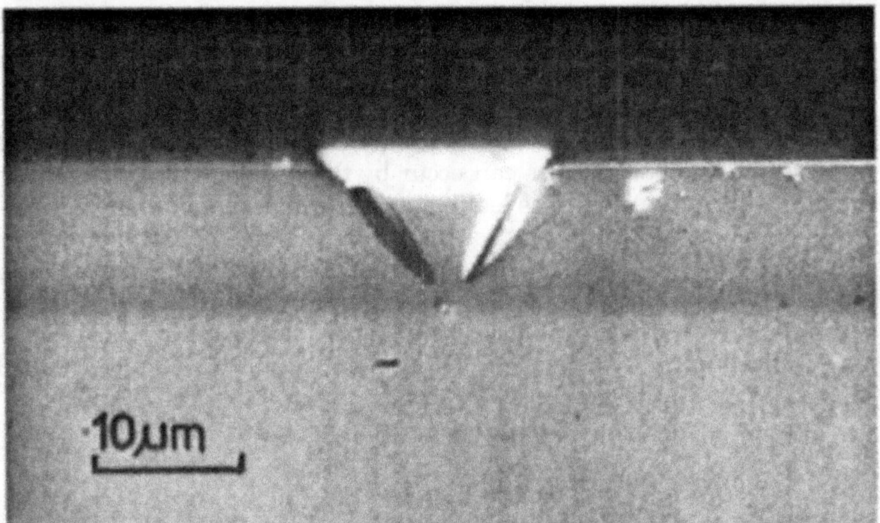

Figure 6.82 Fracture edge after preferential etch of an epitaxial Si layer showing stacking faults having nucleated at the interace between epitaxial layer and substrate (SEM). (*From Kolbesen and Strunk*[31]*; courtesy of G. Franz.*)

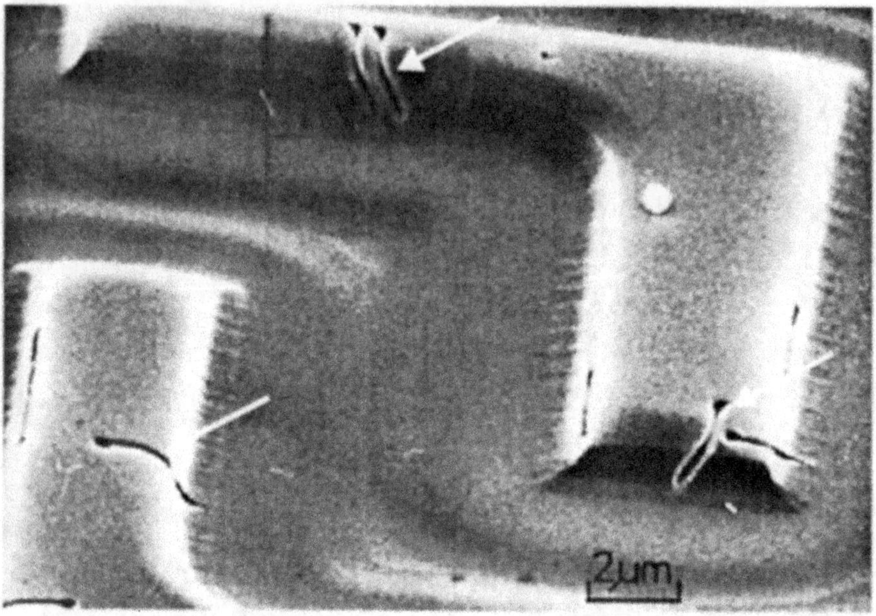

Figure 6.83 Oxidation-induced stacking faults (arrows) that grew from the oxide well into the active regions of bipolar transistors (SEM after preferential etch). (*From Kolbesen and Strunk*[31]*; courtesy of G. Franz.*)

is made by various combinations of gold and aluminum pads and wires. Aluminum bond pads are subject to corrosion in the presence of moisture and halogen species such as chlorine and bromine, leading to the deterioration of the bonds as shown in Fig. 6.86. Also, brittle intermetallics such as Al_2Au and Au_5Al can form by solid-state diffusion, which leads to cracking from subsequent thermal and mechanical stresses. Also, open circuits can occur by the diffusion of aluminum or gold into such intermetallics leading to the formation and eventual coalescence of voids. The plating solution (specially gold plating) often has deliberately added constituents which can be plated and subsequently cause intergranular weakness by segregation, thallium being a prime example. An example of blistering and subsequent cracking due to chlorine contamination between the Kovar leads and the gold-over-nickel plating is shown in Fig. 6.87.

In the case of solder bumps, some common sources of defects are

1. Undercutting of the pad (Fig. 6.88)
2. Broken terminal metal layer between solder and pad (Fig. 6.89)
3. Adhesion of the metallic interfaces can be compromised by the formation of excess intermetallic compounds (Fig. 6.90)

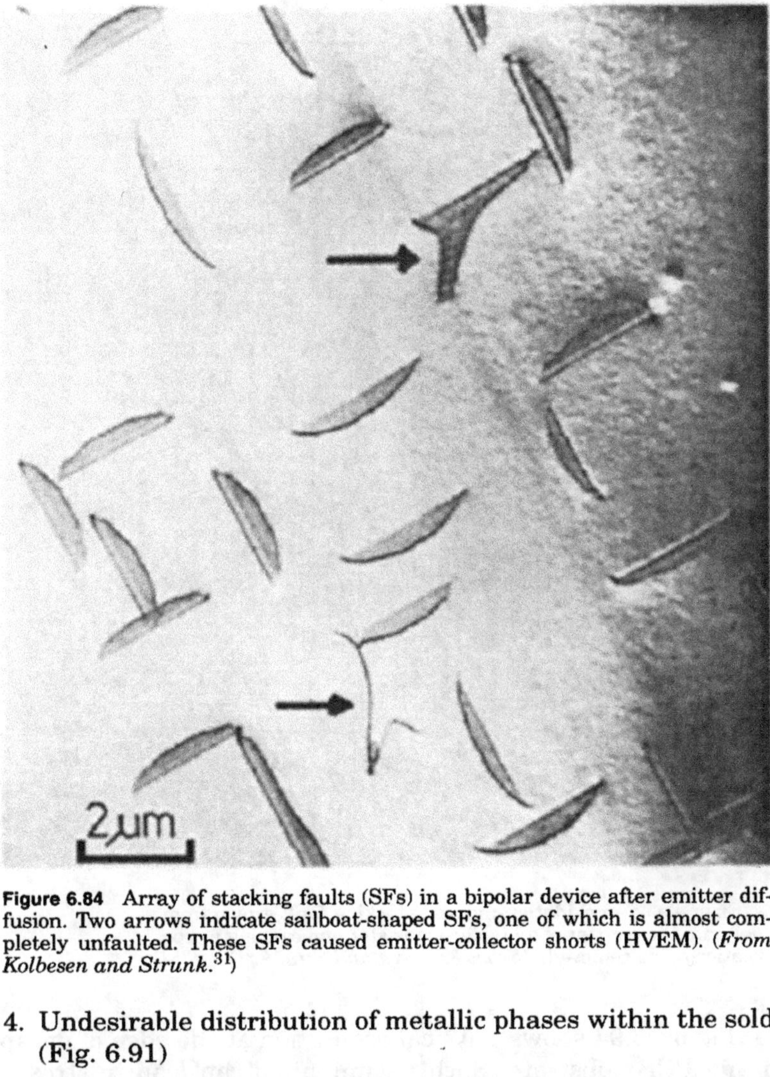

Figure 6.84 Array of stacking faults (SFs) in a bipolar device after emitter diffusion. Two arrows indicate sailboat-shaped SFs, one of which is almost completely unfaulted. These SFs caused emitter-collector shorts (HVEM). (*From Kolbesen and Strunk.*[31])

4. Undesirable distribution of metallic phases within the solder bump (Fig. 6.91)

Plastic packages are nonhermetic and are subject to adsorption of moisture during storage. During assembly onto the PCB by reflow soldering, the package is rapidly heated from ambient to about 200°C, causing rapid conversion of the adsorbed moisture to steam. The result is an explosive delamination/cracking phenomenon known as *popcorning*. This phenomenon is illustrated schematically in Fig. 6.92 while Fig. 6.93 shows evidence of popcorning in a plastic BGA through a C-SAM image.

The encapsulation process is subject to a variety of defects, such as the formation of voids, sweeping of the bond wires, and inadequate

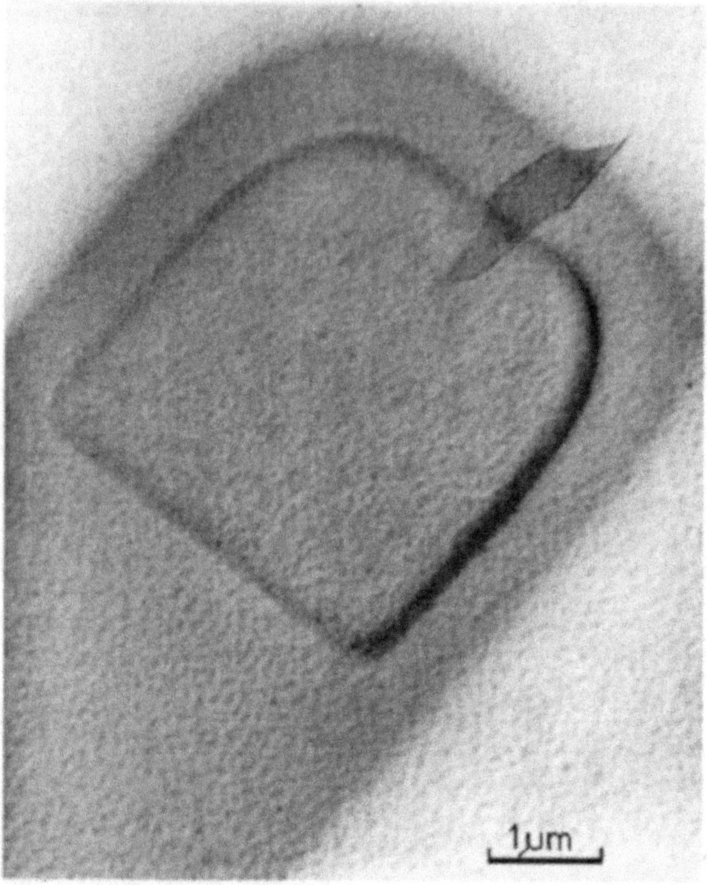

Figure 6.85 Example of an oxidation-induced stacking fault grown from the oxide well into the emitter of a bipolar transistor and causing a pipe (HVEM, isolation oxide removed). (*From Kolbesen and Strunk.*[31])

adhesion. Figure 6.94 shows physical separation at the edge of encapsulation and PCB substrate. Such delamination can lead to stresses which in turn can lead to lifted ball bonds at the die pads (Fig. 6.95) and fractured bond wires (Fig. 6.96).

In terms of the leads used at this level of packaging, Table 6.15 lists the chemical composition of several common alloys used as SMT lead frames. Most package leads are made of relatively stiff alloy 42, which is solder connected. Issues related to lead failure include the stiffness of the material used, the compliance of the lead design, the thermal excursions encountered, the effect on the solderability arising from the sharpness of the knee and toe of the leads, etc.

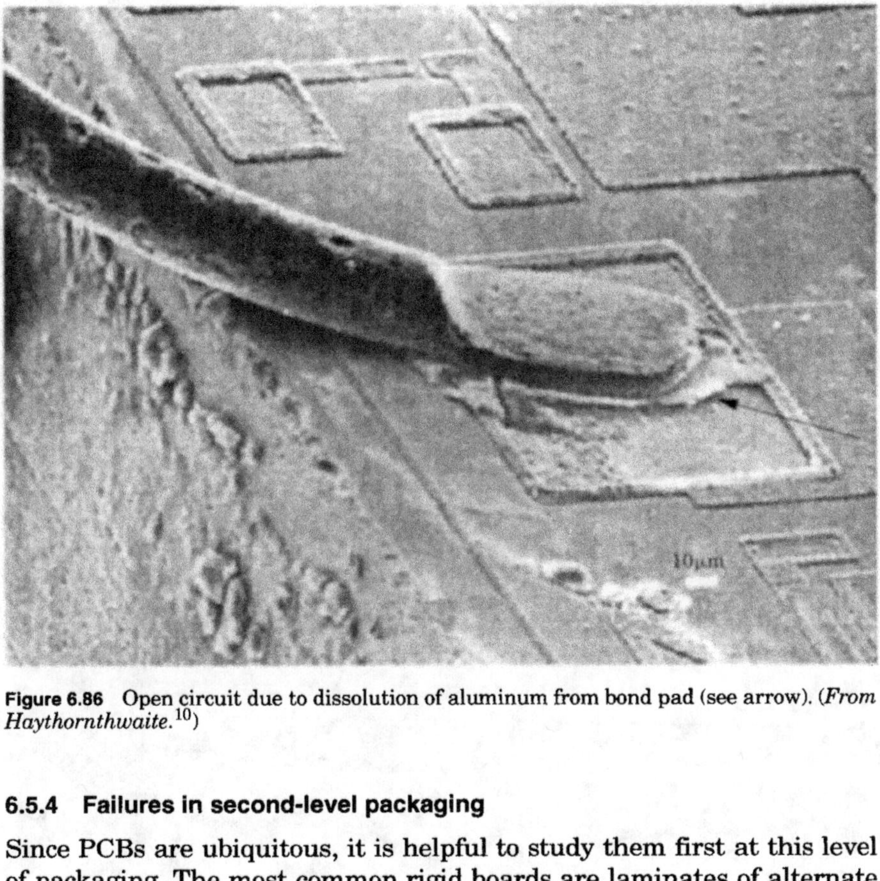

Figure 6.86 Open circuit due to dissolution of aluminum from bond pad (see arrow). (*From Haythornthwaite.*[10])

6.5.4 Failures in second-level packaging

Since PCBs are ubiquitous, it is helpful to study them first at this level of packaging. The most common rigid boards are laminates of alternate layers of resin (insulating) and copper-impregnated (conducting) FR-4 glass epoxy. Electrical contact between the different conducting layers is by three types of plated vias: plated-through-hole vias (30 to 42 mils in diameter) and smaller-diameter (6- to 10-mil) buried and blind vias, all of which are plated with copper (Fig. 6.97). In the case of both PIH and SMT packages, the most common component attachment technique is soldering (lead-tin eutectic alloy) followed by the use of conductive adhesives. The electrical signal and power are distributed throughout the PCB using printed circuit lines or traces. These are formed by either a subtractive- or additive-etch process. Following this step, an insulating mask (nonwettable by solder) is applied over all areas of the PCB with the exception of areas of bare copper areas for subsequent soldering. The solderability of these exposed copper areas is deteriorated by the formation of copper oxides, and various techniques such as tin-lead finishing, use of organic solderability preservatives, hot air solder leveling and use

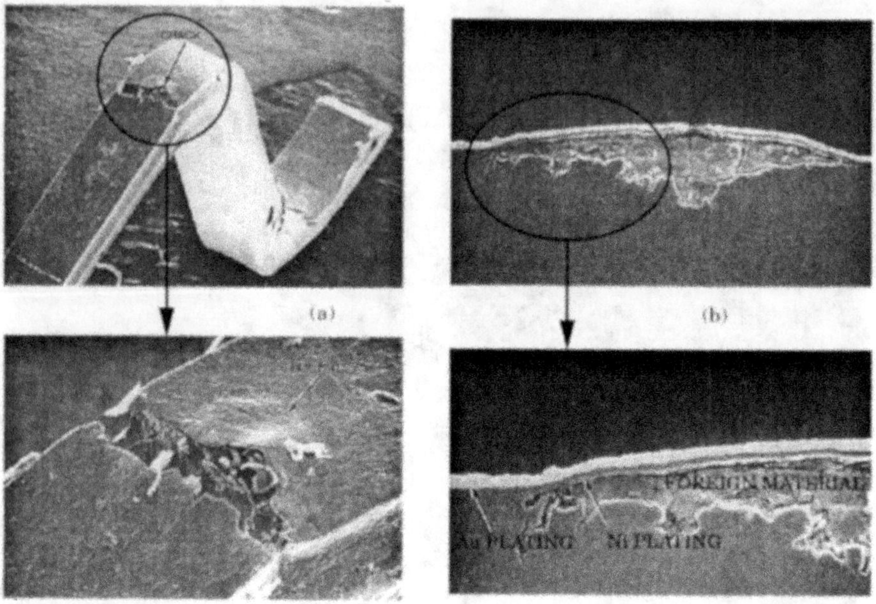

Figure 6.87 Blister and crack defects: (*a*) plan views and (*b*) cross-sectional views. (*From Martin.*[12])

Figure 6.88 Overhang at the edge of scribe line. (*From Lau and Pao.*[32])

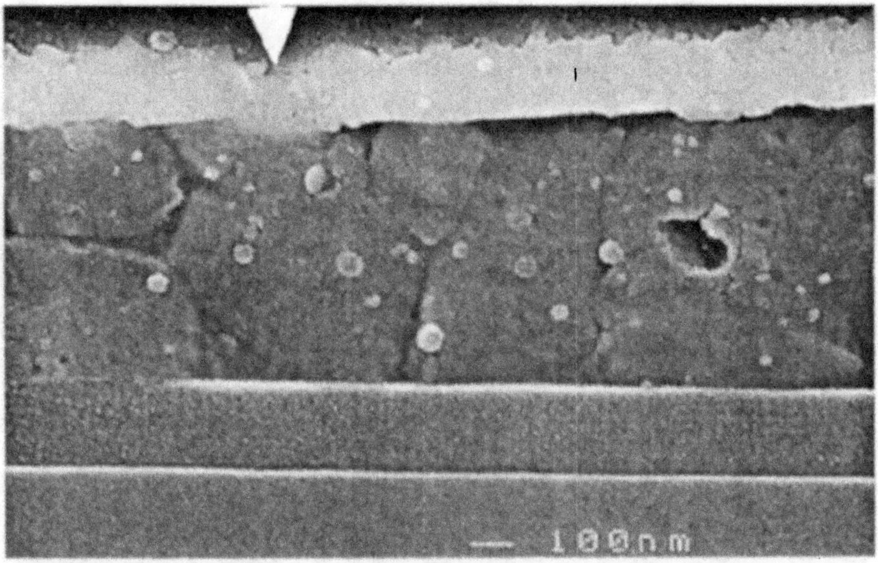

Figure 6.89 A broken terminal metal layer (see arrow). (*From Lau and Pao.*[32])

Figure 6.90 Intermetallic growth (arrow) at interface between terminal metal and solder. (*From Lau and Pao.*[32])

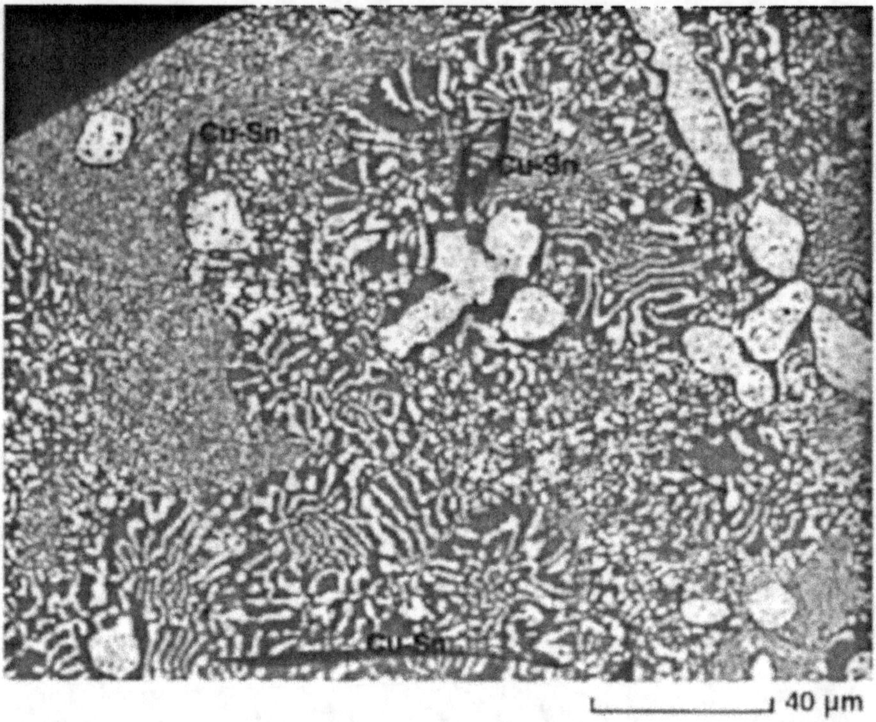

Figure 6.91 A classic micrograph of a noneutectic Sn-Pb, having a composition to the Pb-rich side of the eutectic point. Here, the presence of the Pb-rich dendrite along with the last solidifying eutectic structure can be seen. (*From Woychik.*[33])

of palladium and gold platings are used to alleviate this problem. In addition to rigid PCBs, flexible PCBs and thermoplastic-based molded circuits (two- and three-dimensional) are used at this level of packaging.

The conductive copper traces in the PCBs can have the following defects:

1. Reduction in trace thickness called *pits*
2. Reduction in trace width called *mouse bites*
3. Superfluous copper as shown in Fig. 6.98
4. Bridging between traces and lands as shown in Fig. 6.99
5. Compromised (cut or nicked) conductors as shown in Fig. 6.100

Two of the commonly used laminates for PCBs are designated G-10 and FR-4; in each case the resin and base material used is epoxy and glass fiber, respectively, with the FR-4 having a flame retardant added. If any of the fibers in the long direction (called warp fibers) are

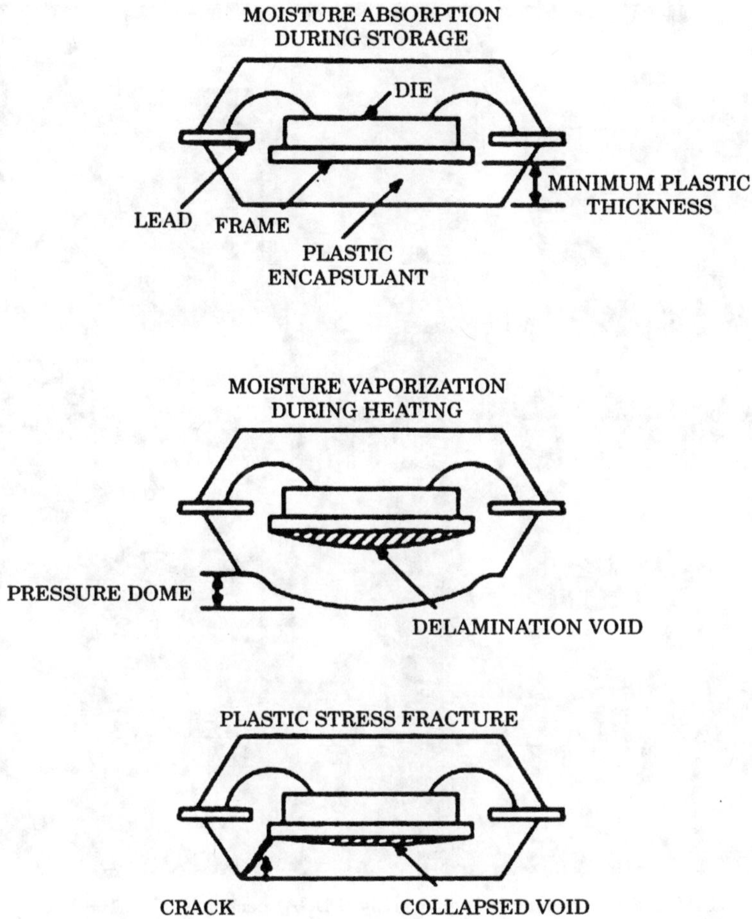

Figure 6.92 Schematic representation of the mechanism of popcorning during soldering. (*From Prough and Pope.*[34])

hollow (trapped air), the copper plating operations in PTH assemblies can lead to a short circuit by the plating of copper on the inside of the fiber. This is known as *wicking* and is shown in Fig. 6.101. The adhesion of epoxy to the glass fiber of PCB laminates is improved by pretreatment of the glass cloth in silane. In spite of this, the fiber-epoxy interface can crack during subsequent manufacturing operations as shown in Fig. 6.102. These cracks can trap copper plating solutions and can also provide pathways for subsequent transport of copper, moisture and ionic contamination. Likewise, there may be delamination between the resin and deposited copper as shown in Fig. 6.103. In the laminate, epoxy starvation can result in loss of adhesion between the fiber and the epoxy. Sometimes the PWB fibers can expand and detach

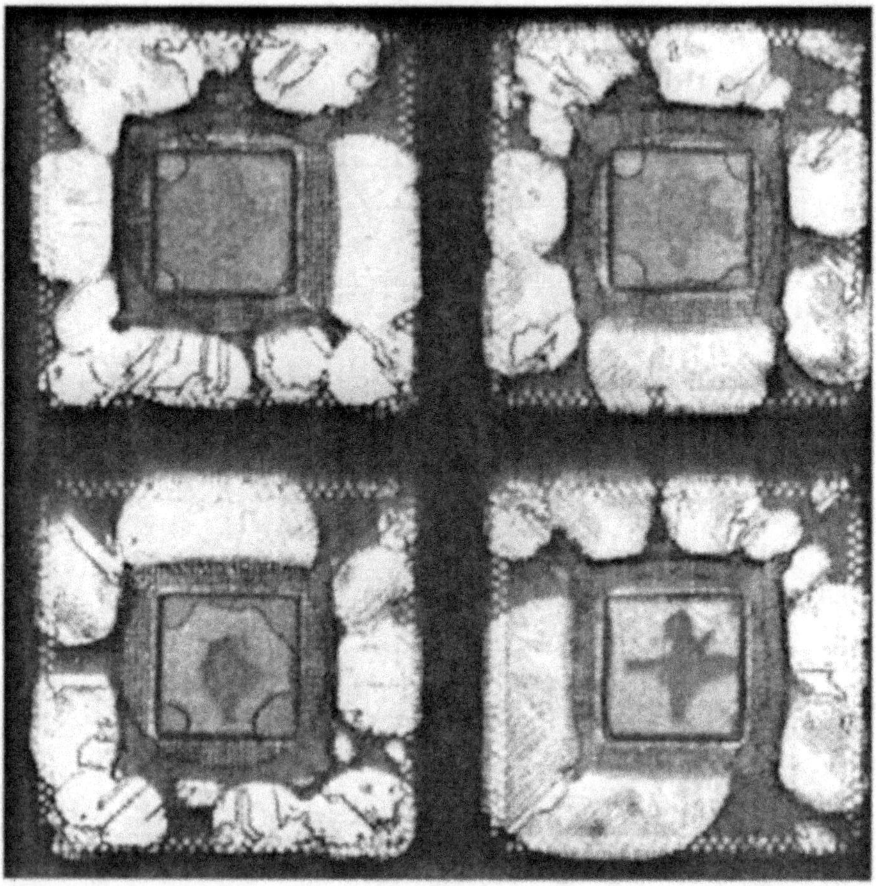

Figure 6.93 C-SAM image of plastic ball grid array (PBGA) packages after 168 hours of exposure at 85°C and 85% relative humidity showing popcorning. (*From Lau and Pao.*[32])

during the rapid volatilization of moisture during soldering. This can appear as white spots or crosses (pluses) called measles. *Crazing* is a term used for merged measling spots. Localized microcrazing around holes is referred to as *haloing*. All of these defects lead to electrical shorts due to loss of electrical insulation by providing pathways for copper migration and moisture adsorption. Figure 6.104 illustrates measling in the PCB in the vicinity of three soldered joints. When the epoxy in multilayer laminates contracts due to thermal stress, laminate voids can form adjacent to inner-layer copper foils or PTH barrels. The occasional fracture of overhangs during etching can result in slivers or particles of metal plating which in turn can cause short circuits or dielectric breakdown.

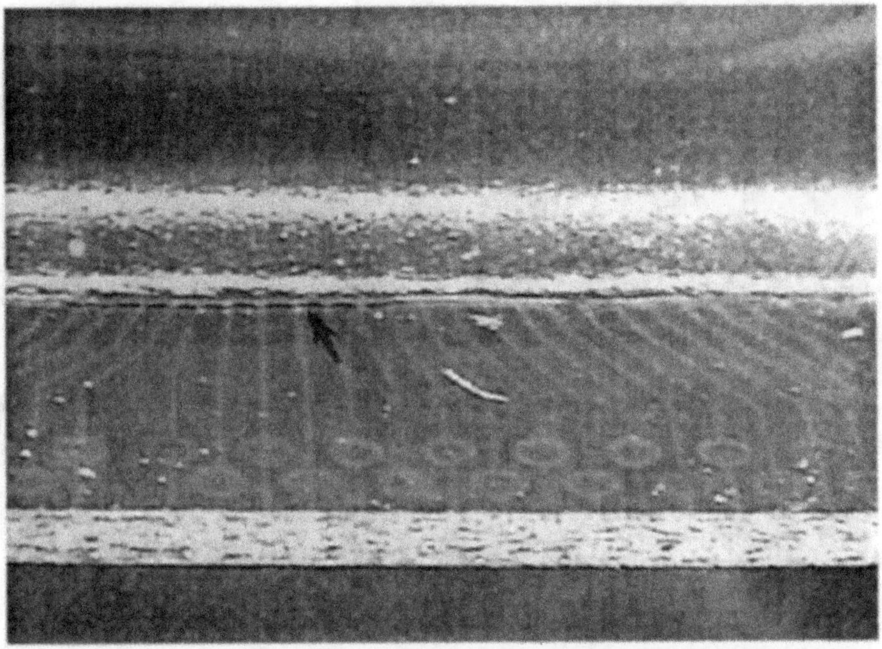

Figure 6.94 Separation (arrow) at the edge of encapsulation and PCB substrate. (*From Parekh et al.*[35])

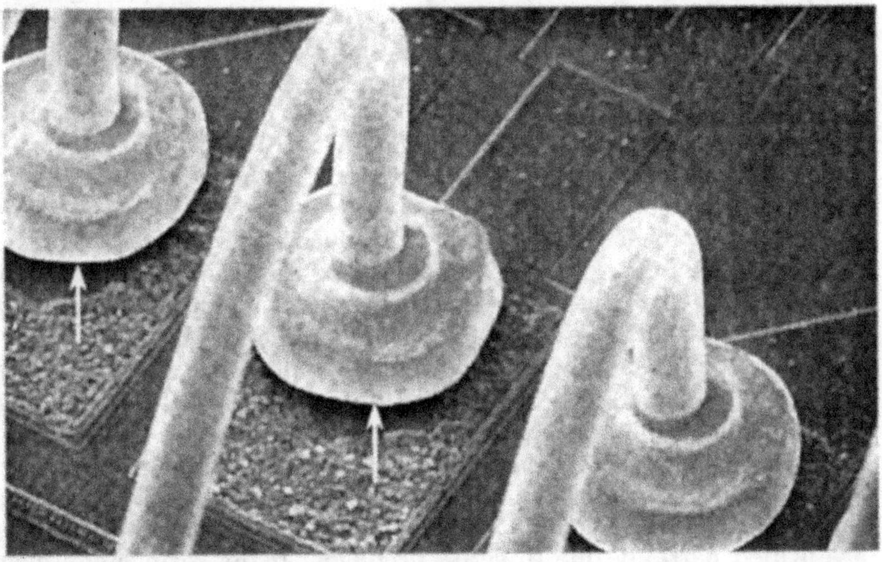

Figure 6.95 SEM image of lifted ball bonds (arrows) from die pads caused by delamination of encapsulation. (*From Parekh et al.*[35])

Figure 6.96 SEM image of broken bond wire caused by delamination of encapsulation. (*From Parekh et al.*[35])

During the plating and etching process, a minute amount of remnant resist can prevent the plated copper from being etched, resulting in a short, while the complementary problem of the copper getting etched at undesired locations can lead to an open circuit. This is especially problematic since the thin remnant copper can occasionally cause the circuit to test "good." The presence of copper nodules, residual copper and blooming between circuit lines can reduce the insulation between lines. Thermal stresses due to thermal cycling or thermal shock can cause the copper traces to crack, as shown in Fig. 6.105. Adhesion problems during plating can be caused by a variety of cleanliness and improper handling problems, such as fingerprints, grease, dust, scratches, etc. Figure 6.106 shows an example of a fingerprint left on a PCB due to careless handling. In addition, improper plating processes can lead to minute gaps in the plating known as *pinholes*.

Plated-through-holes provide electrical contact between different layers in multilayer PWBs. The manufacturing steps for forming PTHs include drilling, deburring, desmearing, electroless copper deposition (20 to 100 μm), and barrel plating with copper followed by tin or tin-lead plating (to assist in subsequent soldering processes). During the drilling process, open circuits can occur when through-holes are drilled partially due to setup problems. If the resin melts over the inner layers during the drilling process, as shown in Fig. 6.107, electrical contact is lost and can lead to subsequent failure of the PWB.

TABLE 6.15 Nominal Compositions of Most Common Lead Frame Metal Alloys Used in Surface-Mount Leaded Packages

Alloy name	Min. Cu (including Ag), wt.%	Fe, wt.%	Sn, wt.%	P, wt.%	Other, wt.%
C 151	99.9				0.10 Zr
C 155	99.75			0.06	0.027–0.10 Ag, 0.80–0.13 Mg
C 194	97.0	2.35		0.03	0.12 Zn, 0.03 Pb max.
C 195	96.0	1.50	0.6	0.03	0.8 Co
Alloy 42		58			42 Ni

SOURCE: From Capillo.[36]

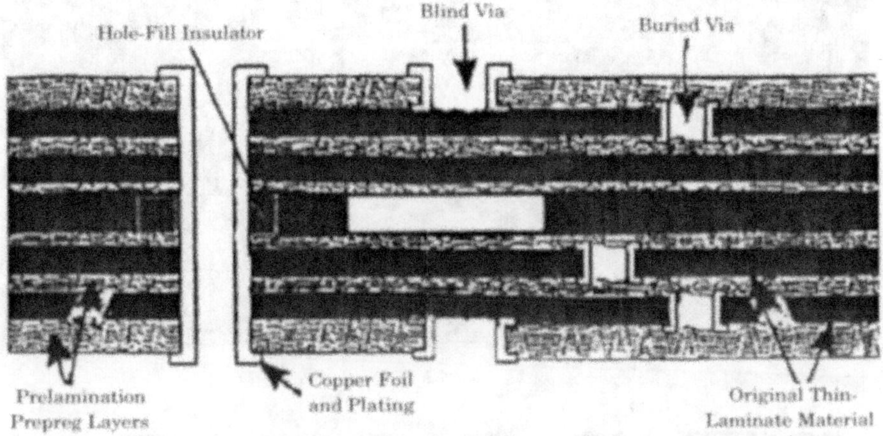

Figure 6.97 Multilayer metal-core board with buried and blind vias. (*From Hinton.*[37])

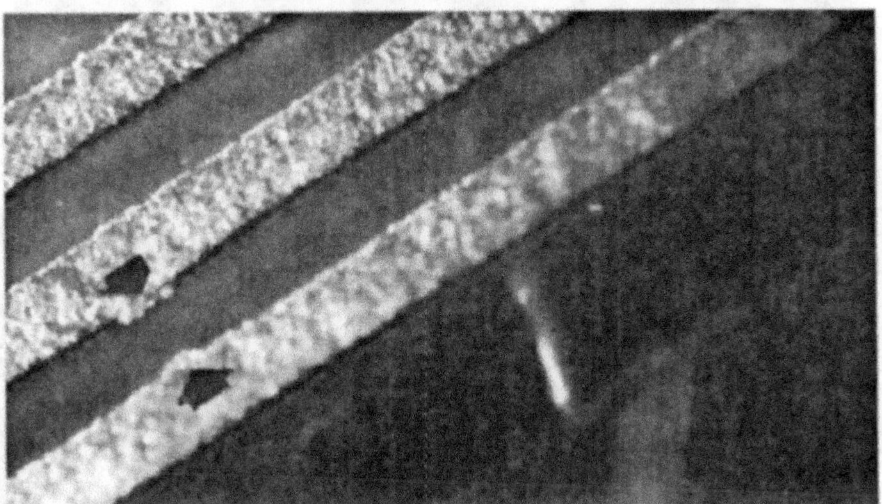

Figure 6.98 Superfluous copper on copper line (arrows). (*From Mansilla.*[38])

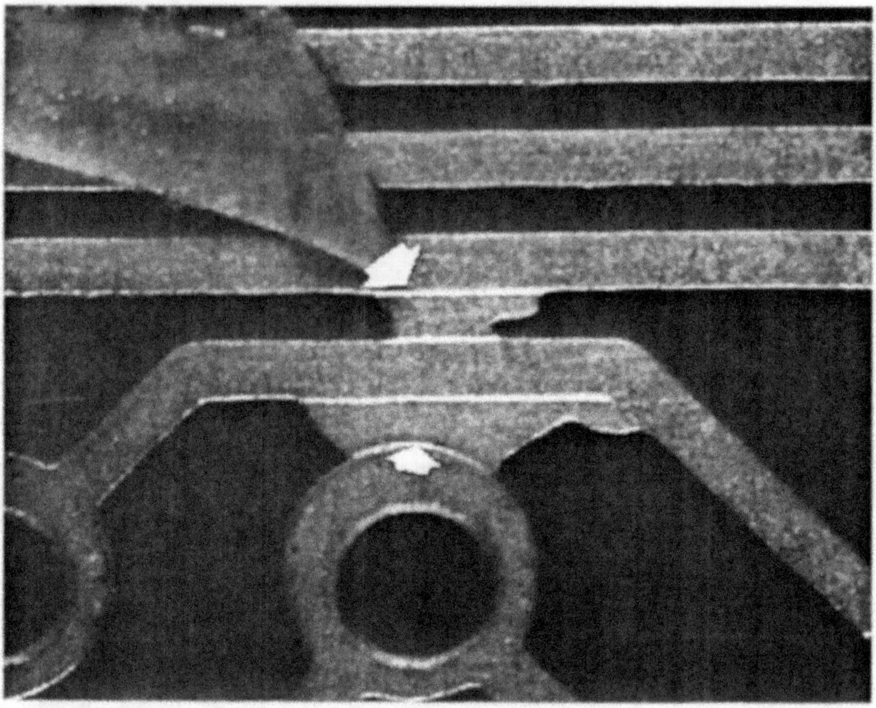

Figure 6.99 Bridging across copper line (arrows). (*From Mansilla.*[38])

Figure 6.100 Cut, nicked, and cracked conductors. (*From Mansilla.*[38])

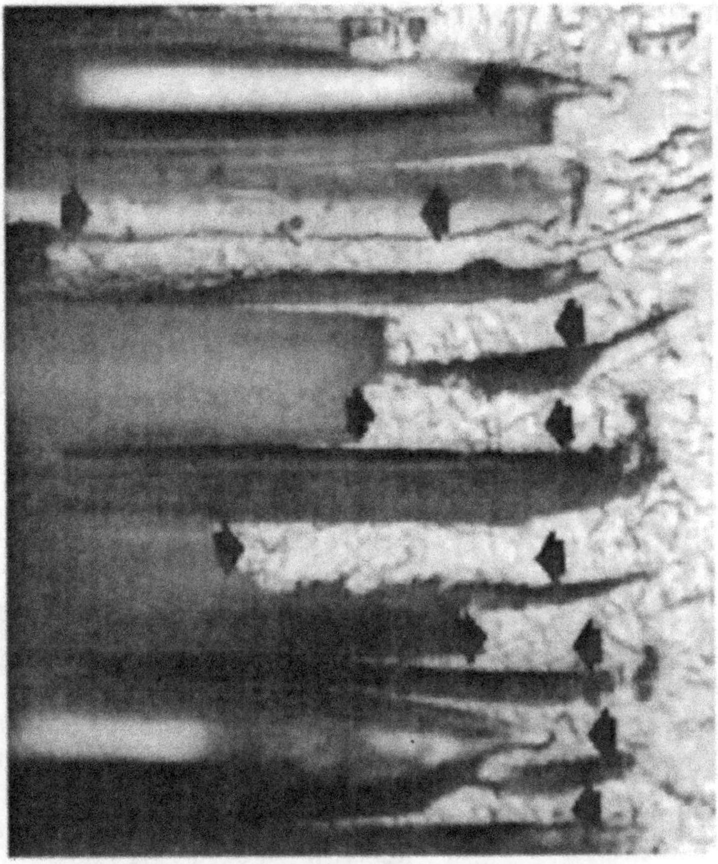

Figure 6.101 Wicking (arrows) of electrodeposited copper along glass fibers. (*From Mansilla.*[38])

Other defects in the PCBs that can be traced back to the drilling operation include electrical shorts due to misregistration between the centerlines of terminal pads and holes, as shown in Fig. 6.108. Poor or improper drilling can also lead to folds and nodules in the plating of through holes (Fig. 6.109). When the inner layer metallization in the board gets flattened due to either a dull drill bit or excessive drill speed, the metal can get cold-worked, leading to future cracking as shown in Fig. 6.110. This effect is known as *nail heading*. Following deburring of the top and bottom surfaces of the PWBs, an etching (chemical or plasma) process is employed to remove any remnant resin debris and smear. Employing the etching process for slightly longer than strictly necessary results in an *etch-back* of the resin as shown schematically in Fig. 6.111. Etch-back generally improves the adhesion of the subsequent metal plating in addition to providing

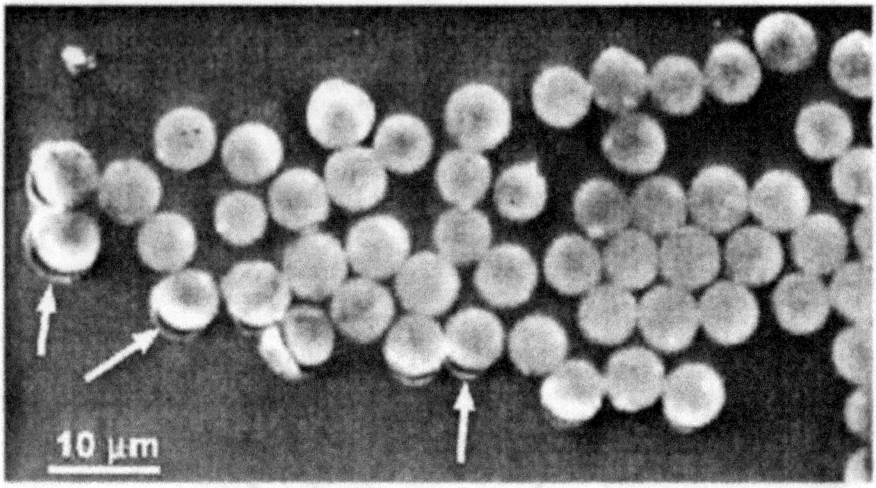

Figure 6.102 Delamination between the resin and the fibers (arrows). (*From Stadterman and Osterman*[39]*; courtesy of CALCE-EPSC.*)

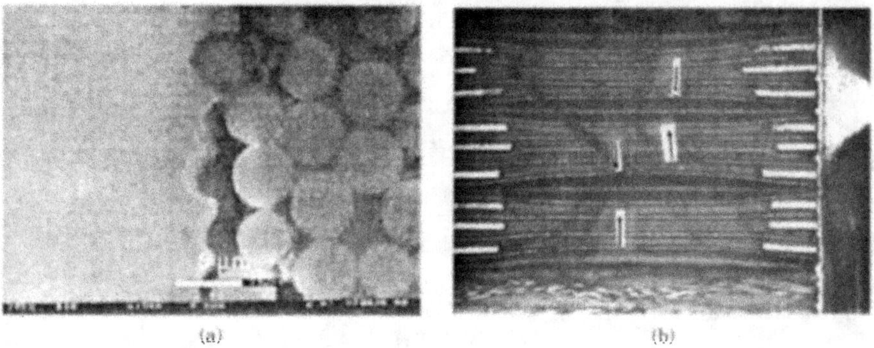

(a) (b)

Figure 6.103 Delamination between the resin and the copper. [(*a*) *From Stadterman and Osterman*[39]*; courtesy of CALCE-EPSC.* (*b*) *From McKeeby and Phillips.*[40]]

more surface area for subsequent metallization. The copper plating in the holes is prone to defects such as uneven thickness (Fig. 6.112) and cracking either during the plating operation or during subsequent soldering operations. An example of a plating crack is shown in Fig. 6.113. If the fiber bundles are delaminated, the plating can *wick* into the PCB along the fibers, as shown schematically in Fig. 6.114. The copper plated in the through-holes can be subjected to tensile stresses during subsequent processes. Columnar grains in this copper plating that are oriented normal to the axis of the holes can thus lead to weakness and rupture (Viswanadham and Singh[9]).

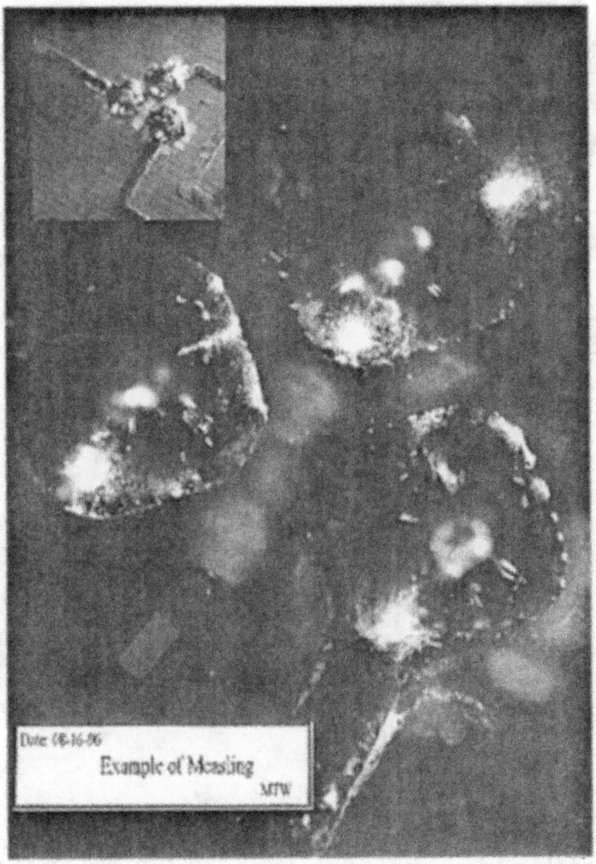

Figure 6.104 Printed wiring board measling surrounding three soldered joints. (*From Vettraino.*[41])

In both PIH and SMT technologies, soldered connections between the PCB and IC packages are subject to failures due to thermal stresses arising out of differences in coefficient of thermal expansion (CTE). While the topic of solderability will be dealt with in a subsequent section, some of the defects associated with soldered joints are discussed here. Cold solder joints, or *frosty joints,* are inherently weak solder joints resulting from soldering done with less-than-adequate heat or on inadequately cleaned surfaces. Improper solder masking and excessive solder paste can cause unintended connections to be formed by a process known as solder bridging as shown in Fig. 6.115. If a component or package has uneven solder wetting forces at its two ends, the resulting unbalanced force can cause either skewing (lateral swing or shifting to one end) of the component or package or *tombstoning,* where the component is flipped

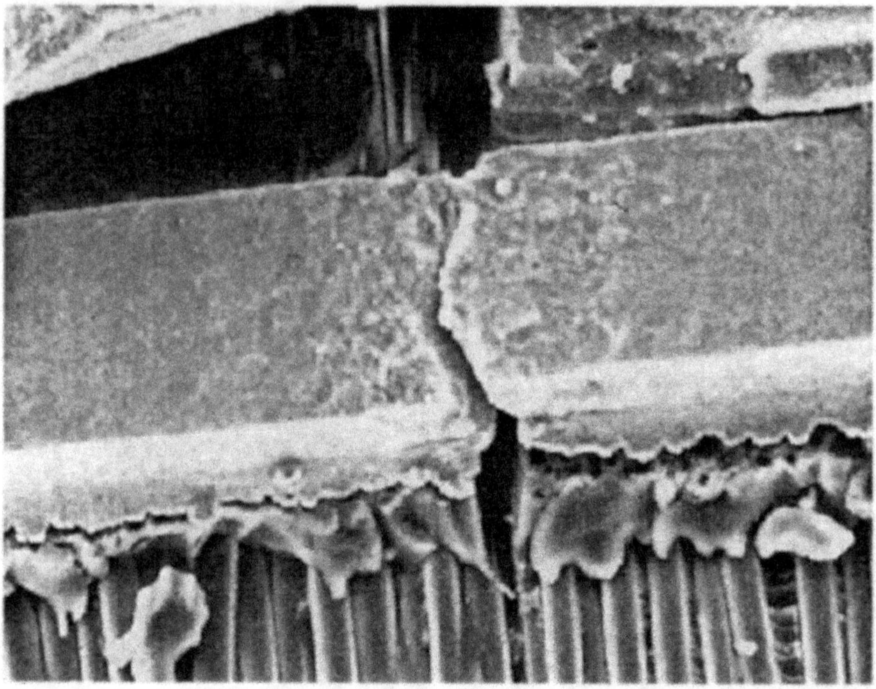

Figure 6.105 Typical copper race on a PCB substrate cracked open due to thermal stress. (*From Parekh et al.*[35])

onto one of its sides into a vertical position. The effect is also known as *drawbridging*, *Manhattan effect*, and *Stonehenge effect*. Figure 6.116 schematically illustrates the phenomena of tombstoning while Fig. 6.117 shows tombstoned chip resistors. In extreme cases, the component or package may actually be cracked by the unbalanced forces.

Electromigration refers to the transport of an ionic species from one conductor to another (separated by a dielectric) caused by an externally imposed electrical potential. Positively charged ions travel to the negative electrode where they are neutralized and deposited usually in the form of dendrites, which can eventually cause short circuits. The three conditions necessary for the phenomena to occur are sufficient moisture, ionic contamination and an applied potential. Figure 6.118 shows examples of dendrite formation. While gold is normally quite noble, it can corrode to a chloride or hydroxide and also can electromigrate.

6.5.5 Failures in third-level packaging

While many different types of hardware are involved in the third level of electronic packaging, connectors are ubiquitous in electronics pack-

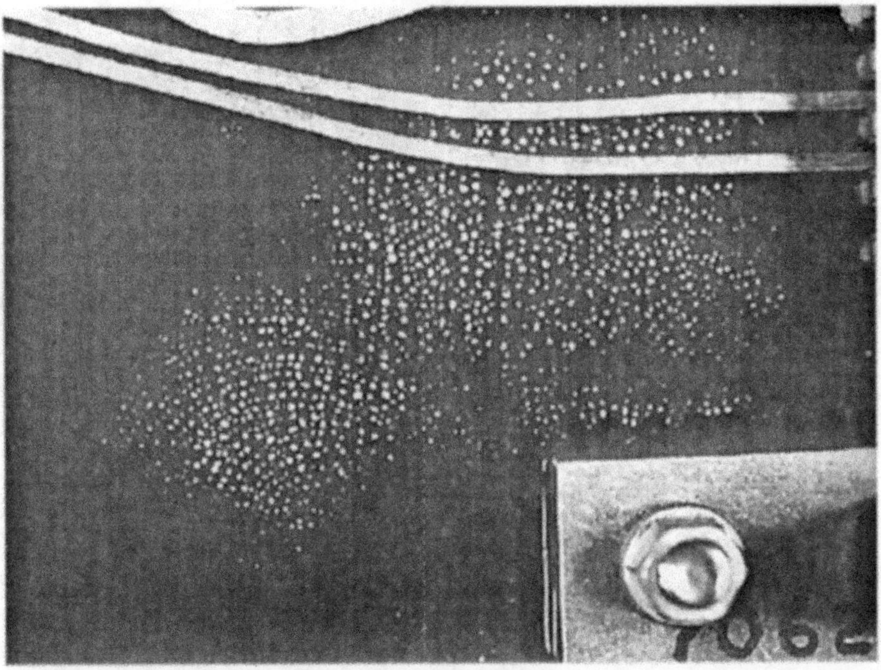

Figure 6.106 Contamination in the form of a fingerprint on a PCB. (*From Blanchard et al.*[42])

aging and thus will be dealt with here. An excellent discussion on the failure mechanisms in connectors can be found in Viswanadham and Singh[9] and this section draws heavily from there.

Connectors serve the dual purposes of providing mechanical and electrical connection between circuits. Some issues associated with the use of connectors and PCB assemblies are

1. Contamination of the contacts with fluxes leading to subsequent corrosion
2. Coating of contacts by insulation during coating of the rest of the PCB
3. Dust and debris in unfilled slots of the connector
4. Overstress of SMT connectors
5. Proper alignment and mating of contacts by application of polarization and guidance mechanism
6. Matching contact metallurgy and application

With respect to the last point, the actual contacts are usually made of base metals such as brass, phosphor bronzes, and beryllium copper. These connectors are then coated appropriately in order to improve

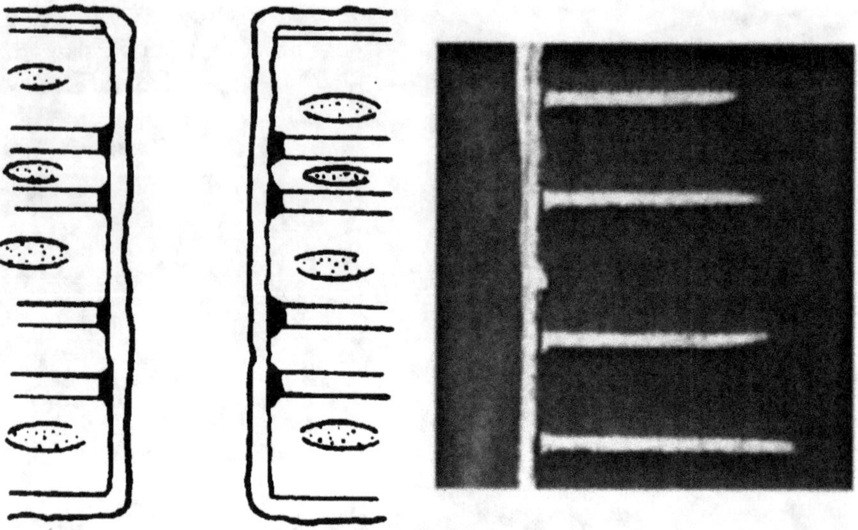

(a) (b)

Figure 6.107 (*a*) Resin smear in plated-through-hole. (*From Brzozowski and Brooks.*[43]) (*b*) Example of resin smear and nail heading. (*From Stadterman and Osterman*[39]; *courtesy of CALCE-EPSC.*)

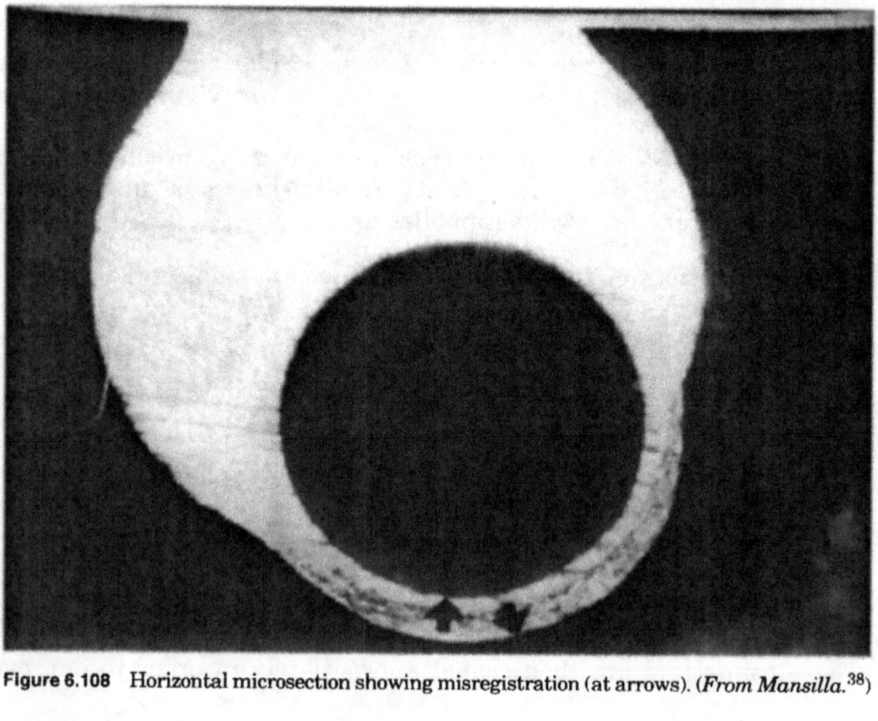

Figure 6.108 Horizontal microsection showing misregistration (at arrows). (*From Mansilla.*[38])

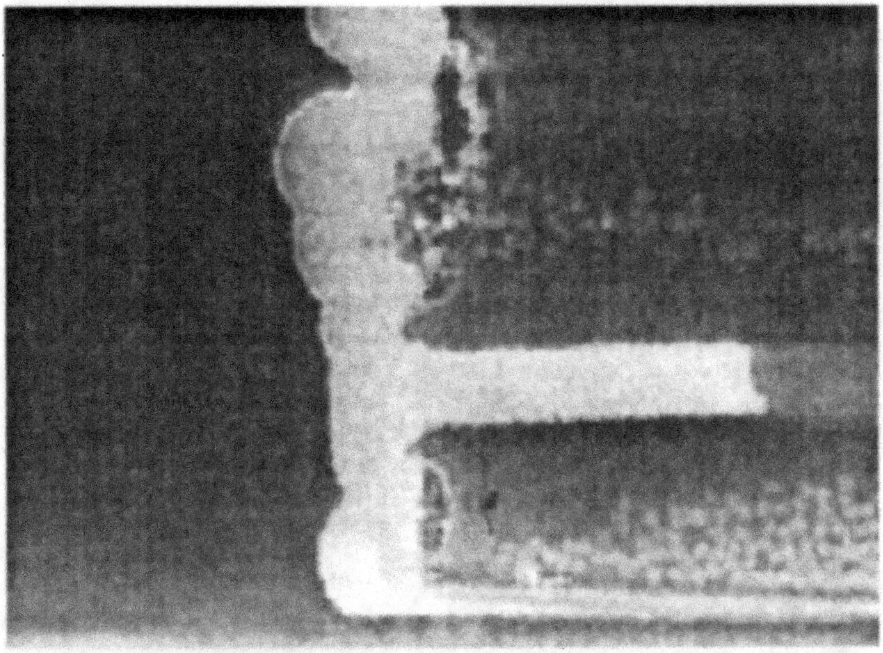

Figure 6.109 Example of plating folds. (*From Stadterman and Osterman*[39]*; courtesy of CALCE-EPSC.*)

various performance criteria such as connection force, current capacity, temperature rise, stress relaxation under temperature, and insertion life. For example, in the case of gold plating, a 1.25- to 2.50-μm nickel underplate is followed by a 0.25- to 2.50-μm gold plate. A lower-cost alternative is palladium coating. The lowest cost alternative is hot dipped or electroplated lead-tin alloy 0.75 to 1.25 mm in thickness. Fretting probability dictates that gold plated contacts and lead-tin contacts not be mated.

Since connectors transmit electrical signals or power between assemblies or subassemblies, durability and low and stable contact resistance are desirable characteristics. The critical part of any connector is the contact. Mechanical failures of connectors can be caused by bent pins and contacts, card warpage, improper shipping, improper tolerances or clearances, debris, and improper fabrication of contacts. The three leading causes of connector failure are excessive current, insulation breakdown and excessive power dissipation. Such failures can result in open circuits in failed components. The alternative result of resistive shorts is more consequential in that it can lead to overheating, melting or electrocution. Either too loose or too tight connections in connectors can result in deformed conductor strands or stripped threads resulting

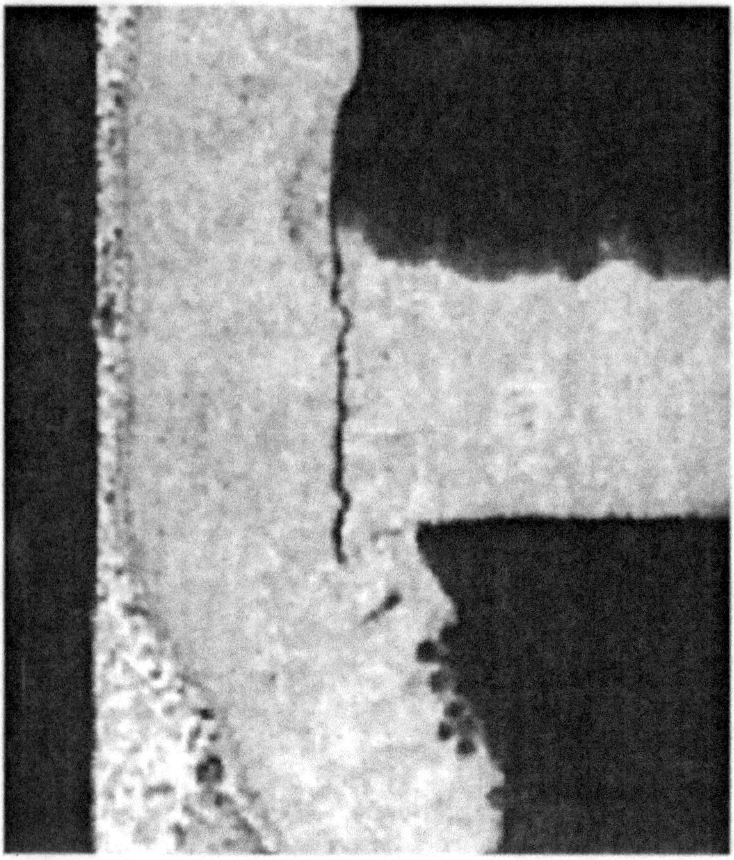

Figure 6.110 Example of nail heading. (*From Stadterman and Osterman*[39]*; courtesy of CALCE-EPSC.*)

in reduced current carrying capacity. Connectors are also subject to normal mechanical failure. An example of tensile overstress failure of 1-mil-diameter aluminum wire used to connect a chip to its package is shown in Fig. 6.119. Some of the other common problems in connectors are discussed below.

Electrical shorting can result from the formation of filamentary single crystals of tin from electroplated tin as shown in Fig. 6.120 and referred to as *whiskering*. One method of alleviating the problem is use of tin-lead plating (93 percent tin–7 percent lead).

Tin plated contacts are subject to fretting corrosion due to relative motion between contact surfaces exposing fresh metal, which gets oxidized as shown in Fig. 6.121. Such relative motion can be caused by vibration, shock, differential thermal expansion and other forms of

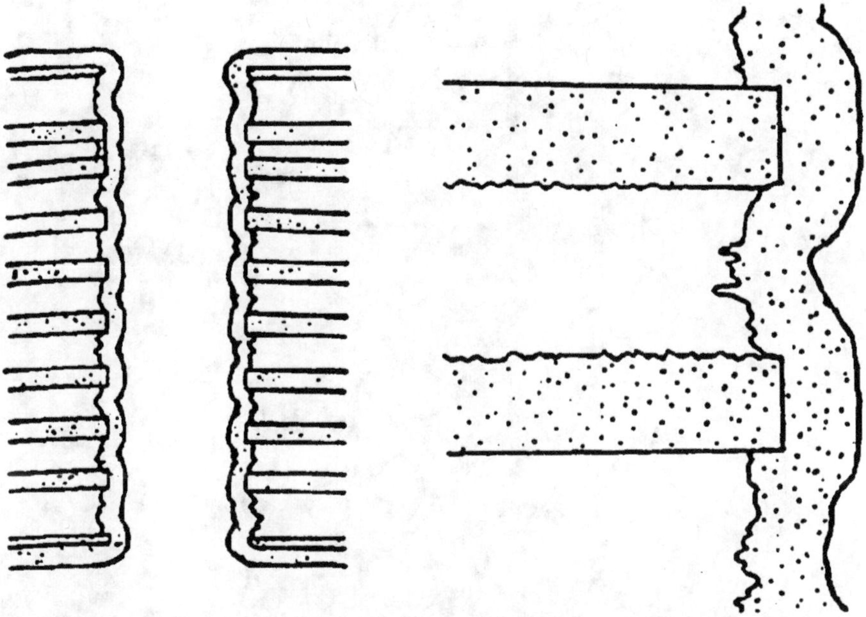

Figure 6.111 Etch back in drilled holes. (*From Brzozowski and Brooks.*[43])

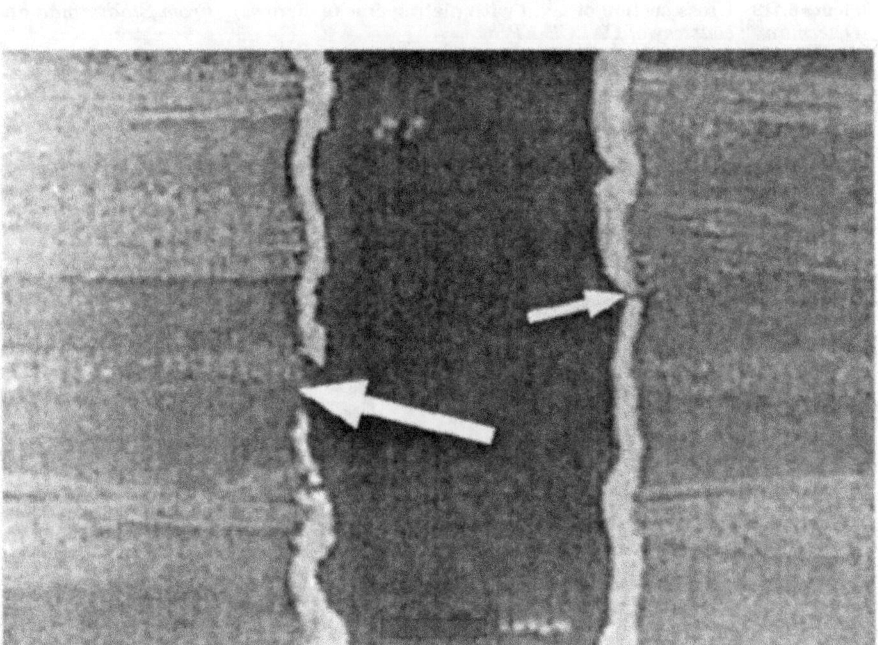

Figure 6.112 Example of plating thickness variation (arrows). (*From Stadterman and Osterman*[39]*; courtesy of CALCE-EPSC.*)

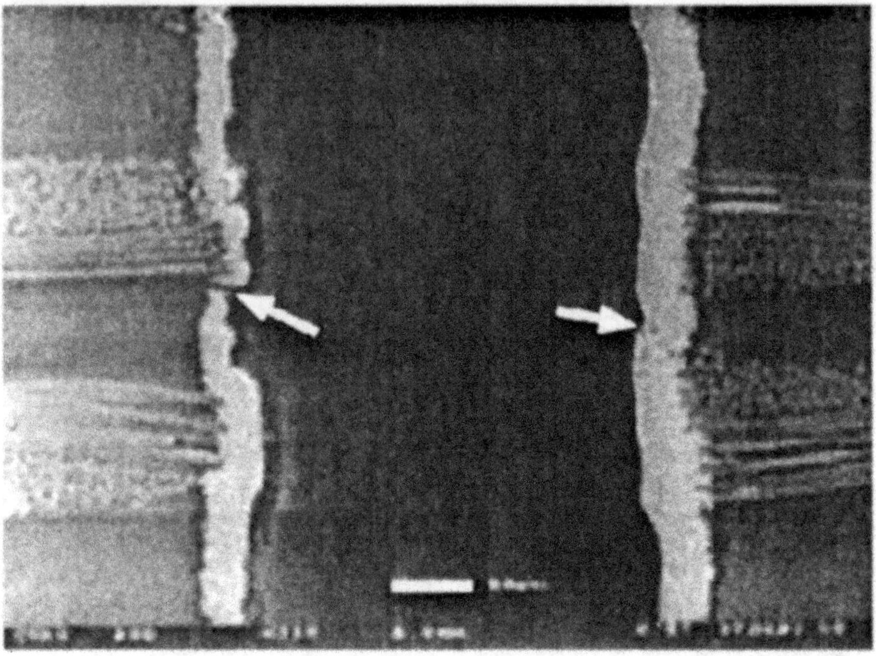

Figure 6.113 Cross section of a via with plating cracks (arrows). (*From Stadterman and Osterman*[39]*; courtesy of CALCE-EPSC.*)

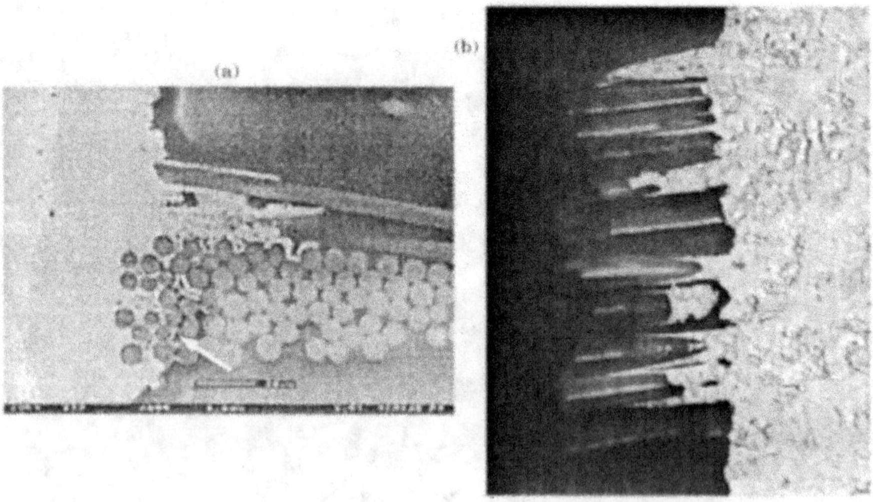

Figure 6.114 (*a*) Example of copper wicking into the fiber bundle. (*From Stadterman and Osterman*[39]*; courtesy of CALCE-EPSC*). (*b*) Example of electroless copper wicking along glass fibers. (*From Goldman.*[44])

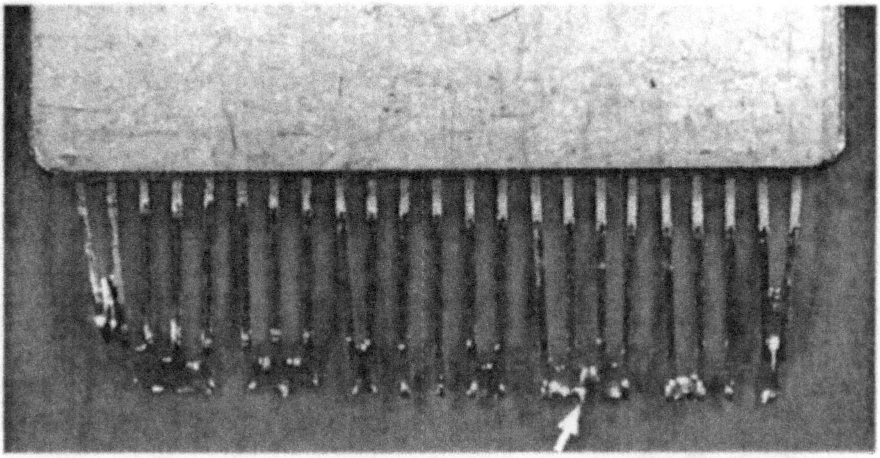

Figure 6.115 Solder bridging (arrow) on SMD leads. (*From Vettraino.*[41])

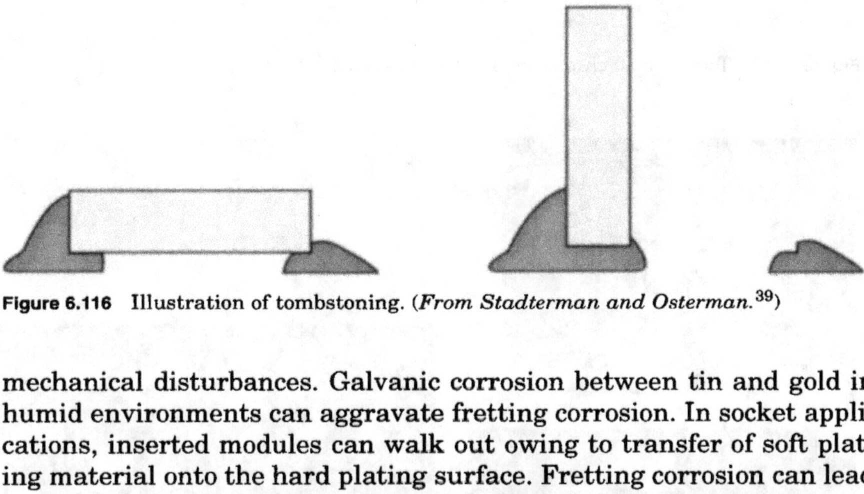

Figure 6.116 Illustration of tombstoning. (*From Stadterman and Osterman.*[39])

mechanical disturbances. Galvanic corrosion between tin and gold in humid environments can aggravate fretting corrosion. In socket applications, inserted modules can walk out owing to transfer of soft plating material onto the hard plating surface. Fretting corrosion can lead to high contact resistance and can be alleviated by the application of lubricants to the contact surface or use of high normal contact force. In the presence of organic vapors, fretting corrosion on palladium-plated contacts can lead to the formation of insulating polymer films by a process known as the frictional polymerization of palladium.

Defects in the plating material of contacts can leave small holes (pores) thereby exposing the base metal to the possibility of corrosion. Such pores can be rendered inactive by the application of a nickel underplate, as shown schematically in Fig. 6.122.

The extended use of connectors under overload conditions will eventually deteriorate their performance. Transient overload conditions such

Figure 6.117 Tombstoned chip resistors. (*From Hinch.*[45])

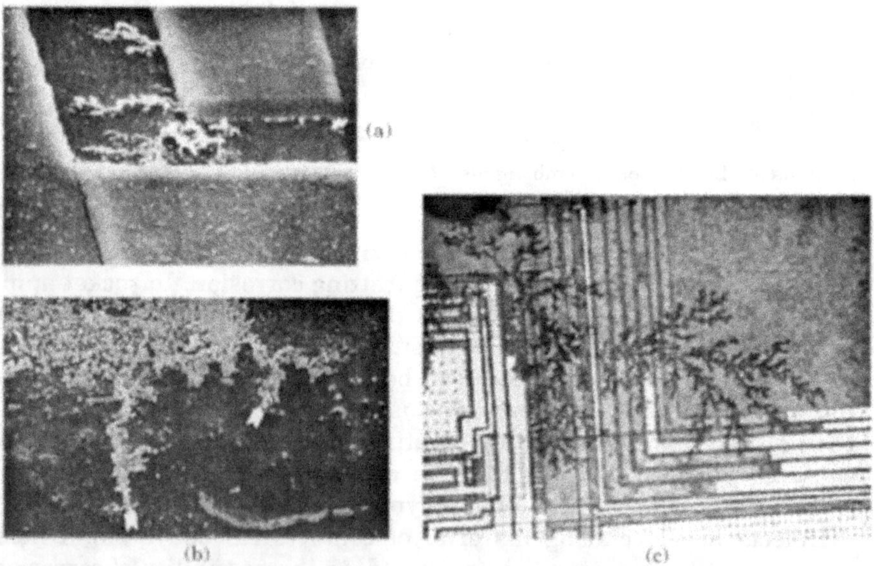

Figure 6.118 Examples of dendrite formation due to electromigration. [(*a*) *From Dicken*[4]; (*b*) *from Silvus*[46]; (*c*) *from Haythornthwaite.*[10]]

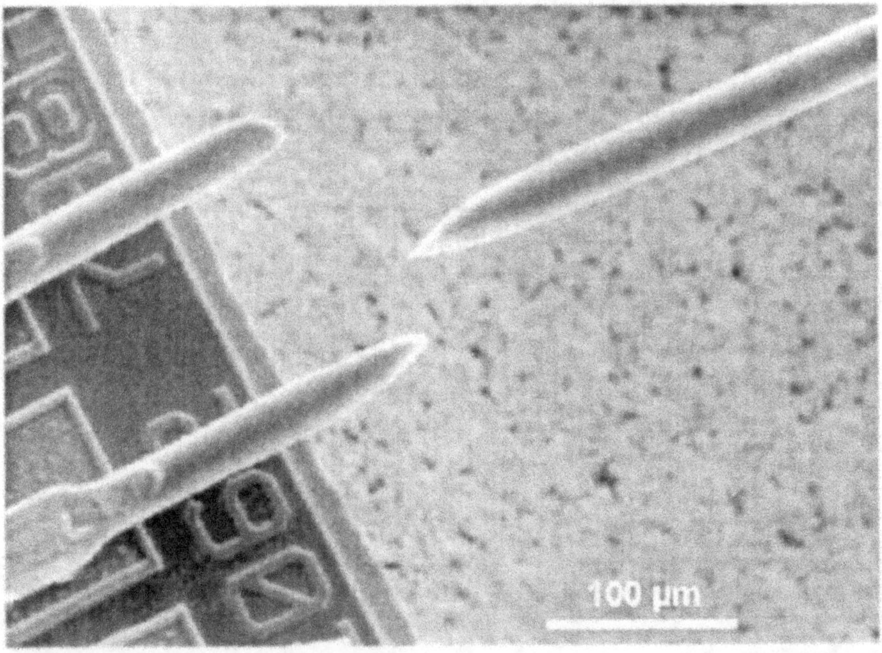

Figure 6.119 One-mil-diameter aluminum wire fractured due to tensile overstress. (*From Haythornthwaite.*[10])

Figure 6.120 SEM micrograph of tin whiskers growing from base of insulated stand-off terminal. (*From Silvus.*[46])

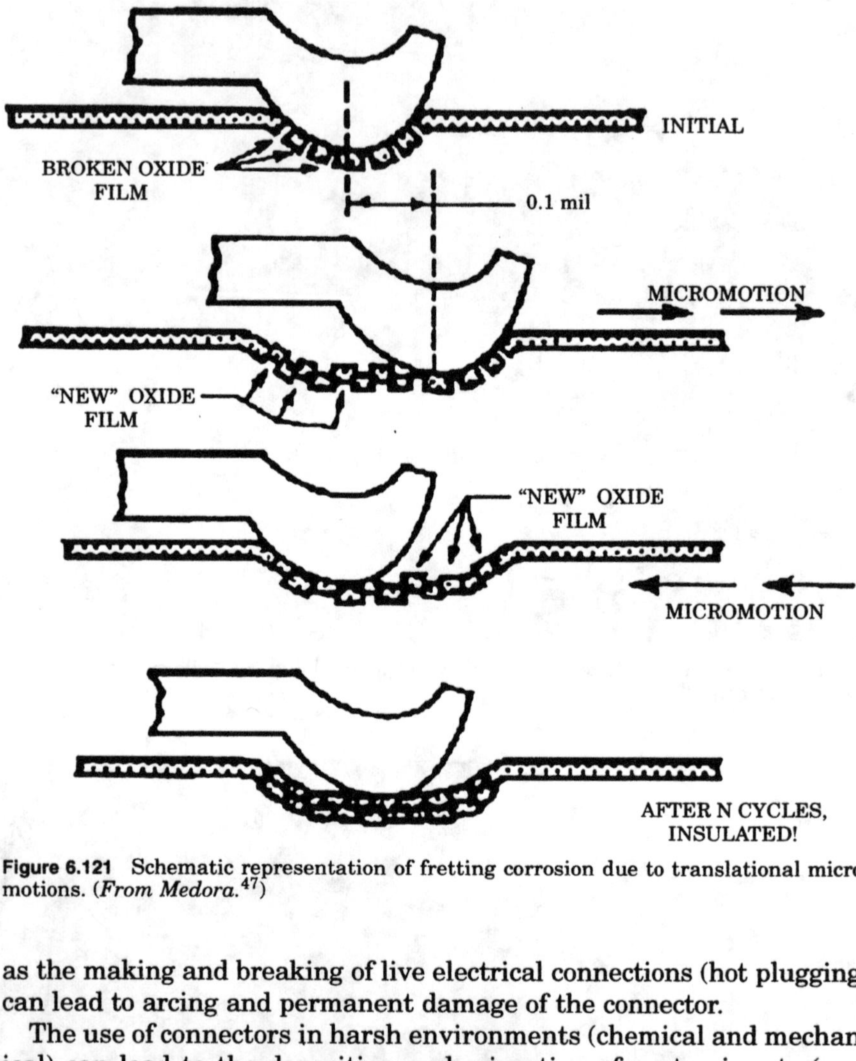

Figure 6.121 Schematic representation of fretting corrosion due to translational micromotions. (*From Medora.*[47])

as the making and breaking of live electrical connections (hot plugging) can lead to arcing and permanent damage of the connector.

The use of connectors in harsh environments (chemical and mechanical) can lead to the deposition and migration of contaminants (e.g., salt spray, carbon, moisture, dirt, and oil), which can cause intermittent or permanent damage.

Another ubiquitous item at this level of packaging is cables formed by application of an electrical insulation on metal wires. The failure of metal wires (generally copper) can be studied using the following failure characteristics (as discussed by Slenski and Galler[48]):

1. Exposure at elevated temperature is evidenced by a microstructure consisting of large, equiaxed grains. An example of this shown for OFHC copper in Fig. 6.123.

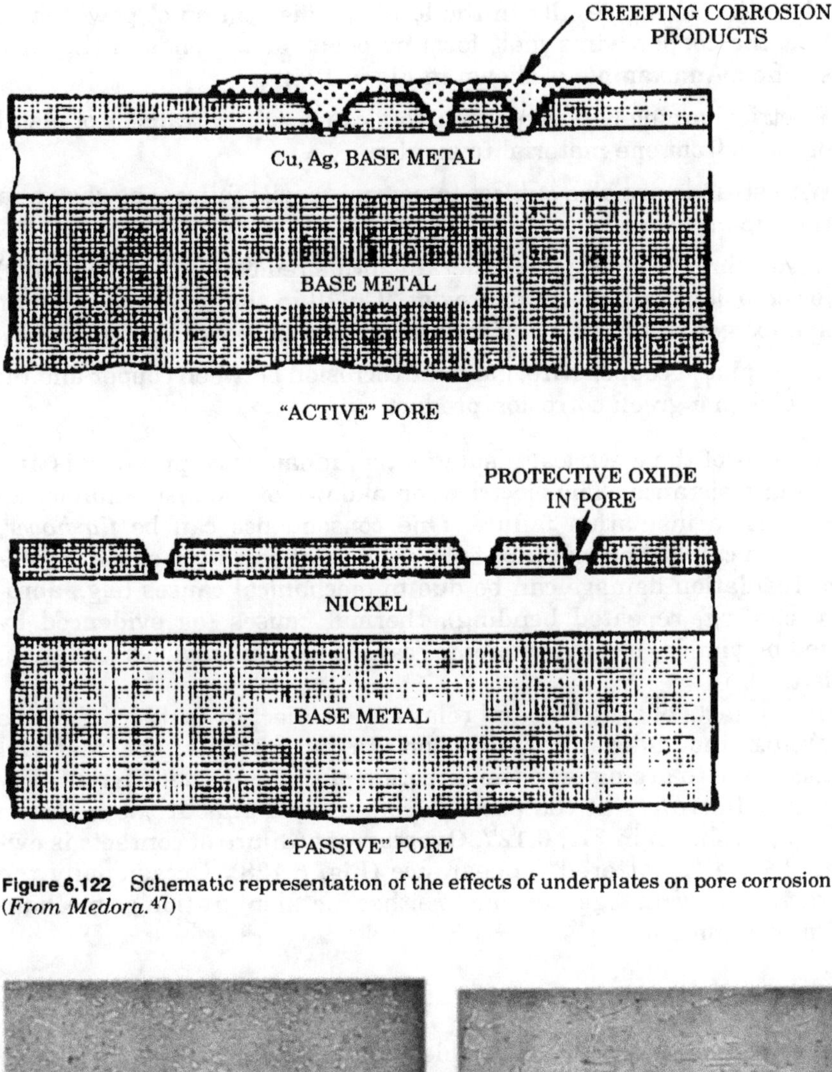

Figure 6.122 Schematic representation of the effects of underplates on pore corrosion. (*From Medora.*[47])

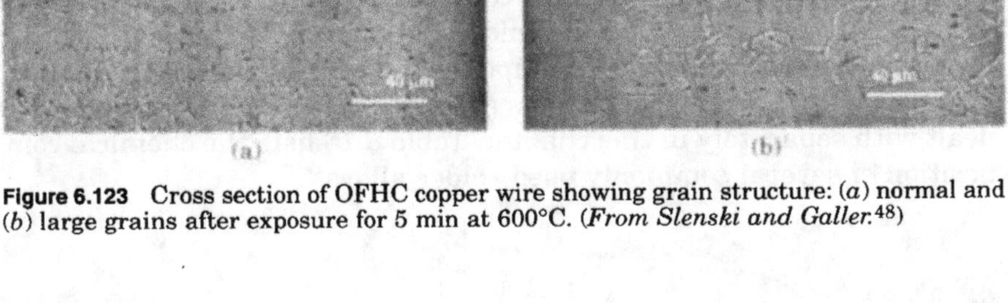

Figure 6.123 Cross section of OFHC copper wire showing grain structure: (*a*) normal and (*b*) large grains after exposure for 5 min at 600°C. (*From Slenski and Galler.*[48])

2. Electrical arcing results in the localized dissipation of power such that the copper wires melt forming beads at the ends of the wire strands. An example is shown in Fig. 6.124.
3. Electrical arcing between conductors can result in transfer of metal droplets from one material to another.
4. Wire strands are also subject to normal tensile failure, as shown in the cup-and-cone fracture in Fig. 6.125.
5. Silver-plated copper wire generally forms red cuprous oxide (Cu_2O) [or occasionally black cupric oxide (CuO)] in the presence of water and oxygen. An example of this *red plague* is shown in Fig. 6.126.
6. In tin-plated copper wire, galvanic corrosion between copper and tin results in a green corrosion product.

In terms of the electrical insulation, a fundamental property is arc-tracking resistance. The electrical breakdown of the insulation material leads to insulation failure. One consequence can be *flashover*, which is a catastrophic failure followed by carbonization of the insulation. Insulation damage can be due to mechanical causes (e.g., abrasion, chafing, repeated bending), thermal causes (as evidenced by glazed polyimide) or environmental causes (e.g., ultraviolet radiation, moisture).

The contacts of switches and relays are subject to contact wear due to arcing and mechanical abrasion. In DC circuits, the material transfer due to arcing is in the direction of conventional current flow resulting in *craters* at the positive contact and *pips* at the negative contact, as shown in Fig. 6.127. Overcurrent failure of contacts is evidenced by a "splattered" appearance (Fig. 6.128). Occasionally the contacts may weld together due to either metal migration or mechanical misalignment.

6.5.6 Solders and solderability

The importance of solders and soldering processes in the manufacture and use of electronic packages cannot be overemphasized. A solder alloy consists of two or more elements that can wet to a surface, most often copper, and then react to form an adhesion layer, and upon solidification of the alloy, produces an interconnection that has good mechanical properties. This connection technique provides both mechanical and electrical connection between components, PWBs and other hardware. The application of soldering spans all levels of packaging and hence is dealt with separately in this chapter. Table 6.16 lists the chemical composition of several commonly used solder alloys.

Figure 6.124 Wire sample exhibiting arcing damage from a 28-Vdc power system. Note balled end and degraded insulation. (*From Slenski and Galler.*[48])

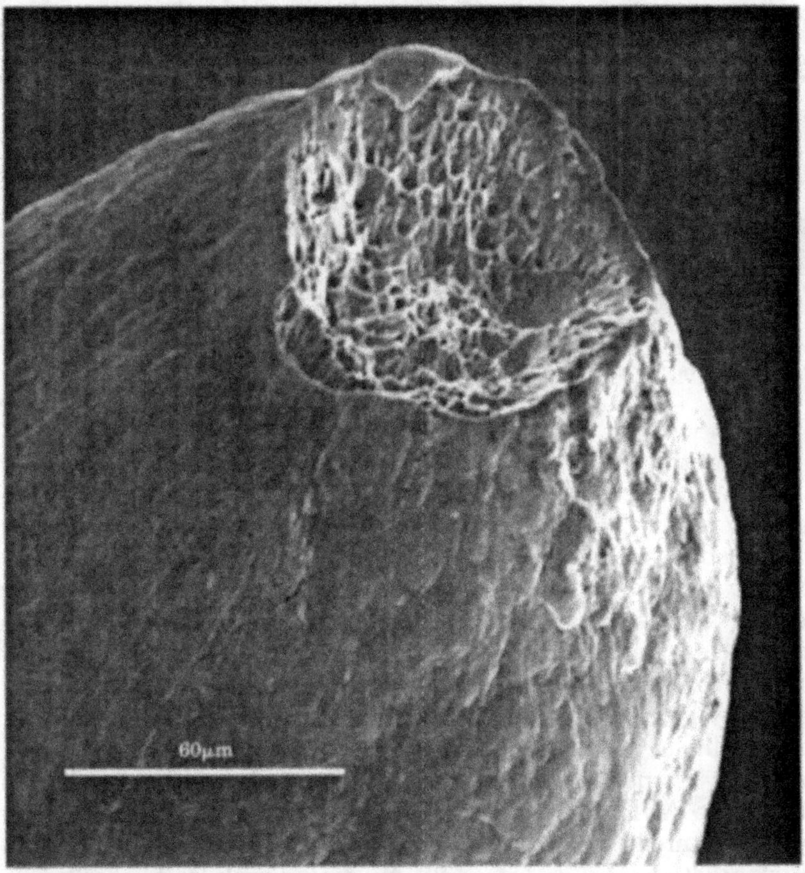

Figure 6.125 SEM micrograph showing the cup-and-cone fracture surface and elongation typically associated with mechanical failure of a copper conductor strand. (*From Slenski and Galler.*[48])

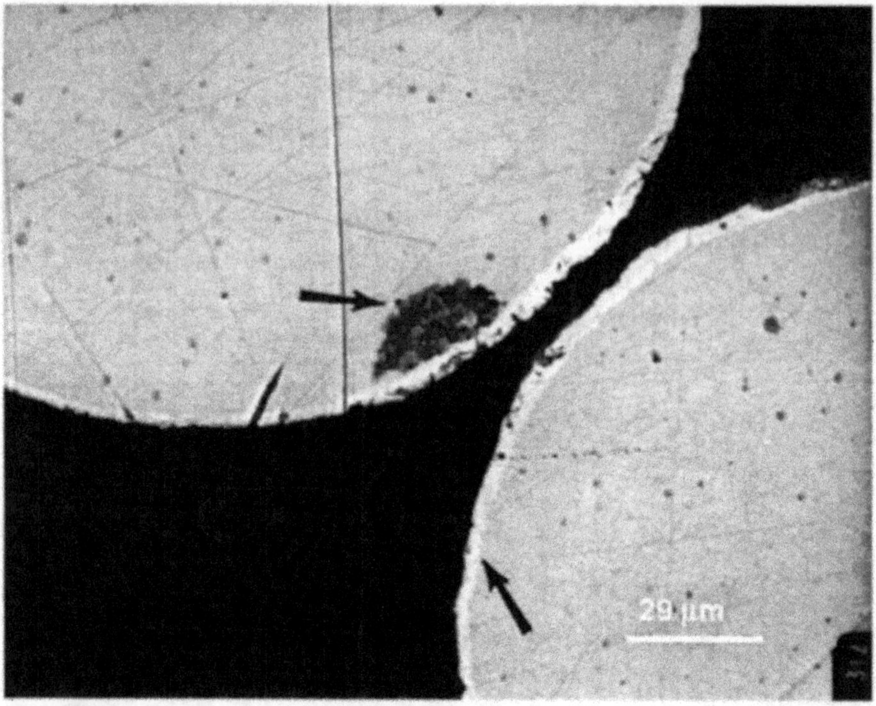

Figure 6.126 Cross section of two wire strands showing corrosion damage attributed to red plague. The copper strands are plated with silver (lower arrow). The dark area (arrow near center) shows corrosion. (*From Slenski and Galler.*[48])

In terms of the solder material, an important property of the material is the ability to properly *wet* the surfaces to be soldered. In a systems sense, solderability refers to the ability to achieve a clean metallic surface on substrates to be joined during a dynamic heating process so that a good wetting of molten solder on the surface of the substrates can be formed. In the case of solder pastes, an additional requirement is the ability to achieve a clean metallic surface on the solder powder so that a complete coalescence of the solder powder particles can be obtained. Solderability is thus dependent on the fluxing efficiency of fluxes used and the quality of the substrate surface (as discussed by Hwang[50]). From a wettability standpoint, substrates can be ranked in the order Sn, Sn/Pb > Cu > Ag/Pd, Ag/Pt > Ni. The component leads to be soldered are generally made of copper, copper alloys, alloy 42 (Fe + ~ 42% Ni), and Kovar (54% Fe + 17% Co + 29% Ni) coated with lead-tin alloys.

In terms of the reliability of soldered joints, one key concern is the formation of intermetallic compounds. The interaction between plated copper and the tin in solder can produce Cu_3Sn and Cu_6Sn_5 inter-

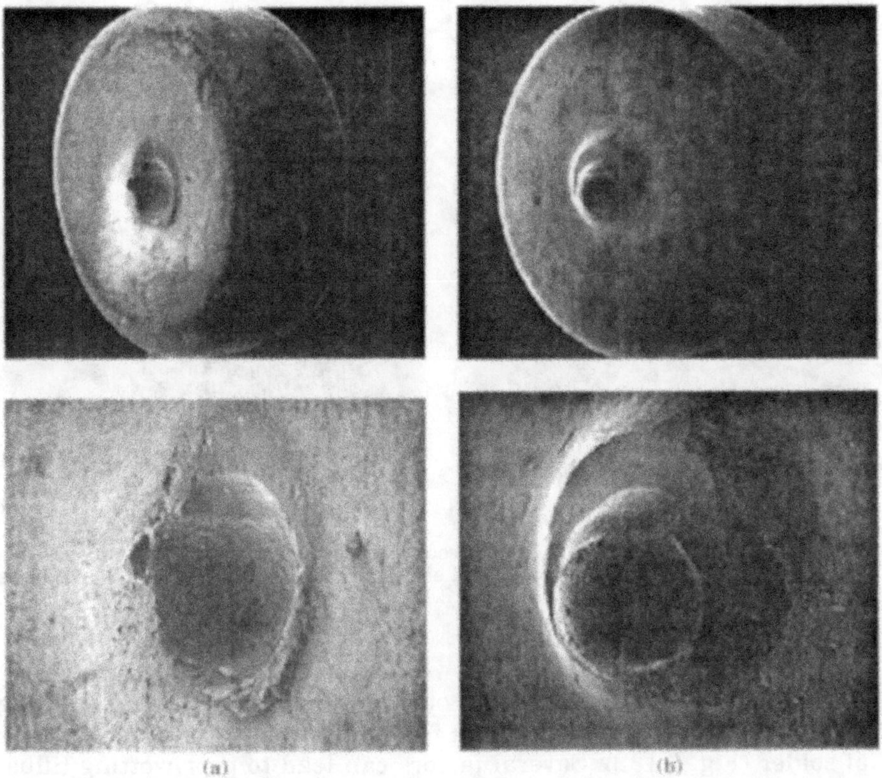

Figure 6.127 (*a*) Positive contact from 10-A circuit breaker cycled 10,000 times with a dc resistive load at rated current. The crater at the center of the contact is shown at higher magnification on the lower photograph. (*b*) Negative contact from a 10-A circuit breaker cycled 10,000 times with a dc resistive load at rated current. The pip at the center of the contact is shown at higher magnification on the lower photograph. (*From Martin et al.*[49])

metallics, as shown in Fig. 6.129. Also, tin and nickel can form Ni_3Sn_4, while gold and tin can form AuSn, $AuSn_2$, and $AuSn_4$. The amount, size, distribution and morphology of these intermetallics can affect the solderability and strength of the soldered joint. While brittle intermetallic compounds contribute to increased joint strength, they lower fatigue strength. Also important is the fact that they form not only during the soldering process but also grow during subsequent processing steps and under certain operating temperatures. The generally accepted thickness of the intermetallic layer is in the range of 1 to 5 mm. Figure 6.130 shows cracks at the interface between the solder and the leads in two different types of joints caused by excessive intermetallic formation.

Nonwetting or poor wetting refers to the effect where the solder coat is discontinuous with rounded areas of base metal showing

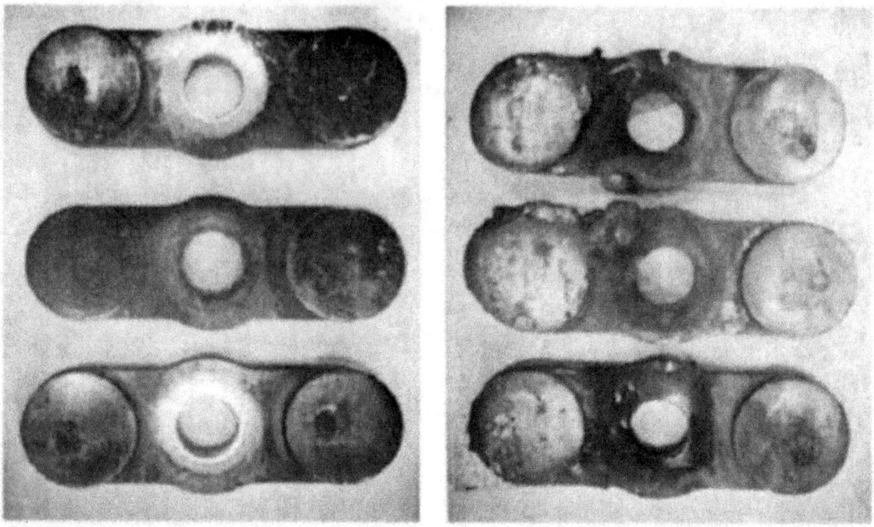

Figure 6.128 Left photo shows a layer of soot on a set of power contactor contacts. This indicates that the contacts were open and away from the arcing. The right photo shows an example of the melting that can occur when the contacts are involved in the arcing. (*From Martin et al.*[49])

through (Fig. 6.131). Dewetting on the other hand refers to pulling back of the solder during cooling forming beads or irregular mounds of solder (Fig. 6.132). Several factors can lead to poor wetting (illustrated in Fig. 6.133):

- Tarnishes, e.g., oxides or sulfides
- Foreign contaminants on the surfaces
- Embedded particles
- Silicone oils
- Solder alloy problems such as low temperature

In some gold-plated leads, the iron and nickel of the base material (alloy 42) can diffuse to the surface through the gold, forming nonwettable oxides and/or hydroxides. An example of such an area on a lead is shown in Fig. 6.134.

In the case of PTH assemblies, poor through-hole filling and decreased solderability of the copper in the hole will adversely affect the quality of PCBs. In addition, rough holes and thin copper plating, laminate voids, copper discontinuity along the periphery of the hole, exposed glass fibers, and drill smear causing folds and fissures in the copper plating can lead to poor solderability and therefore poor PCB

TABLE 6.16 Solder Alloy Compositions (in wt.%)

Composition	Sn	Pb	Sb	Bi, max %	Ag	Cu, max %	Fe, max %	Zn, max %	Al, max %	As, max %	Cd, max %	Others, total max %	Approx. melting range, °C	
													Solidus	Liquidus
Sn 96	Rem	0.10 max	—	—	3.6–4.4	0.2	—	0.005	—	0.05	0.005	—	221	221
Sn 70	69.5–71.5	Rem	0.2–0.5	0.25	—	0.08	0.02	0.005	0.005	0.03	—	0.08	183	193
Sn 63	62.5–63.5	Rem	0.2–0.5	0.25	—	0.08	0.02	0.005	0.005	0.03	—	0.08	183	183
Sn 62	61.5–62.5	Rem	0.2–0.5	0.25	1.75–2.25	0.08	0.02	0.005	0.005	0.03	—	0.08	179	179
Sn 60	59.5–61.5	Rem	0.2–0.5	0.25	—	0.08	0.02	0.005	0.005	0.03	—	0.08	183	191
Sn 50	49.5–51.5	Rem	0.2–0.5	0.25	—	0.08	0.02	0.005	0.005	0.25	—	0.08	183	216
Sn 40	39.5–41.5	Rem	0.2–0.5	0.25	—	0.08	0.02	0.005	0.005	0.02	—	0.08	183	238
Sn 35	34.5–36.5	Rem	1.6–2.0	0.25	—	0.08	0.02	0.005	0.005	0.02	—	0.08	185	243
Sn 30	29.5–31.5	Rem	1.4–1.8	0.25	—	0.08	0.02	0.005	0.005	0.02	—	0.08	185	250
Sn 20	19.5–21.5	Rem	0.80–1.20	0.25	—	0.08	0.02	0.005	0.005	0.02	—	0.08	184	270
Sn 10	9.0–11.0	Rem	0.20 max	0.03	1.7–2.4	0.08	—	0.005	0.005	0.02	—	0.03	268	290
Sn 5	4.5–5.5	Rem	0.50 max	0.25	—	0.08	0.02	0.005	0.005	0.02	—	0.08	308	312
Sb 5	94.0 min	0.20 max	4.0–6.0	—	—	0.08	0.08	0.03	0.03	0.05	0.03	0.03	235	240
Pb 80	Rem	78.5–80.5	0.20–0.50	0.25	—	0.08	0.02	0.005	0.005	0.02	—	0.08	183	277
Pb 70	Rem	68.5–70.5	0.20–0.50	0.25	—	0.08	0.02	0.005	0.005	0.02	—	0.08	183	254
Pb 65	Rem	63.5–65.5	0.20–0.50	0.25	—	0.08	0.02	0.005	0.005	0.02	—	0.08	183	246
Ag 1.5	0.75–1.25	Rem	0.40 max	0.25	1.3–1.7	0.3	0.02	0.005	0.005	0.02	—	0.08	309	309
Ag 2.5	0.25 max	Rem	0.40 max	0.25	2.3–2.7	0.3	0.02	0.005	0.005	0.02	—	0.03	304	304
Ag 5.5	0.25 max	Rem	0.40 max	0.25	5.0–6.0	0.3	0.02	0.005	0.005	0.02	—	0.03	304	380

SOURCE: From Capillo.[36]

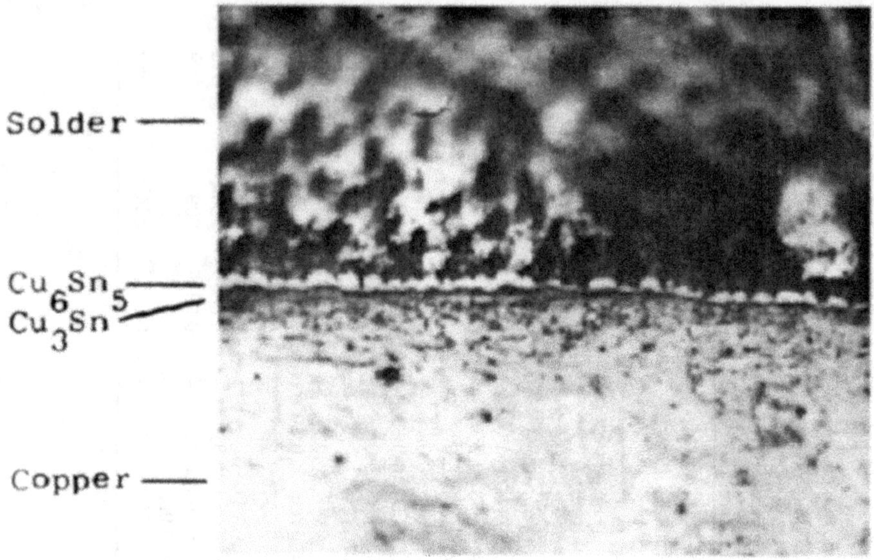

Figure 6.129 Metallographic cross section of soldered joint showing intermetallic compounds. (*From Thwaites.*[51])

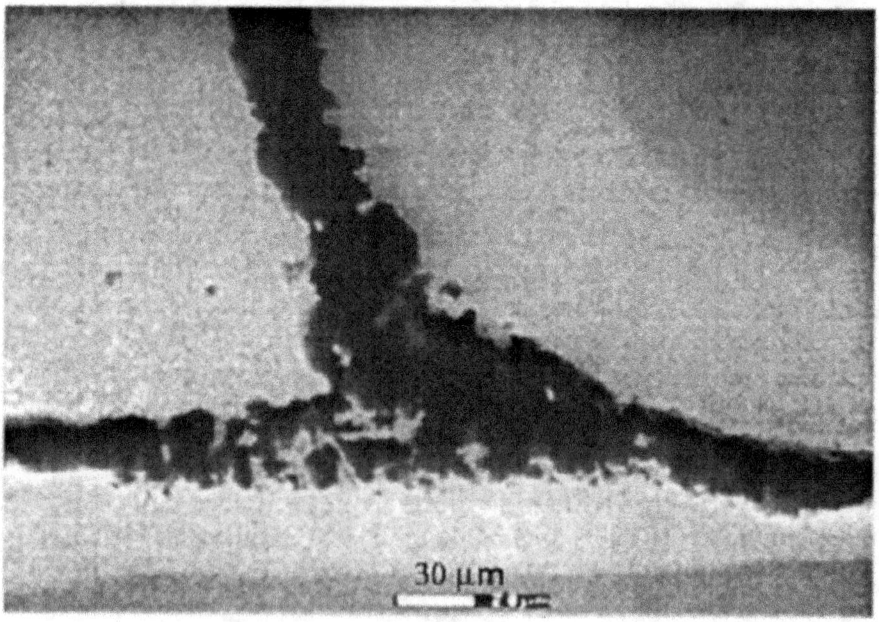

Figure 6.130 Solder separation due to intermetallic formation. (*From Stadterman and Osterman*[39]*; courtesy of CALCE-EPSC.*)

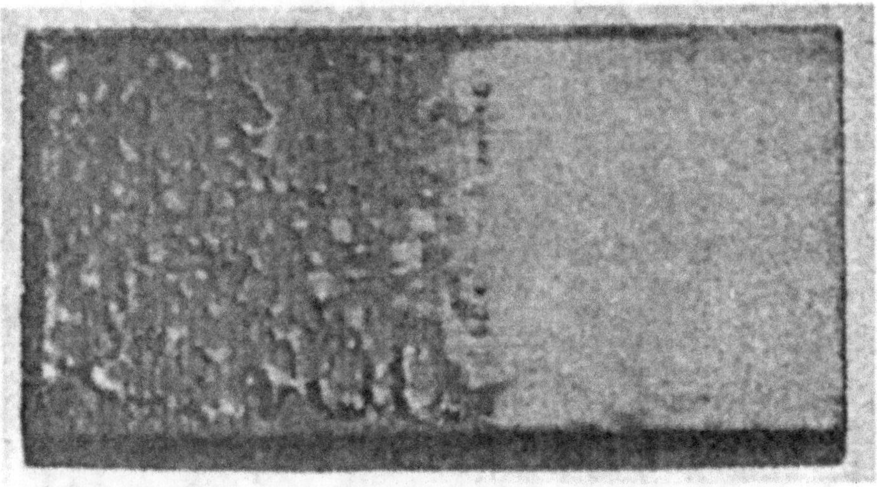

Figure 6.131 An illustration of nonwetting. The portion at the left has been immersed in molten solder. Note the circular patches of base metal visible through the solder. (*From Manko.*[52])

quality. In the case of SMT packages, the voids in the solder result from the high volatile content of the solder pastes. In both PTH and SMT technologies, it appears that a small amount of voids improves the compliance of the joints while higher amounts lead to degradation of joint reliability. In the case of surface-mounted resistors or capacitors, the silver or palladium metallization (adhesion layer) can dissolve in the tin of solder by a process known as leaching. This occurs during the soldering process and causes poor solder fillet in addition to exposure of the underlying ceramic surface. There are two solutions to this problem. The first solution is to use a Sn_{62}-Pb_{36}-Ag_2 alloy but it can be used only in reflow soldering. The silver in the solder alloy is used as a sacrificial element. A second solution used in wave soldering is to use a 50-μin layer of nickel as a barrier layer. Both Ni_3Sn_4 and Ni_3Sn intermetallics can form in this case.

The movement of components during the solidification of the solder can cause a "disturbed solder" joint, such as that shown in Fig. 6.135. When the solder freezes during draining from the wetted surface, a conical defect with a sharp point called an *icicle* may be formed. When such defects form electrical connections, *bridging* is said to have occurred, as shown in Fig. 6.136. Solder webbing occurs due to solder droplets and strings on the bottom surface of PCBs while solder balls form on the top of the board. The outgassing of the epoxy glass laminate and subsequent gas leakage into the hole through plating voids can lead to the formation of a blowhole in the solder, as shown in Fig. 6.137. Some solidification imperfections are only superficial and do not

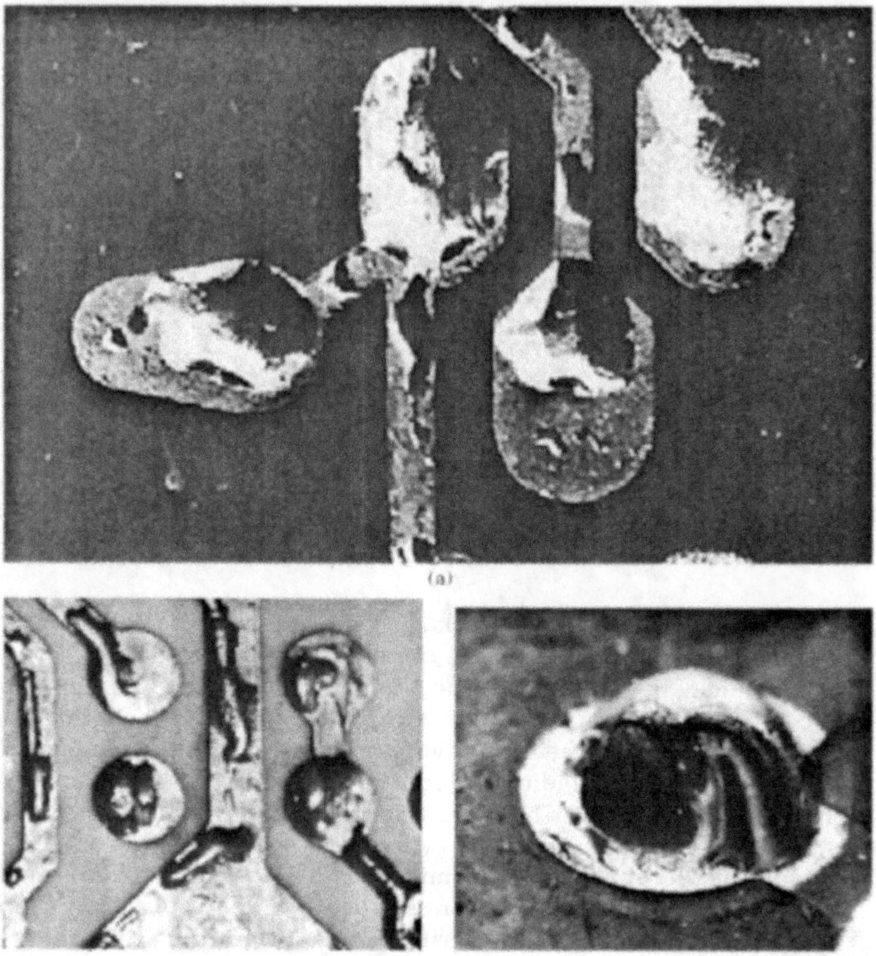

(b)

Figure 6.132 (*a*) Dewetting on a circuit board. (*From Manko.*[52]) (*b*) Commercial example of dewetted soldered joints. (*From Thwaites.*[51])

cause performance deficiencies. Examples are pinholes, freezing lines and alligator skin (grain outline). Grainy or dull solder surfaces can be caused by freezing patterns, impurities in the solder (e.g., aluminum above 0.004 percent and copper above 0.2 percent), intermetallic compounds, dross mixed in solder or insufficient heat fillets. In plated-through-holes, surface tension effects can lead to incomplete hole filling, as shown in Fig. 6.138. Likewise, in SMT packages, if the component lifts, the solder joint can be *starved* (Fig. 6.139) and in the extreme can cause an open joint. Excess solder that obscures fillet con-

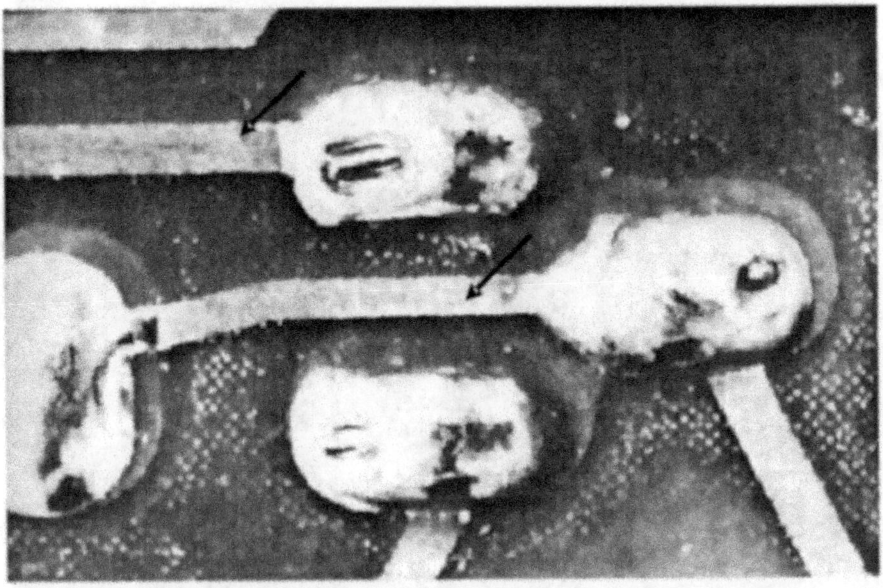

Figure 6.133 Poor wetting on leads (see arrows); only the printed circuit board is well wetted. Note that the solder resist limited the flow to land areas. (*From Manko.*[52])

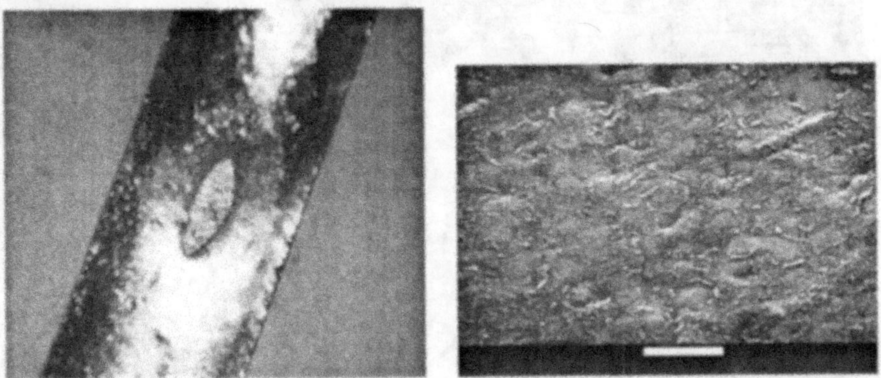

(a) (b)

Figure 6.134 (*a*) Optical micrograph of a nonwetting area on a gold-plated lead. (*b*) SEM micrograph showing "worms" in the nonwetting area. (*From Linn and Wade.*[53])

tours can hide wetting deficiencies or the absence of wire and can short out adjacent conductors (see bridging).

Soldered joints are subject to thermomechanical stresses leading to creep, fatigue and thermal ratcheting. An example of creep damage of a solder bump is shown in Fig. 6.140. In some cases, even room temperature operation can cause creep damage in solder materials. Repeated

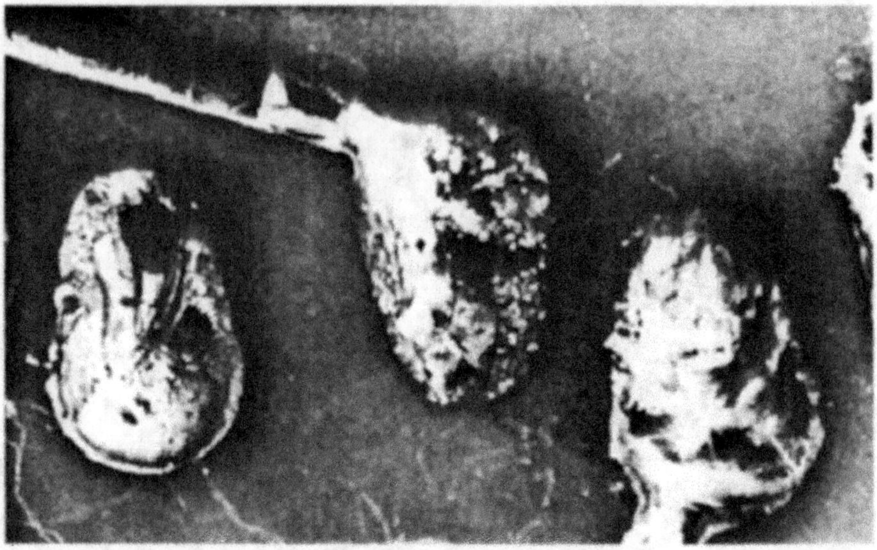

Figure 6.135 A disturbed solder joint. (*From Manko.*[52])

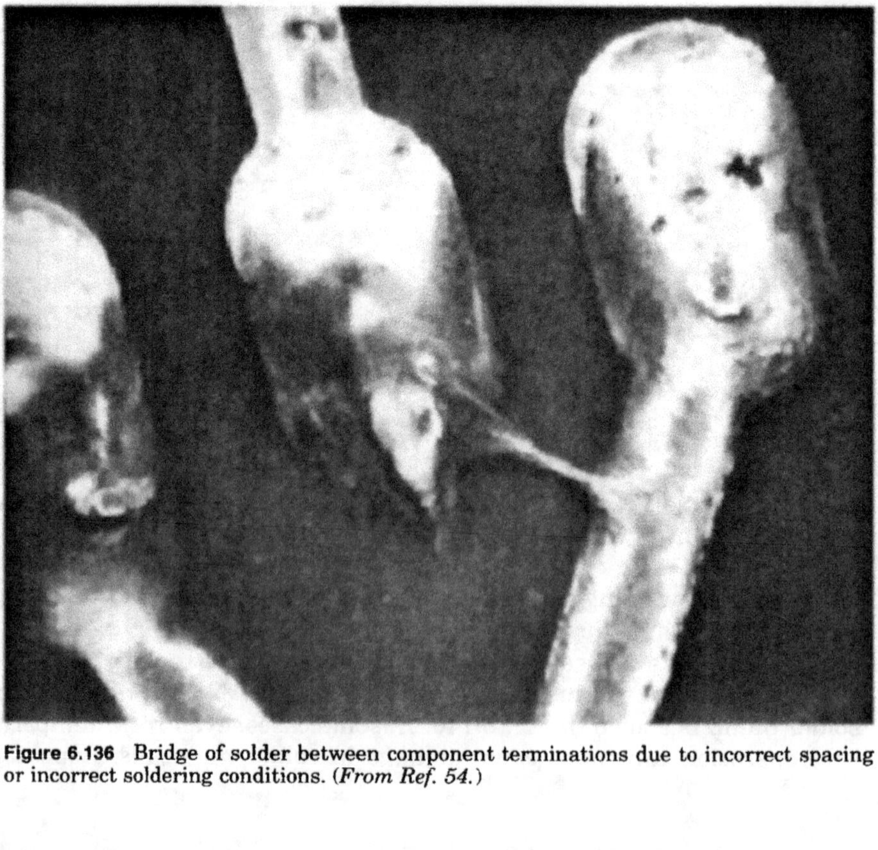

Figure 6.136 Bridge of solder between component terminations due to incorrect spacing or incorrect soldering conditions. (*From Ref. 54.*)

Figure 6.137 A blow hole formed because of outgassing from the epoxy glass. (*From Woodgate.*[55])

mechanical, thermal or electrical stresses can cause fatigue in solder materials. Figure 6.141 shows the effect of different variables on the fatigue properties of lead-tin solder. Table 6.17 shows the fatigue properties of various solder alloys. Figure 6.142 shows a fatigue crack in a gull-wing lead while Fig. 6.143 shows a solder ball broken by fatigue. The progression of a low-cycle fatigue through a PTH joint is shown in Fig. 6.144. Thermal cycling during testing and service of solder joints

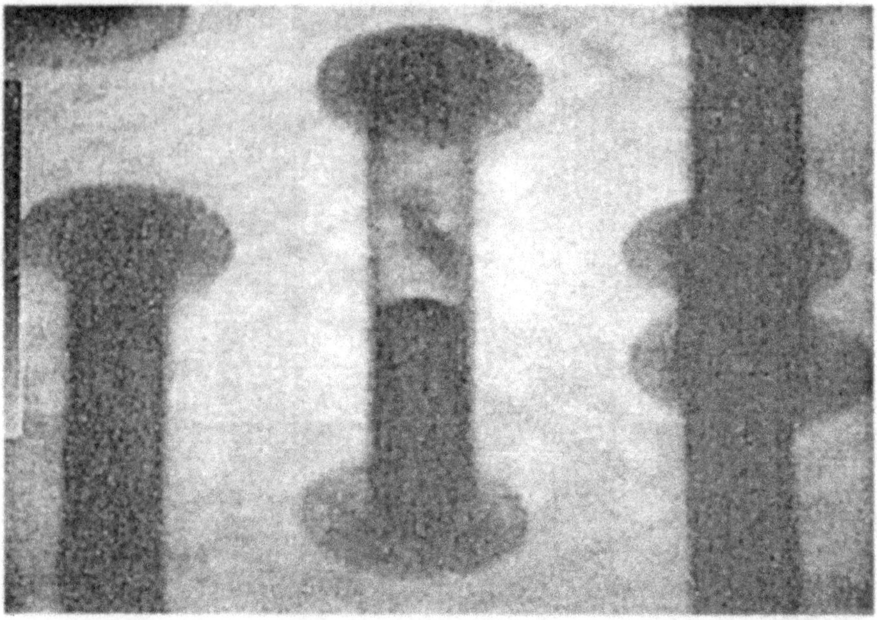

Figure 6.138 Microfocus X-ray radiograph showing incomplete solder filling in a plated-through-hole. (*From Blanchard et al.*[42])

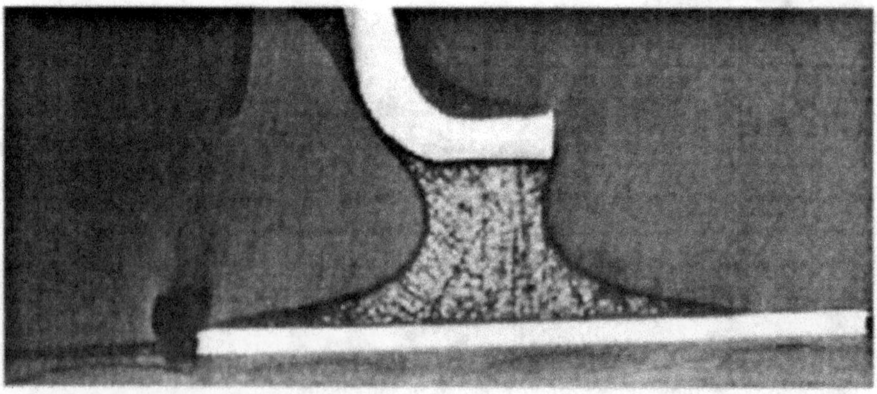

Figure 6.139 Starved solder joint due to component lift. (*From Stadterman and Osterman*[39]; *courtesy of CALCE-EPSC.*)

results in cyclic stresses (due to CTE mismatch) on the solder joint. When these stresses exceed critical values, plastic deformation or fracture can result. An example of this in the soldered joint of a leadless chip resistor is shown in Fig. 6.145. The fractograph of thermally cycled solder in Fig. 6.146 shows both fatigue striations and creep cavitation. Thermal cycling failure of ceramic BGA is shown in Fig. 6.147.

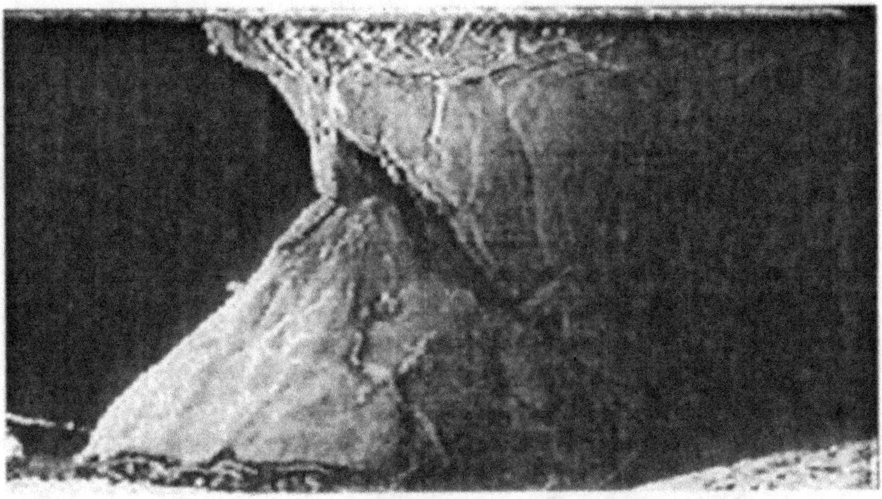

Figure 6.140 Creep fracture of connecting solder bump between two surfaces in a device. (*From Pabbisetty et al.*[13])

Solder joints not only have contaminants in the solder metal itself, but residue from the fluxes used are also entrained on them. Table 6.18 shows allowable contamination levels in the solder alloys. In terms of fluxes, it should be recalled that the purpose of using fluxes is exposure of bare metal by the removal of filmy oxidation films (tarnish) on the surfaces to be soldered. It should thus not be surprising that flux remaining after the soldering operation will corrode components and leads. Residual fluxes from soldering operations can cause corrosion, leakage currents, electrical shorts, coating debonding and insulating contact surfaces. Thus cleaning of soldered joints is an industry unto itself. A note on cleaning of electronic packages in general is also appropriate at this point. During flush-cleaning of assemblies by fluids, particles are mechanically removed, adherent films are dissolved and insoluble matter is displaced. Solvents can be classified into nonpolar and polar solvents, the latter being further subclassified into ionizable and non-ionizable types. Deionized water is nonconductive, its conductivity increasing as the amount of polar contamination increases. The highly conductive electrolytes with large amounts of ionizable polar contaminants can cause problems due to leakage currents over insulators and can lead to galvanic corrosion. During cleaning, the removal of nonpolar dirt is accomplished by "degreasing" with a nonpolar solvent, followed by removal of ionizable contamination using a secondary polar water wash. The details of cleaning processes is beyond the scope of this book and the reader is referred to Manko's[52] excellent treatment of the topic.

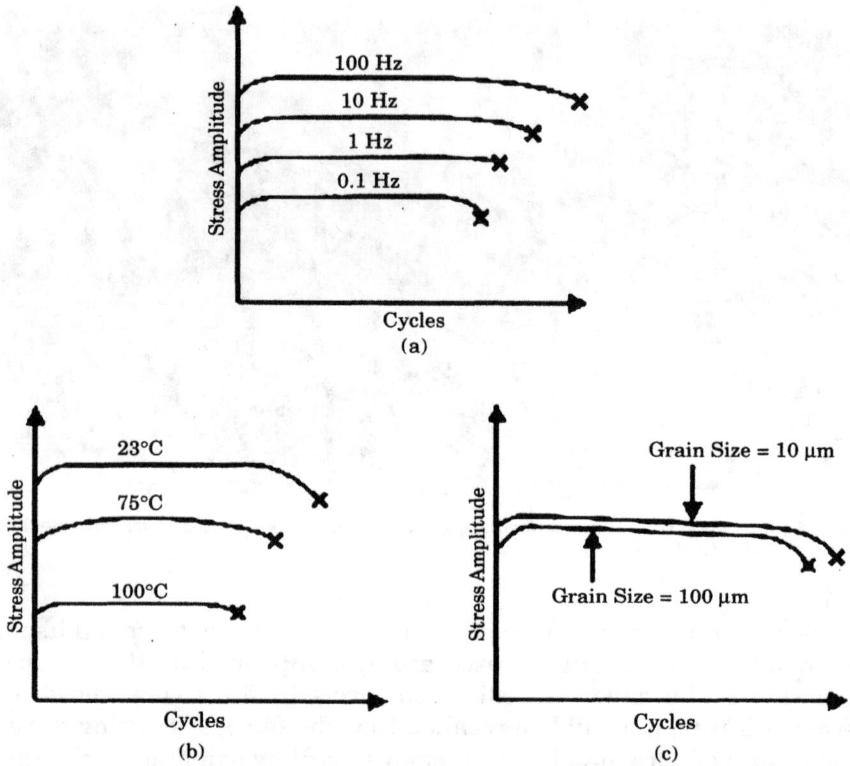

Figure 6.141 Fatigue properties of eutectic Sn-Pb solder showing the effect of (*a*) frequency, (*b*) temperature, and (*c*) grain size on the stress amplitude for constant strain. Note that the curves show an initial increase due to strain hardening, the life increases with increasing frequency approaching a limit at very high frequencies, higher temperatures decrease fatigue life, and fatigue life increases as the grain size decreases. (*From Johnson et al.*[56])

TABLE 6.17 Fatigue Resistance of Solders

Alloy	Fatigue life (relative to Sn 63)
96/4 Sn/Ag	3.03
95/5 Sn/Sb	2.10
62/36/2 Sn/Pb/Ag	1.14
63/37 Sn/Pb	1.00
59/37/4 Sn/Pb/Sb	0.66
40/60 Sn/Pb	0.61

SOURCE: From Prasad.[57]

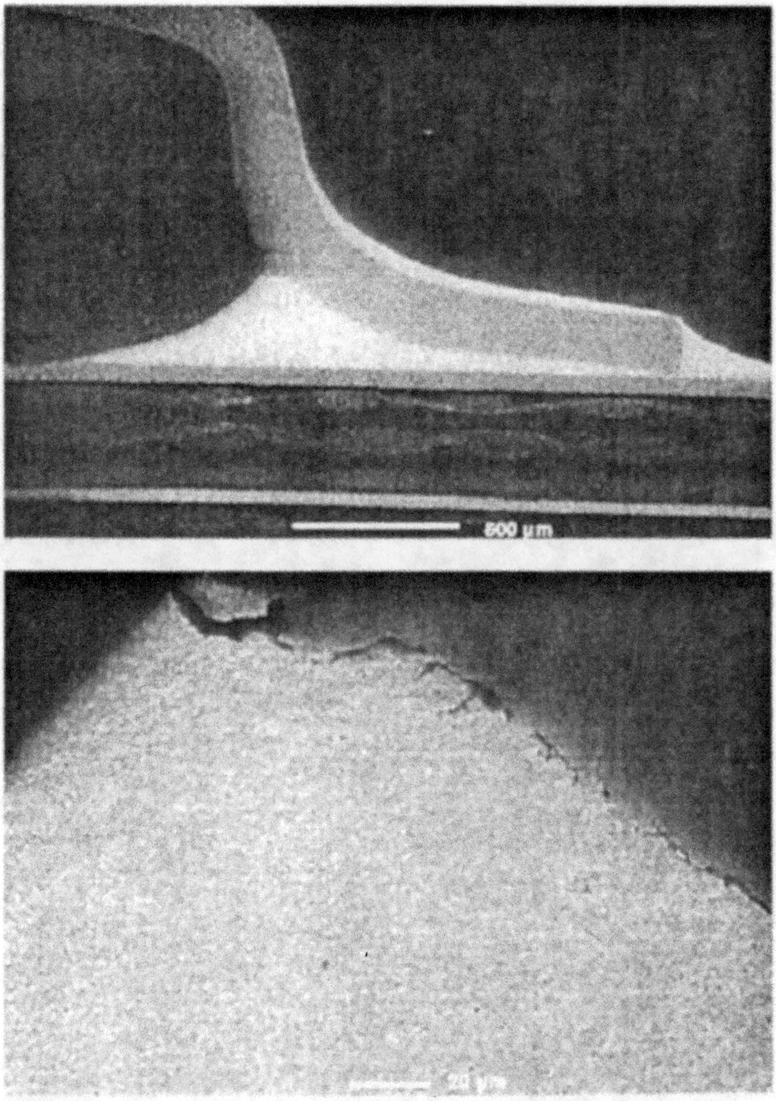

Figure 6.142 Fatigue failure of a gull-wing interconnect. (*From Stadterman and Osterman*[39]*; courtesy of CALCE-EPSC.*)

6.5.7 Lead-free solders

The most commonly used solder in electronics is lead-tin solder (63 percent lead, 37 percent tin). However, with the move toward elimination of lead from solders, several alternative materials are now being used. Table 6.19 shows the relative cost of some elements that can

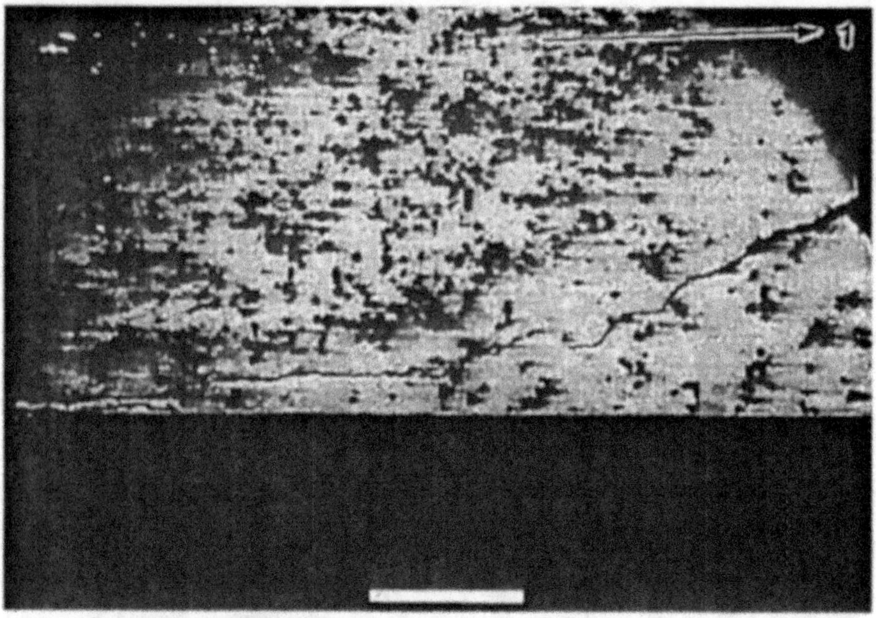

Figure 6.143 Broken solder ball interconnect due to fatigue. (*From Stadterman and Osterman*[39]*; courtesy of CALCE-EPSC.*)

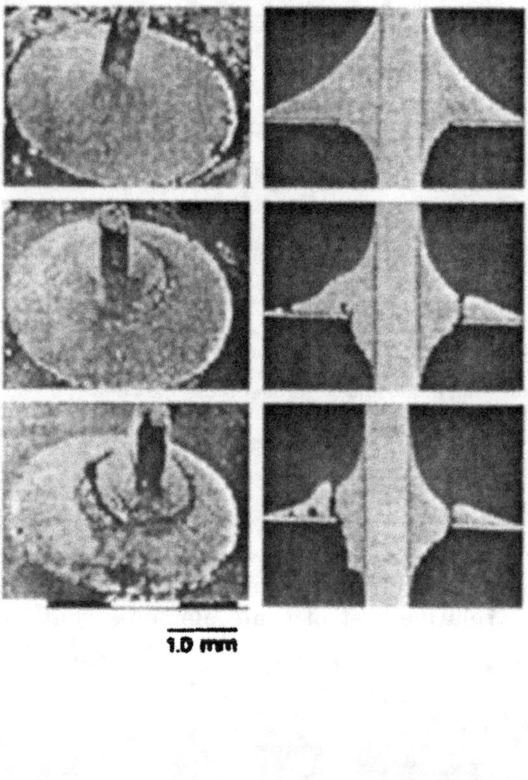

Figure 6.144 Progression of low-cycle fatigue through a single-sided through-hole soldered joint. (*From Vettraino.*[41])

Figure 6.145 SEM micrograph showing creep/fatigue-induced damage near the corner of the resistor: end view after 250 thermal cycles (left); cracking (arrow) in solder microstructure of region A (right). (*From Lau and Pao.*[32])

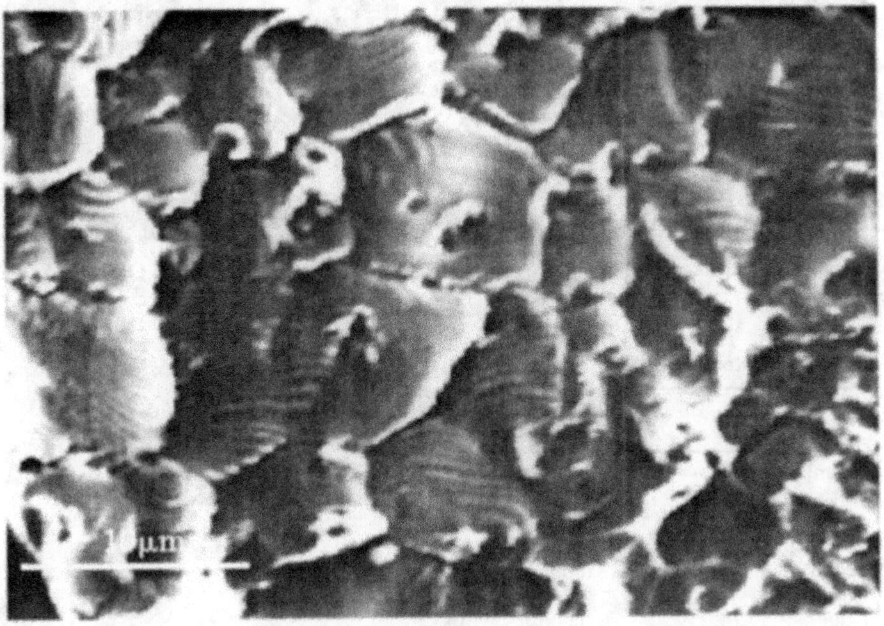

Figure 6.146 An electron microprobe micrograph showing the fatigue striations and creep cavitation of 2512 LCR 63Sn-37Pb solder joint subjected to temperature cycling between –40 and 95°C. (*From Lau and Pao.*[32])

replace lead while Table 6.20 shows characteristics of selected lead-free solder alloys. In general, these alloys exhibit poorer wetting and fatigue characteristics than lead-tin solders. Since the wave soldering temperature is about 45 to 65°C higher than the melting temperature of the solder alloy, the use of the higher melting solder alloys shown in Table 6.20 can lead to cracking of ceramic capacitors during wave soldering.

Figure 6.147 Thermal cycling failure mode of a ceramic ball grid array. (*From Lau and Pao.*[32])

TABLE 6.18 Acceptable Contamination Levels in Eutectic Solder

Metals	Max. allowed level, wt.%
Aluminum	0.006
Antimony	0.500
Arsenic	0.030
Bismuth	0.250
Cadmium	0.005
Copper	0.300
Gold	0.200
Iron	0.020
Silver	0.100
Zinc	0.005
Nickel	0.010

SOURCE: From Prasad.[57]

Tin-bismuth alloys (42 percent tin, 58 percent bismuth) are being used with increasing frequency. Concerns in the use of this alloy are the formation of a low melting ternary eutectic when the alloy is contaminated with lead and its strain rate sensitivity (brittleness under shock loading conditions). In addition, several indium-containing alloys (e.g.,

TABLE 6.19 Alternative Materials to Replace Lead and Their Relative Cost

Replacement elements for lead	Relative cost
Lead (Pb) for reference	1
Antimony	2.2
Bismuth	7.1
Copper	2.5
Indium	194
Silver	212
Tin	6.4
Zinc	1.3

SOURCE: From Prasad.[57]

TABLE 6.20 Examples of Some Lead-Free Solders and Their Properties

Lead-free solder composition, wt.%	Melting point range, °C	Comments
48 Sn, 52 In	118 eutectic	Low melting point, expensive, low strength
42 Sn, 58 Bi	138 eutectic	Established, availability concern of Bi
91 Sn, 9 Zn	199 eutectic	High drossing, corrosion potential
93.5 Sn, 3 Sb, 2 Bi, 1.5 Cu	218 eutectic	High strength, excellent thermal fatigue
95.5 Sn, 3.5 Ag, 1 Zn	218–221	High strength, good thermal fatigue
99.3 Sn, 0.7 Cu	227	High strength and high melting point
95 Sn, 5 Sb	232–240	Good shear strength and thermal fatigue
65 Sn, 25 Ag, 10 Sb	233	Motorola patent, high strength
97 Sn, 2 Cu, 0.8 Sb, 0.2 Ag	226–228	High melting point
96.5 Sn, 3.5 Ag	221	High strength and high melting poin

SOURCE: From Prasad.[57]

an alloy of 77.2 percent tin, 20 percent indium, 2.8 percent silver) offer lead-free alternatives. In addition to the high cost of these materials, these alloys are susceptible to corrosion in a variety of conditions.

6.5.8 Failure of passive components

The failure analysis of passive components in electronic packages such as resistors, capacitors, oscillators and inductors is beyond the scope of this book and the reader is referred to Viswanadham and Singh's[9] excellent treatment of the topic.

References

1. E. A. Amerasekera and F. N. Najm, *Failure Mechanisms in Semiconductor Devices*, 2d ed., Wiley, New York, 1997.
2. L. L. Marsh, R. D. Havens, S. C. Wang, J. A. Malack, and H. B. Ulsh, "Reliability and Testing," in D. P. Seraphim, R. C. Lasky, and C. Y. Li (eds.), *Principles of Electronic Packaging*, McGraw-Hill, New York, 1989.

3. B. P. Richards and P. K. Footner, *The Role of Microscopy in Semiconductor Failure Analysis*, Oxford University Press, New York, 1992.
4. H. K. Dicken, "The Management (?) of Reliability Information," in *Proceedings of Advanced Techniques in Failure Analysis Symposium—1976*, IEEE, New York, 1976.
5. D. D. Dylis, "Overview of Electronic Component Reliability," in P. L. Martin (ed.), *Electronic Failure Analysis Handbook*, McGraw-Hill, New York, 1999.
6. T. W. Lee, "ESD Damage Simulation and Failure Mechanisms," in R. J. Ross, C. Boit, and D. Staab (eds.), *Microelectronic Failure Analysis Desk Reference*, 4th ed., ASM International, Materials Park, Ohio, 1999.
7. T. W. Lee, "Thermomechanical Effects of EOS," in R. J. Ross, C. Boit, and D. Staab (eds.), *Microelectronic Failure Analysis Desk Reference*, 4th ed., ASM International, Materials Park, Ohio, 1999.
8. M. Johnson and D. Pote, "Silicon Precipitate Nodule-Induced Failures of MOSFETs," in *ISTFA '91, Proceedings of the 17th International Symposium for Testing and Failure Analysis*, ASM International, Materials Park, Ohio, 1991.
9. P. Viswanadham and P. Singh, *Failure Modes and Mechanisms in Electronic Packages*, Chapman & Hall, New York, 1998.
10. R. Haythornthwaite, "Case Studies of Metallurgical Failure Mechanisms in Microcircuits," in D. O. Northwood, E. Abramovici, M. T. Shehata, and J. Wylie (eds.), *Microstructural Science*, vol. 26: *Analysis of In-Service Failures and Advances in Microstructral Characterization*, ASM International, Materials Park, Ohio, and International Metallographic Society, Columbus, Ohio, 1999.
11. H. B. Bakoglu, *Circuits, Interconnections, and Packaging for VLSI*, Addison-Wesley Publishing Company, New York, 1990.
12. P. L. Martin, "Semiconductors," in P. L. Martin (ed.), *Electronic Failure Analysis Handbook*, McGraw-Hill, New York, 1999.
13. S. V. Pabbisetty, D. Corum, P. Scott, and K. S. Wills, "Failure Mechanisms in Integrated Circuits," in R. J. Ross, C. Boit, and D. Staab (eds.), *Microelectronic Failure Analysis Desk Reference*, 4th ed., ASM International, Materials Park, Ohio, 1999.
14. C. G. Woychik and R. C. Senger, "Joining Materials an Processes in Electronic Packaging," in D. P. Seraphim, R. C. Lasky, and C. Y. Li (eds.), *Principles of Electronic Packaging*, McGraw-Hill, New York, 1989.
15. J. J. Erickson, "Electronic and Electrical Characterization," in P. L. Martin (ed.), *Electronic Failure Analysis Handbook*, McGraw-Hill, New York, 1999.
16. S. V. Pabbisetty, "Failure Analysis Overview," in R. J. Ross, C. Boit, and D. Staab (eds.), *Microelectronic Failure Analysis Desk Reference*, 4th ed., ASM International, Materials Park, OH, 1999.
17. L. L. Ludwig, Jr., "X-Ray/Radiographic Component Inspection," in P. L. Martin (ed.), *Electronic Failure Analysis Handbook*, McGraw-Hill, New York, 1999.
18. H. Kaplan, "Infrared Thermography," in P. L. Martin (ed.), *Electronic Failure Analysis Handbook*, McGraw-Hill, New York, 1999.
19. L. W. Kessler and J. E. Semmens, "Acoustic Micro Imaging Failure Analysis of Electronic Devices," in P. L. Martin (ed.), *Electronic Failure Analysis Handbook*, McGraw-Hill, New York, 1999.
20. T. E. Adams, "Very High Resolution Reflection Mode Acoustic Microscopy," *Microscopy and Analysis*, The Americas Edition, July 2000.
21. L. Kashar, "Failure Analysis of Electron-Beam Welded Metal Hybrid Packages," in *Proceedings of Advanced Techniques in Failure Analysis Symposium—1976*, IEEE, New York, 1976.
22. J. H. Rose, R. Shuman, T. S. Sriram, and T. Spooner, "Microscopy Methods for Integrated Circuit Interconnect Evaluation," in *ISTFA '93, Proceedings of the 19th International Symposium for Testing and Failure Analysis*, ASM International, Materials Park, Ohio, 1993.
23. P. Tangyunyong, A. Y. Liang, A. W. Righter, D. L. Barton, and J. M. Soden, "Localizing Heat-Generating Defects Using Fluorescent Microthermal Imaging," in *ISTFA '96, Proceedings of the 22nd International Symposium for Testing and Failure Analysis*, ASM International, Materials Park, Ohio, 1993.

24. J. R. Devaney, "Scanning Electron Microscopy and X-Ray Analysis," in P. L. Martin (ed.), *Electronic Failure Analysis Handbook*, McGraw-Hill, New York, 1999.
25. D. E. Newbury, D. C. Joy, P. Echlin, C. E. Fiori, and J. I. Goldstein, *Advanced Scanning Electron Microscopy and X-Ray Microanalysis*, Plenum Press, New York, 1986.
26. L. Rice and W. Chen, "Failure Analysis Using Voltage Contrast and EBIC," in C. Hayzelden, C. Hetherington, and F. Ross (eds.), *Electron Microscopy of Semiconducting Materials and ULSI Devices*, Materials Research Society, Warrendale, Pa., 1998.
27. A. S. Brar and P. B. Narayan, *Materials and Processing Failures in the Electronics and Computer Industry: Analysis and Prevention*, ASM International, Materials Park, Ohio, 1993.
28. "Identification of a Tungsten Particle on a Silicon Wafer," Application Note, by permission of ThermoNoran, Middleton, Wis., 2000.
29. C. Cohn and M. T. Shih, "Packaging and Interconnection of Integrated Circuits," in C. A. Harper (ed.), *Electronic Packaging and Interconnection Handbook*, McGraw-Hill, New York, 2000.
30. J. K. Lumpp, "Hybrid Assemblies," in G. R. Blackwell (ed.), *The Electronic Packaging Handbook*, CRC Press, New York, 2000.
31. B. O. Kolbesen and H. P. Strunk, "Analysis, Electrical Effects, and Prevention of Process-Induced Defects in Silicon Integrated Circuits," in N. G. Einspruch and H. Huff (eds.), *VLSI Electronics Microstructure Science*, Academic Press, New York, 1985.
32. J. H. Lau and Y.-H. Pao, *Solder Joint Reliability of BGA, CSP, Flip Chip, and Fine Pitch SMT Assemblies*, McGraw-Hill, New York, 1997.
33. C. G. Woychik, "Soldering and Cleaning of High Performance Circuit Board Assemblies," in C. A. Harper (ed.), *High Performance Printed Circuit Boards*, McGraw-Hill, New York, 2000.
34. S. D. Prough and D. E. Pope, "Material Development Challenges in High Density Packaging of Advanced VLSI Memory Devices", in A. T. Barfknecht, J. P. Partridge, C. J. Chen, and C.-Y. Li (eds.), *Advanced Electronic Packaging Materials*, Materials Research Society, Pittsburgh, Pa., 1990.
35. K. Parekh, R. Milburn, and K. Georgia, "Package Related Failure Mechanisms in Plastic BGA Packages Used for ASIC Devices," in *ISTFA '96, Proceedings of the 22nd International Symposium for Testing and Failure Analysis*, ASM International, Materials Park, Ohio, 1993.
36. C. Capillo, *Surface Mount Technology*, McGraw-Hill, New York, 1990.
37. P. Hinton, "Fabrication of Rigid Printed Wiring Boards," in M. W. Jawitz (ed.), *Printed Circuit Board Materials Handbook*, McGraw-Hill, New York, 1997.
38. S. S. Mansilla, "Microsection Analysis," in M. W. Jawitz (ed.), *Printed Circuit Board Materials Handbook*, McGraw-Hill, New York, 1997.
39. T. J. Stadterman and M. D. Osterman, "Reliability and Performance of Advanced PWB Assemblies," in C. A. Harper (ed.), *High Performance Printed Circuit Boards*, McGraw-Hill, New York, 2000.
40. D. McKeeby and L. Phillips, "Lamination of Flex and Rigid-Flex Printed Wiring Boards," in M. W. Jawitz (ed.), *Printed Circuit Board Materials Handbook*, McGraw-Hill, New York, 1997.
41. L. G. Vettraino, "Solder Joints," in P. L. Martin (ed.), *Electronic Failure Analysis Handbook*, McGraw-Hill, New York, 1999.
42. R. A. Blanchard, D. Galler, D. Glover, A. Kusko, J. D. Loud, N. K. Medora, and G. J. Mimmack, "Failure Analysis of Printed Wiring Assemblies," in P. L. Martin (ed.), *Electronic Failure Analysis Handbook*, McGraw-Hill, New York, 1999.
43. V. J. Brzozowski and C. T. Brooks, "Printed Wiring Boards," in C. A. Harper and R. N. Sampson (eds.), *Electronic Materials and Processes Handbook*, McGraw-Hill, New York, 1994.
44. P. J. Rice Goldman, "The Desmear/Etchback Processes," in M. W. Jawitz (ed.), *Printed Circuit Board Materials Handbook*, McGraw-Hill, New York, 1997.
45. S. W. Hinch, *Handbook of Surface Mount Technology*, Longman Scientific and Technical, London, 1988.
46. S. Silvus, "Failure Analysis of Passive Components," in R. J. Ross, C. Boit, and D. Staab (eds.), *Microelectronic Failure Analysis Desk Reference*, 4th ed., ASM International, Materials Park, Ohio, 1999.

47. N. K. Medora, "Connection Technology," in P. L. Martin (ed.), *Electronic Failure Analysis Handbook*, McGraw-Hill, New York, 1999.
48. G. Slenski and D. Galler, "Wires and Cables," in P. L. Martin (ed.), *Electronic Failure Analysis Handbook*, McGraw-Hill, New York, 1999.
49. P. Martin, R. A. Blanchard, W. Denson, D. Glover, A. Kusko, and D. Galler, "Switches and Relays," in P. L. Martin (ed.), *Electronic Failure Analysis Handbook*, McGraw-Hill, New York, 1999.
50. J. S. Hwang, "Solder Technologies for Electronic Packaging and Assembly," in C. A. Harper (ed.), *Electronic Packaging and Interconnection Handbook*, McGraw-Hill, New York, 2000.
51. C. J. Thwaites, *Capillary Joining-Brazing and Soft-Soldering*, Research Studies Press, London, 1982.
52. H. H. Manko, *Solders and Soldering: Materials, Design, Production, and Analysis for Reliable Bonding*, 3d ed., McGraw-Hill, New York, 1992.
53. J. H. Linn and W. R. Wade, "Reduced Solderability Effectiveness Due to Base Metal Oxide Formation," in *ISTFA '91, Proceedings of the 17th International Symposium for Testing and Failure Analysis*, ASM International, Materials Park, Ohio, 1991.
54. *Soldering Manual,* 2d ed. rev., American Welding Society, Inc., Miami, Fla., 1978.
55. R. W. Woodgate, *The Handbook of Machine Soldering*, Wiley, New York, 1988.
56. E. A. Johnson, W. T. Chen, and C. K. Lim, "Mechanical Design," in D. P. Seraphim, R. C. Lasky, and C. Y. Li (eds.), *Principles of Electronic Packaging*, McGraw-Hill, New York, 1989.
57. R. P. Prasad, *Surface Mount Technology*, Chapman & Hall, New York, 1997.

Chapter

7

Case Studies

7.1 Introduction

In this chapter we present metallurgical case studies chosen to illustrate the principles of failure analysis. In each study references are made to specific points covered in the previous chapters in order to show how these were used in the analyses.

These studies deal with real failures, in which the scope of the failure analysis was limited by factors such as the time that could be devoted to the problem, the expense of the testing desired, and particularly incomplete information about the failure situation. Thus the reader will find that some of the conclusions are quite tentative, which is a common result of analyses of such real failures.

Microstructural analysis is invariably a part of most metallurgical failure analyses. However, in this book we have made no attempt to review the physical metallurgy of the alloys examined. The reader will find this information in many books and references, such as the two listed in the Bibliography.

7.2 Case A: A Cracked Vacuum Bellows*

7.2.1 Introduction

Tubular metal bellows are useful devices for joining components. One use is to make a vacuum seal for a movable component which must be inserted into a vacuum chamber, such as a shaft that must be moved, either

*The authors acknowledge the work done by David Dellinger on this case study.

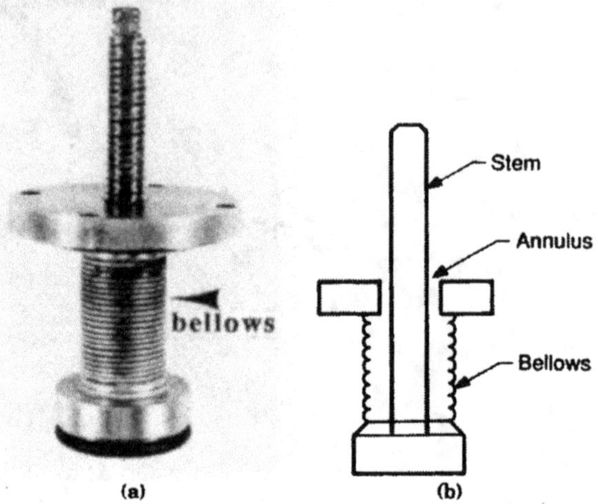

Figure 7.1 (*a*) Bellows used for a vacuum seal. (*b*) Cross-sectional view.

axially or sideways. An example is shown in Fig. 7.1*a*. This device is used to move the seat, with its rubber gasket, down onto a flat surface to provide a vacuum seal. This is accomplished by turning the shaft, which is connected to a pivot joint, as shown in the cross-sectional drawing in Fig. 7.1*b*. Thus the atmosphere on the inside of the bellows never enters the chamber except through the open valve. On the shaft there is no sliding O-ring-type seal, which would be subject to wear and require lubrication.

A bellows very similar to that in Fig. 7.1 was on a vacuum system of an arc melter, which was periodically pumped down to about 10^{-4} torr (mechanical pump only). This was done only occasionally (such as a few times a month). The operators noticed over several months that the time required to attain the desired vacuum had increased and eventually reached an unacceptable length, so that the system had to be examined carefully for a source of leaks. Inspection of the bellows revealed a fine crack across one fin (Fig. 7.2). It was barely discernible by eye and had a greenish deposit at each end of the crack.

7.2.2 Experimental procedure

The surface of the tube was examined in a scanning electron microscope to determine the material from which the tube was made and to obtain chemical analyses of the corrosion deposits, using the energy-dispersive spectrometer (EDS) on the scanning electron microscope. Although the EDS analysis was only qualitative (see Chap. 1), it was

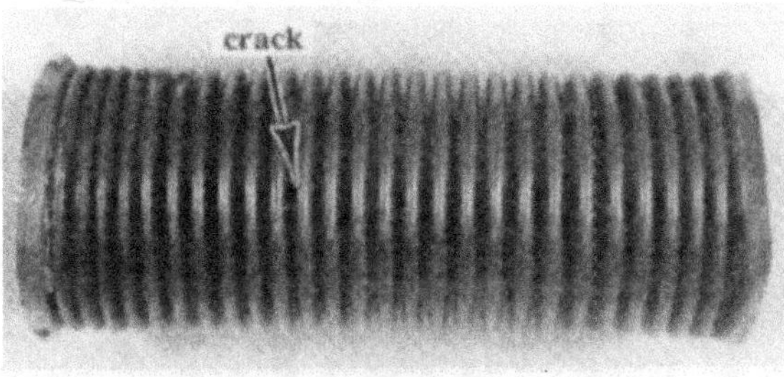

Figure 7.2 Bellows that was analyzed to determine the cause of the fine crack at the location noted by arrow.

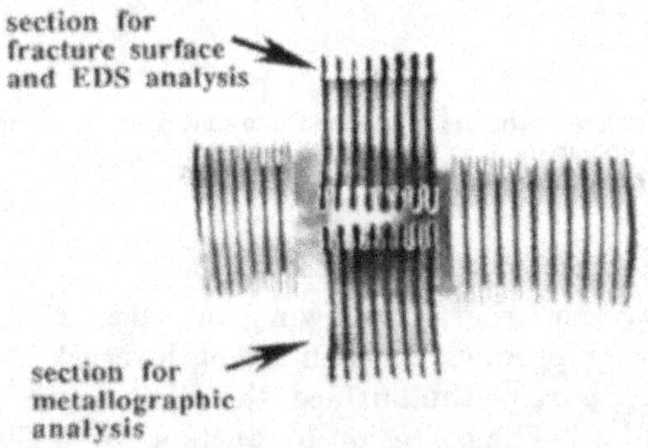

Figure 7.3 Two sections that were removed for analysis.

sufficiently sensitive to distinguish the relative concentration of the elements present.

To examine the inside of the tube, it was sectioned with a thin blade (0.004 mm thick) on a high-speed cutoff wheel using copious water for cooling, followed immediately by rinsing in methanol followed by drying. The sections thus obtained are shown in Fig. 7.3. The section containing the crack was notched on each side of the crack, as shown in Fig. 7.4*a*, and then broken to expose the original crack (Fig. 7.4*b*), which could easily be distinguished from the fresh crack formed while breaking the piece. (The procedure used to break the section stretched the bellows.) One section was examined in the scanning electron microscope. The other section was cut through the crack and one-half embedded in plastic to make a metallographic mount, which allowed examination of the crack itself in cross section. This sample was ground and polished using standard procedures for brass, then etched in an ammonium hydroxide–hydrogen peroxide solution.

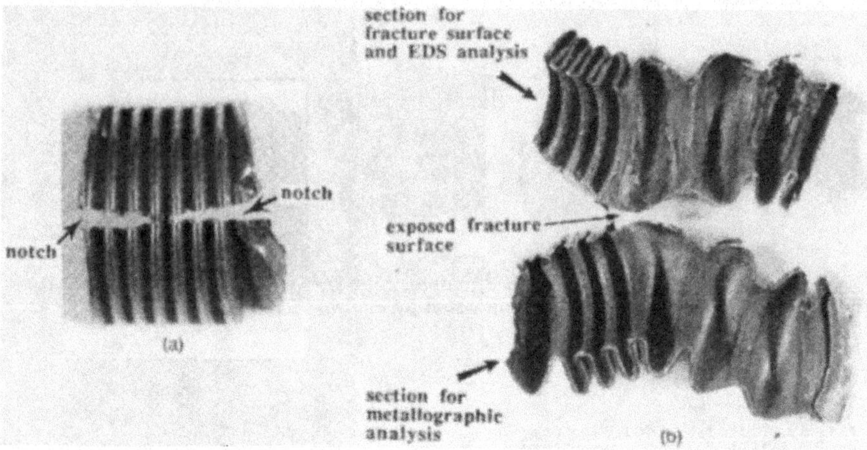

Figure 7.4 (*a*) Section that was notched on each side of the crack, so that upon breaking, the crack surface was exposed. (*b*) Sample in (*a*) after breaking to expose the crack surface.

7.2.3 Results

Figure 7.5 is a scanning electron micrograph showing the crack on the outside of the tube. The corrosion products on each end of the crack are prominent. On the unaffected part of the surface, the EDS analysis revealed only Cu (strong) and Zn. The corrosion products showed Cu, Zn, S, Cl, Ca, and K. Examination of the inside of the bellows showed that the fin adjacent to the one that had the crack contained similar corrosion products at the same location as the crack. On the surface of other fins many cracks were found which had not penetrated through the wall of the tube. The cross-sectional view of the bellows wall in Fig. 7.6 shows such cracks.

The fracture-surface topography was obscured by corrosion products, which were too tenacious to be removed by three replicating tapes (see Sec. 1.4.1). However, EDS analysis revealed on the fresh fracture (caused by breaking the sample) only Cu (strong) and Zn, whereas on the original fracture surface Cu (strong), Cl, Si, and Al were detected. It is to be noted that no Zn was found on the fracture surface.

The microstructure (Fig. 7.7) revealed numerous deformation bands (see Glossary) and bent annealing twins (see Sec. 3.3), showing that the material was in the cold-worked condition (see Brooks,[1] Chap. 1). This appearance was found on the bend in the fins and on the straight sides. Numerous cracks were found across the wall in the bend in many of the fins, such as that noted by the arrow in Fig. 7.7 and shown in Fig. 7.6. Nearer the location of the crack, several of these regions were found

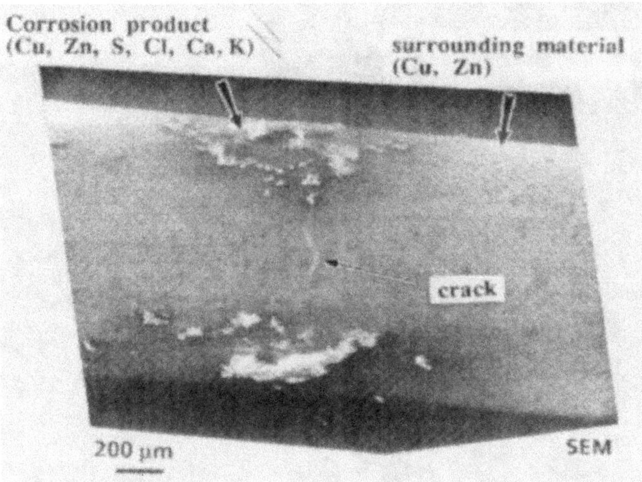

Figure 7.5 Scanning electron micrograph of the crack in the bellows (Fig. 7.2). Corrosion products were on each end, and the elements detected in the EDS analysis of one is given. Also shown are the elements in the EDS analysis from the uncorroded surface, revealing only copper and zinc.

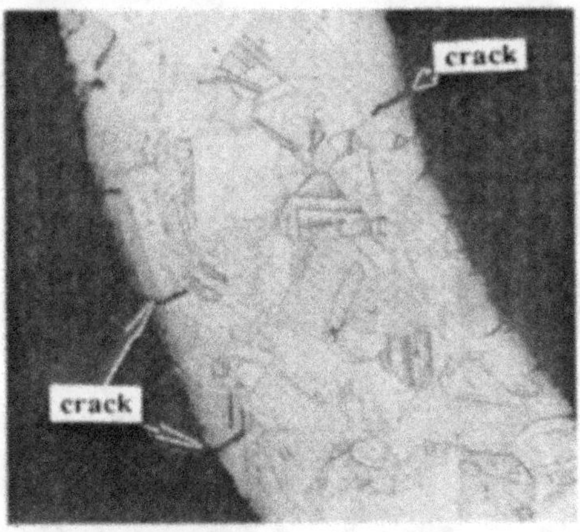

Figure 7.6 Optical micrograph of the cross section of the bellows wall, showing cracks on the surface of the bellows which had not penetrated through the wall thickness.

(Fig. 7.8). The cracks seem to be both intergranular and transgranular, perhaps following the deformation bands in the latter case. Figure 7.9*b* shows an area very near the fracture cross section, and this same area was examined in the scanning electron microscope (Fig. 7.9*a*). The EDS analysis revealed that some of the regions which are shown in relief in Fig. 7.9*a* were relatively low in Zn. Upon examination with an optical microscope (Fig. 7.9*b*) these raised regions had a noticeable copper color to them, compared to the surrounding material.

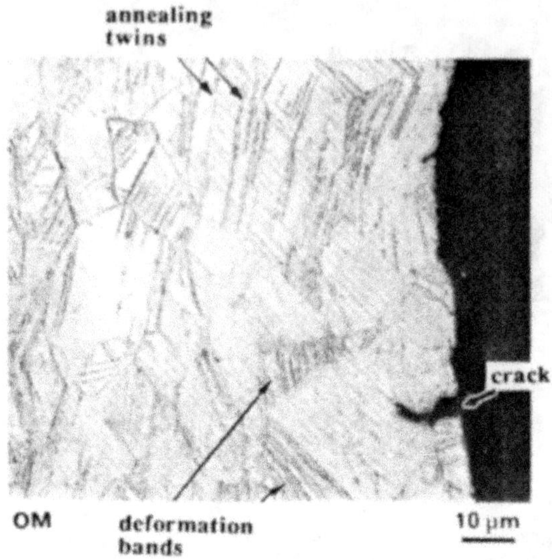

Figure 7.7 Optical micrograph of the cross section of the bellows. Note the deformation bands and bent annealing twins showing that the material was cold-worked.

Figure 7.8 Optical micrograph of the cross section near the crack surface, showing numerous cracks at the outside surface.

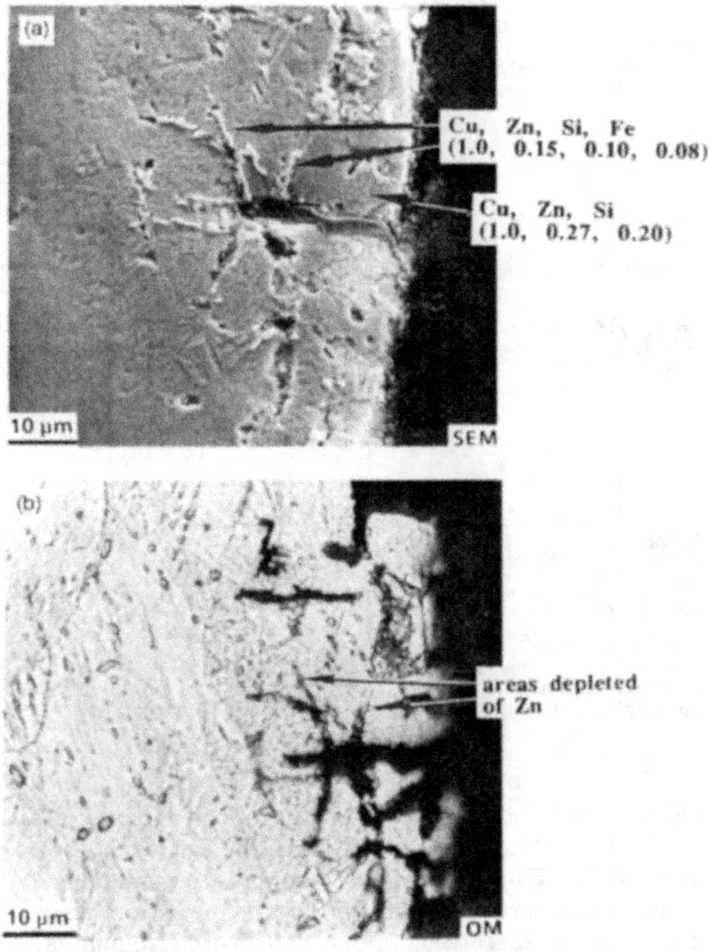

Figure 7.9 (*a*) Scanning electron micrograph showing a crack penetrating the wall. Note that the EDS analysis of the small regions in relief contained less Zn than that of the adjoining matrix. (The analyses are given in terms of the intensity of the elements to that of Cu.) (*b*) Optical micrograph of the same area; the areas noted had a copper color.

7.2.4 Discussion

The construction of bellows involves the formation of fins by the simultaneous application of hydraulic pressure to the inner surface and of axial compression to a tube, such as shown in Fig. 7.10. Clearly at the bend there is severe plastic deformation. However, the microstructure of the straight portion of the tube showed extensive plastic deformation (Fig. 7.7), so the starting tube was already cold-worked to give a strong bellows.

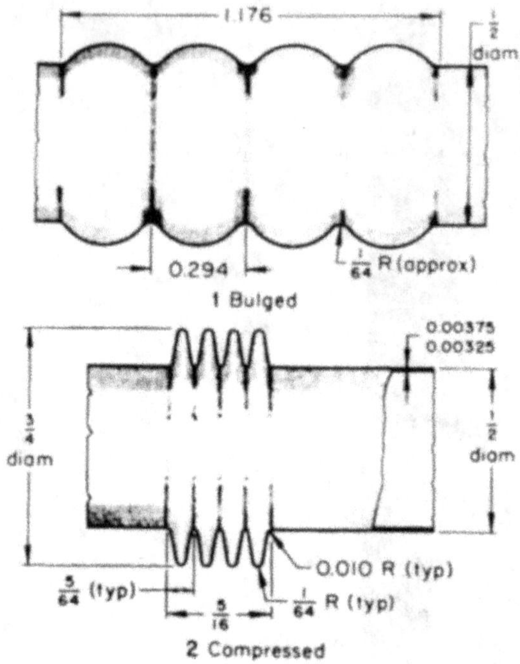

Figure 7.10 Schematic diagram of the process of deforming a tube into a bellows. (*From* Metals Handbook.[2])

The lack of Zn on the crack surface and the copper-depleted grains near the crack surface (Fig. 7.9*a*) show that the corrosion process involved dezincification. This is selective leaching or dealloying,[3–5] a common corrosion mechanism in brasses containing more than about 15 percent Zn.[3–5] The Zn is dissolved, and a copper-rich region is left behind. This usually leads to porosity,[6] although the microstructure in Fig. 7.9 showed no evidence of it. Note in Fig. 7.11 that the zinc-depleted grains show fine, straight annealing twins, with no sign of deformation bands, which indicates that these grains have formed during dezincification. Such fine, twinned grains were found by Polushkin and Shuldener[6] in brass hot-water pipes which had been corroded in service.

A factor that favors dezincification is surface deposits, which deplete the region under the deposit of oxygen and set up an oxygen concentration cell.[5,7] In the case of this failed bellows, the surface on which the crack started was only in contact with air, but if a deposit formed, then such a mechanism could operate. However, the origin of the deposit is not clear. The deposits contained S, Cl, Si, Ca, Al, and K. The outside of the bellows was untarnished, with no evident corrosion except at the crack itself (Fig. 7.5), but the inside surface was corroded. A heavy cor-

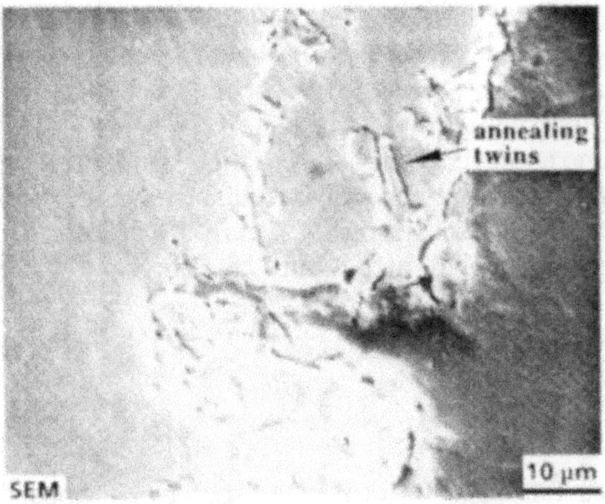

Figure 7.11 Micrograph showing annealing twins in the copper-rich region formed by dezincification.

rosion deposit was found on the inside fin adjacent to the one containing the crack and at the same location. It is speculated that these regions contained flux deposits from the brazing process used to join the bellows to the plates. Fluxes used for brazing brass may contain chlorides (see *Metals Handbook,*[8] p. 1035), an element that is known to promote dezincification.[7] This flux could not be easily removed after brazing because of the small clearance in the annulus (Fig. 7.1*b*). Moisture from air could accumulate in the region because of this and activate the corrosion if it were absorbed by the flux.

Cold-worked brass is susceptible to stress corrosion cracking (season cracking)[4] (see Sec. 3.11), and dezincification can play a key role in such cracking.[5] Transgranular fracture associated with dezincification has been reported in cold-worked brass,[3] which is consistent with the crack appearing to follow the deformation bands in the present study. The many cracks seen on the inside and outside surfaces indicate that the bellows was undergoing generalized stress corrosion cracking, but the process was accelerated at the region of corrosion in which dezincification occurred.

The problem can be avoided by using a brass of no more than 15 percent Zn.[4] If such an alloy will not meet the strength requirements, then another material, such as austenitic stainless steel, will have to be used to make the bellows.

7.2.5 Conclusions

The brass bellows had developed a fine crack through the wall thickness in the cold-worked region of the bend. Other fine cracks were observed which had not penetrated the wall. The crack formation was associated with corrosion, and dezincification occurred in this region. The corrosion was induced by a surface contaminant, probably soldering flux which was not removed during subsequent cleaning of the bellows assembly. The problem may have been aggravated by moisture which condensed in the narrow annulus. The crack probably propagated by stress corrosion cracking (season cracking). Brass containing less than 15 percent Zn usually is not susceptible to this type of attack and cracking.

7.3 Case B: Failure of a Large Air-Conditioning Fan Blade*

7.3.1 Introduction

In some air-cooling systems for large buildings the hot air is passed through a water-cooled chilling tower. The large volume of air requires a relatively large fan, consisting of several blades rotating around a common axis and operating as a turbine fan. A schematic drawing of such a fan is shown in Fig. 7.12. The neck of the blade is the location

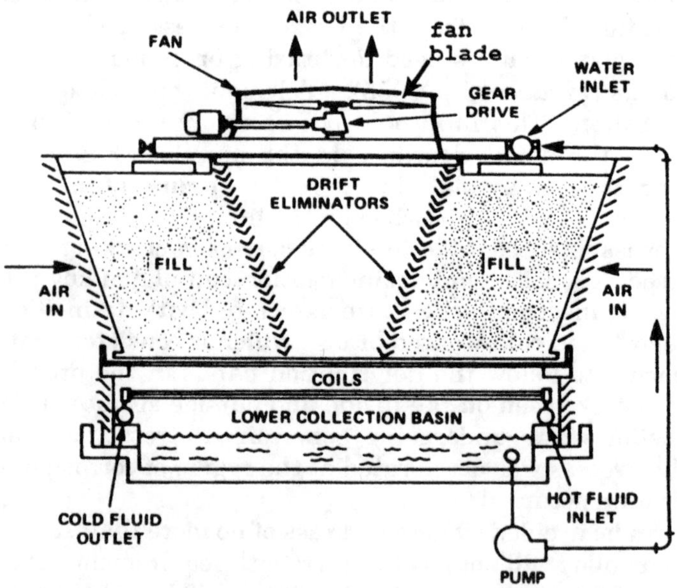

Figure 7.12 Schematic diagram of a centrifugal fan in a cooling tower. (*From* ASHRAE Handbook.[9] *Reprinted by permission of the American Society of Heating, Refrigerating and Air-Conditioning Engineers.*)

*The authors acknowledge the work done by Hwa-Perng Kao on this case study.

that sustains the greatest centrifugal and bending stress, and the subject of this case study is a fan blade which fractured at that location. No history of the blade nor operating conditions were available, except that the blade rotated at about 550 r/min.

7.3.2 Experimental procedure

A drawing of the broken blade is shown in Fig. 7.13. Samples for analysis were cut from the broken blade. The larger pieces were removed using a hack saw, and then smaller pieces were cut using a water-cooled cutoff wheel. After cutting, each piece was immediately washed in water, then acetone, and then dried.

The metallographic samples were mounted in plastic and then ground and polished using standard methods. The etchant used was HF in water, although the microstructure was revealed in the unetched

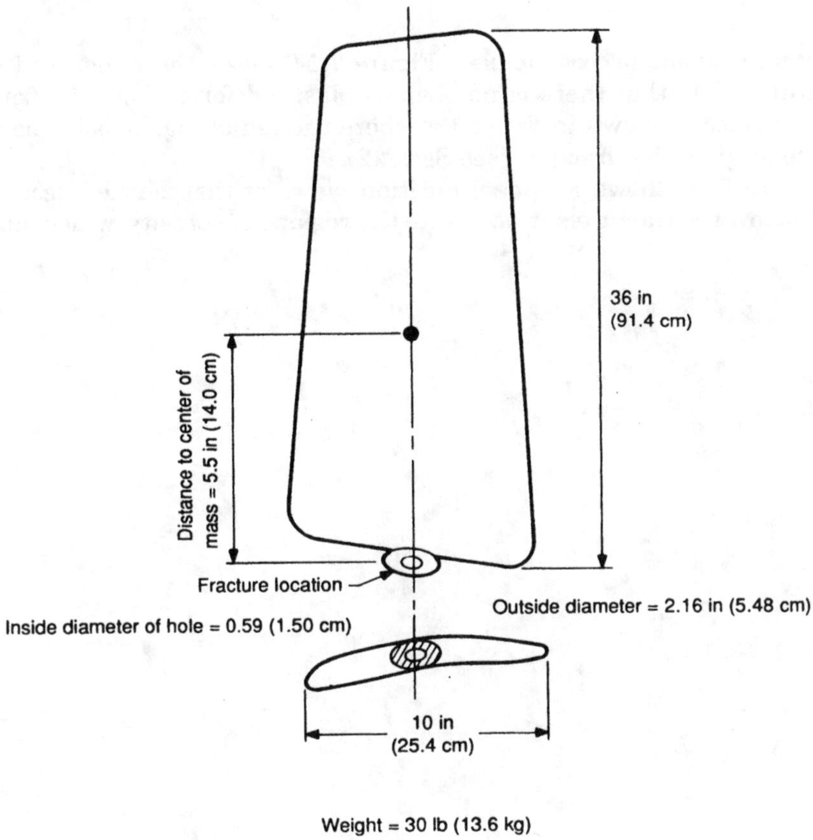

Figure 7.13 Drawing of broken fan blade.

condition. The microstructure was examined using optical microscopy and scanning electron microscopy. The EDS on the scanning electron microscope was used to identify the elements in the alloy and in the phases.

Hardness measurements were made using the Rockwell E scale (accurate to about ±2). Subsize tensile samples were made from the blade and tested to fracture in order to obtain tensile mechanical properties. The tensile strength was accurate to about 3.5 MPa (500 lb/in²) and the elongation at fracture to about 1 percent. Both metallographic and tensile samples were solution heat-treated at 538°C (±5°C) (1000°F) for 12 h, then quenched in water. Some of these were subsequently aged at 155°C (±5°C) (311°F) for 3, 3.5, and 4 h, then air-cooled to 25°C (77°F). These samples allowed a comparison of the effects of heat treatment on the properties and the microstructure.

7.3.3 Results

Fractography and microstructure. Figure 7.14 shows the region of the fracture. Note that there is no obvious plastic deformation. The fracture surface is shown in Fig. 7.15, where the radial marks point back to the origin of the fracture (see Sec. 4.2).

Figure 7.16 shows low-magnification views of the polished section just below the fracture surface. Note the regions of porosity, which sug-

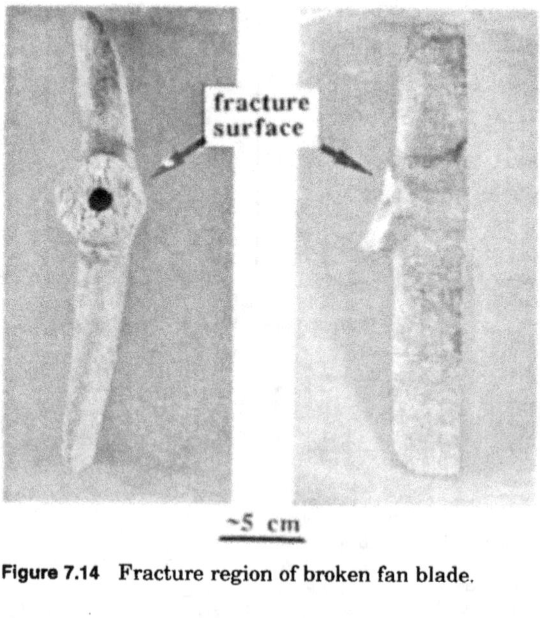

Figure 7.14 Fracture region of broken fan blade.

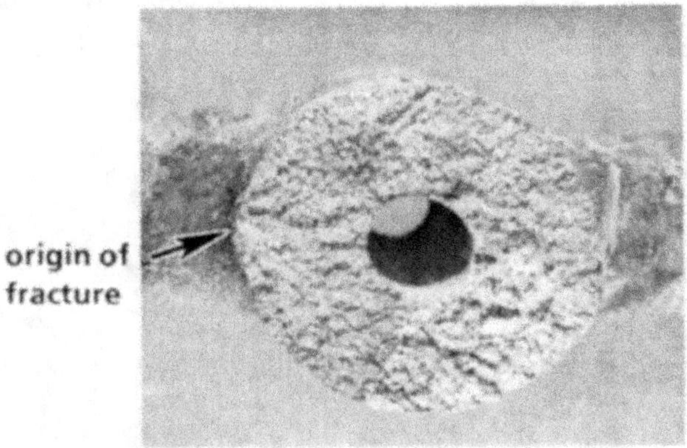

Figure 7.15 Fracture surface, showing radial marks which point to the origin of fracture.

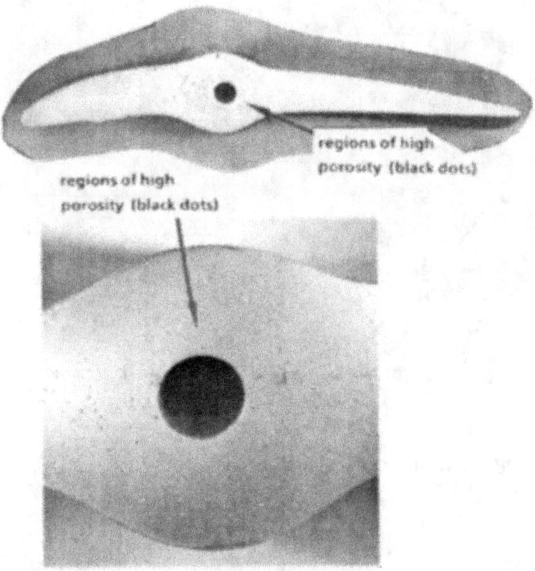

Figure 7.16 Polished section near the fracture surface, showing regions of high porosity in the center.

gest that the blade was made by casting. The spherical areas are due to bubbles of gas that are ejected as the metal freezes and then trapped before they can leave the liquid. These holes are revealed on the fracture surface (Fig. 7.17) and in microstructures, shown in cross section in Fig. 7.18, in a section near the fracture surface. Note that there is also a prominent group of elongated porosity present in this area.

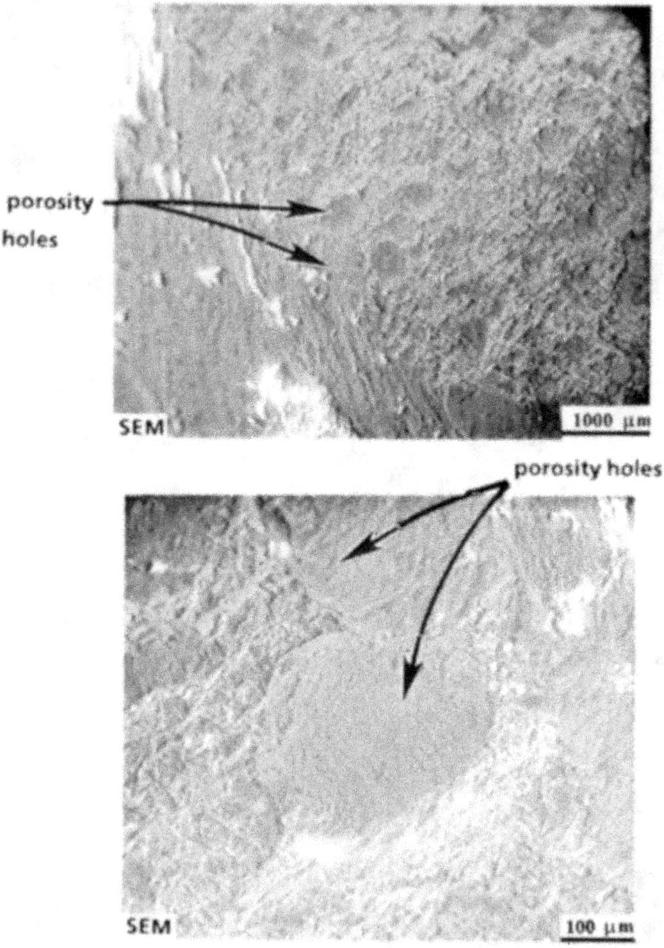

Figure 7.17 Scanning electron micrographs of fracture surface. The spherical holes are where fracture occurred through the holes in Fig. 7.16.

The microstructure in Fig. 7.19 shows interdendritic porosity, which is considerably finer than the spherical holes. Here the dendrites are surrounded by a separate phase. In Fig. 7.20 the phases are identified (based on EDS results). The matrix α is Al, containing a small amount of Si in solid solution. Two types of particles are present—the majority is essentially pure Si, the balance an intermetallic (Fe–Al–Mg–Si) compound. These are regions of a eutectic structure, part of which is α.

These results show that the fan blade was made of an Al casting alloy, a likely candidate being alloy 356 (Table 7.1). Based on the bina-

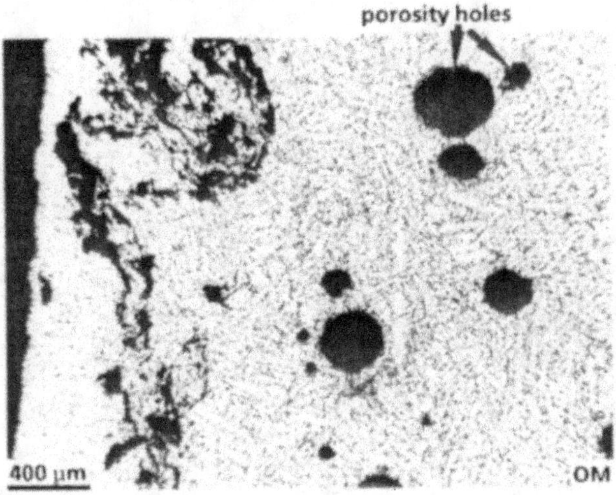

Figure 7.18 Optical micrograph of section near fracture surface, showing porosity.

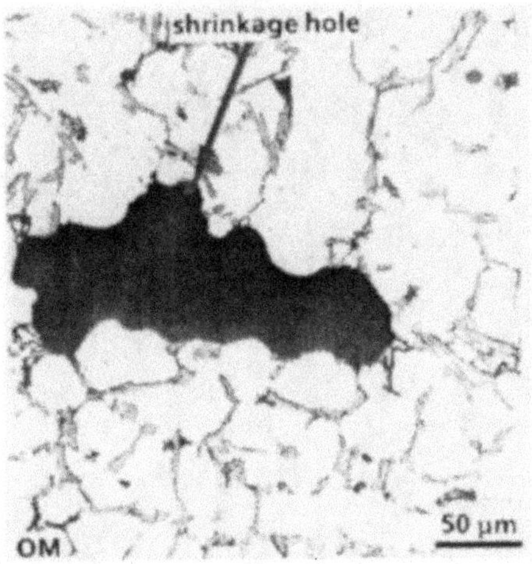

Figure 7.19 Microstructure of region showing interdendritic shrinkage porosity.

ry Al–Si phase diagram (Fig. 7.21), a composition of about 7 percent Si in Al will give the amount of primary α seen in the microstructure (Fig. 7.20*a*) if the alloy is cooled relatively slowly from the liquid to 25°C (77°F), such as would occur for a sand casting.

Note in Fig. 7.20*b* that the particles are relatively sharp. Upon solution-treating a sample, water quenching, then aging for 3 to 4 h at 155°C (311°F), the particles become rounded (Fig. 7.22). Supposedly this morphology enhances tensile mechanical properties.[12]

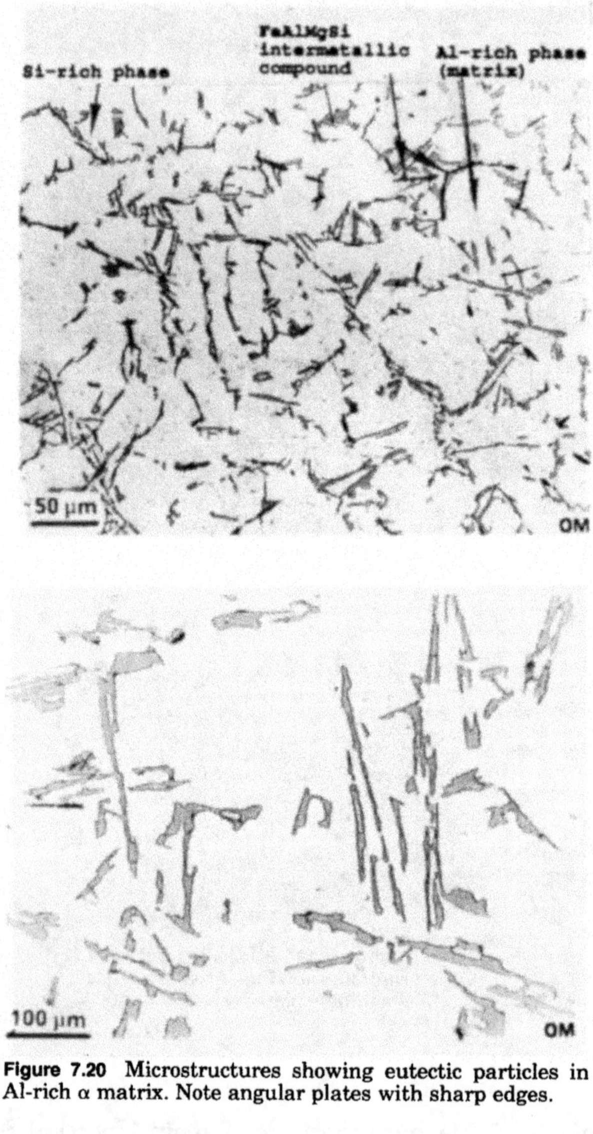

Figure 7.20 Microstructures showing eutectic particles in Al-rich α matrix. Note angular plates with sharp edges.

Mechanical properties. The hardness of the casting ranged from 45 to 55 Rockwell E, the variation being due to the degree of porosity in the area of the hardness indentation. The tensile strength was about 110 MPa (16,000 lb/in^2), and elongation was about 4 percent (Table 7.2). Samples of the blade were solution heat-treated for 12 h at 538°C (1100°F), which as Fig. 7.21 shows, gives the maximum amount of Si dissolved in the α. The hardness was about the same as in the as-received condition (Table

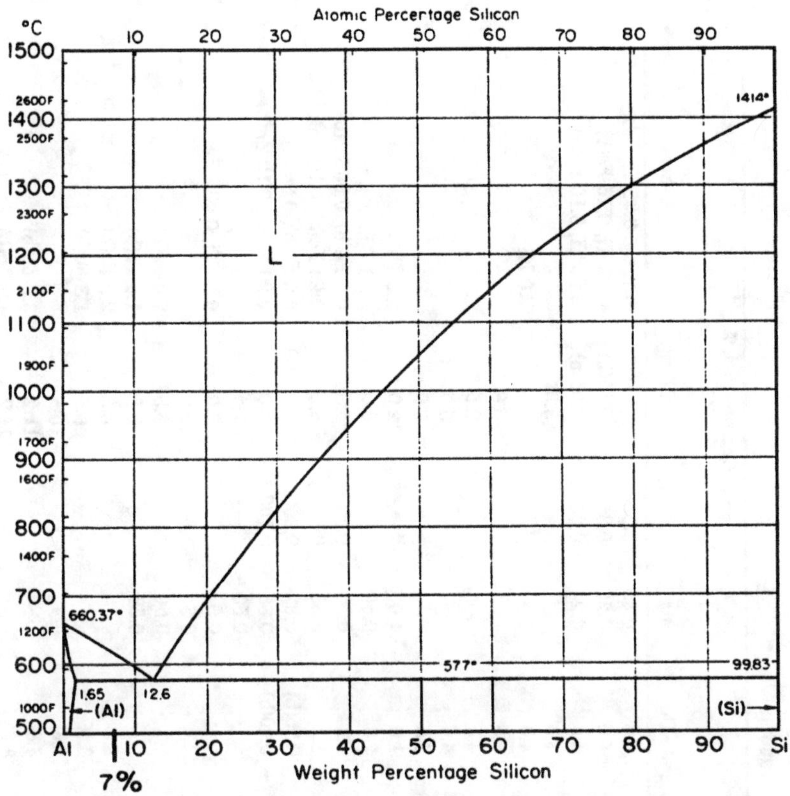

Figure 7.21 Aluminum-silicon phase diagram. The approximate composition of the fan blade alloy is shown at 7% Si. (*From* Metals Handbook.[11])

7.2). Samples were then heated-treated at 155°C (311°F) to induce precipitation hardening,[10] and the results are shown in Fig. 7.23. Note that the hardness increased significantly. The tensile strength is similar to that listed for the heat treatment T6 for this alloy (Table 7.3).

Stress analysis. There are four forces to consider in calculating the stress on the blade—gravitational, lift, drag, and centrifugal. The gravitational force is obtained from the mass of the fan, which was 24 kg (30 lb). The lift force L is given by[14]

$$L = C_L \rho bc \left(\frac{u^2}{2g}\right)$$

where C_L = lift coefficient = 1

ρ = density of air = 1.2 kg/m³ (0.043 lb/in³)

TABLE 7.1 Chemical Composition of Common Aluminum Casting Alloys

Alloys				Composition, %				
AA number	Former AA designation	Former ASTM number	Product(a)	Cu	Mg	Mn	Si	Others
201.0	...	...	S	4.6	0.35	0.35	...	0.7 Ag, 0.25 Ti
206.0	...	...	S or P	4.6	0.25	0.35	0.10(b)	0.22 Ti, 0.15 Fe(b)
A206.0	...	...	S or P	4.6	0.25	0.35	0.05(b)	0.22 Ti, 0.10 Fe(b)
208.0	108	CS43A	S	4.0	...	...	3.0	...
242.0	142	CN42A	S or P	4.0	1.5	...	...	2.0 Ni
295.0	195	C4A	S	4.5	...	...	0.8	...
296.0	B295.0, B195	...	P	4.5	...	...	2.5	...
308.0	A108	SC64A	S or P	4.5	...	...	5.5	...
319.0	319, Allcast	SC64D	S or P	3.5	...	...	6.0	...
336.0	A332.0, A132	SN122A	P	1.0	1.0	...	12.0	2.5 Ni
354.0	354	SC92A	P	1.8	0.50	...	9.0	...
355.0	355	SC51A	S or P	1.2	0.50	0.50(b)	5.0	0.6 Fe(b), 0.35 Zn(b)
C355.0	C355	SC51B	S or P	1.2	0.50	0.10(b)	5.0	0.20 Fe(b), 0.10 Zn(b)
356.0	356	SG70A	S or P	0.25(b)	0.32	0.35(b)	7.0	0.6 Fe(b), 0.35 Zn(b)
A356.0	A356	SG70B	S or P	0.20(b)	0.35	0.10(b)	7.0	0.20 Fe(b), 0.10 Zn(b)
357.0	357	...	S or P	...	0.50	...	7.0	...
A357.0	A357	...	S or P	...	0.6	...	7.0	0.15 Ti, 0.005 Be
359.0	359	SG91A	S or P	...	0.6	...	9.0	...
360.0	360	SG100B	D	...	0.50	...	9.5	2.0 Fe(b)
A360.0	A360	SG100A	D	...	0.50	...	9.5	1.3 Fe(b)
380.0	380	SC84B	D	3.5	...	...	8.5	2.0 Fe(b)
A380.0	A380	SC84A	D	3.5	...	...	8.5	1.3 Fe(b)
383.0	...	SC102A	D	2.5	...	...	10.5	...
384.0	384	SC114A	D	3.8	...	...	11.2	3.0 Zn(b)
A384.0	384	SC114A	D	3.8	...	...	11.2	1.0 Zn(b)

390.0	390	...	D	4.5	0.6	...	17.0	1.3 Zn(b)
A390.0....	A390	...	S or P	4.5	0.6	...	17.0	0.5 Zn(b)
413.0	13	S12B	D	...	...	...	12.0	2.0 Fe(b)
A413.0....	A13	S12A	D	...	...	...	12.0	1.3 Fe(b)
4430......	43	S5B	S	0.6(b)	...	...	5.2	...
A443.0....	43	...	S	0.30(b)	...	...	5.2	...
B443.0....	43	S5A	S or P	0.15(b)	...	...	5.2	...
C443.0....	A43	S5C	D	0.6(b)	...	...	5.2	2.0 Fe(b)
514.0	214	G4A	S	...	4.0	...	...	...
518.0	218	G8A	D	...	8.0	...	...	...
520.0	220	G10A	S	...	10.0	...	...	...
535.0	Almag 35	GM70B	S	...	6.8	0.18	...	0.18 Ti
A535.0....	A218	...	S	...	7.0	0.18	...	...
B535.0....	B218	...	S	...	7.0	...	...	0.18 Ti
712.0	D712.0, D612, 40E	ZG61A	S or P	...	0.6	...	...	5.8 Zn, 0.5 Cr, 0.20 Ti
713.0	613, Tenzaloy	ZC81A,B	S or P	0.7	0.35	...	...	7.5 Zn, 0.7 Cu
771.0	Precedent 71A	ZG71B	S	...	0.9	...	...	7.0 Zn, 0.13 Cr, 0.15 Ti
850.0	750	...	S or P	1.0	...	...	...	6.2 Sn, 1.0 Ni

(a) S = sand casting, P = permanent mold casting, D = die casting. (b) Maximum.

SOURCE: From *Metals Handbook.*[10]

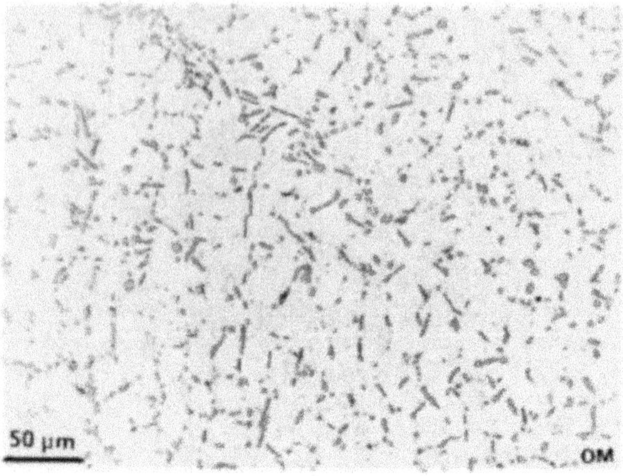

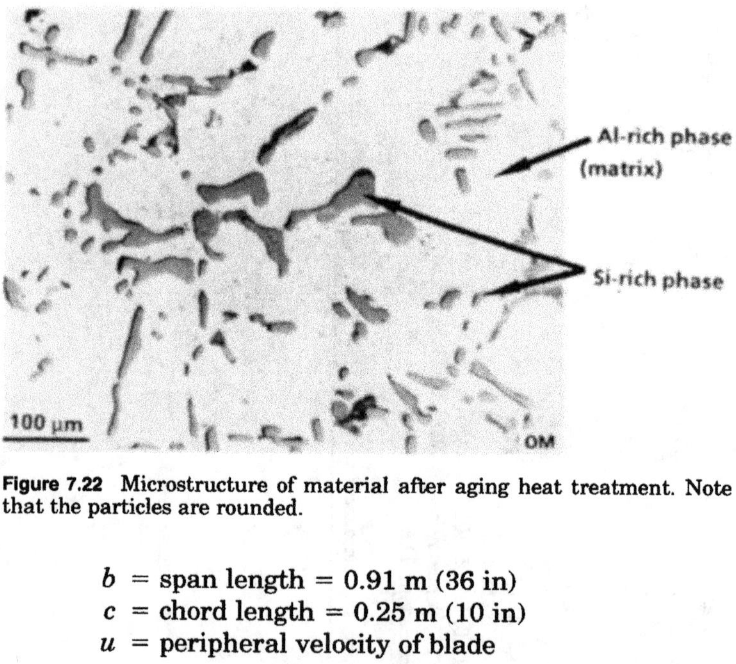

Figure 7.22 Microstructure of material after aging heat treatment. Note that the particles are rounded.

$$b = \text{span length} = 0.91 \text{ m (36 in)}$$
$$c = \text{chord length} = 0.25 \text{ m (10 in)}$$
$$u = \text{peripheral velocity of blade}$$
$$= \left(\frac{\text{r/min}}{60}\right)\pi D = \left(\frac{550}{60}\right)(\pi)(72 \text{ in}) = 53 \text{ m/s (2073 in/s)}$$

Then

$$L = (1)(0.042 \times 10^{-3})(36)(10)(2073^2)\left[\frac{1}{(2)(9.8)(3)(12)}\right]$$
$$= 375 \text{ N (84 lb)}$$

TABLE 7.2 Effect of Heat Treatment on Hardness, Tensile Strength, and Elongation at Fracture for Fan-Blade Material

(a) Results of Tensile Tests

	Heat-treatment Condition			
	As received		Solution treatment + T6	
Sample number	T1	T2	T3	T4
Hardness, Rockwell E	49–55	45–55	77–78	77–78
Tensile strength, lb/in^2	16,200	12,000, low strength, lots of defects	25,900	28,000
Elongation, %	4.1	—	—	2.5
Reduction of area, %	3.6	—	(Broke out of gage)	2.2
Typical Values				
Tensile strength, lb/in^2	16,000		27,000	
Elongation, %	4		3	
Nominal Mechanical Properties of 356-T6				
Tensile strength, lb/in^2	38,000			
Elongation, %	5			

(*b*) Effect of Heat Treatment of Hardness

Heat-treatment condition	Hardness, Rockwell E
Solution treatment (ST): 538°C, 12 h, water-cooled	42, 32, 38, 42, 37
ST + T6 (155°C, 3 h)	77, 77, 79, 74
ST + T6 (155°C, 3.5 h)	73, 76, 78, 75
ST + T6 (155°C, 4 h)	67, 66, 72
As-received, broken blade	45–55

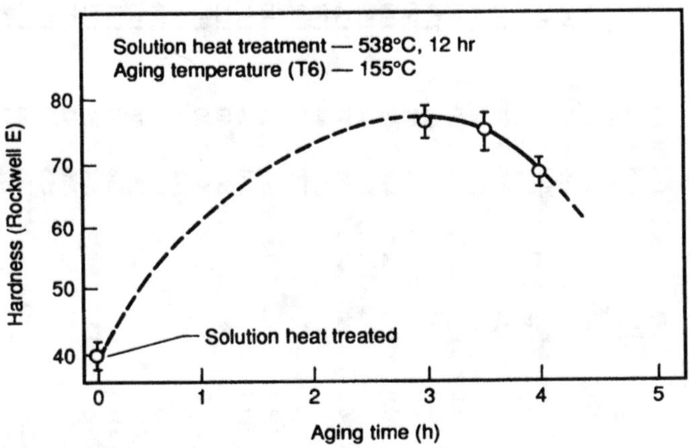

Figure 7.23 Effect of aging at 155°(311°F) on the hardness of the fan blade material.

TABLE 7.3 Typical Mechanical Properties of Common Aluminum Sand Casting Alloys

AA No.	Temper	Tensile strength MPa	Tensile strength ksi	Tensile yield strength(b) MPa	Tensile yield strength(b) ksi	Elongation in 50 mm or 2 in., %	Hardness(c), HB	Compressive yield strength(b) MPa	Compressive yield strength(b) ksi	Shear strength MPa	Shear strength ksi	Endurance limit(d) MPa	Endurance limit(d) ksi	Modulus of elasticity(e) kPa × 10^6	Modulus of elasticity(e) psi × 10^6
201.0	T43	414	60	255	37	17.0	...	...	...	...	...	...	...	...	...
	T6	448	65	379	55	8.0	130	386	56	290	42	...	...	...	...
	T7	469	68	414	60	5.0	...	...	...	...	...	97	14	...	...
A206.0	T4	431	62	264	38	17.0	...	285	41	292	42	...	...	...	...
	T7	436	63	347	50	11.7	...	372	54	257	37	...	...	...	...
213.0	F	207	30	165	24	1.5	85	172	25	165	24	66	9.5	...	...
222.0	T52	241	35	214	31	1.0	100	214	31	172	25	...	...	...	...
	T551	255	37	241	35	<0.5	115	276	40	207	30	59	8.5	74	10.7
	T65	331	48	248	36	<0.5	140	248	36	248	26	62	9	74	10.7
238.0	F	207	30	165	24	1.5	100	207	30	165	24	...	...	...	...
242.0	T571	276	40	234	34	1.0	105	234	34	207	30	72	10.5	71	10.3
	T61	324	47	290	42	0.5	110	303	44	241	35	66	9.5	71	10.3
249.0	T63	476	69	414	60	6.0	...	...	...	...	...	...	...	...	...
	T7	278	62	359	52	9.0	...	414	60	276	40	55	8.0	72	10.5
296.0	T4(f)	255	37	131	19	9.0	75	138	20	207	30	66	9.5	70	10.1
	T6	276	40	179	26	5.0	90	179	26	221	32	69	10	70	10.1
	T7	270	39	138	20	4.5	80	138	20	207	30	63	9	70	10.1
308.0	F	193	28	110	16	2.0	70	117	17	152	22	90	13	...	...
319.0	F	234	34	131	19	2.5	85	131	19	165	24	...	...	...	...
	T6	276	40	186	27	3.0	95	186	27	...	...	...	...	...	...
324.0	F	207	30	110	16	4.0	70	...	...	...	...	...	...	...	...
	T5	248	36	179	26	3.0	90	...	...	...	...	...	...	...	...
	T62	310	45	269	39	3.0	105	...	...	...	...	...	...	...	...
332.0	T5	248	36	193	28	1.0	105	...	...	...	...	...	...	77	11.2
333.0	F	234	34	131	19	2.0	90	131	19	186	27	100	14.5	...	...
	T5	234	34	172	25	1.0	100	172	25	186	27	83	12	...	...
	T6	290	42	207	30	1.5	105	207	30	228	33	103	15	...	...
	T7	255	37	193	28	2.0	90	193	28	193	28	83	12	...	...
336.0	T551	248	36	193	28	0.5	105	193	28	193	28	93	13.5	...	...
	T65	324	47	296	43	0.5	125	296	43	248	36	...	...	...	...

355.0	T51	205	30	165	24	2.0	75	165	24	165	24	...	...	...	...
	T6	290	42	185	27	4.0	90	185	27	235	34	69	10	...	...
	T61	310	45	275	40	1.5	105	275	40	250	36	69	10	...	...
	T7	275	40	205	30	2.0	85	205	30	205	30	69	10	...	...
	T71	250	36	215	31	3.0	85	215	31	185	27	69	10	...	...
C355.0	T61	303	44	234	34	3.0	90	248	36	221	32	97	14	...	...
356.0	F	179	26	124	18	5.0	...	...	...	...	...	...	...	...	...
	T51	186	27	138	20	2.0	...	...	...	...	...	...	...	...	...
	T6	262	38	186	27	5.0	80	186	27	207	30	90	13	72	10.5
	T7	221	32	165	24	6.0	70	165	24	172	25	76	11	72	10.5
A356.0	T61	283	41	207	30	10.0	90	221	32	193	28	90	13	72	10.5
357.0	F	193	28	103	15	6.0	...	...	...	...	...	...	...	...	...
	T51	200	29	145	21	4.0	...	...	...	...	...	...	...	...	...
	T6	359	52	296	43	5.0	100	303	44	241	35	90	13	...	...
A357.0	T6	359	52	290	42	5.0	100	296	43	241	35	103	15	...	...
359.0	T62	345	50	290	42	5.5	...	...	...	...	...	110	16	...	...
A390.0	F	200	29	200	29	<1.0	110	...	...	...	...	...	...	82	11.9
	T5	200	29	200	29	<1.0	110	...	...	...	...	...	...	...	...
	T6	310	45	310	45	<1.0	145	414	60	...	...	117	17	...	...
	T7	262	38	262	38	<1.0	120	359	52	...	...	100	14.5	...	...
443.0	F	159	23	62	9	10.0	45	62	9	110	16	55	8	71	10.3
A444.0	F	165	24	76	11	13.0	44	...	...	...	...	...	...	...	...
	T4	159	23	69	10	21.0	45	76	11	110	16	55	8	...	...
513.0	F	186	27	110	16	7.0	60	117	17	152	22	69	10	...	...
711.0	F	241	35(g)	124	18(g)	8.0(g)	70(g)	...	...	...	...	76	11	76	11.0
850.0	T5	159	23	76	11	12.0	45	76	11	103	15	62	9	71	10.3
851.0	T5	138	20	76	11	5.0	45	76	11	97	14	62	9	71	10.3
852.0	T5	221	32	159	23	5.0	70	159	23	148	21	76	11	71	10.3

(a) Tension and hardness values determined by tests on standard 13-mm (1/2-in.) diam test specimens, without surface machining, each cast in permanent mold. (b) At 0.2% offset. (c) 500-kg (1102-lb) load on 10-mm (0.4-in.) ball. (d) Endurance limits based on 500 million cycles of completely reversed stresses using rotating beam–type machine and specimen. (e) Average of tension and compression moduli; compression modulus is about 2% greater than tension modulus. (f) Properties of T4 approach those of T6 after standing for several weeks at room temperature. (g) Tests made approximately 30 days after casting.

SOURCE: From *Metals Handbook*.[13]

The drag force is negligible since $D \ll L$ (see Wallis[14]). The centrifugal force is given by

$$C = \left(\frac{w}{g}\right)\left(\frac{v^2}{r}\right)$$

where r = distance from center of mass = 0.36 m (14 in)
v = velocity of center of mass = (550/60) (π) (28 in) = 20.5 m/s (806 in/s)
w = mass = 13.6 kg (30 lb)

Then

$$C = 16{,}000 \text{ N (3600 lb)}$$

From these values the stress at the neck was calculated. The centrifugal forces cause a tensile stress σ_c of

$$\sigma_c = \frac{3600}{(\pi)(1.08^2) - (\pi)(0.30^2)} = 7.3 \text{ MPa (1060 lb/in}^2)$$

The bending stress σ_b is caused by the gravitational force and the lift force. The maximum bending occurs at the location at which the fracture originated (see Fig. 7.13). This is given by

$$\sigma_b = \frac{My}{I}$$

where M = moment = (30 + 84) (14) = 180 N · m (1596 in · lb)
I = moment of inertia = (π/4) ($R_o^4 - R_i^4$) = (π/4) ($1.08^4 - 0.30^4$) = 4.40×10^{-7} m^4 (1.07 in^4)
y = R_o = 0.027 m (1.08 in)

So

$$\sigma_b = 11.2 \text{ MPa (1620 lb/in}^2)$$

Thus the tensile stress at point a is 11.2 + 7.3 = 18.5 MPa (2700 lb/in^2).

7.3.4 Discussion

The tensile strength of the broken blade was about 110 MPa (16,000 lb/in^2) (Table 7.2). Since this material is relatively "brittle," as evidenced by the low tensile ductility (Table 7.2), a simple design criterion is to require the tensile strength to be equal to the maximum imposed tensile stress. (Note that in this brittle material, the yield strength is not reported.) However, the loading is both static and variable, the latter involving both variations in the centrifugal stress and machine vibrations. According to Table 7.2, in the aged condition (such as T6) alloy 356 has a fatigue strength (endurance limit) (see Sec. 2.13) of about 60 MPa (8500 lb/in^2), which is about one-third of the tensile strength listed. Thus the fatigue strength of the blade would be about 110/3 = 37 MPa (5300 lb/in^2). The hardness measurements varied by about 50 percent (Table 7.2) due to porosity, and thus the fatigue strength may be as low as $37 \times 0.5 = 19$ MPa (2600 lb/in^2). Due to the roughness of the surface of the casting, this value will be reduced further by about a factor of 2. Thus the expected fatigue strength is about 10 MPa (1300 lb/in^2). This is less than the calculated maximum tensile stress of 19 MPa (2700 lb/in^2), which is too low.

The 356 alloy is precipitation-hardenable,[1,15] but the blade had received no such heat treatment. The measured tensile strength, after aging, of 186 MPa (27,000 lb/in^2) is similar to that reported for this alloy in this condition (Table 7.3). Even if the alloy had been precipitation-hardened, the strength would only have been similar to the applied stress, which is too low a "safety factor" for a dynamic machine part.

7.3.5 Conclusions

The failure was due to the use of material which did not have requisite fatigue strength. Even if the blade had been heat-treated to strengthen it, the safety factor would have been marginal. Thus, consideration should have been given to the use of another material.

7.4 Case C: A Cracked Automobile Flywheel Flex Plate*

7.4.1 Introduction

Automobile parts are subjected to complex vibration loading, and thus fatigue is a commonly observed failure mode in flex plates. This case study involved a cracked automobile flywheel from a 1978 Oldsmobile Custom Cruiser Station Wagon. Figure 7.24 shows the flywheel; the location of the crack is indicated by the arrow. There was no obvious plastic deformation, such as bending. Figure 7.25 shows the relation of such a flywheel to the drive train of an automobile. It serves as a

*The authors acknowledge the work done by David Dellinger on this case study.

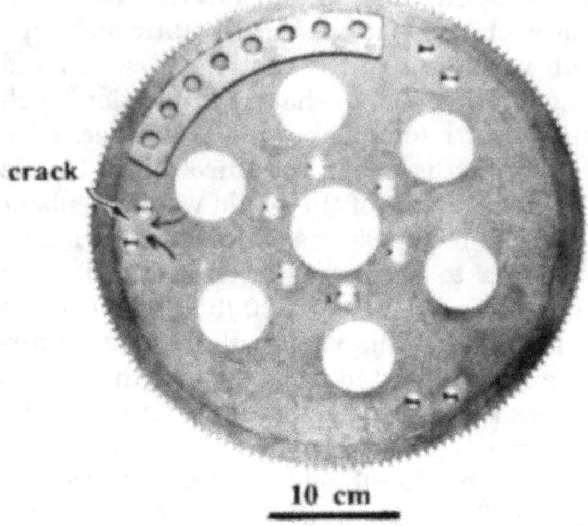

Figure 7.24 Cracked flywheel. The crack is noted by arrow.

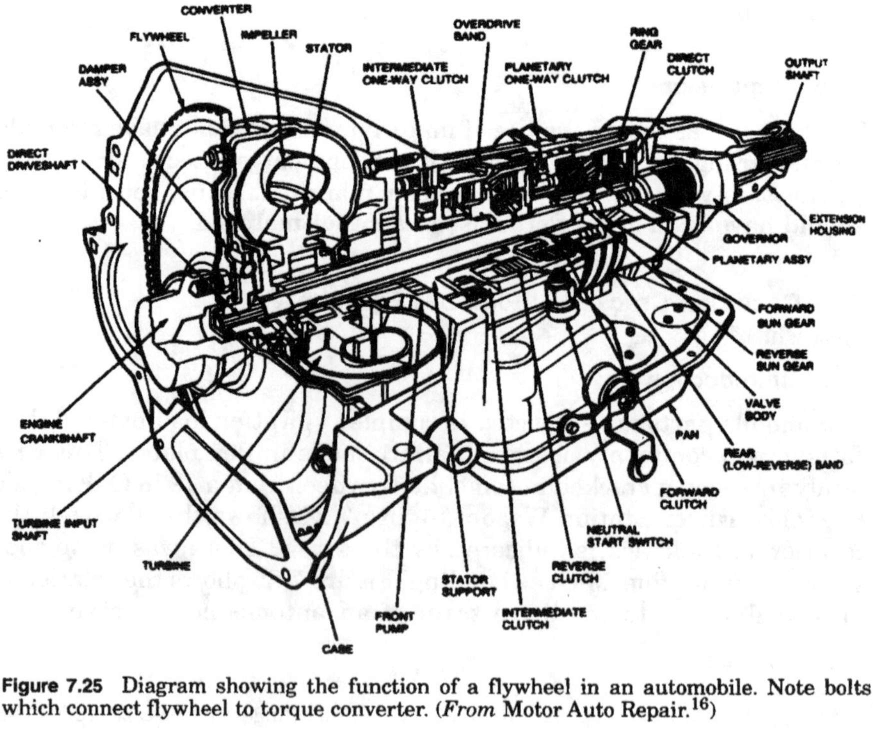

Figure 7.25 Diagram showing the function of a flywheel in an automobile. Note bolts which connect flywheel to torque converter. (*From* Motor Auto Repair.[16])

mechanical link between the engine and the transmission, the torque developed by the engine being transferred to the transmission through the flywheel. The owner stated that there had been no major repairs of the automobile prior to the flywheel failure. The automobile had been driven in excess of 200,000 mi, a reasonable life for such a part.

7.4.2 Experimental procedure

The section containing the crack (Fig. 7.26), and another uncracked section for comparison, were removed by a band saw. From these, samples were cut for metallography. These were prepared by mounting them in plastic and then grinding and polishing. The microstructure was revealed by etching in 2% nitric acid in methanol.

It is seen in Fig. 7.26 that the crack extended from the edge of one hole to just below the adjacent hole. To examine the fracture surface, the sample was cut between the holes, which exposed part of the fracture surface. However, this surface showed extensive rubbing (see Sec. 3.13), so the remaining portion was notched opposite the end of the crack (Fig. 7.27) and broken to expose it. This allowed examination of the entire crack surface in the scanning electron microscope and EDS analysis of it.

7.4.3 Results

Fractography. Low-magnification fractographs are shown in Fig. 7.28. There was no indication of gross plastic deformation. The fracture region extending inward from the left edge through about one-quarter of the thickness is coarse and irregular, and the remainder of the fracture surface is smooth. On the smooth part, conchoidal marks are visible; thus the fracture mode was fatigue (see Sec. 4.5). The coarse fracture surface appears to be the final overload separation area, and this extends to the edge of the hole. Thus the fatigue crack did not originate at the edge of the hole, but somewhere along the right edge. This was confirmed by examination of the fracture surface, which was revealed by breaking the sample. The curvature of the conchoidal marks shows that the origin of the fatigue crack was directly below the other bolt hole (Fig. 7.29). This coincides with one of the concentric grooves surrounding the bolt hole (Fig. 7.30).

The fracture surface near the end which exited at one of the holes was extensively rubbed (but still showed conchoidal marks) (Fig. 7.28*b*), but the other end had minimal rubbing damage, and here fatigue striations were found (Fig. 7.31). (As pointed out in Sec. 3.9, fatigue striations are not often so apparent in steels of complex microstructures. However, in this case the structure consisted mainly of single-phase ferrite of uniform

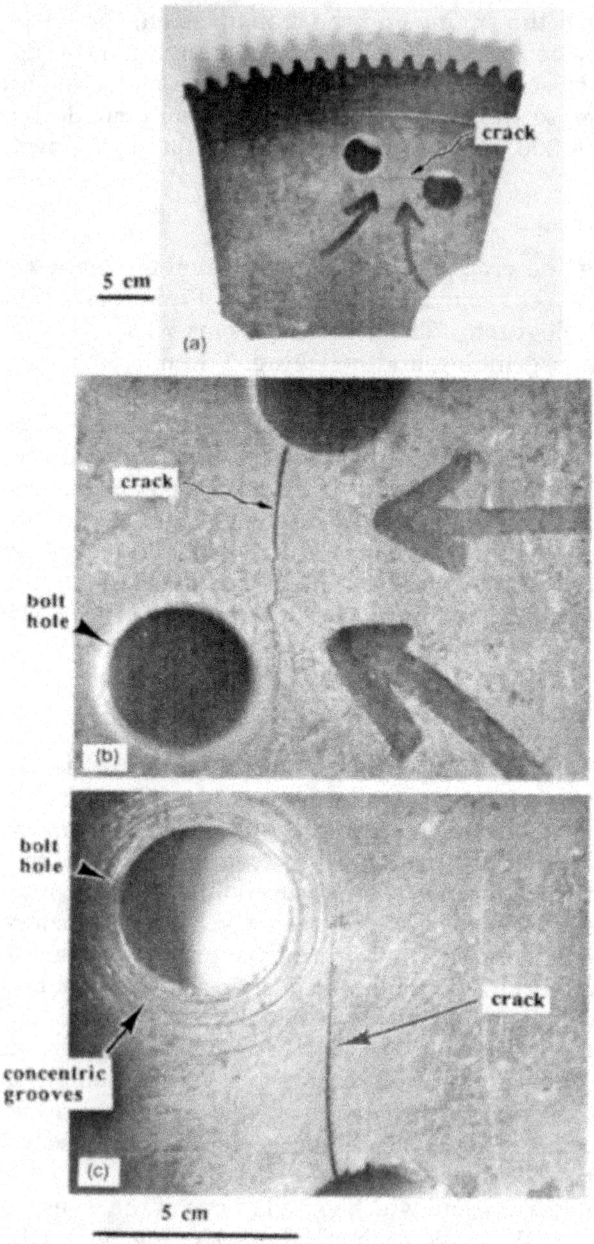

Figure 7.26 Section containing crack.

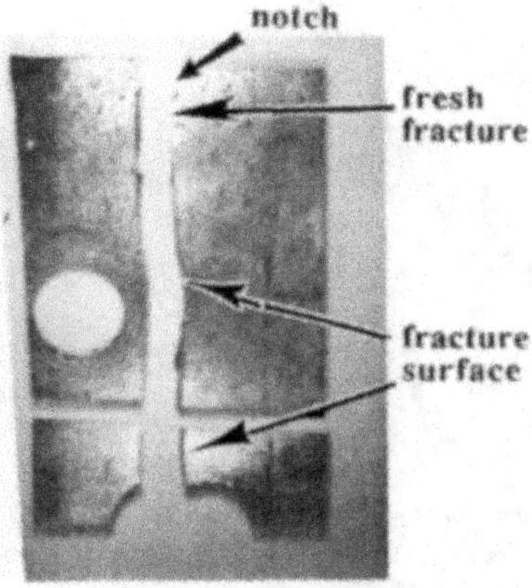

Figure 7.27 Sections cut to reveal fracture surface after breaking the notched part.

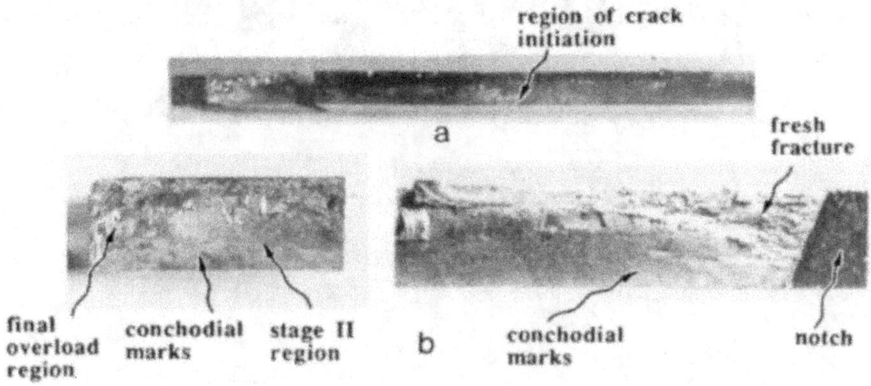

Figure 7.28 Low-magnification scanning electron micrographs of fracture surface.

grain size.) The fracture surface created by fracture after notching, shown in Fig. 7.32, has a dimpled morphology, showing that this region separated by void coalescence (see Sec. 3.5).

Microstructure. The microstructure is shown in Fig. 7.33. EDS analysis revealed Fe, with traces of Mn and Si. The microstructure consisted of a relatively fine ferrite grain size, with a very small amount of pearlite. Thus the plate was probably made of a low-carbon, plain-carbon steel. There was a rather high density of elongated inclusions (Fig. 7.34), and these were also visible on the fracture surface (Fig. 7.35). EDS analysis showed that these were MnS.

7.4.4 Discussion

The flex plate was made of plain-carbon steel having a fine-grain ferrite structure. The service life in excess of 200,000 mi attests to the

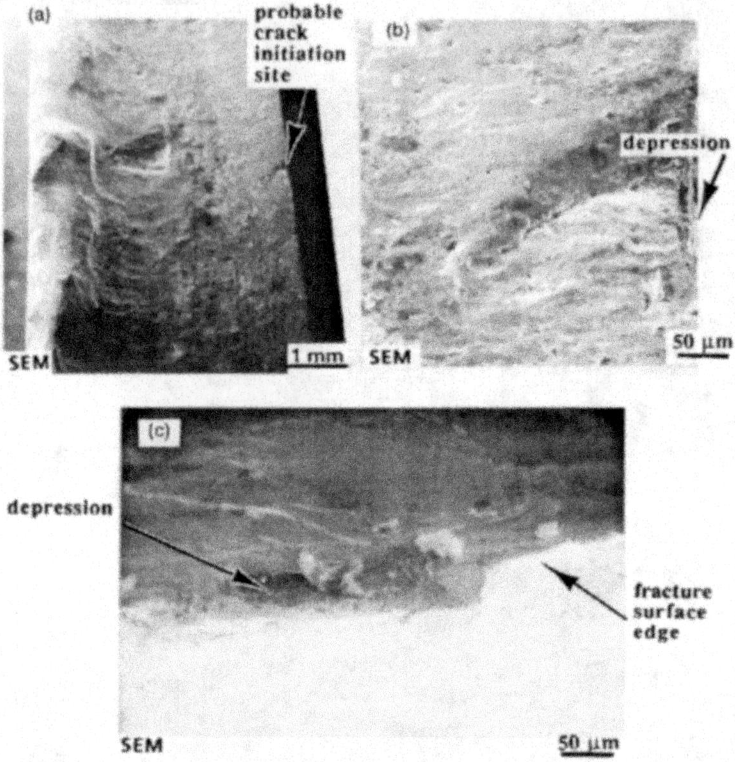

Figure 7.29 Scanning electron micrographs of fracture surface in region of crack origin.

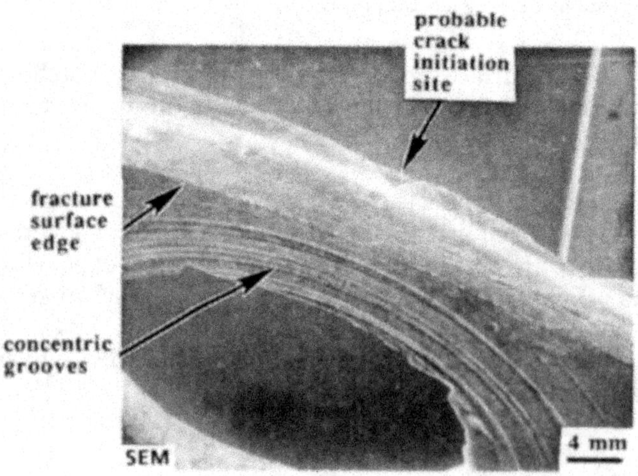

Figure 7.30 Scanning electron micrograph showing damage to the surface from the bolt near the fracture origin.

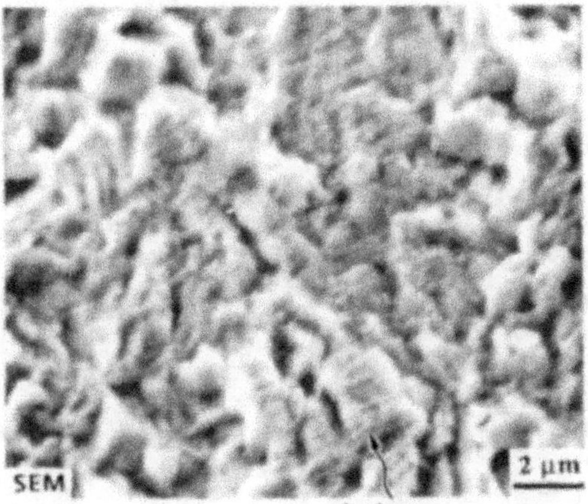

Figure 7.31 Scanning electron micrograph of the fatigue fracture surface showing fatigue striations.

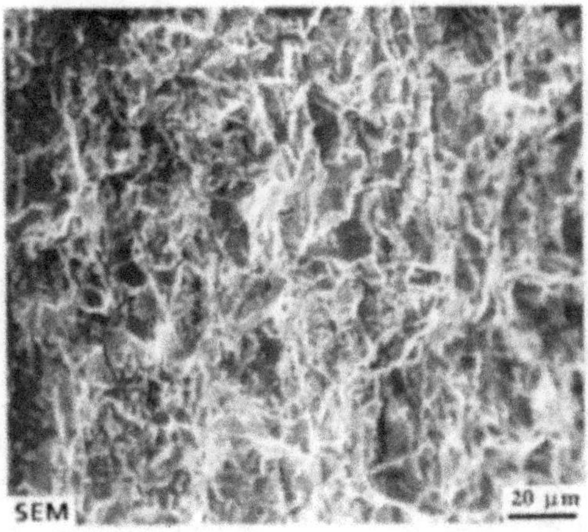

Figure 7.32 Scanning electron micrograph of the fracture surface created by breaking the notched sample.

basic design and choice of material of the plate being sound. In spite of the rather high density of elongated inclusions (Fig. 7.34), they may not have any effect on the fatigue life of such plate material,[17] and cracking does not appear to have originated at the inclusions revealed on the fracture surface (Fig. 7.35). Instead, the origin of the fatigue crack appears to be associated with surface damage due to the bolt or its washer, perhaps when the bolt was tightened during installation.

Although 200,000-mi service is reasonable, a clearer picture of the situation is obtained by estimating the number of cycles of fatigue crack growth. This can be made by dividing the crack length by the fatigue striation spacing (Sec. 3.9). From Fig. 7.28 the distance from the origin to the end of the crack is about 0.5 cm, and from Fig. 7.31, the striation separation is about 0.5 μm. Thus the number of cycles that the fatigue crack grew is about 10^5. On the average, approximately every mile of

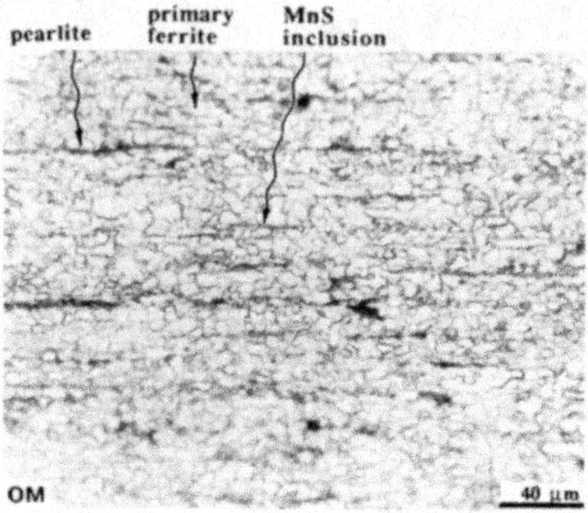

Figure 7.33 Microstructure of the flywheel, showing fine-grain ferrite.

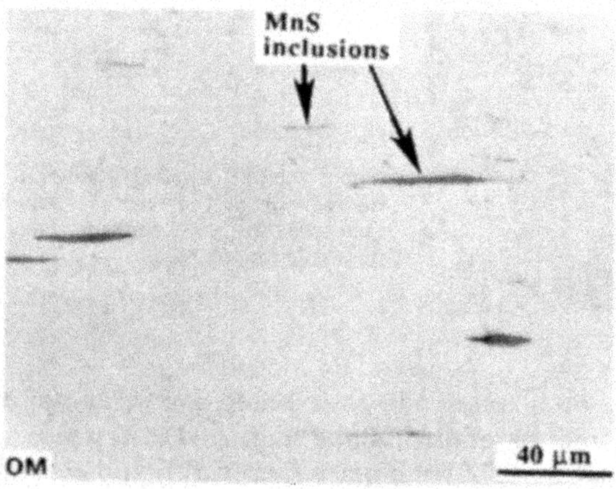

Figure 7.34 Unetched microstructure showing elongated inclusions.

travel the flex plate received a loading sufficiently large to advance the crack one cycle. In fatigue design, the desired fatigue life is usually about 10^6 cycles or greater.

7.4.5 Conclusions

The flex plate cracked in a fatigue mode of fracture. The fatigue crack initiated in a region of the surface damaged by a bolt or its washer. It

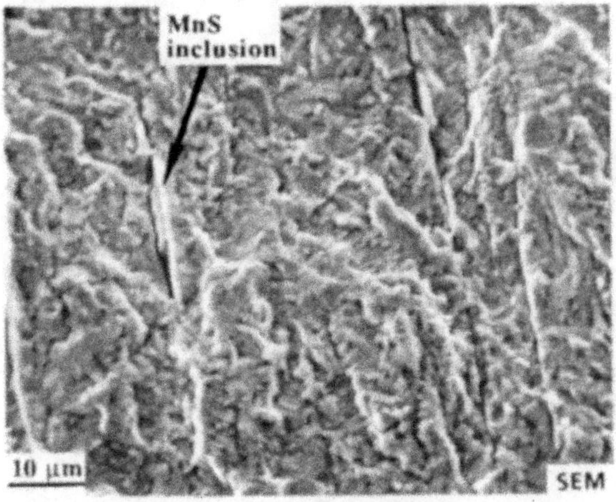

Figure 7.35 Scanning electron micrograph of the fatigue fracture region showing elongated inclusions, identified by EDS as MnS.

should also be remembered that the formation of the fatigue crack in the flex plate may have been symptomatic of excessive vibrations in the automobile from other causes.

7.5 Case D: Failed Welded Railroad Rails*

7.5.1 Introduction

A common method of making railroad rails into very long sections is to weld pieces together by the thermite process. This is illustrated in Fig. 7.36, and the procedure is shown in Fig. 7.37. A mold is placed around the joint between two rails, and a mixture of aluminum and iron oxide powder in a crucible above the mold is ignited. The iron oxide is reduced by the aluminum, forming aluminum oxide. The highly exothermic reaction creates sufficient heat to melt the iron, which then is allowed to flow into the mold. The sensible heat in the molten steel is sufficient to melt part of each rail, and upon cooling, the molten material solidifies to a weld which connects the two rails. The mold is then removed, and the top and sides of the rail are ground to fit the rails on either side. This case study examines fractures which occurred in several thermite welds in railroad rails after only a short service life (for example, 6 months). Figure 7.38 shows a typical broken rail.

*The authors acknowledge the work done by Clayton Crouse on this case study.

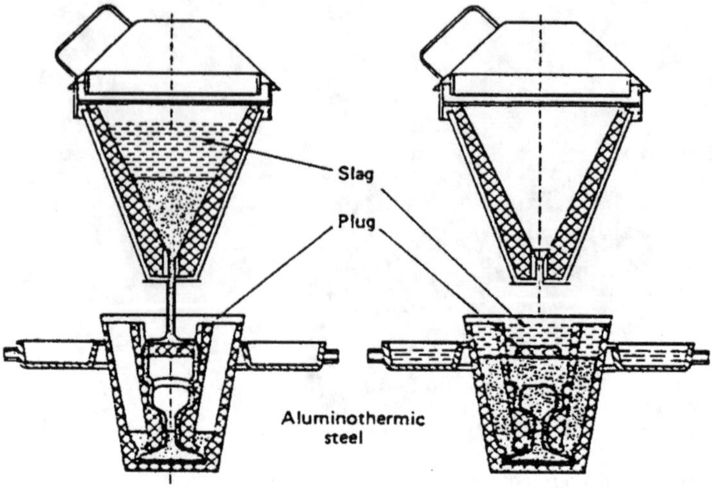

Figure 7.36 Schematic diagram of the process for joining railroad rails by the thermite welding process. (*From H. D. Fricke, in* Metals Handbook,[8] *p. 695.*)

Figure 7.37 Thermite welding process for joining railroad rails. (*From H. D. Fricke, in* Metals Handbook,[8] *p. 695.*)

7.5.2 Experimental procedure

To obtain samples for fractography and metallography, a broken rail was sectioned by a band saw, as shown in Fig. 7.39. The fracture surfaces were cleaned in warm water containing a detergent in an ultrasonic cleaner. A soft brush was also used on the surface after soaking in detergent. After washing in water, the surfaces were cleaned in acetone

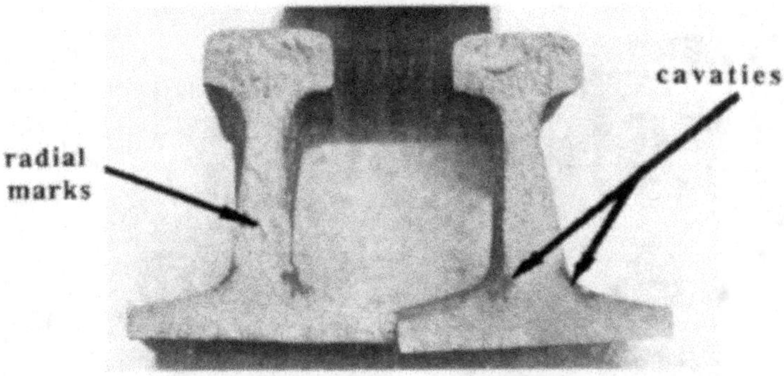

Figure 7.38 Pair of broken rails. Note the radial marks which point to the origin of the fracture.

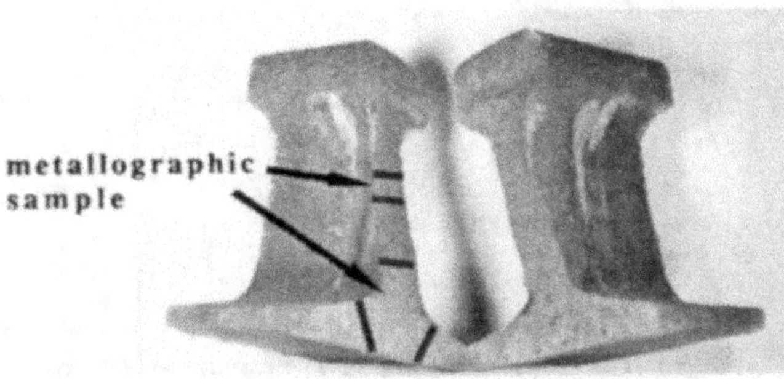

Figure 7.39 Location of samples removed for examination.

agitated by ultrasound, then dried (see Sec. 1.4.1). The metallographic sample was mounted in epoxy, then ground and polished by standard methods. The surface was etched in a solution of 2 percent nitric acid in methanol.

Microhardness measurements using a 500-g load were made on the unetched metallographic sample. The diamond pyramid hardness (DPH) values were accurate to about 10 DPH.

7.5.3 Results

Fractography. Fracture occurred across the weld and not in the adjacent heat-affected zone. The fracture surface in Fig. 7.38 shows radial marks pointing to a fracture origin (Sec. 4.2) on each side of the neck of the bottom web. The crack propagated from these two locations through the cross section of the rail, then to the final failure near the top of the rail. The crack origins contained large cavities, the surface of which consisted of dendrites (Fig. 7.40) (see Sec. 3.10). In these regions

Figure 7.40 Scanning electron micrographs of one of the large cavities at the root of the web, showing dendrites.

cracks could be seen propagating away from the fracture surface. There were similar smaller cavities at various locations on the fracture surface. The fracture surface outside of the cavity areas showed that fracture occurred by cleavage (see Sec. 3.4), as illustrated in Fig. 7.41.

Microstructure. The metallographic sample removed from the web (see Fig. 7.39) allowed examination of the microstructure in cross section, from the fracture surface (area that was molten) through the heat-affected zone and into the unaffected base metal. The locations of microstructures from various areas are shown in Fig. 7.42. The base metal consisted of fine pearlite (Fig. 7.43) with a small amount of primary ferrite. Thus the carbon content is about 0.7 to 0.8 percent. At the surface, decarburization was present, a consequence of fabricating and heat-treating the rail and not associated with the welding process. In the heat-affected zone there was a refinement of the prior austenite

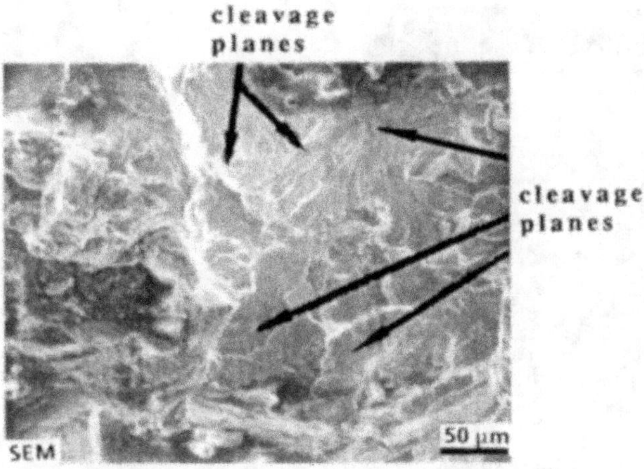

Figure 7.41 Scanning electron micrograph of the fracture surface in the central part of the rail web near the bottom flange. Fracture occurred by cleavage.

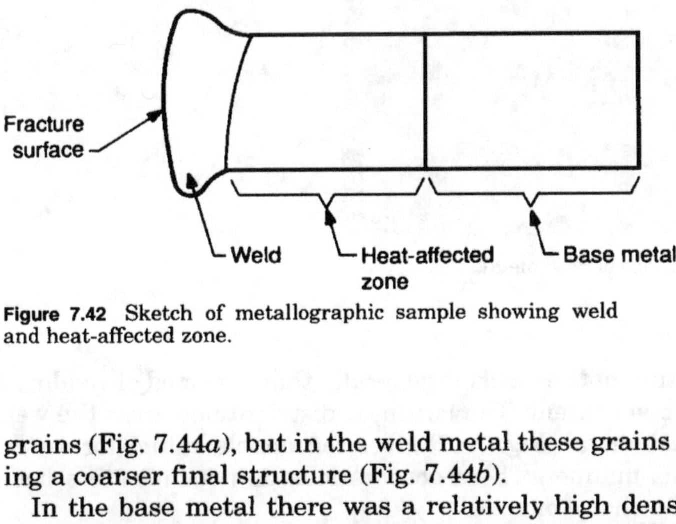

Figure 7.42 Sketch of metallographic sample showing weld and heat-affected zone.

grains (Fig. 7.44*a*), but in the weld metal these grains grew larger, giving a coarser final structure (Fig. 7.44*b*).

In the base metal there was a relatively high density of inclusions elongated parallel to the rail axis (a consequence of hot rolling the rail shape). These were probably MnS inclusions (Fig. 7.45*a*). In the weld zone the inclusions were spherical (Fig. 7.45*b*). Near the edge of the fracture surface, numerous cracks were found (Fig. 7.46).

Hardness measurements. Figure 7.47 shows the hardness distribution across the weld. The DPH values were converted to Rockwell C.[18]

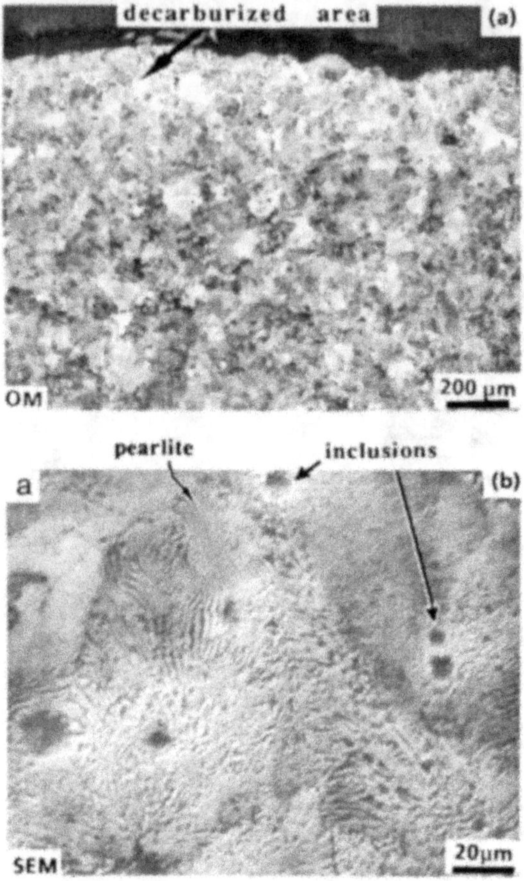

Figure 7.43 Micrographs of base metal.

7.5.4 Discussion

The microstructure of the welds is generally that expected of welding a steel of this carbon content. The hardness distribution across the weld and into the base metal (Fig. 7.47) gives reasonable values for a steel processed in this manner. These observations do not indicate a basic problem with the choice of steel.

The basic cause of the failure of the rails is the presence of the large cavities that formed during solidification, and which acted as crack initiation sites. These cavities formed due to inadequate feeding of the liquid steel during solidification, caused by an improperly designed mold. The railroad company has modified its mold design to alleviate this problem, but field service examination to assess the success of the new design has not been completed.

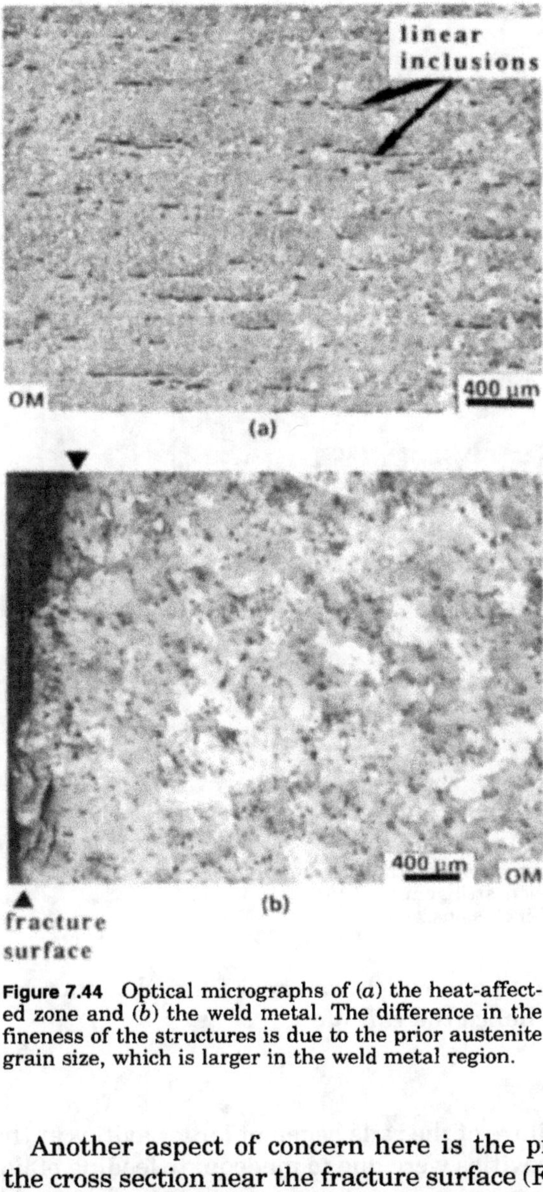

Figure 7.44 Optical micrographs of (*a*) the heat-affected zone and (*b*) the weld metal. The difference in the fineness of the structures is due to the prior austenite grain size, which is larger in the weld metal region.

Another aspect of concern here is the presence of cracks found in the cross section near the fracture surface (Fig. 7.46). These were in the region which had been molten, and most likely are caused by thermal and transformation stresses generated during cooling after solidification. If these cracks existed throughout the weld, then they provided an easy path for the crack to propagate. Establishing the cause of the failure is complicated by the fact that the railroad company had in the previous 2 years changed from a non-heat-treated rail, which had

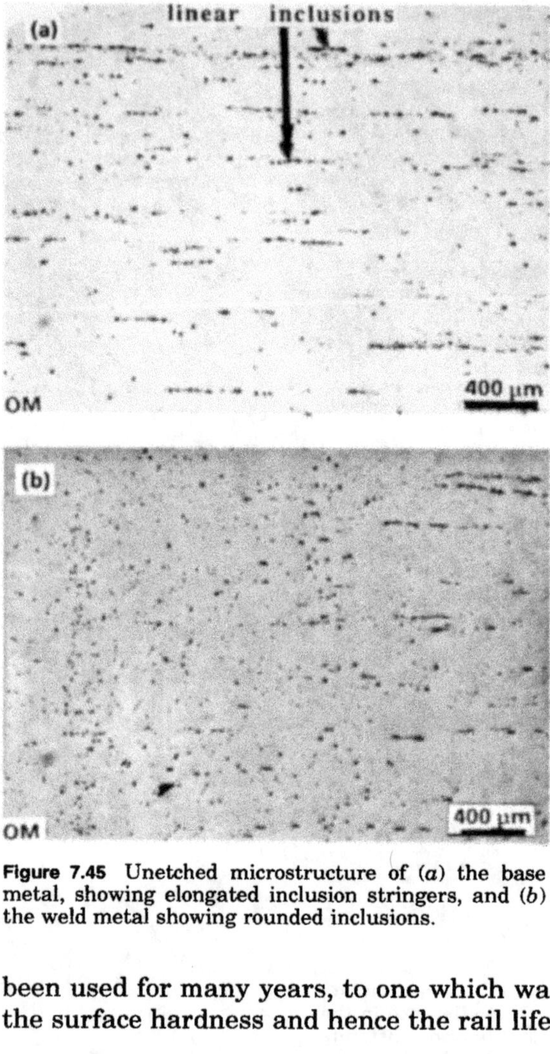

Figure 7.45 Unetched microstructure of (*a*) the base metal, showing elongated inclusion stringers, and (*b*) the weld metal showing rounded inclusions.

been used for many years, to one which was heat-treated to increase the surface hardness and hence the rail life.

7.5.5 Conclusions

The cracks which led to failure of the rails began at large cavities at the bottom of the web. These cavities were due to inadequate feeding of liquid metal. To prevent these required a change in the design of the mold. However, even with proper feeding, the numerous cracks found in the microstructure of the weld are of concern, indicating that limited life may still be a problem. If field testing of the welds using the redesigned molds reveals a continuing fracture problem, then the relation between welding, solidification, and the chemistry of the new rails should be examined carefully.

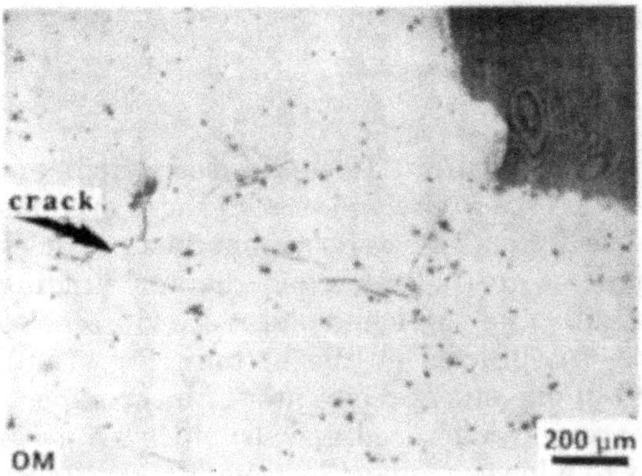

Figure 7.46 Unetched microstructure near the fracture surface showing the presence of numerous cracks.

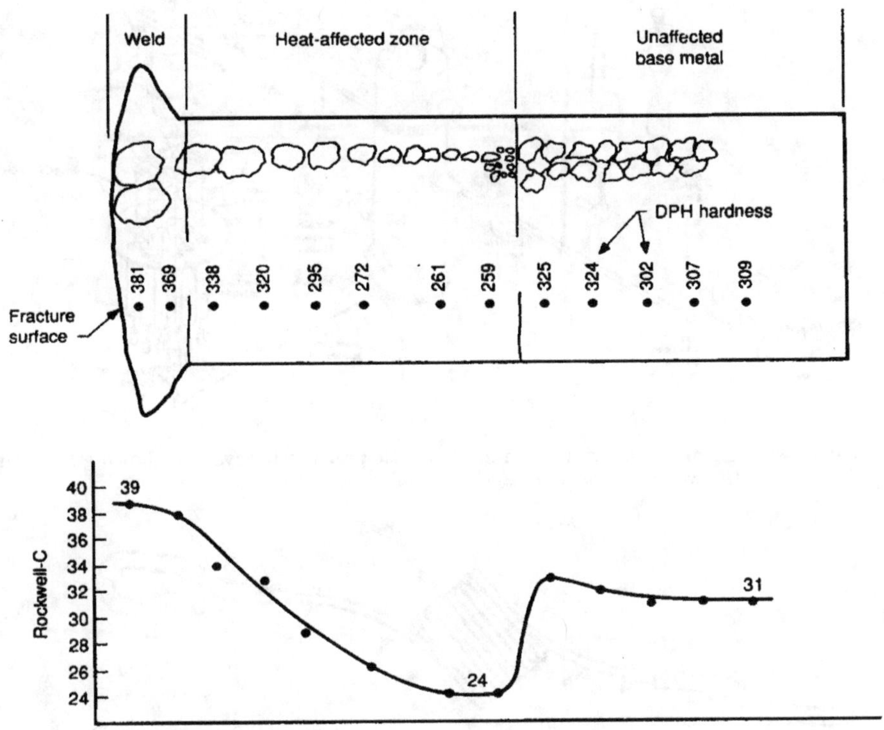

Figure 7.47 Hardness traverse across weld. The locations of these microhardness measurements are approximated.

7.6 Case E: Broken Stainless-Steel Wires from an Electrostatic Precipitator

7.6.1 Introduction

Gas effluents from paper plants contain fine particles which are sometimes removed by an electrostatic precipitator. In these systems thin wires are charged to several thousand volts so that dust particles from flowing gases are attracted to the wires and the surrounding dust collector plates. Periodically, the wires and plates are vibrated to remove these particles into receptacles at the bottom (Fig. 7.48). Several of these wires failed in one such precipitator in an unexpectedly short time. The wires are manufactured in coils which are then straightened by attachment to a lower support by weights (Fig. 7.49).

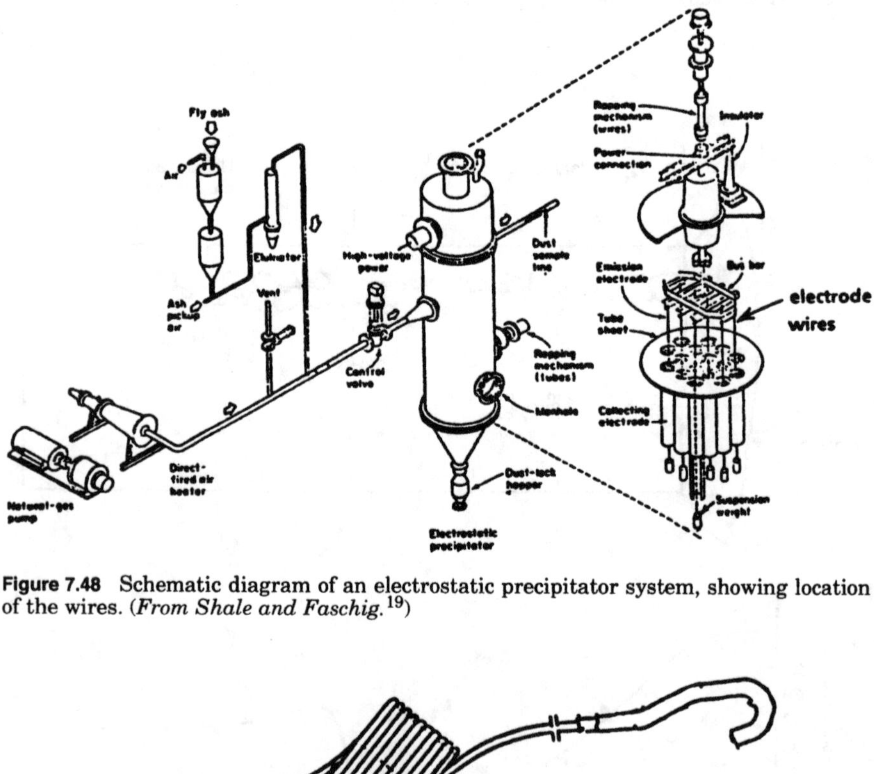

Figure 7.48 Schematic diagram of an electrostatic precipitator system, showing location of the wires. (*From Shale and Faschig.*[19])

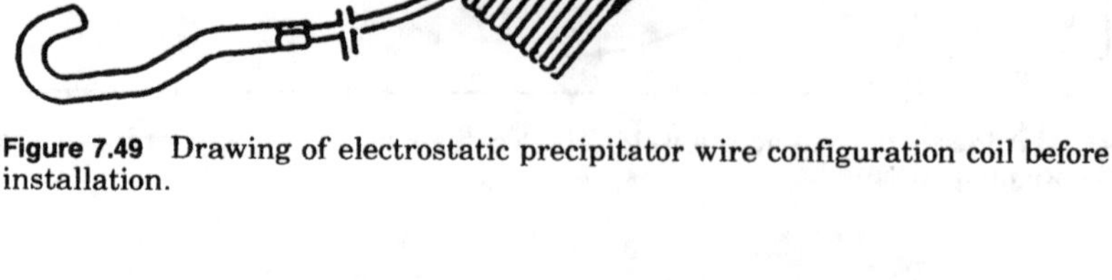

Figure 7.49 Drawing of electrostatic precipitator wire configuration coil before installation.

It was known that the wires were made in Sweden, corresponded to a 316L stainless steel, and had a specification of "3/4 hard."

7.6.2 Experimental procedure

Several broken wires were received. These were cut to a size appropriate for scanning electron microscopy and optical microscopy evaluation. Samples were cleaned in acetone using ultrasonic agitation. Visual examination, macrofractography, and scanning electron fractography were carried out. In addition, microhardness measurements were made on polished metallographic specimens.

7.6.3 Results

Visual examination. Visual examination showed the wire surfaces to be relatively clean, with only superficial corrosion products present. The general plane of the fracture was normal to the wire axis, and there was no obvious indication of plastic deformation.

Hardness measurements. Several microhardness measurements were made and converted to the Rockwell C scale.[20] The average hardness was determined to be 37 Rockwell C (±2).

Fractography. At low magnification the fracture surfaces revealed conchoidal marks (Fig. 7.50) (see Sec. 4.5). From this fractograph the origin of the fatigue fracture is immediately evident. At high magnification (Fig. 7.51), fatigue striations were not seen, but they may have been masked by corrosion products formed on the fracture surface. The final overload region of the fracture surface showed dimples (Fig. 7.52).

A detailed examination of the fracture origin (Fig. 7.50*b*) showed a pit (Fig. 7.53). Such pits were seen not only at the origin of the fracture on several wires, but also along the surface of the wires (Fig. 7.54). These pits had an appearance similar to that seen on stainless steel surfaces in which corrosion pitting occurred.[21]

7.6.4 Discussion

The fractography clearly established that the fracture mode was fatigue. It was also clear that the origin of the fatigue cracks was corrosion pits. Hence the inability of the wires to inhibit pitting corrosion was of concern.

The chemical composition of the steel used to make the wires is given in Table 7.4, which shows the steel to be basically a 316L stainless steel.

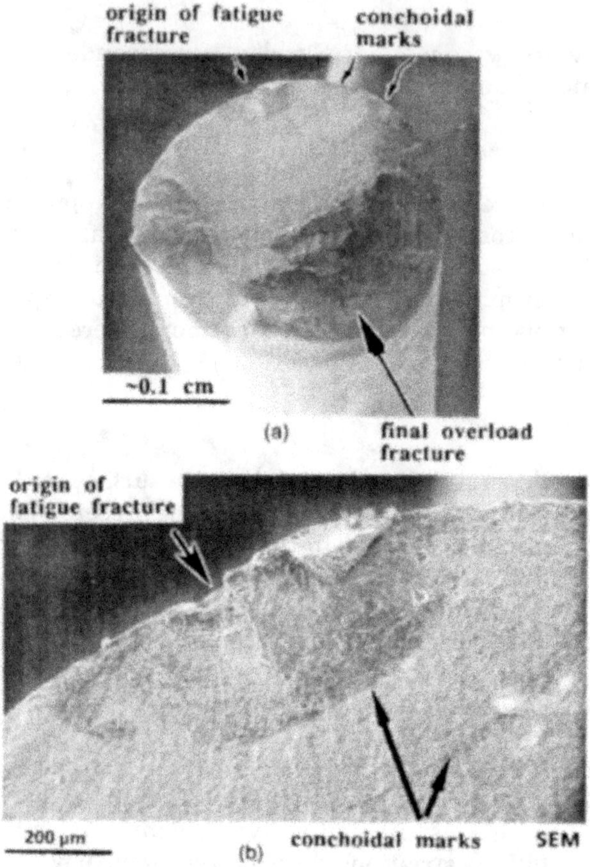

Figure 7.50 Scanning electron micrograph of the fracture surface of one fractured wire, showing that fracture occurred by fatigue.

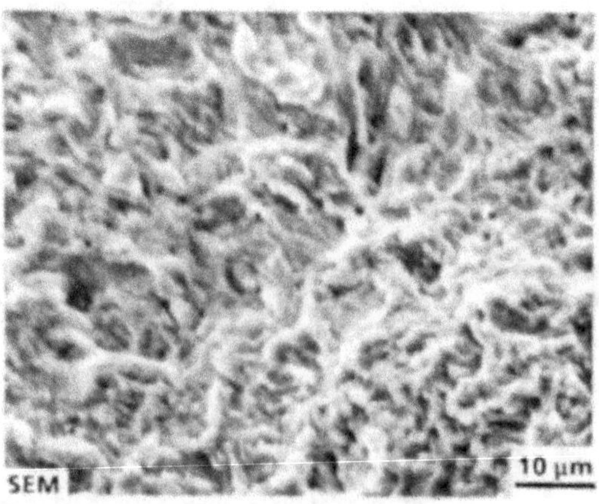

Figure 7.51 Scanning electron micrograph at high magnification of the fatigue fracture propagation region. Note the absence of fatigue striations.

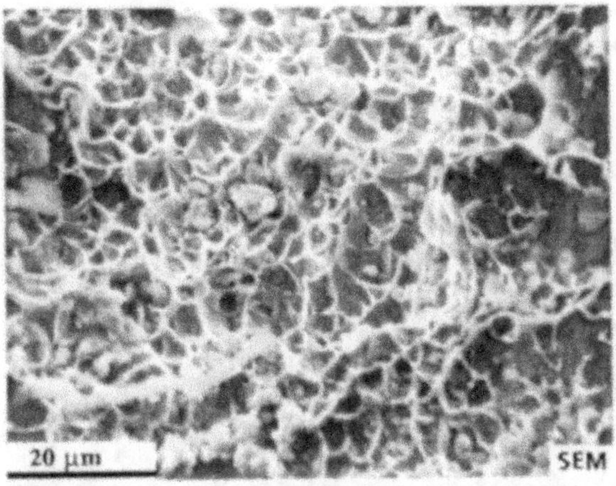

Figure 7.52 Scanning electron micrograph at high magnification of the overload fracture region. The surface shows dimples and some corrosion debris.

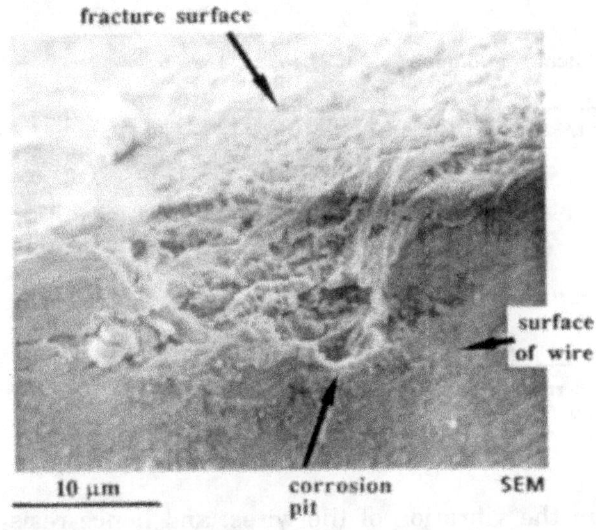

Figure 7.53 Scanning electron micrograph of a corrosion pit at the origin of the fatigue fracture.

The properties of cold-drawn 316 stainless-steel wire are given in Table 7.5. This shows that the hardness in the "3/4 hard" condition should be 29–32 Rockwell C. The measured value of the failed wires (37 Rockwell C) is higher than this and corresponds to the "full hard" condition.

The purpose for specifying that wire in the cold-worked condition be used was not known. The wires are under static load as well as under

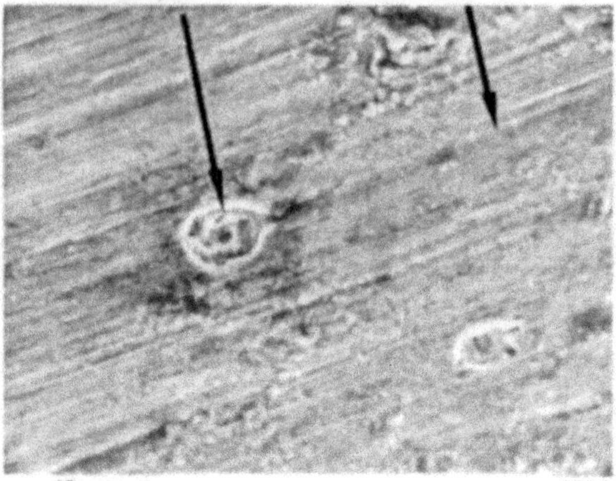

Figure 7.54 Scanning electron micrograph of the surface of a wire showing corrosion pits.

TABLE 7.4 Comparison of Chemical Composition of Swedish and U.S. Steels

Swedish steel SIS 2343, wt.%*	U.S. steel 316 L, wt.%†
0.05C max	0.030C max
1.00Si max	1.00Si max
2.00Mn max	2.00Mn max
0.045P max	0.045P max
0.030S max	0.030S max
16.00–18.00Cr	16.00–18.00Cr
10.50–14.00Ni	10.00–14.00Ni
2.50–3.00Mo	2.00–3.00Mo

*From D. Peckner and I. M. Bernstein, *Handbook of Stainless Steels*, McGraw-Hill, New York, 1977, p. A1-53.
*From *Metals Handbook.*[23]

additional loading from the vibration of the wires, and hence resistance to fatigue fracture is required. Since a common correlation is that the higher the tensile strength, the higher the fatigue strength, then the use of cold-worked stainless steel appears reasonable. However, data (Fig. 7.55)[24] for smooth and notched cold-worked 304 stainless-steel specimens indicate that this steel may be very notch-sensitive in the cold-worked condition. Similar results have been reported for 301 stainless steel (see *Metals Handbook,*[25] p. 765). The fatigue curves in Fig. 7.56[26] for annealed 302 stainless steel show that it is not very notch-sensitive in the annealed condition. Also, data in

TABLE 7.5 Nominal Mechanical Properties of 316 Austenitic Stainless Steel

Condition	Diameter, in	Rockwell C hardness	Yield strength (0.2% offset), 1000 lb/in^2	Tensile strength, 1000 lb/in^2	Elongation (in 2 in), %
¼ hard	0.002–0.020	23–28	100–140	130–170	15–20
	0.021–0.125		100–130	125–165	15–25
	0.126–0.375		80–100	110–130	20–30
½ hard	0.002–0.020	29–32	135–170	165–205	5–17
	0.021–0.125		130–165	160–200	11–18
	0.126–0.375		105–125	135–155	12–20
¾ hard	0.002–0.020	33–37	185–210	205–230	3–10
	0.021–0.125		175–205	200–220	6–12
	0.126–0.375		135–165	165–195	12–20
Full hard	0.002–0.020	38–43	230–260	235–300	1–2
	0.021–0.125		215–245	320–275	2–5
	0.126–0.375		155–185	185–215	3–6

SOURCE: From Ref. 22.

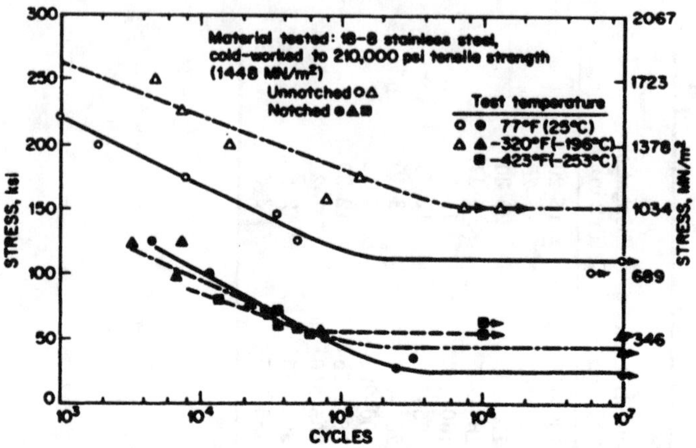

Figure 7.55 Fatigue curves for cold-drawn 304 austenitic stainless steel, showing the notch sensitivity in the cold-worked condition. (*From Spretnak et al.*[24])

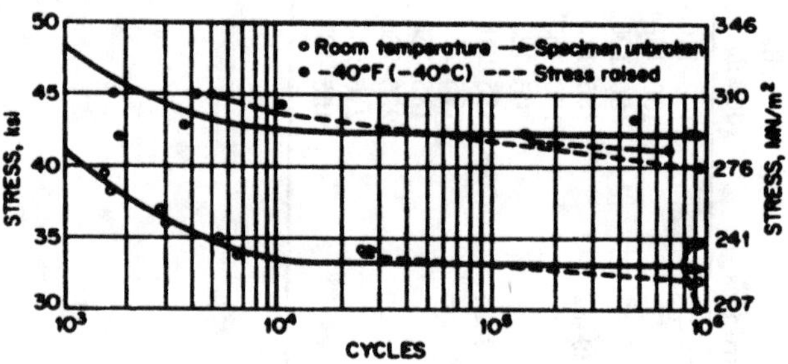

Figure 7.56 Fatigue curves for annealed 302 austenitic stainless steel, showing the lack of notch sensitivity in this condition. (*From Henke.*[26])

Metals Handbook[25] show that with a stress concentration factor K_t of 3, "extra hard" 301 and annealed 304L stainless steel have about the same fatigue strength. Thus the cold-worked 316L steel may have been notch-sensitive.

It thus appears that while the tensile strength was improved by cold working, the material may have been rendered very notch-sensitive. In the chemically aggressive environment of paper-plant effluents the wires were subjected to pitting corrosion. These pits provided the notches for lowering the fatigue strength of the wires.

It is also important to recognize that the fracture of the electrode wires may indicate improper operation of the system, causing excessive vibration and static loading. This aspect was an unknown in this investigation.

7.6.5 Conclusions

The basic cause of the fracture was pitting corrosion, which provided sites for the initiation of fatigue cracks. While the use of material in the cold-worked condition does provide for a better resistance to static overload, the notch sensitivity may be higher for such material. Material with a higher Mo content (such as 317L) is recommended for better resistance to pitting corrosion.

7.7 Case F: Broken Wire Cutters*

7.7.1 Introduction

Diagonal wire cutters are used frequently in a variety of operations, mainly involving electrical wiring. A broken wire cutter was received with the failure location at the point where the handle meets the jaw. Figure 7.57 shows an overall view of the broken cutter (with the failure location indicated). The only visible markings on the broken cutter were "2214" and "DROP FORGED." The manufacturer's trademark was not found, nor was the country of manufacture.

It was learned that the cutters failed while they were being used to cut copper wire of the size normally used in residential house wiring. Wire this size would usually be cut with a larger tool with greater mechanical advantage to reduce the force required by the operator to effect the cut. When received for failure analysis, the cutting edges were in the fully closed position and a screwdriver was used to pry the jaws open. This suggests that the hinge was not functioning properly

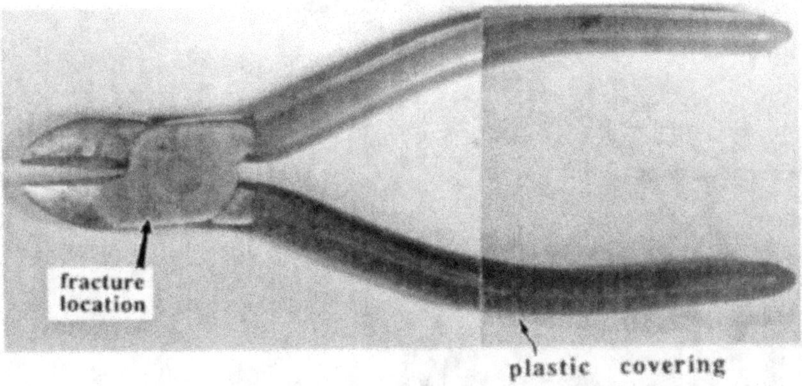

Figure 7.57 Failed wire cutters.

---

*The authors acknowledge the work done by Brian Kruse on this case study.

and this contributed to the excess force required to make the cut when the tool failed.

A survey of manufacturing standards revealed that there were no U.S. standards regarding the manufacture of hand tools. Military standards were not checked. The British Standards Institute and the Japanese Industrial Standards do have standards governing the manufacture of such hand tools.

7.7.2 Experimental procedure

Figure 7.58 shows a low-magnification photograph of the two fracture surfaces and Fig. 7.59 schematically shows the location and orientation of the two fractographs in Fig. 7.58. In the latter, two distinct regions were identified. The outer surface, which is the region of maximum fiber stress during loading, exhibited very little evidence of dimensional change, while some was observed in the central region. In addition, some plastic deformation was evident on the surface that is presumed to be the last point to fail (see Sec. 4.4).

The plastic coating on the handles was cut away to expose the handle area. The chromium plating on the tip of the tool ended at the plastic, and the handles had a sooty substance on them. The tool must have been plated after the plastic was installed, and either the plastic left the residue, or it was left from a heat-treating process.

Following the visual examination, analysis for determining the cause of failure was carried out in three steps. The first step consisted in an

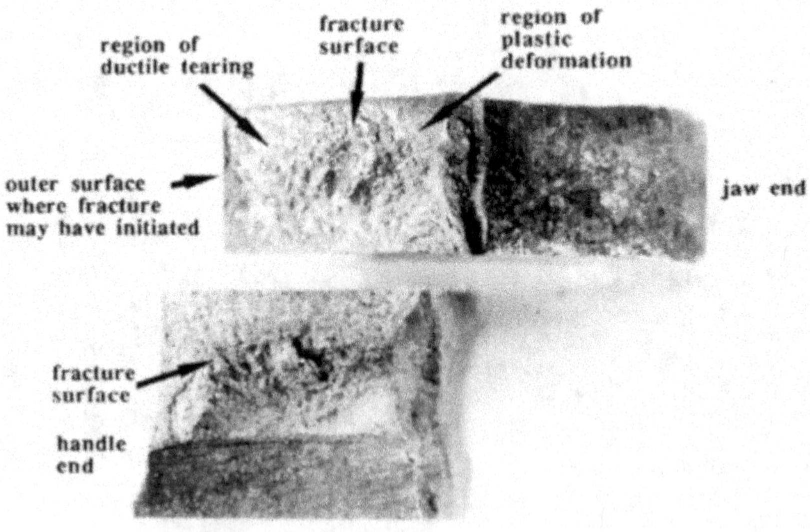

Figure 7.58 Fracture surfaces of the broken pair of cutters. The lower photograph is from the handle and the upper from a location near the hinge pin.

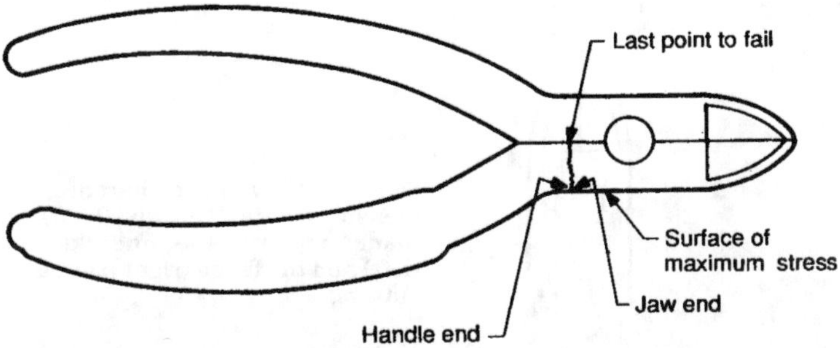

Figure 7.59 Schematic diagram of the wire cutters, showing the location of fracture and the location of the features marked in Fig. 7.58.

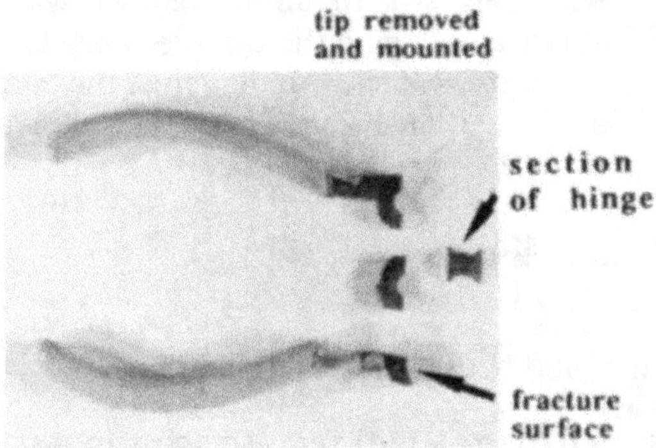

Figure 7.60 Sectioning scheme used for obtaining specimens for microstructural evaluation.

evaluation of the stress levels involved in the wire cutting operations for which the tool was designed. This was followed by a metallographic evaluation of the material used and its heat treatment. Lastly, a detailed fractographic examination was conducted using the scanning electron microscope.

Figure 7.60 shows the failed part after it was sectioned for microstructural evaluation. The tip portion of the cutters was mounted for metallography to expose both a plane parallel to the fracture surface and a cross section of the hinge at its centerline. Rockwell hardness measurements were made using the C scale at the location indicated in Fig. 7.61 on both the failed and an unbroken pair of cutters.

7.7.3 Results

Evaluation of stresses. A pair of unbroken cutters of dimensions similar to those of the broken one was used to measure the force required in a typical cutting operation. The cutters were placed between the compression dies of a Tinius Olsen testing machine and the load required to cut a 10 TW [0.102-in-diameter (0.259-cm)] single-strand insulated copper

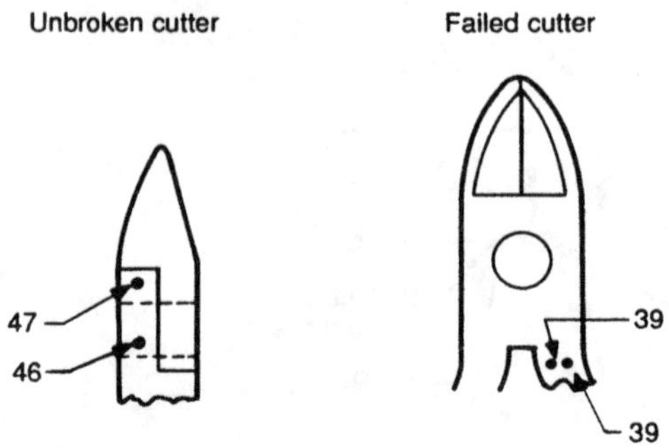

Figure 7.61 Values of hardness measurements (Rockwell C) made on both the unbroken (left) and the failed (right pair of cutters.

wire was measured. A force of 436 N (98 lb) was required to cut the wire, and the cutters appeared to flex elastically during the test (see Sec. 2.2). There was no evident damage to the cutters at the completion of the test. The maximum outer fiber stress imposed on the tool may be approximated by using a simple beam formula,

$$\sigma = \frac{Mc}{I}$$

where M = moment = 6.4 cm × 436 N (2.5 in × 98 lb)
c = 0.34 cm (0.269/2 in)
I = moment of inertia = 0.0112 cm^4 [(0.166 × 0.269^3)/12 in^4]

Hence

$$\sigma = 8.4 \times 10^4 \text{ Pa } (122{,}000 \text{ lb/in}^2)$$

This value now has to be compared to the mechanical properties of the material used.

The results of the hardness measurements are shown in Fig. 7.61. The failed cutters had a somewhat lower hardness. The average value was found to be 40 Rockwell C. This corresponds to an ultimate tensile strength of about 1.24 × 10^9 Pa (180,000 lb/in^2),[13] and a yield strength of about 1.03 × 10^9 Pa (150,000 lb/in^2).[27] It would appear then that the strength was well above the maximum fiber stress calculated. Gross overload may thus be overruled as a possible cause for failure.

Microstructural evaluation. Figure 7.62 is a low-magnification micrograph showing the cross section of the fracture surface from the jaw end. From the graduation in the extent of etching between the outer edge and the center of this micrograph, it appears that the tool was car-

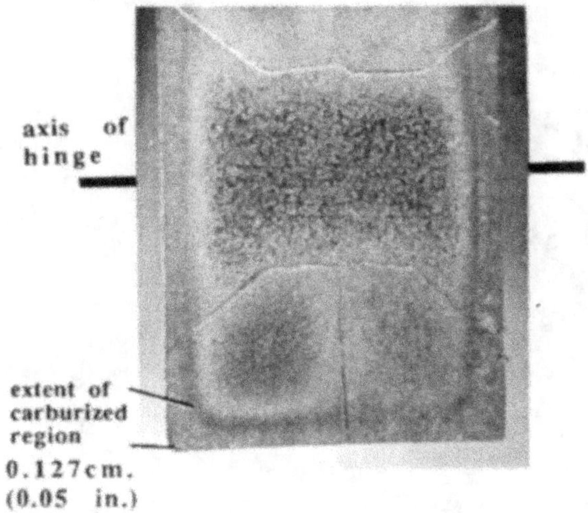

Figure 7.62 Low-magnification micrograph from the fracture region showing the extent of the carburized region.

burized. At higher magnification the microstructure is revealed to be tempered martensite, with undissolved carbides outlining the prior austenite boundaries, as shown in Fig. 7.63. This structure persists to a depth corresponding to about 3 grains (of prior austenite) from the surface, below which the grain-boundary carbides are no longer observed. The prior austenite grain size is estimated to be ASTM 3. There is no evidence of decarburization at the surface. All these points indicate that the material was carburized. In order to confirm this, a section of the handle was removed and austenitized at 980°C (1800°F) for 2 h, then furnace-cooled. The edge-to-center microstructure of this piece is shown at low magnification in Fig. 7.64*a* and clearly reveals the carburized zone. Figure 7.64*b* shows that the center of the specimen was not carburized. This micrograph shows about 10 percent pearlite in a ferrite matrix, indicating a carbon content of approximately 0.1 percent for the steel from which the tool was made. In contrast, the microstructure near the edge (Fig. 7.64*c*) shows pearlite with carbides along the prior austenite boundaries, from which it is deduced that the carbon content of this region is greater than 0.8 percent.

SEM fractography. The fracture surface of the jaw end (Fig. 7.58) was examined in a scanning electron microscope. As shown in the low-magnification fractograph of Fig. 7.65, the outer region is brittle in appearance while the central area shows ductile tearing (see Sec. 3.4) with some secondary cracking. The higher-magnification fractography of Fig. 7.66 is from the outer carburized region where a mixed mechanism composed of

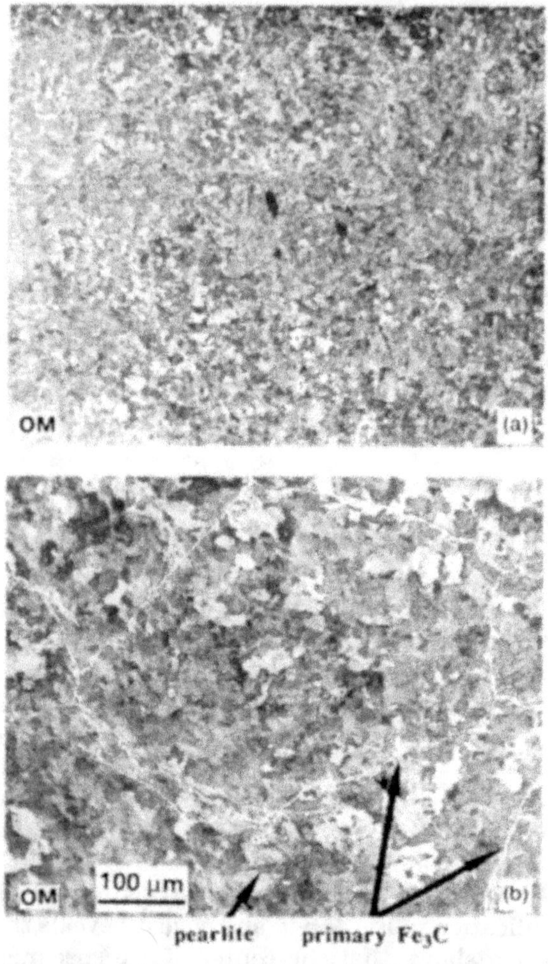

Figure 7.63 (*a*) Low- and (*b*) high-magnification micrographs of the carburized region showing a carbide network along the prior austenite grain boundaries.

cleavage (see Sec. 3.4) and microvoid coalescence (see Sec. 3.5) is evident. The central region shows only dimples, as seen in the high-magnification fractograph of Fig. 7.67, taken adjacent to one of the secondary cracks. The features at the extreme outer edge (region of maximum fiber stress) were masked by the presence of some corrosion products which could not be removed by standard surface-cleaning procedures (see Sec. 1.4.1).

7.7.4 Discussion

The observed hardness of the tool surface near the fracture area correlated to a yield strength that was significantly higher than the imposed

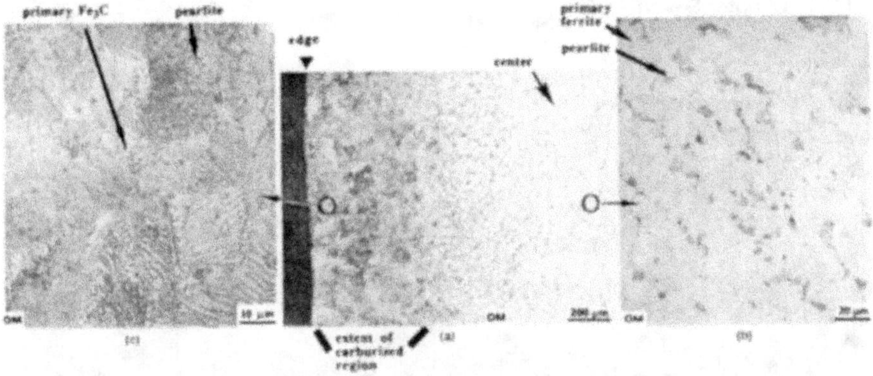

Figure 7.64 (a) Microstructure of the heat-treated specimens. (b) Microstructure from the center showing pearlite and ferrite. (c) The edge shows pearlite and a carbide network along the prior austenite grain boundaries, indicating a carburized structure.

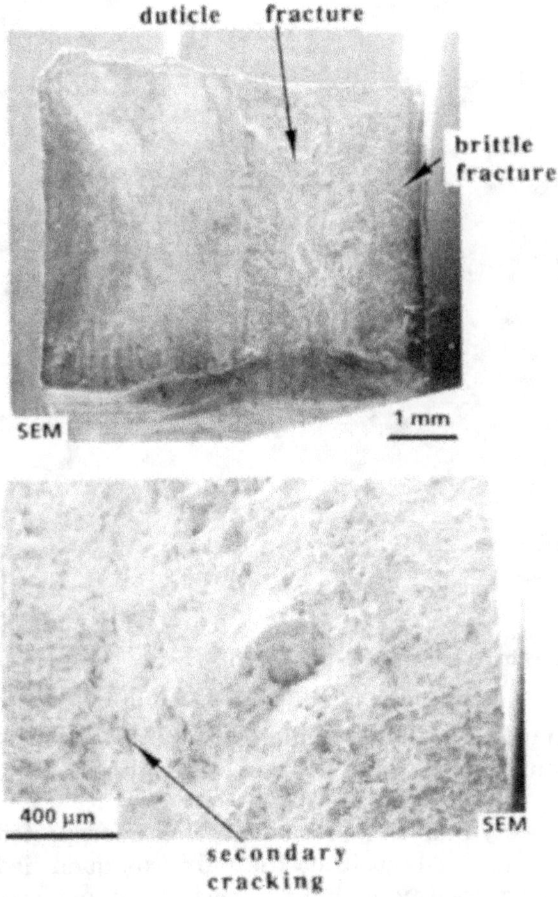

Figure 7.65 Low-magnification fractograph of the fracture surface.

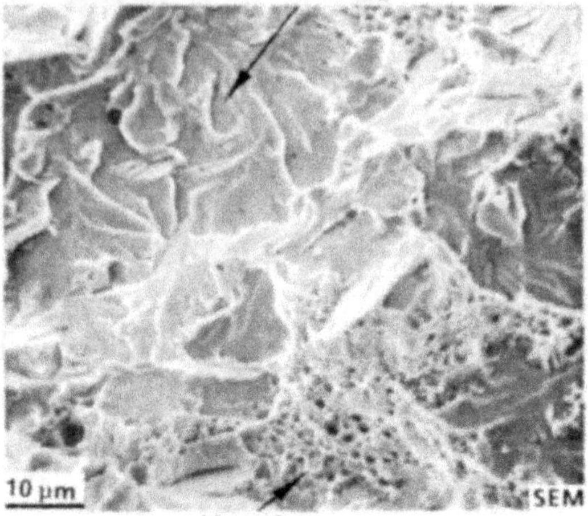

Figure 7.66 SEM fractograph near edge (see Fig. 7.65) showing mixed-mechanism fracture.

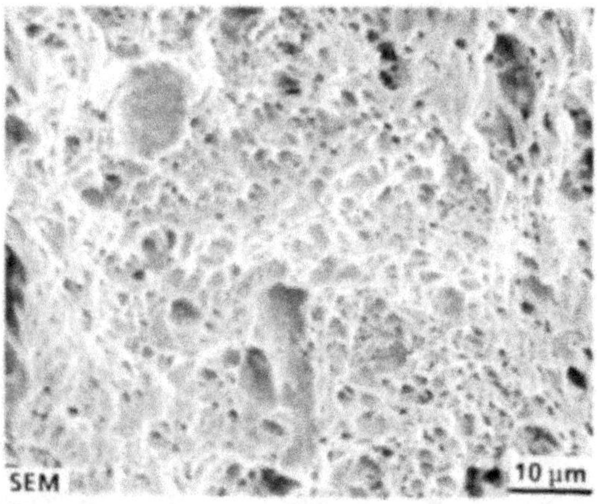

Figure 7.67 SEM fractograph near the secondary crack shown in Fig. 7.65 showing dimples.

stresses. This fact can be misleading, since a microstructure of carbides at prior austenite grain boundaries lowers the fracture toughness of the material.[28]

From the presence of a sooty material on the surface it is suspected that pack carburizing was used. Also, either the part was quenched directly from the carburizing treatment and the carbon content was such that it was beyond the A_{cm} line, thus leading to carbides along the austenite boundaries, or the quench rate was inadequate to suppress the formation of grain-boundary carbides. Since the prior austenite grain size is large (approximately ASTM 3), it is unlikely that the part was reaustenitized. It was in all probability tempered after being quenched directly following carburization.

While the possibility of hydrogen embrittlement (see Sec. 3.11) due to improper plating techniques cannot be completely ruled out, the cleavage fracture features observed at the outer edges are probably the result of the grain-boundary carbide network in this region. If the tool had been carburized prior to its assembly with the hinge pin, there would be a carburized area evident at the hinge pin surface in Fig. 7.62. Thus it is deduced that the tool was carburized after assembly. Uneven thermal expansion of the assembled part during heat treatment could also account for the binding of the hinge pin mentioned as part of the initial inspection.

7.7.5 Conclusions

It appears that the tool failed because of a defective microstructure of a network of primary carbide along prior austenite grain boundaries which lowered the allowable stresses on it. Also the prior austenite grain size was too large. Both of these problems were due to an improper heat treatment.

7.8 Case G: Broken Steel Punch*

7.8.1 Introduction

Pin punches are used to start holes in concrete blocks. A broken $^3/_{32}$-in (0.238-cm) punch (Fig. 7.68) was received with the fracture being quite close to its working point. The punch was known to be made of steel, and it was clearly plated, possibly with nickel. The purpose of the investigation was to determine the cause of failure.

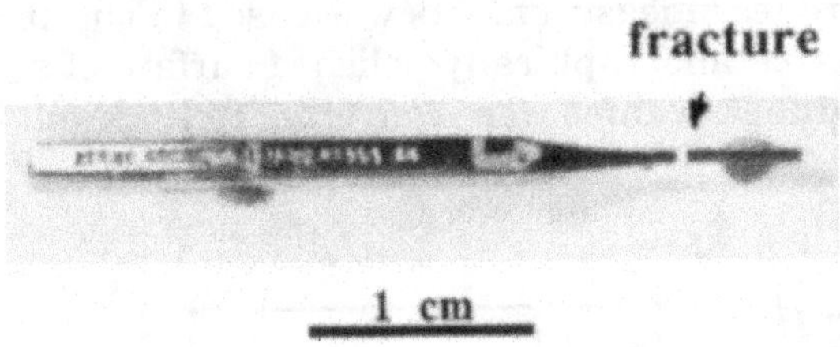

Figure 7.68 Broken punch.

*The authors acknowledge the work done by Bruce D. Cutler on this case study.

In the course of the investigation, the material was determined to be a high-carbon, plain-carbon steel with a tempered martensite structure. There are two well-known causes that can lead to the embrittlement of these steels. Temper embrittlement occurs if the steel is held at or slowly cooled through the 250 to 540°C (482 to 1004°F) temperature range.[29] Tempered martensite embrittlement (TME) occurs after tempering a martensitic structure between 260 and 370°C (500 and 698°F).[29]

TME is aggravated by exposure of the material to hydrogen, as might occur during an electrolytic plating process.[30,31] Higher-strength steels are more susceptible to hydrogen embrittlement.[31] One way to alleviate the problem is to bake the plated components at around 200°C (392°F) for 3 to 4 h. This normally allows the hydrogen to effuse except in cases where the plating itself acts as a diffusion barrier. The steel in this investigation had a relatively high carbon content, and upon quenching from austenite to 25°C (77°F) to harden the material, austenite may have been retained due to the martensite-finish temperature being below 25°C (77°F). During tempering this retained austenite may have decomposed to microconstituents, which are susceptible to hydrogen embrittlement.

7.8.2 Experimental procedure

The first step was a visual examination of the part and photographic documentation of the entire tool (Fig. 7.68). Figure 7.69 shows a schematic representation of how the part was sectioned for the investigation. The fracture surface on section 1 (Fig. 7.69) was examined in a scanning electron microscope to show not only the overall fracture-surface features (low magnification) but also the variation in fine features across the cross section of the fracture surface. In addition, x-ray fluorescent spectroscopic (EDS) analyses were obtained from the bulk material, from an inclusion, and from the plating.

Rockwell C hardness measurements were made on surface *F* of section 6 (Fig. 7.69). Microhardness measurements were used to obtain a hardness profile across the metallographically polished surface of section 3 (surface *C*). Microhardness values were converted to Rockwell C

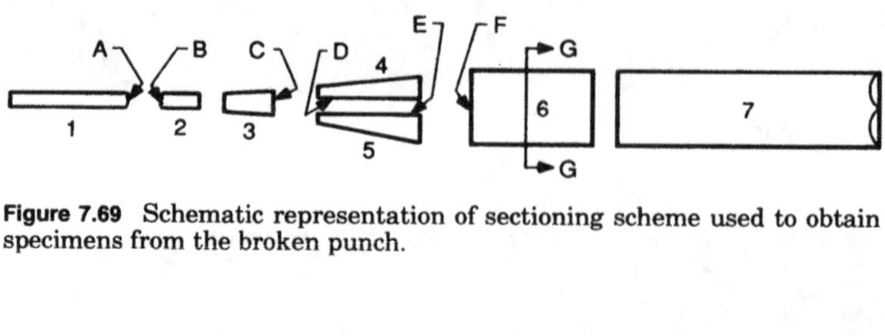

Figure 7.69 Schematic representation of sectioning scheme used to obtain specimens from the broken punch.

values.[18] Microstructural information was obtained by polishing and etching surfaces *D* and *E* of sections 4 and 5, respectively, using standard metallographic procedures.

The determination of the carbon content of the steel was accomplished by austenitizing section 6 at 900°C (1652°F) for 1 h followed by furnace cooling over 24 h. Prior to austenitization most of the plating was ground off. Following austenitization and annealing, section 6 was further sectioned, as shown in Fig. 7.69, to yield surface *G* for metallographic evaluation. From this examination the pearlite and proeutectoid (ferrite or carbide) content of the structure was measured to estimate the carbon content of the steel.

To determine the existence of retained austenite in the material, section 3 was subcooled by immersion in a mixed methanol and dry-ice bath −79°C (−96°F) for approximately 25 min, followed by warming to room temperature in a vacuum desiccator. The microhardness profile across surface *C* of this specimen was then measured and compared to the profile from the as-received specimen (mentioned earlier).

7.8.3 Results

Preliminary information. Very little preliminary information was available as to the exact conditions under which the failure occurred. It was assumed that the tool broke under normal working conditions, which would be impact by a hammer. The package of a similar punch mentioned that the tip was double tempered and the punch itself was nickel-chrome plated. Figure 7.70 shows a closeup of the failed tool at the failure location. There is no evidence of gross plastic deformation or of a pronounced shear lip. Hence the fracture occurred in a brittle manner. The lip visible on the fracture surface (mentioned later) is the plating, which incidentally did not fracture at exactly the same location as the tool itself. It is assumed that the fracture of the plating had no effect on the failure of the tool.

SEM fractography and EDS analysis. Figure 7.71 shows low-magnification (overall) fractographs of the two fracture surfaces. At a slightly higher magnification (Fig. 7.72) these regions are more clearly defined,

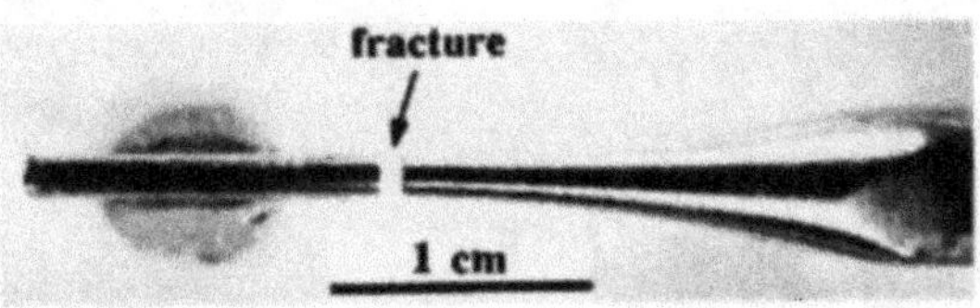

Figure 7.70 Closeup of fracture region of the punch showing lack of gross plastic deformation.

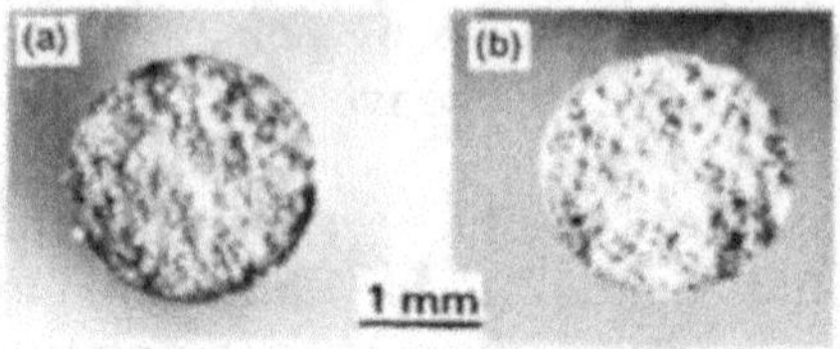

Figure 7.71 Low-magnification photographs of two fracture surfaces of broken punch. (a) Fracture surface of the smaller piece. (b) That of the larger piece.

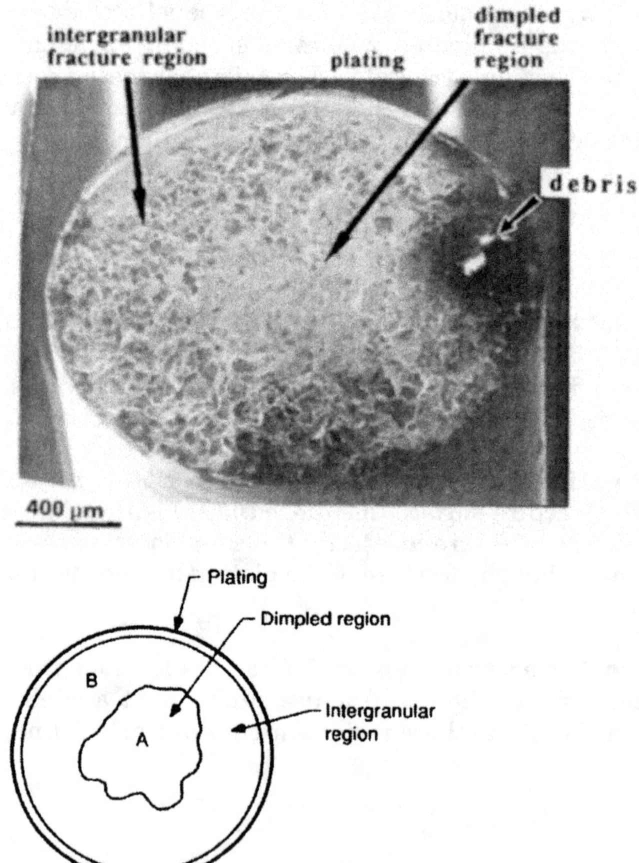

Figure 7.72 Locations of the plating and the dimpled and intergranular areas on the fracture surface of section 1.

and so is the plating. The central dull fibrous region is surrounded by an outer rim of the bright, brittle appearing fracture. Figure 7.73 shows in the center region classic dimple fracture features. In contrast, the stereo-pair fractographs in Fig. 7.74 of the edge region show predominantly a "rock candy" appearance of intergranular fracture, mixed with areas showing dimples.

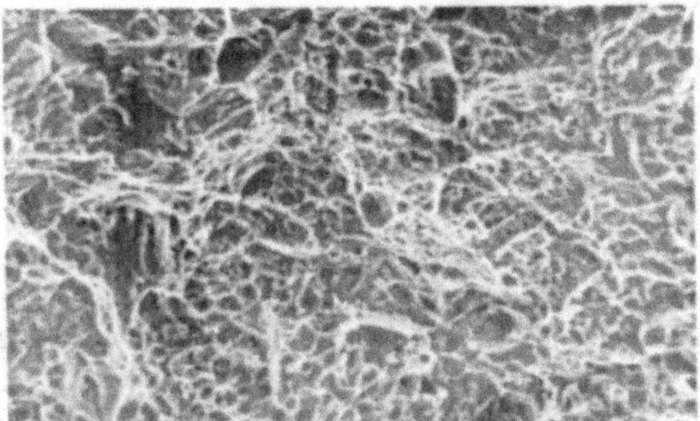

Figure 7.73 Fractograph from center region in Fig. 7.72 showing dimpled fracture features.

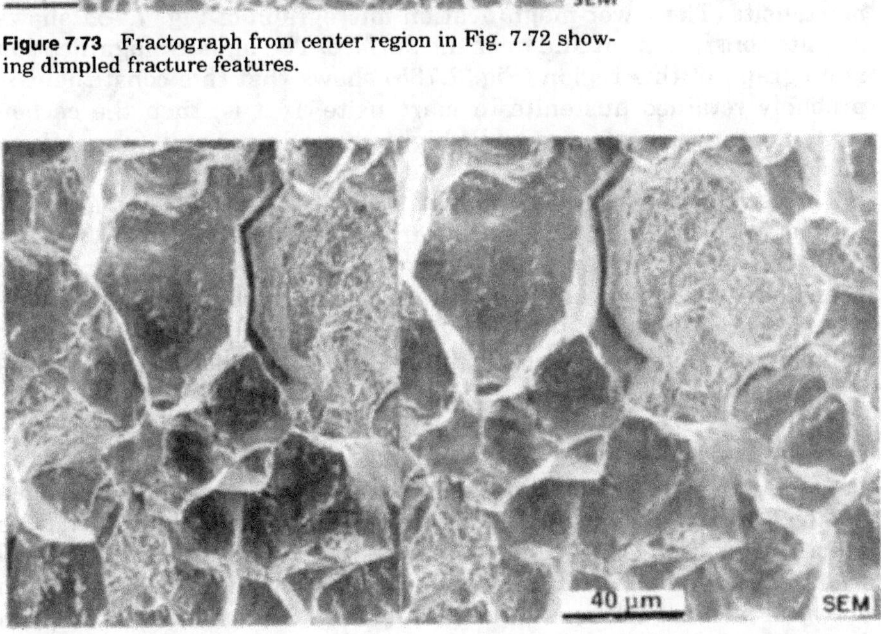

Figure 7.74 Stereo-pair fractographs from region *B* in Fig. 7.72 showing intergranular fracture.

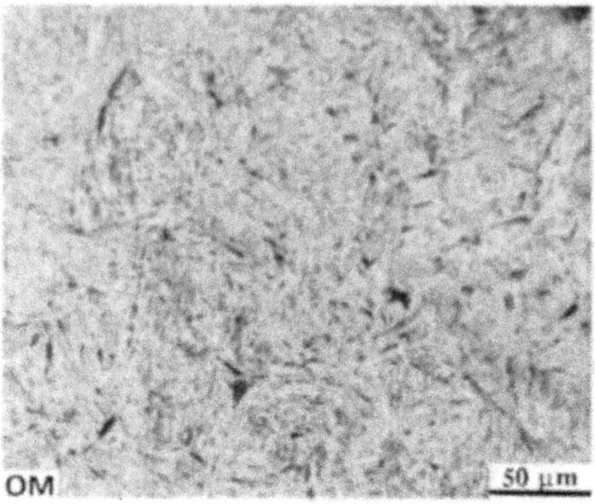

Figure 7.75 Microstructure from central region of punch.

EDS analyses from the bulk material of the fracture surface indicated that the material is plain-carbon steel. The analysis of the plating showed an inner nickel layer and an outer chromium layer.

Microstructural information. Figure 7.75 shows the microstructure from the central region of the punch; it is composed entirely of martensite. The lower-magnification micrograph of Fig. 7.76*a* shows a white constituent at and near the surface. The higher-magnification micrograph of this region (Fig. 7.76*b*) shows that this constituent is probably retained austenite in martensite. If it is, then the carbon content near the surface must be higher than for the bulk (no retained austenite).

The carbon content of the material (bulk) can be estimated from the microstructure in Fig. 7.77, which is from the central region of the heat-treated specimen. In this micrograph the large dark and gray areas are pearlite (about 90 percent), whereas the small bright areas are proeutectoid ferrite (about 10 percent) decorating the prior austenite grain boundaries. From this the carbon content of the material is estimated to be about 0.7 percent.

Hardness measurements. Microhardness readings of the as-received tool obtained from the center of the cross section of the tool averaged about 57 Rockwell C, which is consistent with the values obtained from macrohardness measurements. Readings taken close to the edge, however, varied from 50 to 60 Rockwell C, which could imply variations in the local microstructure encountered by the indenter. After the section had undergone the subcooling treatment mentioned earlier, the readings at both the surface and the center were consistently near 60

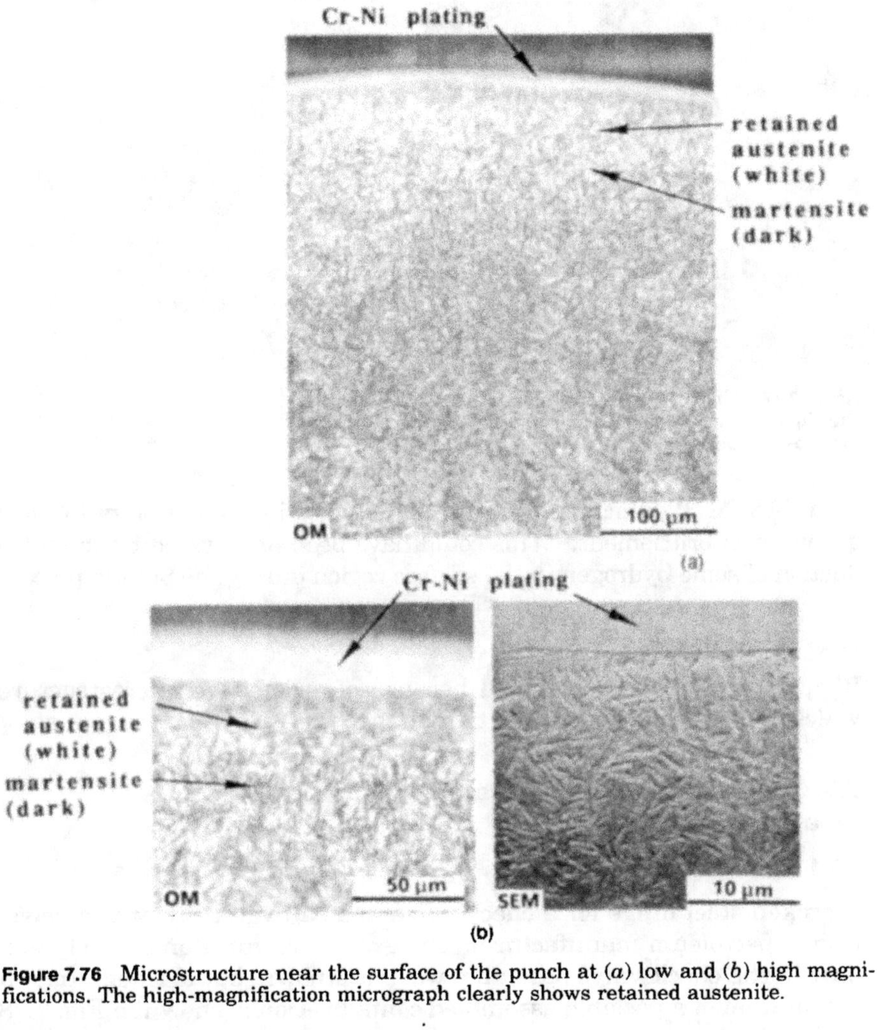

Figure 7.76 Microstructure near the surface of the punch at (*a*) low and (*b*) high magnifications. The high-magnification micrograph clearly shows retained austenite.

Rockwell C, indicating a transformation of retained austenite to martensite at the outer edges of the tool.

7.8.4 Discussion

This tool clearly failed by a mechanism of intergranular fracture along prior austenite boundaries. It appears that the tool was made from high-carbon (about 0.7 percent carbon), plain-carbon steel. The surface was carburized to a higher carbon content, which led to the presence of retained austenite during subsequent quenching. This in itself may not lead to embrittlement of the material. However, the prior austenite grains were

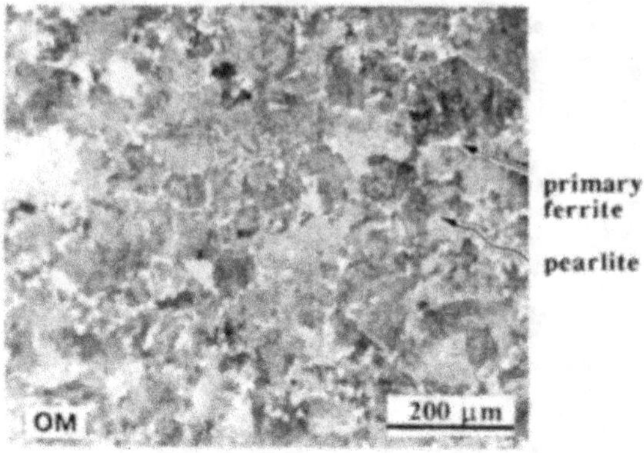

Figure 7.77 Microstructure from the central region of the heat-treated specimen showing pearlite and proeutectoid ferrite.

large (ASTM 3), which is known to be a factor in susceptibility to intergranular embrittlement.[29] This could have been aggravated by the introduction of some hydrogen in the surface region during the plating process.

7.8.5 Conclusions

It appears that the tool failed because of a defective microstructure, which was also very sensitive to hydrogen embrittlement.

7.9 Case H: Broken Stainless-Steel Hinge for a Check Valve

7.9.1 Introduction

A broken steel hinge for a check valve in a saltwater line was received from a petroleum manufacturer. The exact configuration of the broken part in the check-valve assembly was not known, but a schematic representation of a possible assembled configuration is shown in Fig. 7.78. A cover letter accompanying the two broken pieces stated: "These check valves are in service in produce water (saltwater) pipelines. If the pump goes down, these valves close to prevent the high pressure from the discharge line from getting into the low-pressure suction line. When this valve failed, the suction line was pressurized causing considerable damage. Approximate service life 1 year; material is stainless steel; the impact on these valves, when they close, is considerable."

It is to be noted that the following pieces of useful information could not be obtained:

1. The operating environment of the hinge itself, although it was probably the same as for the valve.

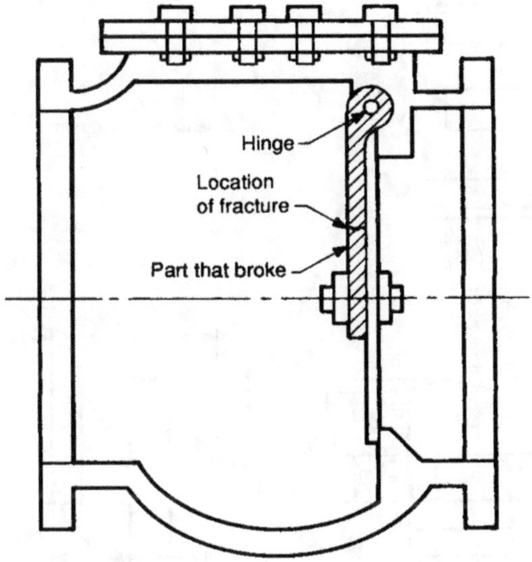

Figure 7.78 Schematic representation of the possible assembled configuration of the hinge and check valve.

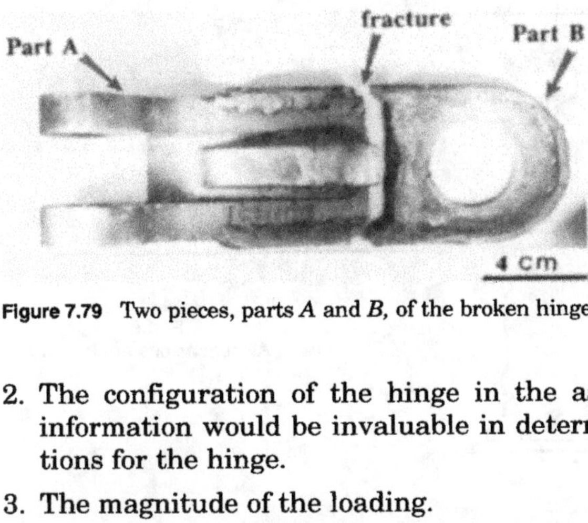

Figure 7.79 Two pieces, parts *A* and *B*, of the broken hinge.

2. The configuration of the hinge in the assembled condition. This information would be invaluable in determining the loading conditions for the hinge.
3. The magnitude of the loading.
4. The grade or class of the stainless steel used.

The broken hinge was received in two parts. Figure 7.79 shows the two parts in their relative operating positions; Figs. 7.80 and 7.81 give the dimensions of the two parts.

7.9.2 Experimental procedure

The first step was a visual examination of the part and photographic documentation of the two pieces. The fracture surfaces of both parts

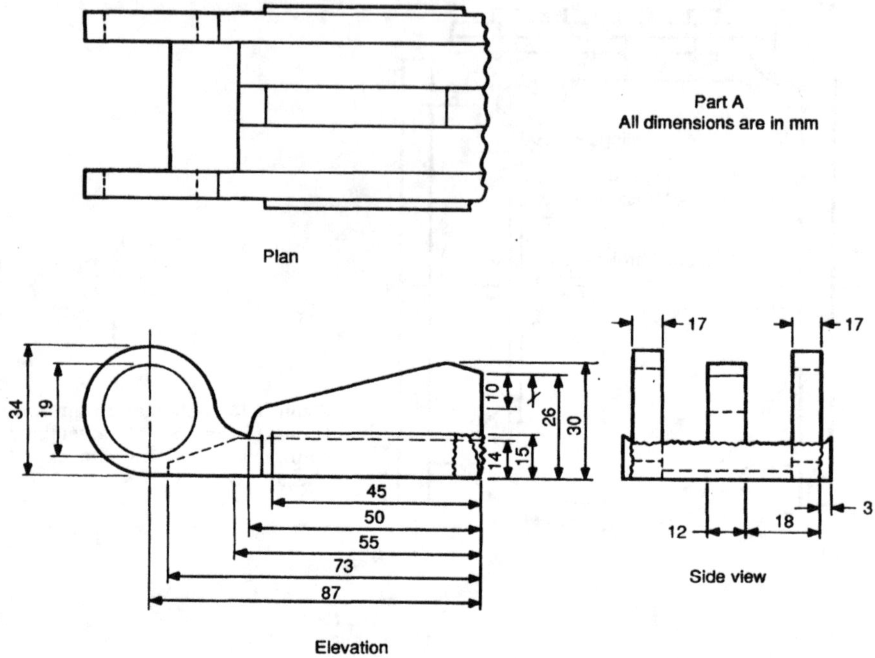

Figure 7.80 Dimensions of part *A* (in millimeters).

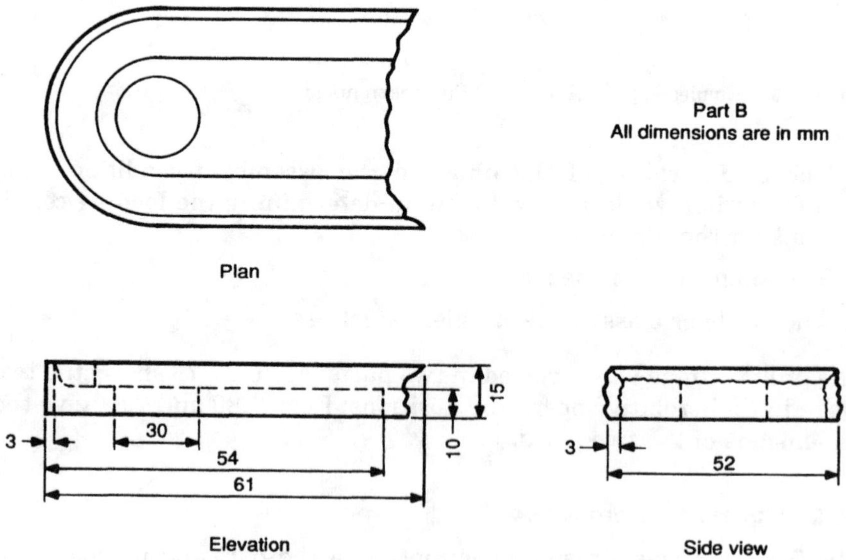

Figure 7.81 Dimensions of part *B* (in millimeters).

were examined carefully, both macroscopically and in a scanning electron microscope. In addition, EDS analyses were obtained from various regions of interest on the fracture surfaces.

Sections were cut from part *A*, as shown schematically in Fig. 7.82, to yield two specimens (called samples 1 and 2) for microhardness measurements, magnetic measurements, and microstructural evaluation. Magnetic measurements were made on sample 1 in an effort to evaluate the level of delta ferrite in the microstructure.[32] Microhardness measurements were also obtained from sample 1. Microstructural

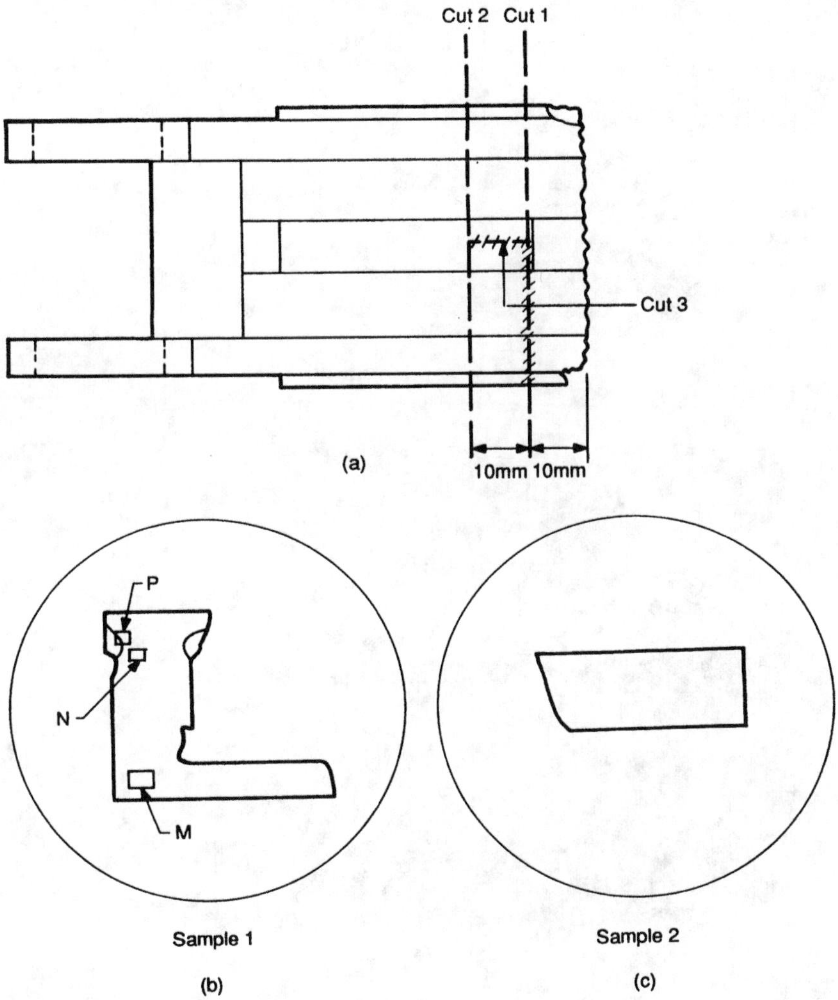

Figure 7.82 (*a*) Schematic representation of the sectioning scheme used to obtain specimens from part *A*. (*b*), (*c*) Two specimens.

information was obtained by polishing and etching (Kalling's etch) samples 1 and 2 using standard metallographic procedures.

7.9.3 Results

Visual inspection. Most of the surface of part *A* appeared to have been painted green. There were three distinct regions in this part: a base plate, a rib on the base plate, and a strap welded on the base. These are identified in Fig. 7.83. One of the filet welds used to join the strap to the base plate is marked in Fig. 7.83*a*. Two stampings were in evidence

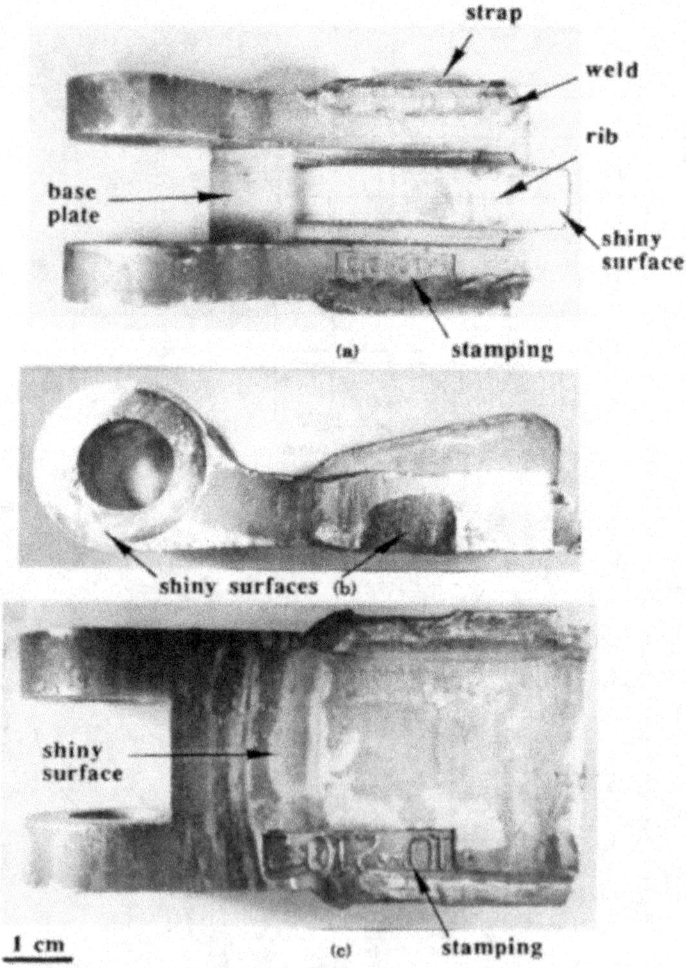

Figure 7.83 Part *A*. (*a*) Plan view. (*b*) Side view. (*c*) Underside. Stampings on part and some shiny surfaces are noted.

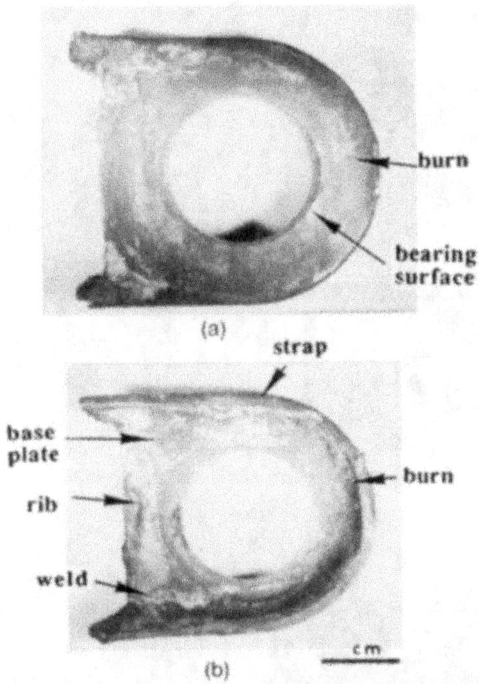

Figure 7.84 Part *B*. (*a*) Top view. (*b*) Underside.

on the base of part *A*—one on the top surface (Fig. 7.83*a*) and one on the undersurface (Fig. 7.83*b*). Some shiny surfaces (rubbed appearance) were in evidence, and these are also marked on Fig. 7.83.

Part *B* was much darker and dirtier in appearance than part *A*. The painting on this part appeared very patchy, flaky, and lighter in color than on part *A*. The overall appearance of part *B* gave the impression that the part had been exposed to high temperature. Figure 7.84 shows the top and underside of this part.

A "bearing surface" is evident around the hole in Fig. 7.84*a*. Figure 7.85 shows two end-on views of part *B*. These clearly indicate bending (plastic deformation) adjacent to the fracture location (see Sec. 4.4).

SEM fractography and EDS analysis. Figure 7.86 shows low-magnification (overall) fractographs of the fracture surfaces of parts *A* and *B* in the as-received condition. As can be seen in Fig. 7.86*a*, the fracture surface of part *A* was rusted in small patches and one-half was covered by a black film. The fracture surface of part *B* (Fig. 7.86*b*) was all but obscured by a black (oxide?) film. Cleaning the fracture surface of *A* was accomplished by repeated use of acetate replicating tape followed by ultrasonically cleaning in a solution of 10 percent oxalic acid. Almost all the rust was removed by this, but the black film persisted (see Sec.

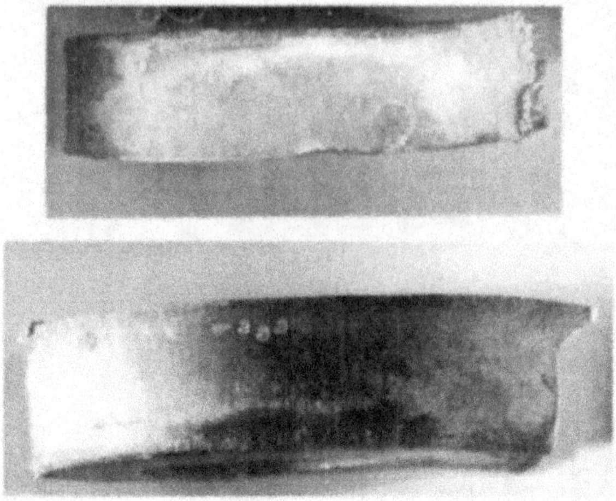

Figure 7.85 Two end views of part *B* showing bending.

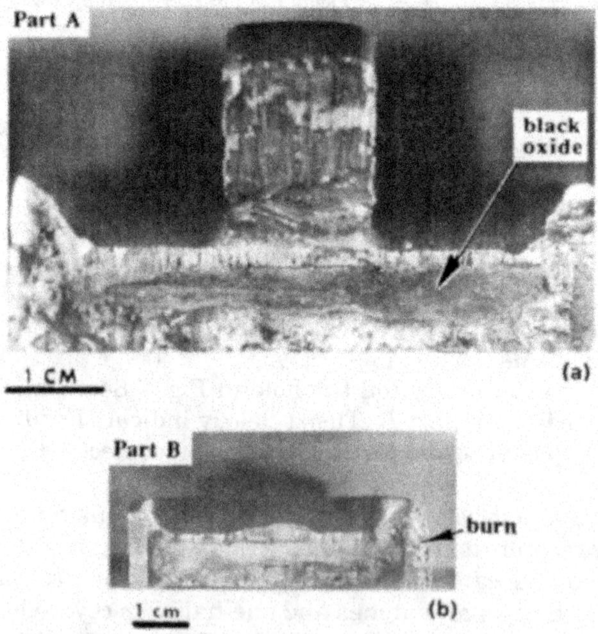

Figure 7.86 Low-magnification fractographs. (*a*) Part *A*. (*b*) Part *B*.

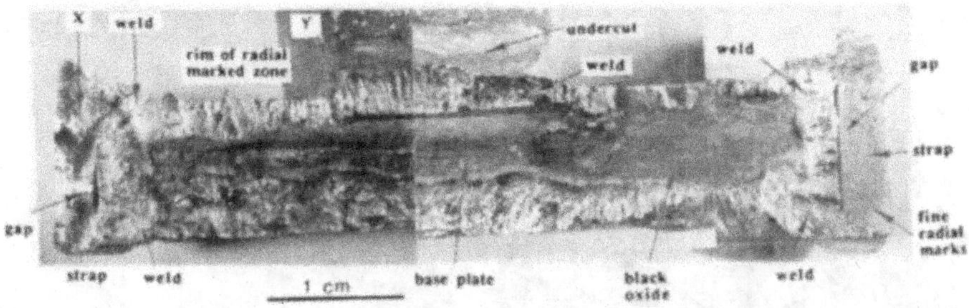

Figure 7.87 Low-magnification fractograph of part *A* showing two distinct regions, one bright and the other dull in appearance.

1.4.1). This obviated the determination of the fracture mechanism by detailed fractography.

Figure 7.87 shows a composite of the fracture surface of part *A*. Two distinct regions can be seen—a bright rim with distinct radial marks and a central dull rectangular region. The radial marks point back toward the central rectangle, which has a machined appearance (see Sec. 4.2). Under a stereo microscope the bright rim consists of very sharp and mirrorlike facets. The strap is seen to be welded on the base plate by two filet welds, one at the top and one at the underside. The latter weld has been ground down. The fracture surfaces of the strap are a bit receded from that of the base plate. The top weld provides a sloped junction between the fracture surfaces of the strap and the base plate. Also the top weld is discontinuous with the strap for a majority of the weld interface with the strap. At the location marked *X* in Fig. 7.87, a small round discontinuity was observed at the weld. The surface of this appeared solidified as opposed to fractured. The surface was dark and around it was a crack in the weld metal which extended to the top surface of the weld. Next to this there was an eye-shaped entrapped slag with a very shiny black appearance. This in turn was surrounded by a rim of bright metal. A partial view of this is shown in Fig. 7.88. The weld between the rib and the base plate of part *A* shows an undercut (Fig. 7.87). As illustrated in Fig. 7.89, at location *Y* marked in Fig. 7.87, a fine crack and radial marks pointing to its origin are seen in the scanning electron micrograph.

EDS analyses were also carried out in the scanning electron microscope on polished and etched metallographic specimens (next section), on both the base plate and the strap. The results are shown in Table 7.6. From these it is surmised that the strap is 347 stainless steel, while the base plate is CF8C (compositions are given in Table 7.7) since some Nb was found in a previous analysis. Table 7.8 shows the results of EDS analyses on the matrix and what was suspected to be delta ferrite in the

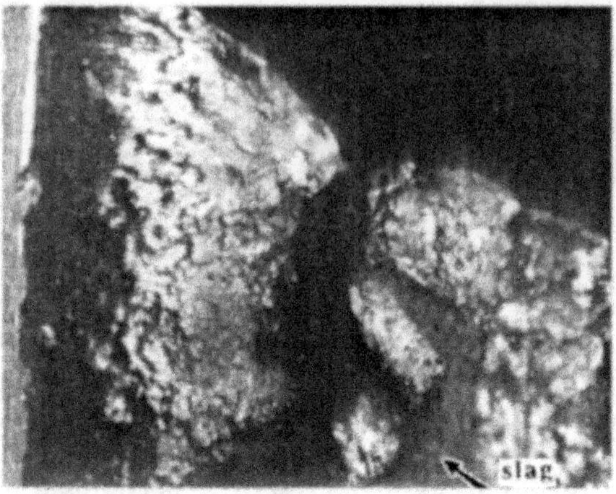

Figure 7.88 Optical photograph of slag found near location X marked in Fig. 7.87.

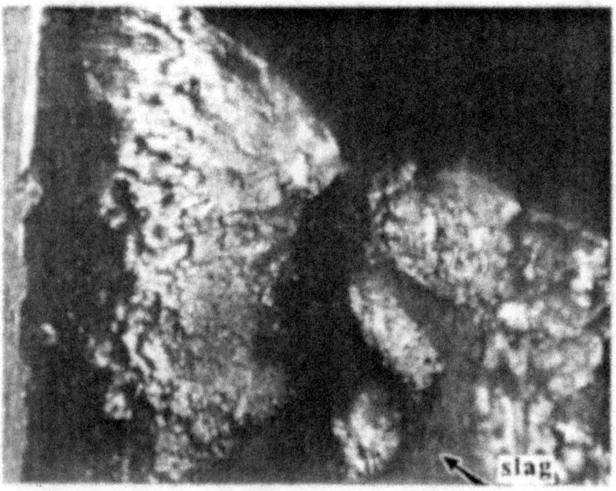

Figure 7.89 Optical photograph of crack found in location Y marked in Fig. 7.87.

microstructure. Indeed the results are consistent with the suspected phase being delta ferrite.

Microstructural information. Figure 7.82 schematically shows samples 1 and 2 and the locations from which they were taken. In the unetched condition, sample 1 shows a large concentration of round inclusions in the base plate and the structure is dendritic. The base plate thus is a casting. In contrast, the strap shows fewer and stringer-type inclusions, which indicates that it is made from rolled stock. One feature of note is

TABLE 7.6 EDS Results from Base Plate and Strap

Base plate			Strap		
keV	Counts	Element	keV	Counts	Element
1.75	51	Si	2.31	87	S
2.34	86	S	5.40	775	Cr
5.40	736	Cr	5.93	203	Cr
5.95	179	Cr	6.40	1745	Fe
6.39	1732	Fe	7.05	251	Fe
7.06	275	Fe	7.50	203	Ni
7.47	192	Ni	8.26	51	Ni
8.25	65	Ni			

TABLE 7.7 Chemical Composition Specification for CF8C Casting and 347 Stainless Steel

CF8C casting*		347 stainless steel†	
Element	wt.%	Element	wt.%
Cr	18.0–21.0	Cr	17.0–19.0
Ni	8.0–11.0	Ni	9.0–13.0
C	0.08 max	C	0.08 max
Mn	1.50 max	Mn	2.00 max
Si	2.00 max	Si	1.00 max
P	0.04 max	P	0.045 max
S	0.04 max	S	0.03 max
Nb	$> 8 \times$ C but < 1.00	Nb + Ta	$> 10 \times$ min. C

*From ASTM A743-79.[33]
†From *Metals Handbook*.[25]

TABLE 7.8 EDS Results from Matrix and Delta Ferrite

Matrix			Ferrite		
keV	Counts	Elements	keV	Counts	Element
1.69	121	Si	5.42	375	Cr
5.43	288	Cr	5.94	127	Cr
5.94	101	Cr	6.42	629	Fe
6.40	717	Fe	7.03	142	Fe
7.05	156	Fe	7.47	65	Ni
7.48	106	Ni			

that the strap lies inclined to the base plate, as illustrated schematically in Fig. 7.90.

Figure 7.91 is an optical micrograph taken from region M in Fig. 7.82b and is from the base plate. The background shows a cored dendritic structure and delta ferrite in the interdendritic areas. Figure

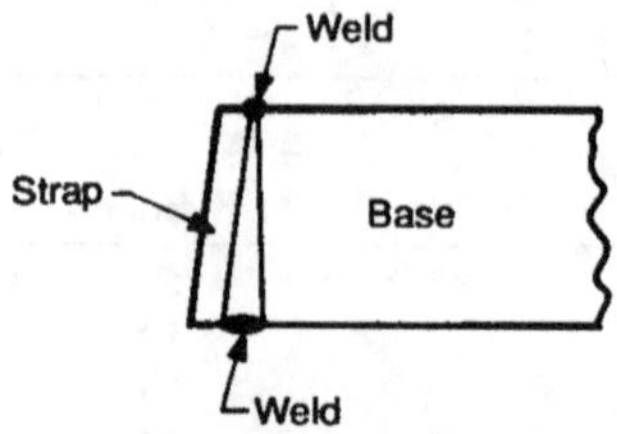

Figure 7.90 Schematic representation of relative inclination between strap and base plate.

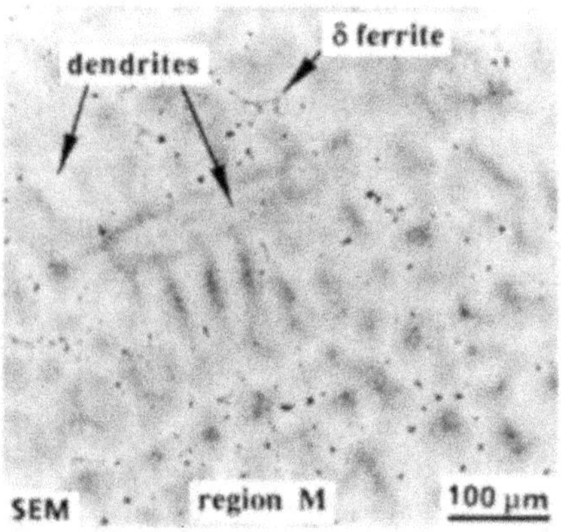

Figure 7.91 Micrograph from region *M* in Fig. 7.82*b*. Note the cored dendritic structure and delta ferrite in the interdendritic areas.

7.92 shows a micrograph from region *N* taken in a scanning electron microscope; the delta ferrite is very evident. A similar micrograph from region *P*, shown in Fig. 7.93, reveals a crack which has propagated through delta ferrite. Note that the weld metal has less delta ferrite than the base plate.

Magnetic measurements. The ferrite content of the materials used in the base plate, weld, and strap was measured using a magnetic gage.[32] Results of readings taken from sample 1 are shown on Fig. 7.94 (locations indicated). At these low ferrite levels the ferrite number is approximately equal to the volume percent ferrite. From these results the range of ferrite contents is deduced to be

Strap: No ferrite

Weld metal: ~1.5%

Heat-affected zone in base plate: ~2.5%

Base plate: ~2.0%

Microhardness measurements. The results of microhardness measurements made on several locations on sample 1 are given in Fig. 7.95.

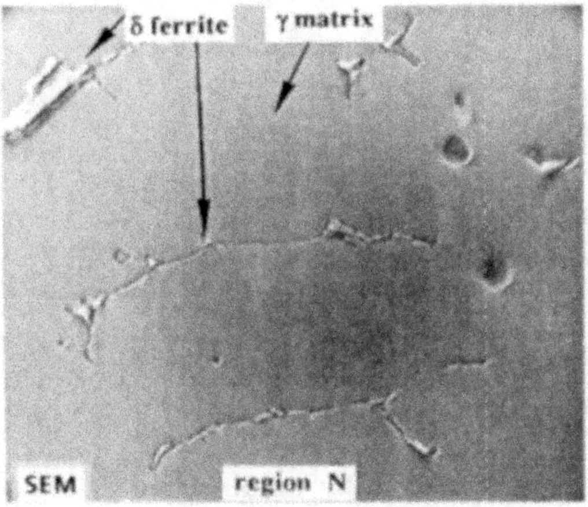

Figure 7.92 Micrograph from region N in Fig. 7.82b showing delta ferrite.

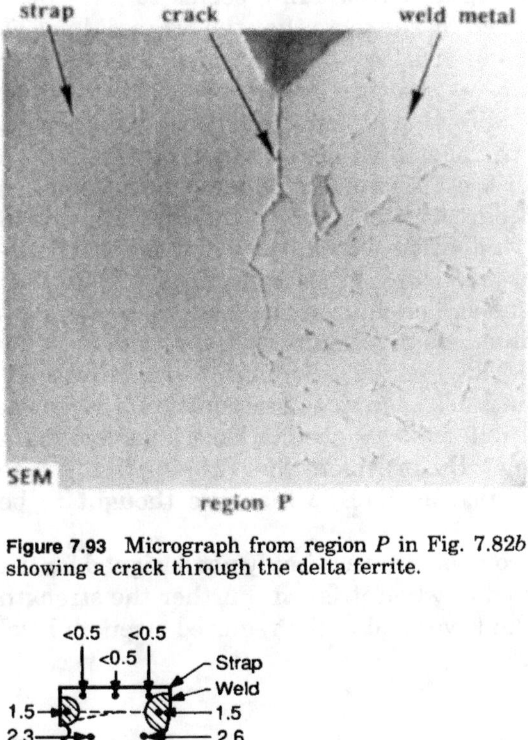

Figure 7.93 Micrograph from region P in Fig. 7.82b showing a crack through the delta ferrite.

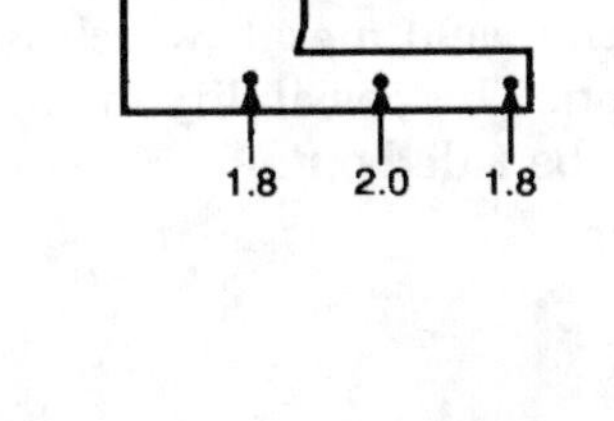

Figure 7.94 Ferrite number of the material at various locations on sample 1 as measured by a magnetic gage.

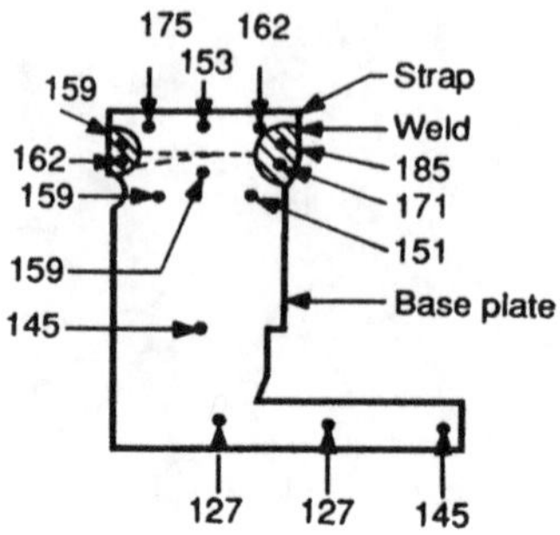

Figure 7.95 Microhardness values at various locations on sample 1.

7.9.4 Discussion

It appears that the part was manufactured by welding two pieces of cast CF8C stainless steel. A strap of 347 stainless steel was then welded on for additional strength. The welding seems to have been done shoddily. This is evidenced by the facts that the strap did not lie flat on the base plate, a crack was found at the weld metal–strap interface, and there is a gap between the weld and strap and base plate (that is, the root of the weld did not penetrate to the point of contact between the strap and the base plate as it should have).

Considering the part under the action of a bending load, as shown in Fig. 7.96, it can be seen that part *A* is stiffened by the rib while part *B* is not. This is a weld region and, as described, the weld also had a discontinuity which can act as an acute stress concentrator. The part seems to have failed under the action of an impact load in excess of its design limits. That the loading was of a bending type is evidenced by the deformation seen in Fig. 7.85 (see Sec. 4.4). The low ferrite content of the weld indicates that the crack seen at the discontinuity is probably a hot crack. Austenitic stainless steels need at least 5 percent delta ferrite to prevent hot cracking,[34] the prevalent theory being that ferrite scavenges tramp elements (such as S, B) which are thought to be responsible for hot cracking.[34]

It is very possible that the component was simply not designed for the type of impact it was subjected to when it failed. Further the strength level of the base plate is at the lower end of the required strength level according to ASTM A743-79.[32]

7.9.5 Conclusions

It appears that the mode of failure was bending overload, probably of the impact type. No clear mechanism of failure was discerned. Except for a slightly higher delta ferrite content in the weld metal (which is desirable), no other material faults were found. The possibility of the improper weld contributing to fracture cannot be ruled out.

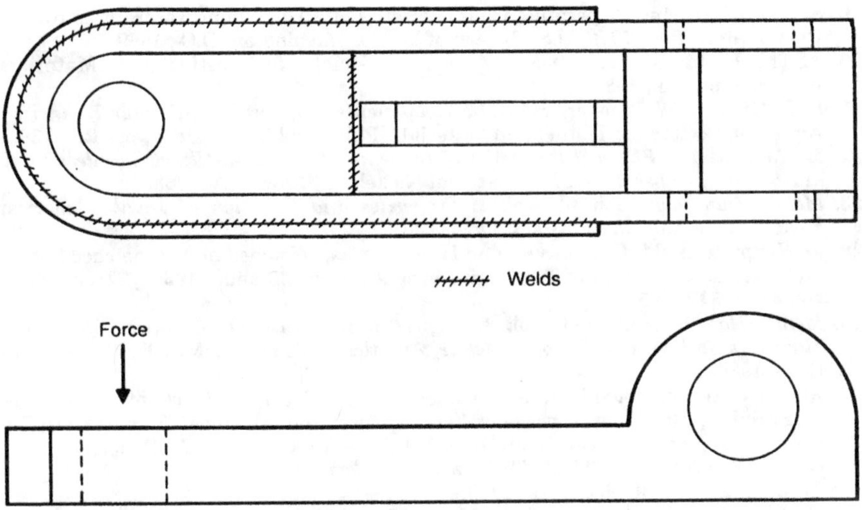

Figure 7.96 Effect of bending loads on the failed part.

References

1. C. R. Brooks, *Heat Treatment, Structure and Properties of Non-Ferrous Alloys,* American Society for Metals, Metals Park, Ohio, 1982.
2. *Metals Handbook,* 8th ed., vol. 4: *Forming,* American Society for Metals, Metals Park, Ohio, 1969, p. 423.
3. R. H. Uhlig, *Corrosion and Corrosion Control,* 2d ed., Wiley, New York, 1971.
4. M. G. Fontana and N. D. Green, *Corrosion Engineering,* 2d ed., McGraw-Hill, New York, 1978.
5. *Metals Handbook,* 9th ed., vol. 13: *Corrosion,* ASM International, Metals Park, Ohio, 1987, p. 131.
6. E. P. Polushkin and H. J. Shuldener, "Corrosion of Yellow Brass Pipes in Domestic Hot-Water Systems—A Metallographic Study," *Trans. AIME,* vol. 161, p. 214, 1945.
7. H. Leidheiser, *The Corrosion of Copper, Tin and Their Alloys,* Wiley, New York, 1971.
8. *Metals Handbook,* 9th ed., vol. 6: *Welding, Brazing and Soldering,* American Society for Metals, Metals Park, Ohio, 1983.
9. *ASHRAE Handbook—Equipment,* American Society for Heating, Refrigeration, and Air-Conditioning Engineers, Atlanta, Ga., 1983, p. 21.7.
10. *Metals Handbook,* 9th ed., vol. 2: *Properties and Selection: Nonferrous Alloys and Pure Metals,* American Society for Metals, Metals Park, Ohio, 1979.
11. *Metals Handbook,* 8th ed., vol. 8: *Metallography, Structure and Phase Diagrams,* American Society for Metals, Metals Park, Ohio, 1973.
12. S. Shivkumar, S. Ricci, Jr., C. Keller, and D. Apelian, "Effect of Solution Treatment Parameters on Tensile Properties of Cast Aluminum Alloys," *J. Heat Treat.,* vol. 8, p. 63, 1990.
13. H. E. Boyer and T. L. Gall (eds.), *Metals Handbook—Desk Edition,* American Society for Metals, Metals Park, Ohio, 1985.
14. R. A. Wallis, *Axial Flow Fans and Ducts,* Wiley, New York, 1983.
15. J. E. Hatch (ed.), *Aluminum—Properties and Physical Metallurgy,* American Society for Metals, Metals Park, Ohio, 1984.
16. *Motor Auto Repair,* 51st ed., Motor, New York, 1987, p. 14-6.
17. R. Kiessling, *Nonmetallic Inclusions in Steel: Part III,* The Metals Society, London, U.K., 1978.
18. ASTM E 140-88, "Standard Hardness Conversion Tables for Metals," American Society for Testing and Materials, Philadelphia, Pa., 1988.

19. Shale and Faschig, "Operating Characteristics of a High-Temperature Electrostatic Precipitator," Rep. 7276, U.S. Bureau of Mines, Washington, D.C., 1969.
20. ASTM E 140-86, vol. 03.01, American Society for Testing and Materials, Philadelphia, Pa., 1987.
21. H. L. Craig, T. W. Crooker, and D. M. Hoeppner (eds.), *Corrosion-Fatigue Technology,* American Society for Testing and Materials, Philadelphia, Pa., 1978, pp. 126, 168.
22. *Mechanical and Physical Properties of Austenitic Chromium-Nickel Stainless Steels at Ambient Temperatures,* International Nickel Co., New York, 1963.
23. *Metals Handbook,* 8th ed., vol. 1: *Properties and Selection of Metals,* American Society for Metals, Metals Park, Ohio, 1961.
24. J. W. Spretnak, M. G. Fontana, and H. E. Brooks, "Notched and Unnotched Tensile and Fatigue Properties of Ten Engineering Alloys at 25 and −196°C," *Trans. ASM,* vol. 43, p. 547, 1951.
25. *Metals Handbook,* 9th ed., vol. 3: *Properties and Selection: Stainless Steels, Tool Materials and Special Purpose Metals,* American Society for Metals, Metals Park, Ohio, 1980.
26. R. H. Henke, as quoted in K. G. Brickner and J. P. Defilippi, "Mechanical Properties of Stainless Steels at Cryogenic Temperature and at Room Temperature," in *Handbook of Stainless Steels,* McGraw-Hill, New York, 1977, p. 20-11 ff.
27. C. R. Brooks, *The Heat Treatment of Ferrous Alloys,* Hemisphere Publ. Corp./McGraw-Hill, New York, 1979.
28. *Republic Alloy Steels,* Cleveland, Ohio, 1968, p. 337.
29. G. V. Van der Voort, "Embrittlement of Steels," in *Metals Handbook,* 10th ed., vol. 1: *Properties and Selection: Irons, Steels, and High-Performance Alloys,* ASM International, Metals Park, Ohio, 1990, p. 689 ff.
30. *Metal Finishing Guidebook,* 43d ed., Metals and Plastics Publ., Hackensack, N.J., 1975, pp. 156–157.
31. S. F. Floreen, "Hydrogen Cracking in Speciality Steels," in R. A. Oriani, J. P. Hirth, and M. Smialowski (eds.), *Hydrogen Degradation of Ferrous Alloys,* Noyes Publ., Park Ridge, N.J., 1985, p. 799 ff.
32. ASTM A799/A-91, "Standard Practice for Steel Castings, Austenitic Alloy, Estimating Ferrite Content Thereof," vol. 01.02, American Society for Testing and Materials, Philadelphia, Pa., 1991.
33. ASTM A743-79, "Standard Specification for Corrosion Resistant Iron-Chromium, Iron-Chromium-Nickel, and Nickel-Base Alloy Castings for General Application," American Society for Testing and Materials, Philadelphia, Pa., 1979.
34. E. Folkhard, *Welding Metallurgy of Stainless Steels,* Springer, New York, 1988, p. 144 ff.

Bibliography

Brooks, C. R.: *Heat Treatment, Structure, and Properties of Non-Ferrous Alloys,* American Society for Metals, Metals Park, Ohio, 1982.

Krauss, G.: *Steels: Heat Treatment and Processing Principles,* ASM International, Metals Park, Ohio, 1990.

Appendix

A

Temperature Conversions*

*Reprinted with permission from *1977 Metal Progress Databook,* American Society for Metals, Metals Park, Ohio, 1977.

Temperature Conversions

The general arrangement of this conversion table was devised by Sauveur and Boylston. The middle columns of numbers (in **boldface** type) contain the temperature readings (°F or °C) to be converted. When converting from degrees Fahrenheit to degrees Celsius, read the Celsius equivalent in the column headed "C". When converting from Celsius to Fahrenheit, read the Fahrenheit equivalent in the column headed "F".

F		C	F		C	F		C	F		C	F		C
.....	**−458**	−272.22	.....	**−308**	−188.89	−252.4	**−158**	−105.56	+17.6	**−8**	−22.22	287.6	**142**	61.11
.....	**−456**	−271.11	.....	**−306**	−187.78	−248.8	**−156**	−104.44	+21.2	**−6**	−21.11	291.2	**144**	62.22
.....	**−454**	−270.00	.....	**−304**	−186.67	−245.2	**−154**	−103.33	+24.8	**−4**	−20.00	294.8	**146**	63.33
.....	**−452**	−268.89	.....	**−302**	−185.56	−241.6	**−152**	−102.22	+28.4	**−2**	−18.89	298.4	**148**	64.44
.....	**−450**	−267.78	.....	**−300**	−184.44	−238.0	**−150**	−101.11	+32.0	**±0**	−17.78	302.0	**150**	65.56
.....	**−448**	−266.67	.....	**−298**	−183.33	−234.4	**−148**	−100.00	+35.6	**+2**	−16.67	305.6	**152**	66.67
.....	**−446**	−265.56	.....	**−296**	−182.22	−230.8	**−146**	−98.89	+39.2	**+4**	−15.56	309.2	**154**	67.78
.....	**−444**	−264.44	.....	**−294**	−181.11	−227.2	**−144**	−97.78	+42.8	**+6**	−14.44	312.8	**156**	68.89
.....	**−442**	−263.33	.....	**−292**	−180.00	−223.6	**−142**	−96.67	+46.4	**+8**	−13.33	316.4	**158**	70.00
.....	**−440**	−262.22	.....	**−290**	−178.89	−220.0	**−140**	−95.56	+50.0	**+10**	−12.22	320.0	**160**	71.11
.....	**−438**	−261.11	.....	**−288**	−177.78	−216.4	**−138**	−94.44	+53.6	**+12**	−11.11	323.6	**162**	72.22
.....	**−436**	−260.00	.....	**−286**	−176.67	−212.8	**−136**	−93.33	+57.2	**+14**	−10.00	327.2	**164**	73.33
.....	**−434**	−258.89	.....	**−284**	−175.56	−209.2	**−134**	−92.22	+60.8	**+16**	−8.89	330.8	**166**	74.44
.....	**−432**	−257.78	.....	**−282**	−174.44	−205.6	**−132**	−91.11	+64.4	**+18**	−7.78	334.4	**168**	75.56
.....	**−430**	−256.67	.....	**−280**	−173.33	−202.0	**−130**	−90.00	+68.0	**+20**	−6.67	338.0	**170**	76.67
.....	**−428**	−255.56	.....	**−278**	−172.22	−198.4	**−128**	−88.89	+71.6	**+22**	−5.56	341.6	**172**	77.78
.....	**−426**	−254.44	.....	**−276**	−171.11	−194.8	**−126**	−87.78	+75.2	**+24**	−4.44	345.2	**174**	78.89
.....	**−424**	−253.33	.....	**−274**	−170.00	−191.2	**−124**	−86.67	+78.8	**+26**	−3.33	348.8	**176**	80.00
.....	**−422**	−252.22	−457.6	**−272**	−168.89	−187.6	**−122**	−85.56	+82.4	**+28**	−2.22	352.4	**178**	81.11
.....	**−420**	−251.11	−454.0	**−270**	−167.78	−184.0	**−120**	−84.44	+86.0	**+30**	−1.11	356.0	**180**	82.22
.....	**−418**	−250.00	−450.4	**−268**	−166.67	−180.4	**−118**	−83.33	+89.6	**+32**	±0.00	359.6	**182**	83.33
.....	**−416**	−248.89	−446.8	**−266**	−165.56	−176.8	**−116**	−82.22	+93.2	**+34**	+1.11	363.2	**184**	84.44
.....	**−414**	−247.78	−443.2	**−264**	−164.44	−173.2	**−114**	−81.11	+96.8	**+36**	+2.22	366.8	**186**	85.56
.....	**−412**	−246.67	−439.6	**−262**	−163.33	−169.6	**−112**	−80.00	+100.4	**+38**	+3.33	370.4	**188**	86.67
.....	**−410**	−245.56	−436.0	**−260**	−162.22	−166.0	**−110**	−78.89	+104.0	**+40**	+4.44	374.0	**190**	87.78
.....	**−408**	−244.44	−432.4	**−258**	−161.11	−162.4	**−108**	−77.78	107.6	**42**	5.56	377.6	**192**	88.89
.....	**−406**	−243.33	−428.8	**−256**	−160.00	−158.8	**−106**	−76.67	111.2	**44**	6.67	381.2	**194**	90.00
.....	**−404**	−242.22	−425.2	**−254**	−158.89	−155.2	**−104**	−75.56	114.8	**46**	7.78	384.8	**196**	91.11
.....	**−402**	−241.11	−421.6	**−252**	−157.78	−151.6	**−102**	−74.44	118.4	**48**	8.89	388.4	**198**	92.22
.....	**−400**	−240.00	−418.0	**−250**	−156.67	−148.0	**−100**	−73.33	122.0	**50**	10.00	392.0	**200**	93.33

.....	**−398**	−238.89	−414.4	**−248**	−155.56	−144.4	**−98**	−72.22	125.6	**52**	11.11	395.6	**202**	94.44
.....	**−396**	−237.78	−410.8	**−246**	−154.44	−140.8	**−96**	−71.11	129.2	**54**	12.22	399.2	**204**	95.56
.....	**−394**	−236.67	−407.2	**−244**	−153.33	−137.2	**−94**	−70.00	132.8	**56**	13.33	402.8	**206**	96.67
.....	**−392**	−235.56	−403.6	**−242**	−152.22	−133.6	**−92**	−68.89	136.4	**58**	14.44	406.4	**208**	97.78
.....	**−390**	−234.44	−400.0	**−240**	−151.11	−130.0	**−90**	−67.78	140.0	**60**	15.56	410.0	**210**	98.89
.....	**−388**	−233.33	−396.4	**−238**	−150.00	−126.4	**−88**	−66.67	143.6	**62**	16.67	413.6	**212**	100.00
.....	**−386**	−232.22	−392.8	**−236**	−148.89	−122.8	**−86**	−65.56	147.2	**64**	17.78	417.2	**214**	101.11
.....	**−384**	−231.11	−389.2	**−234**	−147.78	−119.2	**−84**	−64.44	150.8	**66**	18.89	420.8	**216**	102.22
.....	**−382**	−230.00	−385.6	**−232**	−146.67	−115.6	**−82**	−63.33	154.4	**68**	20.00	424.4	**218**	103.33
.....	**−380**	−228.89	−382.0	**−230**	−145.56	−112.0	**−80**	−62.22	158.0	**70**	21.11	428.0	**220**	104.44
.....	**−378**	−227.78	−378.4	**−228**	−144.44	−108.4	**−78**	−61.11	161.6	**72**	22.22	431.6	**222**	105.56
.....	**−376**	−226.67	−374.8	**−226**	−143.33	−104.8	**−76**	−60.00	165.2	**74**	23.33	435.2	**224**	106.67
.....	**−374**	−225.56	−371.2	**−224**	−142.22	−101.2	**−74**	−58.89	168.8	**76**	24.44	438.8	**226**	107.78
.....	**−372**	−224.44	−367.6	**−222**	−141.11	−97.6	**−72**	−57.78	172.4	**78**	25.56	442.4	**228**	108.89
.....	**−370**	−223.33	−364.0	**−220**	−140.00	−94.0	**−70**	−56.67	176.0	**80**	26.67	446.0	**230**	110.00
.....	**−368**	−222.22	−360.4	**−218**	−138.89	−90.4	**−68**	−55.56	179.6	**82**	27.78	449.6	**232**	111.11
.....	**−366**	−221.11	−356.8	**−216**	−137.78	−86.8	**−66**	−54.44	183.2	**84**	28.89	453.2	**234**	112.22
.....	**−364**	−220.00	−353.2	**−214**	−136.67	−83.2	**−64**	−53.33	186.8	**86**	30.00	456.8	**236**	113.33
.....	**−362**	−218.89	−349.6	**−212**	−135.56	−79.6	**−62**	−52.22	190.4	**88**	31.11	460.4	**238**	114.44
.....	**−360**	−217.78	−346.0	**−210**	−134.44	−76.0	**−60**	−51.11	194.0	**90**	32.22	464.0	**240**	115.56
.....	**−358**	−216.67	−342.4	**−208**	−133.33	−72.4	**−58**	−50.00	197.6	**92**	33.33	467.6	**242**	116.67
.....	**−356**	−215.56	−338.8	**−206**	−132.22	−68.8	**−56**	−48.89	201.2	**94**	34.44	471.2	**244**	117.78
.....	**−354**	−214.44	−335.2	**−204**	−131.11	−65.2	**−54**	−47.78	204.8	**96**	35.56	474.8	**246**	118.89
.....	**−352**	−213.33	−331.6	**−202**	−130.00	−61.6	**−52**	−46.67	208.4	**98**	36.67	478.4	**248**	120.00
.....	**−350**	−212.22	−328.0	**−200**	−128.89	−58.0	**−50**	−45.56	212.0	**100**	37.78	482.0	**250**	121.11
.....	**−348**	−211.11	−324.4	**−198**	−127.78	−54.4	**−48**	−44.44	215.6	**102**	38.89	485.6	**252**	122.22
.....	**−346**	−210.00	−320.8	**−196**	−126.67	−50.8	**−46**	−43.33	219.2	**104**	40.00	489.2	**254**	123.33
.....	**−344**	−208.89	−317.2	**−194**	−125.56	−47.2	**−44**	−42.22	222.8	**106**	41.11	492.8	**256**	124.44
.....	**−342**	−207.78	−313.6	**−192**	−124.44	−43.6	**−42**	−41.11	226.4	**108**	42.22	496.4	**258**	125.56
.....	**−340**	−206.67	−310.0	**−190**	−123.33	−40.0	**−40**	−40.00	230.0	**110**	43.33	500.0	**260**	126.67
.....	**−338**	−205.56	−306.4	**−188**	−122.22	−36.4	**−38**	−38.89	233.6	**112**	44.44	503.6	**262**	127.78
.....	**−336**	−204.44	−302.8	**−186**	−121.11	−32.8	**−36**	−37.78	237.2	**114**	45.56	507.2	**264**	128.89
.....	**−334**	−203.33	−299.2	**−184**	−120.00	−29.2	**−34**	−36.67	240.8	**116**	46.67	510.8	**266**	130.00
.....	**−332**	−202.22	−295.6	**−182**	−118.89	−25.6	**−32**	−35.56	244.4	**118**	47.78	514.4	**268**	131.11
.....	**−330**	−201.11	−292.0	**−180**	−117.78	−22.0	**−30**	−34.44	248.0	**120**	48.89	518.0	**270**	132.22
.....	**−328**	−200.00	−288.4	**−178**	−116.67	−18.4	**−28**	−33.33	251.6	**122**	50.00	521.6	**272**	133.33
.....	**−326**	−198.89	−284.8	**−176**	−115.56	−14.8	**−26**	−32.22	255.2	**124**	51.11	525.2	**274**	134.44
.....	**−324**	−197.78	−281.2	**−174**	−114.44	−11.2	**−24**	−31.11	258.8	**126**	52.22	528.8	**276**	135.56
.....	**−322**	−196.67	−277.6	**−172**	−113.33	−7.6	**−22**	−30.00	262.4	**128**	53.33	532.4	**278**	136.67
.....	**−320**	−195.56	−274.0	**−170**	−112.22	−4.0	**−20**	−28.89	266.0	**130**	54.44	536.0	**280**	137.78
.....	**−318**	−194.44	−270.4	**−168**	−111.11	−0.4	**−18**	−27.78	269.6	**132**	55.56	539.6	**282**	138.89
.....	**−316**	−193.33	−266.8	**−166**	−110.00	+3.2	**−16**	−26.67	273.2	**134**	56.67	543.2	**284**	140.00
.....	**−314**	−192.22	−263.2	**−164**	−108.89	+6.8	**−14**	−25.56	276.8	**136**	57.78	546.8	**286**	141.11
.....	**−312**	−191.11	−259.6	**−162**	−107.78	+10.4	**−12**	−24.44	280.4	**138**	58.89	550.4	**288**	142.22
.....	**−310**	−190.00	−256.0	**−160**	−106.67	+14.0	**−10**	−23.33	284.0	**140**	60.00	554.0	**290**	143.33

F		C	F		C	F		C	F		C	F		C
557.6	292	144.44	870.8	466	241.11	1832.0	1000	537.78	3398.0	1870	1021.1	4964.0	2740	1504.4
561.2	294	145.56	874.4	468	242.22	1850.0	1010	543.33	3416.0	1880	1026.7	4982.0	2750	1510.0
564.8	296	146.67	878.0	470	243.33	1868.0	1020	548.89	3434.0	1890	1032.2	5000.0	2760	1515.6
568.4	298	147.78	881.6	472	244.44	1886.0	1030	554.44	3452.0	1900	1037.8	5018.0	2770	1521.1
572.0	300	148.89	885.2	474	245.56	1904.0	1040	560.00	3470.0	1910	1043.3	5036.0	2780	1526.7
575.6	302	150.00	888.8	476	246.67	1922.0	1050	565.56	3488.0	1920	1048.9	5054.0	2790	1532.2
579.2	304	151.11	892.4	478	247.78	1940.0	1060	571.11	3506.0	1930	1054.4	5072.0	2800	1537.8
582.8	306	152.22	896.0	480	248.89	1958.0	1070	576.67	3524.0	1940	1060.0	5090.0	2810	1543.3
586.4	308	153.33	899.6	482	250.00	1976.0	1080	582.22	3542.0	1950	1065.6	5108.0	2820	1548.9
590.0	310	154.44	903.2	484	251.11	1994.0	1090	587.78	3560.0	1960	1071.1	5126.0	2830	1554.4
593.6	312	155.56	906.8	486	252.22	2012.0	1100	593.33	3578.0	1970	1076.7	5144.0	2840	1560.0
597.2	314	156.67	910.4	488	253.33	2030.0	1110	598.89	3596.0	1980	1082.2	5162.0	2850	1565.6
600.8	316	157.78	914.0	490	254.44	2048.0	1120	604.44	3614.0	1990	1087.8	5180.0	2860	1571.1
604.4	318	158.89	917.6	492	255.56	2066.0	1130	610.00	3632.0	2000	1093.3	5198.0	2870	1576.7
608.0	320	160.00	921.2	494	256.67	2084.0	1140	615.56	3650.0	2010	1098.9	5216.0	2880	1582.2
611.6	322	161.11	924.8	496	257.78	2102.0	1150	621.11	3668.0	2020	1104.4	5234.0	2890	1587.8
615.2	324	162.22	928.4	498	258.89	2120.0	1160	626.67	3686.0	2030	1110.0	5252.0	2900	1593.3
618.8	326	163.33	932.0	500	260.00	2138.0	1170	632.22	3704.0	2040	1115.6	5270.0	2910	1598.9
622.4	328	164.44	935.6	502	261.11	2156.0	1180	637.78	3722.0	2050	1121.1	5288.0	2920	1604.4
626.0	330	165.56	939.2	504	262.22	2174.0	1190	643.33	3740.0	2060	1126.7	5306.0	2930	1610.0
629.6	332	166.67	942.8	506	263.33	2192.0	1200	648.89	3758.0	2070	1132.2	5324.0	2940	1615.6
633.2	334	167.78	946.4	508	264.44	2210.0	1210	654.44	3776.0	2080	1137.8	5342.0	2950	1621.1
636.8	336	168.89	950.0	510	265.56	2228.0	1220	660.00	3794.0	2090	1143.3	5360.0	2960	1626.7
640.4	338	170.00	953.6	512	266.67	2246.0	1230	665.56	3812.0	2100	1148.9	5378.0	2970	1632.2
644.0	340	171.11	957.2	514	267.78	2264.0	1240	671.11	3830.0	2110	1154.4	5396.0	2980	1637.8
647.6	342	172.22	960.8	516	268.89	2282.0	1250	676.67	3848.0	2120	1160.0	5414.0	2990	1643.3
651.2	344	173.33	964.4	518	270.00	2300.0	1260	682.22	3866.0	2130	1165.6	5432.0	3000	1648.9
654.8	346	174.44	968.0	520	271.11	2318.0	1270	687.78	3884.0	2140	1171.1	5450.0	3010	1654.4
658.4	348	175.56	971.6	522	272.22	2336.0	1280	693.33	3902.0	2150	1176.7	5468.0	3020	1660.0
662.0	350	176.67	975.2	524	273.33	2354.0	1290	698.89	3920.0	2160	1182.2	5486.0	3030	1665.6
665.6	352	177.78	978.8	526	274.44	2372.0	1300	704.44	3938.0	2170	1187.8	5504.0	3040	1671.1
669.2	354	178.89	982.4	528	275.56	2390.0	1310	710.00	3956.0	2180	1193.3	5522.0	3050	1676.7
672.8	356	180.00	986.0	530	276.67	2408.0	1320	715.56	3974.0	2190	1198.9	5540.0	3060	1682.2
676.4	358	181.11	989.6	532	277.78	2426.0	1330	721.11	3992.0	2200	1204.4	5558.0	3070	1687.8
680.0	360	182.22	993.2	534	278.89	2444.0	1340	726.67	4010.0	2210	1210.0	5576.0	3080	1693.3
683.6	362	183.33	996.8	536	280.00	2462.0	1350	732.22	4028.0	2220	1215.6	5594.0	3090	1698.9
687.2	364	184.44	1000.4	538	281.11	2480.0	1360	737.78	4046.0	2230	1221.1	5612.0	3100	1704.4
690.8	366	185.56	1004.0	540	282.22	2498.0	1370	743.33	4064.0	2240	1226.7	5702.0	3150	1732.2
694.4	368	186.67	1007.6	542	283.33	2516.0	1380	748.89	4082.0	2250	1232.2	5792.0	3200	1760.0
698.0	370	187.78	1011.2	544	284.44	2534.0	1390	754.44	4100.0	2260	1237.8	5882.0	3250	1787.7
701.6	372	188.89	1014.8	546	285.56	2552.0	1400	760.00	4118.0	2270	1243.3	5972.0	3300	1815.5
705.2	374	190.00	1018.4	548	286.67	2570.0	1410	765.56	4136.0	2280	1248.9	6062.0	3350	1843.3
708.8	376	191.11	1022.0	550	287.78	2588.0	1420	771.11	4154.0	2290	1254.4	6152.0	3400	1871.1
712.4	378	192.22	1040.0	560	293.33	2606.0	1430	776.67	4172.0	2300	1260.0	6242.0	3450	1898.8
716.0	380	193.33	1058.0	570	298.89	2624.0	1440	782.22	4190.0	2310	1265.6	6332.0	3500	1926.6

719.6	382	194.44	1076.0	580	304.44	2642.0	1450	787.78	4208.0	2320	1271.1	6422.0	3550	1954.4
723.2	384	195.56	1094.0	590	310.00	2660.0	1460	793.33	4226.0	2330	1276.7	6512.0	3600	1982.2
726.8	386	196.67	1112.0	600	315.56	2678.0	1470	798.89	4244.0	2340	1282.2	6602.0	3650	2010.0
730.4	388	197.78	1130.0	610	321.11	2696.0	1480	804.44	4262.0	2350	1287.8	6692.0	3700	2037.7
734.0	390	198.89	1148.0	620	326.67	2714.0	1490	810.00	4280.0	2360	1293.3	6782.0	3750	2065.5
737.6	392	200.00	1166.0	630	332.22	2732.0	1500	815.56	4298.0	2370	1298.9	6872.0	3800	2093.3
741.2	394	201.11	1184.0	640	337.78	2750.0	1510	821.11	4316.0	2380	1304.4	6962.0	3850	2121.1
744.8	396	202.22	1202.0	650	343.33	2768.0	1520	826.67	4334.0	2390	1310.0	7052.0	3900	2148.8
748.4	398	203.33	1220.0	660	348.89	2786.0	1530	832.22	4352.0	2400	1315.6	7142.0	3950	2176.6
752.0	400	204.44	1238.0	670	354.44	2804.0	1540	837.78	4370.0	2410	1321.1	7232.0	4000	2204.4
755.6	402	205.56	1256.0	680	360.00	2822.0	1550	843.33	4388.0	2420	1326.7	7322.0	4050	2232.2
759.2	404	206.67	1274.0	690	365.56	2840.0	1560	848.89	4406.0	2430	1332.2	7412.0	4100	2260.0
762.8	406	207.78	1292.0	700	371.11	2858.0	1570	854.44	4424.0	2440	1337.8	7502.0	4150	2287.7
766.4	408	208.89	1310.0	710	376.67	2876.0	1580	860.00	4442.0	2450	1343.3	7592.0	4200	2315.5
770.0	410	210.00	1328.0	720	382.22	2894.0	1590	865.56	4460.0	2460	1348.9	7682.0	4250	2343.3
773.6	412	211.11	1346.0	730	387.78	2912.0	1600	871.11	4478.0	2470	1354.4	7772.0	4300	2371.1
777.2	414	212.22	1364.0	740	393.33	2930.0	1610	876.67	4496.0	2480	1360.0	7862.0	4350	2398.8
780.8	416	213.33	1382.0	750	398.89	2948.0	1620	882.22	4514.0	2490	1365.6	7952.0	4400	2426.6
784.4	418	214.44	1400.0	760	404.44	2966.0	1630	887.78	4532.0	2500	1371.1	8042.0	4450	2454.4
788.0	420	215.56	1418.0	770	410.00	2984.0	1640	893.33	4550.0	2510	1376.7	8132.0	4500	2482.2
791.6	422	216.67	1436.0	780	415.56	3002.0	1650	898.89	4568.0	2520	1382.2	8222.0	4550	2510.0
795.2	424	217.78	1454.0	790	421.11	3020.0	1660	904.44	4586.0	2530	1387.8	8312.0	4600	2537.7
798.8	426	218.89	1472.0	800	426.67	3038.0	1670	910.00	4604.0	2540	1393.3	8402.0	4650	2565.5
802.4	428	220.00	1490.0	810	432.22	3056.0	1680	915.56	4622.0	2550	1398.9	8492.0	4700	2593.3
806.0	430	221.11	1508.0	820	437.78	3074.0	1690	921.11	4640.0	2560	1404.4	8582.0	4750	2621.1
809.6	432	222.22	1526.0	830	443.33	3092.0	1700	926.67	4658.0	2570	1410.0	8672.0	4800	2648.8
813.2	434	223.33	1544.0	840	448.89	3110.0	1710	932.22	4676.0	2580	1415.6	8762.0	4850	2676.6
816.8	436	224.44	1562.0	850	454.44	3128.0	1720	937.78	4694.0	2590	1421.1	8852.0	4900	2704.4
820.4	438	225.56	1580.0	860	460.00	3146.0	1730	943.33	4712.0	2600	1426.7	8942.0	4950	2732.2
824.0	440	226.67	1598.0	870	465.56	3164.0	1740	948.89	4730.0	2610	1432.2	9032.0	5000	2760.0
827.6	442	227.78	1616.0	880	471.11	3182.0	1750	954.44	4748.0	2620	1437.8	9122.0	5050	2787.7
831.2	444	228.89	1634.0	890	476.67	3200.0	1760	960.00	4766.0	2630	1443.3	9212.0	5100	2815.5
834.8	446	230.00	1652.0	900	482.22	3218.0	1770	965.56	4784.0	2640	1448.9	9302.0	5150	2843.3
838.4	448	231.11	1670.0	910	487.78	3236.0	1780	971.11	4802.0	2650	1454.4	9392.0	5200	2871.1
842.0	450	232.22	1688.0	920	493.33	3254.0	1790	976.67	4820.0	2660	1460.0	9482.0	5250	2898.8
845.6	452	233.33	1706.0	930	498.89	3272.0	1800	982.22	4838.0	2670	1465.6	9572.0	5300	2926.6
849.2	454	234.44	1724.0	940	504.44	3290.0	1810	987.78	4856.0	2680	1471.1	9662.0	5350	2954.4
852.8	456	235.56	1742.0	950	510.00	3308.0	1820	993.33	4874.0	2690	1476.7	9752.0	5400	2982.2
856.4	458	236.67	1760.0	960	515.56	3326.0	1830	998.89	4892.0	2700	1482.2	9842.0	5450	3010.0
860.0	460	237.78	1778.0	970	521.11	3344.0	1840	1004.4	4910.0	2710	1487.8	9932.0	5500	3037.7
863.6	462	238.89	1796.0	980	526.67	3362.0	1850	1010.0	4928.0	2720	1493.3	10022.0	5550	3065.5
867.2	464	240.00	1814.0	990	532.22	3380.0	1860	1015.6	4946.0	2730	1498.9	10112.0	5600	3093.3

Appendix B

Metric Conversion Factors*

To convert from	To	Multiply by	To convert from	To	Multiply by
angstrom	m	1.0000×10^{-10}(a)	hp(e)	W	7.4570×10^{2}
atm	Pa	1.0133×10^{5}	hp(f)	W	7.4600×10^{2}
Btu(b)	J	1.054×10^{3}	in.	m	2.5400×10^{-2}
Btu(b)/ft²·h	W/m²	3.1525	in.²	m²	6.4516×10^{-4}
Btu(b)/ft²·h·°F	W/m²·K	5.6745	in.³	m³	1.6387×10^{-5}
Btu(b)·ft/h·ft²·°F	W/m·K	1.7296	in. of Hg(g)	Pa	3.3864×10^{3}
Btu(b)/ft²·s	W/m²	1.135×10^{4}	in. of water(c)	Pa	2.4908×10^{2}
Btu(b)·in./ft²·h·°F	W/m·K	1.4413×10^{-1}	K	°C	$t_{°C} = t_K - 273.15$
Btu(b)·in./s·ft²·°F	W/m·K	5.1887×10^{2}	kgf	N	9.80665(a)
Btu(b)/lbm·°F	J/kg·K	4.1840×10^{3}	kgf/mm²	Pa	9.80665×10^{6}(a)
cal(b)	J	4.1840 (a)	ksi	MPa	6.8948
cal(b)/cm·s·°C	W/m·K	4.1840×10^{2}(a)	ksi	Pa	6.8948×10^{6}
cal(b)/g	J/kg	4.1840×10^{3}(a)	ksi$\sqrt{in.}$	MPa$\sqrt{m}$	1.089
cal(b)/g·°C	J/kg·K	4.1840×10^{3}(a)	lb(h)	kg	4.5359×10^{-1}
circ mil	m²	5.0671×10^{-10}	lb/in.³	kg/m³	2.7680×10^{4}
°C	K	$t_K = t_{°C} + 273.15$	lbf	N	4.4482
degree	rad	1.7453×10^{-2}	lbf·in.	N·m	1.1298×10^{-1}
dyne/cm²	Pa	1.0000×10^{-1}(a)	lbf·ft	N·m	1.3558
°F	°C	$t_{°C} = (t_{°F} - 32)/1.8$	MPa$\sqrt{m}$	MNm$^{-3/2}$	1.0000(a)
°F	K	$t_K = (t_{°F} + 459.67)/1.8$	µin.	m	2.5400×10^{6}(a)
ft	m	3.0480×10^{-1}	mil	m	2.5400×10^{-5}(a)
ft²	m²	9.2903×10^{-2}	N/m²	Pa	1.0000(a)
ft³	m³	2.8317×10^{-2}	oersted	A/m	79.578
ft of water(c)	Pa	2.9890×10^{3}	oz/ft²	kg/m²	3.0515×10^{-1}
ft²/h (thermal diffusivity)	m²/s	2.58064×10^{-5}(a)	psi	Pa	6.8948×10^{3}
ft·lbf	J	1.3558	°R	K	$t_K = t_{°R}/1.8$
ft·lbf/s	W	1.3558	ton(j)	kg	9.0718×10^{2}
ft/s	m/s	3.0480×10^{-1}	ton(k)	kg	1.0160×10^{3}
gauss	T	1.0000×10^{-4}(a)	ton/in.²	Pa	1.3786×10^{4}
gallon(d)	m³	3.7854×10^{-3}	tonne	kg	1.0000×10^{3}(a)
g/cm³	kg/m³	1.0000×10^{3}(a)	torr	Pa	1.3332×10^{2}
g/cm³	Mg/m³	1.0000(a)	Ω/circ mil·ft	Ω·m	1.6624×10^{-9}

(a) Exactly. (b) Thermochemical. (c) At 4 °C (39.2 °F). (d) U.S. liquid. (e) Mechanical (1 hp = 550 ft·lbf/s). (f) Electrical. (g) At 0 °C (32 °F). (g) Avoirdupois. (j) Short; equal to 2000 lbm. (k) Long; 2240 lbm.

*Reprinted with permission from H. E. Boyer and T. L. Gall (eds.), *Metals Handbook—Desk Edition,* American Society for Metals, Metals Park, Ohio, 1985.

Appendix

B

Metric Conversion Factors

Appendix

C

Converting Common Units from the English to the Metric (SI) System*

---

*Reprinted with permission from *1977 Metal Progress Databook,* American Society for Metals, Metals Park, Ohio, 1977.

The International System of Units (SI for short) is a modernized version of the metric system. It is built upon seven base units and two supplementary units. Derived units are related to base and supplementary units by formulas in the right-hand column. Symbols for units with specific names are given in parentheses. The information in this Data Sheet, adapted from the revised "Metric Practice Guide," Standard E380 ASTM, includes a selected list of factors for converting U. S. customary units to SI units.

Metric Units and Conversion Factors

Quantity	Unit	Formula
	Base Units	
length	metre (m)	
mass	kilogram (kg)	
time	second (s)	
electric current	ampere (A)	
thermodynamic temperature	kelvin (K)	
amount of substance	mole (mol)	
luminous intensity	candela (cd)	
	Supplementary Units	
plane angle	radian (rad)	
solid angle	steradian (sr)	
	Derived Units	
acceleration	metre per second squared	m/s^2
activity (of a radioactive source)	disintegration per second	(disintegration)/s
angular acceleration	radian per second squared	rad/s^2
angular velocity	radian per second	rad/s
area	square metre	m^2
density	kilogram per cubic metre	kg/m^3
electric capacitance	farad (F)	A • s/V
electric conductance	siemens (S)	A/V
electric field strength	volt per metre	V/m
electric inductance	henry (H)	V • s/A
electric potential difference	volt (V)	W/A
electric resistance	ohm (Ω)	V/A
electromotive force	volt (V)	W/A
energy	joule (J)	N • m
entropy	joule per kelvin	J/K
force	newton (N)	$kg \cdot m/s^2$
frequency	hertz (Hz)	(cycle)/s
illuminance	lux (lx)	lm/m^2
luminance	candela per square metre	cd/m^2
luminous flux	lumen (lm)	cd • sr
magnetic field strength	ampere per metre	A/m
magnetic flux	weber (Wb)	V • s
magnetic flux density	tesla (T)	Wb/m^2
magnetomotive force	ampere (A)	—
power	watt (W)	J/s
pressure	pascal (Pa)	N/m^2
quantity of electricity	coulomb (C)	A • s
quantity of heat	joule (J)	N • m
radiant intensity	watt per steradian	W/sr
specific heat	joule per kilogram-kelvin	J/kg • K
stress	pascal (Pa)	N/m^2
thermal conductivity	watt per metre-kelvin	W/m • K
velocity	metre per second	m/s
viscosity, dynamic	pascal-second	Pa • s
viscosity, kinematic	square metre per second	m^2/s
voltage	volt (V)	W/A
volume	cubic metre	m^3
wavenumber	reciprocal metre	(wave)/m
work	joule (J)	N • m

Metric Conversion Factors

To convert from	To	Multiply by
atmosphere (760 mm Hg)	Pa	$1.013\ 25 \times 10^5$
Btu (International Table)	J	$1.055\ 056 \times 10^3$
Btu (International Table)/hour	W	$2.930\ 711 \times 10^{-1}$
calorie (International Table)	J	4.186 800*
centipoise	Pa • s	$1.000\ 000^* \times 10^{-3}$
centistoke	m^2/s	$1.000\ 000^* \times 10^{-6}$
circular mil	m^2	$5.067\ 075 \times 10^{-10}$
degree Fahrenheit	°C	$tC = (tF - 32)/1.8$
foot	m	$3.048\ 000^* \times 10^{-1}$
foot²	m^2	$9.290\ 304^* \times 10^{-2}$
foot³	m^3	$2.831\ 685 \times 10^{-2}$
foot-pound-force	J	1.355 818
foot-pound-force/minute	W	$2.259\ 697 \times 10^{-2}$
foot/second²	m/s^2	$3.048\ 000^* \times 10^{-1}$
gallon (U.S. liquid)	m^3	$3.785\ 412 \times 10^{-3}$
horsepower (electric)	W	$7.460\ 000^* \times 10^2$
inch	m	$2.540\ 000^* \times 10^{-2}$
inch²	m^2	$6.451\ 600^* \times 10^{-4}$
inch³	m^3	$1.638\ 706 \times 10^{-5}$
inch of mercury (60 F)	Pa	$3.376\ 85 \times 10^3$
inch of water (60 F)	Pa	$2.488\ 4 \times 10^2$
kilogram-force/centimetre²	Pa	$9.806\ 650^* \times 10^4$
kip (1000 lbf)	N	$4.448\ 222 \times 10^3$
kip/inch² (ksi)	Pa	$6.894\ 757 \times 10^6$
ounce (U.S. fluid)	m^3	$2.957\ 353 \times 10^{-5}$
ounce-force (avoirdupois)	N	$2.780\ 139 \times 10^{-1}$
ounce-mass (avoirdupois)	kg	$2.834\ 952 \times 10^{-2}$
ounce-mass/ft²	kg/m^2	$3.051\ 52 \times 10^{-1}$
ounce-mass/yard²	kg/m^2	$3.390\ 575 \times 10^{-2}$
pint (U.S. liquid)	m^3	$4.731\ 765 \times 10^{-4}$
pound-force (lbf avoirdupois)	N	4.448 222
pound-mass (lbm avoirdupois)	kg	$4.535\ 924 \times 10^{-1}$
pound-force/inch² (psi)	Pa	$6.894\ 757 \times 10^3$
pound-mass/inch³	kg/m^3	$2.767\ 990 \times 10^4$
pound-mass/foot³	kg/m^3	$1.601\ 846 \times 10$
quart (U.S. liquid)	m^3	$9.463\ 529 \times 10^{-4}$
ton (short, 2000 lbm)	kg	$9.071\ 847 \times 10^2$
torr (mm-Hg)	Pa	$1.333\ 22 \times 10^2$
watt-hour	J	$3.600\ 000^* \times 10^3$
yard	m	$9.144\ 000^* \times 10^{-1}$
yard²	m^2	$8.361\ 274 \times 10^{-1}$
yard³	m^3	$7.645\ 549 \times 10^{-1}$

*Exact

Multiplication Factors	Prefix	SI Symbol
$1\ 000\ 000\ 000\ 000 = 10^{12}$	tera	T
$1\ 000\ 000\ 000 = 10^9$	giga	G
$1\ 000\ 000 = 10^6$	mega	M
$1\ 000 = 10^3$	kilo	k
$100 = 10^2$	hecto*	h
$10 = 10^1$	deka*	da
$0.1 = 10^{-1}$	deci*	d
$0.01 = 10^{-2}$	centi*	c
$0.001 = 10^{-3}$	milli	m
$0.000\ 001 = 10^{-6}$	micro	μ
$0.000\ 000\ 001 = 10^{-9}$	nano	n
$0.000\ 000\ 000\ 001 = 10^{-12}$	pico	p
$0.000\ 000\ 000\ 000\ 001 = 10^{-15}$	femto	f
$0.000\ 000\ 000\ 000\ 000\ 001 = 10^{-18}$	atto	a

*To be avoided where possible

Appendix

D

Rockwell C and B Hardness Numbers for Steel*

*Reprinted with permission from H. E. Boyer and T. L. Gall (eds.), *Metals Handbook—Desk Edition,* American Society for Metals, Metals Park, Ohio, 1985.

Rockwell C-scale hardness No.	Vickers hardness No.	Brinell hardness No., 3000-kg load, 10-mm ball: Standard ball	Brinell hardness No., 3000-kg load, 10-mm ball: Tungsten carbide ball	Rockwell hardness No.: A scale, 60-kg load Brale indenter	Rockwell hardness No.: B scale, 100-kg load, 1/16-in.-diam ball	Rockwell hardness No.: D scale, 100-kg load, Brale indenter	Rockwell superficial hardness No., superficial Brale indenter: 15N scale, 15-kg load	Rockwell superficial hardness No., superficial Brale indenter: 30N scale, 30-kg load	Rockwell superficial hardness No., superficial Brale indenter: 45N scale, 45-kg load	Knoop hardness No., 500-g load and greater	Shore Scleroscope hardness No.	Tensile strength (approx), 1000 psi	Rockwell C-scale hardness No.
Rockwell C-Scale Hardness Numbers													
68	940	...	...	85.6	...	76.9	93.2	84.4	75.4	920	97	...	68
67	900	...	...	85.0	...	76.1	92.9	83.6	74.2	895	95	...	67
66	865	...	...	84.5	...	75.4	92.5	82.8	73.3	870	92	...	66
65	832	...	(739)	83.9	...	74.5	92.2	81.9	72.0	846	91	...	65
64	800	...	(722)	83.4	...	73.8	91.8	81.1	71.0	822	88	...	64
63	772	...	(705)	82.8	...	73.0	91.4	80.1	69.9	799	87	...	63
62	746	...	(688)	82.3	...	72.2	91.1	79.3	68.8	776	85	...	62
61	720	...	(670)	81.8	...	71.5	90.7	78.4	67.7	754	83	...	61
60	697	...	(654)	81.2	...	70.7	90.2	77.5	66.6	732	81	...	60
59	674	...	(634)	80.7	...	69.9	89.8	76.6	65.5	710	80	351	59
58	653	...	615	80.1	...	69.2	89.3	75.7	64.3	690	78	338	58
57	633	...	595	79.6	...	68.5	88.9	74.8	63.2	670	76	325	57
56	613	...	577	79.0	...	67.7	88.3	73.9	62.0	650	75	313	56
55	595	...	560	78.5	...	66.9	87.9	73.0	60.9	630	74	301	55
54	577	...	543	78.0	...	66.1	87.4	72.0	59.8	612	72	292	54
53	560	...	525	77.4	...	65.4	86.9	71.2	58.6	594	71	283	53
52	544	(500)	512	76.8	...	64.6	86.4	70.2	57.4	576	69	273	52
51	528	(487)	496	76.3	...	63.8	85.9	69.4	56.1	558	68	264	51
50	513	(475)	481	75.9	...	63.1	85.5	68.5	55.0	542	67	255	50
49	498	(464)	469	75.2	...	62.1	85.0	67.6	53.8	526	66	246	49
48	484	(451)	455	74.7	...	61.4	84.5	66.7	52.5	510	64	238	48
47	471	442	443	74.1	...	60.8	83.9	65.8	51.4	495	63	229	47
46	458	432	432	73.6	...	60.0	83.5	64.8	50.3	480	62	221	46
45	446	421	421	73.1	...	59.2	83.0	64.0	49.0	466	60	215	45
44	434	409	409	72.5	...	58.5	82.5	63.1	47.8	452	58	208	44
43	423	400	400	72.0	...	57.7	82.0	62.2	46.7	438	57	201	43
42	412	390	390	71.5	...	56.9	81.5	61.3	45.5	426	56	194	42
41	402	381	381	70.9	...	56.2	80.9	60.4	44.3	414	55	188	41
40	392	371	371	70.4	...	55.4	80.4	59.5	43.1	402	54	182	40
39	382	362	362	69.9	...	54.6	79.9	58.6	41.9	391	52	177	39
38	372	353	353	69.4	...	53.8	79.4	57.7	40.8	380	51	171	38
37	363	344	344	68.9	...	53.1	78.8	56.8	39.6	370	50	166	37
36	354	336	336	68.4	(109.0)	52.3	78.3	55.9	38.4	360	49	161	36
35	345	327	327	67.9	(108.5)	51.5	77.7	55.0	37.2	351	48	157	35
34	336	319	319	67.4	(108.0)	50.8	77.2	54.2	36.1	342	47	153	34
33	327	311	311	66.8	(107.5)	50.0	76.6	53.3	34.9	334	46	149	33
32	318	301	301	66.3	(107.0)	49.2	76.1	52.1	33.7	326	44	145	32
31	310	294	294	65.8	(106.0)	48.4	75.6	51.3	32.5	318	43	141	31
30	302	286	286	65.3	(105.5)	47.7	75.0	50.4	31.3	311	42	138	30
29	294	279	279	64.7	(104.5)	47.0	74.5	49.5	30.1	304	41	135	29
28	286	271	271	64.3	(104.0)	46.1	73.9	48.6	28.9	297	40	131	28
27	279	264	264	63.8	(103.0)	45.2	73.3	47.7	27.8	290	39	128	27
26	272	258	258	63.3	(102.5)	44.6	72.8	46.8	26.7	284	38	125	26
25	266	253	253	62.8	(101.5)	43.8	72.2	45.9	25.5	278	38	122	25
24	260	247	247	62.4	(101.0)	43.1	71.6	45.0	24.3	272	37	119	24
23	254	243	243	62.0	100.0	42.1	71.0	44.0	23.1	266	36	117	23
22	248	237	237	61.5	99.0	41.6	70.5	43.2	22.0	261	35	114	22
21	243	231	231	61.0	98.5	40.9	69.9	42.3	20.7	256	35	112	21

Rockwell B-scale hardness No.	Vickers hardness No.	Brinell hardness No., 10-mm-diam ball, 500-kg load	Brinell hardness No., 10-mm-diam ball, 3000-kg load	Rockwell hardness No., A scale, 60-kg load, Brale indenter	Rockwell hardness No., C scale, 150-kg load, Brale indenter	Rockwell hardness No., F scale, 60-kg load, 1/16-in.-diam ball	Rockwell superficial hardness No., 1/16-in.-diam ball, 15T scale, 15-kg load	Rockwell superficial hardness No., 1/16-in.-diam ball, 30T scale, 30-kg load	Rockwell superficial hardness No., 1/16-in.-diam ball, 45T scale, 45-kg load	Knoop hardness No., 500-g load and greater	Shore Scleroscope hardness No.	Tensile strength (approx), 1000 psi	Rockwell B-scale hardness No.
Rockwell B-Scale Hardness Numbers													
98	228	189	228	60.2	(19.9)	...	92.5	81.8	70.9	241	34	107	98
97	222	184	222	59.5	(18.6)	...	92.1	81.1	69.9	236	33	104	97
96	216	179	216	58.9	(17.2)	...	91.8	80.4	68.9	231	32	102	96
95	210	175	210	58.3	(15.7)	...	91.5	79.8	67.9	226	...	99	95
94	205	171	205	57.6	(14.3)	...	91.2	79.1	66.9	221	31	97	94
93	200	167	200	57.0	(13.0)	...	90.8	78.4	65.9	216	30	94	93
92	195	163	195	56.4	(11.7)	...	90.5	77.8	64.8	211	...	92	92
91	190	160	190	55.8	(10.4)	...	90.2	77.1	63.8	206	29	90	91
90	185	157	185	55.2	(9.2)	...	89.9	76.4	62.8	201	28	88	90
89	180	154	180	54.6	(8.0)	...	89.5	75.8	61.8	196	27	86	89
88	176	151	176	54.0	(6.9)	...	89.2	75.1	60.8	192	...	84	88
87	172	148	172	53.4	(5.8)	...	88.9	74.4	59.8	188	26	82	87
86	169	145	169	52.8	(4.7)	...	88.6	73.8	58.8	184	26	81	86
85	165	142	165	52.3	(3.6)	...	88.2	73.1	57.8	180	25	79	85
84	162	140	162	51.7	(2.5)	...	87.9	72.4	56.8	176	...	78	84
83	159	137	159	51.1	(1.4)	...	87.6	71.8	55.8	173	24	76	83
82	156	135	156	50.6	(0.3)	...	87.3	71.1	54.8	170	24	75	82
81	153	133	153	50.0	...	...	86.9	70.4	53.8	167	...	73	81
80	150	130	150	49.5	...	...	86.6	69.7	52.8	164	23	72	80
79	147	128	147	48.9	...	...	86.3	69.1	51.8	161	...	70	79
78	144	126	144	48.4	...	...	86.0	68.4	50.8	158	22	69	78
77	141	124	141	47.9	...	...	85.6	67.7	49.8	155	22	68	77
76	139	122	139	47.3	...	...	85.3	67.1	48.8	152	...	67	76
75	137	120	137	46.8	...	99.6	85.0	66.4	47.8	150	21	66	75
74	135	118	135	46.3	...	99.1	84.7	65.7	46.8	148	21	65	74
73	132	116	132	45.8	...	98.5	84.3	65.1	45.8	145	...	64	73
72	130	114	130	45.3	...	98.0	84.0	64.4	44.8	143	20	63	72
71	127	112	127	44.8	...	97.4	83.7	63.7	43.8	141	20	62	71
70	125	110	125	44.3	...	96.8	83.4	63.1	42.8	139	...	61	70
69	123	109	123	43.8	...	96.2	83.0	62.4	41.8	137	19	60	69
68	121	107	121	43.3	...	95.6	82.7	61.7	40.8	135	19	59	68
67	119	106	119	42.8	...	95.1	82.4	61.0	39.8	133	19	58	67
66	117	104	117	42.3	...	94.5	82.1	60.4	38.7	131	...	57	66
65	116	102	116	41.8	...	93.9	81.8	59.7	37.7	129	18	56	65
64	114	101	114	41.4	...	93.4	81.4	59.0	36.7	127	18	...	64
63	112	99	112	40.9	...	92.8	81.1	58.4	35.7	125	18	...	63
62	110	98	110	40.4	...	92.2	80.8	57.7	34.7	124	...	...	62
61	108	96	108	40.0	...	91.7	80.5	57.0	33.7	122	17	...	61
60	107	95	107	39.5	...	91.1	80.1	56.4	32.7	120	...	...	60
59	106	94	106	39.0	...	90.5	79.8	55.7	31.7	118	...	...	59
58	104	92	104	38.6	...	90.0	79.5	55.0	30.7	117	...	...	58
57	103	91	103	38.1	...	89.4	79.2	54.4	29.7	115	...	...	57
56	101	90	101	37.7	...	88.8	78.8	53.7	28.7	114	...	...	56
55	100	89	100	37.2	...	88.2	78.5	53.0	27.7	112	...	...	55

(a) For carbon and alloy steels in the annealed, normalized, and quenched-and-tempered conditions; less accurate for cold worked condition and for austenitic steels. The values in **boldface type** correspond to the values in the joint SAE-ASM-ASTM hardness conversions as printed in ASTM E140, Table 2. The values in parentheses are beyond normal range and are given for information only.

Appendix

E

The Relations between ASTM Grain Size and Average Grain "Diameter"*

*Adapted from *1966 Book of ASTM Standards,* part 31, American Society for Testing and Materials, Philadelphia, Pa., 1966;

ASTM micro-grain size number	Calculated "diameter" of average grain		ASTM micro-grain size number	Calculated "diameter" of average grain	
	mm	in.		mm	in.
		X10^{-3}			X10^{-3}
00	0.508	20.0	7.5	0.027	1.05
0	0.359	14.1	-	0.025	0.984
0.5	0.302	11.9	8.0	0.0224	0.884
1.0	0.254	10.0	-	0.0200	0.787
-	0.250	9.84	8.5	0.0189	0.743
1.5	0.214	8.41	9.0	0.0159	0.625
-	0.200	7.87	-	0.0150	0.591
-	0.180	7.09	9.5	0.0134	0.526
2.0	0.179	7.07	10.0	0.0112	0.442
2.5	0.151	5.95	-	0.0100	0.394
-	0.150	5.91	10.5	0.00944	0.372
3.0	0.127	5.00	-	0.00900	0.354
-	0.120	4.72	-	0.00800	0.315
3.5	0.107	4.20	11.0	0.00794	0.313
-	0.099	3.90	-	0.00700	0.276
4.0	0.090	3.54	11.5	0.00667	0.263
4.5	0.076	2.97	-	0.00600	0.236
-	0.070	2.76	12.0	0.00561	0.221
5.0	0.064	2.50	-	0.00500	0.197
-	0.060	2.36	12.5	0.00472	0.186
5.5	0.053	2.10	-	0.00400	0.158
-	0.050	1.97	13.0	0.00397	0.156
6.0	0.045	1.77	13.5	0.00334	0.131
-	0.040	1.58	-	0.00300	0.118
6.5	0.038	1.49	14.0	0.00281	0.111
-	0.035	1.38	-	0.00250	0.098
7.0	0.032	1.25			
-	0.030	1.18			

Appendix

F

Comments on Magnification Markers

It is customary to indicate the magnification on a photograph of a microstructure (micrograph), but when the photograph size is changed (such as enlarged), the original magnification must be corrected. A method of avoiding this is to place on the micrograph a bar or marker that has the correct dimension for the magnification of the photograph. Then if the magnification of the photograph is changed, the marker also changes dimensions to maintain the correct dimension at the new magnification.

For example, consider a micrograph at 100×. Then a marker 1 cm long would correspond to a dimension of 1/100 = 0.01 cm. It is common to express this in micrometers. Since a micrometer is equal to 10^{-4} cm, then the 0.01-cm dimension could also be labeled as 100 μm (0.01 × 10,000). If the micrograph is enlarged 2.5 times, then the marker becomes 2.5 × 1 = 2.5 cm long, and it is still labeled as 100 μm.

In general,

$$M = \frac{x}{y \times 10^{-4}}$$

where M is the magnification, x is the marker length in centimeters, and y is the marker value in micrometers on the micrograph.

At very high magnification, the marker value may be less than 1. For example, if the magnification is 40,000×, then 1 cm would correspond to 0.25 μm. In such cases, the marker corresponding to a certain value in angstroms (Å) may be used. 1 Å is equal to 10^{-8} cm, so in this case the marker dimension y in angstroms would be obtained from the equation as $40{,}000 = 1/(y \times 10^8)$. Then $y = 2500$ Å.

For the SI system the nanometer (nm) is used. 1 nm equals 0.1 Å, so for the preceding case, the marker would be labeled 25,000 nm.

Such markers are very useful because the actual sizes of microstructural features can be determined. For example, if the marker is labeled 2 μm and is 1 cm long, and a particle is present which is about 0.5 cm long in the micrograph, then its *actual* dimension in the sample is about 1 μm.

Appendix

G

Acronyms Used in Electronic Failure Analysis

AE	atomic emission
AES	Auger electron spectroscopy
AF	atomic fluorescence
AFM	atomic force microscopy/microscope
AMI	acoustic microimaging
BGA	ball grid array
BSE	back-scattered electron
CFC	chlorofluorocarbon
CL	cathodoluminescence
C-SAM	C-mode scanning acoustic microscopy
CSP	chip scale packaging
CTE	coefficient of thermal expansion
CVD	chemical vapor deposition
DIP	dual inline package
DIT	differential infrared thermography
DMA	dynamic mechanical analysis
DRAM	dynamic random access memory
DSC	differential scanning calorimetry
EBIC	electron beam–induced current
ECL	emitter-coupled logic
EDX	energy-dispersive x-ray
EDXA	energy-dispersive x-ray analysis

EOS	electrical overstress
ESD	electrostatic discharge
FAA	flame atomic absorption
FBSOA	forward bias safe operating area
FTIR	Fourier transform infrared
GFAA	graphite furnace atomic absorption
GOI	gate oxide integrity
HF-SAM	high-frequency scanning acoustic microscopy
HPLC	high-pressure liquid chromatography
IC	integrated circuit
ICP-E	inductively coupled plasma—emission
ICP-MS	inductively coupled plasma—mass spectrometry
I/O	input/output
IP	inner plane
IR	infrared
IRM	infrared microscopy/microscope
JDM	Joule displacement microscopy
LCC	leadless chip carrier
LEEM	low-energy electron microscopy
LM	light microscopy
MCM	multichip module
MELF	metal electrode leadless face (SMT component)
MFM	magnetic force microscopy
MOS	metal oxide semiconductor
MOSFET	metal oxide semiconductor field-effect transistor
nMOS	metal oxide semiconductor with n-type channel for “on”
OM	optical microscope/microscopy
OP-AMP	operational amplifier
PCB	printed-circuit board
PDM	photo displacement microscopy
PGC	pyrolysis-gas chromatography
PGC-MS	pyrolysis-gas chromatography–mass spectroscopy
PIH	pin-in-hole (also called PTH)
PIN	p-i-n diode (I is for intrinsic)
PLCC	plastic leaded chip carrier
PTH	plated-through-hole (also called PIH)
PWB	printed wiring board

RBS	Rutherford back-scattered spectroscopy
RBSOA	reverse bias safe operating area
RI	radiographic inspection
SAM	scanning acoustic microscopy
SCM	single-chip module
SCR	silicon controlled rectifier
SE	secondary electron
SEAM	scanning electron acoustic microscopy
SEM	scanning electron microscope/microscopy
SEM-BSE	SEM back-scattered electron
SEM-CL	SEM cathodoluminescence
SEM-EBIC	SEM electron beam–induced current
SEM-VC	SEM voltage contrast
SLAM	scanning laser acoustic microscopy
SLM	scanning light microscopy
SLM-OBIC	SLM optical beam–induced current
SMT	surface mount technology
SNFLM	scanning near field light microscopy
SOG	spin-on-glass
SPM	scanning probe microscopy
SRAM	static random access memory
STM	scanning tunneling microscope/microscopy
STP	standard thermal profile
TAB	tape-automated bonding
TEM	transmission electron microscope/microscopy
TGA	thermogravimetric analysis
TMA	thermomechanical analysis
VMOS	V-shaped groove metal oxide semiconductor
WDS	wavelength-dispersive spectrometry
XPS	x-ray photoelectron spectroscopy
XRF	x-ray fluorescence

Glossary*

abrasion The process of grinding or wearing away through the use of abrasives; a roughening or scratching of a surface due to *abrasive wear.*

abrasive wear The removal of material from a surface when hard particles slide or roll across the surface under pressure. The particles may be loose or may be part of another surface in contact with the surface being abraded. Compare with *adhesive wear.*

adhesive wear The removal or displacement of material from a surface by the welding together and subsequent shearing of minute areas of two surfaces that slide across each other under pressure. Compare with *abrasive wear.*

alligatoring The longitudinal splitting of flat slabs in a plane parallel to the rolled surface. Also called fishmouthing.

alligator skin See *orange peel.*

ambient Surrounding; usually used in relation to temperature, as "ambient temperature" surrounding a certain part or assembly.

annealing twin A *twin* formed in a crystal during recrystallization.

anode The electrode of an electrolytic cell at which oxidation occurs. Contrast with *cathode.*

arrest lines (marks) See *beach marks.*

asperity In *tribology,* a protuberance in the small-scale topographical irregularities of a solid surface.

axial Longitudinal, or parallel, to the axis or centerline of a part. Usually refers to axial compression or axial tension.

axial strain Increase (or decrease) in length resulting from a stress acting parallel to the longitudinal axis of a test specimen.

banded structure A segregated structure consisting of alternating, nearly parallel bands of different composition, typically aligned in the direction of primary hot working.

beach marks Macroscopic (visible) progression marks on a fracture surface that indicate successive position of the advancing crack front. The classic appearance is of irregular elliptical or semielliptical rings, radiating outward from one or more origins. Beach marks (also known as clamshell marks, tide marks, or arrest marks) are typically found on service fractures where the part

*Reprinted with permission from *Metals Handbook,* 9th ed., vol. 11: *Failure Analysis and Prevention,* American Society for Metals, Metals Park, Ohio, 1986.

is loaded randomly, intermittently, or with periodic variations in mean stress or alternating stress. Not to be confused with *striations,* which are microscopic and form differently.

breaking stress See *rupture stress.*

Brinell hardness number HB A number related to the applied load and to the surface area of the permanent impression made by a ball indenter, computed from

$$\mathrm{HB} = \frac{2P}{\pi D(D - \sqrt{D^2 - d^2})}$$

where P is applied load, kgf; D is diameter of ball, mm; and d is mean diameter of impression, mm.

Brinell hardness test A test for determining the hardness of a material by forcing a hard steel or carbide ball of specified diameter into it under a specified load. The result is expressed as the *Brinell hardness number.*

brinelling Damage to a solid bearing surface characterized by one or more plastically formed indentations brought about by overload. This term is often applied in the case of rolling-element bearings. See also *false brinelling.*

brittle Permitting little or no plastic (permanent) deformation prior to fracture.

brittle crack propagation A very sudden propagation of a crack with the absorption of no energy except that stored elastically in the body. Microscopic examination may reveal some deformation not noticeable to the unaided eye. Contrast with *ductile crack propagation.*

brittle erosion behavior Erosion behavior having characteristic properties (e.g., little or no plastic flow, the formation of cracks) that can be associated with *brittle fracture* of the exposed surface. The maximum volume removal occurs at an angle near 90°, in contrast to approximately 25° for *ductile erosion behavior.*

brittle fracture Separation of a solid accompanied by little or no macroscopic plastic deformation. Typically, brittle fracture occurs by rapid crack propagation with less expenditure of energy than for *ductile fracture.*

brittleness The tendency of a material to fracture without first undergoing significant *plastic deformation.* Contrast with *ductility.*

buckle (1) An indented valley in the surface of a sand casting due to expansion of the molding sand. (2) A local waviness in a metal bar or sheet, usually transverse to the direction of rolling.

buckling A compression phenomenon that occurs when, after some critical level of load, a bulge, bend, bow, kink, or other wavy condition is produced in a beam, column, plate, bar, or sheet product form.

bulk modulus See *bulk modulus of elasticity.*

bulk modulus of elasticity *K* The measure of resistance to change in volume; the ratio of hydrostatic stress to the corresponding unit change in volume. This elastic constant can be expressed by

$$K = \frac{\sigma_m}{\Delta} = \frac{-p}{\Delta} = \frac{1}{\beta}$$

where K is bulk modulus of elasticity, σ_m is hydrostatic or mean stress tensor, p is hydrostatic pressure, and β is compressibility. Also known as bulk modulus, compression modulus, hydrostatic modulus, and volumetric modulus of elasticity.

carbon flotation Segregation in which free graphite has separated from the molten iron. This defect tends to occur at the upper surfaces of the cope of the castings.

casting shrinkage See *liquid shrinkage, shrinkage cavity, solidification shrinkage,* and *solid shrinkage.*

catastrophic wear Rapidly occurring or accelerating surface damage, deterioration, or change of shape caused by wear to such a degree that the service life of a part is appreciably shortened or its function is destroyed.

caustic cracking A form of *stress-corrosion cracking* most frequently encountered in carbon steels or iron-chromium-nickel alloys that are exposed to concentrated hydroxide solutions at temperatures of 200 to 250°C (400 to 480°F). Also known as caustic embrittlement.

caustic embrittlement See *caustic cracking.*

cavitation The formation and rapid collapse within a liquid of cavities or bubbles that contain vapor or gas or both. Cavitation caused by severe turbulent flow often leads to *cavitation damage.*

cavitation damage The degradation of a solid body resulting from its exposure to cavitation. This may include loss of material, surface deformation, or changes in properties or appearance

cavitation erosion See *cavitation damage.*

centerline shrinkage Shrinkage or porosity occurring along the central plane or axis of a cast metal section.

chafing fatigue Fatigue initiated in a surface damaged by rubbing against another body. See also *fretting.*

Charpy test An impact test in which a V-notched, keyhole-notched, or U-notched specimen, supported at both ends, is struck behind the notch by a striker mounted at the lower end of a bar that can swing as a pendulum. The energy that is absorbed in fracture is calculated from the height to which the striker would have risen had there been no specimen and the height to which it actually rises after fracture of the specimen. Contrast with *Izod test.*

chevron pattern A fractographic pattern of radial marks (shear ledges) that looks like nested letters V; sometimes called a herringbone pattern. Chevron patterns are typically found on brittle fracture surfaces in parts whose widths are considerably greater than their thicknesses. The points of the chevrons can be traced back to the fracture origin.

chill A white iron structure that is produced by rapid solidification.

chord modulus A slope of the chord drawn between any two specific points on a stress-strain curve. See also *modulus of elasticity.*

clamshell marks See *beach marks.*

cleavage (1) Fracture of a crystal by crack propagation across a crystallographic plane of low index. (2) The tendency to cleave or split along definite crystallographic planes.

cleavage crack A crack that extends along a plane of easy *cleavage* in a crystalline material.

cleavage fracture A fracture, usually of a polycrystalline metal, in which most of the grains have failed by *cleavage,* resulting in bright reflecting facets. It is one type of *crystalline fracture* and is associated with low-energy brittle fracture. Contrast with *shear fracture.*

cleavage plane A characteristic crystallographic plane or set of planes in a crystal on which *cleavage fracture* occurs easily.

cold shot A small globule of metal that solidified prematurely and is embedded in but not entirely fused with the surface of the casting.

cold shut A discontinuity on or immediately beneath the surface of a casting, caused by the meeting of two streams of liquid metal that failed to merge. A cold shut may have the appearance of a crack or seam with smooth, rounded edges.

columnar structure A coarse structure of parallel, elongated grains formed by unidirectional growth that is most often observed in castings. This results from diffusional growth accompanied by a solid-state transformation.

composite material A heterogeneous, solid structural material consisting of two or more distinct components that are mechanically or metallurgically bonded together, such as a wire or filament of a high-melting substance embedded in a metal or nonmetal matrix.

compression modulus See *bulk modulus of elasticity.*

compressive Pertaining to forces on a body or part of a body that tend to crush, or compress, the body.

compressive strength The maximum *compressive stress* a material is capable of developing. With a brittle material that fails in compression by fracturing, the compressive strength has a definite value. In the case of ductile, malleable, or semiviscous materials (which do not fail in compression by a shattering fracture), the value obtained for compressive strength is an arbitrary value dependent on the degree of distortion that is regarded as effective failure of the material.

compressive stress A stress that causes an elastic body to deform (shorten) in the direction of the applied load. Contrast with *tensile stress.*

contact fatigue Cracking and subsequent pitting of a surface subjected to alternating hertzian stresses such as those produced under rolling contact or combined rolling and sliding. The phenomenon of contact fatigue is encountered most often in rolling-element bearings or in gears, where the surface stresses are high due to the concentrated loads and are repeated many times during normal operation.

corrosion The chemical or electrochemical reaction between a material, usually a metal, and its environment that produces a deterioration of the material and its properties. See also *corrosion fatigue, crevice corrosion, denickelification, dezincification, erosion-corrosion, exfoliation, filiform corrosion, fretting corrosion, galvanic corrosion, general corrosion, graphitic corrosion, impingement attack, interdendritic corrosion, intergranular corrosion, internal oxidation, oxidation, parting, pitting, poultice corrosion, rust, selective leaching, stray-current corrosion, stress-corrosion cracking,* and *sulfide stress cracking.*

corrosion fatigue Cracking produced by the combined action of repeated or fluctuating stress and a corrosive environment at lower stress levels or fewer cycles than would be required in the absence of a corrosive environment.

corrosive wear *Wear* in which chemical or electrochemical reaction with the environment is significant.

crack extension Δa An increase in crack size. See also *crack length, effective crack size, original crack size,* and *physical crack size.*

crack length (depth) a In *fatigue* and *stress-corrosion cracking,* the *physical crack size* used to determine the crack growth rate and the *stress-intensity factor.* For a compact-type specimen, crack length is measured from the line connecting the bearing points of load application. For a center-crack tension specimen, crack length is measured from the perpendicular bisector of the central crack. See also *crack size.*

crack mouth opening displacement (CMOD) See *crack opening displacement.*

crack opening displacement (COD) On a K_{Ic} specimen, the opening displacement of the notch surfaces at the notch and in the direction perpendicular to the plane of the notch and the crack. The displacement at the tip is called the crack tip opening displacement (CTOD); at the mouth, it is called the crack mouth opening displacement (CMOD).

crack plane orientation An identification of the plane and direction of a fracture in relation to product geometry. This identification is designated by a hyphenated code, the first letter(s) representing the direction normal to the crack plane and the second letter(s) designating the expected direction of crack propagation.

crack size a A lineal measure of a principal planar dimension of a crack. This measure is commonly used in the calculation of quantities descriptive of the stress and displacement fields. In practice, the value of crack size is obtained from procedures for measurement of *physical crack size, original crack size,* or *effective crack size,* as appropriate to the situation under consideration. See also *crack length (depth).*

crack tip opening displacement (CTOD) See *crack opening displacement.*

crack tip plane strain A stress-strain field near a crack tip that approaches *plane strain* to the degree required by an empirical criterion.

creep The time-dependent strain occurring under stress. The *creep strain* occurring at a diminishing rate is called primary or transient creep; that occurring at a minimum and almost constant rate, secondary or steady-rate creep; that occurring at an accelerating rate, tertiary creep.

creep rate The slope of the creep-time curve at a given time determined from a cartesian plot.

creep-rupture strength The stress that will cause fracture in a creep test at a given time in a specified constant environment. Also known as stress-rupture strength.

creep strain The time-dependent total strain (extension plus initial gage length) produced by applied stress during a creep test.

creep strength The stress that will cause a given *creep strain* in a creep test at a given time in a specified constant environment.

creep stress The constant load divided by the original cross-sectional area of the specimen.

crevice corrosion Localized *corrosion* of a metal surface at, or immediately adjacent to, an area that is shielded from full exposure to the environment because of close proximity between the metal and the surface of another material.

cross direction See *transverse direction.*

crush An indentation in a casting surface due to displacement of sand into the mold cavity when the mold is closed.

crystalline fracture A pattern of brightly reflecting crystal facets on the fracture surface of a polycrystalline metal, resulting from *cleavage fracture* of many individual crystals. Contrast with *fibrous fracture* and *silky fracture*; see also *granular fracture.*

cup fracture (cup-and-cone fracture) A mixed-mode fracture, often seen in tension test specimens of a ductile material, where the central portion undergoes plane-strain fracture and the surrounding region undergoes plane-stress fracture. One of the mating fracture surfaces looks like a miniature cup; it has a central depressed flat-face region surrounded by a shear lip. The other fracture surface looks like a miniature truncated cone.

cupping The condition sometimes occurring in heavily cold-worked rods and wires, in which the outside fibers are still intact and the central zone has failed in a series of cup-and-cone fractures.

cut A raised rough surface on a casting due to erosion by the metal stream of part of the sand mold or core.

cycle *N* In fatigue, one complete sequence of values of applied load that is repeated periodically.

cyclic load (1) Repetitive loading, as with regularly recurring stresses on a part, that sometimes leads to fatigue fracture. (2) Loads that change value by following a regular repeating sequence of change.

cyclic stressing See *cyclic load.*

decarburization Loss of carbon from the surface layer of a carbon-containing alloy due to reaction with one or more chemical substances in a medium that contacts the surface.

deformation A change in the form of a body due to stress, thermal change, change in moisture, or other causes. Measured in units of length.

deformation bands Bands produced within individual grains during cold working which differ variably in orientation from the matrix.

deformation curve See *stress-strain diagram.*

dendrite A crystal with a treelike branching pattern. Dendrites are most evident in cast metals slowly cooled through the solidification range.

denickelification *Corrosion* in which nickel is selectively leached from nickel-containing alloys. Most commonly observed in copper-nickel alloys after extended service in fresh water. See also *selective leaching.*

depletion Selective removal of one component of an alloy, usually from the surface or preferentially from grain-boundary regions. See also *selective leaching.*

deposit attack See *poultice corrosion.*

deposit corrosion See *poultice corrosion.*

dezincification *Corrosion* in which zinc is selectively leached from zinc-containing alloys. Most commonly found in copper-zinc alloys containing less than 85% Cu after extended service in water containing dissolved oxygen. See also *selective leaching.*

diamond pyramid hardness test See *Vickers hardness test.*

dimpled rupture fracture A fractographic term describing *ductile fracture* that occurs through the formation and coalescence of microvoids (dimples) along the fracture path. The fracture surface of such a ductile fracture appears dimpled when observed at high magnification and usually is most clearly resolved when viewed in a scanning electron microscope.

distortion Any deviation from an original size, shape, or contour that occurs because of the application of stress or the release of *residual stress.*

ductile crack propagation Slow crack propagation that is accompanied by noticeable *plastic deformation* and requires energy to be supplied from outside the body. Contrast with *brittle crack propagation.*

ductile erosion behavior Erosion behavior having characteristic properties (i.e., considerable *plastic deformation*) that can be associated with *ductile fracture* of the exposed solid surface. A characteristic ripple pattern forms on the exposed surface at low values of angle of attack. Contrast with *brittle erosion behavior.*

ductile fracture Fracture characterized by tearing of metal accompanied by appreciable gross *plastic deformation* and expenditure of considerable energy. Contrast with *brittle fracture.*

ductility The ability of a material to deform plastically before fracturing. Measured by *elongation* or *reduction of area* in a tension test, by height of *cupping* in a cupping test, or by the radius or angle of bend in a bend test. Contrast with *brittleness*; see also *plastic deformation.*

dynamic Moving, or having high velocity. Frequently used with high strain rate ($>0.1\ s^{-1}$) testing of metal specimens. Contrast with *static.*

effective crack size a_e The *physical crack size* augmented for the effects of crack tip plastic deformation. Sometimes the effective crack size is calculated from a measured value of a physical crack size plus a calculated value of a plastic-zone adjustment. A preferred method for the calculation of effective crack size compares compliance from the secant of a load-deflection trace with the elastic compliance from a calibration for the type of specimen.

elastic constants The factors of proportionality that relate elastic displacement of a material to applied forces. See also *bulk modulus of elasticity, modulus of elasticity, Poisson's ratio,* and *shear modulus.*

elastic deformation A change in dimensions directly proportional to and in phase with an increase or decrease in applied force.

elasticity The property of a material by virtue of which deformation caused by *stress* disappears upon removal of the stress. A perfectly elastic body completely recovers its original shape and dimensions after release of stress.

elastic limit The maximum *stress* a material is capable of sustaining without any permanent *strain* (deformation) remaining upon complete release of the stress. See also *proportional limit.*

elastic strain See *elastic deformation.*

elongation A term used in mechanical testing to describe the amount of extension of a test piece when stressed. See also *elongation, percent* and *stress.*

elongation, percent The extension of a uniform section of a specimen expressed as percentage of the original gage length:

$$\text{Elongation, \%} = \frac{L_f - L_o}{L_o} \times 100$$

where L_o is original gage length and L_f is final gage length. See also *elongation.*

embrittlement The severe loss of ductility and/or toughness of a material, usually a metal or alloy.

endurance limit The maximum *stress* below which a material can presumably endure an infinite number of *stress cycles.* If the stress is not completely reversed, the value of the mean stress, the minimum stress, or the *stress ratio* also should be stated. Compare with *fatigue limit.*

erosion Destruction of materials by the abrasive action of moving fluids, usually accelerated by the presence of solid particles carried with the fluid. See also *erosion-corrosion.*

erosion-corrosion A conjoint action involving *corrosion* and *erosion* in the presence of a moving corrosive fluid, leading to the accelerated loss of material.

exfoliation *Corrosion* that proceeds laterally from the sites of initiation along planes parallel to the surface, generally at grain boundaries, forming corrosion products that force metal away from the body of the material, giving rise to a layered appearance.

failure A general term used to imply that a part in service (1) has become completely inoperable, (2) is still operable but is incapable of satisfactorily performing its intended function, or (3) has deteriorated seriously, to the point that it has become unreliable or unsafe for continued use.

false brinelling Damage to a solid bearing surface characterized by indentations not caused by *plastic deformation,* resulting from overload but thought to be due to other causes such as *fretting corrosion.* See also *brinelling.*

fatigue The phenomenon leading to *fracture* under repeated or fluctuating stresses having a maximum value less than the ultimate tensile strength of the material. See also *fatigue failure, high-cycle fatigue, low-cycle fatigue,* and *ultimate strength.*

fatigue crack growth rate *da/dN* The rate of crack extension caused by constant-amplitude fatigue loading, expressed in terms of crack extension per cycle of load application.

fatigue failure Failure that occurs when a specimen undergoing fatigue completely fractures into two parts or has softened or been otherwise significantly reduced in stiffness by thermal heating or cracking. Fatigue failure generally occurs at loads which applied statically would produce little perceptible effect. Fatigue failures are progressive, beginning as minute cracks that grow under the action of the fluctuating stress.

fatigue life The number of *stress cycles* that can be sustained prior to failure under a stated test condition.

fatigue limit The maximum *stress* that presumably leads to fatigue fracture in a specified number of *stress cycles.* If the stress is not completely reversed, the value of the mean stress, the minimum stress, or the *stress ratio* also should be stated. Compare with *endurance limit.*

fatigue notch factor K_f The ratio of the *fatigue strength* of an unnotched specimen to the fatigue strength of a notched specimen of the same material and condition; both strengths are determined at the same number of *stress cycles.*

fatigue notch sensitivity q An estimate of the effect of a notch or hole of a given size and shape on the fatigue properties of a material; measured by $q = (K_f - 1)/(K_t - 1)$, where K_f is the *fatigue notch factor* and K_t is the *stress-concentration factor.* A material is said to be fully notch-sensitive if q approaches a value of 1.0; it is not notch-sensitive if the ratio approaches 0.

fatigue ratio The *fatigue limit* under completely reversed flexural stress divided by the *tensile strength* for the same alloy and condition.

fatigue strength The maximum *stress* that can be sustained for a specified number of *stress cycles* without failure, the stress being completely reversed within each cycle unless otherwise stated.

fatigue striation See *striation.*

fatigue wear *Wear* of a solid surface caused by *fracture* arising from material fatigue.

fiber (1) The characteristic of wrought metal that indicates directional properties. It is revealed by etching of a longitudinal section or is manifested by the fibrous or woody appearance of a fracture. It is caused chiefly by extension of the constituents of the metal, both metallic and nonmetallic, in the direction of working. (2) The pattern of *preferred orientation* of metal crystals after a given deformation process, usually wiredrawing. See also *texture.*

fiber-reinforced composite A material consisting of two or more discrete physical phases, in which a fibrous phase is dispersed in a continuous matrix phase. The fibrous phase may be macro-, micro-, or submicroscopic, but it must retain its physical identity so that it could conceivably be removed from the matrix intact.

fiber stress Local *stress* through a small area (a point or line) on a section where the stress is not uniform, as in a beam under a bending load.

fibrous fracture A gray and amorphous *fracture* that results when a metal is sufficiently ductile for the crystals to elongate before fracture occurs. When a fibrous fracture is obtained in an impact test, it may be regarded as definite evidence of the toughness of the metal. See also *crystalline fracture* and *silky fracture.*

fibrous structure (1) In forgings, a structure revealed as laminations, not necessarily detrimental, on an etched section or as a ropy appearance on a *fracture.* (2) In wrought iron, a structure consisting of slag fibers embedded in ferrite. (3) In rolled steel plate stock, a uniform, lamination-free, fine-grained structure on a fractured surface.

filiform corrosion *Corrosion* that occurs under some coatings in the form of randomly distributed threadlike filaments.

fisheye A discontinuity found on the fracture surface of a weld in steel that consists of a small pore or inclusion surrounded by an approximately round, bright area.

fishmouthing See *alligatoring.*

flake A short, discontinuous internal crack in ferrous metals attributed to stresses produced by localized transformation and hydrogen-solubility effects during cooling after hot working. In fracture surfaces, flakes appear as bright, silvery areas with a coarse texture. In deep acid-etched transverse sections, they appear as discontinuities that are usually in the midway to center location of the section. Also termed hairline cracks and shatter cracks.

flow Movement (slipping or sliding) of essentially parallel planes within an element of a material in parallel directions; occurs under the action of *shear* *stress.* Continuous action in this manner, at constant volume and without disintegration of the material, is termed *yield, creep,* or *plastic deformation.*

flow lines Texture showing the direction of metal flow during hot or cold working. Flow lines often can be revealed by etching the surface or a section of a metal part.

fluting A type of *pitting* in which cavities occur in a regular pattern, forming grooves or flutes. Fluting is caused by *fretting* or by electric arcing.

fold A defect in metal, usually on or near the surface, caused by continued fabrication of overlapping surfaces.

fractography Descriptive explanation of a fracture process, especially in metals, with specific reference to photographs of the fracture surface. Macrofractography involves low magnification (<25×); microfractography high magnification (>25×).

fracture The irregular surface produced when a piece of metal is broken. See also *crystalline fracture, fibrous fracture, granular fracture, intergranular fracture, silky fracture,* and *transgranular fracture.*

fracture mechanics See *linear-elastic fracture mechanics.*

fracture stress See *rupture stress.*

fracture test Test in which a specimen is broken and its fracture surface is examined with the unaided eye or with a low-power microscope to determine such factors as composition, grain size, case depth, or discontinuities.

fracture toughness A generic term for measures of resistance to the extension of a crack. The term is sometimes restricted to results of fracture mechanics tests, which are directly applicable in fracture control. However, the term commonly includes results from simple tests of notched or precracked specimens not based on fracture mechanics analysis. Results from tests of the latter type are often useful for fracture control, based on either service experience or empirical correlations with fracture mechanics tests. See also *stress-intensity factor.*

fretting *Wear* that occurs between tight-fitting surfaces subjected to oscillation at very small amplitude. This type of wear can be a combination of *oxidative wear* and *abrasive wear.* See also *fretting corrosion.*

fretting corrosion The deterioration at the interface between contacting surfaces as the result of *corrosion* and slight oscillatory slip between the two surfaces.

fretting fatigue Fatigue fracture that initiates at a surface area where *fretting* has occurred.

galling A condition whereby excessive friction between high spots results in localized welding with subsequent *spalling* and a further roughening of the rubbing surfaces of one or both of two mating parts.

galvanic corrosion Accelerated *corrosion* of a metal because of an electrical contact with a more noble metal or nonmetallic conductor in a corrosive electrolyte.

gas hole A hole in a casting or weld formed by gas escaping from molten metal as it solidifies. Gas holes may occur individually or in clusters, or may be distributed throughout the solidified metal.

gas porosity Fine holes or pores within a metal that are caused by entrapped gas or by evolution of dissolved gas during solidification.

general corrosion A form of deterioration that is distributed more or less uniformly over a surface. See also *corrosion.*

glide See *slip.*

grain An individual crystal in a polycrystalline metal or alloy, including twinned regions or *subgrains* if present.

grain boundary An interface separating two grains at which the orientation of the lattice changes from that of one grain to that of the other. When the orientation change is very small, the boundary is sometimes referred to as a *subboundary structure.*

grain-boundary corrosion Same as *intergranular corrosion*; see also *corrosion* and *interdendritic corrosion.*

grain flow Fiberlike lines on polished and etched sections of forgings caused by orientation of the constituents of the metal in the direction of working during forging. Grain flow produced by proper die design can improve required *mechanical properties* of forgings.

granular fracture A type of irregular surface produced when metal is broken that is characterized by a rough, grainlike appearance, rather than a smooth or fibrous one. It can be subclassified as *transgranular* or *intergranular.* This type of fracture is frequently called *crystalline fracture*; however, the inference that the metal broke because it "crystallized" is not justified, because all metals are crystalline in the solid state. See also *fibrous fracture* and *silky fracture.*

graphitic corrosion Deterioration of gray cast iron in which the metallic constituents are selectively leached or converted to corrosion products, leaving the graphite intact; it occurs in relatively mild aqueous solutions and in buried pipe and fittings. The term "graphitization" is commonly used to identify this form of *corrosion,* but is not recommended because of its use in metallurgy for the decomposition of carbide to graphite.

hairline crack See *flake.*

hardness A measure of the resistance of a material to surface indentation or abrasion; may be thought of as a function of the *stress* required to produce some specified type of surface deformation. There is no absolute scale for hardness; therefore, to express hardness quantitatively, each type of test has its own scale of arbitrarily defined hardness. Indentation hardness can be measured by *Brinell, Knoop, Rockwell, Scleroscope,* and *Vickers hardness tests.*

Hartmann lines See *Lüders lines.*

heat-affected zone That portion of the base metal that was not melted during brazing, cutting, or welding, but whose *microstructure* and *mechanical properties* were altered by the heat.

herringbone pattern See *chevron pattern.*

high-cycle fatigue *Fatigue* that occurs at relatively large numbers of cycles. The arbitrary, but commonly accepted, dividing line between high-cycle fatigue and *low-cycle fatigue* is considered to be about 10^4 to 10^5 cycles. In practice, this distinction is made by determining whether the dominant component of the *strain* imposed during cyclic loading is elastic (high cycle) or plastic (low cycle), which in turn depends on the properties of the metal and on the magnitude of the nominal *stress.*

Hooke's law A material in which *stress* is linearly proportional to *strain* is said to obey Hooke's law. This law is valid only up to the *proportional limit,* or the end of the straight-line portion of the *stress-strain diagram.* See also *modulus of elasticity.*

hot crack See *solidification shrinkage crack.*

hot tear A crack or fracture formed before completion of solidification because of hindered contraction. A hot tear is frequently open to the surface of the casting and thus exposed to the atmosphere. This may result in *oxidation, decarburization,* or other metal-atmosphere reactions at the tear surface.

hydrogen blistering The formation of blisters on or below a metal surface from excessive internal hydrogen pressure. Hydrogen may be formed during cleaning, plating, corrosion, etc.

hydrogen damage A general term for the embrittlement, cracking, blistering, and hydride formation that can occur when hydrogen is present in some metals.

hydrogen embrittlement A condition of low ductility or hydrogen-induced cracking in metals resulting from the absorption of hydrogen. See also *hydrogen-induced delayed cracking.*

hydrogen-induced delayed cracking A term sometimes used to identify a form of *hydrogen embrittlement,* in which a metal appears to fracture spontaneously under a steady stress less than the *yield stress.* There is usually a delay between the application of stress (or exposure of the stressed metal to hydrogen) and the onset of cracking. Also referred to as static fatigue.

hydrostatic modulus See *bulk modulus of elasticity.*

impact energy The amount of energy required to fracture a material, usually measured by means of an *Izod test* or *Charpy test.* The type of specimen and test conditions affect the values and therefore should be specified.

impact load An especially severe shock load such as that caused by instantaneous arrest of a falling mass, by shock meeting of two parts (in a mechanical hammer, for example), or by explosive impact, in which there can be an exceptionally rapid buildup of stress.

impact strength See *impact energy.*

impingement attack *Corrosion* associated with turbulent flow of liquid. May be accelerated by entrained gas bubbles. See also *erosion-corrosion.*

inclusion A particle of foreign material in a metallic matrix. The particle is usually a compound (such as an oxide, sulfide, or silicate), but may be of any substance that is foreign to (and essentially insoluble in) the matrix. Inclusions are usually considered undesirable, although in some cases—such as in free-machining metals—manganese sulfides, phosphorus, selenium, or tellurium may be deliberately introduced to improve machinability.

intercrystalline See *intergranular.*

intercrystalline corrosion See *intergranular corrosion.*

intercrystalline cracking See *intergranular cracking.*

interdendritic corrosion Corrosive attack that progresses preferentially along interdendritic paths. This type of attack results from local differences in composition commonly encountered in alloy castings. See also *corrosion.*

interface The boundary between two contacting parts or regions of parts.

intergranular Between crystals or grains. Also termed intercrystalline. Contrast with *transgranular.*

intergranular corrosion *Corrosion* occurring preferentially at grain boundaries, usually with slight or negligible attack on the adjacent grains. See also *interdendritic corrosion.*

intergranular cracking Cracking or fracturing that occurs between the grains or crystals in a polycrystalline aggregate. Contrast with *transgranular cracking.*

intergranular fracture Brittle fracture of a metal in which the fracture is between the grains, or crystals, that form the metal. Contrast with *transgranular fracture.*

intergranular stress-corrosion cracking *Stress-corrosion cracking* in which the cracking occurs along grain boundaries.

internal oxidation (1) The formation of isolated particles of *corrosion* products beneath the metal surface. This occurs as the result of preferential oxidation of certain alloy constituents by inward diffusion of oxygen, nitrogen, sulfur, etc. Also called subsurface corrosion. (2) Preferential in situ oxidation of certain components of phases within the bulk of a solid alloy accomplished by diffusion of oxygen into the body. This is commonly used to prepare electrical contact materials.

intracrystalline See *transgranular.*

intracrystalline cracking See *transgranular cracking.*

Izod test A type of impact test in which a V-notched specimen, mounted vertically, is subjected to a sudden blow delivered by the weight at the end of a pendulum arm. The energy required to break off the free end is a measure of the impact strength or toughness of the material. Contrast with *Charpy test.*

***J* integral** A mathematical expression; a line or surface integral that encloses the crack front from one crack surface to the other, used to characterize the *fracture toughness* of a material having appreciable plasticity before fracture. The J integral eliminates the need to describe the behavior of the material near the crack tip by considering the local stress-strain field around the crack front; J_{Ic} is the critical value of the J integral required to initiate growth of a preexisting crack.

Knoop hardness number HK A number related to the applied load and to the projected area of the permanent impression made by a rhombic-based pyramidal diamond indenter having included edge angles of 172° 30′ and 130° 0′ computed from the equation

$$\text{HK} = \frac{P}{0.07028d^2}$$

where P is applied load, kgf; and d is long diagonal of the impression, mm. In reporting Knoop hardness numbers, the test load is stated.

Knoop hardness test An indentation hardness test using calibrated machines to force a rhombic-based pyramidal diamond indenter having specified edge angles, under specified conditions, into the surface of the material under test and to measure the long diagonal after removal of the load.

lamination (1) A type of discontinuity with separation or weakness generally aligned parallel to the worked surface of a metal. (2) In electrical components such as motors, a blanked piece of electrical sheet that is stacked up with several other identical pieces to make a stator or rotor.

lap (1) A surface imperfection on worked metal caused by folding over a fin overfill or similar surface condition, then impressing this into the surface by subsequent working without welding it. (2) A flat surface that holds an abrasive for polishing operations.

leaching See *selective leaching.*

linear-elastic fracture mechanics A method of fracture analysis that can determine the *stress* (or load) required to induce fracture instability in a structure containing a cracklike flaw of known size and shape. See also *stress-intensity factor.*

liquid metal embrittlement The decrease in *ductility* of a metal caused by contact with a liquid metal.

liquid shrinkage The reduction in volume of liquid metal as it cools to the liquidus.

longitudinal direction That direction parallel to the direction of maximum elongation in a worked material. See also *normal direction* and *transverse direction.*

low-cycle fatigue *Fatigue* that occurs at relatively small numbers of cycles ($<10^4$ cycles). Low-cycle fatigue may be accompanied by some plastic, or permanent, deformation. Compare with *high-cycle fatigue.*

Lüders lines Elongated surface markings or depressions, often visible to the unaided eye, that form along the length of a tension specimen at an angle of approximately 45° to the loading axis. Caused by localized *plastic deformation,* they result from discontinuous (inhomogeneous) yielding. Also known as Lüders bands, Hartmann lines, Piobert lines, or stretcher strains.

macroscopic Visible at magnifications at or below 25×.

macroshrinkage Isolated, clustered, or interconnected voids in a casting that are detectable macroscopically. Such voids are usually associated with abrupt changes in section size and are caused by feeding that is insufficient to compensate for *solidification shrinkage.*

macrostructure The structure of metals as revealed by macroscopic examination of a specimen. The examination may be carried out using an as-polished or a polished and etched specimen.

magnification The ratio of the length of a line in the image plane (for example, ground glass or a photographic plate) to the length of the same line in the object. Magnifications are usually expressed in linear terms and in units called diameters.

malleability The characteristic of metals that permits *plastic deformation* in compression without fracture. See also *ductility.*

matrix The continuous or principal phase in which another constituent is dispersed.

maximum strength See *ultimate strength.*

mechanical (cold) crack A crack or fracture in a casting resulting from rough handling or from thermal shock, such as may occur at shakeout or during heat treatment.

mechanical properties The properties of a material that reveal its elastic and inelastic (plastic) behavior when force is applied, thereby indicating its suitability for mechanical (load-bearing) applications. Examples are *elongation, fatigue limit, hardness, modulus of elasticity, tensile strength,* and *yield strength.* Compare with *physical properties.*

mechanical twin A *twin* formed in a crystal by simple shear under external loading.

metal penetration An imperfection on the surface of a casting caused by the penetration of molten metal into voids between refractory particles of the mold.

microcrack A crack of microscopic proportions. Also termed microfissure.

microfissure See *microcrack.*

microporosity Extremely fine *porosity* in castings.

microscopic Visible only at magnifications above 25×.

microshrinkage A casting imperfection consisting of interdendritic voids. Microshrinkage results from contraction during solidification where the opportunity to supply filler material is inadequate to compensate for shrinkage. Alloys with wide ranges in solidification temperature are particularly susceptible.

microstructure The structure of metals and alloys as revealed after polishing and etching a specimen, at magnifications greater than 25×.

misrun A casting not fully formed because of solidification of metal before the mold is filled.

mode One of the three classes of crack (surface) displacements adjacent to the crack tip. These displacement modes are associated with stress-strain fields around the crack tip and are designated I, II, and III. See also *crack tip plane strain* and *crack opening displacement.*

modulus of elasticity *E* (1) The measure of rigidity or stiffness of a metal; the ratio of stress, below the *proportional limit,* to the corresponding strain. In terms of the *stress-strain diagram,* the modulus of elasticity is the slope of the stress-strain curve in the range of linear proportionality of stress to strain. Also known as Young's modulus. (2) For materials that do not conform to *Hooke's law* throughout the elastic range, the slope of either the tangent to the stress-strain curve at the origin or at low stress, the secant drawn from the origin to any specified point on the stress-strain curve, or the chord connecting any two specific points on the stress-strain curve is usually taken to be the modulus of elasticity. In these cases, the modulus is referred to as the *tangent modulus, secant modulus,* or *chord modulus,* respectively.

modulus of rigidity See *shear modulus.*

modulus of rupture Nominal stress at fracture in a bend test or torsion test. In bending, modulus of rupture is the bending moment at fracture divided by the section modulus. In torsion, modulus of rupture is the torque at fracture divided by the polar section modulus. See also *modulus of rupture in bending* and *modulus of rupture in torsion.*

modulus of rupture in bending S_b The value of maximum tensile or compressive stress (whichever causes failure) in the extreme fiber on a beam loaded to failure in bending, computed from

$$S_b = \frac{Mc}{I}$$

where M is maximum bending moment, computed from the maximum load and the original moment arm; c is initial distance from the neutral axis to the extreme fiber where failure occurs; and I is initial moment of inertia of the cross section about the neutral axis. See also *modulus of rupture.*

modulus of rupture in torsion S_s The value of maximum shear stress in the extreme fiber of a member of circular cross section loaded to failure in torsion, computed from

$$S_s = \frac{Tr}{J}$$

where T is maximum twisting moment, r is original outer radius, and J is polar moment of inertia of the original cross section. See also *modulus of rupture.*

necking (1) Reduction of the cross-sectional area of metal in a localized area by stretching. (2) Reduction in the diameter of a portion of the length of a cylindrical shell or tube.

neutron embrittlement Embrittlement resulting from bombardment with neutrons, usually encountered in metals that have been exposed to a neutron flux in the core of a reactor. In steels, neutron embrittlement is evidenced by a rise in the ductile-to-brittle transition temperature. See also *radiation damage.*

nominal strength See *ultimate strength.*

normal direction That direction perpendicular to the plane of working in a worked material. See also *longitudinal direction* and *transverse direction.*

notch See *stress concentration.*

notch acuity Relates to the severity of the *stress concentration* produced by a given notch in a particular structure. If the depth of the notch is very small compared with the width (or diameter) of the narrowest cross section, acuity may be expressed as the ratio of the notch depth to the notch root radius. Otherwise acuity is defined as the ratio of one-half the width (or diameter) of the narrowest cross section to the notch root radius.

notch brittleness Susceptibility of a material to *brittle fracture* at points of *stress concentration.* For example, in a notch tension test, the material is said to be notch brittle if the notch strength is less than the tensile strength of an unnotched specimen. Otherwise it is said to be notch ductile.

notch depth The distance from the surface of a test specimen to the bottom of the notch. In a cylindrical test specimen, the percentage of the original cross-sectional area removed by machining an annular groove.

notch rupture strength The ratio of applied load to original area of the minimum cross section in a stress-rupture test of a notched specimen.

notch sensitivity A measure of the reduction in strength of a metal caused by the presence of *stress concentration.* Values can be obtained from static, impact, or fatigue tests.

notch strength The maximum load on a notched tension-test specimen divided by the minimum cross-sectional area (the area at the root of the notch). Also called notch tensile strength.

notch tensile strength See *notch strength.*

orange peel A surface roughening in the form of a pebble-grained pattern where a metal of unusually coarse grain is stressed beyond its elastic limit. Also known as pebbles and alligator skin.

original crack size a_o The *physical crack size* at the start of testing.

oxidation (1) A reaction in which there is an increase in valence resulting from a loss of electrons. Contrast with *reduction.* (2) A *corrosion* reaction in which the corroded metal forms an oxide; usually applied to reaction with a gas containing elemental oxygen, such as air.

oxidative wear A type of *wear* resulting from the sliding action between two metallic components that generates oxide films on the metal surfaces. These oxide films prevent the formation of a metallic bond between the sliding surfaces, resulting in fine wear debris and low wear rates.

parting The selective *corrosion* of one or more components of a solid-solution alloy.

pebbles See *orange peel.*

physical crack size a_p The distance from a reference plane to the observed crack front. This distance may represent an average of several measurements along the crack front. The reference plane depends on the specimen form, and it is normally taken to be either the boundary or a plane containing either the load line or the centerline of a specimen or plate.

physical properties Properties of a metal or alloy that are relatively insensitive to structure and can be measured without the application of force; for example, density, electrical conductivity, coefficient of thermal expansion, magnetic permeability, and lattice parameter. Does not include chemical reactivity. Compare with *mechanical properties.*

pinhole A small, rounded hole just below the surface of a casting, sometimes visible only after machining. Such holes, often localized, have bright interior surfaces.

Piobert lines See *Lüders lines.*

pitting (1) *Corrosion* of a metal surface, confined to a point or small area, that takes the form of cavities. (2) In *tribology,* a type of wear characterized by the presence of surface cavities formed by processes such as fatigue, local adhesion, or cavitation.

plane strain The stress condition in *linear elastic fracture mechanics* in which there is zero strain in a direction normal to both the axis of applied *tensile stress* and the direction of crack growth (i.e., parallel to the crack front); most nearly achieved in loading thick plates along a direction parallel to the plate surface. Under plane-strain conditions, the plane of fracture instability is normal to the axis of the principal tensile stress.

plane-strain fracture toughness K_{Ic} The crack extension resistance under conditions of *crack tip plane strain.* See also *stress-intensity factor.*

plane stress The stress condition in *linear elastic fracture mechanics* in which the stress in the thickness direction is zero; most nearly achieved in loading very thin sheet along a direction parallel to the surface of the sheet. Under plane-stress conditions, the plane of fracture instability is inclined 45° to the axis of the principal *tensile stress.*

plane-stress fracture toughness K_c The value of the crack-extension resistance at the instability condition determined from the tangency between the *R curve* and the critical crack-extension force curve of the specimen. See also *stress-intensity factor.*

plastic deformation The permanent (inelastic) distortion of metals under applied stresses that strain the material beyond its *elastic limit.*

plowing In *tribology,* the formation of grooves by *plastic deformation* of the softer of two surfaces in relative motion.

Poisson's ratio The absolute value of the ratio of the transverse strain to the corresponding axial strain, in a body subjected to uniaxial stress; usually applied to elastic conditions.

polycrystalline Comprising an aggregate of more than one crystal and usually a large number of crystals.

pore (1) A small void in the body of a metal. (2) A minute cavity in a powder metallurgy compact, sometimes intentional. (3) A minute perforation in an electroplated coating.

porosity Fine holes or pores within a metal.

poultice corrosion A term used in the automotive industry to describe the *corrosion* of vehicle body parts due to the collection of road salts and debris on ledges and in pockets that are kept moist by weather and washing. Also called deposit attack or deposit corrosion.

preferred orientation A condition of a polycrystalline aggregate in which the crystal orientations are not random, but rather exhibit a tendency for alignment with a specific direction in the bulk material, commonly related to the direction of working. See also *fiber* and *texture.*

primary creep The first, or initial, stage of *creep,* or time-dependent deformation.

principal stress (normal) The maximum or minimum value of the normal stress at a point in a plane considered with respect to all possible orientations of the considered plane. On such principal planes the shear stress is zero. There are three principal stresses on three mutually perpendicular planes. The state of stress at a point may be (1) uniaxial, a state of stress in which two of the three principal stresses are zero; (2) biaxial, a state of stress in which only one of the three principal stresses is zero; or (3) triaxial, a state of stress in which none of the principal stresses is zero. Multiaxial stress refers to either biaxial or triaxial stress.

proportional limit The maximum *stress* at which *strain* remains directly proportional to stress; the upper end of the straight-line portion of the stress-strain or load-elongation curve. See also *elastic limit.*

quasicleavage fracture A fracture mode that combines the characteristics of *cleavage fracture* and *dimpled rupture fracture.* An intermediate type of fracture found in certain high-strength metals.

quenching crack A crack formed as a result of thermal stresses produced by rapid cooling from a high temperature.

radial marks Lines on a fracture surface that radiate from the fracture origin and are visible to the unaided eye or at low magnification. Radial marks result from the intersection and connection of brittle fractures propagating at different levels. Also known as shear ledges. See also *chevron pattern.*

radiation damage A general term for the alteration of properties of a material arising from exposure to ionizing radiation (penetrating radiation), such as x rays, gamma rays, neutrons, heavy-particle radiation, or fission fragments in nuclear fuel material. See also *neutron embrittlement.*

ratchet marks Lines on a fatigue fracture surface that result from the intersection and connection of fatigue fractures propagating from multiple origins. Ratchet marks are parallel to the overall direction of crack propagation and are visible to the unaided eye or at low magnification.

rattail A shallow, indented, and irregular line on a casting surface due to sand expansion.

***R* curve** A plot of crack-extension resistance as a function of stable crack extension, which is the difference between either the *physical crack size* or the *effective crack size* and the *original crack size. R* curves normally depend on specimen thickness and, for some materials, on temperature and strain rate.

reduction (1) In cupping and deep drawing, a measure of the percentage decrease from blank diameter to cup diameter, or of diameter reduction in redrawing. (2) In forging, rolling, and drawing, either the ratio of the original to final cross-sectional area or the percentage decrease in cross-sectional area. (3) A reaction in which there is a decrease in valence resulting from a gain in electrons. Contrast with *oxidation.*

reduction in area See *reduction of area.*

reduction of area The difference between the original cross-sectional area of a tension specimen and the smallest area at or after fracture as specified for the material being tested. Also known as reduction in area.

residual stress Stress present in a body that is free of external forces or thermal gradients.

river pattern A characteristic pattern of cleavage steps running parallel to the local direction of crack propagation on the fracture surfaces of grains that have separated by *cleavage.*

rock candy fracture A fracture that exhibits separated-grain facets; most often used to describe an *intergranular fracture* in a large-grained metal.

Rockwell hardness number HR A number derived from the net increase in the depth of impression as the load on an indenter is increased from a fixed minor load to a major load and then returned to the minor load. Rockwell hardness numbers are always quoted with a scale symbol representing the penetrator, load, and dial used.

Rockwell hardness test An indentation hardness test using a calibrated machine that utilizes the depth of indentation, under constant load, as a measure of hardness. Either a 120° diamond cone with a slightly rounded point or a 1.6- or 3.2-mm-diameter (1/16- or 1/8-in) steel ball is used as the indenter.

Rockwell superficial hardness number See *Rockwell hardness number* and *Rockwell superficial hardness test.*

Rockwell superficial hardness test Same as *Rockwell hardness test,* except that smaller minor and major loads are used.

rupture stress The stress at failure. Also known as breaking stress or fracture stress.

rust A *corrosion* product consisting primarily of hydrated iron oxide. A term properly applied only to ferrous alloys.

sand hole A pit in the surface of a sand casting resulting from a deposit of loose sand on the surface of the mold.

scab A raised and rough area on the surface of a casting due to sand being dislodged from the surface of the mold.

Scleroscope hardness number HSc or HSd A number related to the height of rebound of a diamond-tipped hammer dropped on the material being tested. It is measured on a scale determined by dividing into 100 units the average rebound of the hammer from a quenched (to maximum hardness) and untempered AISI W-5 tool steel test block.

Scleroscope hardness test A dynamic indentation hardness test using a calibrated instrument that drops a diamond-tipped hammer from a fixed height onto the surface of the material being tested. The height of rebound of the hammer is a measure of the hardness of the material.

scoring In *tribology,* a severe form of wear characterized by the formation of extensive grooves and scratches in the direction of sliding.

scratching In *tribology,* the mechanical removal and/or displacement of material from a surface by the action of abrasive particles or protuberances sliding across the surfaces. See also *plowing.*

scuffing A form of *adhesive wear* that produces superficial scratches or a high polish on the rubbing surfaces. It is observed most often on inadequately lubricated parts.

seam An unfused fold or lap that appears as a crack on a metal surface.

season cracking Cracking resulting from the combined effects of corrosion and internal stress. A term usually applied to *stress-corrosion cracking of brass.*

secant modulus The slope of the secant drawn from the origin to any specified point on a stress-strain curve. See also *modulus of elasticity.*

secondary creep See *creep.*

segregation Nonuniform distribution of alloying elements, impurities, or phases.

selective leaching *Corrosion* in which one element is preferentially removed from an alloy, leaving a residue (often porous) of the elements that are more resistant to the particular environment. See also *decarburization, denickelification, dezincification,* and *graphitic corrosion.*

sensitization In austenitic stainless steels, the precipitation of chromium carbides, usually at grain boundaries, on exposure to temperatures of about 550 to 850°C (1000 to 1550°F), leaving the grain boundaries depleted of chromium and therefore susceptible to preferential attack by a corroding (oxidizing) medium.

shatter crack See *flake.*

shear bands Bands in which *deformation* has been concentrated inhomogeneously in sheets that extend across regional groups of grains. Only one system is usually present in each regional group of grains, different systems being present in adjoining groups. The bands are noncrystallographic and form on planes of maximum *shear stress* (55° to the compression direction). They carry most of the deformation at large strains.

shear fracture A *ductile fracture* in which a crystal (or a polycrystalline mass) has separated by sliding or tearing under the action of shear stresses. See also *shear stress.*

shear ledges See *radial marks.*

shear lip A narrow, slanting ridge along the edge of a fracture surface. The term sometimes also denotes a narrow, often crescent-shaped, fibrous region at the edge of a fracture that is otherwise of the cleavage type, even though this fibrous region is in the same plane as the rest of the fracture surface.

shear modulus *G* The ratio of *shear stress* to the corresponding *shear strain* for shear stresses below the *proportional limit* of the material. Values of shear modulus are usually determined by torsion testing. Also known as modulus of rigidity.

shear strain The tangent of the angular change, due to force, between two lines originally perpendicular to each other through a point in a body.

shear strength The maximum *shear stress* that a material is capable of sustaining. Shear strength is calculated from the maximum load during a shear or torsion test and is based on the original dimensions of the cross section of the specimen.

shear stress (1) A *stress* that exists when parallel planes in metal crystals slide across each other. (2) The stress component tangential to the plane on which the forces act. Also known as tangential stress.

shock load The sudden application of an external force that results in a very rapid buildup of *stress*—for example, piston loading in internal combustion engines.

shrinkage See *casting shrinkage.*

shrinkage cavity A void left in cast metals as a result of *solidification shrinkage.* Shrinkage cavities occur in the last metal to solidify after casting.

silky fracture A metal fracture in which the broken metal surface has a fine texture, usually dull in appearance. Characteristic of tough and strong metals. Contrast with *crystalline fracture* and *granular fracture.*

slant fracture A type of fracture appearance, typical of *plane-stress* fractures, in which the plane of metal separation is inclined at an angle (usually about 45°) to the axis of the applied stress.

slip *Plastic deformation* by the irreversible shear displacement (translation) of one part of a crystal relative to another in a definite crystallographic direction and usually on a specific crystallographic plane. Sometimes called glide. See also *flow.*

S-N curve A plot of stress (S) against the number of cycles to failure (N). The stress can be the maximum stress (S_{max}) or the alternating stress amplitude (S_a). The stress values are usually nominal stress, i.e., there is no adjustment for stress concentration. The diagram indicates the *S-N* relationship for a specified value of the mean stress (S_m) or the stress ratio (A or R) and a specified probability of survival. For N a log scale is almost always used. For S a linear scale is used most often, but a log scale is sometimes used. Also known as *S-N* diagram.

S-N diagram See *S-N curve.*

solidification shrinkage The reduction in volume of metal from beginning to end of solidification.

solidification shrinkage crack A crack that forms, usually at elevated temperature, because of the internal (shrinkage) stresses that develop during solidification of a metal casting. Also termed hot crack.

solid shrinkage The reduction in volume of metal from the solidus to room temperature.

spalling The cracking and flaking of particles out of a surface.

static Stationary or very slow. Frequently used in connection with routine tension testing of metal specimens. Contrast with *dynamic.*

static fatigue See *hydrogen-induced delayed cracking.*

steady-rate creep See *creep.*

strain The unit of change in the size or shape of a body due to force.

strain hardening An increase in hardness and strength caused by *plastic deformation* at temperatures below the recrystallization range. Also known as work hardening.

stray-current corrosion *Corrosion* caused by electric current from a source external to the intended electric circuit, for example, extraneous current in the earth.

stress The intensity of the internally distributed forces or components of forces that resist a change in the volume or shape of a material that is or has been subjected to external forces. Stress is expressed in force per unit area and is calculated on the basis of the original dimensions of the cross section of the specimen. Stress can be either direct (tension or compression) or shear. Usually expressed in pounds per square inch (lb/in^2) or megapascals (MPa).

stress amplitude One-half the algebraic difference between the maximum and minimum *stress* in one cycle of a repetitively varying stress.

stress concentration A change in contour or a discontinuity that causes local increases in *stress* in materials under load. Typical are sharp-cornered grooves or notches, threads, fillets, holes, etc. Also called stress raiser.

stress-concentration factor K_t A multiplying factor for applied *stress* that allows for the presence of a structural discontinuity such as a notch or hole; K_t equals the ratio of the greatest stress in the region of the discontinuity to the nominal stress for the entire section. Also known as theoretical stress-concentration factor.

stress-corrosion cracking SCC A cracking process that requires the simultaneous action of a corrodent and sustained *tensile stress.* This excludes corrosion-reduced sections that fail by fast fracture. It also excludes intergranular or transgranular corrosion, which can disintegrate an alloy without applied or residual stress. See also *corrosion.*

stress cycle The smallest segment of the stress-time function that is repeated periodically.

stress-intensity factor K A scaling factor used in *linear-elastic fracture mechanics* to describe the intensification of applied *stress* at the tip of a crack of known size and shape. At the onset of rapid crack propagation in any structure containing a crack, the factor is called the critical stress-intensity factor, or the *fracture toughness.* Various subscripts are used to denote different loading conditions or fracture toughnesses:

K_c Plane-stress fracture toughness. The value of stress intensity at which crack propagation becomes rapid in sections thinner than those in which plane-strain conditions prevail.

K_I Stress-intensity factor for a loading condition that displaces the crack faces in a direction normal to the crack plane. Also known as the opening mode of deformation.

K_{Ic} Plane-strain fracture toughness. The minimum value of K_c for any given material and condition, which is attained when rapid crack propagation in the opening mode is governed by plane-strain conditions.

K_{Id} Dynamic fracture toughness. The fracture toughness determined under dynamic loading conditions; it is used as an approximation of K_{Ic} for very tough materials.

K_{Iscc} Threshold stress intensity for stress-corrosion cracking when loading conditions meet plane-strain requirements.

K_Q Provisional value for plane-strain fracture toughness.

K_{th} Threshold stress intensity for stress-corrosion cracking. A value of stress intensity characteristic of a specific combination of material, material condition, and corrosive environment above which stress-corrosion crack propagation occurs and below which the material is immune from stress-corrosion cracking.

ΔK The range of the stress-intensity factor during a fatigue cycle.

stress raiser See *stress concentration.*

stress ratio *A* or *R* The algebraic ratio of two specified *stress* values in a *stress cycle.* Two commonly used stress ratios are (1) the ratio of the alternating stress amplitude to the mean stress, $A = S_a/S_m$, and (2) the ratio of the minimum stress to the maximum stress, $R = S_{min}/S_{max}$.

stress-rupture strength See *creep-rupture strength.*

stress-strain curve See *stress-strain diagram.*

stress-strain diagram A graph in which corresponding values of *stress* and *strain* are plotted against each other. Values of stress are usually plotted vertically (ordinate or y axis) and values of strain horizontally (abscissa or x axis). Also known as deformation curve and stress-strain curve.

stretcher strains See *Lüders lines.*

striation A fatigue fracture feature often observed in electron micrographs that indicates the position of the crack front after each succeeding cycle of *stress.* The distance between striations indicates the advance of the crack front across that crystal during one *stress cycle,* and a line normal to the striation indicates the direction of local crack propagation. Not to be confused with *beach marks,* which are much larger (macroscopic) and form differently.

stringer In wrought materials, an elongated configuration of microconstituents or foreign material aligned in the direction of working. The term is commonly associated with elongated oxide or sulfide inclusions in steel.

subboundary structure (subgrain structure) A network of low-angle boundaries, usually with misorientations less than 1° within the main grains of a microstructure.

subgrain A portion of a crystal or *grain,* with an orientation slightly different from the orientation of neighboring portions of the same crystal.

subsurface corrosion See *internal oxidation.*

sulfidation The reaction of a metal or alloy with a sulfur-containing species to produce a sulfur compound that forms on or beneath the surface of the metal or alloy.

sulfide stress cracking SSC Brittle failure by cracking under the combined action of *tensile stress* and *corrosion* in the presence of water and hydrogen sulfide.

tangential stress See *shear stress.*

tangent modulus The slope of the stress-strain curve at any specified *stress* or *strain.* See also *modulus of elasticity.*

temper brittleness Brittleness that results when certain steels are held within, or are cooled slowly through, a certain range of temperatures below the transformation range. The brittleness is manifested as an upward shift in ductile-to-brittle transition temperature, but only rarely produces a low value of *reduction of area* in a smooth-bar tension test of the embrittled material.

tensile strength In *tension testing,* the ratio of the maximum load to the original cross-sectional area. See also *ultimate strength*; compare with *yield strength.*

tensile stress A *stress* that causes two parts of an elastic body, on either side of a typical stress plane, to pull apart. Contrast with *compressive stress.*

tensile testing See *tension testing.*

tension The force or load that produces *elongation.*

tension testing A method of determining the behavior of materials subjected to uniaxial loading, which tends to stretch the metal. A longitudinal specimen of known length and diameter is gripped at both ends and stretched at a slow, controlled rate until rupture occurs. Also known as tensile testing.

tertiary creep See *creep.*

texture In a polycrystalline aggregate, the state of distribution of crystal orientations. In the usual sense, it is synonymous with *preferred orientation,* in which the distribution is not random. See also *fiber.*

theoretical stress-concentration factor See *stress-concentration factor.*

thermal fatigue Fracture resulting from the presence of temperature gradients that vary with time in such a manner as to produce cyclic stresses in a structure.

thermal shock The development of a steep temperature gradient and accompanying high stresses within a structure.

thermal stresses Stresses in metal resulting from nonuniform temperature distribution.

tide marks See *beach marks.*

torsion A twisting action applied to a shaftlike or cylindrical member. The twisting may be either reversed (back and forth) or unidirectional (one way).

torsional stress The *shear stress* on a transverse cross section resulting from a twisting action.

transcrystalline See *transgranular.*

transcrystalline cracking See *transgranular cracking.*

transgranular Through or across crystals or grains. Also called intracrystalline or transcrystalline.

transgranular cracking Cracking or fracturing that occurs through or across a crystal or grain. Also called transcrystalline cracking. Contrast with *intergranular cracking.*

transgranular fracture Fracture through or across the crystals or grains of a metal. Also called transcrystalline fracture or intracrystalline fracture. Contrast with *intergranular fracture.*

transient creep See *creep* and *primary creep.*

transverse direction Literally, "across," usually signifying a direction or plane perpendicular to the direction of working. In rolled plate or sheet, the direction across the width is often called long transverse; the direction through the thickness, short transverse.

tribology The science concerned with the design, friction, lubrication, and wear of contacting surfaces that move relative to each other.

tuberculation The formation of localized *corrosion* products that appear on a surface as knoblike prominences (tubercules).

twin Two portions of a crystal with a definite orientation relationship; one may be regarded as the parent, the other as the twin. The orientation of the twin is a mirror image of the orientation of the parent across a twinning plane or an orientation that can be derived by rotating the twin portion about a twinning axis. See also *annealing twin* and *mechanical twin.*

twin bands Bands across a crystal grain, observed on a polished and etched section, where crystallographic orientations have a mirror-image relationship to the orientation of the matrix grain across a composition plane that is usually parallel to the sides of the band.

ultimate strength The maximum *stress* (tensile, compressive, or shear) a material can sustain without fracture, determined by dividing maximum load by the original cross-sectional area of the specimen. Also known as nominal strength or maximum strength.

Vickers hardness number HV A number related to the applied load and the surface area of the permanent impression made by a square-based pyramidal diamond indenter having included face angles of 136°, computed from

$$\mathrm{HV} = 2P \sin \frac{\alpha/2}{d^2} = \frac{1.8544P}{d^2}$$

where P is applied load, kgf; d is mean diagonal of impression, mm; and α is face angle of indenter, 136°.

Vickers hardness test An indentation hardness test employing a 136° diamond pyramid indenter (Vickers) and variable loads, enabling the use of one hardness scale for all ranges of hardness—from very soft lead to tungsten carbide. Also known as diamond pyramid hardness test.

volumetric modulus of elasticity See *bulk modulus of elasticity.*

Wallner lines A distinct pattern of intersecting sets of parallel lines, usually producing a set of V-shaped lines, sometimes observed when viewing brittle fracture surfaces at high magnification in an electron microscope. Wallner lines are attributed to interaction between a shock wave and a brittle crack front propagating at high velocity. Sometimes Wallner lines are misinterpreted as fatigue striations.

wear Damage to a solid surface, generally involving progressive loss of material, due to relative motion between that surface and a contacting surface or substance.

wear rate The rate of material removal or dimensional change due to *wear* per unit of exposure parameter—for example, quantity of material removed (mass, volume, thickness) in unit distance of sliding or unit time.

whiskers (1) Metallic filamentary growths, often microscopic in size, that attain very high strengths. (2) Oxide whiskers, such as sapphire, which because of their strength and inertness at high temperatures are used as reinforcements in metal-matrix composites.

work hardening See *strain hardening.*

yield Evidence of *plastic deformation* in structural materials. See also *creep* and *flow.*

yield point The first *stress* in a material, usually less than the maximum attainable stress, at which an increase in *strain* occurs without an increase in stress. Only certain metals—those that exhibit a localized, heterogeneous type of transition from elastic to plastic deformation—produce a yield point. If there is a decrease in stress after yielding, a distinction may be made between upper and lower yield points. The load at which a sudden drop in the flow curve occurs is called the upper yield point. The constant load shown on the flow curve is the lower yield point.

yield strength The stress at which a material exhibits a specified deviation from proportionality of stress and strain. The specified deviation is usually 0.2 percent for most metals. Compare with *tensile strength.*

yield stress The stress level of highly ductile materials, such as structural steels, at which large *strains* take place without further increase in *stress.*

Young's modulus See *modulus of elasticity.*

References to Glossary

Merriman, A. D., *A Dictionary of Metallurgy,* Pitman, 1958.

"Standard Definitions of Terms Relating to Corrosion and Corrosion Testing," G 15, *Annual Book of ASTM Standards,* vol. 03.02, ASTM, Philadelphia, Pa., 1984, pp. 133–137.

"Standard Definitions of Terms Relating to Fatigue Testing and the Statistical Analysis of Fatigue Data," E 206, *Annual Book of ASTM Standards,* vol. 03.01, ASTM, Philadelphia, Pa., 1984, pp. 340–345.

"Standard Definitions of Terms Relating to Methods of Mechanical Testing," E 6, *Annual Book of ASTM Standards,* vol. 03.01, ASTM, Philadelphia, Pa., 1984, pp. 119–129.

"Standard Terminology Relating to Erosion and Wear," G 40, *Annual Book of ASTM Standards,* vol. 03.02, ASTM, Philadelphia, Pa., 1984, pp. 239–246.

"Standard Terminology Relating to Fracture Testing," E 616, *Annual Book of ASTM Standards,* vol. 03.01, ASTM, Philadelphia, Pa., 1984, pp. 671–684.

Index

ABOUT THE AUTHORS

CHARLIE R. BROOKS is Professor of Metallurgical Engineering and Alumni Distinguished Service Professor Emeritus at the University of Tennessee. He has received several teaching awards, including the M. E. Brooks Distinguished Faculty Award of the College of Engineering and the Albert Easton White Teaching Award from ASM International. He is also a Fellow of ASM International.

ASHOK CHOUDHURY is Commercialization Manager for the Office of Technology Transfer at Oak Ridge National Laboratory, Oak Ridge, Tennessee. He is the recipient of several teaching awards and technology transfer awards, and a co-recipient of an R&D 100 Award.

www.ingramcontent.com/pod-product-compliance
Lightning Source LLC
LaVergne TN
LVHW010620300926
849162LV00001B/17
* 9 7 8 0 0 7 1 3 5 7 5 8 6 *